HANDBOOK OF CITIES AND NETWORKS

RESEARCH HANDBOOKS IN URBAN STUDIES

In this urban century the need for innovative and rigorous research on the challenges and opportunities facing our cities has never been so pressing. This timely series brings together critical and thought-provoking contributions on key topics and issues in urban research from a range of social science perspectives. Comprising specially commissioned chapters from leading academics these comprehensive *Research Handbooks* feature cutting-edge research and are written with a global readership in mind. Equally useful as reference tools or high-level introductions to specific topics, issues, methods and debates, these *Research Handbooks* will be an essential resource for academic researchers and postgraduate students.

Titles in the series include:

Handbook of Urban Segregation
Edited by Sako Musterd

Handbook of Megacities and Megacity-regions
Edited by Danielle Labbé and André Sorensen

Handbook on City and Regional Leadership
Edited by Markku Sotarauta and Andrew Beer

Handbook of Cities and Networks
Edited by Zachary P. Neal and Céline Rozenblat

Handbook of Cities and Networks

Edited by

Zachary P. Neal

Associate Professor, Department of Psychology, Michigan State University, USA

Céline Rozenblat

Professor of Urban Geography, Faculty of Geosciences and Environment, Université de Lausanne, Switzerland

RESEARCH HANDBOOKS IN URBAN STUDIES

Edward Elgar PUBLISHING

Cheltenham, UK • Northampton, MA, USA

Published by
Edward Elgar Publishing Limited
The Lypiatts
15 Lansdown Road
Cheltenham
Glos GL50 2JA
UK

Edward Elgar Publishing, Inc.
William Pratt House
9 Dewey Court
Northampton
Massachusetts 01060
USA

A catalogue record for this book
is available from the British Library

Library of Congress Control Number: 2021936871

This book is available electronically in the **Elgar**online
Geography, Planning and Tourism subject collection
http://dx.doi.org/10.4337/9781788114714

MIX
Paper from
responsible sources
FSC® C013604

ISBN 978 1 78811 470 7 (cased)
ISBN 978 1 78811 471 4 (eBook)

Typeset by Servis Filmsetting Ltd, Stockport, Cheshire
Printed and bound by CPI Group (UK) Ltd, Croydon, CR0 4YY

Contents

List of contributors viii

PART I THEORETICAL CONCEPTIONS OF CITIES AND NETWORKS

1 The levels and scales of urban networks 2
Céline Rozenblat and Zachary P. Neal

2 From networks of cities to systems of cities 16
Denise Pumain

3 Complex networks and fundamental urban processes 41
Luís M.A. Bettencourt

PART II CITIES AND NETWORKS IN HISTORY

4 Settlement networks and sociocultural evolution 63
Elizabeth Bogumil and Christopher Chase-Dunn

5 Sizing up Roman urbanism 88
J.W. Hanson

6 Associational life in the rounding out of dynamic cities: an in-depth methodology and application to Newcastle city-region in the nineteenth century 107
Michael Barke and Peter J. Taylor

PART III METHODS AND MODELS OF CITY–NETWORK INTERACTIONS

7 The structure of urban networks 127
Marc Barthelemy

8 Modeling the co-evolution of cities and networks 166
Juste Raimbault

9 Ties through place: socio-material network analyses in urban studies 194
Meg Bartholomew and Alasdair Jones

10 Theory and method in macro-comparative social network analysis 215
Matthew C. Mahutga and Robert Clark

11 The role of proximity and distance in inter-urban networks 239
Thomas Sigler, Kirsten Martinus and Petr Matous

12 About being in the middle: conceptions, models and theories of centrality in urban studies 252
Michiel van Meeteren

PART IV NETWORK PROCESSES WITHIN CITIES

13 The city of opportunity: designing Cities4People 273
Karima Kourtit, Peter Nijkamp and Tigran Haas

14 Community organizing and interorganizational network changes in a justice system reform coalition in Chicago 293
Brian D. Christens and Daniel G. Cooper

15 Racial/ethnic residential segregation and urban spatial networks in the United States 313
Kathryn Freeman Anderson and Joseph Galaskiewicz

16 Urban mobility and segregation examined through networked travel activity 331
Susan A. Burtner and Alan T. Murray

17 The impact of urban social life on sexual networks and health 350
Patrick Janulis and Michelle Birkett

18 Modeling dissemination of health information and beliefs in urban social networks 366
Sara S. Metcalf, Harvey D. Palmer, Qiuyi Zhang and Mary E. Northridge

19 Conflict resolution and opinion pooling in city planning 389
Michael Batty

20 Up close and impersonal: locative media and the changing nature of the networked individual in the city 409
Darryl A. Pieber and Anabel Quan-Haase

21 Far away ties, never so close: the geographical spread of social support resources for mobile individuals 427
Romina Cachia and Isidro Maya-Jariego

PART V NETWORK PROCESSES BETWEEN CITIES

22 Is maritime transport an urban network? The interplay between global container flows and urban hierarchies 449
César Ducruet

23 Unravelling the forces underlying urban industrial agglomeration 472
Neave O'Clery, Samuel Heroy, François Hulot and Mariano Beguerisse-Díaz

24 Cities, networks, polycentrism: examining the place of polycentrism in spaces of flows 493
Kathy Pain and Shuai Shi

25 Intracity and intercity networks of multinational firms, 2010–2019 511
Céline Rozenblat

26 Uneven ties! The imposition of inequality through interscalar networks 557
Ronald Wall and Umakrishnan Kollamparambil

27 Research progress of Chinese city networks 585
Fenghua Pan, Cheng Fang and Xiande Li

28 The GaWC perspective on global-scale urban networks 601
Ben Derudder and Peter J. Taylor

29 Global cities, centripetal wealth transfer and uneven development 618
Christof Parnreiter

Index 633

Contributors

Kathryn Freeman Anderson, University of Houston, USA.

Michael Barke, Northumbria University, United Kingdom.

Marc Barthelemy, Institute for Theoretical Physics, France.

Meg Bartholomew, Woods Bagot, United Kingdom.

Michael Batty, University College London, United Kingdom.

Mariano Beguerisse-Díaz, University of Oxford, United Kingdom.

Luís M.A. Bettencourt, University of Chicago, USA.

Michelle Birkett, Northwestern University, USA.

Elizabeth Bogumil, University of California Riverside, USA.

Susan A. Burtner, University of California Santa Barbara, USA.

Romina Cachia, University of Malta.

Christopher Chase-Dunn, University of California Riverside, USA.

Brian D. Christens, Vanderbilt University, USA.

Robert Clark, University of Oklahoma, USA.

Daniel G. Cooper, Chicago Metropolitan Planning Council, USA.

Ben Derudder, Katholieke Universiteit Leuven, Belgium.

César Ducruet, Centre National de la Recherche Scientifique, France.

Cheng Fang, Beijing Normal University, People's Republic of China.

Joseph Galaskiewicz, University of Arizona, USA.

Tigran Haas, KTH Royal Institute of Technology Stockholm, Sweden.

J.W. Hanson, University of Reading, United Kingdom.

Samuel Heroy, University of Oxford, United Kingdom.

François Hulot, University of Oxford, United Kingdom.

Patrick Janulis, Northwestern University, USA.

Alasdair Jones, London School of Economics, United Kingdom.

Umakrishnan Kollamparambil, University of the Witwatersrand, South Africa.

Karima Kourtit, Jheronimus Academy of Data Science, the Netherlands.

Xiade Li, Shangai Normal University, People's Republic of China.

Matthew C. Mahutga, University of California Riverside, USA.

Kirsten Martinus, University of Western Australia.

Petr Matous, University of Sydney, Australia.

Isidro Maya-Jariego, Universidad de Sevilla, Spain.

Michiel van Meeteren, Loughborough University, United Kingdom.

Sara S. Metcalf, State University of New York Buffalo, USA.

Alan T. Murray, University of California Santa Barbara, USA.

Zachary P. Neal, Michigan State University, USA.

Peter Nijkamp, Jheronimus Academy of Data Science, the Netherlands.

Mary E. Northridge, New York University, USA.

Neave O'Clery, University of Oxford, United Kingdom.

Kathy Pain, University of Reading, United Kingdom.

Harvey D. Palmer, State University of New York Buffalo, USA.

Fenghua Pan, Beijing Normal University, People's Republic of China.

Christof Parnreiter, University of Hamburg, Germany.

Darryl A. Pieber, Western University, Canada.

Denise Pumain, University Paris I Panthéon-Sorbonne, France.

Anabel Quan-Haase, Western University, Canada.

Juste Raimbault, University College London, United Kingdom.

Céline Rozenblat, Université de Lausanne, Switzerland.

Shuai Shi, University of Reading, United Kingdom.

Thomas Sigler, The University of Queensland, Australia.

Peter J. Taylor, Northumbria University, United Kingdom.

Ronald Wall, University of the Witwatersrand, South Africa.

Qiuyi Zhang, State University of New York Buffalo, USA.

PART I

THEORETICAL CONCEPTIONS OF CITIES AND NETWORKS

1. The levels and scales of urban networks

Céline Rozenblat and Zachary P. Neal

The urban networks discussed in this volume, and that appear in the literature more broadly, are characterized by significant diversity. This is perhaps not a surprise as the study of urban networks is necessarily interdisciplinary, drawing on theoretical foundations from geography, economics, psychology and sociology, and on methodological tools including ethnographic and qualitative methods from sociology, and quantitative methods from mathematics and physics. However, although the flexibility of network models to capture a wide range of urban phenomena is a key strength of the approach and a source of intellectual diversity, it can also be a source of confusion. Different fields and different research questions require studying different types of urban networks, often defined in very different ways, which obscures their commonalities.

In this introductory chapter, we sketch a framework for integrating the diversity of urban networks by situating them along the dimensions of level and scale. These two dimensions define, respectively, the aggregation and spatial scope of the nodes, and therefore provide critical parameters for defining an urban network. In some instances, a network's level and scale are defined implicitly by the research question, but we contend there is still value in being explicit about level and scale. Similarly, although a great deal of past research on urban networks has explored only specific intersections of level and scale (for example, networks of people at the local scale, or networks of cities at the global scale), we contend that exploring urban networks with different combinations of levels and scales offers opportunities for new insights that the reader will find in this volume. We begin by describing the level/scale framework in general, then discuss the case of economic urban networks as an extended example, and use the framework to explore commonalities among the diverse urban networks in this volume. We conclude by discussing ways that levels and scales can be made more explicit in urban networks, and the potential benefits for studying urban networks at multiple levels and scales.

1. DEFINING URBAN NETWORKS

Networks are distinguished from other types of data by their focus not on individual cases, but on relationships. Therefore, defining any network, including urban networks, begins by specifying the relationship that the network is intended to capture. That is, in network terms, what are the edges? The relationships or edges that serve as the building blocks of an urban network, giving it a particular structure, might include physical flows, such as the movement of people in a transportation network, intangible flows, such as the movement of information in communication networks, or the physical infrastructures that facilitate such flows (for example, a street network). However, they can also include affective relationships (for example, friendship), biological relationships (for example, kinship), relationships of similarity (for example, co-location). Defining the relationship that the network is intended to capture, already defines to a significant extent what type of network it is. Therefore, and since there are so many types of relationships that an urban network might aim to capture, clarity and precision about the specific relationship an urban network does capture is critical for clearly defining the network.

1.1 Levels of Aggregation

After the relationships or edges of an urban network have been defined, the nodes – that is, the objects participating in these relationships and between which the edges exist (or not) – must be defined. Butts (2009, p. 414) recommends that 'nodes should be defined so as to include all distinct entities that are capable of participating in the relationship under study'. Following this recommendation, avoiding any aggregation of distinct entities, defines nodes at what we call the micro level. However, nodes can also be aggregated and defined at the higher meso level or macro level. These aggregations of smaller nodes into larger nodes is sometimes essential because, as Pavé (2006, p. 45) explains, 'properties of processes having no significance at lower levels appear at each level [through a process known as] emergence'. It is from this perspective that Berry proposed, in his 1964 seminal paper 'Cities as systems within systems of cities', that several levels of systems characterize cities: cities are themselves spaces of interactions of actors and institutions (cities as systems) and are positioned into spaces of other cities with which they interact (systems of cities).

Thus, the level of an urban network characterizes the extent to which the nodes represent aggregations of the smallest distinct entities capable of participating in the relationship under study to larger systems (parts of cities, cities or systems of cities) that transform their individual actions and interactions. There are many ways that micro-level nodes might be aggregated; for example, individuals can be aggregated into groups with common interests or with common demographic characteristics, while individual firms can be aggregated into groups by sector or by value chain. Importantly, the aggregation of low-level nodes (for example, people) into higher-level nodes (for example, demographic groups) also implies aggregating their edges (for example, person-to-person friendships are aggregated as group-to-group friendships). During this aggregation, although the basic definition of the edges themselves remains constant (that is, friendship), the nature of the relationship does not necessarily retain the same significance after aggregation owing to the non-linearity in the aggregation process. Following our example, friendship between persons is not necessarily the same type of relationship as friendship between groups (that is, the criteria for measuring friendship and the consequences in behavior are different for individuals and for groups).

The challenge of defining the level of a network mirrors the 'modifiable areal unit problem' (MAUP) in geography (Openshaw and Taylor 1979), where distinct spatial locations can be aggregated into spatial units in different ways (for example, defining a 'city' as a legal municipality versus a metropolitan area). As Wolf et al. (2020, p. 2) explain, 'Place characteristics are contingent on where place boundaries are drawn'. Therefore, as is the case with the MAUP, defining an urban network's nodes at different levels can yield very different networks, with different structures, capable of capturing different processes and of answering different research questions. Therefore, clarity and precision about the level of (aggregation of) the nodes in an urban network is critical for clearly defining the network.

1.2 Scales of Inclusion

After the edges and nodes of an urban network have been defined, finally it is necessary to define which of these nodes (and their edges) should be included in or excluded from the network. In the social network literature, the specification of which nodes are part of a network is known as the 'boundary specification problem' (Laumann et al. 1989). Since in the context of urban networks the notion of a network 'boundary' is often specifically spatial, this becomes a spatial-scale specification problem. An urban network may be defined at the local scale to include only the nodes within a local area, or at the global scale to include all nodes regardless of their location, or at any spatial scale between these extremes (for example, regional, national, and so on). The selection of a particular scale directly defines the size of the network (that is, the number of nodes and edges it includes), and can significantly impact its structure as it implies the inclusion of some nodes and edges and the exclusion of others. However, still more substantively, as Wolf et al. (2020, p. 2) explain, 'Characteristics of a geographical process are contingent on the scale at which the process is measured'. That is, for example, trade processes may operate in radically different ways among a local network of cities within a region versus among a global network of cities. Therefore, clarity and precision about the spatial scale of (inclusion of) the nodes in an urban network is critical for clearly defining the network.

2. ECONOMIC URBAN NETWORKS: AN EXAMPLE

The framework sketched in the previous section is intentionally abstract, allowing it to accommodate the diversity of urban networks. However, a concrete example serves to illustrate how edges, levels and scales might be defined in practice, and how different definitions can yield different urban networks. Here, we focus on the broad example of economic urban networks.

2.1 Defining the Relationship

In an urban economic network, the relationship under study is the exchange of capital. Exchanges of capital can take many forms, including exchanges of financial capital (that is, money), physical capital (that is, goods), intellectual capital (that is, information) and human capital (that is, labor). By defining an urban network in which the edges capture exchanges of capital, before we have provided any other detail about the network, it is already clear that it is an economic network.

2.2 Defining the Levels

The smallest distinct entities which can exchange capital are individual people, which serve as the nodes in an economic urban network defined at the micro level. Defining the network at this level may be useful for understanding, for example, how neighbors provide assistance to one another (for example, borrowing a lawnmower), but for other research questions it is more useful to conceptualize capital exchanges as occurring between aggregations of people. For example, studying the movement of capital between firms (that is, between people grouped into firms to which they contribute their labor) provides an urban economic network at the meso level that may be useful for understanding supply and production chains. Aggregating further, studying the movement of capital between entire cities (that is, between firms grouped into cities) provides an urban economic network at the macro level that may be useful for understanding the organization of command and control centers in the world economy.

Notably, although the micro-, meso- and macro-level definitions of an urban economic network differ in their aggregation of the nodes, they are all defined by the same relationship: the exchange of capital between these nodes. However, the specific nature of the relevant capital exchanges may depend on the level. For example, individuals may exchange their working (human) capital and relatively small amounts of money, while firms exchange large quantities of goods and services in trade and supplying networks, and cities exchange labor pools and billions of dollars in financial products.

2.3 Defining the Scales

An urban economic network might be studied at several different spatial scales, each of which involves the inclusion of different sets of nodes in the network. For example, a micro-level urban economic network in which the nodes are individuals might be studied at the local scale, including all individuals in a given neighborhood in the network. In contrast, a macro-level urban economic network in which the nodes are cities might be studied at the global scale, including all the cities in the world.

2.4 Combining Levels and Scales

A great deal of the existing research on urban networks has focused on only some potential combinations of level and scale. For example, Neal's (2013) framework combines levels and scales to define micro-, meso- and macro-urban networks, which correspond to the shaded cells along the diagonal in Table 1.1. However, each combination of level and scale potentially defines a unique urban economic network, and each of these networks may be useful for understanding different urban economic processes.

Table 1.1 Combinations of level and scale to define classic urban economic networks

Level (aggregation of linkages and nodes)		Geographical scale (network boundary)		
		1 – Local	2 – Regional	3 – Global
A – Macro	Cities	City as a whole entity	Central place system (e.g., Christaller 1933)	World city network (e.g., Friedmann 1986 [1995])
B – Meso	Firms	Clusters (Porter 2000)	Regional economic integration and functional polycentrism (e.g., Scott 2001; Boschma 2005)	Global inter-firm network (e.g., Cohen 1981)
C – Micro	People	A neighborly economic support network (e.g., Wellman 1979)	A local labor market (e.g., Granovetter 1974)	The global economic elite (Sassen 1991)

A network defined at the micro level and local scale (cell C1 in Table 1.1) may capture a network of exchanges of resources among the neighbors in a particular neighborhood, thereby providing insight into how neighbors provide one another economic assistance. At a larger scale, a micro-level regional-scale network (cell C2) might capture employer–employee relationships in a metropolitan area, and thereby capture the exchanges that constitute a local labor market (Granovetter 1974). At a still larger scale, a micro-level global-scale network (cell C3) would include economic exchanges between all people everywhere, and thus would be simultaneously very detailed but also impractical to collect and analyze. A notable exception is Sassen's (1991) study of economic exchanges among a global sample of individuals, which provides insight into how economically influential individuals interact on the global stage.

A network defined at the meso level and local scale (cell B1 in Table 1.1) could capture the alliances and complementarities among firms in a particular district within a city, which may be useful for understanding industrial clusters that facilitate competitive advantage (for example, Porter 2000). At a larger scale, a meso-level regional-scale network (cell B2) may focus on the inter-firm exchanges of information throughout the region that promote innovation through proximity (for example, Boschma 2005) and create locally specialized systems of production (for example, Scott 2001). Finally, at a still larger scale, a meso-level global-scale network (cell B3) would include all the supplier–buyer or command–control relationships between firms around the world, allowing for exploration of the new international division of labor (for example, Cohen 1981).

An urban economic network with nodes defined at the macro level may not exist at the local scale (cell A1 in Table 1.1) because in such a network the nodes are whole cities, but studies considering the city as a whole entity, without scrutinizing the internal interactions between actors, may be viewed as falling within the scope of this category. A macro-level regional-scale network, capturing the economic relations between cities in a region, lies at the heart of economic-base and central-place theories (for example, Christaller 1933). Finally, a macro-level global-scale network, capturing the specialized economic links of all the world's cities, is the focus of research on the world city network (for example, Friedmann 1986 [1995]).

3. FINDING COMMONALITIES THROUGH SCALES AND LEVELS

The confusion between levels and scales arises because the space–time scope (scale) usually characterizes nested levels of organization. However, as explained previously, levels depend on the size and nature of the relationships (links) between individuals (nodes) that are considered in the network, while the scale is the boundary of a set of interacting nodes. Disentangling scales from levels makes clear the diversity of network approaches which attribute different roles and properties to cities in their contribution to network formation.

We explore this two-dimensional perspective in Chapters 2 to 29 of this volume. The explicit deployment of the levels supporting each chapter will underline emergence mechanisms appearing with the combination of multiple and dense interactions (section 3.1 in this chapter). After the levels of the study are identified, the definition of scales depends on the network's context; one or more scales can be adopted (section 3.2). Thus, the comparison and the classification of the levels and scales of the networks presented in this volume clarifies the diversity of city conceptualizations among these studies, enriching the mutual constitutive relationship between cities and networks.

3.1 Levels

The definition of levels starts with the initial lower level where the basic networks are considered (section 3.1.1). Upper levels underline these emergence mechanisms (bottom up or top down), that are the core of the different social sciences' contributions to the science of city (section 3.1.2). Micro/macro or micro/meso/macro suppose some nested levels, but non-nested levels are also possible, as well as multidimensional levels considering several criteria of aggregation (section 3.1.3).

3.1.1 Lower micro level of relations

The lower micro level is the basic frame of urban networks. Nodes can be individuals, actors or firms, with more or less specification on their interactions. For many chapter authors (Bettencourt, Chapter 3; Metcalf et al., Chapter 18; Bartholomew and Jones, Chapter 9; Christens and Cooper, Chapter 14; and Cachia and Maya-Jariego, Chapter 21), individual persons develop specific connectivity according to their socioeconomic status, which leads to individuals' activity spaces inside each city. For Janulis and Birkett (Chapter 17), individuals develop their networks according to their sexual preferences, norms and constraints, and their partners, which extend the conditions to define the basic networks. These micro networks link individuals and create their social proximity. This social proximity is not the goal per se for Batty (Chapter 19), but proximity serves to channel opinion influences from one person to another. More implicitly, the city is a set of interacting actors for Hanson (Chapter 5), Raimbault (Chapter 8), Pumain (Chapter 2) or van Meeteren (Chapter 12), but not studied per se. They use the (bounded) rationality of actors to explain the formation of urban processes, but without defining the individual properties and their relationship.

For Bogumil and Chase-Dunn (Chapter 4), diverse networks as merchant trade, intermarriage, political–military alliances and conflicts, and more generally communication explain the history of cities' settlements, while Parnreiter (Chapter 29) focuses on contemporary trade networks between producer service firms and their clients. The networks between firms are privileged in the inter-city networks whether the links would be ownership relations (Rozenblat, Chapter 25), greenfield investments (Wall and Kollamparambil, Chapter 26) or mergers and acquisitions between firms (Pain and Shi, Chapter 24). For O'Clery et al. (Chapter 23), firms are considered by their activity and linkages rely on co-located activities, so nodes are not the firms, but their activities.

Geographical space as location or movement can also constitute the basic frame at the micro level of the city. For Barthelemy (Chapter 7), interacting position nodes in the city constitute the micro-level framework. At the inter-city level, Ducruet (Chapter 22) considers maritime traffic as the initial network. Interestingly, other authors mix the movement and the social interaction between actors (Kourtit et al., Chapter 13; Anderson and Galaskiewicz, Chapter 15; and Burtner and Murray, Chapter 16). Thus, spatial and social movement combine very explicitly to produce urban life. Going a step further, Pieber and Quan-Haase (Chapter 20) propose that individuals interact with space through digital media. For them, direct social relationships disappear and are replaced by the indirect relationship of each individual with places. Here we are in the classic geographic debate, where we wonder if the basic urban framework is constituted by the social relationships in space or by the relationships between society and space. It is interesting that this old question arises with the new digital media and their implication for the relationship of inhabitants to their cities.

A final category of chapters considers the lower level as an already aggregated collection of individuals (which we could term meso level). In this category we can distinguish two sub-categories. First, some authors' approaches start at a unique aggregated level (for example, communities of actors formed by cognitive, institutional, organizational and social distances; Sigler et al., Chapter 11), while others start at an even more aggregated level of cities as a whole (Mahutga and Clark, Chapter 10; and Raimbault). Second, some approaches are rooted in the mix of two levels because the relationships are defined by belonging to a group. In particular is the instance of the interlocking model which assumes that linkages exist between members of each group. This volume presents different examples of the application of this model: inside cities with the bipartite network formed by people and associations in Newcastle city-region in the nineteenth century (Barke and Taylor, Chapter 6); intra or inter-firm networks (Pan et al., Chapter 27; Derudder and Taylor, Chapter 28; and Pain and Shi for London).

3.1.2 The meso and upper nested levels

Collective properties that define intrinsically the mutual formation of cities and networks emerge from network processes occurring in cities at a lower-level definition. Most of the studies presented in this volume use three levels (micro, meso and macro) or more that are often nested.

A typical example is the set of approaches described by Pumain, where individual interactions (micro level) create cities or city-regions (meso level) interacting with other cities in systems of cities (macro level), but numerous other schemes are possible. Barke and Taylor define the city using two nested meso levels: the city as a set of associations within which members interact by co-membership, and the city as interacting groups. Thus, these two meso levels constitute the capacity of Newcastle to specialize in coal exportation and to become an industrial powerhouse in relation to other cities at the macro level. Similarly, Pain and Shi, and Rozenblat, studying polycentric city regions, also consider two meso levels: local centers interacting to form the city region that is itself situated among other cities with which it interacts at the macro level.

Other examples mix social relationships and space to define the meso level: for Kourtit et al., public spaces represent the meso level where actors interplay; they name them 'platforms of interactions' or the 'microcosmic city'. Thus the 'smart city' (macro level) is a network of microcosmic constituents. For Burtner and Murray, the meso level is the aggregation of people by common ethnicity and neighborhood. Accordingly, Bettencourt considers cities as groups formed by social transactions and diffusion, or by infrastructure networks. These mixed approaches support the view that social and physical properties interact to form the city.

3.1.3 Non-nested levels

When multiple levels are defined, they are not necessarily nested. In their chapter, Bartholomew and Jones suggest that cities create a double embeddedness: social relations are embedded in a local structure of other relationships, which in turn is embedded in geographic space. Thus, they define four meso levels according to different types of ties: avocational associations, institutional ties, familiar strangers and eponymous relationships. These meso levels are not nested, but in a way they are arranged hierarchically by the intensity of the relationships they imply, from closer relationships for eponymous relationships to more distant relationships for avocational associations.

3.1.4 The city in inter-level interactions

Upper levels in all cases are assumed to create emergent processes between basic initial relationships. Direct and indirect collective mechanisms rise from the interacting multiple social networks. Each individual participates in several networks, transforming his or her behavior in a particular way for future interactions. The change of individual behaviors could transform the cities, since cities, with their dense social networks, are a strong path of diffusion for a new way of life and new practices (Hägerstrand 1952; Cliff et al. 1992). In turn, the previous cities' structure support or slow down individual actions. Dynamics of multiple networks at micro levels are embedded in the collective trend they create in a structuration way where individual actions and collective structure are intertwined (Giddens 1984). Assuming levels is like considering them as regulators of the maintenance of the systems forming cities. In transition stages, individual behaviors and their levels of emergence transform in new ways of regulation and the binding of actions.

Levels can be seen as solely social groups or solely as space, or a mix of both. In most of the chapters presented in this volume, the city is a system comprising a set of social groups or a set of places that are all assumed to have a consistent intensity and nature for the networks building them. Thus, a city represents some evolving structural properties that contribute to shape social practices of networking through these socio-spatial elements, and the city itself is transformed by the effect of the evolution of these networks.

Consequently, cities are the macro level if we consider only what happens inside the city, or cities are only an intermediate meso level if we assume that each city is a part of a system of cities that constitutes the macro level.

3.2 Scales

By the same diversity with which cities are assumed to constitute a particular level of emergent processes, cities are also situated among various scales, depending on the questions addressed. Remember that the scales are considered here as the extent of nodes and edges included in the network. Most of the time, scales are geographical, but some examples show scales that are not necessarily geographic or that go beyond a simple geographical scope.

3.2.1 Geographical scales

Most of the chapters adopt some nested geographical scales, while varying first according to their intra- or inter-urban perspective. For intra-urban approaches, most of the chapters have the lower scale as a neighborhood (scale 1) and the larger scale as a whole city (scale 3; Anderson and Galaskiewicz; Burtner and Murray; and Metcalf et al.), with sometimes an intermediate scale in between, such as district (scale 2; Kourtit et al.).

For inter-urban approaches, cities' boundaries constitute the lower scale (scale 1) which are situated inside hinterlands/regions/kingdoms/states/countries (scale 2), which is itself in the global system space (Pumain; Bogumil and Chase-Dunn; Hanson, Mahutga; Sigler et al.; and van Meeteren).

A third set of chapters broaden the scales from intra-urban to inter-urban (Bettencourt; Christens and Cooper; Pain and Shi; and Rozenblat). Here the lower scale is a part of the city (neighborhood or locality) and the city appears at a second embedding scale, itself interacting with other cities at higher scales that could comprise countries, which themselves are interacting in the global space.

3.2.2 Non-nested geographical scales

Geographical scales are not necessarily nested. For example, neighborhoods, third places, venues and activity spaces formed by the movement of actors (Bartholomew and Jones; and Janulis), or sets of locations with amenities (Batty), are distinct ways to qualify different spaces that are not nested spaces. In addition, Bartholomew and Jones propose the scale 'online' which concerns only intra-urban relationships, despite its worldwide dimension. So, the scales are used here to delineate the set of nodes and their relationships with a freer range of different meanings. The freedom of these authors is also visible when they address the different levels formed by social proximity, without any space, as a general premise of organization of the cities.

3.2.3 Non-geographical scales

The variety of scales is extended outside geographical space by O'Clery et al. when they apply, for each city, three types of distances between their firms, to define the network boundaries, following the evolutionary economy approaches of relatedness: (1) occupations with similar skill requirements using occupational data, (2) shared customers and suppliers using an input–output matrix, and (3) research and development proximity using patent citation data. Each city is seen as a whole, and networks according to the three 'proximities' are built between located firms (micro level) forming industrial clusters (meso level), together constituting the economy of each city. The unique geographical scale is thus the city (here US metropolitan statistical areas, MSAs), while other metrics are used to vary the scale in non-geographic dimensions.

4. MULTI-LEVEL AND MULTISCALE APPROACHES OF CITIES AND NETWORKS

The set of 28 chapters underlines the variety of possible definitions of levels and scales of networks when referring to cities. Disentangling levels and scales permits going beyond the assumption that larger levels correspond to larger scales.

4.1 Privileging the Diagonal of Levels and Scales?

As we observed at the beginning of this chapter, it is often assumed in network or system analyses that levels and scales would vary together privileging the diagonal of a table crossing both criteria. While this may often be a useful approach to city networks, many of the chapters in this volume illustrate that more nuanced approaches are possible.

Table 1.2 Levels and scales of the 'City of opportunity' (Kourtit et al.)

Level (aggregation of nodes and linkages)		Geographical scales (network boundary)		
		The city is considered within different geographical scales		
The city is considered as a level of organization		1 – Neighborhood	2 – District	3 – The whole city
A – Macro	'Smart city' – network of microcosmic constituents	Identification of critical transitions and vulnerabilities at the local and district scales of urban spaces		XXQ city – 'city of opportunity': decentralized, interconnected urban areas by network linkages and new information technology/city management
B – Meso	Public spaces where actors interact (platforms of interactions) – 'microcosmic city'	'Active space' ties in a capability approach where is built the 3H-city triangle (happiness, health and home)		Environmental resources: infrastructures and places
C – Micro	Interacting actors	Micro-behavioral, 'opportunity-orientated' and 'bounded rationality' perspective at local and neighborhood level. Physical and virtual proximity		Collection of big data

The conceptualization proposed by Kourtit et al. illustrates a classical intra-urban approach (Table 1.2). The city's inhabitants are seen as interacting actors (cell C1 in Table 1.2) for which the interactions could be intensified in 'active spaces' that are defined very locally and that emerge at a meso level (cell B1) in the '3H-city triangle' (happiness, health and home). This emergence is evaluated thanks to the collection of big data at the scale of the whole city (cell C3). What Kourtit et al. term the 'city of opportunity' (XXQ city) or 'smart city', is the network of these platforms of interactions emerging at the macro level of the whole city scale (cell A3). The city management of this process is based on resources situated in the active spaces of the whole city at the meso level (cell B3), while the identification of critical transitions and vulnerabilities is conducted by the whole city's government at the local and district scales of urban spaces (cell A2). Thus, the district scale does not seem active at the meso level of 'active spaces' (cell B2), as the district scale only participates in the general evaluation of their contribution to the whole XXQ city (cell A3). Then the diagonal is not complete, which applies to most of the chapters presented on intra-urban approaches.

Shifting now to an inter-urban approach, Bogumil and Chase-Dunn's chapter 'Settlement networks and sociocultural evolution' is an exemplary construction on levels and scales (Table 1.3). The authors define the basic linkages as the trade between merchants, intermarriage, political-military alliances and conflicts, and communication, all often implying institutions' political power or control. Before settlements became sedentary, every interaction remained isolated and fragmented in camps and migration circuits (cell C2 in Table 1.3) that occupied a larger regional space than cities. A meso

Table 1.3 Levels and scales of the 'settlement networks and sociocultural evolution' (Bogumil and Chase-Dunn)

Level (aggregation of nodes and linkages)		Geographical scales (network boundary) The city is considered within different geographical scales		
The city is considered as a level of organization		1 – Local	2 – Regions/states	3 – International/global
A – Macro	Settlements in core/semi-peripheral/peripheral hierarchies		Rise of networks that emerged to link settlements => early instance of semi-peripheral development/hilly flanks/chiefdoms	Global hierarchy of cities => multi-centric political structure of interstate networks
B – Meso	Settlements and institutions	First sedentary villages	Effect on cities' size and differentiation City-states	Large city-regions
C – Micro	Merchants' trades, intermarriage, political-military alliances and conflicts, and communication		Camps and migration circuits	

level emerged from the aggregation of relations created in sedentary settlements (cell B1). Interestingly, the portion of space is reduced in local spaces during this transition. Then an early instance of core–periphery relationships emerged from the interactions between settlements at the region/state scale (cell A2), in limited areas as there were no long-distance interactions. This favored some cities and not others, making the size of cities grow unevenly and attributing some of them functions of capitals and city-states (cell B2). At the global scale, the encounters of several systems having uneven development, increased the sociocultural complexity of the core/periphery model creating a 'multi-centric political structure of interstate networks' (cell A3). At the meso level, a few cities benefitted from this global integration creating large city regions (cell B3). Bogumil and Chase-Dunn specify that higher levels of complexity cannot emerge directly out of low levels, since scale expansion from local to regional and from regional/state to global participated together with levels, in this emergence. However, this scaling-up and levelling-up shift was not necessarily in the diagonal of the 3×3 matrix of levels and scales.

Finally, Bartholomew and Jones's chapter on 'Ties through place: socio-material network analyses in urban studies', where both social (levels) and spatial (scales) contribute in different ways to identity formation and social cohesion, offers an example that has neither nested levels nor nested scales (Table 1.4). The complementary scales and levels are described as multiple ways that intersections may occur across spaces and between the different types of ties, where unplanned encounters would occur. They have no hierarchy, but act as complementary social and spatial dimensions. Not all the cells of Table 1.4

Table 1.4 Levels and scales of the 'Ties through place: socio-material network analyses in urban studies' (Bartholomew and Jones)

Types of ties to form communities	Types of proximities to define spaces of interaction			
	The city is considered within different spaces			
	Neighborhood	Third places	Activity spaces	Online relationship
Avocational associations	Socializing frequency is distance dependent			
Institutional ties		It is the interactions that occur in institutional settings, such as schools, that lead to greater tie generation		
Familiar strangers		Frequency of meeting others with similar interests increases with the density of locations serving that interest in a given area	Friends are more likely to share urban space than random user pairs, and they also share more locations	
Eponymous relationships				

are described, since many social forms of community operate regardless of the types of space, and types of spaces create platforms of interactions whatever the type of ties. Online relationships are a particular form of social tie, which is not spatial, but still also contributes to the cohesion of the city or of parts of the city.

4.2 The City as a Level or a Scale?

The three examples serve to enlighten the general model sketched at the beginning of this chapter and to disentangle how scales and levels operate together, thereby illustrating, that the relationship between cities and networks involves numerous aspects of levels and scales. Cities form different types of communities with uneven contents, multiple natures of relationship, creating cohesions and communities by social, economic, cognitive, political, spatial, digital and other forms of proximity.

To the question 'Is the city a level or a scale?', we can answer that the city is both a level and a scale: it represents a social/economic/cognitive process that emerges in specific spaces, and the city can eventually be considered as emerging from its relationships with other cities that constitute a system of cities. It is both a top-down and bottom-up complex process evolving in relation to the transformations of social relationships and global trends. The urban space is strongly dependent on what is considered to be the basic constituent of this emergence, namely, the nodes and linkages of the networks. This is surely why cities' boundaries are so difficult to delineate in a universal way: rooted in sociocultural evolving relationships, it reflects the specific places of the organization of the society. As with every evolving system, cities' boundaries continuously move according to the transformation of its components, of its levels and of its scales. Scales of cities depend on the networks' scope of each city, leading some authors, following Lefevre (1970), to argue that urbanization would be generalized over the planet in a planetary urbanization (Katsikis 2018).

4.3 Role of Digital Networks in Cities (Levels, Scales)?

With the increasing role of digital networks, the city continues to evolve but also to reinvent social relationships in space. Several chapters deal with this new scale and how levels and spatial urban scales transform under the effects of these new communication modes that occupy increasingly more aspects of citizens' life. Digital networks both create a new space but also new communities, renewing the basic constitution of cities. Cities are not disappearing with these digital communications but, instead, will gain new strength and new imaginary symbols. Technology, culture and identity have new ways to convey citizen expressions and claims, mixing online social networks, locational digital tools for meeting or repulsion, and reviving traditional human feelings with new tools. Democracy, climate change and global issues of human development, health and well-being are some of the main issues that must be solved by networks and cities for the future.

5. PRESENTATION OF THE STRUCTURE OF THE VOLUME

This volume explores the state of the art of the intertwining of cities and network relationships. The individual chapters represent an interdisciplinary set of collaborations introducing a large variety of approaches, in levels and scales of networks, often going outside the traditional co-variations of levels and scales we discussed previously. The chapters are grouped in five parts:

Part I – Theoretical conceptions of cities and networks
Part II – Cities and networks in the history
Part III – Methods and models of city–network interactions
Part IV – Network processes within cities
Part V – Network processes between cities.

This clean grouping belies the diversity and depth of the chapters as each chapter often addresses multiple issues (for example, theory and method, or intra-urban and inter-urban) and could have appeared in multiple parts. However, each chapter contributes to some understanding of the relationships between cities and networks, constituting a kaleidoscopic contribution which reflects the complexity of these relationships. We hope that the volume will be useful for scholars of all disciplines who would like to delve deeper into one or several dimensions of cities' and networks' mutual constitution, and that this plurality of approaches will open large avenues for further multidisciplinary studies.

ACKNOWLEDGEMENT

We thank all the cited chapters' authors who reviewed this chapter, helping to make this introduction more consistent with their own chapters. However, the responsibility of the content developed in this introductory chapter remains with the authors.

REFERENCES

Berry, B.J. (1964), 'Cities as systems within systems of cities', *Papers of the Regional Science Association*, **13** (1), 146–63.
Boschma, R. (2005), 'Proximity and innovation: a critical assessment', *Regional Studies*, **39** (1), 61–74.
Butts, C.T. (2009), 'Revisiting the foundations of network analysis', *Science*, **325** (5939), 414–16.
Christaller, W. (1933), Die zentralen Orte in Süddeutschland: eine ökonomisch-geographische Untersuchung über die Gesetzmässigkeit der Verbreitung und Entwicklung der Siedlungen mit städtischen Funktionen. University Microfilms. English translation 1966, *Central Places in Southern Germany*, trans. C.W. Baskin, Englewood Cliffs, NJ: Prentice Hall.
Cliff, A., A. Pred and T. Hägerstrand (1992), Classics in human geography revisited: Hägerstrand, T. (1967), 'Innovation diffusion as a spatial process. Chicago: University of Chicago Press. Translation and postscript by Allan Pred', *Progress in Human Geography*, **16** (4), 541–4.
Cohen, R.B. (1981), 'The new international division of labor', in M. Dear and A.J. Scott (eds), *Multinational Corporations, and the Urban Hierarchy, Urbanisation and Urban Planning in Capitalist Society*, London: Methuen, pp. 309–44.
Friedmann, J. (1986), 'The world city hypothesis', repr. 1995 in P.L. Knox and P.J. Taylor (eds), *World Cities in a World-System*, Cambridge: Cambridge University Press, pp. 317–31.
Granovetter, M. (1974), *Getting a Job: A Study of Contacts and Careers*, Cambridge, MA: Harvard University Press.
Giddens A. (1984), *The Constitution of Society: Outline of the Theory of Structuration*, Berkeley and Los Angeles, CA: University of California Press.
Hägerstrand, T. (1952), *The Propagation of Innovation Waves*, London: Royal University of Lund, Department of Geography.
Katsikis N. (2018), 'Visualizing the planetary urban', in J. Harrison and M. Hoyler (eds), *Doing Global Urban Research*, London: Sage, pp. 12–33.
Laumann, E.O., P.V. Marsden and D. Prensky (1989), 'The boundary specification problem in network analysis', in L.C. Freeman, D.R. White and A. Kimball Romney (eds), *Research Methods in Social Network Analysis*, Fairfax, VA: George Mason University Press, pp. 61–87.
Lefévre, H. (1970), *Le manifeste differentialiste*, Paris: Gallimard.
Neal, Z.P. (2013), *The Connected City. How Networks Are Shaping the Modern Metropolis*, New York: Routledge.
Openshaw, S. and P.J. Taylor. (1979), 'A million or so correlation coefficients: three experiments on the modifiable areal unit problem', in N. Wrigley (ed.), *Statistical Applications in the Spatial Sciences*, London: Pion, pp. 127–44.
Pavé, A. (2006), 'Biological and ecological systems hierarchical organisation', in D. Pumain (ed.), *Hierarchy in Natural and Social Sciences*, Dordrecht: Springer, pp. 39–70.
Porter, M.E. (2000), 'Location, competition, and economic development: local clusters in a global economy', *Economic Development Quarterly*, **14** (1), 15–34.
Sassen, S. (1991), *The Global City*, New York: Princeton University Press.
Scott, A. (2001), *Global City-Regions: Trends, Theory, Policy*, Oxford: Oxford University Press.
Wellman, B. (1979), 'The community question: the intimate networks of East Yorkers', *American Journal of Sociology*, **84** (5), 1201–31.
Wolf, L.J., S. Fox, R. Harris, R. Johnston, K. Jones, D. Manley et al. (2020), 'Quantitative geography III: future challenges and challenging futures', *Progress in Human Geography*, May, doi:10.1177/0309132520924722.

2. From networks of cities to systems of cities

Denise Pumain

1. INTRODUCTION

Philosophy then social sciences took an interest in defining what a city is and finding ways for improving its political organization, shaping its esthetics or developing its performance. Many theories flourished that provide a vision about what 'a' city is or how it should be. For example, economists constructed, since the first suggestions by Marshall (1890), a theory of 'the' city as a means of offering agglomeration economies that provides a subset of diverse externalities favorable to enhance the functioning and development of firms (Fujita et al. 1999). Among all the social sciences, geographers were the first, followed by historians and archaeologists, to engage in searching for a theory that is not that of 'the' city but which readily admits that cities form in networks connecting cities of differentiated size and function (Peris et al. 2018).

Building a theory is a long and complex process. Ideally starting with observation, measurement, following a formalization in logical, mathematical or computing models, then validation through many empirical tests on international comparisons, theory also interferes with available data, technologies, current scientific paradigms and ideological backcloth of the time during the course of history. Many controversies and personal or political rivalries take their place in what is by no means a linear and regularly constructive process. We provide here a very simplified historical view by reducing the emergence of urban theories to three main steps: first, the observation of their relationships with territories and trade networks; second, how they integrated observation of socio-economic changes within concepts of systems dynamics and co-evolution; and third, how they may take further inspiration from the recent availability of not yet fully explored data about real-time interaction networks at all geographical scales.

2. TERRITORIAL RATHER THAN RETICULAR PRINCIPLE

The first axiom of geographical urban theories is that cities are never isolated, but are organized in systems of interconnected cities, over territories of various large dimensions (Berry 1964; Rozenblat et al. 2018; Pumain 2020). Based on observations of cities over the very long term, whatever the forms of political and economic organization of societies, it seems established that cities never appear in isolation, but always (or with a very high statistical frequency) as sets of cities connected by different types of relationships and exchanges (Renfrew and Poston 1979; Fletcher 1986; Liu 1996). Trade and/or violent predation have enabled cities to accumulate wealth in a more efficient way than villages were able to after the 'Neolithic revolution' (storage and then the invention of agriculture), which had already multiplied population densities by a factor of 100 compared with hunter–gatherer populations (Marcus and Sabloff 2008).

Cities differ from villages. Even if their definition may vary according to cultural areas and countries of the world, they are generally characterized by the importance of the non-agricultural activities that developed there. Cities have freed themselves from the constraints of resource limitations and uncertainties in their immediate environment (their site) by establishing connections with other sites in order to find complementary resources and reduce the effects of local hazards (Reymond 1981). In this way, they build up a 'geographical situation', defined by accessibility to other sites and other types of resources, which enables them to resist the consequences of natural hazards (earthquakes, climatic accidents and epidemics) or conflicts (wars and predations) occurring on their site and in their neighborhood. Most important, thanks to higher densities, interactions generated by proximity within and between cities increase the probability of innovations appearing through the exchange and hybridization of local inventions, and give cities powers for capturing and accumulating resources far greater than those villages provide.[1] This may explain why the urbanization process has profoundly transformed the settlement of human societies on the Earth's surface with the progress of technologies and complexity of societies since the emergence of the first cities around 10000 years ago. From multiplying the probabilities of fruitful interactions and widening their range, the spatial organization of cities including multilevel networks is a partly endogenous engine for ensuring urban development (Schmitt 2014).

In the spatial patterns of emergence of these sets of cities, two organizing principles are classically identified, each corresponding to one of the two main definitions of promoting spatial interactions through geographical proximity and accessibility: the territorial principle, which is based on neighborhood proximity, measured in kilometers; and the reticular principle, defines proximity by the capacity to access communications in a network and is measured by indices of connectedness. Whatever the historical era or the geographical area, these two principles are always closely associated in all practices related to proximity. Close neighborhoods develop with networks of roads that allow for pathways and routes between resources and habitats, and long-distance networks are only effective if they link territories between which there is an exchange of resources, population and information. Depending on the place and time in history, particular types of relationships may have prevailed in the formation of cities and the systems they build (Pirenne 1925; Vance 1970; Bird 1973; Braudel 1979). However, some urban theories choose to insist on the role of networks, while others choose to leave it implicit (Pumain 1992; Camagni 1993; Pumain and Robic 1996).

2.1 Cities and Territory: The Implicit Network

Cities participated in the spatial organization of territories, that is, parts of the Earth's surface that were dominated by a sub-group of human population that used its resources and designed its specific linkages in the organization of society. Within the ancient representations in relation to military strategy, political domination, organization of agricultural markets and distribution of goods, cities are part of a territorial coverage. The city is considered here in a process of paving the geographical space. The city controls a region, disposes of it or serves it; it dominates a territory.

In ancient Europe and Asia, cities were defensive places protected by walls that were, together with other signs of urbanity (such as wealth, beauty and hygiene), one of the signs of the grandeur of the place. As early as the seventeenth century in France, the defensive system changed scale and was transferred from the locality to the kingdom (Pumain and Robic 1996). Vauban built a series of fortifications along the borders, classified in three orders: the fortified squares surround the *pré carré*, a continuous territory whose limits were defined. The theoretical representation of a system of military squares organized to protect the whole territory was formulated later in the first half of the eighteenth century, including a spatial model of the spacing between places, the hierarchy of sizes and their depth arrangement. The speed at which troops moved, the time it took to transport food and ammunition, and the duration of resistance to siege, all of which are themselves inducers of seat size, governed the framework of this system. The organization of military space tends to control the organization of civil space. For a number of theorists, the hierarchy of war zones is in line with that of the kingdom's cities. The communication routes, essential to the functioning of the strategic system, are subject to particular control by the engineers. Under the Restoration, strategic thinking became even more precise by underlining the problems of interdependence and hierarchy between the points of the territory. Moreover, it is in military art that the notion of 'network' (the 'plan of the permanent network of fortifications', 1821, quoted by Guillerme 1986, p. 31) appears at the same time.

Similar to the art of war, reflection on political or administrative domination has generated a theoretical representation of the city as the center of a paving of space. The capital is, par excellence, the city that ensures territorial control: the center of concentration of power, the place of demonstration, and the node of information systems that cover the country from the center to the periphery. Synonymous with asymmetrical relations, it is the metropolis, the head, whose impulses are transmitted in small parts by a series of capitals. This view is well illustrated by Marco Polo's description of the Empire of the Great Khan according to the main cities and political articulations (Deluz 1989). He accompanied his description of the cities with a qualifier that roughly corresponds to their rank in the administrative hierarchy of China.

No political theory of cities was formulated during the French Revolution; the constituents' debates on departmentalization focused primarily on the size of the constituencies and paid little attention to cities, in an anti-urban attitude that prolonged the attitude of the eighteenth century (Ozouf-Marignier 1989). The representations of the city brought into play are of two types: first, the cultural, traditional representation, which emphasizes urbanity, the quality of the elites gathered as much as the beauty and healthiness of the place. A more novel functionalist representation emerges at the same time which, by emphasizing the role and activities of the city, takes into account the city's relationship with its environment, through jurisdiction and centrality. This structure of vernacular representations and representations of the constituents is consistent with the interpretations by Bernard Lepetit (1988) of the spatial structures identifiable in the eighteenth century, cities are for the most part locally anchored; they constitute a whole without systematic dependencies and with a low degree of hierarchy.[2] It is with the Saint-Simonians and the construction of railways at the beginning of nineteenth century that this perspective changes.

2.2 Urban Hierarchy as Described with Central Place Theory

Reflection on the city as a market serving a specific catchment area or rural 'fallout', which in turn supplies the city and supports its growth, is very old and prevailed long before achieving the overall formalization of the service of a territory by cities. The latter is rooted in theories from, for example, the economist Cantillon (1755 [1952]), or von Thünen's (1826) theory of the isolated state (1826). Von Thünen situates the organizing role of the city by showing the dependence of the agricultural territory on its accessibility to the market center, responsible for the concentric structure of land use around it, but his model does not consider the urban repercussions nor the relationships established between neighboring sub-systems.

During the twentieth century, the emergence of the notion of services (and 'tertiary' activities according to Colin Clark's 1940 classification) encouraged the flowering of research on areas of urban influence. Many empirical studies took place during the interwar period, whether in an academic framework in Europe (Great Britain, France and Germany, in particular) or in the framework of applied research (the USA); the bibliography accompanying the theory of central places proposed by Robert Dickinson (1934) testifies to this abundance of research on the service areas of American cities.

In a similar way, the major systematic studies of regional urban networks carried out by European geographers, even until the 1960s, still give priority to the study of city-to-rural relationships rather than city-to-city relationships. The idea of a total coverage of the territory by cities underlies the descriptions that geographers made in the middle of the twentieth century, despite their use in French of the expression 'urban network'. At that time, the 'network' of cities was almost never studied as a set of nodes and lines. In a perspective inspired by Marxism, which characterized the city as a predator, that is, a landowner taking benefit from the rent, the question of rural levies underpins empirical investigation, through the study of land ownership. Thus, in many works, a representation of the city rooted in a rural space guides the conception of the coverage of the territory by cities. However, it was less a question of land rent than of commercial exchange and consumption of services that was to give rise to the theorization of groups of cities as hierarchies of central places. Even if Saint-Simonian Jean Reynaud (1841), 'discovered' by Robic (1982), invented the theory, it is the geographer Walter Christaller (1933) who a century later stands out as a symbol.

Christaller's central places theory aims to explain the number, size and spacing of cities (Christaller 1933). Its major postulate is that cities organize the territory to which they provide services in a generally regular spatial and hierarchical pattern of centers characterized by different levels of urban functions. The example of the medical service supports this demonstration. Centrality is defined by the exercise, in a city, of an activity intended for an external clientele, located in the complementary region, contiguous to the city. The spatial range of the service, based on the transportation cost that consumers agree to add to the cost of the good requested, and the threshold of appearance, the volume of clientele necessary for the survival of the service supply, regulate the spacing of the centers. Low-cost or frequently used goods have smaller ranges than more expensive or less commonplace goods. Goods and services of comparable scope are grouped in the same centers, and centers at a given level have all the functions of lower-level centers, plus a few more. From circular, the service area becomes hexagonal to account for the vicinity of other centers at the same level without overlap. The number of centers and their arrangement may differ slightly according to three different criteria. The market principle provides the best possible accessibility to the centers for customers. The transport principle leads to a smaller number of centers arranged in a way that limit the length of communication routes needed to link the centers. The administrative principle characterizes the inclusion of an entire lower-level administrative district in the upper-level administrative district, instead of sharing the clientele between competing centers as in the previous two principles. On the maps drawn up by Christaller for southern Germany or for the whole of Europe, as on the diagrams of his models, the boundary between urban catchment areas is always strongly drawn. The complementary regions (or market areas) are disjointed and constitute a complete paving of the territory. In his first applications of the gravity model, Reilly (1931) sought to calculate the location of this boundary. It was only much later that their probabilistic (Huff 1963) or blurred (Ponsard 1977) character, or even the possible discontinuities of the areas of influence, were insisted upon.

The urban hierarchy as conceived in central place theory derives from the functional levels rather than from the spatial organization of the nodes. The hierarchy of cities is thus a different notion from the concept of hierarchy used to describe the organization of institutions, linked to the exercise of a specific power or attributions. The urban hierarchy is mainly a differentiation of functional levels. This notion is based on the recurrent observation in empirical studies of the presence of specific facilities, shops or services in specific cities, and that these centers have several levels, graded by the frequency of appearance, the rarity of use and the spatial scope of these functions, by the size of their zones of influence, and by the diversity of the functions gathered. Generally, but not always, the rarity and diversity of functions are greater the larger the size of a city. This essential property of urban networks has been recognized as fundamental since its first mention in the literature. Jean Reynaud (1841) based his description of what he named the 'system of cities' on the existence of several levels of services. While the inhabitants of villages group together under the impulse of 'sociability', craftsmen concentrate in one of these villages, 'not only because of the pleasure they have in being together', but also 'because of the continual need they have for each other' (Reynard 1841, p. 678), what we would now term agglomeration economies. Placed at the center of a hexagon in which it serves several villages, this village of artisans is itself, together with others, part of a higher-level center for the use of less frequently used services. The author believes that three or four levels of centers between the simple village and the capital of a large country can be detected.

The two major principles of central place theory – central supply of services and grouping of services at the same level in the same centers – are the origin of a nested hierarchy of levels of functions, the design of which has greatly guided the theorization and analysis of the regional groups of cities that French-speaking geographers will term 'urban networks' in the 1960s. In the formulations of Christaller (1933), in the experimentations carried out by Berry (1967), after Lösch (1940) in the Iowa plains, in the mathematical models proposed by Beckman and McPherson (1970), in the many works on regional urban networks and in tests of the theory quoted by Berry and Pred (1965), or Parr (1985), urban hierarchy is almost never deduced from the analysis of city-to-city relationships, from the observation of flows or exchanges between cities or, even and especially, from direct measures of dependence, domination or 'command' (despite the sporadic use of this latter term in the literature). Indeed, the ordered levels of the urban hierarchy are deduced from the degree of rarity and spatial range of their activities. The facts of polarization, that is, the organization of flows centered on cities, are most often used only to measure the extent of areas of influence and to help identify levels among centers. At most, a few analyses mention the concentration of decision-making power in particular higher levels of the hierarchy, for example, as regards banking power. These allusions are more frequent in connection with administrative organization and comparisons between urban networks that are generally centralized or dominated by a particular metropolis or capital city.

However, the nesting of the levels, as shown by observations, is not as perfect as in theory. There are overlaps of competence, equipment is lacking in some centers where it should normally be and contrarily is present in centers that are too small. Brian Berry (1967), Richard Preston (1991), Reginald Golledge (2002) and many others have discussed this 'vagueness' in the hierarchy of central locations. Seeing the urban hierarchy as a set of centers emanating from a territory that they frame and serve cannot fit with all observable urban networks. For example, specialized cities, such as major seaports or mining and steelmaking centers, were sometimes considered in the literature of that time as off the grid. The specialization of a limited number of cities in activities that are not trivial services to the population is, however, a characteristic property of urban networks, admittedly of secondary importance but just as universal as their hierarchical organization (Paulus and Vacchiani-Marcuzzo 2015). In Lösch's models, postulated by the economic base theory (Hoyt 1961) as well as by that of growth poles (Pumain and Torre 2020), specialization has nevertheless remained, until recently, poorly integrated into the territorial theory of urban networks, perhaps owing to the lack of an explicit formulation of the nature and form of the relationships that are established between cities. It is only recently that the improved availability of data on a variety of inter-urban flows enabled mapping of the multilevel spatial organization of French cities in different types of sub-systems according to the range and nature of their interactions (Berroir et al. 2017). In section 3.3 of this chapter we see how taking into account the dynamics of urban systems and modeling the genesis of their structure through their socio-economic interactions necessitate a description of the inter-urban networks that do not belong to the common customer services observed in central place theory, but are generated by more specialized urban activities.

In summary, these formalizations of the 'urban network' are partial in the theoretical field they address, that is, the service of the population living in the territory, but also in their formal orientation, which neglects the interrelationships between the nodes of the network. The ambiguity of using the expression 'urban network' in the work inspired by central place theory is apparent. Although Christaller (1933) names one of his models a 'transport model', its purpose is only to see which effects an economy in the construction of traffic routes has on the spatial pattern of cities. The model does not specify the possible hierarchy of roads by traffic, nor the intensity of the relationships. It is also revealing that the concept of centrality, which could connote a position in a network, was explicitly defined by Christaller (1933) as the ability of a city to offer goods and services to an external clientele located in the 'complementary region' adjacent to the city. The criticisms addressed to central place theory today largely concern its reference to the interpretative form of 'territorial coverage', which leads to explaining the hierarchical and spatial form of an urban network by its function of controlling and serving a territory. As the majority of the population now lives in cities, some people believe that it is an aberration to conceive a theory where the entire urban organization depend on the services provided to rural populations in complementary regions. Other critics question an oversimplified representation in this theory of the behavior of actors, close to that of homo economicus minimizing transport costs for each product. Several authors have endeavored to study the consequences of more complex behaviors, such as that of multiple-purpose purchasing trips, on the spatial forms of urban networks, and more generally to make the foundations of the theory more explicit at the individual level, that is, those relating to the behavior of households and businesses (Beguin 1988; Fujita 1989). Moreover, the presence of specialized functions, whose logic of location is a-hierarchical, may explain why, instead of distinct levels as expected from central place theory, there is always an almost continuous statistical distribution of city sizes (Beguin and Thisse 1979). Finally, central place theory is criticized for its static character; the hierarchy of cities and the models of spatial organization correspond to a situation of equilibrium in which a territory is shared between competing centers.

However, in the previous century, both Jean Reynaud and the geographer Kohl (1841), whom Christaller quotes, had strongly emphasized the interdependence between traffic routes and cities as trade nodes. The conception of the city as a node in a communications network is another very old representation of cities.

2.3 Cities as Stops on Long-distance Routes

The reticular principle is advocated for explaining the historical emergence of subsets of cities where interdependencies have been built up, less in connection with the resources of a local or regional territory than on the enormous profits derived from the exchange of rare products over very long distances. These exchanges have spatial ranges that are out of all proportion to the smaller scopes of the most frequent spatial interactions permitted by the state of transport technology at a given time (Renfrew 1975). Cities thus also formed in networks of stops along routes, based on exchanges of expensive and coveted products, such as the many cities along the 'Silk Roads'. They form urban networks staggered along traffic routes, the location of which often corresponds with major accidents of geographic relief that have from the outset facilitated passage or navigation (Mackinder 1890; Pirenne 1925). There are numerous examples, as the Phoenician trading posts (1200 BC) including Tyre in Lebanon, and towns in Cyprus, Crete, Sicily, Malta, Cadiz in Spain, or Carthage, or the Venetian trading posts of the thirteenth to fifteenth centuries also spread around the Mediterranean. The Hanseatic towns of the twelfth to seventeenth centuries along the Baltic Sea were an example of multilateral network creation through trading contracts, whereas the cities of triangular trade associated with the European colonial empires from the sixteenth to the nineteenth century formed a network between Africa, North American and Europe. More recently, the chain of ports along the maritime container route in Asia (Ducruet 2008; Ducruet et al. 2018) or even to the nodes of the world financial markets that complete the daily transaction time in only three major stops in very powerful economic centers (New York, Tokyo and London) are examples of how functional networks can shape urban patterns. In these networks of cities, the trade links build interdependencies that ensure the development of cities for as long as the exclusivity and profitability of these specialized exchanges lasts.

The effect of the reticular principle also appears in what has been characterized as the constitutive duality of urban networks in developing countries having undergone colonization (Taaffe et al. 1963). A former, generally regular, network of central places founded by local markets and polities according to the territorial principle has been superimposed by a network of disproportionately large urban centers, usually seaports, whose development is associated with an economy of a much wider scale (for example, three of India's four largest metropolises are of this type (Swerts et al. 2018)). The exchanges associated with these exceptional urban developments differ from those of the former networks of continental or maritime stops along roads by the strong qualitative and quantitative asymmetries of center–periphery relationships. These cases result in a macrocephaly characteristic of many former colonized countries (Moriconi-Ebrard 1993).

When countries were colonized and settled by a majority European population, such as the USA (Bretagnolle et al. 2018), or even a lesser fraction, as in South Africa (Baffi et al. 2018) for instance, the origin of cities is frequently associated with the development of seaports and then roads or railways networks. In these New World countries, the development of local market and administrative centers, according to the principles of central place theory, occurred historically along pioneer fronts at the same time as, or even after, the generation of the first major cities, through the effect of long-distance trade and political relationships. The territorial and reticular principles are always mixed, to a greater or lesser extent, in the genesis of urban networks. However, although Jean Reynaud or Walter Christaller envisaged particular modalities of the evolution of urban networks, posterity has hardly retained a mention of these hypotheses, probably because their conceptions were not sufficiently based on the relational properties of cities, which are studied by theories that assume another form of representation of urban networks.

2.4 Emergence of Network Thinking through Interurban Competition

It is remarkable that despite the general acknowledgement of the role of networks in urban development, very little attempt at formalizing the relationship appeared before the end of the twentieth century. Lalanne (1863)'s attempt at formalizing the relationship between the size of urban centers, their distances and the newly designed railway lines is an exception A few analytical studies discussed the direction of causality between urban growth and the development of railways (see, for instance, relating to France, the alternative results in Pumain 1982b; Bretagnolle 2003; Mimeur et al. 2018; and Raimbault 2018). Physical or technical networks alone cannot explain urban developments, and only multivariate approaches including urban complexity could help in further developing the contribution of networks to urban theory.

We must mention here a precursor remarkable as much for his erudition as for the complexity of his reasoning, which was forgotten for centuries. Giovanni Botero (1588 [1948]) introduced a very comprehensive vision while exploring the 'causes of the beauty and greatness of cities' (Gaudin and Pumain 2000; Pumain and Gaudin 2002). He provided a formalized verbal description of their multi-dimensional capacities and mutual adaptation in a context of fierce inter-urban competition. Giovanni Botero listed a series of factors conducive to urban growth: authority, strength, the withdrawal of population and wealth from neighboring regions, and utility, whether in the convenience of the site, the fertility of the surrounding area or the ease of access (by land and sea). Through a systematic comparison carried out since antiquity and on a global scale, drawing examples from Brazil and China as well as from Europe and Africa, Botero emphasizes that each of these properties is necessary, but not sufficient, for the development of a large city. By showing that natural dispositions are reinforced by human artifice, he is led to outline which specific activities, in competition with other cities, are the basis for the size of cities and under what conditions these activities function well. These are the residence of princely or religious powers ('cresce a proporzione del dominio' – 'it [the city] . . . grows in proportion to power' – Botero 1948, p. 402); the exercise of justice, the relationship with which increases with the level of jurisdiction, knowledge and the necessary accompanying freedoms; industrial production or the freeing of commerce from taxes; and the development of traffic routes. It then sets out the conditions for what is currently named the 'sustainability' of urban growth, in terms that recall the current formulation of dynamic models. Beyond the universal trend towards growth in human numbers, Botero (1948, p. 405) links the growth of a city to its potential capacity (*'dalla nutrivita d'esse città'*), the latter being itself determined by its attractiveness, and limited by random disturbances (famines, plagues, wars, and so on) and by competition from other cities. The growth of a city is thus dependent on the specialization of its trade in a product sought after by its quantity or quality (termed later in the literature 'comparative advantage'). It tends to be self-replicating – *le cause della produzione delle cose et della conservazione loro sonn l'istesse* ('the causes of production are the same as the causes of conservation') (Botero 1948, p. 405) – and lasts as long as justice, peace and abundance reign. The city, however, is a brake on its own growth, since its large size causes increasing supply difficulties.

Expressed in these terms, Botero's theory of urban growth is reminiscent of the current formulation of systemic theories of cities. Can it be said that all it lacks are the equations for a dynamic model transcription?

The notion of a system of interdependent cities has only been explicitly formulated very gradually. We can quote here the development of organicist views by a few major European geographers at the beginning of twentieth century (Pumain and Robic 1996). Harold Mackinder (1902), when examining the evolution of British cities, describes three successive spatial and functional modes of operation: first, the function of local crossroads; second, mining or metallurgical towns based on the exploitation of localized resources; and third, cities that specialize in more diversified functions (tourism, residence, transport, and so on). Most importantly, and differing from previous generations of cities, their location does not correspond to a site of resources, but instead to a situation within an urban ensemble. Consequently, the expression 'ensemble of cities' is no longer adequate: it is a system of interrelationships that is involved, creating 'federations of cities'. Without being systematic, the explanation of this evolutionary conception, inspired by a theory of the evolution of species, is based on technical modernity ('steam and electricity are changing our geographical conceptions'; Mackinder 1902, p. 330). It underlines the change in scale of spatial organizations and their increasing complexity. Thus, urban forms are changing so that they are no longer based on the spatial continuity of the built environment, but on coordinated systems of settlement, specializations and complementarities. In an article devoted to the evolution of cities, Elisée Reclus (1895) has explored this urban structuring around the London conurbation, which projects not only ring-shaped suburbs around the city, but also specialized satellites. Only the spatial integration linked to the space–time contraction allows this organization (Bretagnolle et al. 1998). A few years later, Patrick Geddes (1915) believed that this new form of settlement organization should be named 'conurbation', a term he created to designate the new city-regions that broke the old city–country opposition by creating vast, interconnected entities comprising several centers linked by high-density communication routes. Perhaps less acute in France, where the notion of agglomeration is sufficient to describe new forms of urbanization, geographical thinking at the beginning of the twentieth century nevertheless theorized the emergence of new urban regional forms. Vidal de la Blache notes the modern trend towards metropolization (the concentration of populations and activities in large cities) and the regrouping of cities, ports and industrial centers. According to him, these new phenomena of 'concentration' break with the classic constraint of urban spacing. To interpret them, he constructs a notion which, borrowed from Mackinder, does not depart from the idea of the crossroads. Vidal de la Blache's 'nodality' is conceived as a local emergence, linked to a particular gathering of functions. Through this, the city brings together regional activities, and links this productive space with its markets. The nodality of a city is the 'reunion of all the auxiliaries required by commercial and industrial life' (Vidal de la Blache 1911 [1913], p. 8). It is not measured by the population, or still less by the number of civil servants, and would require a distinction of the activities that fall within its scope. Typical of the dynamic 'regional city' and the industrial metropolis, nodality is a characteristic of modern economic functioning.

While Vidal de la Blache's work is the seed of the idea that would be expressed currently as the emergence of a level of organizational complexity, stemming from the life sciences, several biological metaphors are explicitly associated with the formulation of phenomena of interdependence between cities, whether as organic comparisons (the nervous system in Geddes, for example) or the assimilation between the workings of life and economic and social processes. The organicist metaphor proves to be particularly appropriate to these conceptions of the city, which see the city as a participant in a totality. This organicist metaphor is recurrent; it has never completely disappeared. The notion of urban function used by geographers or town planners comes from this perspective, just as clearly as more obvious propositions, such as statements that cities are born, live and die, or even have a soul, an identity.

3. SYSTEMS OF CITIES, COMPLEXITY AND CO-EVOLUTION

The inspiration for conceptualizing sets of interdependent cities shifted from biology towards physics during the second half of twentieth century. The expression 'system of cities' appears as early as 1841, in text by Jean Reynaud. However, the theoretical content of this notion has evolved significantly. In the nineteenth century, the use of this term denotes a relatively rigid organization that has a purpose and a function. An example would be the blood system or the nervous system. The substance of the notion is to emphasize the interdependencies between parts of the system, which are related by common development or regulation, and which form a whole that is more than the sum of its parts. Models from physics or cybernetics were disseminated later in the social sciences about settlement systems. Jean Gottmann (1947) recalled the temptation of a social thermodynamics, which Gibbs's work at Yale in the 1910s gave rise to, and suggested that we should take inspiration from it in order to establish a methodology for analyzing interdisciplinary crossroads.

The real transfer of the concept took place during the 1960s. In 1964, Berry, with his famous expression 'cities as systems within systems of cities', emphasized the two relevant levels of organization and observation: on the one hand, the city, considered as a set of neighborhoods, institutions or actors in systemic relations, and on the other, the system of cities, organized by the principles of central place theory. Alan Pred (1973) made historical analyses of innovation diffusion and growth in systems of cities; he also scrutinized the evolution of variegated urban interactions in developed countries (Pred 1977). He identifies the notion of 'city system' by applying a very general definition as:

> a set, national or regional, of interdependent cities in such a way that any significant change in the economic activities, occupational structure, income or population of one of the cities directly or indirectly brings about any change in the economic activities, occupational structure, income or population of one or more other elements of the whole. (Pred 1977, p. 13)

However, subsequent empirical research on how city systems evolve was to clarify this proposal and in particular the notion of 'significant change'. Indeed, it is not just any local transformation that is likely to affect the other cities in the system.

We recall here briefly how an evolutionary theory of urban systems (Pumain 1997) emerged from that long research history. Formalizing the evolution of cities within a theory that is embedding most of the empirical observations about their organization and evolution required three logical steps: first, linking the distribution of city sizes with models of urban growth; second, linking the transformation of socio-economic profiles of interdependent cities with the diffusion of innovations and scaling laws (Pumain 2004); and third, using simulation models for reconstructing these observations and testing the validity of the theoretical hypotheses on a diversity of cases.

3.1 Distribution of City Sizes and Urban Growth

Recurrent observations about sets of cities over historical times brought evidence of a paradoxical persistency of a specific structural identity of a group of cities over time, despite the vagaries of demographic and economic growth.

> While the urban population increases in quantity and proportion to the total population, while the number of cities and their size continue to grow, while this process is accompanied by very significant changes in productive systems and in the living and housing conditions of populations, we observe a very large persistency in the relative positions of cities in the hierarchy of sizes and a persistence of functional specialization among urban areas. (Pumain 1982a, p. 7)

Whereas according to Walter Christaller (1933, p. 82) the statistical regularities in the distribution of city sizes coined by Auerbach (1913) and Zipf (1941) were merely 'playing with numbers', an abundant literature emerged after them in connection with systemic theories and complex systems (Cottineau 2017). Zipf's (1941, pp. 417–41) 'rank size rule' raised several statistical variants that were often tested on data that lacked rigor in the definition and delineation of urban entities. The diversity of the results obtained led to misunderstandings and uninformed discussions, which I have reported on several occasions in subsequent writings (Pumain 2006, 2012).

However, few authors at that time had considered the model of settlement size hierarchy proposed by Gibrat (1931), according to a lognormal distribution, which has the advantage of being testable as generative process. The model links the shape of the distribution of city sizes to the process of randomly distributing the growth between cities. Relying on harmonized databases of all cities and towns (defined as urban agglomerations or urban functional areas) in a territory, observed at short intervals over a long period of time thanks to census data, made it possible to verify both the degree of adequacy of a theoretical model of urban hierarchies and the plausibility of the growth hypotheses predicted by Gibrat's model. The process of proportional growth (added quantities at short time intervals are proportional to the acquired size) thus gives rise to a lognormal distribution of city sizes if urban growth rates (added quantity divided by size) are randomly distributed whatever the city size and not correlated with previous rates. Brian Robson (1973) had tested this model successfully on the evolution of urban agglomerations in England and Wales during the nineteenth century. He mentioned, however, a few slight deviations between empirical observations and the pure stochastic model, which I also observed when testing it on French data from the nineteenth and twentieth centuries (Pumain 1982a). Gabaix and Iohannides (2004) tested it on US data and many other authors have done so since.

Although not perfect, the good quality of fit of the model made it a satisfying first approximation for summarizing the urban growth process. The contribution to urban theory was thus double: the model provides, first, a statistical explanation of the shape of urban hierarchies and, second, a better understanding of the persistency of their spatial pattern. This explanation could be admitted by geographers, but not by specialists of urban economics for whom the rank-size rule remained a 'mystery' (Krugman 1996), mainly because economists wanted to derive it from a process that would be defined at the level of individual firms and persons.

3.2 Inferring Co-evolution from Comparative Socio-economic Change

Admitting that a simple stochastic process of urban growth can explain the distribution of city sizes, with its strong inequalities and its persistence over time, is a step forward in knowledge. It is a morphogenetic explanation, as the morphology of the system of cities is thus related to the dynamics of this system. However, from the viewpoint of geographical theory it is frustrating since in this stochastic statistical model intentional individual initiatives disappear because their result is random and cities are supposed to be independent from each other – which they are not. This also means taking the search for an explanation a step forward: why do cities grow all together in that way? What justifies that all well-connected cities in the same territory have on average the same growth rate over the long term, despite the large fluctuations observed in this growth spatially and over shorter periods?

Considering this, leads to searching for an explanation by investigating how migrations (that represent the attractiveness of cities) relate to the jobs cities offer to the population, or how investments select cities where they locate. During particular periods, the fastest growing cities, which are the most attractive, are those where job-creating and income-generating activities are developed. Geographers have a long tradition of classifying cities according to their activities and skills, and thus defining urban functions or functional specializations (Berry and Smith 1972; Pumain and Saint-Julien 1976). Multivariate analyses provide a synthetic description of the most frequent associations and exclusions of activities in cities. Among contemporary cities, a major difference still separates two types of cities, those with predominantly manufacturing bases and those orientated towards services, inherited from the highly distinctive specializations created with the

Industrial Revolution of the nineteenth century. A secondary differentiating factor represents traces of recent (after 1950) changes improving technologies and practices (termed knowledge-intensive activities) among industrial and tertiary specializations in France as in the USA (Paulus and Vacchiani-Marcuzzo 2015) and in China (Swerts 2016). Looking at the relative stability or slow evolution of that fundamental structure over a few decades, we found a surprising parallelism in the trajectories of individual cities: the qualitative and quantitative changes occurring in the social and economic composition of urban populations were almost the same in every city (Pumain and Saint-Julien 1978; Paulus and Vacchiani-Marcuzzo 2015). This incremental and distributed process of socio-economic change contributes to a type of simple homothetic translation of the former hierarchical and functional structure of the urban system.

A constantly renewed flow of innovations modifies the socio-economic profile of cities. Many socio-economic changes, specific to a city, do not affect the structure of the system, since they are not sufficiently persistent from one period to the next; they are simple fluctuations that do not alter the relative place of cities in the system if they occur when the system is on a stable trajectory. The change in structure occurs when one of these fluctuations persists and increases to such an extent that it permanently alters this position, most often for several cities at the same time, often associated with an innovation wave and the development of a new specialization. In these cases, there is a bifurcation in the history of the system. This evolution is comparable to that described in physics as self-organizing open systems (Allen 1997; Wilson 2012). Behind the structuring of these systems, according to an architecture that is identifiable because it is persistent, there is no network operator, nor an explicit political or economic purpose, which could be described as a function to be optimized. That is, systems of cities are not in equilibrium at a given moment, they transform under the action of internal and external dynamic processes. Their structure is the product of this evolution.

The evolutionary theory of urban systems (Pumain 1997) includes in a unified process the organizational duality between the hierarchical organization of the systems of cities controlling and serving the territory, on the one hand, and their networking for the exercise of various specialized functions, on the other. It links the structure of systems of cities with the process of diffusion of innovations. When a city (usually a large one) adopts an innovation, it is very quickly imitated in other cities and this process filters down the urban hierarchy. Cities' adaptive power to change depends on their relative position in the system, based on resources, both site related and accumulated. The Swedish geographer Torsten Hägerstrand (1966, p. 40) formalized that observation and termed this process a 'hierarchical diffusion of innovation'. A large number of authors have illustrated the link between the speed of adaptation to change and previous accumulation (summarized by the size of cities, their financial power, and skills of their inhabitants and institutions) and illustrated the hierarchical diffusion process of innovations in urban systems (Pred 1973, 1977; Robson 1973).

Compared with the simplistic description of urban growth provided by Gibrat's model, these results attain several gains in explanation. First, it is not because they are statistically independent entities that cities grow at roughly the same rate, but because they share the same type of transformation of their economic and social profiles, from being part of a territory under common laws and societal regulations and processes. This is likely to happen owing to the multiple connections ensuring a rapid communication of information between cities. Second, Gibrat's model, when completed by including growth impulses from innovation waves and their hierarchical diffusion within the system of cities (Tornqvist 1973; Favaro and Pumain 2011), helps to better understand the recurrent deviations that were observed between empirical data on urban growth and Gibrat's stochastic model, that is, a slight positive correlation between growth rates and city size (at the beginning of each innovation wave) and a persistency of high or low growth rates in the same cities over some periods of time. This also provides an understanding of the bifurcations in the trajectories of individual cities that are created by the major functional specialization, successively amplifying urban growth then urban decline according to the stage in the corresponding product cycles.

We were able to use that knowledge for developing an interpretation, which is better anchored in geographical urban theory, of the new frame of analysis proposed under the label of 'scaling laws' by physicists (Bettencourt et al. 2007). Instead of imagining an urbanization process where largest cities would simply be a replication of smaller ones constrained by universal power laws,[3] we explain the level and observed variations in values of the exponents of urban scaling laws according to the stage of development of the considered urban attributes in the urban system (Pumain et al. 2006, 2009). Exponents of non-linear relationships between an urban attribute and city size are above 1 during the first stage of adoption of innovation in the urban system, when they are captured by and concentrate in the largest cities, and below 1 when they withdraw in smaller towns of cheaper land rents and wages after their stage of wide diffusion all over the system. Over time, the exponent values evolve according to this theory (Paulus 2004; Paulus and Pumain 2011; Finance and Swerts 2020). Scaling laws are thus convenient instruments for expressing urban centralities at different aggregated levels of urban settlements (Cottineau et al. 2017) or for detecting urban agglomeration economies (Cottineau et al. 2019).

Thus, starting from an interpretation of a temporal multivariate analysis, we have made a discovery that was later confirmed with a series of different investigations. To summarize that discovery: size inequalities and qualitative socio-economic differences between cities are traces of their co-evolution (which is an interactive and proactive adaptation with feedbacks to the innovations they create) within interconnected systems of cities. The expansive trend that characterizes these systems in a quantitative way (growth in population and wealth) and qualitatively (through technological and societal innovation) is an essential driving force of their structure and evolution. Elements of this theoretical explanation already existed in urban geography, especially in the theory of hierarchical diffusion of innovations (Hägerstrand 1952). The novelty also comes from the simulation models which enable testing the hypothesis of the theory.

3.3 Reconstructing Evolutionary Trajectories for Testing the Evolutionary Theory of Urban Systems

The evolutionary theory of urban systems includes hypotheses on the generating processes that explain their structure and possibly allow predicting their future evolution. This conception can be tested on empirical data about past evolution, using mathematical and computational concepts, and models proposed by the science of complex systems that has been developing in recent decades. The principle is that interactions between cities explain their trajectory and the structure of the system of cities. Multi-agent models are best suited to represent the exchanges between cities, differentiated according to their functions and spatial distribution. We briefly summarize this research process, which originated in creating a whole family of Simpop models.[4]

The past 30 years enriched the theories of self-organization in physics (Prigogine 1996; Haken 1977), evolution of living species in biology and adaptive cognitive systems in economy or social networks (Arthur 1994; Allen 1997; Weidlich 2006). Our representations of system dynamics changed, especially by emphasizing the conditions of emergence of new structures from local interactions between adapting individuals. The main epistemological questions have therefore shifted from the autonomy of systems relative to their environment, towards the identification of attractors governing their dynamics and, finally, to their capacity for innovation within a context of uncertain and changing rules of social interactions. Compared with these earlier representations of self-organization in models, the notion of emergent properties refers to a more explicit modelling of individual behaviour and interactions, usually in agent-based models or in multi-agent systems. Emergent properties appear at a higher level of aggregation than the original description of the system. They consist most of times in the formation of a structure at the macro level (which may arise in an unexpected way from the individuals and their interactions as they are described in the model), and more rarely consist in the appearance of a true novelty, which was not expected but can be confirmed afterwards in empirical testing. Ideally, the latter would be the prediction of a social innovation.

Multi-agents systems (MAS) are especially useful as simulation tools for modelling dynamics, when it is essentially explained by the heterogeneity of individual features and their interaction. They enable the modeller to associate qualitative and quantitative rules, and to integrate several levels of organization as well as diverse time scales and dynamics relationships. They are used in natural, social or cognitive sciences, as artificial laboratories (*in silico*) for observing the behaviour of agents at an individual or collective level, and analysing the evolution of the structures which emerge at a macro level (Ferber 1995). They appear as a reasonably promising technique for simulating geographic worlds, mainly because of their ability to consider the environment of a system, their acceptance of a wide conceptual diversity of agents (allowing for multi-level analysis) and their flexibility regarding interaction rules, especially in spatial relationships.

Multi-agents systems are much more flexible than differential equations are for simulating spatial and evolving interactions, including quantitative and qualitative effects (Pumain and Sanders 2013). Through the definition of rules at individual level, they can reproduce the circulation of information between cognitive and decision-making agents. They simulate at the upper level the emergence of collective or aggregated structures, which can be tested statistically. The rules can be adapted for varying space and time scales of interaction during the course of history. Multi-agents systems do not solve the problem of choosing a 'good' theoretical representation. A first application to geography was the Simpop model (Bura et al. 1996; Sanders et al. 1997), which used MAS for simulating the emergence of a functionally differentiated system of towns and cities from an initial generally homogeneous rural settlement system, over a period of some 2000 years. This model differs from most MAS in two aspects. First, its agents are immobile, as they represent places in geographical space, even if their interactions are affected by the technical innovations in the communication systems and the increasing level of urban resources that modify the relative situation of elementary cities in the network of their relational space.

Second, as Simpop is a model of interactions between territories (aggregated geographical objects), the behaviour of these agents is not reducible to the behaviour of individual persons, for instance, as in models of cognitive economy. Compared with the difficulty of representing spatial interactions with parameters such as in mathematical models using systems of differential equations, MAS models enable to differentiate different types of spatial interactions processes. Within the Simpop model, inter-urban interactions for market activities are freely competing but constrained by a distance decay model of the gravity type, while administrative functions levy taxes in fixed spatial subdivisions. Specialized activities develop interactions according to specific types of networks that may reach long-distance and cross-national boundaries. These longer-distance and more freely distributed interactions are linked to the location of resources for manufacturing and to resort amenities for tourism, for instance.

Andersson et al. (2003) use an algorithm generating 'scale-free' networks. This corresponds to a class of growing networks whose node degrees are power-law distributed because of a rule of preferential attachment – according to a growth process similar in its consequences to Gibrat's model (Albert and Barabasi 2002). In Andersson et al.'s model, the nodes of the network represent pieces of land which over time become increasingly connected by edges representing exchanges of goods and services. (The result of this trade is simulated by a trade benefit or financial investment directed from one node to another.) The model proceeds by adding new links between already developed nodes, with a probability of this occurring that is proportional to the relative size of the node in the total number of nodes, and by selecting new nodes. The mean probability of developing existing nodes is significantly higher than that relating to the development of new nodes. Spatial rules are added to specify this selection process, according to hypotheses about a distance–decay interaction model. Thus, this model is designed to simulate the urbanization process on the intra-urban scale, or the formation of urban hierarchies, or both. In any event, the concept of 'scale-free' networks, or 'small worlds' (Watts 2004), seems well adapted to the simulation of urban systems, since they reveal the hierarchical structure that emerges in progressively constructed networks.

What are interurban interactions in the Simpop models? As our intention is to simulate the development of urban systems that include a large number of towns and cities, it would be unrealistic to think of interactions as real flows of exchanged goods, people or information. The interactions that are simulated in the model are not these first-order inter-urban exchanges, but more abstract, second-order interactions, which represent an interpretation of the effect of concrete flows on the relative dynamics of cities. For example, urban functions are essential attributes of the cities. These do not give an exhaustive representation of the economic profiles of the cities, since they are attributed only to the cities that have developed a major specialization in a particular sector of activity during the corresponding innovation cycle. Similarly, the exchange of products and resources among cities on the 'market place' (cities selling and buying according to their level of supply and demand) does not reflect the totality of the urban economy, only the specialized part of the inter-urban market, which is likely to give rise to urban growth differentials.

Our model is a geographic model, in that spatial interactions are supposed to reflect the power of cities in relation to range of influence of their activities and support for new developments from their access to generally extended markets. Three types of spatial interactions are distinguished for reflecting the most frequent types of inter-urban exchanges, linked to different constraints. Proximity-constrained interactions are representative of many activities for which the distance between supply and demand is an essential constraint. They are the rule for all central place functions, whatever their level and range, and even if that spatial range is increasing over time; under that rule, the probability of exchanges are distributed according to a model of gravity type. Territorially constrained interactions are limiting a city's influence within boundaries, regional or national. They correspond to all types of administrative or political activities and the interaction rule is modulated according to the nature of the activity; for instance, a capital can levy taxes in an exhaustive way on all cities belonging to its region or state, whereas for other activities this rule can attribute only a preference for a territorial market. Interactions within specialized networks are free from distance constraints, even though exploring them for developing new markets along this line may have differential costs according to the distance. Long-distance trade, maritime transport, part of tourism activities or manufacturing industry are following this type of spatial interaction rule.

The results of experiments with this model provided further advances in explaining the evolution of systems of cities, proving the importance of urban networks in generating more urban growth. The observed historical major dynamic trend leads to an increasing hierarchization: the inequalities in city sizes are becoming larger than those that would be predicted with Gibrat's model, contradicting the theories predicting counter-urbanization (Berry 1976). A new discovery that global cities may have existed since the Middle Ages also emerged, since the cities that are at the head of urban hierarchies always expand a significant part of their interactions beyond the boundaries of their own system (Bretagnolle and Pumain 2010). We have verified in further studies, including in Brazil, Russia, India, China and South Africa (the BRICS countries), that metropolization and simplification from below of urban hierarchies (leading to 'shrinking cities') are common dynamics trends observed in several systems of cities in the world (Pumain et al. 2015, para. 16).

The research program GeoDiverCity[5] paved the way for developing a new method of building simulation models which takes into account not only the general urban dynamics for reconstructing common stylized facts but also the major historical features that distinguish regions of the world. The MARIUS model designed by Clementine Cottineau and Paul Chapron, with the help of the OpenMOLE simulation platform, intended to reconstruct the trajectories of cities within the boundaries of the territorial system of the former USSR (Cottineau 2014). In parallel, the authors invented an incremental method for model building: from a hierarchy of factors explaining the differential urban growth that was revealed by statistical analysis of the observed trajectories of individual cities. They implemented, first, the simplest and most generic model of urban growth (that is, a Gibrat's stochastic model without spatial interaction), and then introduced a more sophisticated mechanism of constraints on urban growth as well as some specific environmental conditions (Cottineau et al. 2015). At each step, the computed deviation between the expected shape of urban hierarchy and typology of urban trajectories and the computed shape and typology helped to measure the retro-predictive capability of the model. For instance, it was important to introduce not only resource location for generating functional specialization, but also the political decision of large investments in these urban areas. A new geographical modelling framework is thus proposed, which operates at different levels of precision, according to an adjustable focus evaluating the degree of specificity of a particular evolution in a generic dynamic, for a given granularity of description (Cottineau et al. 2015).

The OpenMOLE simulation platform allows the integration of sophisticated model validation processes, including genetic algorithms and access to parallel grid computing. We have been able to acquire more certainty in the quality of parameter estimation for a generic model, SimpopLocal (Schmitt 2014), dedicated to the emergence of a system of cities some 4000 thousand years after the Neolithic period, with the help of the simProcess platform developed by Sébastien Rey-Coyrehourcq (2015). Compared to the few hundred simulations that were available when estimating with trial-and-error by-hand methods in the former experiments with the first Simpop models, it was possible to develop an almost exhaustive exploration of the space of parameters, through some 500 million replications of the same model (Schmitt et al. 2015). The theoretical hypotheses, thus tested and assumed to lead to the emergence of a system of cities based on interactions propagating innovations between population centers, could be considered not only as sufficient but also as necessary to reconstruct an urban hierarchy and a correct diversity of urban trajectories. A new procedure for the rigorous exploration of the behaviors of a model, the 'calibration profile', was developed for that precise purpose (Reuillon et al. 2015). This outcome is important for the humanities and social sciences because it opens the door to providing proof when answering questions about complex social processes.

4. TOWARDS MULTI-LEVEL URBAN NETWORKS

While concluding a book that we edited on world urbanization (Rozenblat et al. 2018) we made a plea for 'multi-scalar urban policies'. The available urban knowledge still appears too fragmented among a variety of disciplines for efficiently undertaking enough inspiring policies. It is also not yet connecting well the accumulated knowledge from the past with the new urban sciences (Batty 2013). It also remains difficult to establish sound reasoning at levels of individual cities and at the level of systems of cities. The construction of simulation models including spatial interactions and networks has made considerable progress, particularly in relation to technical validation (Pumain and Reuillon 2017), but remains limited by the lack of precise information on the interactions between the different levels of urban systems. Until very recently the lack of data observed at individual level hampered a realistic consideration of inter-individual relationships in urban modeling.

Probabilistic reasoning in statistical and simulation models takes into account the material impossibility of knowing at any given moment the exact position and state of all the elements that make up the system, and denotes the existence of several levels of observation in the system. With regard to cities, the micro level corresponds to that of the individual or collective stakeholders that make up the social environment of a city, who are connected and mutually informed of their actions, and who compete with those of other cities to capture the profits associated with the exploitation of innovations. The routes and relationships that unite the cities in a system are also social networks through which information flows. Some of these networks organize hierarchically, *de jure* (administration) or de facto (distribution), while others do not, and the possibility of their capture by any city in the system is not excluded a priori. Successful specialization at a given moment is the result of exploiting the initial advantage conferred by the early adoption of an economic, social or political innovation that is bound to develop. The actors know about structure of the urban network (hierarchical form and specializations) and can use it to comply with it or to challenge it. In spite of the diversity of the stakeholders' personalities, motivations and actions, the aggregation of their behavior produces regularities that can be represented at the level of the evolution of a city as well as that of the system of cities, with a random factor, a degree of unpredictability of varying importance corresponding to the ignorance in which we find ourselves as regards decisions taken by the stakeholders.

There is still a lack of research into the form taken by local social networks (within a city) compared with that of cities as a whole. In particular, it is not clear what makes a local synergy, a particular form of social relationships in a city, effective, which some believe would facilitate its adaptation to change. The probable effect of these social interactions is made explicit with a real success for specific networks, such as financial linkages between firms and their subsidiaries, which are located within one city and between different cities (Rozenblat 2010). They are possible sources of agglomeration economies and network economies. That type of data greatly enhances knowledge about the functioning of the emerging system of cities at the global scale.

Progress in this direction is now anticipated from several directions; the expanded computing capacity and new algorithms enable the statistical treatment and visualization of the structure and evolution of very large networks to be handled more precisely (Rozenblat and Melançon 2013). Concepts and methods for social network analysis develop rapidly, as demonstrated in previous works developing applications for cities (Pflieger and Rozenblat 2010; Neal 2013) and in several chapters of this volume. Sophisticated simulation methods help in identifying the key processes that connect the communication networks and the development of systems of cities (Raimbault 2018). The technical concerns for developing smart cities appear in a variety of research about better-informed communication systems, at local (Keyvan-Ekbatani et al. 2012) or global scale (Graham 1999). Massive data extracted from smartphones traces, and exchanges in social networks collected from the web (such as Twitter, Facebook or Weibo; see, for instance, Fang et al. 2020), are promising sources of research but currently often lack enough substance in the attributes they inform to be used efficiently as proxies for measuring inter-urban interactions. Many lessons about the importance and variety of urban interactions may be learned from the gigantic experiment of their drastic reduction worldwide and at all geographical scales during the Covid-19 epidemics. A first exploration of these data is being made which relies on a simulation model that considers spatial interactions of a fractal form at all geographical levels (Abbasi et al. 2020).

We illustrate the evolutionary functioning of implicit or materialized inter-urban networks in Figure 2.1 according to three major historical eras separated by two major transitions. About 6000 years ago, when cities emerged some 3000 years after agriculture was invented, urban growth started through sharing technical and societal inventions with close settlements. That enabled much higher densities and social diversification within cities through the multiplication of social interactions facilitated by higher densities, but their development was hampered by many hazards in their environment owing to natural catastrophes or war predations and destructions.[6] Prepared long before but accelerating after 1800, with the Industrial Revolution emerges a period of apparently free development of urban networks sustained by an enormous gain in energetic power and diversification of manufacturing production that was accompanied by expanding urbanization all

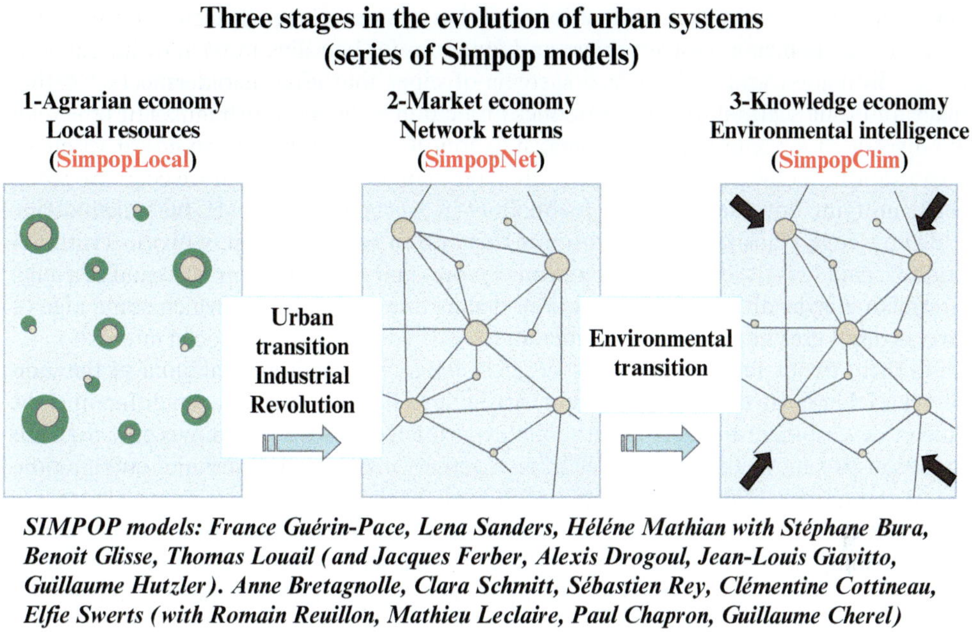

Figure 2.1 Inter-urban networks within evolutionary theory of systems of cities

over the world. Growth in individual income and life expectancy characterize this period, during which cities become the habitat of the majority of humanity and the way of living a reference for all. However, that huge development was made at the expense of planetary resources that are not priced at the level of their real value. Metaphorically, we estimate that three to six planets would be necessary to ensure raising the consumption level of the world population to that of Europe or the USA. Moreover, environmental degradation and climate change have raised concerns about the necessary ecological transition needed to achieve more sober consumption and better care of the quality of interactions between human activities and natural resources. To succeed in this transition, the existing systems of cities fortunately have the diversity of ready-built networks. They supply their numerous and various channels for conveying top-down incentives, from international or national agencies, for adopting new regulations, as well as circulating bottom-up locally generated initiatives and sharing adapted alternative views for developing cities that will be smart for their inhabitants.

5. CONCLUSION

Systems of cities are a form of adaptation to change generated by human societies, an invention that is generally not perceived as technical, but which is the result of historical processes of self-organization. Several disciplines in social sciences conceptualize each aspect of their intrinsic complexity. Economically, urbanization is reflected in the accumulation of capital, knowledge and wealth in attracting firms in multiplying networks that accompany technological creation and innovation. Geographically, urbanization articulates settlements of diverse sizes and functions for handling most daily interactions within diversified and hierarchized systems of cities that efficiently connect at higher space and time scales different economies and cultures. Socially, urbanization promotes the diversification and increasing sophistication of institutional and political forms of organization, including intensification and refinement of the social division of labor, partly unifying but also deepening inequalities in access to wealth, well-being, education and knowledge. Culturally, urbanization, often held to be synonymous with the term 'civilization', catalyzes the evolution of collective mores and cognitive representations through intermixing, hybridization and education, opening up prospects for better knowledge of diversity and greater acceptance of the other.

All these trends seem to represent progress for humankind, as translated at the individual level in the historical elevation of synthetic indicators, such as average income, the human development index or healthy life expectancy. However, we now know that this evolution also marks the irruption of human activity in energy balances and the functioning of the Earth's ecological system, which leads to the proposal of a new stratum, the Anthropocene, in the geological ages. Cities have accompanied and sustained the current over-consumption of planetary resources. The increasing scarcity of available materials and energy sources, the disasters caused by climate changes that have been under way for at least two centuries, and a large variety of global threats, also seem to be a consequence of the success of urbanization.

As revealed by the Covid-19 pandemic, and the consequent lockdown that stopped temporarily a large proportion of social interactions for more than half the world's population, cutting the networks that connect people within and between cities reduces not only the propagation of a pandemic but also the economic efficiency of cities. Thus, networks must be a significant part of urban theories explaining their functioning and development.

NOTES

1. For a more complete and concrete definition, using the example of the Roman Empire, see Chapter 5 in this volume.
2. Lepetit gave a chapter about the equipment of paved roads all over France in the eighteenth century the title 'l'impensable réseau' (the unthinkable network).
3. This is not a well-adapted description since systems of cities are not ergodic (Pumain 2012).
4. Simpop website: https://web.archive.org/web/20181224152427/simpop.parisgeo.cnrs.fr/ (accessed 22 March 2021).
5. European Research Council (ERC) advanced grant GeoDiverCity 2011–2016 (PI Denise Pumain).
6. Chapter 4 in this volume documents well the urban settlements of that period.

REFERENCES

Abbasi, M., A.L. Bollini, J.L.B. Castillo, A. Deppman, J.P. Guidio, P.T. Matuoka, et al. (2020), 'Fractal signatures of the COVID-19 spread', *Chaos, Solitons & Fractals*, **140** (November), art. 110119.

Albert, R. and A.L. Barabási (2002), 'Statistical mechanics of complex networks', *Reviews of Modern Physics*, **74** (1), 47–97.

Allen, P. (1997), *Cities and Regions as Self-Organizing Systems: Models of Complexity*, Amsterdam: Gordon and Breach.

Andersson, C., A. Hellervik, K. Lindgren, A. Hagson and J. Tornberg (2003), 'The urban economy as a scale-free network', *Physical Review E*, **68** (3), 036124.

Arthur, W.B. (1994), *Increasing Returns and Path Dependence in the Economy*, Ann Arbor, MI: University of Michigan Press.

Auerbach, F. (1913), 'Das Gesetz der Bevölkerungskonzentration', *Petermans Mitteilungen*, **59** (1), 74–6.

Baffi, S., I. Turok and C. Vacchiani-Marcuzzo (2018), 'The South African urban system', in C. Rozenblat, D. Pumain and E. Velasquez (eds), *International and Transnational Perspectives on Urban Systems*, Singapore: Springer, pp. 285–314.

Batty, M. (2013), *The New Science of Cities*, Cambridge, MA: MIT Press.

Beckmann, M.J. and J.C. MacPherson (1970), 'City size distribution in a central place hierarchy', *Journal of Regional Science*, **10** (1), 25–33.

Beguin, H. (1988), 'La région et les lieux centraux', in C. Ponsard (ed.), *Analyse économique spatiale*, Paris: PUF.

Beguin, H. and J. Thisse (1979), 'An axiomatic approach to geographical space', *Geographical Analysis*, **11** (4), 325–41.

Berroir, S., N. Cattan, F. Dobruszkes, M. Guérois, F. Paulus and C. Vacchiani-Marcuzzo (2017), 'French urban systems: a relational approach', *Cybergeo, European Journal of Geography*, (807), doi:10.4000/cybergeo.35587.

Berry, B.J.L. (1964), 'Cities as systems within systems of cities', *Papers of the Regional Science Association*, **13** (1), 147–63.

Berry, B.J.L. (1967), *Geography of Market Centers and Retail Distribution*, Englewood Cliffs, NJ: Prentice-Hall.

Berry, B.J.L. (1976), *Urbanization and Counter-Urbanization*, Beverly Hills, CA: Sage.

Berry, B.J.L. and A. Pred (1965), *Central Place Studies: A Bibliography of Theory and Applications*, Philadelphia, PA: Regional Science Research Institute.

Berry, B.J.L. and K.B. Smith (1972), *City Classification Handbook*, New York: Wiley Interscience.

Bettencourt, L.M., J. Lobo, D. Helbing, C. Kühnert and G.B. West (2007), 'Growth, innovation, scaling, and the pace of life in cities', *Proceedings of the National Academy of Sciences*, **104** (17), 7301–06.

Bird, J.H. (1973), 'Of central places, cities and seaports', **58** (2), *Geography*, 105–18.

Botero, G. (1588), *Delle cause della grandezza et magnificenza delle città. Tre libri*, new edition 1598 in *La Ragion di Stato*, 7th edn, Venice, and 1948, L. Firpo (ed.), Turin: Unione Tipografico Editrice Torinese, pp. 341–409.

Braudel, F. (1979), *Civilisation matérielle, économie et capitalisme (XVe–XVIIIe siècle)*, 3 vols, Paris: Armand Colin.

Bretagnolle, A. (2003), 'Vitesse des transports et sélection hiérarchique entre les villes françaises', in *Données urbaines*, vol. 4, Paris: Anthropos, pp. 309–23.

Bretagnolle, A. and D. Pumain (2010), 'Simulating urban networks through multiscalar space-time dynamics (Europe and United States, 17th–20th centuries)', *Urban Studies*, **47** (13), 2819–39.

Bretagnolle, A., F. Paulus, S. Kim and C. Vacchiani-Marcuzzo (2018), 'The US urban system', in C. Rozenblat, D. Pumain and E. Velasquez (eds), *International and Transnational Perspectives on Urban Systems*, Singapore: Springer, pp. 69–98.

Bretagnolle, A., D. Pumain and C. Rozenblat (1998), 'Space-time contraction and the dynamics of urban systems', *Cybergeo: European Journal of Geography*, (61), doi:10.4000/cybergeo.373.

Bura, S., F. Guérin-Pace, H. Mathian, D. Pumain and L. Sanders (1996), 'Multi-agent systems and the dynamics of a settlement system', *Geographical Analysis*, **28** (2), 161–78.

Camagni, R.P. (1993), 'From city hierarchy to city network: reflections about an emerging paradigm', in T.R. Lakshmanan and P. Nijkamp (eds), *Structure and Change in the Space Economy: Festschrift in Honor of Martin J. Beckmann*, New York: Springer-Verlag, pp. 66–87.

Cantillon, R. (1755), *Essai sur la nature du commerce en général*, London: Fletcher Gyles, repr. 1952, Paris: INED.

Christaller, W. (1933), *Die Zentralen Orte in Suddeutschland: eine Ökonomisch-Geographische Untersuchung Über die Gesetz Massigkeit der Verbreitung und Entwicklung der Siedlungen mit Städitschen Funktionen*, Jena: Fischer Verlag.

Clark C. (1940), *The Conditions of Economic Progress*, London: Macmillan.

Cottineau C. (2014), 'L'évolution des villes dans l'espace post-soviétique. Observation et modélisations', PhD dissertation, Université Paris 1 – Panthéon-Sorbonne.

Cottineau, C. (2017), 'MetaZipf. A dynamic meta-analysis of city size distributions', *PLOS One*, **12** (8), e0183919.

Cottineau, C., P. Chapron and R. Reuillon (2015), 'Growing models from the bottom up. An evaluation-based incremental modelling method (EBIMM) applied to the simulation of systems of cities', *Journal of Artificial Societies and Social Simulation*, **18** (4), 9, doi:10.18564/jasss.2828.

Cottineau, C., O. Finance, E. Hatna, E. Arcaute and M. Batty (2019), 'Defining urban clusters to detect agglomeration economies', *Environment and Planning B: Urban Analytics and City Science*, **46** (9), 1611–26.

Cottineau, C., E. Hatna, E. Arcaute and M. Batty (2017), 'Diverse cities or the systematic paradox of urban scaling laws', *Computers, Environment and Urban Systems*, **63** (May), 80–94.

Deluz, C. (1989), 'Villes et organisation de l'espace: la Chine de Marco Polo', in M. Bourin (ed.), *Villes, bonnes villes, cités et capitales. Mélanges offerts à Bernard Chevalier*, Tours: University of Tours, pp. 161–8.
Dickinson, R.E. (1934), 'The metropolitan regions of the United States', *Geographical Review*, **24** (2), 278–91.
Ducruet, C. (2008), 'Typologie mondiale des relations ville-port', *Cybergeo, European Journal of Geography*, (417), doi: 10.4000/cybergeo.17332.
Ducruet, C., S. Cuyala and A. El Hosni (2018), 'Maritime networks as systems of cities: the long-term interdependencies between global shipping flows and urban development (1890–2010)', *Journal of Transport Geography*, **66** (January), 340–55.
Fang, C., X. Yu, X. Zhang, J., Fang and H. Liu (2020), 'Big data analysis on the spatial networks of urban agglomeration', *Cities*, **102** (July), art. 102735.
Favaro, J.-M. and D. Pumain (2011), 'Gibrat revisited: an urban growth model including spatial interaction and innovation cycles', *Geographical Analysis*, **43** (3), 261–86.
Ferber, J. (1995), *Les systèmes multi-agents: vers une intelligence collective*, Paris: InterEditions.
Finance, O. and E. Swerts (2020), 'Scaling laws in urban geography. Linkages with urban theories, challenges and limitations', in D. Pumain (ed.), *Theories and Models of Urbanisation*, Cham: Springer, pp. 67–96.
Fletcher, R. (1986), 'Settlement archaeology: world-wide comparison', *World Archaeology*, **18** (1), 59–83.
Fujita, M. (1989), *Urban Economic Theory: Land Use and City Size*, Cambridge: Cambridge University Press.
Fujita, M., P. Krugman and A.J. Venables (1999), *The Spatial Economy: Cities, Regions and International Trade*, New York: Wiley.
Gabaix, X. and Y.M. Iohannides (2004), 'The evolution of city size distributions', in V. Henderson and J.F. Thisse, *Handbook of Regional and Urban Economics*, Amsterdam: North Holland, pp. 2341–78.
Gaudin, J.P. and D. Pumain (2000), 'Quelques métaphores au miroir des analyses spatiales: réseaux de villes et réseaux de pouvoir', *Revue européenne des Sciences sociales*, **38** (117), 105–19.
Geddes, P. (1915), *Cities in Evolution. An Introduction to the Town Planning Movement and to the Study of Civics*, London: William and Norgate
Gibrat, R. (1931), *Les inégalités économiques*, Paris: Sirey.
Golledge, R.G. (2002), 'The nature of geographic knowledge', *Annals of the Association of American geographers*, **92** (1), 1–14.
Gottmann J. (1947), 'De la méthode d'analyse en géographie humaine', *Annales de géographie*, **56** (301), 1–12
Graham, S. (1999), 'Global grids of glass: on global cities, telecommunications and planetary urban networks', *Urban Studies*, **36** (5–6), 929–49.
Guillerme, A. (1986), 'L'émergence du concept de réseau 1820-1830', *FLUX Cahiers scientifiques internationaux Réseaux et Territoires*, **2** (5), 30–47.
Hägerstrand, T. (1952), *The Propagation of Innovation Waves*, London: Royal University of Lund, Department of Geography.
Hägerstrand, T. (1966), 'Aspects of the spatial structure of social communication and the diffusion of information', *Papers of the Regional Science Association*, **16** (1), 27–42.
Haken, H. (1977), 'Synergetics', *Physics Bulletin*, **28** (9), 412–14.
Hoyt, H. (1961), 'The utility of the economic base method in calculating urban growth', *Land Economics*, **37** (1), 51–8.
Huff, D.L. (1963), 'A probabilistic analysis of shopping center trade areas', *Land Economics*, **39** (1), 81–90.
Keyvan-Ekbatani, M., A. Kouvelas, I. Papamichail and M. Papageorgiou (2012), 'Exploiting the fundamental diagram of urban networks for feedback-based gating', *Transportation Research Part B: Methodological*, **46** (10), 1393–403.
Kohl, J.G. (1841), *Der Verkehr und die Ansiedelungen der Menschen in ihrer Abhängigkeit des Gestaltung der Erdoberfläche*, Dresden and Leipzig: Arnold.
Krugman, P. (1996), 'Confronting the mystery of urban hierarchy', *Journal of the Japanese and International economies*, **10** (4), 399–418.
Lalanne, L. (1863), 'Essai d'une théorie des réseaux de chemin de fer, fondée sur l'observation des faits et sur les lois primordiales qui président au groupement des populations', *Comptes rendus des séances de l'Académie des sciences*, no. 57, 2nd semester, 206–10.
Lepetit, B. (1988), *Les Villes dans la France moderne (1740-1840)*, Paris: Albin Michel.
Liu, L. (1996), 'Settlement patterns chiefdom variability and the development of early states in North China', *Journal of Anthropological Archaeology*, **15** (3), 237–88.
Lösch, A. (1940), *Die räumliche Ordnung der Wirtschaft*, Iena: G. Fischer.
Mackinder, H.J. (1890), 'The physical basis of political geography', *Scottish Geographical Magazine*, **6** (2), 78–84.
Mackinder, H.J. (1902), *Britain and the British Isles*, London: W. Heinemann.
Marcus, J. and J.A. Sabloff (eds) (2008), *The Ancient City. New Perspectives on Urbanism in the Old and New World*, Santa Fe, NM: SAR Press.
Marshall, A. (1890), *Principles of Economics*, London: Macmillan.

Mimeur, C., F. Queyroi, A. Banos and T. Thévenin (2018), 'Revisiting the structuring effect of transportation infrastructure: an empirical approach with the French railway network from 1860 to 1910', *Historical Methods: A Journal of Quantitative and Interdisciplinary History*, **51** (2), 65–81.
Moriconi-Ebrard, F. (1993), *L'urbanisation du monde depuis 1950*, Paris: Anthropos.
Neal, Z.P. (2013), *The Connected City: How Networks are Shaping the Modern Metropolis*, New York: Routledge.
Ozouf-Marignier, M.-V. (1989), *La formation des départements. La représentation du territoire français à la fin du 18è siècle*, Paris: Editions des Hautes Etudes en Sciences Sociales.
Parr, J.B. (1985), 'A note on the size distribution of cities over time', *Journal of Urban Economics*, **18** (2), 199–212.
Paulus, F. (2004), 'Coévolution dans les systèmes de villes: croissance et spécialisation des aires urbaines françaises de 1950 à 2000', PhD dissertation, Université Paris I – Panthéon-Sorbonne, Paris.
Paulus, F. and D. Pumain (2011), 'Salaire et hiérarchie urbaine', in D. Pumain and M.-F. Mattei (eds), *Données urbaines*, vol. 6, Paris: Anthropos, pp. 205–16.
Paulus, F. and C. Vacchiani-Marcuzzo (2015), 'Knowledge economy and competitiveness: economic trajectories of French cities since the 1960s', in A. Cusinato and A. Philippopoulos-Mihalopoulos (eds), *Knowledge Creating Milieus in Europe. Firms, Cities, Territories*, Heidelberg: Springer, pp. 157–70.
Peris, A., E. Meijers and M. van Ham (2018), 'The evolution of the systems of cities literature since 1995: schools of thought and their interaction', *Networks and Spatial Economics*, **18** (3), 533–54.
Pflieger, G. and C. Rozenblat (2010), 'Introduction. Urban networks and network theory: the city as the connector of multiple networks', *Urban Studies*, **47** (13), 2723–35.
Pirenne, H. (1925), *Les villes du Moyen-Âge*, Paris: PUF.
Ponsard, C. (1977), 'Hiérarchie des places centrales et graphes-flous', *Environment and Planning A*, **9** (11), 1233–52.
Pred, A. (1973), *Urban Growth and the Circulation of Information: The United States System of Cities, 1790-1840*, Cambridge, MA: Harvard University Press.
Pred, A. (1977), *City Systems in Advanced Societies*, London: Hutchison.
Preston, R.E. (1991), 'Central place theory and the Canadian urban system', in Tr. Bunting and P. Filion (eds), *Canadian Cities in Transition*, Toronto: Oxford University Press, pp. 148–77.
Pumain, D. (1982a), *La dynamique des villes*, Paris: Economica.
Pumain, D. (1982b), 'Chemin de fer et croissance urbaine', *Annales de Géographie*, **91** (507), 529–50.
Pumain, D. (1992), 'Urban networks versus urban hierarchies?', *Environment and Planning A*, **24** (10), 1377–9.
Pumain, D. (1997), 'Vers une théorie évolutive des villes', *L'Espace Géographique*, **26** (2), 119–34.
Pumain, D. (2004), 'Scaling laws and urban systems', Working Paper No. 04-02-002, Santa Fe Institute Santa Fe, NM.
Pumain, D. (2006), 'Alternative explanations of hierarchical differentiation in urban systems', in D. Pumain (ed.), *Hierarchy in Natural and Social Sciences*, Dordrecht: Springer, pp. 169–222.
Pumain, D. (2012), 'Urban systems dynamics, urban growth and scaling laws: the question of ergodicity', in J. Portugali, H. Meyer, E. Stolk and E. Tan (eds), *Complexity Theories of Cities Have Come of Age: An Overview with Implications to Urban Planning and Design*, Heidelberg and Berlin: Springer, pp. 93–106.
Pumain, D. (ed.) (2020), *Theories and Models of Urbanisation*, Cham: Springer.
Pumain, D. and J.-P. Gaudin (2002), 'Systèmes de villes et pouvoir. L'analyse de Giovanni Botero à l'époque de la Renaissance', *Cybergeo, European Journal of Geography*, (227), doi:10.4000/cybergeo.1836.
Pumain, D. and M.C. Robic (1996), 'Théoriser la ville', in P.H. Derycke, J.M. Huriot and D. Pumain (eds), *Penser la ville, théories et modèles*, Paris: Anthropos, pp. 107–61.
Pumain, D. and R. Reuillon (2017), *Urban Dynamics and Simulation Models*, Cham: Springer.
Pumain, D. and T. Saint-Julien (1976), 'Fonctions et hiérarchie dans les villes françaises', *Annales de géographie*, **85** (470), 385–439.
Pumain, D. and T. Saint-Julien (1978), *Les dimensions du changement urbain*, Paris: CNRS.
Pumain, D. and L. Sanders (2013), 'Theoretical principles in inter-urban simulation models: a comparison', *Environment and Planning A*, **45** (9), 2243–60.
Pumain, D. and A. Torre (2020), 'Regional science: economy and geography in France and French-speaking countries', *Papers in Regional Science*, **99** (2), 1–21.
Pumain, D., F. Paulus, C. Vacchiani-Marcuzzo and J. Lobo (2006), 'An evolutionary theory for interpreting urban scaling laws', *Cybergeo: European Journal of Geography*, (343), doi:10.4000/cybergeo.2519.
Pumain, D., F. Paulus and C. Vacchiani-Marcuzzo (2009), 'Innovation cycles and urban dynamics', in D. Lane, D. Pumain, S. van der Leeuw and G.B. West (eds), *Complexity Perspective in Innovation and Social Change*, Dordrecht: Springer, pp. 237–60.
Pumain, D., E. Swerts, C. Cottineau, C. Vacchiani-Marcuzzo, A. Ignazzi, A. Bretagnolle, et al. (2015), 'Multilevel comparison of large urban systems', *Cybergeo, European Journal of Geography*, (706), doi:10.4000/cybergeo.26730.
Raimbault, J. (2018), 'Modeling the co-evolution of cities and networks', arXiv preprint, arXiv:1804.09430.

Reclus, E. (1895), 'The evolution of cities', *Contemporary Review*, **67** (2), 246–64.
Reilly, W.J. (1931), *The Law of Retail Gravitation*, New York: Knickerbocker Press.
Renfrew, C. (1975), 'Trade as action at a distance: question of integration and communication', in A. Sabloff and C.C. Lamberg-Karlovsky (eds), *Ancient Civilization and Trade*, Albuquerque, NM: University of New Mexico Press, pp. 3–59.
Renfrew, C. and T. Poston (1979), 'Discontinuities in the endogenous change of settlement pattern', in C. Renfrew and K.L. Cooke (eds), *Transformations: Mathematical Approaches to Culture Change*, Amsterdam: Academic Press, pp. 437–61.
Reuillon, R., C. Schmitt, R. De Aldama and J.B. Mouret (2015), 'A new method to evaluate simulation models: the calibration profile (CP) algorithm', *Journal of Artificial Societies and Social Simulation*, **18** (1), 12.
Rey-Coyrehourcq, S. (2015), 'Une plateforme intégrée pour la construction et l'évaluation de modèles de simulation en géographie', PhD dissertation, Paris 1-Panthéon-Sorbonne.
Reymond, H. (1981), 'Une problématique théorique', in H. Isnard, J.-B. Racine and H. Reymond, *Problématiques de la géographie*, Paris: PUF, pp. 163–262.
Reynaud, J. (1841), 'Villes', in C. Gosselin (ed.), *Encyclopédie nouvelle*, vol. 8, Paris: Charles Furne, pp. 670–87.
Robic, M.-C. (1982), 'Cent ans avant Christaller, une théorie des lieux centraux', *L'Espace Géographique*, **11** (1), 5–12.
Robson, B. (1973), *Urban Growth, an Approach*, London: Methuen.
Rozenblat, C. (2010), 'Opening the black box of agglomeration economies for measuring cities' competitiveness through international firm networks', *Urban Studies*, **47** (13), 2841–65.
Rozenblat, C. and G. Melançon (eds) (2013), *Methods for Multilevel Analysis And Visualisation of Geographical Networks*, Dordrecht: Springer Netherlands.
Rozenblat, C., D. Pumain and E. Velasquez (eds) (2018), *International and Transnational Perspectives on Urban Systems*, Singapore: Springer Nature.
Sanders, L., D. Pumain, H. Mathian, F. Guérin-Pace and S. Bura (1997), 'SIMPOP: a multiagent system for the study of urbanism', *Environment and Planning B: Planning and Design*, **24** (2), 287–305.
Schmitt, C. (2014), 'Modélisation des systèmes de peuplement: de SimpopLocal à SimpopNet', PhD thesis, Université Paris I-Panthéon-Sorbonne, Paris.
Schmitt, C., S. Rey-Coyrehourcq, R. Reuillon and D. Pumain (2015), 'Half a billion simulations, evolutionary algorithms and distributed computing for calibrating the SimpopLocal geographical model', *Environment and Planning B*, **42** (2), 300–315.
Swerts, E. (2016), 'The rising stars of urban growth in China', *GeoDiverCity*, blog, 25 May, accessed 29 January 2020 at https://geodivercity.parisgeo.cnrs.fr/blog/2016/05/the-rising-stars-of-urban-growth-in-china/.
Swerts, E., E. Denis and P. Mukhopadhyay (2018), 'Diffuse urbanization and mega-urban regions in India: between reluctant and restrictive urbanism?', in C. Rozenblat, D. Pumain and E. Velasquez (eds), *International and Transnational Perspectives on Urban Systems*, Singapore: Springer, pp. 237–62.
Taaffe, E., R. Morrill and P. Gould (1963), 'Transport extension in underdeveloped countries', *Geographical Review*, **53** (4), 503–29.
Tornqvist, G.E. (1973), 'Systems of cities and information flows', *Human Geography*, **38**, 1–85.
Vance, J. E. (1970), *The Merchant's World: The Geography of Wholesaling*, Englewood Cliffs, NJ: Prentice-Hall.
Vidal de la Blache, P. (1911), 'De la relativité des divisions régionales', *Athéna*, repr. 1913 in *Les divisions régionales de la France*, Paris: Félix Alcan, pp. 3–14.
Von Thünen, J.H. (1826), *Der Isolierte Staat in Beziehung auf Landwirtschaft und Naturalökonomie*, Hamburg: Friedrich Perthes.
Watts, D.J. (2004), *Small Worlds: The Dynamics of Networks between Order and Randomness*, Princeton, NJ: Princeton University Press.
Weidlich, W. (2006), *Sociodynamics: A Systematic Approach to Mathematical Modelling in the Social Sciences*, North Chelmsford, MA: Courier Corporation.
Wilson, A. (2012), *Catastrophe Theory and Bifurcation (Routledge Revivals): Applications to Urban and Regional Systems*, 2nd edn, London and New York: Routledge.
Zipf, G.K. (1941), *National Unity and Disunity*, Bloomington, IN: Principia Press.

3. Complex networks and fundamental urban processes

Luís M.A. Bettencourt

1. INTRODUCTION

Complex networks provide a particularly compelling theoretical and modeling approach to cities and urban systems. While cities are made of people, places and organizations, it is the social, economic and physical connections among these entities that start to reveal how and why cities exist, are sustained and grow (Hall 1998; Jacobs 1985). These connections also play a role in structuring individual entities themselves, for example, via firm interactions in markets, the role of institutions in shaping behavior and the general tendency for creating knowledge and labor specialization so characteristic of professional occupations in cities.

Thus, a shift in perspective away from places and individuals, towards emphasizing various types of interaction networks has been advocated as one of the cornerstones for constructing a more integrated and general science of cities (Batty 2017; Bettencourt 2021), for emphasizing the importance of human ecological factors (Sampson 2012) and, even, as a general philosophical stance for capturing the complexity and fluidity of human social phenomena (Deleuze et al. 2012). Empirical and theoretical developments now underway are also starting to show how quantitative network properties and concepts constitute a natural target for developing general urban theory (Bettencourt 2013), beyond the descriptive statistics of data.

The origin of these insights stems from a number of unifying themes now made apparent by both universal urbanization (UN-HABITAT 2016) (most nations worldwide are now urban) and the growing availability of data on many different but integrated facets of urban life.

A number of general realizations, which naturally form a basis for urban theory, follow from these comparative analyses and are increasingly modeled via network concepts. Human population agglomerations, from temporary camps in small-scale human societies (Ortman et al. 2014, 2015, 2016; Hamilton et al. 2018) to today's megacities (Bettencourt 2013; Bettencourt et al. 2007a; Glaeser 2012; Kennedy 2011) are driven primarily by human socioeconomic interactions. These interactions are necessary owing to strong complementarities between people, from finding a mate to the maintenance and development of shared knowledge (culture), to strong interdependence of resource exchange and the division of labor and knowledge in modern cities (Bettencourt 2014a; Bettencourt et al. 2014; Duranton and Puga 2004; Puga 2010; Pumain et al. 2006; Youn et al. 2016).

All these interactions are facilitated by spatial proximity (Alonso 1977; Fujita et al. 2001; Glaeser 2012; Hall 1998), and thus can be aided and structured by built environments and associated flows, costs of movement and land use, as well as by social and cultural opportunities (Fischer 1995). Although cities have always been described partially in these terms, over the past few years comparative quantitative analysis of cities throughout space, time and at varying socioeconomic levels of development have distilled frameworks for modeling cities based on networks (Batty 2017; Bettencourt 2013, 2014a, 2021). In this chapter, I briefly review its main results, predictions and open questions for near future research.

The chapter is organized in three parts, which are interdependent but can, to a large extent, be treated separately and have been the subject of different strands in the literature.

The first, and in my view most fundamental, strand deals with the conceptualization of human settlements as socioeconomic networks. I argue, in section 2, that socioeconomic interactions are primary to infrastructure and space since this is the sequence observed historically in developing towns and cities (Bresson and Rendall 2016; Kintigh et al. 2014; Kostof and Tobias 1999; Ortman et al. 2015, 2014) and because adopting such a perspective enables us to better understand and predict the socioeconomic performance of cities (Bettencourt 2013). Associated with socioeconomic connectivity are the most important set of general phenomena displayed by people in urban societies, known as network effects (Bettencourt 2013, 2014a; Ciccone 2002; Duranton and Puga 2004; Schlapfer et al. 2014). These effects state that the properties of cities are not proportional to numbers of people or space occupied, say, but to the connections between them, which are naturally a superlinear function of the of number of connected elements. This emergent phenomenon enables, for example, cities to display increasing returns to population scale for economic outputs (Bettencourt et al. 2007a, 2007b), and creates the centripetal forces that cause population agglomeration in the first place (Berry 1964; Fujita et al. 2001).

I briefly review the main facts about urban socioeconomic networks in cities, including how they grow with city size, and how they mediate productivity through general mechanisms for the division of labor and knowledge. I also discuss what we know about these networks' statistics, and issues of change in socioeconomic connectivity associated with economic development and inequality through the life course, as well as their connection to more macroscopic theories of economic growth (Jones and Romer 2009; Lucas 1988; Romer 1989).

It is in this light that the role of infrastructure networks and space (section 3) also become more evident, as mediating costs of movement and structuring land uses that enable interaction networks to grow and become dedicated to higher-value activities (Alonso 1977; Bettencourt 2013). There is a general logic to the infrastructure networks of cities, which leads to particular predictable aspects of their topology and patterns of use (Brelsford et al. 2018), including average energy dissipation and congestion (Barthélemy and Flammini 2006; Bettencourt 2013). This local connection between socioeconomic networks in cities and corresponding infrastructure also enables us to analyze spatial data in cities to detect when the two networks – social and infrastructural – are well matched, and potentially diagnose infrastructure deficits (Brelsford et al. 2018) (and surpluses, for example, in shrinking cities).

Section 4 describes flows between cities in relation to networks (Batty 2017; Levy 2010; Molloy et al. 2011); see Figure 3.1. These flows are both causes and effects of the properties of each city, as posited by the first law of geography (Miller 2004). I discuss how these flows structure the distribution of city sizes in an urban system, and when these properties signal selection or randomness of choices in migration (Bettencourt and Zünd

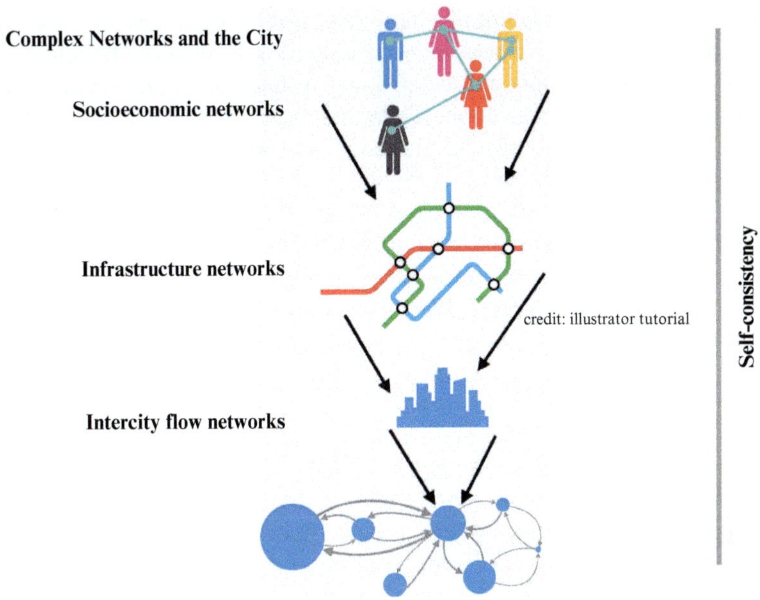

Note: There are many encapsulated levels of interaction and exchange that describe the dynamics of cities. Network dynamics at different scales must be self-consistent and produce a set of interdependent urban general phenomena (see Figure 3.2 in section 3.5 of this chapter).

Figure 3.1 Complex networks and the city

2020; Gabaix 1999; Gibrat 1931). In section 5, I finish with a discussion of open questions and potential directions for future research at the interface between network theory and urban science and its applications.

2. SOCIOECONOMIC NETWORKS IN THE CITY

The essential character of cities is that all people must depend on others for almost everything they do (Bettencourt 2014a; Smith and Krueger 2003). These interdependencies are mediated by vast complex networks of social exchange, economic markets, civic channels and more. From housing, to urban services, such as water, sanitation, safety and emergency protection, to all manner of consumer goods and services, life in the city requires interactions with many other people and organizations to achieve functions in our life that are complementary to our own primary occupation (Bettencourt et al. 2014; Duranton and Puga 2004; Park et al. 2010; Wirth 1938). Even when people do not have a primary job, such as children or the retired, the expectation remains that life in the city provides all manner of supporting activities and services, including schools and health services.

This strong interdependence is only possible as social transactions among many people (mostly strangers) become possible and more goods and services can be exchanged safely, reliably and affordably (Bettencourt 2014a; Fujita et al. 2001). This idea can be thought of as a relative decrease in transaction or coordination costs (Besbris et al. 2015; North

1992; Williamson 1979), making links in networks of services and exchange possible, and indeed favorable relative to situations of autarky and self-reliance (Bettencourt 2014a). This process of urbanization can be associated with a state transition in functional network structures from small-scale local connectivity associated with subsistence (Bettencourt 2014a), to urbanization, when widespread, larger network degree becomes common, and the formation of large (dominant) connected components emerge at city and national levels.

This general situation of greater connectivity associated with living in cities has been described by many authors as the starting point for entire urban disciplines. In sociology, Robert Park described the condition of living in cities as a series of adaptations of 'human behavior in the urban environment' (Park et al. 2010). Following this line of enquiry, a great deal of urban sociology has since adopted the perspective of 'human ecology', which Park defined as 'the science which seeks to isolate [social forces] and to describe the *constellations of persons and institutions* which the cooperation of these forces produces' (Park et al. 2010, p. 2, emphasis added). Thus, as in other parts of science, while patterns of interaction are form, we must also seek function, in the forces, or processes, whereby people and institutions become interconnected in cities (Jacobs 1985; Park et al. 2010).

In urban economics, foundational ideas by Alfred Marshall (Fujita et al. 2001; Marshall 1890) in the context of his study of industrial districts (not so much cities), have been distilled by contemporary authors in relation to several features of spatial proximity that make economic activity more efficient and inventive (Bettencourt 2021; O'Sullivan 2012). Marshall emphasized the co-location of producers and consumers (later the basis for models of new economic geography; Fujita et al. 2001; Krugman 1991), the production of thick labor markets that support specialization, and the sharing and generation of information often described in the contemporary literature as knowledge spillovers (Acs et al. 1994; Anselin et al. 1997; Griliches 1991). Other authors in urban economics have emphasized the importance of sharing space and institutions, including physical infrastructure and public goods, which can be more necessary and transformative in the presence of large numbers of people and firms (Duranton and Puga 2004). Similar arguments have been core to economic theory since Adam Smith (Smith and Krueger 2003), but the role of cities as the environments where large and diverse markets can take place and be sustained despite high apparent costs has only very gradually become evident (Jacobs 1985; Lucas 1988).

Most recently, in the context of the search for a theory of endogenous economic growth, a number of authors have highlighted the importance of cities as highly connected environments, where human capital (education and skills) is encouraged to form and concentrate (Lucas 1988) and where knowledge, the non-rival good at the basis of growth in these theories (Jones and Romer 2009; Romer 1989), can be created and elaborated. Robert Lucas, emphasized the value of social and economic proximity for growth (Lucas 1988) and encouraged economists to learn from the insights of Jane Jacobs (Lucas 1988). More recently, Jones and Romer attempted to estimate the scaling of complex social networks underlying economic growth, as an open problem in improving underlying theory (Jones and Romer 2009).

Starting from a different perspective, much closer to early urban sociology, social psychologists emphasized the experience of living in cities as being of predictable behavioral adaptations to the condition of cognitive overload (Milgram 1969), where the number of choices, possibilities and solicitations becomes overwhelming, so that individuals must segment their lives in functions and time, and learn to cope with novelty and challenges (Park et al. 2010; Wirth 1938).

In all these general approaches, the emphasis is on interactions and connections of many different kinds (social, economic, employment, learning and co-location), but not on rigid arrangements or specific forms of social or economic connectivity. Indeed, a great deal of the conceptual emphasis on the nature of social networks in cities across disciplines is on opportunity, possibility, incompleteness, fluidity and adaptation (Milgram 1969; Park et al. 2010; Wirth 1938). This dynamic approach to networks is thus imperative for coping with incomplete information and change that keeps urban environments dynamic and creative, though also imperfect, rather than optimized into specific rigid structures. We return to what the study of these dynamics can do for network science in the discussion in section 5.

What, then, can be said in general of the social and economic networks that make up cities? First, that there is a general statistical quantitative trend for people in larger cities to have larger, and functionally more diverse, socioeconomic networks. This expectation has been expressed by many earlier authors (Herrera-Yagüe et al. 2015; Park et al. 2010; Wirth 1938), and has been confirmed both by a number of traditional surveys and, more recently, via cell phone network analysis (Andris and Bettencourt 2014; Schlapfer et al. 2014), in a number of different nations.

The average increase in connectivity per person, $k(N,t)$ with city population size, N, and time, t, is predicted by urban scaling theory to follow a scaling relationship of the form (Bettencourt 2013; Bettencourt et al. 2007a; Schlapfer et al. 2014):

$$k(N,t) = K_0(t)N(t)^{\delta}e^{\xi_i(t)}, \tag{3.1}$$

where $K_0(t)$ is a size-independent but time dependent prefactor and $\xi_i(t)$ are residuals for each city away from the average scaling pattern of connectivity. What is interesting is that the exponent (or elasticity, in the language of economics), $\delta \simeq \frac{1}{6} < 1$, is a general number common to most cities and is less than unity (Andris and Bettencourt 2014; Bettencourt 2013; Bettencourt et al. 2014). This means that even as people have on average more connections in larger cities, this increase is smaller than the corresponding increase in population, which is naturally proportional to $N(t)$ (exponent $= 1$). This means that each individual's network reaches less and less of the city as a whole, so that larger cities express a pattern of higher specialization at the individual level, within a global pattern of greater connectivity at the city level. The limitations of individual connectivity growth relative to that of the city at large create the phenomenon that social horizons (the fraction of the city that someone can experience directly through their ego network) recede with city size, allowing cities to contain more subcultures (Fischer 1995), more moral regions (Park et al. 2010) and, therefore, more specializations than smaller and more unified communities (Bettencourt et al. 2014; Feldman and Audretsch 1999; Wirth 1938).

Second, despite evident average results, these measures are observed to be variable among people. For example, a useful general description of the statistical distribution of network degree in cell phone networks is a lognormal (Schlapfer et al. 2014). Technically, these studies measure the average degree (number of connections, as well as number of calls and call duration). In all instances, these quantities are observed to scale superlinearly with city size and manifest approximately lognormal statistics.

This type of distribution, as well as power-law distributions of degree, follow from processes in which degree grows multiplicatively, as the consequence of new connections introduced with a probability proportional to the number of connections a person has (Clauset et al. 2009). A simple mechanism for this to occur is that we meet most new people through someone else we already know (Gabaix 1999; Gibrat 1931; Schlapfer et al. 2014) with some finite probability. This insight is contained in simple algorithms of preferential attachment (Barabási 1999; Price 1976; Simon 1955), but is also a much more general property of exponential growth processes, where the growth rate is itself a stochastic quantity (Gabaix 1999; Gibrat 1931; Montroll and Shlesinger 1982).

Third, the main function of the expansion of connectivity in cities is associated with the integration of the division of labor and knowledge in cities. It is observed, for example, that the number of unique occupations and business types tends to increase with city size, and that the relative abundance of these types is a predictable function of urban population sizes (Bettencourt et al. 2014; Youn et al. 2016). The number and abundance of occupations is clearly limited by the opportunities to integrate functions back together in relation to someone's life: if we cease to perform an essential function, such as growing food, we need to obtain it via a connection with others, such as from a grocery store or restaurant. This is observed in the data for the number of occupations in US metropolitan areas, which scale with population size to the extent afforded by social connectivity (Bettencourt et al. 2014). Thus, connectivity, specialization and occupational diversity are all facets of the same urban phenomenon whereby knowledge can be made deeper (for example, better cooking) through individual specialization, and obtain greater value and impact via its integration through networks with the labor and knowledge of others (Arrow 1971; Bettencourt 2014a; Bettencourt et al. 2014; Puga 2010; Smith and Krueger 2003). Consequently, this expanding socioeconomic connectivity of larger cities is not random, but instead reflects strong social and economic complementarities in knowledge and labor.

These effects of expanding connectivity and functional complementarity can have surprising consequences for processes that spread by social connections, whether they are contagious diseases or information. The key parameter for epidemic processes, which transcends all detailed models, is the basic reproductive number R_0 (Anderson and May 2010). It accounts for the number of secondary cases triggered by the presence of an index (initial) case and is therefore closely associated with the early growth rate for an epidemic process in a susceptible population. For a contagion network, the reproductive number is related to the statistics of degree, k, as:

$$R_0 = \lambda \frac{\langle k^2 \rangle}{\langle k \rangle} = \lambda \langle k \rangle \left(1 + \frac{\sigma_k^2}{\langle k \rangle^2} \right), \tag{3.2}$$

where λ is the infection probability per contact, $\langle \ldots \rangle$ denotes expectation values over the population and σ_k^2 is the network degree of variance. For a lognormal degree distribution, we write the degree average and variance as:

$$\langle k \rangle = e^{\mu + \frac{\sigma^2}{2}}, \sigma_k^2 = (e^{\sigma^2} - 1)\, e^{2\mu + \sigma^2}, \tag{3.3}$$

with the parameters $\mu = \langle \ln k \rangle$, $\sigma^2 = \langle (\ln k - \langle \ln k \rangle)^2 \rangle$. This results in a simple and elegant expression for the reproductive number:

$$R_0(N_i) = \lambda e^{\sigma^2} \langle k(N_i) \rangle \simeq \lambda k_0 \, e^{\sigma^2} N_i^{\delta}, \tag{3.4}$$

where, in the last equality, we introduced the scaling of network degree with city population size. We see therefore that, in general, the reproductive number is expected to be a function of city size N_i, and specifically to be larger in bigger cities (Chowell et al. 2007; Dalziel et al. 2018; Schlapfer et al. 2014). How much larger, depends on the behavior of the log-variance, σ^2, and whether this parameter is city-size dependent, an issue that can generate statistical corrections (beyond mean-field) to the simplest expectations from urban scaling theory (Bettencourt 2013), which predict $\delta \simeq \frac{1}{6}$. This result was recently confirmed during the early stages of the Covid-19 pandemic in the USA (Stier et al. 2020).

A larger reproductive rate for spreading processes in networks in larger cities has two important consequences for both health and processes of communication and innovation (Acs 2004; Bettencourt et al. 2007a; Dalziel et al. 2018; Feldman and Audretsch 1999). First, the most important meaning of the reproductive rate, R_0, is as a threshold for an epidemic signal to propagate through a population (Anderson and May 2010; Bettencourt et al. 2006). For $R_0 < 1$, an introduction will die off because it is damped in transmission, while for $R_0 > 1$, the process will be amplified and become an epidemic in that it can be transmitted quickly to almost everyone in the population: it becomes a chain reaction. Since, as we have seen, we expect R_0 to increase with city size, we should expect large cities to be more susceptible to contagious diseases and the spread of information, both formal and in relation to culture, fashions and behaviors. This is anecdotally true. As R_0 may be locally larger in more densely interacting communities within cities, this may make these communities the natural reservoirs of specific contagious diseases, an issue especially important currently relative to the persistence of vector-borne diseases in pockets of large cities, even in off-season periods, when we may expect $R_0 < 1$, owing to environmental factors, such as temperature.

Second, the growth rate, and thus the speed of diffusion, of a signal is faster in larger cities, with epidemics and information ripping through larger communities at a higher pace than in smaller towns. This makes efforts to contain epidemics harder in larger cities, but may facilitate certain processes for the spread of information and innovations.

Finally, the size of an epidemic outbreak is also related to the reproductive number. In complex epidemic models this needs to be computed numerically, but for a simple susceptible-infected model (Anderson and May 2010) we can integrate the dynamics and write the explicit expression:

$$\frac{S_0}{S_\infty} - 1 = e^{R_0\left(1 - \frac{S_\infty}{N_i}\right)} - 1 \simeq R_0\left(1 - \frac{S_\infty}{N_i}\right), \tag{3.5}$$

where S_0 is the initial population of susceptibles (before the outbreak) and S_∞ the (smaller) final population of susceptibles. We see therefore that a larger R_0 leads to more extensive epidemics, in the relative sense that more of the susceptible population is affected (Stier et al. 2020). Thus, with an R_0 that is larger in larger cities, we should expect epidemics to be more extensive, which on the one hand involves larger fractions of the population, and on the other may create crowd immunity to future outbreaks if exposure triggers long-lasting defenses.

The dynamics of epidemic processes in populations highlight the importance of connectivity, in general, and of its dependence on city characteristics, in particular. However, considerably more complicated dynamics of human development, economic growth, and so on require knowledge of the detailed dynamics of these quantities and their connections to networks, which is still very much a work in progress.

The general observations and studies using large contemporary datasets, such as from cell phones, do not always lead to a rich enough empirical picture of the individuals involved and their use of associated socioeconomic connectivity (Schlapfer et al. 2014; Wang et al. 2015; Zhong et al. 2015). A strand of literature based on traditional social surveys has shed some light into these issues and provides complementary starting points for potential generalizations.

For example, Claude Fischer measured the ego (individual) networks of people in locations in the Bay Area, with different characters more or less urban (Fischer 1982). He also associated these places with a number of different issues of interest to social scientists, including mood, social involvement and values. The strength of this work is its richness in terms of what type of networks (kin, neighbors or professional) people form, depending on the character of their lives (more or less urban) but also their own personal status, in relation to employment, age, education and so on.

In general, Fischer (1982) and other similar studies (Carrasco et al. 2008a, 2008b; Wellman and Berkowitz 1988) find qualitatively different personal networks in rural and urban areas, with the former involving more kin and neighbors, and the latter involving more work relations and (non-kin related) friends. This structural transformation does not seem to be associated with a degradation of psychological condition in larger cities (on the contrary) or less social and civic involvement, but it does express qualitative changes in the condition of living in cities and that urban environments tend to be more attractive to younger professionals, for example. Other large sociological surveys, have focused on issues of collective action and efficacy related not so much to the rural–urban divide, but to how well local neighborhood communities can handle challenges and cultivate trust and agency (Sampson 2012). This research is intimately connected to the idea of neighborhood effects, and the realization that local environments in cities are important in shaping people's behaviors and opportunities, especially for children and young adults (Chetty et al. 2017; Sampson 2012; Sampson and Laub 1990, 1995; Sharkey and Faber 2014). Thus, better-connected neighborhoods, both in internal structure of residents and their connections to civic and political organizations, tend to do better in relation to their socioeconomic indicators and ability to create positive change. A similar association between the connectivity of places and socioeconomic status has also been observed at the national level, using telephone networks (Eagle et al. 2010). They are also related to the observed connection between the economic complexity of cities and nations (Hausmann and Hidalgo 2011), their larger and more diverse trade flows, and their higher levels of economic wealth (Hidalgo and Hausmann 2009).

Thus, we see that there is a general expectation, playing out at different scales, between larger connectivity and the socioeconomic status of individuals, communities and cities. Connectivity in these contexts, is both cause and effect of an expanded set of opportunities, ideas and influence (Berry and Garrison 1958a; Bonacich 1987) that dynamically integrates agents with networked processes of information and resource flows. These ideas transcend time, space and forms of political and economic organization. There is a need, however, for a better understanding of dynamical mechanisms by which the extent and types of connectivity mediate learning processes and resource flows that, on aggregate over time and populations, result in human development and economic growth.

3. URBAN INFRASTRUCTURE NETWORKS

The built spaces and engineering systems that develop to support socioeconomic activity in cities can also be described as networks (Barthelemy 2016; Bettencourt 2013). These systems are spatial and physical (roads, pipes and subways), occupying some of the volume taken up by cities, while also creating spaces where socioeconomic activity and movement take place not only for people but also for materials, goods and information. Their characteristics modulate how the city is used by people and organizations, including how large the city can be spatially (Alonso 1977; Anas et al. 1998; Bettencourt 2013), where network effects are concentrated in space and time (Wang et al. 2018, 2015; Zhong et al. 2015), congestion and associated energy dissipation (Bettencourt 2013; Depersin and Barthelemy 2018), the value of rents and the shape of the city as a three-dimensional space (Alonso 1977; Anas et al. 1998; Glaeser 2008).

Although these networks and spaces have been the primary focus of urban planning, they are not a feature of all cities: very early cities (Ortman et al. 2014, 2015), and even poor neighborhoods currently (Brelsford et al. 2018, 2019), typically have only incipient or non-existing infrastructural networks, especially those that mediate services, such as sanitation, water, electrical power and larger-scale traffic or emergency services (Brelsford et al. 2017). Keeping these situations in mind is instructive, as they provide us with accidental historical experiments for characterizing the functions of physical networks in cities and the deficits associated with their absence.

The physical structure of most cities can be understood in terms of a simple repeating topological (not typically geometric) pattern. Cities are made up of two interconnected spaces with different character: places, made up of buildings and public spaces, where most life and work take place, and access networks, which connect places with each other and to other cities, and consist of street networks as well as railroads, pipes, electrical cables, optical fiber and so on.

The reason for dissecting the spaces of cities into places and access networks (Brelsford et al. 2018) is that their topological structure becomes very simple and a number of rigorous statements (theorems) can then be asserted about their general properties. First, places in cities are aggregated as city blocks, which are sets of places surrounded by street networks. Each place typically requires physical adjacency (a terminal connection) to the street network, so that each place can be reached from any other place traveling over the access network. These properties have a general topology (but variable geometry, or detailed shape): the space of places can be seen as a set of islands in a surface spanned by the street network, which has a universal topological number, the Euler characteristic $\chi = 1 - n_b$, with n_b the number of city blocks in a city. This topological invariant is a discrete measure of city size, different from population or total area. The two are proportional if the average population (or area) in each city block is a well-defined number independent of city size, but usually this is not the case.

The structure of city blocks repeats over space, setting many of the properties of the corresponding access network, which shares with it the same topology

(Brelsford et al. 2018). As a whole, this structure of street networks can be analyzed as a particularly simple type of network, known as a planar graph (Brelsford et al. 2018). One area of intense research activity has been the statistical analysis of street patterns, after these have been represented as complex networks (graphs). City streets are often abstracted from the real access surface to form a primal graph, where nodes represent intersections and edges represent centerlines (Buhl et al. 2006; Cardillo et al. 2006; Chan et al. 2011; Porta et al. 2006a). Alternatively, the access system has also been represented as a dual graph, where nodes represent streets, and edges represent street intersections (Jiang 2007; Jiang et al. 2014; Kalapala et al. 2006; Porta et al. 2006b). The authors of reference (Porta et al. 2006a) argue for particular advantages of the primal approach.

After these graphs have been constructed from emerging urban and regional data, many standard complex network properties have been measured, including statistical distributions of street lengths (Courtat et al. 2011; Jiang 2007), several measures of centrality (Chan et al. 2011; Crucitti et al. 2006; Lämmer et al. 2006; Masucci et al. 2009, 2014), and the distribution of block areas (Lämmer et al. 2006; Masucci et al. 2009; Strano et al. 2012).

In most instances, except for intensively planned layouts (Crucitti et al. 2006; Louf and Barthelemy 2014; Strano et al. 2012), these statistical distributions are very broad (power-laws or lognormals) suggesting the large variability of street patterns and block areas and, in some instances, even their apparent scale independence (power laws), at least over some ranges.

Although combinations of these measures have been used to attempt to classify urban shape classes (Crucitti et al. 2006; Louf and Barthelemy 2014; Marshall 2005), the large variability makes the geometric classification of cities somewhat arbitrary, depending on quantities chosen and given clustering cutoffs between categories. This has been known more qualitatively through many attempts to distinguish city geometries (Alexander 1965; Hillier 2005, 2015) based on street patterns.

In addition to this repeating structure of city blocks with diverse sizes and geometries, a city's street network typically shows a degree of hierarchy, with wider, faster main roads (ultimately highways) providing cross-city access, while smaller streets and lanes enable more local circulation and set the character of neighborhoods. The final ingredient of infrastructural urban networks is that they are decentralized (Bettencourt 2013; Samaniego and Moses 2008), meaning that, unlike the vascular system of an animal or a river network, it is possible to travel from point to point in the network without having to go through a unique center (the animal's heart).

These general characteristics can be brought together into a self-consistent calculation of how street networks scale with population size, in their total area, width, flows and associated congestion (Bettencourt 2013; Depersin and Barthelemy 2018). The simplest predicted scaling exponents are $1 - \delta \simeq \frac{5}{6}$ for road surface (and built-up area), $\delta \simeq \frac{1}{6}$ for average width, and $1 + \delta \simeq 7/6$, for energy dissipation (and associated measures of congestion) (Bettencourt 2013). The build-up area of cities tends to scale similar to its access network area, which also has been experiencing spatial expansion and loss of density, as transportation technologies have become cheaper relative to incomes, as predicted by general urban theory (Alonso 1977; Angel et al. 2011; Bettencourt 2013; O'Sullivan 2012; Seto et al. 2011).

Since infrastructure networks, despite their geometric variability, share general topological and functional properties in their relationship with places and socioeconomic activity, we can also use these insights to assess the deficit (and possibly the surplus) of infrastructure given the other quantities. These methods rely on the topological network properties of places within blocks and whether they are universally connected to infrastructure meaning that all buildings have accesses. These methods generate algorithmic optimization strategies that start by analyzing maps, extracting city blocks and associated access network topological characteristics and, if necessary, propose extending existing street networks (growing the accesses graph) until the topology associated with universal connectivity is reached at minimal cost (Brelsford et al. 2018, 2019).

Another interesting application of network analysis in cities, typically using cell-phone data or other types of individual geolocation, deals with the detailed spatialization of socioeconomic activity. This enables us to identify people's activity spaces (Carrasco et al. 2008a; Wang et al. 2015; Zhong et al. 2015) and analyze when these overlap with each other and with points of interest, including intentional functions designed into the built environment (libraries, parks, recreation spots, and so on). These observations allow us to expand insights of social network surveys about where different types of interactions take place. For example, it is typical that downtown areas concentrate most large, mixed forms of connectivity, and that individuals who spend a great deal of time in these environments (statistically younger people and professionals) tend to display larger and more diverse social networks (Wang et al. 2015). Residential neighborhoods however, and especially suburbs, can be more segregated and can be more dedicated to non-economic activities (Sampson 2012). Ideas of neighborhood effects and, more generally, of opportunity in cities are thereby being transformed through a rich combination of mapping socioeconomic interactions over the built spaces of cities (Wang et al. 2018).

In all these ways, network structures revealing the relationships between places, socioeconomic activity and access networks can become the basis for new approaches to urban planning and policy that seek to develop infrastructure and amenities that are well matched and supportive of socioeconomic life and the realization of personal and institutional opportunities for well-being and development (Bettencourt 2014b, 2019, 2021).

4. INTERCITY FLOWS

In geography, over the past decades, there have been two distinct general arguments motivating modeling cities. The first starts with cities' internal structure, as we have done here in the previous sections, and from this perspective subsequently considers the interactions between towns and cities. The second approaches cities in the reverse order: it considers the flows of people goods and information between sites, and assumes that these external flows are the main forces determining the characteristics of cities (Batty 2017; Berry 1964; Berry and Garrison 1958a; Christaller 1966; Fujita et al. 2001; Pred 1977; Swerts and Pumain 2013). A seemingly different but related question deals with the conditions under which a set of cities can be considered independent versus when they become interrelated and 'act as one', forming an urban system (Batty 2006; Berry and Garrison 1958a; Christaller 1966; Bettencourt and Zünd 2020).

These questions create a classic example of a bottom-up versus a top-down dilemma. As in any complex system, the preference for one approach over the other is moot: both effects are important and must be taken together as a self-consistent set of properties of cities. In this section, I try to show how these network connections, of different quantities and at different scales, are interrelated and form a consistent whole.

The best studied case of intercity flows, and naturally the place to start, deals with migration, that is, the long-term movement of people changing their place of residence from one city to another (Bettencourt and Zünd 2020; Greenwood 1997; Molloy et al. 2011; Simini et al. 2012).

Intuitively, the question of the urban system hinges on how strong the flows of people between its constituent cities are. At the extreme limit of no migration (and no other flows), cities would be completely independent from each other, and their dynamics would be the result of only internal forces, which would be idiosyncratic across places. However, if intercity flows are strong, then their populations will mix, thus bridging different places. Their identity and characteristics, including their skills, knowledge and culture, will become part of several cities and, over time, a collective property of the full set of cities in such a network of exchange, which then form an urban system.

Demographers and ecologists know the quantitative implication of strong inter-population flows for their development and growth (Caswell 2001; Cohen 1979). Their models, similar to the model for human populations in an urban system, are based on processes of births, death and migration. The main results, known as a set of population ergodic theorems, is that if migration flows between cities form a connected graph then, at a sufficiently long time set by the strength of the flows, the set of cities will act as a system (Batty 2017; Bettencourt and Zünd 2020; Caswell 2001). This means, in the simplest formulation, that there is a dominant eigenvector of the flow matrix to which the distribution of relative city sizes converges, and that all cities will share the same population growth rate (Gibrat's law) (Gabaix 1999; Gibrat 1931; Sutton 1997). Where the flows are stochastic, the situation becomes less stringent but there are still guarantees about the statistics of city size variables and their rates of change.

There is an important particular simplification of this picture often associated with urban systems, and specifically with their relative size distribution statistics. When flows in and out of each city are balanced and vanish on the net (as occurs under the standard gravity law of migration) (Greenwood 1975; Stewart 1941; Wilson 2013; Zipf 2012), only random fluctuations of these flows away from zero drive the system dynamics. The resulting effect is that city sizes will now behave effectively as random multiplicative processes and, under some boundary conditions on the behavior of the smallest places, the distribution of city sizes may become very broad and approach the rank-size rule (Zipf's law) (Gabaix 1999; Saichev et al. 2010). This is a very special limiting behavior of the growth dynamics of a system of cities. Its necessary conditions are typically not observed exactly, so that the size distribution (Bettencourt and Zünd 2020) (measured in population or other quantities, such as built-up area or gross domestic product) tends to only approximately resemble Zipf's law. Departures from Zipf's law are most common for small cities (Eeckhout 2004), which Zipf's law would predict to be much more common than typically observed. For example, according to Zipf's law, if the largest city in the system has a population of 20 million (as in the USA, Brazil, China or India) then there should be a thousand cities of size 20000 and a million towns of size 20, which is counterfactual. Moreover, whenever there is a tangible, persistent reason for a city to grow faster or slower than others, Zipf's law cannot apply. These scenarios tend to not be universal as they depend on particular aspects of each place and time, but they are also common and indeed intentional, involving stated policy efforts and expenditures, such as in China currently (Bettencourt and Zünd 2020; Monkkonen et al. 2016; Rozenblat et al. 2018; Zhao 2003).

These results generalize promptly to other flow networks, including those of money, goods and information. The exchanges of these quantities between cities are typically less tracked by official statistical bureaus. In the USA, there is a long-standing commodity survey (Essletzbichler 2015), which tracks the transportation of physical goods between counties and metropolitan areas. There are also estimates for the value of exports and imports originating in US metropolitan areas (Marchio and Parilla 2018), though these are largely imputed based on national data. In other nations, to my knowledge the situation is even more uncertain, with very little to no empirical information on flows of goods, services, energy or information between cities, other than migration.

A related set of network properties for the flows between cities deals with their functions relative to city size, which we have discussed previously to some extent. There are general theoretical expectations, developed in the context of central place theory (Berry and Garrison 1958b; Christaller 1966) and its extensions, and in the context of general systems theory (Odum 2007) that the economic activity of cities, and their energetic and resource needs, are interconnected by a system of flows with specific characteristics. These ideas also frame concepts of industrial ecology (Bettencourt and Brelsford 2015) towards mapping urban metabolism (Kennedy et al. 2011; Warren-Rhodes and Koenig 2001), defined as comprehensive input–output models of urban areas, for various inflows and outflows of resources, which becomes a centerpiece of analysis for making cities more sustainable in relation to their impact on external environments.

While there is currently no universal theory to predict the nature and magnitude of these inter-city flows, concepts of central place theory and general systems theory emphasize that cities of different sizes develop complementary functions, as do people inside cities. In the standard picture, larger cities develop typically as seats of political governance, financial and innovation services, centers for assembly and coordination for simpler goods and products. As a consequence, larger cities require inflows of energy and resources from smaller places. These places tend to then focus on resource extraction, agriculture and other forms of primary production that can be traded up the urban hierarchy to larger cities. This type of relative functional differentiation by city size constitutes an urban hierarchy (Batty 2006; Rozenblat et al. 2018), where non-rival flows of informational quantities from larger cities can service many smaller places in exchange for rival physical goods, resources and energy.

Moreover, these exchanges can have a significance that goes beyond trade. In general systems theory, this system of flows is finely balanced so that information generated in larger cities controls and coordinates (in the sense of engineering theory) the functioning of the entire urban hierarchy, creating a self-reinforcing, non-linear dynamics that encourages the efficient production of basic resources and energy in smaller places that, in turn, flow up to larger cities, which continue to research and coordinate the properties of such production and flows (Odum 2007). This integration at the level of the system of cities is then capable of ensuring coordinated economic growth across the entire urban system, creating value in small cities owing to increases in (labor) production efficiencies of basic goods and energy, while at the same time enabling both the growth of cities and the expansion of their socioeconomic networks, thus also creating new knowledge and value, especially in research, product development and new services.

Although these ideas have been with us for some time, their development into more quantitative predictive theory has been handicapped by the lack of both empirical evidence and interdisciplinary theory at all the necessary scales of the problem. More empirical work needs to be done to test their applicability and generality to urban systems (Bettencourt 2021). It should nevertheless be clear that their development constitutes a promising integrated framework for understanding processes of economic growth, human development and material flows in urbanizing human societies, cutting across scales, from individuals to cities and entire nations and spanning diverse quantities from energy and resources to labor and information.

5. DISCUSSION

Networks are idealized quantitative structures that enable us to model, in a common formalism, many different types of connections, structures and flows. We have seen how social networks, economic transactions, physical roads, pipes and cables, as well as migrating people and intercity trade, create structures of connectivity that are relevant to cities and urban systems and can all be modelled by networks or graphs (Newman et al. 2006).

The type of connections discussed here constitute a general description of cities, both as spaces and as dynamical flows. This description leads to a number of statistical quantities that display empirical regularities across time, space and city sizes, and that have become the focus of integrated urban theory (Bettencourt 2021).

In this chapter, I presented a road map for understanding the interconnected multi-scale structure and main properties of cities and urban system in terms of network quantities. It has not been my intention to produce an extensive review of what is now a very large literature on network models and studies of urban data. Instead, I highlighted significant, high-level unifying themes that, taken together, constitute the basis of a unifying theory of cities and urban systems. From this perspective, it also becomes possible to identify a number of extant problems, which, in my view, will be key for future empirical and theoretical developments.

I highlight three themes. The issue of predictability, the related issue of historic contingency, and the contention that the type of network theory described here will constitute the basis for a predictive statistical dynamics of urbanizing human societies, capable of better accounting for processes of social change, economic growth and human development.

Let us first deal with the issue of predictability: are network analyses of cities mere descriptive statistics, or are they part of a theory capable of making generalizable statements to new situations? Ultimately, this is an empirical question but, in my view, there is plenty of evidence already that many aspects of cities and urban systems are systematic and predictable, in that they can be measured in many urban systems at different times in history, geographies and levels of development, leading to the same essential results, not only in terms of rough statistical regularities but also the same numbers (Bettencourt 2013, 2021; Ortman et al. 2015). These phenomena include city size distributions, urban hierarchy, features of migration flows, the structure of urban infrastructure

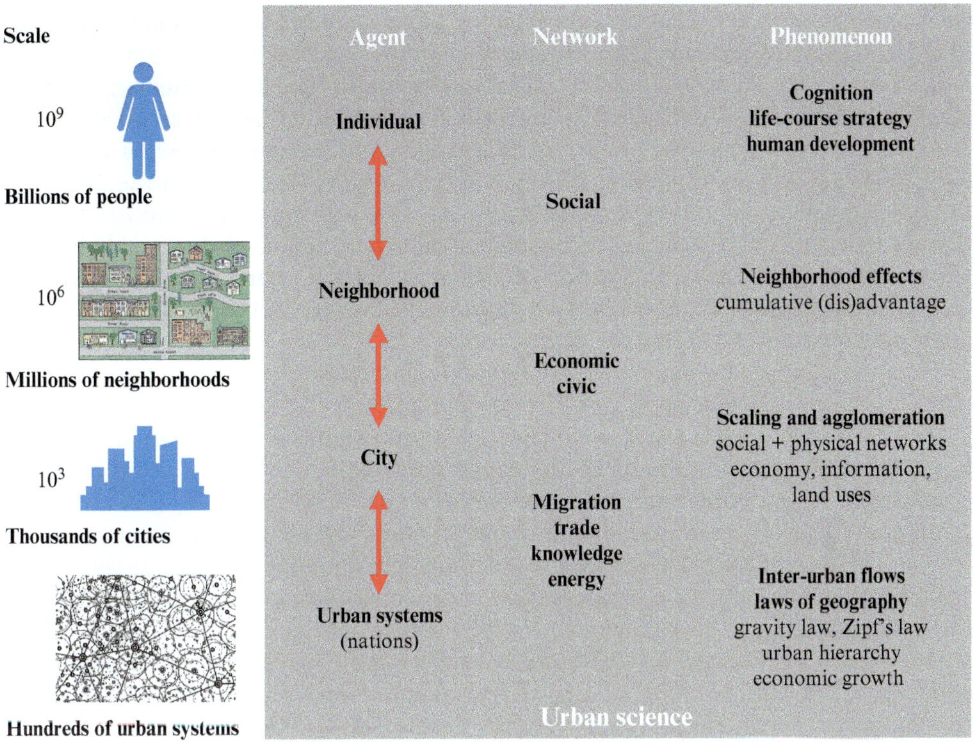

Note: Urban science requires the simultaneous quantitative and predictive modeling of many different networks in cities, at different scales.

Figure 3.2 Quantitative and predictive modeling of different networks in cities

networks, socioeconomic networks, scaling relations and the specific values of their exponents, neighborhood effects, and individual trajectories of cumulative advantage and disadvantage, among others (see Figure 3.2). These regularities constitute the basis for an integrated body of knowledge capable of predictions and generalizations, both qualitative and quantitative. That cities are complex systems does not make them close to what Jane Jacobs termed disorganized complexity (including chaos, sand piles and the like), but rather what she insisted is a problem in organized complexity, describing cities metaphorically as 'organisms that are replete with unexamined, but obviously intricately interconnected, and surely understandable, relationships' (Jacobs 1961, pp. 438–9). Here, we have seen that it is by measuring and understanding these relationships as networks, that we can indeed make them understandable.

Related to the issue of predictability is always the question of historical contingency. All complex systems, including cities, are historical systems (Bairoch 1991; Bettencourt 2020; Bettencourt et al. 2010; Mumford 1989), in that evolutionary dynamics takes place over time and iterates on both intentional choices (selection) and accidents. This is a feature of urban dynamics at all scales, from individual lives to the history of nations as urban systems. Evolutionary dynamics is predictable but in a very different way from models of classical physics or, indeed, of microeconomics, which are deterministic and constructed in order to maximize specific simple idealized quantities, such as an action or utility.

By contrast, modeling evolutionary dynamics requires population thinking, a fundamentally statistical approach and an emphasis on information (Andersen 2004; Frank 2013; Price 1970). Population and temporal averaging produce properties that can often be predicted in relation to both structural features of networks and associated rates of change. Ingredients for this type of modeling are both a strong sense of constraints on the dynamics of the system, and a population statistical approach that highlights the importance of choices and information (Bettencourt 2020). The first issue is already well developed in urban theory, while the second remains a work in progress, raising many interesting questions for future research.

The issue of emphasis on constraints versus evolutionary population dynamics also maps well to models that apply at different time scales, with the former being relevant in the short term and the latter over the longer term. Many of the classical models of cities, embedding the concept of spatial equilibrium, express urban dynamics in terms of constraints. For example, in urban economics and urban scaling theory (Alonso 1977; Bettencourt 2013; Glaeser 2008; O'Sullivan 2012), the budget constraint allocates urban incomes to transportation costs, rents and other consumption, thereby defining the spatial extent of the city via movement costs and incomes. These ideas, supplemented by the basic embedding of social interaction in build spaces, enable the calculation of urban scaling exponents (Bettencourt 2013, 2020; Bettencourt et al. 2014; Ortman et al. 2014), and, when supplemented with models of consumption and production, lead to most models of urban economics (Anas et al. 1998; Glaeser 2008; O'Sullivan 2012).

Note how little an individual's life or free will is constrained by these assumptions. In effect, given that people balance their bank accounts and live in space and time, all else remains free to choose. Thus, it is critical to keep in mind that predictability of a few average urban quantities does not imply the predictability of most others or any substantial loss of human agency or individuality.

This should be obvious, but it is worth repeating. Since urbanites must make a living and obtain urban services on a daily basis, this leaves the manner in which all this is done – which job, home, where, how – largely unspecified. The existence of networks with some average properties specify only a few numbers within a universe of possibilities that are exponential on the number of choices that any individual makes every day. Thus, form and descriptive statistics are not sufficient as a basis for modeling because any observed network only stands in for a general process that requires networks with particular average properties but says very little about their detailed connections and embeddedness in specific people, places and times.

It then becomes a more difficult question of how people's individual exploratory and learning behavior accumulates over time in a population and can lead to the most fascinating properties of cities as environments that create such fast change in human societies (Bettencourt 2020; Jones and Romer 2009; Lucas 1988). This, in my view, remains the most important open question for a dynamic network theory of cities. Many very important questions hinge on understanding processes of economic growth, human development and urban sustainability from the perspective of networks.

REFERENCES

Acs, Z.J. (2004), 'Innovation and the growth of cities', in R. Capello and P. Nijkamp (eds), *Contributions to Economic Analysis*, **266** (1), 635–58.

Acs, Z.J., D.B. Audretsch and M.P. Feldman (1994), 'R & D spillovers and recipient firm size', *Review of Economic and Statistics*, **76** (2), 336–40.

Alexander, C. (1965), 'A city is not a tree', *Architectural Forum*, **122** (1), 58−62.

Alonso, W. (1977), *Location and Land Use: Toward a General Theory of Land Rent*, Cambridge, MA: Harvard University Press.

Anas, A., R. Arnott and K.A. Small (1998), 'Urban spatial structure', *Journal of Economic Literature*, **36** (3), 1426–64.

Andersen, E.S. (2004), 'Population thinking, Price's equation and the analysis of economic evolution', *Evolutionary and Institutional Economics Review*, **1** (1), 127–48.

Anderson, R.M. and R.M. May (2010), *Infectious Diseases of Humans: Dynamics and Control*, repr., Oxford: Oxford University Press.

Andris, C. and L. Bettencourt (2014), 'Development, information and social connectivity in Côte d'Ivoire', *Infrastructure Complexity*, **1**, art. 1, doi:10.1186/s40551-014-0001-4.

Angel, S., J. Parent, D.L. Civco, A. Blei and D. Potere (2011), 'The dimensions of global urban expansion: estimates and projections for all countries, 2000–2050', *Progress in Planning*, **75** (2), 53–107.

Anselin, L., A. Varga and Z. Acs (1997), 'Local geographic spillovers between university research and high technology innovations', *Journal of Urban Economics*, **42** (3), 422–48.

Arrow, K.J. (1971), 'The economic implications of learning by doing', in F.H. Hahn (ed.), *Readings in the Theory of Growth*, London: Palgrave Macmillan, pp. 131–49.

Bairoch, P. (1991), *Cities and Economic Development: From the Dawn of History to the Present*, trans. C. Baider, Chicago, IL: University of Chicago Press.

Barabási, A.L. and R. Albert (1999), 'Emergence of scaling in random networks', *Science*, **286** (5439), 509–12.

Barthelemy, M. (2016), *The Structure and Dynamics of Cities: Urban Data Analysis and Theoretical Modeling*, Cambridge: Cambridge University Press.

Barthélemy, M. and A. Flammini (2006), 'Optimal traffic networks', *Journal of Statistical Mechanics: Theory and Experiment*, art. L07002, doi:10.1088/1742-5468/2006/07/L07002.

Batty, M. (2006), 'Hierarchy in cities and city systems', in D. Pumain (ed.), *Hierarchy in Natural and Social Sciences*, Dordrecht: Springer Netherlands, pp. 143–68.

Batty, M. (2017), *The New Science of Cities*, Cambridge, MA and London: MIT Press.

Berry, B.J.L. (1964), 'Cities as systems within systems of cities', *Papers in Regional Science*, **13** (1), 147–63.

Berry, B.J.L. and W.L. Garrison (1958a), 'Recent developments of central place theory', *Papers in Regional Science*, **4** (1), 107–20.

Berry, B.J.L. and W.L. Garrison (1958b), 'A note on central place theory and the range of a good', *Economic Geography*, **34** (4), 304–11.

Besbris, M., J.W. Faber, P. Rich and P. Sharkey (2015), 'Effect of neighborhood stigma on economic transactions', *Proceedings of the National Academy of Science*, **112** (16), 4994–8.

Bettencourt, L.M.A. (2013), 'The origins of scaling in cities', *Science*, **340** (6139), 1438–41.

Bettencourt, L.M.A. (2014a), 'Impact of changing technology on the evolution of complex informational networks', *Proceedings of the IEEE*, **102** (12), 1878–91.

Bettencourt, L.M.A. (2014b), 'The kind of problem a city is: new perspectives on the nature of cities from complex systems theory', in D. Offenhuber and C. Ratti (eds), *Decoding the City*, Basle: Birkhauser Verlag, pp. 168–79.

Bettencourt, L.M.A. (2019), 'Designing for complexity: the challenge to spatial design from sustainable human development in cities', *Technology Architecture + Design*, **3** (1), 24–32, doi:10.1080/24751448.2019.1571793.

Bettencourt, L.M.A. (2020), 'Urban growth and the emergence statistics of cities', *Science Advances*, **6** (34), eaat8812, doi:10.1126/sciadv.aat8812.

Bettencourt, L.M.A. (2021), *Introduction to Urban Science: Evidence and Theory of Cities as Complex Systems*, Cambridge, MA: MIT Press.

Bettencourt, L.M.A. and C. Brelsford (2015), 'Industrial ecology: the view from complex systems', *Journal of Industrial Ecology*, **19** (2), 195–7.

Bettencourt, L.M.A. and D. Zünd (2020), 'Demography and the emergence of universal patterns in urban systems', *Nature Communications*, **11**, art. 4584, doi:10.1038/s41467-020-18205-1.

Bettencourt, L.M.A., A. Cintrón-Arias, D.I. Kaiser and C. Castillo-Chávez (2006), 'The power of a good idea: quantitative modeling of the spread of ideas from epidemiological models', *Physica A: Statistical Mechanics and its Applications*, **364** (May), 513–36, doi:10.1016/j.physa.2005.08.083.

Bettencourt, L.M.A., J. Lobo and D. Strumsky (2007b), 'Invention in the city: increasing returns to patenting as a scaling function of metropolitan size', *Research Policy*, **36** (1), 107–20.
Bettencourt, L.M.A., J. Lobo, D. Helbing, C. Kühnert and G.B. West (2007a), 'Growth, innovation, scaling, and the pace of life in cities', *Proceedings of the National Academy of Science*, **104** (17), 7301–6.
Bettencourt, L.M.A., J. Lobo, D. Strumsky and G.B. West (2010), 'Urban scaling and its deviations: revealing the structure of wealth, innovation and crime across cities', *PLoS ONE*, **5** (11), e13541, doi:10.1371/journal.pone.0013541.
Bettencourt, L.M.A., H. Samaniego and H. Youn (2014), 'Professional diversity and the productivity of cities', *Scientific Reports*, **4** (1), 1–16.
Bonacich, P. (1987), 'Power and centrality: a family of measures', *American Journal of Sociology*, **92** (5), 1170–82.
Brelsford, C., J. Lobo, J. Hand and L.M.A. Bettencourt (2017), 'Heterogeneity and scale of sustainable development in cities', *Proceedings of the National Academy of Science*, **114** (34), 8963–8.
Brelsford, C., T. Martin and L.M. Bettencourt (2019), 'Optimal reblocking as a practical tool for neighborhood development', *Environment and Planning B: Urban Analytics and City Science*, **46** (2), 303–21.
Brelsford, C., T. Martin, J. Hand and L.M.A. Bettencourt (2018), 'Toward cities without slums: topology and the spatial evolution of neighborhoods', *Science Advances*, **4** (8), eaar4644, doi:10.1126/sciadv.aar4644.
Bresson, A. and S. Rendall (2016), *The Making of the Ancient Greek Economy: Institutions, Markets, and Growth in the City-States*, Princeton, NJ: Princeton University Press.
Buhl, J., J. Gautrais, N. Reeves, R.V. Solé, S. Valverde, P. Kuntz, et al. (2006), 'Topological patterns in street networks of self-organized urban settlements', *European Physical Journal B: Physics of Condensed Matter*, **49** (4), 513–22.
Cardillo, A., S. Scellato, V. Latora and S. Porta (2006), 'Structural properties of planar graphs of urban street patterns', *Physical Review E*, **73** (6), 066107, doi:10.1103/PhysRevE.73.066107.
Carrasco, J.A., B. Hogan, B. Wellman and E.J. Miller (2008a), 'Collecting Social network data to study social activity-travel behavior: an egocentric approach', *Environment and Planning B: Planning and Design*, **35** (6), 961–80.
Carrasco, J.A., B. Hogan, B. Wellman and E.J. Miller (2008b), 'Agency in social activity interactions: the role of social networks in time and space', *Tijdschrift voor Economische en Sociale Geografie*, **99** (5), 562–83.
Caswell, H. (2001), *Matrix Population Models: Construction, Analysis, and Interpretation*, 2nd edn, Sunderland, MA: Sinauer Associates.
Chan, S.H.Y., R.V. Donner and S. Lämmer (2011), 'Urban road networks — spatial networks with universal geometric features? A case study on Germany's largest cities', *European Physical Journal B*, **84** (4), 563–77.
Chetty, R., D. Grusky, M. Hell, N. Hendren, R. Manduca and J. Narang (2017), 'The fading American dream: trends in absolute income mobility since 1940', *Science*, **356** (6336), 398–406.
Chowell, G., H. Nishiura and L.M.A. Bettencourt (2007), 'Comparative estimation of the reproduction number for pandemic influenza from daily case notification data', *Journal of the Royal Society Interface*, **4** (12), 155–66.
Christaller, W. (1966), *Central Places in Southern Germany*, Upper Saddle River, NJ: Prentice-Hall.
Ciccone, A. (2002), 'Agglomeration effects in Europe', *European Economic Review*, **46** (2), 213–27.
Clauset, A., C.R. Shalizi and M.E.J. Newman (2009), 'Power-law distributions in empirical data', *SIAM Review*, **51** (4), 661–703.
Cohen, J.E. (1979), 'Ergodic theorems in demography', *Bulletin (New Series) of the American Mathematical Society*, **1** (2), 275–95.
Courtat, T., C. Gloaguen and S. Douady (2011), 'Mathematics and morphogenesis of cities: a geometrical approach', *Physical Review E*, **83** (3), 036106, doi:10.1103/PhysRevE.83.036106.
Crucitti, P., V. Latora and S. Porta (2006), 'Centrality measures in spatial networks of urban street', *Physical Review E*, **73** (3), 036125, doi:10.1103/PhysRevE.73.036125.
Dalziel, B.D., S. Kissler, J.R. Gog, C. Viboud, O.N. Bjørnstad, C.J.E. Metcalf, et al. (2018), 'Urbanization and humidity shape the intensity of influenza epidemics in U.S. cities', *Science*, **362** (6410), 75–9.
Deleuze, G., F. Guattari, M. Foucault, R. Hurley, M. Seem and H.R. Lane (2012), *Anti-Oedipus: Capitalism and Schizophrenia*, London and New York: Continuum.
Depersin, J. and M. Barthelemy (2018), 'From global scaling to the dynamics of individual cities', *Proceedings of the National Academy of* Sciences, **115** (10), 2317–22.
Duranton, G. and D. Puga (2004), 'Micro-foundations of urban agglomeration economies', in J.V. Henderson and J.-F. Thisse (eds), *Handbook of Regional and Urban Economics*, Amsterdam: Elsevier, pp. 2063–117.
Eagle, N., M. Macy and R. Claxton (2010), 'Network diversity and economic development', *Science*, **328** (5981), 1029–31.
Eeckhout, J. (2004), 'Gibrat's law for (all) cities', *American Economic Review*, **94** (5), 1429–51.
Essletzbichler, J. (2015), 'Relatedness, industrial branching and technological cohesion in US metropolitan areas', *Regional Studies*, **49** (5), 752–66.

Feldman, M.P. and D.B. Audretsch (1999), 'Innovation in cities: science-based diversity, specialization and localized competition', *European Economic Review*, **43** (2), 409–29.

Fischer, C.S. (1982), *To Dwell among Friends: Personal Networks in Town and City*, Chicago, IL: University of Chicago Press.

Fischer, C.S. (1995), 'The subcultural theory of urbanism: a twentieth-year assessment', *American Journal of Sociology*, **101** (3), 543–77.

Frank, S.A. (2013), 'Natural selection. VII. History and interpretation of kin selection theory', *Journal of Evolutionary Biology*, **26** (6), 1151–84.

Fujita, M., P.R. Krugman and A.J. Venables (2001), *The Spatial Economy: Cities, Regions, and International Trade*, Cambridge, MA: MIT Press.

Gabaix, X. (1999), 'Zipf's law for cities: an explanation', *Quarterly Journal of Economics*, **4** (3), 739–67.

Gibrat, R. (1931), *Les inégalités économiques; applications: aux inégalités des richesses, à la concentration des entreprises, aux populations des villes, aux statistiques des familles, etc., d'une loi nouvelle, la loi de l'effect proportionnel*, Paris: Recueil Sirey.

Glaeser, E.L. (2008), *Cities, Agglomeration, and Spatial Equilibrium*, Oxford: Oxford University Press.

Glaeser, E.L. (2012), *Triumph of the City: How Our Greatest Invention Makes Us Richer, Smarter, Greener, Healthier, and Happier*, New York: Penguin Books.

Greenwood, M.J. (1975), 'Research on internal migration in the United States: a survey', *Journal of Economic Literature*, **13** (2), 397–433.

Greenwood, M.J. (1997), 'Internal migration in developed countries', in M. Rosenzweig and O. Stark (eds), *Handbook of Population and Family Economics*, Amsterdam: Elsevier, pp. 647–720.

Griliches, Z. (1991), 'The search for R&D spillovers', Working Paper No. 3768, National Bureau of Economic Research, Cambridge, MA.

Hall, P. (1998), *Cities in Civilization*, New York: Pantheon Books.

Hamilton, M.J., B. Buchanan and R.S. Walker (2018), 'Scaling the size, structure, and dynamics of residentially mobile hunter-gatherer camps', *American Antiquity*, **83** (4), 701–20.

Hausmann, R. and C.A. Hidalgo (2011), 'The network structure of economic output', *Journal of Economic Growth*, **16** (4), 309–42.

Herrera-Yagüe, C., C.M. Schneider, T. Couronné, Z. Smoreda, R.M. Benito, P.J. Zufiria, et al. (2015), 'The anatomy of urban social networks and its implications in the searchability problem', *Scientific Reports*, **5** (1), 1–13.

Hidalgo, C.A. and R. Hausmann (2009), 'The building blocks of economic complexity', *Proceedings of the National Academy of Sciences*, **106** (26), 10570–75.

Hillier, B. (2005), 'The art of place and the science of space', *World Architecture*, **185** (1), 96–102.

Hillier, B. (2015), *Space is the Machine: A Configurational Theory of Architecture*, London: Space Syntax.

Jacobs, J. (1961), *The Death and Life of Great American Cities*, New York: Random House.

Jacobs, J. (1985), *Cities and the Wealth of Nations: Principles of Economic Life*, New York: Vintage Books.

Jiang, B. (2007), 'A topological pattern of urban street networks: universality and peculiarity', *Physica A: Statistical Mechanics and its Applications*, **384** (2), 647–55.

Jiang, B., Y. Duan, F. Lu, T. Yang and J. Zhao (2014), 'Topological structure of urban street networks from the perspective of degree correlations', *Environment and Planning B: Planning and Design*, **41** (5), 813–28.

Jones, C. and P. Romer (2009), 'The new Kaldor facts: ideas, institutions, population, and human capital', Working Paper No. 15094, National Bureau of Economic Research, Cambridge, MA.

Kalapala, V., V. Sanwalani, A. Clauset and C. Moore (2006), 'Scale invariance in road networks', *Physical Review* E, **73** (2), 026130, doi:10.1103/PhysRevE.73.026130.

Kennedy, C., S. Pincetl and P. Bunje (2011), 'The study of urban metabolism and its applications to urban planning and design', *Environmental Pollution*, **159** (8–9), 1965–73.

Kennedy, C.M. (2011), *The Evolution of Great World Cities: Urban Wealth and Economic Growth*, Toronto: University of Toronto Press.

Kintigh, K.W., J.H. Altschul, M.C. Beaudry, R.D. Drennan, A.P. Kinzig, T.A. Kohler, et al. (2014), 'Grand challenges for archaeology', *Proceedings of the National Academy of Sciences*, **111** (3), 879–80.

Kostof, S. and R. Tobias (1999), *The City Shaped: Urban Patterns and Meanings Through History*, Boston: MA: Little, Brown.

Krugman, P. (1991), 'Increasing returns and economic geography', *Journal of Political Economy*, **99** (3), 483–99.

Lämmer, S., B. Gehlsen and D. Helbing (2006), 'Scaling laws in the spatial structure of urban road networks', *Physica A: Statistical Mechanics and its Applications*, **363** (1), 89–95.

Levy, M. (2010), 'Scale-free human migration and the geography of social networks', *Physica A: Statistical Mechanics and its Applications*, **389** (21), 4913–17.

Louf, R. and M. Barthelemy (2014), 'A typology of street patterns', *Journal of The Royal Society Interface*, **11** (101), 20140924, doi:10.1098/rsif.2014.0924.

Lucas, R.E. (1988), 'On the mechanics of economic development', *Journal of Monetary Economics*, **22** (1), 3–42.

Marchio, N. and J. Parilla (2018), 'Export Monitor 2018: exports increase in most of metro America in 2017 amidst rising uncertainty in national trade policy', Brookings Institution, Washington, DC, accessed 6 October 2019 at https://www.brookings.edu/research/export-monitor-2018/.
Marshall, A. (1890), *Principles of Economics*, London: Macmillan.
Marshall, S. (2005), *Streets & Patterns*, London and New York: Spon.
Masucci, A.P., D. Smith, A. Crooks and M. Batty (2009), 'Random planar graphs and the London street network', *European Physical Journal B*, **71** (2), 259–71.
Masucci, A.P., K. Stanilov and M. Batty (2014), 'Exploring the evolution of London's street network in the information space: a dual approach', *Physical Review E*, **89** (1), 012805, doi:10.1103/PhysRevE.89.012805.
Milgram, S. (1969), 'The experience of living in cities: a psychological analysis', invited address to the Division of General Psychology at the Annual Meeting of the American Psychological Association, Washington, DC, September.
Miller, H.J. (2004), 'Tobler's first law and spatial analysis', *Annals of the Association of American Geographers*, **94** (2), 284–9.
Molloy, R., C.L. Smith and A.K. Wozniak (2011), 'Internal migration in the United States', Working Paper No. 17307, National Bureau of Economic Research, Cambridge, MA.
Monkkonen, P., A. Comandon and J. Zhu (2016), 'Economic segregation in transition China: evidence from the 20 largest cities', *Urban Geography*, **38** (7), 1039–61.
Montroll, E.W. and M.F. Shlesinger (1982), 'On 1/f noise and other distributions with long tails', *Proceedings of the National Academy of Sciences*, **79** (10), 3380–83.
Mumford, L. (1989), *The City in History: Its Origins, its Transformations, and its Prospects*, San Diego, New York and London: Harcourt.
Newman, M.E.J., A.-L. Barabási and D.J. Watts (eds) (2006), *The Structure and Dynamics of Networks*, Princeton, NJ: Princeton University Press.
North, D.C. (1992), 'Institutions, transaction costs and economic growth', *Economic Inquiry*, **25** (3), 419–28.
O'Sullivan, A. (2012), *Urban Economics*, 8th edn, New York: McGraw-Hill/Irwin.
Odum, H.T. (2007), *Environment, Power, and Society for the Twenty-First Century: The Hierarchy of Energy*, New York: Columbia University Press.
Ortman, S.G., A.H.F. Cabaniss, J.O. Sturm and L.M.A. Bettencourt (2015), 'Settlement scaling and increasing returns in an ancient society', *Science Advances*, **1** (1), e1400066, doi:10.1126/sciadv.1400066.
Ortman, S.G., A.H.F. Cabaniss, J.O. Sturm and L.M.A. Bettencourt (2014), 'The pre-history of urban scaling', *PLoS ONE*, **9** (2), e87902, doi:10.1371/journal.pone.0087902.
Ortman, S.G., K.E. Davis, J. Lobo, M.E. Smith, L.M.A. Bettencourt and A. Trumbo (2016), 'Settlement scaling and economic change in the Central Andes', *Journal of Archaeological Science*, **73** (September), 94–106.
Park, R.E., E.W. Burgess, R.D. McKenzie and M. Janowitz (2010), *The City: Suggestions for Investigation of Human Behavior in the Urban Environment*, Chicago, IL: University of Chicago Press.
Porta, S., P. Crucitti, and V. Latora (2006a), 'The network analysis of urban streets: a primal approach', *Environment and Planning B: Planning and Design*, **33** (5), 705–25.
Porta, S., P. Crucitti, and V. Latora (2006b), 'The network analysis of urban streets: a dual approach', *Physica A: Statistical Mechanics and its Applications*, **369** (2), 853–66.
Pred, A. (1977), *City Systems in Advanced Economies: Past Growth, Present Processes, and Future Development Options*, New York: Wiley.
Price, D.D.S. (1976), 'A general theory of bibliometric and other cumulative advantage processes', *Journal of the American Society for Information Science*, **27** (5), 292–306.
Price, G.R. (1970), 'Selection and covariance', *Nature*, **227** (August), 520–21.
Puga, D. (2010), 'The magnitude and causes of agglomeration economies', *Journal of Regional Science*, **50** (1), 203–19.
Pumain, D., F. Paulus, C. Vacchiani-Marcuzzo and J. Lobo (2006), 'An evolutionary theory for interpreting urban scaling laws', *Cybergeo: European Journal of Geography*, **343**, doi:10.4000/cybergeo.2519.
Romer, P. (1989), 'Endogenous technological change', Working Paper No. 3210, National Bureau of Economic Research, Cambridge, MA.
Rozenblat, C., D. Pumain and E. Velasquez (eds) (2018), *International and Transnational Perspectives on Urban Systems*, Singapore: Springer Singapore.
Saichev, A.I., Y. Malevergne and D. Sornette (2010), *Theory of Zipf's law and beyond*, Heidelberg and New York: Springer.
Samaniego, H. and M.E. Moses (2008), 'Cities as organisms: allometric scaling of urban road networks', *Journal of Transport and Land Use*, **1** (1), 21–39.
Sampson, R.J. (2012), *Great American City: Chicago and the Enduring Neighborhood Effect*, Chicago, IL: University of Chicago Press.
Sampson, R.J. and J.H. Laub (1990), 'Crime and deviance over the life course: the salience of adult social bonds', *American Sociological Review*, **55** (5), 609–27.

Sampson, R.J. and J.H. Laub (1995), *Crime in the Making: Pathways and Turning Points through Life*, Cambridge, MA: Harvard University Press.

Schlapfer, M., L.M.A. Bettencourt, S. Grauwin, M. Raschke, R. Claxton, Z. Smoreda, et al. (2014), 'The scaling of human interactions with city size', *Journal of The Royal Society Interface*, **11** (98), 20130789, doi:10.1098/rsif.2013.0789.

Seto, K.C., M. Fragkias, B. Güneralp and M.K. Reilly (2011), 'A meta-analysis of global urban land expansion', *PLoS ONE*, **6** (8), e23777, doi:10.1371/journal.pone.0023777.

Sharkey, P. and J.W. Faber (2014), 'Where, when, why, and for whom do residential contexts matter? Moving away from the dichotomous understanding of neighborhood effects', *Annual Review of Sociology*, **40**, 559–79, doi:10.1146/annurev-soc-071913-043350.

Simini, F., M.C. González, A. Maritan and A.-L. Barabási (2012), 'A universal model for mobility and migration patterns', *Nature*, **484** (7392), 96–100.

Simon, H.A. (1955), 'On a class of skew distribution functions', *Biometrika*, **42** (3–4), 425–40.

Smith, A. and A.B. Krueger (2003), *The Wealth of Nations*, repr., New York: Bantam Classics.

Stewart, J.Q. (1941), 'The "gravitation," or geographical drawing power, of a college', *Bulletin of the American Association of University Professors (1915-1955)*, **27** (1), 70–75.

Stier, A., M.G. Berman and L. Bettencourt (2020), 'COVID-19 Attack Rate Increases with City Size', Mansueto Institute for Urban Innovation Research Paper No. 19, University of Chicago, IL, accessed 29 March at https://ssrn.com/abstract=3564464.

Strano, E., V. Nicosia, V. Latora, S. Porta and M. Barthélemy (2012), 'Elementary processes governing the evolution of road networks', *Scientific Reports*, **2** (1), 1–8.

Sutton, J. (1997), 'Gibrat's legacy', *Journal of Economic Literature*, **35** (1), 40–59.

Swerts, E. and D. Pumain (2013), 'A statistical approach to territorial cohesion: the Indian city system', *L'Espace Géographique*, **42** (1), 75–90.

UN-HABITAT (2016), *World Cities Report 2016: Urbanization and Development – Emerging Futures*, New York: United Nations.

Wang, Q., N.E. Phillips, M.L. Small and R.J. Sampson (2018), 'Urban mobility and neighborhood isolation in America's 50 largest cities', *Proceedings of the National Academy of Sciences*, **115** (30), 7735–40.

Wang, Y., C. Kang, L.M.A. Bettencourt, Y. Liu and C. Andris (2015), 'Linked activity spaces: embedding social networks in urban space', in M. Helbich, J. Jokar Arsanjani and M. Leitner (eds), *Computational Approaches for Urban Environments*, Cham: Springer International, pp. 313–36.

Warren Rhodes, K. and A. Koenig (2001), 'Escalating trends in the urban metabolism of Hong Kong: 1971–1997', *AMBIO: A Journal of the Human Environment*, **30** (7), 429–38.

Wellman, B. and S.D. Berkowitz (eds) (1988), *Social Structures: A Network Approach, Structural Analysis in the Social Sciences*, Cambridge and New York: Cambridge University Press.

Williamson, O.E. (1979), 'Transaction-cost economics: the governance of contractual relations', *Journal of Law and Economics*, **22** (2), 233–61.

Wilson, A.G. (2013), *Entropy in Urban and Regional Modelling*, London: Routledge.

Wirth, L. (1938), 'Urbanism as a way of life', *American Journal of Sociology*, **44** (1), 1–24.

Youn, H., L.M.A. Bettencourt, J. Lobo, D. Strumsky, H. Samaniego and G.B. West (2016), 'Scaling and universality in urban economic diversification', *Journal of The Royal Society Interface*, **13** (114), 20150937, doi:10.1098/rsif.2015.0937.

Zhao, Y. (2003), 'The role of migrant networks in labor migration: the case of China', *Contemporary Economic Policy*, **21** (4), 500–511.

Zhong, C., M. Schläpfer, S.M. Arisona, M. Batty, C. Ratti and G. Schmitt (2015), 'Revealing centrality in the spatial structure of cities from human activity patterns', *Urban Studies*, **54** (2), 437–55.

Zipf, G.K. (2012), *Human Behavior and the Principle of Least Effort: An Introduction to Human Ecology*, Mansfield Centre, CT: Martino.

PART II

CITIES AND NETWORKS IN HISTORY

4. Settlement networks and sociocultural evolution

Elizabeth Bogumil and Christopher Chase-Dunn

INTRODUCTION

Anthropologists and historical comparative social scientists have seen value in understanding not only how different locations and cultures are distinct but also where and when sociocultural complexity and hierarchy have emerged over the past 12000 years. Knowledge of where and when the sizes of human settlements[1] have changed over time is useful for testing aspects of general models of sociocultural evolution.[2] This chapter examines the roles of settlement networks in sociocultural evolution using the comparative world-systems perspective. The first section focuses on settlements, and the interaction networks built among them, before the emergence of states. The second section describes how interaction networks were involved in the rise and fall of large settlements and polities and in the emergence of settlements that were larger than any that had existed before. The third section discusses the early emergence of cities and states in Mesopotamia, Egypt, the Indus River Valley, the Yellow River Valley, Mesoamerica and the Andes, and the expansion of empires and their building of empire cities. The fourth section considers the emergence and spread of semiperipheral capitalist city-states, and their roles in the construction of commodity trade networks and the spread of commercializations networks in world-systems in which state power and tributary accumulation remained the main instruments of reproduction. Then we consider how the emergence of the Europe-centered capitalist world-system drove and was driven by the establishment of core world cities and dependent colonial cities. We conclude by providing an overview of the implications of research on the evolution of settlement systems for understanding the contemporary global system of city-regions and possible futures for the world city system.

Anthropological and World Historical Frameworks of Comparison

The emergence of sedentism, the growth of settlements and interaction networks among settlements and polities,[3] are fundamental processes of social change that required the invention of institutions that could facilitate cooperation and enable polities to effectively compete with one another for resources, including territory. The study of settlement-size distributions – the relative sizes of interacting settlements within polities and in networks of independent interacting polities – provides another important window for viewing and comprehending the emergence and growth of human organizational complexity and hierarchy.

An anthropological and world historical framework of comparison allows us to study the evolution of small-scale (prestate) polities as well as the emergence and growth of cities and the rise of more hierarchical of settlement-size distributions. Ethnographers and archaeologists have studied small-scale nomadic polities of foragers (hunter-gatherers). All humans lived in small-scale nomadic polities until about 12000 years ago. Nomadic foragers lived in temporary camps, but they tended to follow yearly migration circuits that brought them back to the same seasonal locations, and so they too had 'settlement systems' in that they utilized geographical space in a patterned way. Around 11 000 years ago, nomadic foragers in the Levant (present-day Lebanon) began staying longer in winter hamlets. This was the beginning of sedentism. Termed the 'Natufian culture' by archaeologists, these first sedentary people gathered natural stands of grain that grew in the rain-watered valleys of the Levant and stored this food so that they could remain in one location through the winter (Bar-Joseph and Belfer-Cohen 1991). These were the first villagers on Earth. They were not horticulturalists. They were sedentary hunter-gatherers, and polities of this kind continued to exist in some locations, such as California, until they were colonized by the expanding modern world-system in the eighteenth and nineteenth centuries of our era. An anthropological framework of comparison enables us to analyze the settlements systems of nomads and the emergence of sedentism, and the roles that nomads and small-scale polities and interaction networks have played in the emergence of larger settlement systems based on horticulture and agriculture. However, in order to comprehend the long-term processes of the development of complexity and hierarchy we need to combine the study of small-scale systems with a world historical framework that examines regional systems in which cities (settlements with at least 50 000 residents) emerged. This occurred in Mesopotamia and Egypt in the early Bronze Age, and these civilizations also invented writing; so now we have both archaeological and documentary evidence for studying them.

The Comparative Evolutionary World-systems Perspective

Chase-Dunn and Hall (1997) reconfigured the world-system perspective that had emerged to study the modern global system (Shannon 1992; Hall et al. 2013) to compare spatially small world-systems with medium-sized regional systems and the current global world-system. They defined world-systems as systemic interaction networks that link settlements and polities in reciprocal interaction webs (networks) that condition the reproduction and change of local social structures.[4] The word 'world' here refers to the world of real interactions (trade, warfare, communications and intermarriage) that reproduce the social structures and institutions of human groups. Worlds were small when transportation and communication technologies imposed a tyranny of distance that constrained the consequences of interaction to extend relatively short distances. These were the small social worlds in which people lived (see also Chase-Dunn and Mann 1998; Chase-Dunn and Lerro 2016).

Chase-Dunn and Hall (1997) noted that different types of systemic interaction often have different spatial scales in small and medium-sized world-systems. Bulk goods (food) networks were usually smaller than political or military networks in which independent polities were making war and security alliances with each another. These political or military networks were smaller than prestige-goods trade networks in which goods that were very valuable relative to their weight were traded in down-the-line[5] exchange networks or carried by long-distance traders. The spatial bounding of world-systems must choose a focal locale and then use the principal of fall-off to determine the distances from that point to the outer edge of the interaction system (Renfrew 1975, 1977). This is usually, at most, two or three indirect (non-contiguous) links. The important point here is that the networks in which individual settlements are linked with other settlements typically cross the borders between independent polities (and so they are international).

Once one has spatially bounded the interaction networks starting from a focal settlement, then the issue of core–periphery relationships can be examined. The modern core–periphery hierarchy is often discussed as the Global North and the Global South. World-systemists see the Global South as comprising small, poor and relatively powerless peripheral countries in Africa, Asia and Latin America and larger semiperipheral countries (that is, Mexico, Brazil, China, Indonesia, India and Russia), or smaller countries at middle levels of economic development (that is, Taiwan, South Korea, Israel and South Africa) that are in the modern semiperiphery. The core comprises most of the national societies in Europe and North America, but also Japan and Australia. However, core–periphery relationships may also be studied in earlier world-systems.[6] Chase-Dunn and Hall (1997) distinguish between core–periphery differentiation, which exists when polities with different degrees of population density[7] systemically interact with one another, and core–periphery hierarchy which exists when some polities exploit and dominate other polities. Chase-Dunn and Hall (1997, ch. 5) also propose that the phenomenon of semiperipheral development is an important cause of sociocultural evolution in which polities in semiperipheral positions are often the agents of increases in the scale and complexity of world-systems.

Human Settlements before the Rise of States

The anthropological framework enables us to study the settlement systems of nomads. Nomads do not wander aimlessly across the face of the Earth. They move to the most easily obtainable and desirable food, water and other necessities. They harvest nature, but nature is seasonal and the harvesting depletes some of the resources that are being taken. The hunting of big game depletes the local supply and so it is easier to move the camp than to travel long distances to hunt. Camps are the settlements of nomadic peoples, and seasonal migration routes are their patterned settlement systems. Camps and migration circuits are the patterned ways in which nomads utilize space in socially constructed ways (Dodgshon 1987). Archaeologists have discovered that seasonal migration circuits became smaller as the population density of North American nomadic foragers rose (Fagan 1991; Nassany and Sassaman 1995) in the transition from Paleoindian big-game hunters to Archaic hunters, who were still nomads but were hunting smaller types of game (see Figure 4.1). The Archaic nomadic bands developed spear-point styles that were distinctive from those of their neighbors, whereas Paleo spear points were much more similar across wide regions of North America. Nassaney and Sassaman (1995) suggest that these regional tool-kit distinctions mean that the Archaic bands were forming distinct local and regional cultural identities.

The shorter migration circuits were a step on the way to sedentism in which nomadic bands move down the food chain from hunting large game to relying on smaller game (which depletes more slowly) and more vegetable foraging. Population density (the number of people per area of land) increased and resource use became more diversified. Both nomadic and sedentary foragers are known to have used fire to increase the growth of food-producing plants and grazing areas attractive to game. This type of activity has been termed 'proto-agriculture' (Bean and Lawton 1976).

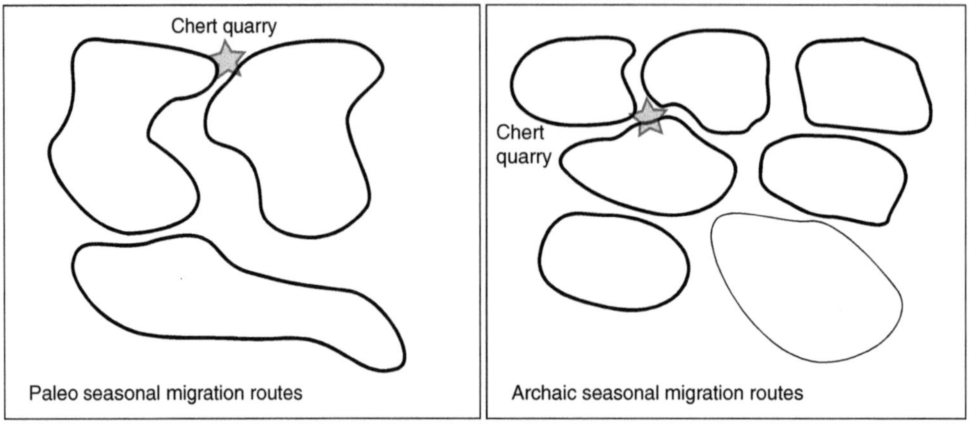

Figure 4.1 Paleo and Archaic migration circuits

The First Villagers

It is commonly believed that all hunter-gatherers were nomadic and that sedentism emerged with horticulture during the 'Neolithic revolution', but this is wrong. Sedentism emerged before horticulture, and sedentary foraging societies survived into recent centuries in some ecologically abundant locations, such as California and the Pacific Northwest. Some hunter-gatherers in prime environments worked out how to exploit less vulnerable natural resources, such as seeds, tubers, small game and fish. They were able to live for a substantial part of the year in semi-permanent winter villages without depleting the environment. These Mesolithic diversified foragers moved to other locations during the summer for seasonal hunting or gathering. Some of the population of the winter villages moved to seasonal camps during the spring and summer. Thus, the emergence of sedentism that began with the transition to smaller seasonal circuits continued in small steps until the winter village became the permanent home of most of the population. The transition from nomadism to sedentism was a matter of seasonal camps becoming occupied for increasingly longer periods of time, and with some of the population remaining, while others went off to other locations during special seasons. The earliest sedentary societies were of diversified foragers in locations in which nature was bountiful enough to allow hunter-gatherers to feed themselves all year without migrating. These early villagers continued to interact with still-nomadic peoples in both trade and warfare. This was the first instance of core–periphery differentiation.

In many Mesolithic world-systems the largest villages comprised only about 250 people. In other regions there were larger villages, and regions with different population densities were often in systemic interaction with one another (Chase-Dunn and Mann 1998). Settlement-size hierarchies emerged when a village at a crucial location, often located at the confluence of two streams, became the home place of important personages and the location of larger non-residential ritual spaces, such as sweat lodges. Sedentary foragers developed long-distance trading networks, and the shift from nomadism to sedentism can be understood as a transition from a system in which people moved to resources to a system in which resources were moved to people.

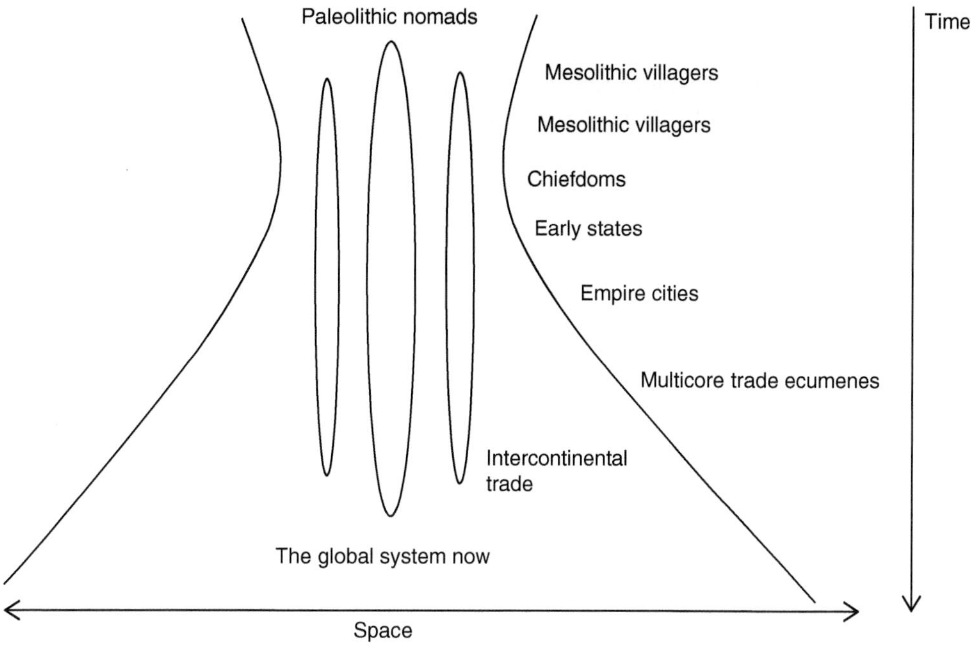

Figure 4.2 Spatial networks shrank and then expanded with the emergence of sedentism

The spatial network aspects of this transition are interesting. As mentioned previously, in the description of the emergence of smaller seasonal migration circuits and regionally differentiated tool styles, nomadic systems went from spatially very large (in that people used larger territories on a part-time basis) to smaller, and to very small with the emergence of sedentism. However, the settlement systems of sedentary peoples began again to get larger if we consider the rise of trade networks that emerged to link settlements that were distant from one another (see Figure 4.2). Exchange between settlements and polities enabled the emergence of greater population density, since villagers were able to obtain needed food during periods of scarcity without resorting to raiding neighboring settlements for supplies. These reciprocal exchange networks continued to grow, though they also occasionally shrank, repeating a pattern that Chase-Dunn and Hall (1997) have termed 'oscillation'. Eventually, the systemic interaction networks extended globally (Earth-wide), and thus were named 'globalization'. So, networks first shrank with the coming of sedentism, and then expanded again to eventually become completely global with the arrival of intercontinental oceanic voyaging.

Sedentism also increased the population growth rate, as village life was far more conducive to child-rearing than was the life of nomads. Families living in permanent villages could afford to have more closely timed children than nomadic peoples could manage (McNeill and McNeill 2003). This speeded up the repeated emergence of population pressure on resources, which was a major driver of sociocultural evolution (Kirch 1984, 2017; Johnson and Earle 1987).

The Hilly Flanks

Sedentism arose in multiple locations throughout the world, which had no known interaction or communication with each other. This is an example of parallel or convergent evolution in which similar circumstances led to similar developmental outcomes. It was in regions adjacent to those that first developed sedentary diversified foraging that Neolithic horticulture first emerged. In the Levant it was neighbors of the Natufians who occupied less naturally fecund upland valleys who first went to the trouble of planting fields of grain (Moore 1982).

When the nomadic foragers in neighboring areas tried to emulate the sedentary lifestyle of the Mesolithic villagers, they found that their less abundant natural stands of grain were quickly eaten up. To increase the productivity of their lands, they experimented with planting some of the seeds that they had gathered. This intervention in nature's production of food resources increased the productivity of land, but required more labor. It allowed more people to be fed per area of land, once again increasing population density (Hayden et al. 1981). This emergence of horticulture in the hilly flanks near the earlier emergence of sedentary foragers was an early instance of semiperipheral development in which jealous hill-dwellers were willing to work harder to catch up with neighboring village-living lowlanders. The emergence of horticulture was a continuation of the trend from large nomadic circuits, to smaller circuits, to sedentism and then to intensified production. Horticulture allowed villages to be larger in population size. As with sedentism, horticulture emerged independently in at least eight world regions – another instance of parallel evolution.

The next steps in intensification of production and the emergence of yet larger settlements occurred with the formation of chiefdoms on the Susiana Plain (in what is now Iran) that used small-scale irrigation systems to increase the productivity of agriculture. The organization and coordination of irrigation projects is favored by the emergence of hierarchy within polities, as has been shown in a recent study of 155 small-scale polities in Austronesia (Sheehan et al. 2018). However, the first chiefdoms on Earth were in a region in the Fertile Crescent not far from where sedentism and horticulture had first emerged. The rise of chiefdoms produced the first two-tiered, settlement-size hierarchies.

SETTLEMENT-SIZE HIERARCHIES

The spatial aspect of population density is one of the most fundamental variables for understanding the constraints and possibilities of human social organization. The settlement-size distribution – the relative population sizes of the settlements within a region – is an important and easily ascertained aspect of all sedentary social systems. Also, the functional differences among settlements are a basic feature of the division of labor that links households and communities into larger polities and interpolity networks. The emergence of social hierarchies is often related to, and inferred from, the size hierarchy of a set of settlements. The building of monumental architecture in larger settlements has been closely associated with the emergence of more hierarchical social structures: complex chiefdoms and early states.

The spatial relationships among settlements in a region, and their relative sizes, can be

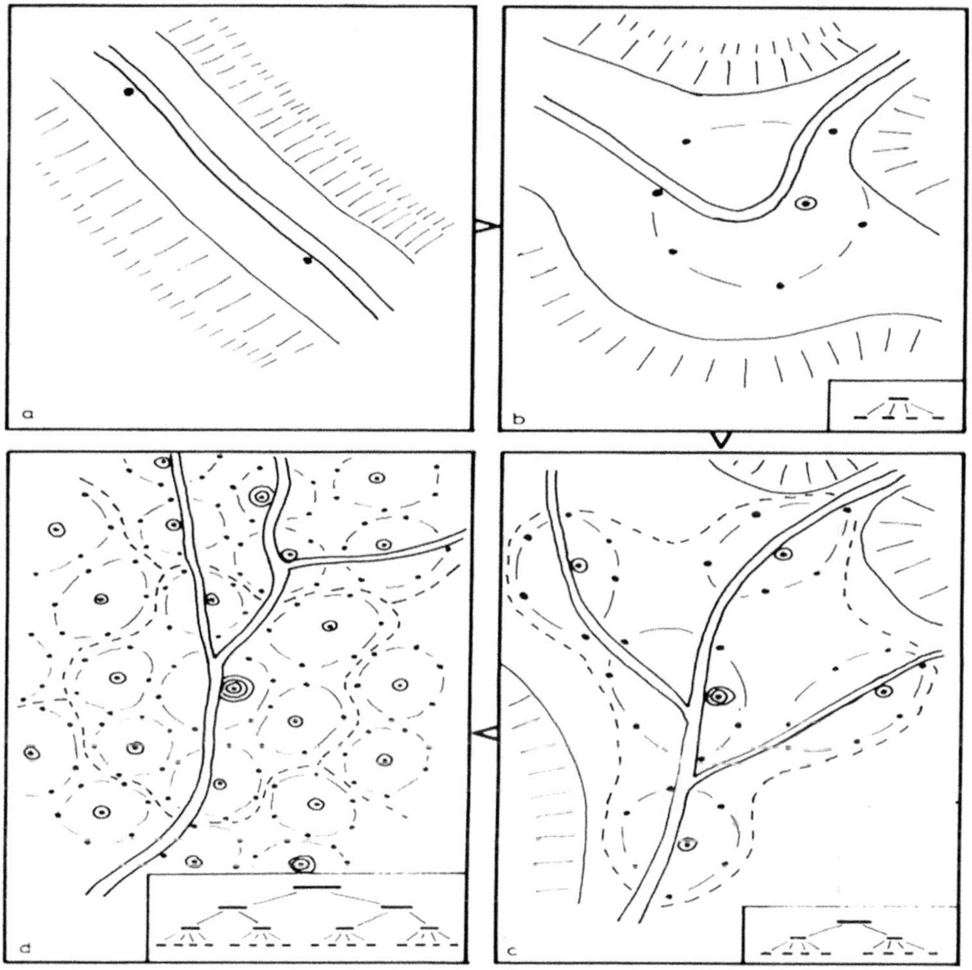

Source: Nissen (1988, p. 42).

Figure 4.3 *Settlement-size hierarchies*

inferred from archeological evidence, and this is an empirically useful pattern that allows us to compare preliterate systems with those for which we have documentary evidence. Figure 4.3 shows Hans Nissen's drawings of one-tiered (a), two-tiered (b), three-tiered (c) and four-tiered (d) settlement-size hierarchies.

Settlement-size distributions are often graphed to show the relative population sizes of the settlements in a region. Figure 4.4 shows a steep settlement-size distribution in which the largest settlement is substantially larger than the second largest, and a flat distribution in which all the settlements are about the same size. Urban geographers suggest that a spatial-size hierarchy is often related to the distribution of functions across settlements and transportation costs (Christaller 1966). Production of goods and services that can easily be distributed across a whole region from a central place will tend to be located

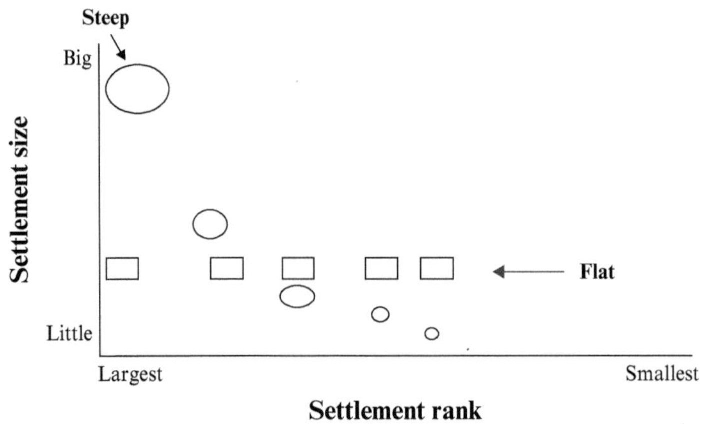

Figure 4.4 Flat and steep settlement-size distributions

in the largest settlement, whereas products that cannot easily be stored or transported will be produced locally in the smaller settlements. The range of goods creates the space economy. This approach was developed to describe market societies, but is also relevant for understanding settlement systems in which exchange was organized as reciprocity, as transportation costs and labor time had to be considered in the effort to be generous.

Urban geographers contend that there is a tendency for settlement-size hierarchies to approximate a rank-size or lognormal size distribution. A rank-size distribution exists when the second largest settlement is half the size of the largest, the third largest is one-third the size of the largest, and so on. A lognormal distribution is similar in shape. It exists when the ranked population sizes of settlements in a region fall on a straight line when the population sizes have been transformed to a logarithmic scale. Urban primacy is said to exist when the largest city in a region is larger than would be expected based on the rank-size or lognormal distributions – larger than twice the size of the second largest settlement. Empirical studies have shown that many settlement-size hierarchies do approximate the rank-size rule, but some are flat (as with a one-tiered size distribution) and many are primate (for example, Chase-Dunn 1985).

THE RISE OF CITIES AND STATES

The original gardening (horticulture) in the Levant spread both west into the valley of the Nile and east toward Mesopotamia. Larger villages and towns emerged in rain-watered regions, and population growth led to the migration of farmers away from the original heartland of gardening. Horticultural techniques also diffused from group to group and were combined with the domestication of pigs, sheep and goats. Domestication of animals and the use of dairy products and meat from domesticated animals eventually made it possible to move a few notches back up the food chain, reversing some of the descent that had followed the depletion of big game. Still-nomadic hunter-gatherers traded with the Neolithic towns, and new forms of specialized pastoral nomadism developed based on the herding of domesticated animals in areas that were adequate for pasture but not for farming.

The simple model here is that technological development (planting and animal husbandry) increased population density, and this facilitated the emergence of larger settlements and social hierarchies. McNeill and McNeill (2003) noted that state formation in East Asia followed the spread of rice cultivation (for example, from China, to Korea, to Japan). They contend that storable and tradable grain is far more conducive to state formation than are crops that are not easily stored (such as yams).[8] However, there is evidence from the Chesapeake region of indigenous North America that the adoption of planting of storable food does not always immediately lead to greater complexity and hierarchy (Chase-Dunn and Hall 1999).

The arrival of maize-planting in the Chesapeake region allowed the formerly Mesolithic diversified foragers living in large villages to redisperse into widely spread farmsteads, and to reduce the intensity of their trading and ritual symbolization of group identity and social hierarchy. So, increasing productivity can, under some conditions, lead to deconcentration and less social hierarchy. The ability to produce a surplus does not automatically lead to hierarchy formation. Surplus production must be possible in order to support non-producers, but hierarchy formation tends to occur when population pressures have led to increasing levels of conflict. In the example of the Chesapeake, the arrival of maize reduced population pressure, and so those hierarchies and larger villages that had already emerged went into decline. It was not until population pressure had returned after a period of population growth and increasing competition for land that villages and hierarchies grew again. New hierarchies emerged to regulate access to scarce resources and to reduce the intensity of conflict (Chase-Dunn and Hall 1997, ch. 6).

As villages eventually grew larger, so did trade networks. Craft specialists began producing for export and importing raw materials. Trade networks expanded and intensified, but not permanently. All networks exhibited a pattern of expansion and contraction (oscillation), and these waves were punctuated by occasional upward sweeps that connected much larger regions. Thus, the waves of globalization and deglobalization that have been shown to have occurred in the past 150 years (Chase-Dunn et al. 2000), are only the most recent and largest instances of a much older pattern of exchange network expansion and contraction.

To the Flood Plain

According to Nissen (1988, ch. 3) the first three-tiered settlement system in Southwest Asia emerged on the Susiana Plain (in what is now in Iran adjacent to the Mesopotamian flood plain) during the Ubaid period (5500–4000 BCE). This would indicate the presence of complex chiefdoms, and Wright (1986) notes the importance of the existence of complex chiefdoms in a region as the necessary organizational prerequisite for the emergence of pristine states (states that emerge in a context in which there have been no previous states). That is, the first states did not emerge directly from egalitarian societies. Evidence from Uqair, Eridu and Ouelli shows that there were also Ubaid sites on the Lower Mesopotamian flood plain that were as large as the sites on the Susiana Plain. The Early Ubaid phase at Tell Ouelli shows remarkably complex architecture as early as anything on the Susiana Plain. Thus, there was an inter-regional exchange and warfare network of chiefdoms based on a mix of rain-watered and small-scale irrigated agriculture. This was the context in which cities and states first emerged on Earth.

In the next period (Uruk or Late Chalcolithic from 4000–3100 BCE) the first true city (Uruk) grew up on the floodplain of lower Mesopotamia, and other cities of similar large size soon emerged in adjacent locations. Uruk had a peak population of about 50000. Surrounding these unprecedentedly large settlements were smaller towns and villages that formed the first four-tiered settlement systems (Adams 1966). Other cities soon emerged on the floodplain, constituting a world-system of city-states. For seven centuries after the emergence of Uruk, the Mesopotamian world-system was an interactive network of city-states competing with one another for land, glory and control of the complicated transportation routes that linked the Mesopotamian floodplain with the peoples and natural resources of adjacent regions.

Mesopotamia was the original heartland of civilization, understood as the combination of irrigated agriculture, writing, cities and states. States also emerged later in the Uruk period on the Susiana Plain (Wright 1998) and these also developed four-tiered settlement systems (Flannery 1998, p. 17). This was an instance of uneven development; the transition from an inter-regional inter-chiefdom system to an inter-city-state system that emerged first in Mesopotamia and then spread to the adjacent Susiana plain.

Cities and states grew larger with the development of social complexity, but they did not grow smoothly. Instead, there were cycles of growth and decline and sequences of uneven development in all the regions of the world in which cities and states emerged. The invention of new techniques of power, production and exchange ultimately made possible the more complex and hierarchical societies that emerged. The processes of uneven development by which smaller and newer semiperipheral settlements overcame and transformed larger and older settlements has been a fundamental aspect of sociocultural evolution since the invention of sedentary life. Also, sociocultural evolution is analogous to ecological succession. Higher levels of complexity cannot emerge directly out of low levels. Production of surplus and the formation of chiefdoms must form the institutional soil out of which state formation can grow. However, states do not form automatically. They are the products of human innovation in a situation in which they have been made possible by earlier developments, and in which they can solve problems that are posed under current conditions. Thus, there is a degree of historical contingency and agency in the process, and this is evident in the pattern of semiperipheral development, since it is not all semiperipheries that transform the systems of which they are a part.

Sedentary–Nomadic Coevolution

Sedentary societies interacted with still-nomadic societies from the beginning of sedentism. The original core–periphery division of labor was between sedentary and nomadic foragers. As sedentary societies developed horticulture, they also domesticated plants and animals, especially in ancient Southwest Asia and Egypt. Domesticated wheat and barley retain their grains longer and are easier to harvest than are the wild varieties. Foragers had lived with dogs for millennia, but goats, sheep, pigs, cattle, donkeys, horses and cattle were domesticated by either sedentary peoples, or by nomadic peoples who became pastoralists in interaction with sedentary farmers. Archaeologists have studied farmer-forager interactions in many contexts (for example, Gregg 1988, 1991). Nomadism co-evolved with sedentism, a process most famously described for East and Central Asia by Owen Lattimore (1940 [1951]) where Central Asian steppe nomads became specialists in the herding of sheep and horses, trading meat and animals with farmers along the steppe frontier. Nomadic pastoralists played an important role in the formation of sedentary states and empires because they not only supplied meat and animal products, but they soon underwent political evolution as well. The Central Asian steppe nomad confederacies were able to mobilize large cavalries to attack agrarian empires, and they pillaged and extracted tribute in a process most clearly described by Thomas Barfield (1989). It was also former nomadic pastoralists who formed sedentary semiperipheral states on the edges of old core regions, and who were often the protagonists of empire formation when they conquered adjacent core states. Thus, the dynamics of sedentary–nomadic relationships played an important role in the evolution of the tributary modes of accumulation – the uses of institutionalized coercion to structure a system of hierarchical exploitation and domination.

Power and Size: Cities and Empires

What is the relationship between the sizes of settlements and power in intergroup relations? Under what circumstances does a society with greater population density have power over adjacent societies with lower population density, and when might this relationship not hold? Population density is often assumed to be a sensible proxy for relative societal power. Indeed, Chase-Dunn and Hall (1997) employ high relative population density as a major indicator of core status within a world-system. However, Chase-Dunn and Hall are careful to distinguish between core–periphery differentiation and core–periphery hierarchy. Only the latter constitutes actively employed intersocietal domination or exploitation, and Chase-Dunn and Hall warn against inferring power directly from differences in population density.

In many world-systems, military superiority is a key dimension of intersocietal relations. Military superiority is generally a function of population density and the proximity of a large and coordinated group of combatants to contested regions. The winner of a confrontation is the group that can quickly bring the larger number of combatants together. This general demographic basis of military power has been transformed by military technology, including transportation technologies. Factors including better weapons, better training in the arts of war, faster horses, better boats, greater solidarity among soldiers and their leaders, as well as advantageous terrain, alter the simple correlation between population size and military power.

Ironically, George Modelski's (2003) important study of the growth of world cities completely ignored the growth of states and empires, though Modelski is himself an astute scholar of international relations and geopolitical power. Modelski contended that cities were the most important driving force of world-system evolution and that we may conveniently ignore states and empires. The relationship between political power and settlements itself evolved over the millennia, so that analysis of the relationship between size and power is necessary to understand what happened.

The most important general exception (from a comparative evolutionary perspective) to the size–power relationship is the phenomenon of semiperipheral development mentioned previously. The pattern of uneven development by which formerly more complex polities lost their place to less-developed polities has taken several forms, depending on the institutional terrain on which interpolity competition was occurring. Less relatively dense semiperipheral marcher chiefdoms often conquered older core chiefdoms to create larger paramount chiefly polities (Kirch 1984). Similarly, semiperipheral marcher states, usually recently settled peripheral peoples out on the edge of an old region of core states, frequently were the agents of a new core-wide empire based on conquest (Mann 1986; Turchin 2003).

Another exception is the phenomenon of semiperipheral capitalist city-states – polities in the spaces between tributary empires and states that specialized in long-distance trade and commodity production.[9] Although these were rarely the largest cities within the world-systems dominated by tributary empires, they played a transformational role in the expansion of production for exchange and commodification in the ancient and classical world-systems. Also, less-dense semiperipheral Europe was the locus of a virile form of capitalism that condensed in a region that was home to many unusually proximate semiperipheral capitalist city-states. This development, and the military technology that emerged in the competitive and capitalist European interstate system, made it possible for less-dense Europe to achieve global predominance over the more densely populated older core regions of Afroeurasia (Chase-Dunn and Hall 1997; Chase-Dunn and Lerro 2016). The more recent hegemonic ascent of formerly semiperipheral national states such as England and the United States are further examples of the phenomenon of semiperipheral development.

The phenomenon of semiperipheral development does not totally undermine the proposition that societal power and demographic size are likely to be correlated. What it implies is that this correlation can be overcome by other factors, and that these processes are not entirely random. Denser core polities are regularly overcome or out-competed by less-dense semiperipheral polities, but it does not follow that all semiperipheral or peripheral regions have this advantage. On the contrary, in most world-systems most low-density polities are subjected to the power of higher-density polities. Semiperipheral development is an important exception to this general rule.

Why should a settlement system have a steeper size distribution when there is a greater concentration of power? The answer is that large settlements, and especially large cities, require greater concentrations of resources to support their large populations. Therefore, population size has itself been suggested as an indicator of power (Morris 2010, 2013; Taagepera 1978). However, these resources may be obtainable locally and the settlement-size hierarchy may simply correspond to the geographical distribution of resources. People cluster near oases in a desert environment. In this instance, it is not the political or economic power of the central settlement over surrounding areas that produces a centralized settlement system, but the geographical distribution of necessary or desirable resources (water). In many systems, however, we have reason to believe that relations of power, domination and exploitation do affect the distribution of human populations in space. Many large cities are as large as they are because they can draw upon far-flung regions for food and raw materials. If a city can use political or military power, or economic power, to acquire resources from surrounding cities, it will be able to support a larger population than the dominated cities can, and this will produce a hierarchical settlement-size distribution.

The effect can also go the other way. Some cities can dominate others because they have larger populations. Great population size makes possible the assembly of large armies or navies, and this may be an important factor creating or reinforcing steep settlement-size distributions.

The relationship between power and settlement systems is contingent on technology as well as political and economic networks manifested through their institutions. Thus, the relationship between urban growth and decline sequences and the growth or decline sequences of empires varies across different systems or in the same regional system over time as new institutional developments emerge. We know that the development of new techniques of power, as well the integration of increasing larger regions into systems of interaction, production and trade, facilitate the emergence of increasingly larger polities and cities. Thus, there has been a secular trend at the global level and within regions between settlement sizes and polity sizes over the past six millennia.

Studies of the relationship between the rise and fall of empires and the growth or decline phases of the largest cities in the same regions have found differences in the temporal relationship between the growth and decline of largest cities and largest empires. Partial correlations that take out the long-term trend show that the medium-term relationship between city and empire growth is significantly positive in Mesopotamia (2800–650 BCE), South Asia (1800 BCE–1500 CE) and Europe (430 BCE–1800 CE), but not in Egypt, West Asia and East Asia (Chase-Dunn et al. 2005: table 5.2). In the regions where there are significant correlations, this is sometimes owing to the big empires building their own big capital cities, but at other times a big city appears that is outside the largest empire. This suggests that regions go through general phases of expansion and contraction in which cities and empires grow and then decline, and this supposition is confirmed by the finding in all regions of high partial correlations between the growth or decline phases of largest and second largest cities (Chase-Dunn et al. 2005: table 5.3). Surprisingly, there is also a similar set of significant positive partial correlations in all five regions studied between the growth or decline phases of largest and second largest empires (Chase-Dunn et al. 2005: table 5.4). This latter is unexpected as territorial growth is a zero-sum game among adjacent empires, and yet the medium-term temporal correlations are positive, indicating that empires get larger and smaller together within regions. This is strong evidence that regions experience cycles of growth and decline that affect both cities and states.

We should also note an important finding reported in Gilbert Rozman's (1973) comparison of the rise of cities in Japan and China. Rozman notes that, in China, early dynasties built large primate capitals in a context of agricultural villages, but it took 2000 years for this to evolve into a lognormal city-size distribution with middle-sized cities performing intermediary roles linking the capital with the villages. In Japan the process started later, but the emergence of middle-sized cities occurred a great deal more quickly because the Japanese adopted the Chinese organizational and institutional feature that facilitated the emergence of middle-sized cities.

To better understand the timing and location of scale changes in the sizes of cities, George Modelski (2003) improved Tertius Chandler's (1987) compendium of the population sizes of the largest cities since the Bronze Age.[10] Studies of the population sizes of the largest cities in world regions show that there have been cycles in which the largest city in each region increased in size and then decreased in size, and these cycles were punctuated by occasional upsweeps in which the largest city increased to a size that was substantially bigger than that of the earlier size peaks.[11] From studies of the growth and decline of large cities, Modelski (2003) and Fletcher (1995) both noticed the emergence of size ceilings before an upsweep to the next larger ceiling. Fletcher contended that, in addition to resource constraints on the sizes of settlements, large settlements require the invention of institutions that enable large numbers of people to live close to one another. Settlement sizes increased when one polity in a region figured out how to make a large city work, and then other polities in adjacent regions copied, and sometimes improved upon, these techniques, and so their cities were able to catch up in size with the earlier large city. A size ceiling was reached owing to resource and institutional constraints.

The SetPol Research Working Group has identified those instances in which the scale of cities significantly changed (upsweeps and downsweeps) (Inoue et al. 2015) and has begun testing the hypothesis that these scale changes were caused by semiperipheral marcher states (Inoue et al. 2016). In our studies of upsweeps and non-core (peripheral and semiperipheral) marcher states (Chase-Dunn 1988; Inoue et al. 2016) we examined four regional world-systems (Mesopotamia, Egypt, East Asia and South Asia) as well as the expanding central political or military network that is designated by David Wilkinson's (1987) temporal and spatial bounding of state systems since the Bronze Age. We found that nine of the 18 urban upsweeps we studied were produced by non-core marcher state conquests, and eight directly followed and were caused by upsweeps in the territorial sizes of polities (Inoue et al. 2015). Whereas about half of the upsweep events were caused by one or other form of non-core development, there were a significant number of upsweep events in which the causes seem to be substantially internal to the polity that carried out the upsweep (Inoue et al. 2016). Another important finding is that collapses, in which the largest settlements in a region fall below the level of the earlier trough, and stay down for a long period, are rare. Individual cites collapse, but systems of cities rarely do.

THE MODERN WORLD CITY NETWORK

The problem of sustainable urbanization is crucial for the human encounter with the consequences of our ballooning environmental footprint (Gruebler and Buettner 2013). Over half the human population of the Earth now lives in very large cities, and these have spread rapidly over the land as population densities within cities have decreased (suburbanization) and cities have become interconnected into huge city-regions. The size hierarchy of world cities has been flattening as megacities in the non-core countries have caught up, in overall population size, with the global cities of the core. This flattening has important implications for theories of urban growth, globalization and the future of global inequalities. This section considers the conceptualization of world cities and city-regions, and the idea of a global system of cities. We consider the global city-size distribution and the implications of its flattening for the question of the limits of settlement size, and the problems of how to spatially bound cities and city-regions. We further discuss the emergence of low-density and multicentric cities.

The role of city systems in the reproduction and transformation of human social institutions was altered by the emergent predominance of capitalist accumulation. Whereas most of the important cities of agrarian tributary states were centers of control and coordination for the extraction of labor and resources from vast empires by means of institutionalized coercion, the most important cities in the modern world have increasingly supplemented the coordination of force with the manipulations of money and the production of commodities. Many of those who studied the early modern world-system noted the importance of cities as primary nodes through which goods and services flowed and the importance of finance capital and its relationship with state power in the core (Braudel 1984; Arrighi 1994).

The long rise of capitalism was promoted by semiperipheral capitalist city-states, usually maritime coordinators of trade protected by naval power. The Italian city-states of Venice and Genoa are perhaps the most famous of these, but the Phoenician city-states of the Mediterranean exploited a similar interstitial niche within a larger system dominated by tributary empires. These capitalist trading city-states rewired, expanded and intensified the existing networks of exchange. The niche pioneered by capitalist city-states expanded and became more predominant in the guise of core capitalist nation-states in a series of transformations from Venice and Genoa to the Dutch Republic (led by Amsterdam) and eventually the *Pax Britannica* coordinated by the great world city of the nineteenth century, London (Chase-Dunn and Willard 1994). Thus did capitalism move from the semiperiphery to the core, constituting a world-system in which the logic of profit-making had become more important than the logic of tribute and taxation. In 1900 CE London was still the largest world city, but New York was coming up fast (see Figure 4.5).

Within London, the political and financial functions were spatially separated: empire in Westminster and money in the City. In the twentieth-century hegemony of the United

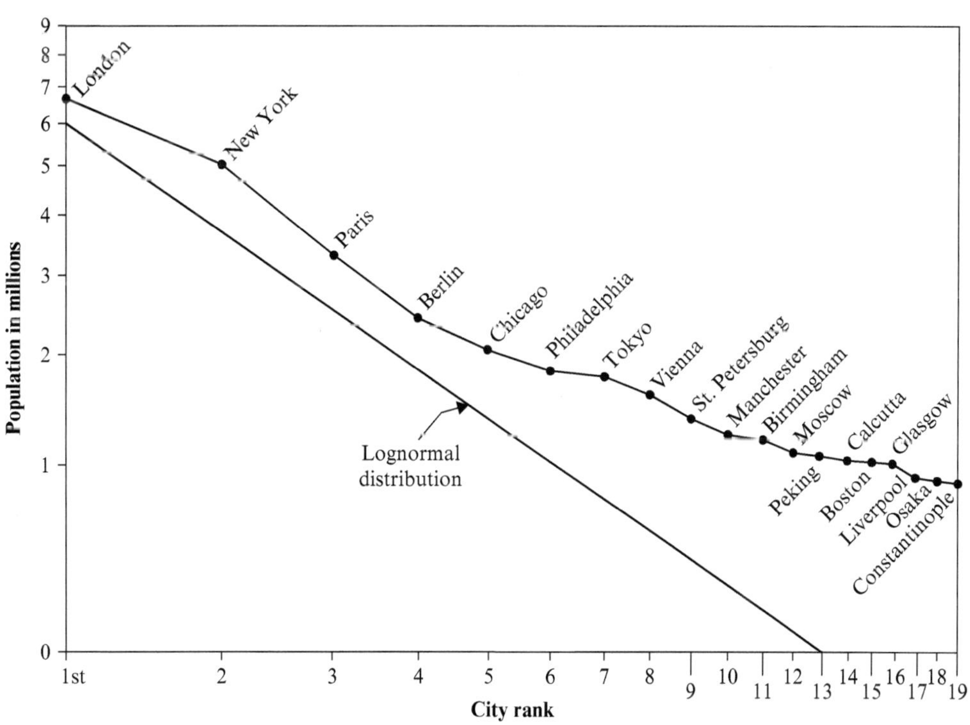

Figure 4.5 The world city-size distribution in 1900 CE

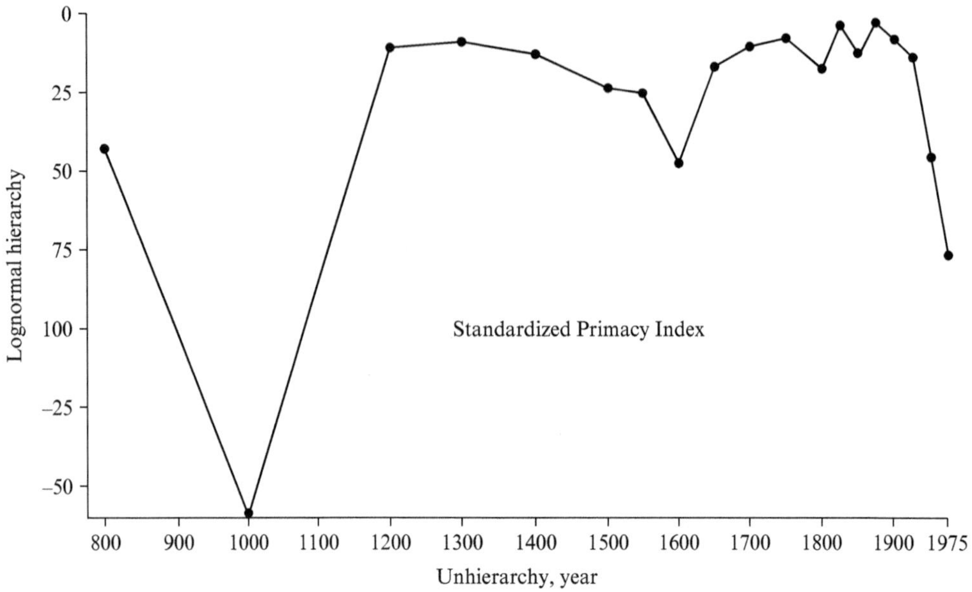

Figure 4.6 Changes in the city-size distribution of the Europe-centered system, 800 to 1975 CE

States these global functions became located in separate cities (Washington, DC, and New York). Thus, the role of cities in world-systems changed greatly as capitalism became the predominant mode of accumulation over the past 500 years. In the tributary world-systems the biggest cities were the capitals of empires based on their ability to extract resources using institutionalized coercion (armies, bureaucracies, and so on). Capitalist cities existed, but they were in the semiperipheral interstices between the large tributary empires. With the rise of Europe, capitalist cities became the most important cities in the whole world-system. Amsterdam, London and New York have been the global centers of the modern world-system.

The relationship between power and size continued to operate (until recently) in the modern system. Figure 4.6 displays changes in the settlement-size distribution of the largest cities in the European-centered world-system since 800 CE (Chase-Dunn and Willard 1994). The city-size distribution of interstate systems is almost always flatter than the size distributions of settlements within a single polity, since the multicentric political structure of interstate networks affects the size distribution of settlements (for example, Figure 4.5). Figure 4.6 uses the Standardized Primacy Index, a measure of deviation from the lognormal rule (Walters 1985). The Europe-centered city system was never steeper than the lognormal distribution and it was occasionally much flatter. The periods of flatness mainly correspond with times of political decentralization in which there was an absence of a hegemonic core power (Chase-Dunn and Willard 1994). The descent into city-size flatness that began in the second half of the twentieth century has continued since then.

WORLD CITIES AND THE GLOBAL SETTLEMENT SYSTEM

According to the theorists of global capitalism it was during the 1970s that the world-system entered a new period in which the system's structure was qualitatively altered by the rise of the neoliberal globalization project (Castells 1996; Robinson 2014). This new model of development displaced the Keynesian national development model that had been hegemonic since World War II. Neoliberalism attacked, and tried to dismantle, the institutions of reformed capitalism that had been built in response to the world revolution of 1917: the welfare state, labor unions that had succeeded in obtaining middle-class incomes for primary sector workers, progressive tax policies, controls on currency manipulations and capital flows, and public education institutions that had expanded access to formerly excluded groups. The neoliberals used advantages stemming from new information and communication technologies to financialize and globalize the world economy, with the export of manufacturing jobs from the core to the semiperiphery and periphery, and the rise of global cities exercising economic power over the global economy (Sassen 1991). John Friedmann's (1986) 'world city hypothesis' identified a set of world cities and the contradictory relationships of globalized production and management and the territorialized political networks of the international system of states.

Although the global city network in recent decades is hierarchical, with world cities at the top (Friedmann 1986; Sassen 2001), according to Alderson and Beckfield (2007) this hierarchy does not exactly mirror the tripartite core–periphery hierarchy since the world-city network became partially decoupled from the geopolitical network of states over the course of twentieth-century globalization. Contrary to the neo-liberal claim that there is now a level playing field (Friedman 2006), the structure of the world-city system network has not become less hierarchical across the era of globalization (Alderson and Beckfield 2007). Indeed, the world-city network became more unequal when some cities, such as New York, London and Tokyo, gained more control over the finances of the world economy relative to other megacities, such as those in Latin America (Timberlake and Smith 2012).

Saskia Sassen and others have further developed the global city hypotheses. Global cities acquired new functions beyond acting as centers of international trade and banking. They became: (1) concentrated control locations in the world economy that use advanced telecommunication facilities; (2) important centers for finance and specialized producer service firms; (3) coordinators of state power; (4) sites of innovative post-Fordist forms of industrialization and production; and (5) markets for the products and innovations produced (Brenner 2004; Sassen 2006). These structural shifts in the functioning of cities 'impacted both the international economic activity and urban form where major cities concentrate control over vast resources, while financial and specialized service industries have restructured the urban social and economic order' (Sassen 1991, p. 4). During the 1990s, New York became specialized in equity trading, London in currency trading and Tokyo in large bank deposits (Slater 2004).

Beaverstock et al. (1999) and Taylor (2003) used Sassen's focus on producer services to classify 55 alpha, beta and gamma world cities based on the presence of office networks and headquarters of accountancy, advertising, banking or finance, and law firms (see Figure 4.7).

The hierarchy of global cities shown in Figure 4.7 provides an important insight into the world settlement system that has emerged in the period of neoliberal global capitalism.

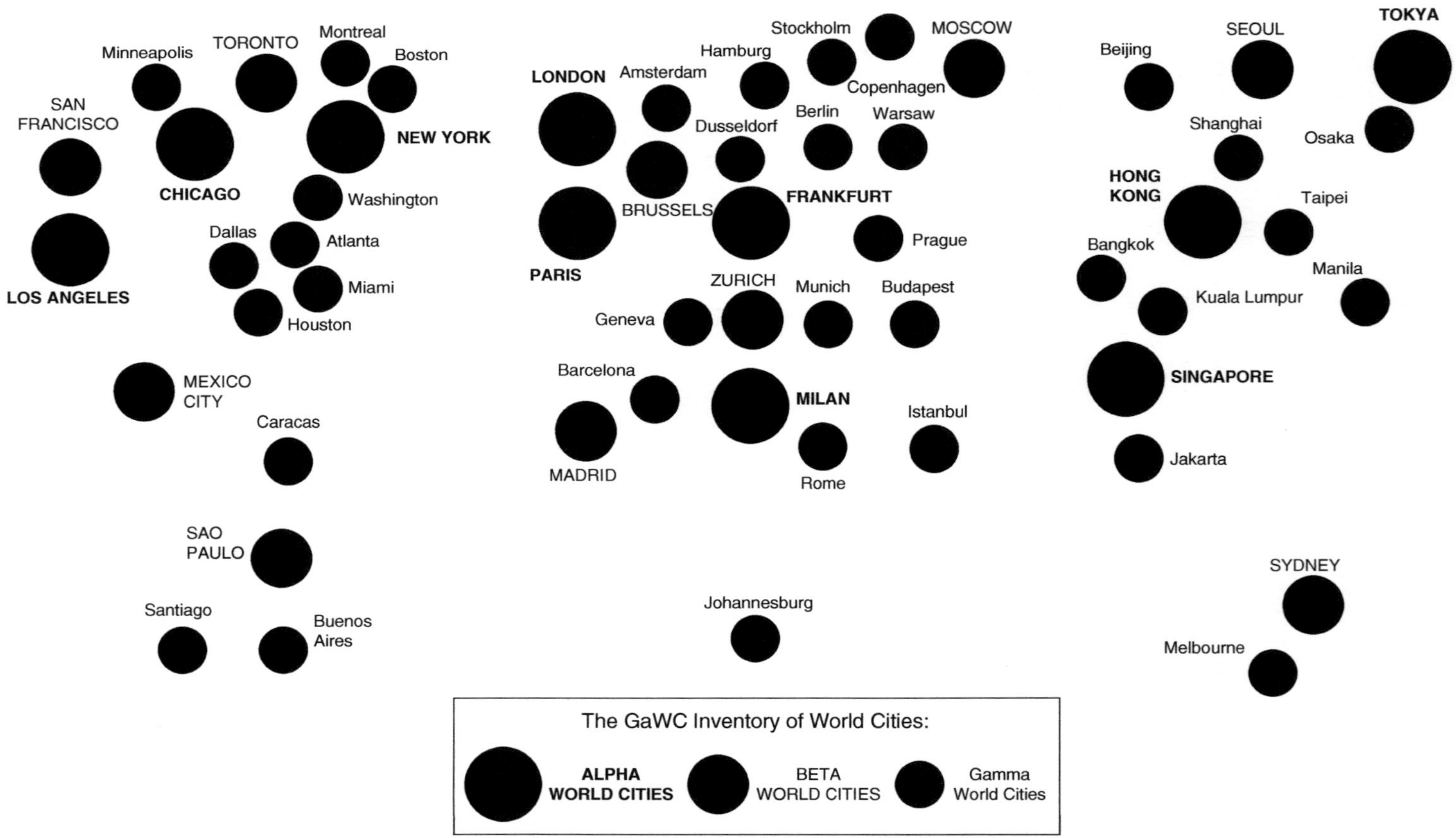

Figure 4.7 *Alpha, beta and gamma world cities according to Beaverstock et al. (1999)*

Although the world city-size distribution based on the population sizes of the largest cities has flattened since 1950, this has not produced a flat globalized world in which a core–periphery hierarchy no longer exists. Instead, the world cities of the Global North continue to concentrate the command and control functions of the world economy, while joined to some extent by megacities in the rising countries of the semiperiphery in Asia and Latin America and including a third tier of megacities in peripheral countries – huge metropolises containing vast numbers of recent migrants from the countryside who subsist in informalized slums lacking regular employment and urban services (Davis 2006).[12]

A large amount of the research on the global city system and affiliated networks has been based on case studies of individual cities that seek to identify the processes leading to their emergence and positioning within the larger system. Janet Abu-Lughod (1999) traced the developmental histories of New York City, Chicago and Los Angeles through their upward mobility in the world city system (see also Taylor and Lang 2005). While these US metropoles share similar characteristics with other world cities, they have substantial differences in geography, original economic functions, transportation and political history that make the stories of each fascinating instances of globalization and local peculiarities (see also Davis 1992).

CITY-REGIONS

Another phenomenon of recent urbanization is the emergence of city-regions, large areas in which big cities are located rather closely to one another and intervening areas are mainly suburbanized. Urban geographers have noted that populations in the rural areas and small towns of core countries are thinning, and people are concentrating in these city-regions (Simmonds and Hack 2000; Scott 2001; Segbers 2007). The city-region phenomenon is made plain by observing global maps that show city lights at night produced from satellite and shuttle images (see Figure 4.8).

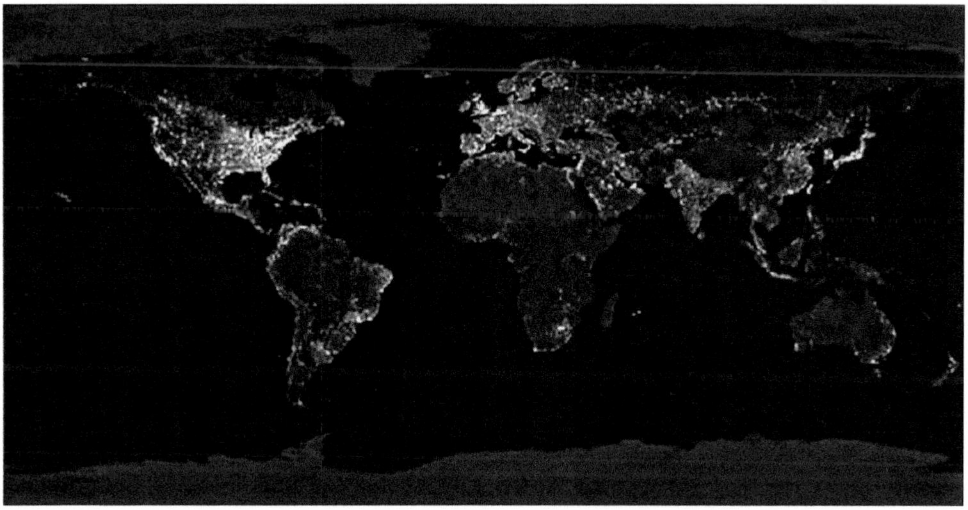

Figure 4.8 City lights photographed from satellites and shuttles

All the continents have city-regions, but the largest are those found in the eastern half of the United States and the western portion of Europe, with several other regions also displaying this phenomenon. These city-regions are linked together spatially by overlapping suburban areas.

The flattening of the global city-size distribution since 1950, shown in Figure 4.6, has several causes. As in earlier periods, the city-size distribution became less hierarchical (flatter) during times in which a global hegemon was in decline. The cities of challenging powers grow, and the cities of the former hegemon cease to grow as fast. The recent descent into flatness of the world city-size distribution is partly due to the declining hegemony of the United States and the rise of challengers with very large cities. Many of the world's largest cities are now in the semiperiphery and the periphery. Changes in the city-size distribution reflect to some extent changes in the distribution of economic and political or military power. The flattening since the 1950s is partly owing to the rapid growth of large cities in Japan and China. However, megacities in the Global South have also grown significantly during the period of recent globalization. These megacities in non-core countries are very different versions, despite their large size. The neoliberal structural adjustment policies of the International Monetary Fund, which were part of the globalization project, imposed austerity on countries of the Global South and encouraged large-scale export agriculture that drove small farmers from the land into the 'planet of slums' (Davis 2006). Another factor that accounts for cities in the Global South catching up with core cities in population size has to do with differences in the demographic transition. Most core countries have achieved a replacement fertility rate, but most of the Global South still has a higher fertility rate and faster population growth, boosting the growth of megacities.

Roland Fletcher (personal communication) contends that contemporary institutional and infrastructural inventions only allow for megacities to function at maximum populations of around 20 million and this serves as a type of ceiling which has allowed cities in the Global South to catch up in population size with the largest cities in the core states. Fletcher's notion of an upper limit on the size of large cities may also be part of the explanation for the emergence of city-regions not gigacities.

The size and density of city-regions are related to global differences in the level of economic development. The global city-region size hierarchy is related to differential economic and political or military power. Countries in the periphery have succeeded in producing very large megacities, but their city-regions are not as large and dense as those in the core (see Figure 4.8).

The Future City of Humans

Over half the 7.6 billion people on Earth now live in cities. The low-density suburban sprawl that has taken over the process of urban growth in the core is immensely expensive in resource use. Urbanization has a huge direct effect on the environment, as cities absorb heat from the sun and then release it, and humans use huge amounts of energy in cities, which contributes to global warming. The urban heat island, in which urban regions are hotter than surrounding areas owing to energy usage and because building materials absorb more heat from the sun (Sailor 2011), is an important phenomenon that is contributing to warming.

Although many core cities have deindustrialized, large cities in the semiperiphery have industrialized and are now the new sites of intense labor struggles (Tilly 1995; Silver 2003). The global reserve army of labor (rural people still not employed in the formal economy) is still large but the long-run tendency for wages, materials and taxes to rise will probably resume and continue, eventually causing a crisis for capitalist accumulation (Wallerstein 2003). The expansion of the sharing economy during the neoliberal crackdown on wages will also challenge the logic of capitalism (Mason 2015).

Peter Taylor (2003) contends that globalization has decreased the importance of nation-states and increased the importance of cities, and that this may be a good thing as cities are more easily governable by communities of citizens. Human settlement systems have been strongly involved in the processes of sociocultural evolution for thousands of years as cybernetic nodes of innovation, and in the processes of uneven development that have led to the rise and fall of states, empires and modern hegemons. Regarding the latter, we can expect that new forms of governance relevant to solutions of the emergent problems of the twenty-first century may be invented and implemented in the cities of the Global South, especially in Brazil, India, Mexico and China. Curitiba, in Brazil, has already successfully demonstrated a new form of sustainable urbanism that will become increasingly relevant as the natural resources that have been the basis of urban sprawl become further depleted (Rabinowitz and Leitman 1996). The pink-tide left populist regimes that emerged in Latin America in reaction to neoliberal globalization (Rodrik 2018) were important supporters of transnational global justice movements that have contested neoliberal global governance and pushed from below toward a new type of globalization. The cities of the Global South remain fertile spaces for finding solutions for our increasingly urbanized planet.

The world settlement system has become a network of large cities and city-regions that contain over half the human population. There are immense inequalities between those powerful global cities that control global financial and military apparatuses that affect everyone, and the huge cities of the Global South that face growing problems of poverty, disease and insecurity. The modern world-system has, over the past several centuries, exhibited cycles of globalization (Chase-Dunn et al. 2000), the rise and fall of hegemonic core powers, and upward trends in population growth and economic development. The massive global inequalities that emerged during the nineteenth century have not been reduced despite the rapid economic development of India and China (Bornschier 2010).

The three main challenges of the twenty-first century are:

- addressing huge environmental issues and moving in the direction of sustainable development;
- addressing the issue of huge inequalities between the global North and South; and
- restructuring global governance to prevent a recurrence of warfare among the great powers and to improve the capacity for managing the other two challenges.

All the large cities of the world will be affected by the timing and rapidity of these emerging crises, and so a plan for restructuring the global system to more effectively meet these challenges is greatly needed. The strengthening and democratization of the United Nations would enhance global cooperation for meeting these challenges and for supporting the empowerment of the peoples of the world.

NOTES

1. The term 'settlement' includes camps, hamlets, villages, towns and cities. Settlements are spatially bounded for comparative purposes as the contiguous built-up area.
2. The term 'evolution' still requires explanation. Here we are discussing sociocultural evolution, not biological evolution. Social science can discover the causes of changes in the degree of complexity and hierarchy of human societies without taking a position on whether or not this has been a good or a bad thing. Ideas about progress and regress are important matters of values but they need not be settled in order to know the causes of patterned social change.
3. We use the term 'polity' to generally denote a spatially-bounded realm of sovereign authority such as a band, tribe, chiefdom, state or empire.
4. The important insight is that all human polities have systemic interactions with their neighbors, so it does not make sense to study them one at time.
5. In down-the-line exchange, goods move from group to group to group.
6. Core, semiperiphery and periphery are relational concepts whose empirical content varies across systems.
7. The size of the largest settlements is one fairly reliable indicator of population density and of core-periphery differentiation. Core polities have larger settlements. Peripheral polities are nomadic or have smaller settlements.
8. Anderson's (2019, p. 27) study of climate change and dynastic cycles in East Asia says: 'Overall, in the last analysis, the dominance of wet-rice agriculture allowed a centralized imperial agrarian regime to dominate even if it had a rather small land base, such as the Kanto Plain in Japan and the Red River Valley in Vietnam.'
9. We define capitalism generally as the accumulation of wealth by means of the production and distribution of commodities and financial services. When the rulers of a specialized trading state obtain most of their wealth from trade and the production of commodities it is a capitalist state.
10. Modelski (2003) used population cut-offs that increase with time to designate world cities. During the ancient era (3000–1000 BCE) world cities must have had a population of 10 000 or more. During the classical era (1000 BCE to 1000 CE) they must have had 100 000 people, and in the modern era (1000 CE to the present) they must have had at least 1 million.
11. These swings and sweeps are being studied by the Settlements and Polities (SetPol) Research Working Group at the Institute of Research on World-Systems at the University of California-Riverside. The project website is at http://irows.ucr.edu/research/citemp/citemp.html (accessed 17 February 2021).
12. Another important hypothesis of the global cities literature is based on Saskia Sassen's (1991) observations about class polarization and the casualization of work within globalizing cities. The research of Gareth Stedman Jones (1971) on Irish immigration into London's East End in the mid-nineteenth century shows that a similar process of peripheralization of the core and precarious labor was occurring during the middle of *Pax Britannica*.

REFERENCES

Abu-Lughod, J.L. (1999), *New York, Chicago, Los Angeles: America's Global Cities*, Minneapolis, MN: University of Minnesota Press.

Adams, R.M. (1966), *The Evolution of Urban Society: Early Mesopotamia and Prehispanic Mexico*, Chicago, IL: Aldine.

Alderson, A.S. and J. Beckfield (2007), 'Globalization and the world city system: preliminary results from a longitudinal data set', in P.J. Taylor, B. Deruder, P. Saey and F. Witlox (eds), *Cities in Globalization: Practices, Policies and Theories*, London: Routledge Taylor and Francis, pp. 21–35.

Anderson, E.N. (2019), *Mandate of Heaven*, Berlin: Springer Verlag.

Arrighi, G. (1994), *The Long Twentieth Century: Money, Power, and the Origins of Our Time*, London: Verso.

Bar-Yosef, O. and A. Belfer-Cohen (1991), 'From sedentary hunter-gatherers to territorial farmers in the Levant', in S.A. Gregg (ed.), *Between Bands and States*, Carbondale, IL: Southern Illinois University, pp. 181–202.

Barfield, T.J. (1989), *The Perilous Frontier: Nomadic Empires and China*, Cambridge, MA: Basil Blackwell.

Bean, L.J. and H. Lawton (1976), 'Some explanations for the rise of cultural complexity in native California with comments on proto-agriculture and agriculture', in L.J. Bean and T.C. Blackburn (eds), *Native Californians: A Theoretical Retrospective*. Socorro, NM: Ballena Press, pp. 19–48.

Beaverstock, J.V., R.G. Smith and P.J. Taylor (1999), 'A roster of world cities', *Cities*, **16** (6), 445–58.

Bornschier, V. (2010), 'On the evolution of inequality in the world system', in C. Suter (ed.), *Inequality Beyond Globalization*, New Brunswick, NJ: Transaction, pp. 39–64..
Braudel, F. (1984), *Civilization and Capitalism 15th-18th Century, Volume III: The Perspective of the World*, London: Collins.
Brenner, N. (2004), *New State Spaces: Urban Governance and the Rescaling of Statehood*, New York: Oxford University Press.
Chandler, T. (1987), *Four Thousand Years of Urban Growth*, Lewiston, NY: St. David's University Press.
Castells, M. (1996), *The Rise of the Network Society*, Cambridge, MA: Blackwell.
Chase-Dunn, C. (1985), 'The coming of urban primacy in Latin America', *Comparative Urban Research*, **11** (1–2), 14–31.
Chase-Dunn, C. (1988), 'Comparing world systems: toward a theory of semipheral development', *Comparative Civilizations Review*, **19** (19), 29–66.
Chase-Dunn, C. and T.D. Hall (1997), *Rise and Demise: Comparing World-Systems*, New York: Routledge.
Chase-Dunn, C. and T.D. Hall (1999), 'The Chesapeake world-system: complexity, hierarchy and pulsations of long range interactions in prehistory', paper presented at the American Sociological Association 1999 Conference, Chicago, IL, 6–10 August, accessed 17 February 2021 at http://wsarch.ucr.edu/archive/papers/c-d&hall/asa99b/asa99b.htm.
Chase-Dunn, C. and B. Lerro (2016), *Social Change: Globalization from the Stone Age to the Present*, New York: Routledge.
Chase-Dunn, C. and K.M. Mann (1998), *The Wintu and Their Neighbors: A Small World-System in Northern California*, Tucson, AZ: University of Arizona Press.
Chase-Dunn, C. and A. Willard (1994), 'Cities in the central political-military network since CE 1200', *Comparative Civilizations Review*, **30** (Spring), 104–32.
Chase-Dunn, C., A. Alvarez and D. Pasciuti. (2005), 'Power and size; urbanization and empire formation in world-systems', in C. Chase-Dunn and E.N. Anderson (eds), *The Historical Evolution of World-Systems*, New York: Palgrave, pp. 92–112.
Chase-Dunn, C., Y. Kawano and B. Brewer (2000), 'Trade globalization since 1795: waves of integration in the world-system', *American Sociological Review*, **65** (1), 77–95.
Christaller, W. (1966), *Central Places in Southern Germany*, Englewood Cliffs, NJ: Prentice-Hall.
Davis, M. (1992), *City of Quartz: Excavating the future in Los Angeles*, New York: Vintage.
Davis, M. (2006), *Planet of Slums*, London: Verso.
Dodgshon, R.A. (1987), *The European Past: Social Evolution and Spatial Order*, London: Macmillan.
Fagan, B.M. (1991), *Ancient North America*, New York: Thames and Hudson.
Flannery, K.V. (1998), 'The ground plans of archaic states', in G. Feinman and J. Marcus (eds), *Archaic States*, Santa Fe, NM: School of American Research, pp. 15–57.
Fletcher, R. (1995), *The Limits of Settlement Growth: A Theoretical Outline*, Cambridge: Cambridge University Press.
Friedmann, J. (1986), 'The world city hypothesis', *Development and Change*, **17** (1): 69–83.
Friedman, T.L. (2006), *The World is Flat: A Brief History of the Twenty-first Century*, New York: Farrar, Straus and Giroux.
Gregg, S.A. (1988), Foragers *and Farmers: Population Interaction and Agricultural Expansion in Prehistoric Europe*, Chicago, IL: University of Chicago Press.
Gregg, S.A. (ed.) (1991), *Between Bands and States*, Carbondale, IL: Southern Illinois University.
Gruebler, A. and T. Buettner (2013), 'Urbanization past and future', in A. Gruebler and D. Fisk (eds), *Energizing Sustainable Cities*, New York: Routledge, pp. 17–30.
Hall, T.D., C. Chase-Dunn and H. Inoue (2013), 'Wallerstein, Immanuel', in R.J. McGee and R.L. Warms (eds), *Theory in Social and Cultural Anthropology*, vol. 2, Thousand Oaks, CA: Sage, pp. 909–12.
Hayden, B., S. Bowdler, K.W. Butzer, M.N. Cohen, B. Druss, R.C. Dunnell, et al. (1981), 'Research and development in the Stone Age: technological transitions among hunter-gatherers', *Current Anthropology*, **22** (5), 519–48.
Inoue, H., A. Álvarez, E.N. Anderson, K. Lawrence, T. Neal, D. Khutkyy, et al. (2016), 'Comparing world-systems: empire upsweeps and non-core marcher states since the Bronze Age', accessed 17 February 2021 at https://irows.ucr.edu/papers/irows56/irows56.htm.
Inoue, H., A. Álvarez, E.N. Anderson, A. Owen, R. Álvarez, K. Lawrence, et al. (2015), 'Urban scale shifts since the Bronze Age: upsweeps, collapses and semiperipheral development', *Social Science History*, **39** (2), 175–200.
Inoue, H., A. Álvarez, E.N. Anderson, K. Lawrence, T. Neal, D. Khutkyy, et al. (2016), 'Comparing world-systems: empire upsweeps and non-core marcher states since the Bronze Age', paper presented at the Pacific Sociological Association 2016 Conference, Oakland, CA, 30 March–2 April.
Johnson, A.W. and T. Earle (1987), *The Evolution of Human Societies: From Foraging Tribe to Agrarian State*, Stanford, CA: Stanford University Press.

Jones, G.S. (1971), *Outcast London: A Study in the Relationship between Classes in Victorian Society*, Oxford: Clarendon.
Kirch, P.V. (1984), *The Evolution of Polynesian Chiefdoms*, Cambridge: Cambridge University Press.
Kirch, P.V. (2017), *On the Road of the Winds: An Archaeological History of the Pacific Islands before European Conquest*, Berkeley, CA: University of California Press.
Lattimore, O. (1940), *Inner Asian Frontiers of China*, New York: American Geographical Society, 2nd edn 1951, Boston: Beacon Press.
Mann, M. (1986), *The Sources of Social Power: A History of Power from the Beginning to A.D. 1760*, Cambridge: Cambridge University Press
Mason, P. (2015), *Postcapitalism*, New York: Farrer, Straus and Giroux.
McNeill, J. and W.H. McNeill (2003), *The Human Web*, New York: Norton.
Modelski, G. (2003), *World Cities: -3000 to 2000*, Washington, DC: Faros 2000.
Moore, A.M.T. (1982), 'The first farmers in the Levant', in R.J. Braidwood, C. Young, P.E.L. Smith and P. Mortenson (eds), *The Hilly Flanks and Beyond: Essays on the Prehistory of Southwestern Asia*, Chicago, IL: University of Chicago Press, pp. 91–112.
Morris, I. (2010), *Why the West Rules – For Now*, New York: Farrer, Straus and Giroux.
Morris, I. (2013), *The Measure of Civilization*, Princeton, NJ: Princeton University Press.
Nassaney, M.S. and K.E. Sassaman (eds) (1995), *Native American Interactions: Multiscalar Analyses and Interpretations in the Eastern Woodlands*, Knoxville, TN: University of Tennessee Press.
Nissen, H.J. (1988), *The Early History of the Ancient Near East, 9000-2000 B.C.*, Chicago, IL: University of Chicago Press.
Rabinovitch, J. and J. Leitman (1996), 'Urban planning in Curitiba', *Scientific American*, **274** (3), 46–53.
Renfrew, C.R. (1975), 'Trade as action at a distance: questions of integration and communication', in J.A. Sabloff and C.C. Lamborg-Karlovsky (eds), *Ancient Civilization and Trade*, Albuquerque, NM: University of New Mexico Press, pp. 3–59.
Renfrew, C.R. (1977), 'Alternative models for exchange and spatial distribution', in T. J. Earle and T. Ericson (eds), *Exchange Systems in Prehistory*, New York: Academic Press, pp. 71–90.
Robinson, W.I. (2014), *Global Capitalism and the Crisis of Humanity*, Cambridge: Cambridge University Press.
Rodrik, D. (2018), 'Populism and the economics of globalization', *Journal of International Business Policy*, **1** (1–2), 12–33.
Rozman, G. (1973), *Urban Networks in Ching China and Tokugawa Japan*, Princeton, NJ: Princeton University Press.
Sailor, D.J. (2011), 'A review of methods for estimating anthropogenic heat and moisture emissions in the urban environment', *International Journal of Climatology* **31** (2), 189–99.
Sassen, S. (1991), *Global Cities*, Princeton, NJ: Princeton University Press.
Sassen, S. (2001), *The Global City: New York, London, and Tokyo*, Princeton, NJ: Princeton University Press.
Sassen, S. (2006), *Territory, Authority, Rights: From Medieval to Global Assemblages*, Princeton, NJ: Princeton University Press.
Scott, A.J. (ed.) (2001), *Global City-Regions: Trends, Theory, Policy*, New York: Oxford University Press.
Segbers, K. (ed.) (2007), *The Making of Global City Regions: Johannesburg, Mumbai, Sao Paulo and Shanghai*, Baltimore, NJ: Johns Hopkins University Press.
Shannon, T.R. (1992), *An Introduction to the World-systems Perspective*, Boulder, CO: Westview Press.
Sheehan, O., J. Watts, R.D. Grey and Q.D. Atkinson (2018), 'Coevolution of landesque capital intensive agriculture and sociopolitical hierarchy', *Proceedings of the National Academy of Sciences of the United States of America (PNAS)*, **115** (14), 3628–33.
Silver, B.J. (2003), *Forces of Labor: Workers' Movements and Globalization since 1870*, Cambridge: Cambridge University Press.
Simmonds, R. and G. Hack (eds) (2000), *Global City Regions: Their Emerging Forms*, London: Spon.
Slater, E. (2004), 'The flickering global city', *Journal of World-Systems Research*, **10** (3), 591–608.
https://www.nationalgeographic.com/photography/proof/2017/09/yiwu-china-commodities-market/.
Taagepera, R. (1978), 'Size and duration of empires: growth-decline curves, 3000 to 600 BC', *Social Science Research*, **7** (2), 180–96.
Taylor, P.J. (2003), *World City Network*, London: Routledge.
Taylor, P. and R.E. Lang. (2005), 'US cities in the "World City Network"', report, February, Brookings Institution, Washington, DC.
Tilly, C. (1995), 'Globalization threatens labor's rights', *International Labor and Working-Class History*, **47** (Spring), 1–23.
Timberlake, M. and D.A. Smith (2012), 'Global cities and world city networks', in S. Babones and C. Chase-Dunn (eds), *Routledge Handbook of World-Systems Analysis*, London: Routledge, pp. 247–55.
Turchin, P. (2003), *Complex Population Dynamics: A Theoretical/Emperical Synthesis*, Princeton, NJ: Princeton University Press.

Wallerstein, I. (2003), *World-Systems Analysis*, Durham, NC: Duke University Press.
Walters, P.B. (1985. 'Systems of cities and urban primacy: problems of definition and measurement', in M. Timberlake (ed.), *Urbanization in the World-Economy*, Orlando, FL: Academic Press, pp. 63–120.
Wilkinson, D. (1987), 'Central civilization', *Comparative Civilizations Review*, **17** (17), 31–59.
Wright, H.T. (1986), 'The evolution of civilization', in D.J. Meltzer (ed.), *American Archaeology, Past and Future: A Celebration of the Society for American Archaeology, 1935-1985*, Washington, DC: Smithsonian Institutional Press, pp. 323–65.
Wright, H.T. (1998), 'Uruk states in Southwestern Iran', in G. Feinman and J. Marcus (eds), *Archaic States*, Santa Fe, NM: School of American Research, pp. 173–97.

5. Sizing up Roman urbanism

J. W. Hanson

> Cities gleam with radiance and charm, and the whole earth has been beautified like a garden.
> (Aelius Aristides, *To Rome*, 99)

INTRODUCTION: RHETORICAL QUESTIONS

An interesting window onto the urban life of the ancient world comes from a speech delivered by the orator and author Aelius Aristides in Rome in AD 154 before the imperial household. This was his moment, and his address was intended to win favour in the court by glorifying the achievements of Rome. The text is important, nonetheless, since it gives us an impression of what contemporaries understood, or at least what they chose to emphasize, about the singular changes to the ancient world that came with imperial rule. The orator not only relates that there was a dramatic increase in the numbers of cities ('when were there ever so many cities?'), but also that, while old sites continued to flourish, new sites were also founded ('some newly founded, others increased under and by you'; Aelius Aristides, *To Rome*, 93–4). Intriguingly, the *rhetor* (teacher of rhetoric) also calls up an image of cities lying cheek by jowl ('did ever a man who lived then travel across the country as we do, counting the cities by days, and sometimes riding on the same day through two or even three cities, as if he was passing through one only?'; Aelius Aristides, *To Rome*, 93–4). The sense throughout is not only of greater wealth, affluence and prosperity, but also of the embellishment of cities through new amenities and comforts, new-found confidence in their ability to endure, and the explosion of urban life. Yet, what do we know about the urbanism of the Roman Empire? To what extent was the development of cities the cause or the result of the new connectedness and integration that must have come with the expansion of the empire?

In general, scholarship on ancient urbanism has been defined by two trends (Hanson 2016). The first is that we have never had a reliable idea of how many cities there were, how they were distributed across space and time, or how many individuals lived in them. We have never had a comprehensive list of sites or complete maps of their distribution. Although N.J.G. Pounds (1969) attempted to create a tentative map of cities in the ancient world in the second century AD (Figure 5.1), which has been reproduced several times (including in the Very Short Introduction to Classics series (Beard 1995)), his work was seriously hampered by his reliance on a mixture of ancient sources, such as Pausanias's *Guide to Greece*, Pliny's *Natural History* and Caesar's *Commentaries*, which cannot be expected to be consistent. This means that his work not only misses out half the empire, but also gives no sense as to how cities developed or how large they eventually became. Although there has been much discussion about the sizes of ancient cities, we have never had an indisputable method for estimating their numbers of residents. We have some idea about a couple of the most important sites, such as Rome, but we have almost no

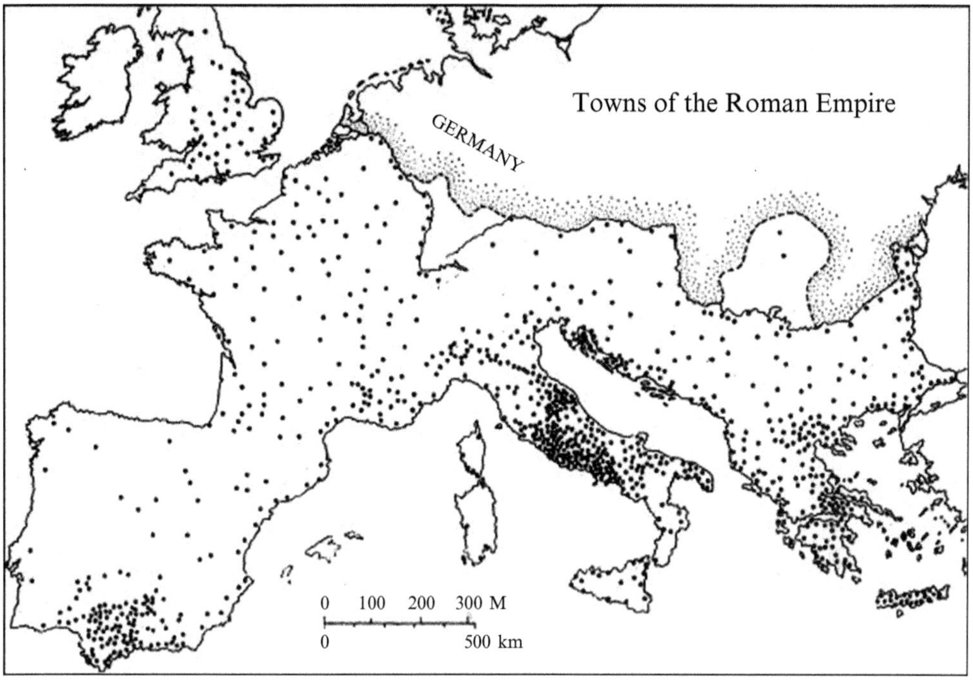

Source: After Pounds (1969) and Scheidel (2007).

Figure 5.1 The cities of the north-west in the second or third centuries AD

information about other settlements in the ancient world. This means it has been virtually impossible to discuss with confidence the size of the overall urban population, the urbanization rate or the total population.

It is striking that discussion about the urbanism of the ancient world has generally been kept separate from the rest of urban studies (Hanson 2016). The dominant approach to urbanism has been to either treat the city as an abstract concept or to concentrate on case studies, which has often led to an over-fixation on the amount of variation between cities, the nature of their design and amenities, or the reasons for their divergent experiences. It has therefore been difficult to contextualize individual settlements within the urban system, since most research has tended to treat them in isolation, instead of as components in wider systems, hierarchies and networks – even though the latter has been an important theme of scholarship on more recent and contemporary urbanism for some time.

The question of the structure and scale of ancient urbanism is not a sterile issue, since we have often assumed that it has had a considerable impact on the development of modern Europe, including offering the template for the revival of urbanism in the medieval and early modern periods, and its expansion in more recent times (Hanson 2016). Moreover, ancient urbanism has been seen as the main source for the stock of cities and towns in our own times, many of which remained important centres in their own right and continued to have significant roles in wider systems, networks and hierarchies. The lack of detailed information about ancient urbanism therefore leaves us with two questions for which we as yet have only inadequate answers: how does it compare with the urbanism of other periods and places? If we can establish this, why did one fail, while the other has been so successful?

As a consequence, although we have always strongly suspected that urbanism was a defining feature of the ancient world, we have not had a complete or reliable overview of its structure or scale until recent times. This has inhibited attempts to contextualize individual cities with the urban system, investigate how they functioned in wider systems, networks and hierarchies, or examine their impact on later development. During the past decade, however, there has been increasing interest in attempting to quantify the amount of urbanism in the ancient world, sparked, at least in part, by recognition that the scale of urbanism experienced in our times might well be one of the greatest challenges facing us in the twenty-first century (each generation gains the historiography it deserves). This initiative would not be feasible without reference to the vast amount of data about the ancient world collected in the field over the past 100 years or so, or new ways of working with them. In this chapter, I review the recent research that has made it feasible to discuss some of the basic aspects of the urbanism of the Roman Empire, including the numbers of settlements, the sizes of their inhabited areas and estimated numbers of residents, and the nature of urban hierarchies and networks, before examining future directions.

DEFINING URBANISM

It is notoriously difficult to define urbanism, owing not only to the complexity of urbanism as a phenomenon, but also because cities can be seen in a number of ways, including as a number of people living in the same place, a physical space, an official status, or a mixture of all three (Laurence et al. 2011, p. 1).

The ancient sources that have come down to us give us two important insights into how we should define urbanism (Hanson 2016). The first is that cities encompassed both the settlement and its hinterland, and the second is that they did not just comprise people, but were also marked out by their monuments and status. The most famous indication of ancient views about urbanism is Pausanias's account of Panopeus in his *Description of Greece*, where he is dismissive of this place's claim to be an urban centre:

> It is twenty stades to Panopeus, a city of the Phocians, if one can give the name of city to those who possess no government offices, no gymnasium, no theatre, no agora, no water descending to a fountain, but live in bare shelters just like mountain cabins, right on a ravine. (Pausanias, *Description of Greece*, X.4.1)

So far so good. But how can we reconcile these insights with the evidence to which we have access?

To address these issues, we might take our lead from recent work in urban studies and attempt to come up with a working definition of urbanism by concentrating on sites that are more likely to have engaged in secondary and tertiary activities rather than in primary activities (Hanson 2016). This can be gauged by whether they had a specific size population (such as 1000, 5000 or more individuals) or had specific functions (such as historical, social, cultural, religious, political, administrative, juridical and economic roles). Although we do not have direct evidence for these features, we can approximate them by looking at the size of inhabited areas, monuments and civic status. This provides us with a number of criteria, which include not only whether sites conform to thresholds of 10 or 50 hectares (which are used to estimate the numbers of inhabitants in settlements), but also whether they had particular monuments, such as public spaces, associated public buildings, urban grids, leisure and entertainment structures, and religious, sanitary and defensive structures, and whether they had particular statuses, such as roles as provincial capitals, *conventus*, *metropolis* or *nome* capitals, *coloniae*, *municipia*, *civitates* and *poleis*, or various other rights and privileges. These features do not necessarily coincide with each other, so there is a small number of sites that do not meet the criteria for size, but nonetheless have significant monuments or civic status. Inevitably there will be differences of opinion about exactly which sites should be included or excluded from such a catalogue. In the following section, I have therefore not tried to use this working definition to arrive at a hard and fast list of cities, but to light upon a basic catalogue that can then be used to explore the overall nature of urbanism, including how it changed over time or by region, and how large cities might have been, both individually and collectively.

Since most scholars have used varying thresholds, such as 1000 or 5000 individuals, for the ancient world and either 5000 or 10000 individuals for the medieval and early modern periods and either 10000 or 100000 for the industrial and post-industrial periods, it is necessary to use the same thresholds in estimating the size of the urban population, urbanization rate and total population for the sake of comparing the ancient world with more recent times (Chandler 1987; Bairoch 1988). I have therefore followed the convention of dividing sites into different classes and then using these different classes to estimate the size of the urban population and compare it to other places and periods.

I have also restricted myself to a conventional definition of the limits of the Roman Empire, that is, the region covered by Italy and the provinces at their maximum extent in AD 117 and the period between the first century BC and the third century AD.

STRUCTURE AND SCALE

The Numbers of Cities

> This country [the Roman Empire] has more than four hundred smaller cities and towns.
> (Yu Huan, *The Weilue* – a Chinese historian writing between AD 239 and 265)

To understand the structure and scale of ancient urbanism, I have created a new catalogue of cities in the Roman Empire between the first century BC and the third century AD (Hanson 2016). Given the amount of material that needed to be collated, this attempt would not have been feasible without drawing on recent big data approaches, synthesizing existing sets of data and then using a series of methods to establish patterns and trends. The approach has been termed 'the shotgun method' by one recent scholar, discussing the urbanism of the Greek world in the Classical period (Hansen 2006). As he explains:

> to study ancient history is like hunting hares. The hunter uses a shotgun instead of a rifle. His weapon does not hit the bull's eye and is not constructed for big game, but the spreading out of the pellets to cover a broader field is very efficient when used against smaller animals. Similarly, the quantifications used presented by the ancient historian are never precise, but within certain limits they can provide extremely valuable information about ancient societies. (Hansen 2006, p. 1)

There is no need to describe how this catalogue was created in detail, as it has been described at length elsewhere, except to say that I attempted to use a combination of textual and archaeological sources to draw up a list of cities, using ancient geographical sources, such as Strabo, Pliny and Ptolemy, and modern atlases and gazetteers (Hanson 2016). I then combed the existing literature relating to these sites for information about the size of their inhabited areas, monuments, and status, which was used to gauge whether sites conformed to the criteria discussed previously. I then attempted to locate these sites in space, searching for recognizable features in Google Earth, and to date them, either based on evidence for their original foundation (which can be found in changes to their names or the construction of particular features and the laying out of urban grids) or when they first reached a specific size or acquired the other attributes mentioned previously. This catalogue has now been made available online.[1]

This material offers us a radical new view of the urbanism of the Roman Empire. This evidence suggests that there were at least 1388 sites with urban characteristics in the Roman Empire (Figure 5.2), although the exact figure is subject to change based on

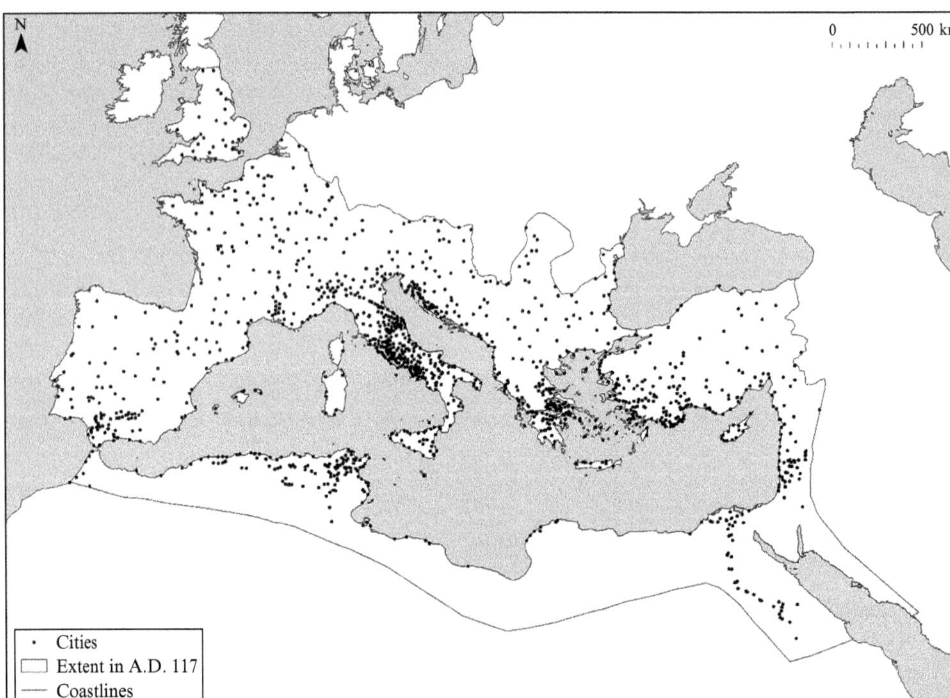

Source: After Hanson (2016).

Figure 5.2 The cities of the Roman world in the Imperial period

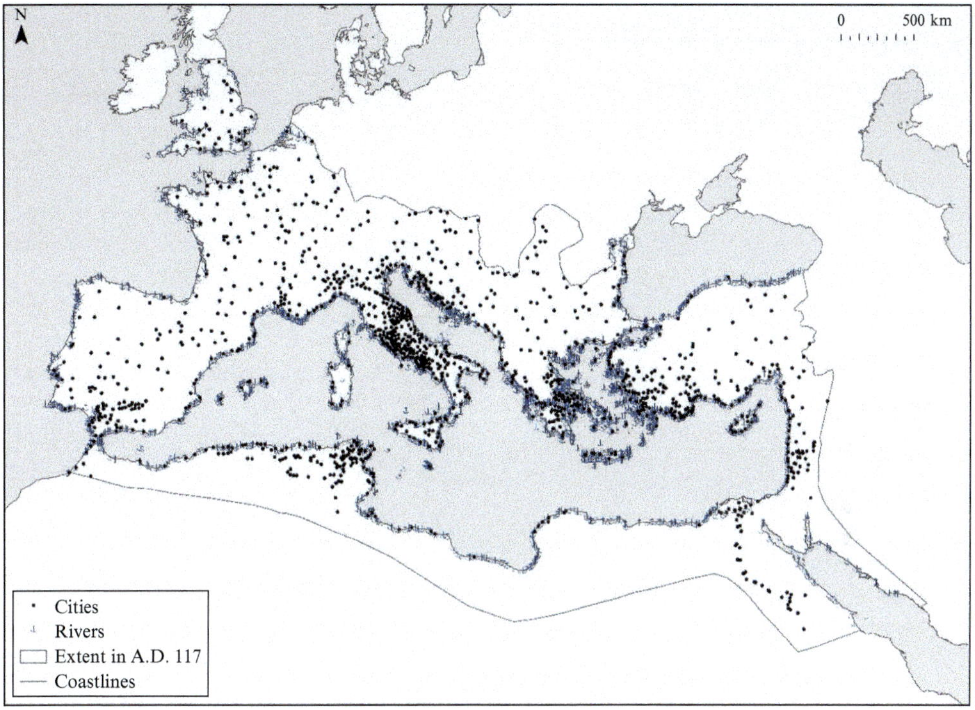

Source: Data from ancientportsantiques.com (accessed 31 August 2018).

Figure 5.3 Harbours in the ancient world

new discoveries. The experience of each region was not even, however, since these sites were more concentrated in Italy, Greece and the east, as well as North Africa and the north-west, than in the rest of the ancient world. These were the areas most congenial to settlement and that offered the best transportation links through sea or river networks, or where roads were most developed.

This is borne out by a basic analysis of the distances between sites and sea, river and road networks, which reveals that the overwhelming majority of sites had access to at least one of these networks (Figures 5.3–5.5). Around 32 per cent of cities are within 5 kilometres of known harbours, 26 per cent within the same distance of navigable rivers and, unsurprisingly, 78 per cent within the same distance of documented roads (this means that 88 per cent of cities have access to one of these networks). It should also not escape our notice, that the largest sites are concentrated on maritime, fluvial or terrestrial networks (or where different types of route intersect), suggesting that a high premium was placed on communication, with sea, river and road systems.

The most striking aspect of this catalogue, however, is the breath-taking proliferation and dispersion of cities that it reveals, rising from 822 in 100 BC to a peak of 1362 in AD 200, before dropping slightly to 1350 in AD 300, and spreading from the Italian peninsular, the Greek mainland and islands, and the Middle East (Asia Minor, the Levant and Egypt) to North Africa, the Balkans and the north-western Europe (Gaul and Germany,

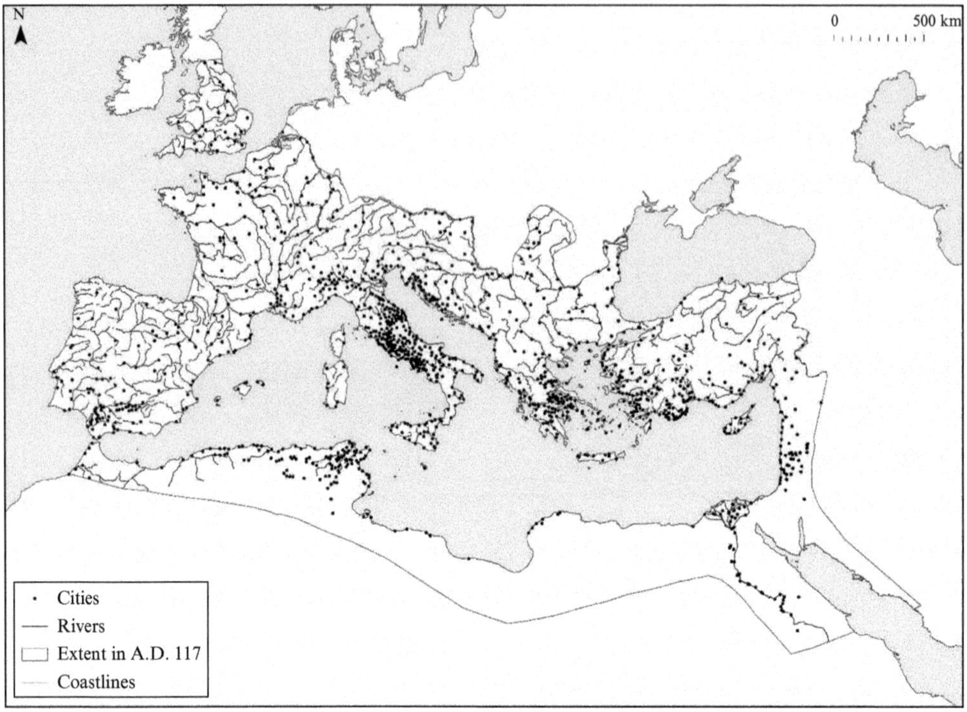

Source: After Hanson (2016).

Figure 5.4 Rivers in the ancient world

the Iberian peninsula and Britain). This suggests a remarkable number of new foundations each year, almost two per annum – not much by the standards of modern nations such as China, but remarkable for a civilization that did not rely on widespread use of technologies underpinned by fossil fuels.

The increase in the numbers of cities was also uneven and was most marked in the first century BC, before levelling off throughout the first, second and third centuries AD, and was mainly found in the newer additions to the empire, shifting between North Africa, the Balkans, Gaul and Germany, the Iberian peninsula and then Britain. There was very little turnover in the overall stock of cities, however, and they seem to have been remarkably resilient to sackings and natural disasters (with the exception of Pompeii and the other cities in the shadow of Vesuvius), although their numbers of residents must have fluctuated. The expansion of the Roman Empire often preceded changes in their number and distribution, which suggests that the development of the state was directly related to the increase in the extent of urbanism. It is remarkable, however, that only a limited number of cities were deliberately founded through the deduction of *coloniae* or the establishment of forts, which suggests that most of these changes were not the result of colonization or military activity, and that they must have been a direct or indirect result of the other changes that accompanied imperial rule.

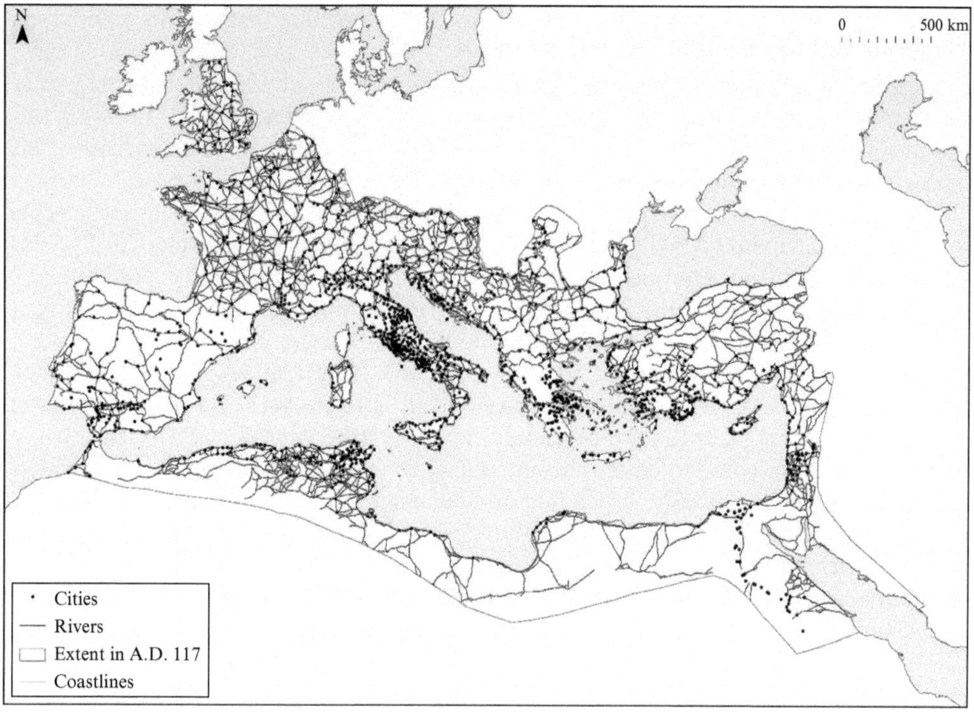

Source: After Hanson (2016).

Figure 5.5 *Roads in the ancient world*

The Numbers of Inhabitants

> No longer are savage islands dreaded, nor their rocky shores feared; everywhere are houses, and inhabitants, and settled government, and civilized life. What most frequently meets our view, and sometimes occasions complaint, is our teeming population: our numbers are burdensome to the world.
>
> (Tertullian, *On the Soul*, 30.3–4)

The most common forms of evidence that have been used to estimate the populations of ancient settlements are textual sources, inscriptions and papyri (Hanson 2016; Hanson and Ortman 2017). These are notoriously difficult to use, however, because they usually refer only to specific groups, such as the numbers of individuals who served on councils or called themselves citizens, as well as those who received various types of handouts or could be taxed or raised as an army. As a consequence, it is nearly always necessary to extrapolate from the figures in the sources to obtain the total population. There are recurrent issues concerning the exact ratio of men to women and children, as well as the share of slaves, foreigners and other groups. This exercise is also complicated because many of the numbers referred to both the settlement and its immediate hinterland or the surrounding countryside. A typical example of this is a stray remark by Galen about Pergamum to the effect that 'if then our citizens amount to as many as 40 000, likewise if you add their wives and slaves, you will find yourself admitting that you are richer than 120 000 folks' (Galen, *On the Natural Faculties*, V.49). If we assume that each of these citizens had two children, this would give us a figure of about 200 000. However, since the area of Pergamum, even with the later extension, was no more than 220 hectares, this would suggest 909 people per hectare, which is evidently too large to be believable – if his figure has any basis, Galen must be referring to both the site and its hinterland.

An alternative source of evidence for the numbers of residents in ancient settlements is archaeological material (Wilson 2011; Hanson 2016; Hanson and Ortman 2017). This material includes: the numbers of beneficiaries of different kinds of euergetism; the amount of water that was supplied by aqueducts; the numbers of seats in entertainment structures, such as theatres, amphitheatres and hippodromes or circuses. Unfortunately, this material has also been shown to be problematic. The number of beneficiaries of different types of euergetism seems to have had little to do with the size of the total population and was likely to have been conditioned by other factors, such as the expense that the benefactor could tolerate. The amount of water that was supplied by aqueducts must have varied enormously, depending on how much of it was used for non-domestic versus domestic use and how much was obtained through other means, such as wells and cisterns, which have much lower textual and archaeological visibility. The number of seats in entertainment structures is unreliable, since many of these structures were designed to accommodate both the residents of the immediate settlement and those of its nearest neighbours. This is illustrated by the riot reported by Tacitus between the people of Pompeii and Nuceria in AD 59, which is depicted in the fresco from the Casa della Rissa nell'Anfiteatro in Pompeii (Tacitus, *Annals*, XIV, 17). These issues are most obvious when we compare the relative sizes of entertainment structures in different cities. Although the largest amphitheatre in the ancient world was boasted by the capital (according to recent estimates, the Colosseum could have accommodated around 65 000 individuals), many of the others are to be found in less expected places: Capua, for example (Rose 2005).

In the past decade or so, there has been increasing interest in estimating the numbers of residents in ancient settlements by measuring the sizes of their inhabited areas and multiplying the results by a range of densities (Hansen 2006; Wilson 2011; Hanson 2016; Hanson and Ortman 2017). The strength of this method is that it enables us to control and manipulate our estimates of the population of each settlement. It also enables us to establish a range of figures for the densities of each settlement, which can then be used to explore the upper and lower limits of what might have been possible for human numbers (Wilson 2011). Another asset of the method is that it allows us to keep separate the evidence for the inhabited areas of each settlement and the evidence for the range of population densities that are applied to it. The evidence for the inhabited areas of each settlement is directly derived from maps, plans and satellite images, which means that it can be easily subjected to verification or falsification by other scholars. It is also relatively easy to obtain, so it can be generated from the existing stock of information, expanded with new evidence or updated in the light of new information. In a similar fashion, it is possible to base the range of population densities on whatever evidence is available, and this can then be compared with what is known about settlements from other civilizations to judge its appropriateness. Most importantly, this method allows us to estimate the populations of a substantially larger number of sites than other methods do. As Wilson has noted, although the area × density method is not perfect, it is the only method that allows us to correct the existing bias towards the most well-known sites and for which there is sufficient evidence for a large enough sample to enable us to explore overall patterns and trends, such as the size of the whole urban population, the urbanization rate and the total population (Wilson 2011).

I have therefore used a number of features to measure the size of the inhabited area of each site, including the area enclosed by walls, the extents of urban grids, the locations of monumental structures, the sizes of residential zones, the situation of cemeteries and, even, the character of natural features, such as changes in relief, rivers and coastlines, working from a combination of existing maps and plans (Hanson 2016). This allows us to estimate the sizes of the inhabited areas of 885 of the 1388 sites in the catalogue (*circa* 64 per cent) (Figure 5.6).

One of the weaknesses of this method, however, is that it has often been necessary to use a single figure or a wide range of figures for densities to estimate their numbers of inhabitants. Most scholars have suggested that the population densities of ancient cities would have ranged between 100 and 400 people per hectare (Wilson 2011). An issue with these ranges, however, is that they do not take sufficient account of known differences between the domestic architecture of sites, such as the incidence of apartment blocks.

There is now a body of theory which suggests that there is a consistent set of

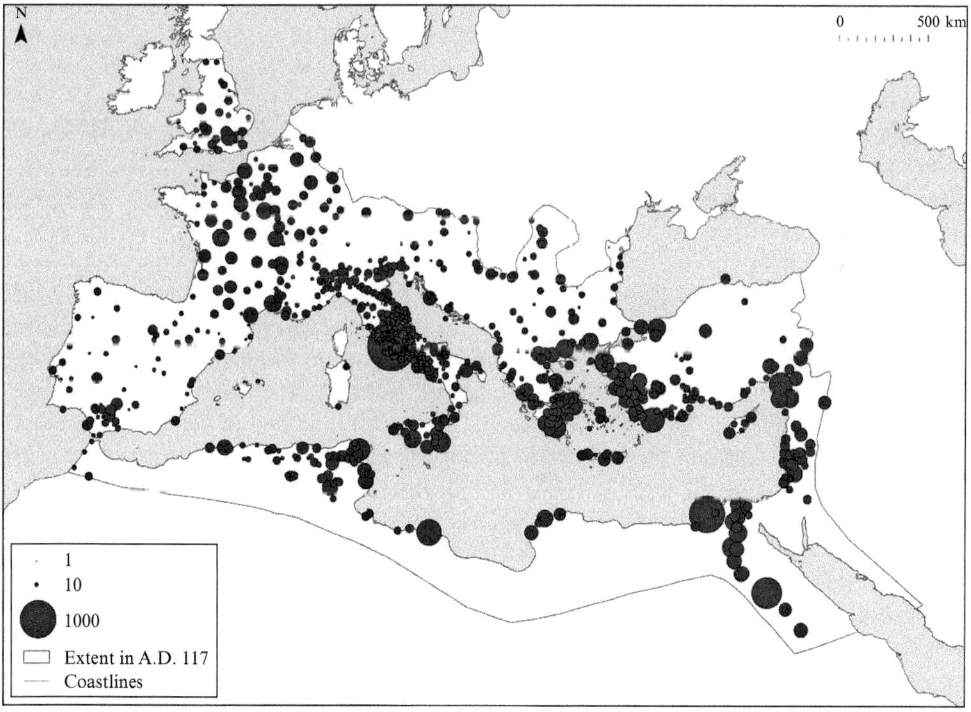

Source: After Hanson and Ortman (2017).

Figure 5.6 The estimated sizes of the inhabited areas of cities in the Roman world during the Imperial period

relationships between the numbers of residents in contemporary settlements and some of their most important traits, including densification (Hanson and Ortman 2017). To establish whether this was also true of the ancient world, I attempted to collect enough evidence for the numbers of residential units in particular neighbourhoods in a selection of sites and then to explore the relationships between their inhabited areas, estimated numbers of inhabitants and densities. To do this, I took a sample of sites, counted the numbers of residential structures in specific neighbourhoods, suggested a figure for the average size of each household (which I assumed ranged between three and seven and averaged about five), estimated the number of occupants in these units, calculated the size of each zone and then divided my figures for the number of occupants in each neighbourhood by the size of this neighbourhood to generate new estimates for their densities. The best examples of this are Ostia, Pompeii and Herculaneum (Hanson and Ortman 2017). At Ostia, there is evidence for 3153 structures, including 22 houses and 184 apartments of different kinds, which were divided into 3131 flats. At Pompeii, there is evidence for 1115 structures, including 498 houses, 468 *tabernae* (shops), 133 apartments and 52 that resist categorization. At Herculaneum, there is evidence for 80 structures, including 20 houses, 20 *tabernae*, 19 apartments and 21 structures that cannot be classified. This gives us evidence for a total of 52 sites, which are scattered throughout the settlement hierarchy and dotted throughout the Greek and Roman world from the fourth century BC to the fourth century AD. The results suggest that there is indeed a strong relationship between the inhabited areas, densities and numbers of inhabitants in these sites. This can be expressed as a simple equation, which allows us to predict the populations of some of the most important settlements with remarkable accuracy and precision. This provides us with a new tool that can be used to estimate the populations of the other settlements in the ancient world, as long as we have evidence for their inhabited areas (it should be noted that this can also be applied to urban and rural sites) (Figure 5.7).

These results confirm many of our suspicions about the scale of some of the most important sites. The area of Rome was about 1783 hectares (based on the Augustan *regiones*), which suggests a population density of 518 people per hectare, and a total population of 923000, which is nearly identical to the figure of 1 million that is normally suggested (Scheidel 2007). The area of Alexandria was around 972 hectares (based on the urban grid and walls), which suggests a population of about 411000 (at 422 people per hectare), which is also in line with the figure of half a million cited by scholars (Scheidel 2007). The area of Antioch was about 399 hectares (based on the urban grid and walls), suggesting a population of about 125000 (at 313 people per hectare), which also compares well with figures suggested by scholars (Scheidel 2007). Meanwhile, at the other end of the spectrum, the area of Pompeii was at least 60 hectares, suggesting a population of about 10000 (at 166 people per hectare), which is identical to the figure of 10000 suggested by others (Wallace-Hadrill 1994).

One of the most interesting things about these results, however, is that there is a much larger number of intermediate-sized sites than expected, which is reflected in a surprisingly large average of about 54 hectares, suggesting about 8500 residents (at 159 people per hectare). Moreover, although many of the largest sites are concentrated in both the Italian peninsular and the east, there are also significant numbers in some areas of North Africa and the north-west, especially Gaul and Germany, challenging the assumption that there was an exceptional amount of urbanism in the heartlands compared with the

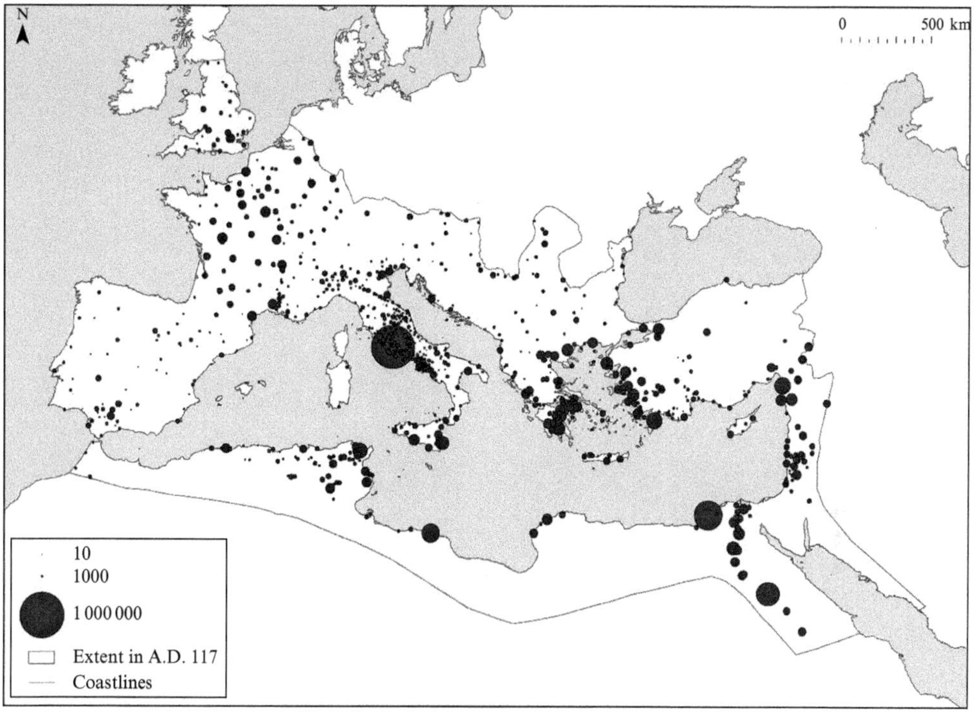

Source: After Hanson and Ortman (2017).

Figure 5.7 The estimated numbers of inhabitants in cities in the Roman world during the Imperial period

rest of the empire. These differences can be most readily appreciated by looking at a cartogram of the total number of residents in each site according to squares of 10 000 square kilometres (Figure 5.8).

These results should give us pause for thought, since current thinking suggests that cities of this scale must have had an important effect on their hinterlands, which might have led to considerable changes in the structure and organization of rural areas (Scheidel 2007; Wilson 2011; Hanson 2016). Since these settlements would have exerted considerable demand for both necessity and luxury commodities, they must have relied on the intensive exploitation of their immediate hinterlands and the extensive importation of various items from further afield. This must have relied on a considerable amount of investment in rural areas, through changes in organization and technology, and the development of an organized and efficient transport system, as well as widespread monetarization to mediate these exchanges. This has been well documented in the case of Rome, where the rise of the capital seems to have led to the development of a halo of *villae* in its immediate hinterland, geared towards specialist commodities, such as violets and roses, but there is no reason why other sites could not have had a similar effect, albeit at a lower level. In addition, since it is almost certain that urban areas would have had lower natality and higher mortality than rural areas, they also must have relied on a substantial amount of

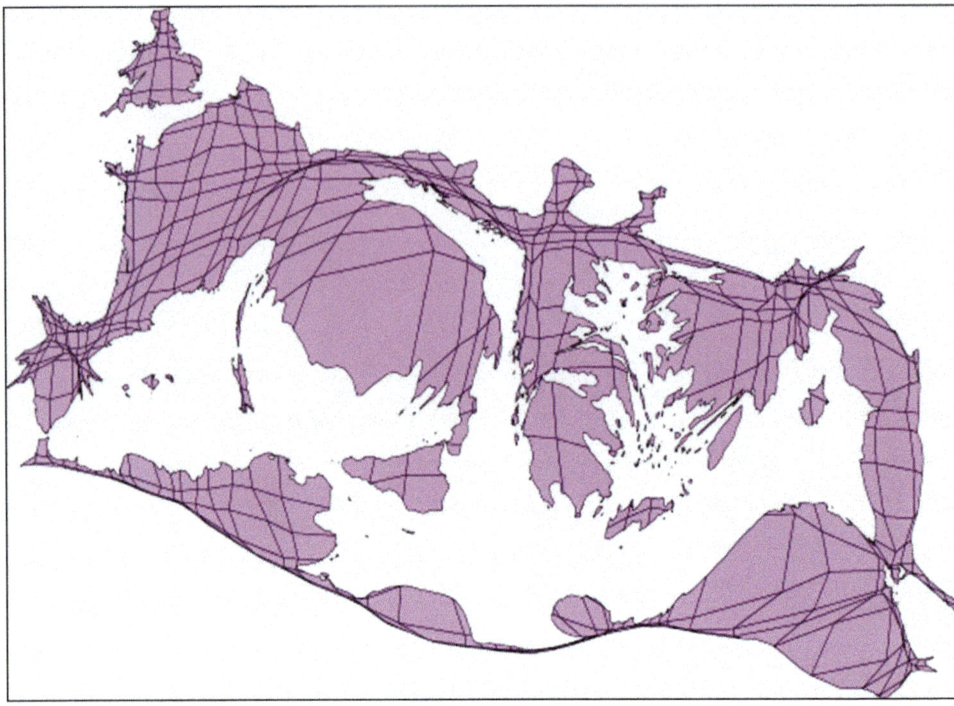

Figure 5.8 The estimated size of the urban population over time, represented as a cartogram

migration to maintain their sizes, let alone increase them. If true, this also suggests that urban areas must have had relatively high wages and standards of living to attract new migrants, although there is no compelling evidence for this as yet.

A useful concept for understanding these relationships is the extent of span of control, or the ratio between a given unit and the numbers of units subordinate to it, such as the number of managers and the numbers of staff subordinate to them in a given organization (Hanson 2016). A narrow span of control has been seen as being motivated by security and military considerations, such as maintaining law and order, while a wide span of control has been seen as being more suited to financial concerns, such as collecting taxes and tribute. Interestingly, when we calculate the ratio for ancient cities, although we would expect the optimum figure to be about six, the actual span of control between Rome and provincial capitals was about 40, and between them and *coloniae* and *municipia* about 30, suggesting that their main role was connected with the collection of taxes and tribute, rather than security and defence.

Note that, although the largest sites, such as Rome and Alexandria, tend to have the largest numbers and therefore the greatest diversities of monuments and statuses, there is no obvious relationship between their size and specific types of structures, although there is one with rights and privileges (provincial capitals, and *conventus*, *metropolis* and *nome* capitals tend to have the largest average size, followed by *coloniae*, and then *municipia*, as we would expect). This reminds us that there was not necessarily a standard 'kit' for

Table 5.1 The urban population, total population and urbanization rate of the Roman Empire

	All	5000 threshold	1000 threshold
Count of estimates	885	757	410
Sum of estimates (millions)	9.8	9.7	8.8
Total number of cities	1388	1388	1388
Proportion of catalogue (%)	64	64	64
Extrapolation	15.3	15.2	13.7
Total population (millions)	65.5	65.5	65.5
Urbanization rate (%)	23	23	21

Table 5.2 Survival rates for ancient cities

Bin	Number in AD 200	Number in AD 2000	Percentage
Up to 10000	179	74	41
10000 to 100000	225	92	41
100000 to 1000000	6	4	67

ancient cities, and how much their overall design and amenities were the result of local responses to the ideal of urban life (Laurence et al. 2011; Hanson 2016).

In addition, we can also suggest new figures for the size of the urban population by totting up the estimates that we have for all sites (Table 5.1) (Hanson 2016; Hanson and Ortman 2017). This suggests that there must have been a total of at least 9 million residents in cities over the threshold of 5000 individuals. As I have argued elsewhere (Hanson 2016; Hanson and Ortman 2017), although the most important settlements are more likely to have been found and recorded, they are also more likely to have been obliterated by later developments, meaning that the chances of having information about different classes of settlement should be relatively even. If this is correct, we can assume that the sites that we have evidence for represent a relatively random sample and we can scale it up accordingly. To test this assumption, we can look at how many of the sites with over 5000 residents in AD 150 have at least the same number in AD 2000, using modern censuses (Table 5.2).[2] This shows that the chances of sites surviving to modern times are about even for most size bands, although this exercise is very dependent on sample size. As a consequence, since the figure is based on a sample of 885 sites out of a total of 1388 sites (about two-thirds), it is difficult to escape the conclusion that there must have been at least 14 million residents in cities over the threshold of 5000 individuals (which is almost double the figure of 7–9 million suggested by some scholars).

As Lo Cascio (2009) has noted, if the urban population really is larger than normally thought, then either the urbanization rate must have been much higher than 10 per cent or the total population must have been much higher than 59–72 million, both of which have been suggested before. This suggests that either the urbanization rate must have been closer to 25–30 per cent or that the total population must have been close to 150 million. There is something to be said for both scenarios, since one makes sense with recent estimates of the urbanization rate of the Greek world, while the other works well with recent estimates for the gross population of modern Europe (Bairoch 1988; Hansen 2006). Either way, these results have significant implications for how we think about the ancient world, since they suggest that large numbers of its inhabitants lived in urban communities, even if they were outnumbered by those living in rural communities, which must in turn have implications for how we think about the overall structure and performance of the ancient economy.[3]

That is as far as the available evidence allows us to go. We might be able to improve on these figures in the future, however, by investigating the relationships between the numbers of residents and various other material remains, such as inscriptions, coins or ceramics, as these are either more abundant (allowing us to estimate the numbers of residents in an even larger selection of sites) or that can be dated more readily (allowing us to track change over time).

Hierarchies and Networks

Another way of dissecting urban systems is to examine their rank-size distributions (Hanson 2016). These methods are useful, since we would expect the rank-size distribution of a normal urban system to follow something similar to a power law, so that it forms a straight line with a slope of close to –1 when shown on a doubly logarithmic graph. Deviations from this archetypical distribution towards more concave or convex ones have been interpreted as either the result of different levels of connectivity and integration or the result of the erroneous amalgamation or division of urban systems. The rank-size relationship is therefore a useful tool, not only for assessing relative levels of development, but also for confirming whether we are dealing with a single system or not. One of the big questions with the interpretation of these types of distributions, however, is what effect missing values will have on the overall shape of the distribution. In particular, since this exercise is based on ranking data, we might expect missing values to artificially compress the distribution. To account for this, I have attempted to create a more realistic distribution by artificially adjusting the ranks of sites to mimic the inclusion of missing values at regular intervals throughout the catalogue (since we have evidence for about two-thirds of these sites, I have added one for every two) (Figure 5.9). We can then investigate the shape of the distribution using Drennan and Peterson's method, which is based on measuring the area between the log-normal line and the observed rank-size distribution (Drennan and Peterson 2004). This gives us an index where zero reflects a log-normal distribution and increasingly large positive and negative values reflect increasingly amounts of convexity and concavity respectively. This reveals a distribution that is virtually linear. The largest sites have a slope of almost exactly –1, with an index value of just under 0.1, suggesting that it was not the result of the pooling or partitioning of systems, and that the urban system should be seen as a single, indivisible entity (and therefore that individual regions must have been dependent on the rest of the system). These results are also consonant with recent studies of ancient markets, which have suggested that, although the different quarters of the ancient world were not tied together as tightly as modern markets, they nevertheless functioned as a part of a comprehensive Mediterranean market (Temin 2012).

Although it would be interesting to investigate these dependencies in more detail, such

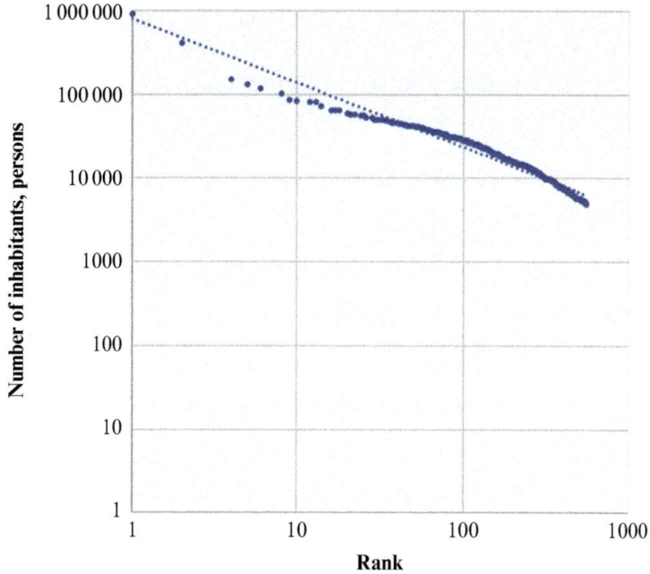

Note: Ranks have been adjusted to mimic the inclusion of missing values in the distribution.

Figure 5.9 The urban hierarchy of the Roman Empire, with rank against estimated numbers of inhabitants

as by investigating how connected and integrated sites were through sea, river or road networks using network analysis, we are still severely limited by the nature of the data that is available. Although we can side-step these issues to an extent by using the kinds of measures of urban potential that have been used elsewhere, these are also disappointing, since they show that there is no correlation between the sizes of cities and their individual scores, meaning that it is difficult to relate these networks to the development of urbanism (de Vries 1984; Bretagnolle et al. 1998; Bosker et al. 2008). The nature of current research therefore raises an important challenge for classical archaeologists and ancient historians to continue to find robust measures of the movement of individuals and commodities, such as through a combination of computational models and the assemblages of mass data-sets of containers and other small finds.

CONCLUSIONS

In this chapter, I have attempted to outline a new orthodoxy about the structure and scale of urbanism in the Roman Empire. This account suggests that there were 1388 cities in the Roman world between the first century BC and the first century AD. Although there are fewer sites with urban characteristics than we have become accustomed to, they covered larger inhabited areas, suggesting that they had larger numbers of residents, than we would expect (as well as more monuments and more civic status). There was considerable growth in the number of cities over time, but this process was not even and was greatest in the first century BC, before slowing down throughout the first and second centuries AD. Cities did spread from east to west, however, this development was uneven and was most marked in North Africa, the Balkans, Gaul and Germany, and Spain in the first century BC, North Africa, the Balkans, and Britain in the first century AD and North Africa and the Balkans in the second century AD.

In addition, this new account reveals that there was a significant number of medium-sized sites and that the average size was relatively high, even without accounting for missing sites. However, although the largest cities were concentrated in both Italy and the east, there were substantial cities in parts of North Africa and the north-west, especially in Gaul and Germany. The number of individuals that lived in these cities was considerable by any standards and could have amounted to as many as 14 million (using the threshold of 5000 individuals), suggesting that either the urbanization rate must have been much higher than 10 per cent or the total population must have been much higher than 59–72 million (it is not easy to say which).

Any assessment of how urbanism compares with other periods or places drives home the sheer extent of urbanism. As always, the closest parallel is the Greek world, for which we now have exceptional data based on a combination of Hansen and Nielsen's (2004) inventory of urban centres, which includes a total of 1035 sites, and Hansen's (2006) work on the sizes of these sites, including estimates for some 233 sites. This yields an estimate for the urban population of 3.3 million, implying an urbanization rate of between 33 and 41 per cent, according to estimates for the total population of between eight and 10 million. Meanwhile, when we compare these data to Bairoch's estimates for Europe in more recent times, these results show that the size of the urban population might not have been surpassed again until at least the eighteenth century (Table 5.3) (Bairoch 1988). This material supports the view that there was a gradual rise in the extent of urbanism throughout the Greek and Roman world, which was taken up again during the medieval and early modern periods, marking out the ancient world as one of a series of efflorescences, which themselves were part of an upward trajectory of undulating expansion and contraction.

As I have argued elsewhere (Hanson 2016), if modern theories about the links between urbanization and economic development are correct, this scale of urbanism must reflect considerable economic development, in both gross domestic product and per capita

Table 5.3 The estimated number of cities, urban population, total population and urbanization rate of Europe, AD 800 to AD 1900

Date	Number of cities	Urban population	Total population	Urbanization rate (%)
800	77	2.2	32.0	6.9
1000	111	3.8	39.0	9.7
1300	242	7.9	75.0	10.5
1500	241	8.2	76.0	10.7
1600	294	10.9	95.0	11.5
1700	330	12.6	102.0	12.3
1800	–	18.6	154.0	12.1
1900	–	108.3	285.0	38.0

Source: Based on Bairoch (1988, tables 7.1, 8.2 and 13.2).

terms, which was caused by a mixture of conquest (that is, an influx of material wealth) and the more intrinsic changes accompanying imperial rule (that is, opportunities for invention or innovation, investment, trade, and so on). The most important cause of economic development might have been cities themselves, owing to their ability to foster specialization and encourage changes in the organization and technological development of rural areas. The sheer scale of urbanism also created areas of intense demand, which can only have been met by a combination of an efficient system for the extraction of necessity and luxury commodities from the immediate hinterlands of settlements, and an effective system for importing items from further afield. The growth of urbanism, however, might also have led many cities to overshoot their ecological carrying capacities, so that they depended on their ability to offset deficits in one region with surpluses in another, essentially linking heterogeneous environments through trade (Dermody et al. 2014). If this is right, this might help us to understand why the system came under so much strain towards the end of this period, since any disruption to this delicate nexus of exchanges would have cut off urban areas from their traditional areas of supply, causing a collapse of living standards, while guaranteeing that rural areas experienced a sudden loss of demand, leading to decreasing investment. This seems to be in contrast to the medieval and early modern periods, when the system was more able to respond to changes in the structure of supply and demand.

FUTURE DIRECTIONS

What has been lacking in this chapter is a more nuanced understanding of the dynamics of settlements, including the mechanisms that might have led to growth and decline, suggesting a manifesto for future work. There is reason to be optimistic about the agenda, however, since the availability of independent estimates for the numbers of residents in sites should allow us to start to unpack the links between the size of their inhabited areas and various aspects of their urban form. This is demonstrated by recent work, which has not only established a consistent relationship between the size of settlements and numbers of residential units, but also with the sizes of communal spaces, such as *fora* and *agorae*, street networks, and numbers of craft and trade associations, suggesting models for densification, interaction, and even specialization or diversification (Hanson and Ortman 2017; Hanson et al. 2017, 2019). This, therefore, raises the possibility that we might continue to use these theories to sharpen our thought about the relationships between their residents and the built environment, enabling us to bring some order to the great mass of material that we have for ancient settlements. This could have significant implications for our views of urbanism in non-modern settings.

ACKNOWLEDGEMENT

This chapter represents the fruits of work done during my time as a research associate at the University of Colorado, Boulder, and before that as a doctoral student at the University of Oxford.

NOTES

1. http://oxrep.classics.ox.ac.uk/databases/ (accessed 15 May 2019).
2. http://www.cgeh.nl/data (accessed 31 January 2018).
3. Incidentally, these results can also be used to shed some light on the fraught debate about whether a low count of 3.5 million or a high count of 12.5 million is right for the total population of Italy. Although neither of these are satisfactory, it is unlikely that the urbanization rate of the region would have been lower than the empire as a whole, which tends to make an inflated rate more reasonable.

REFERENCES

Bairoch, P. (1988), *Cities and Economic Development: From the Dawn of History to the Present*, London: Mansell.

Beard, M. (1995), *Classics: A Very Short Introduction*, Oxford: Oxford University Press.

Bosker, M., S. Brakman, H. Garretsen, H. de Jong and M. Schramm (2008), 'Ports, plagues and politics: explaining Italian city growth 1300–1861', *European Review of Economic History*, **12** (1), 97–131.

Bretagnolle, A., D. Pumain and C. Rozenblat (1998), 'Space-time contraction and the dynamics of urban systems', *Cybergeo: European Journal of Geography*, doc. 61, 10 July, 1–12, doi:10.4000/cybergeo.373.

Chandler, T. (1987), *Four Thousand Years of Urban Growth: An Historical Census*, Lewiston, NY and Lampeter: Edwin Mellen Press.

De Vries, J. (1984), *European Urbanization: A.D. 1500 to A.D. 1800*, London: Methuen.

Dermody, B.J., R.P.H. van Beek, E. Meeks, K. Klein Goldewijk, W. Scheide, Y. van der Velde, et al. (2014), 'A virtual water network of the Roman world', *Hydrology and Earth System Sciences*, **18** (December), 5025–40.

Drennan, R.D. and C.E. Peterson (2004), 'Comparing archaeological settlement systems with rank-size graphs: a measure of shape and statistical confidence', *Journal of Archaeological Science*, **31** (5), 533–49.

Hansen, M.H. (2006), *The Shotgun Method: The Demography of the Ancient Greek City-State Culture*, Columbia, MO and London: University of Missouri Press.

Hansen, M.H. and T.H. Nielsen (2004), *An Inventory of Archaic and Classical Poleis: An Investigation Conducted by the Copenhagen Polis Centre for the Danish National Research Foundation*, Oxford: Oxford University Press.

Hanson, J.W. (2016), *An Urban Geography of the Roman World, 100 BC to AD 300*, Oxford: Archaeopress.

Hanson, J.W. and S.G. Ortman (2017), 'A systematic method for estimating the populations of Greek and Roman settlements', *Journal of Roman Archaeology*, **30** (November), 301–24.

Hanson, J. W., S.G. Ortman and J. Lobo (2017), 'Urbanization and the division of labour in the Roman Empire', *Journal of the Royal Society Interface*, **14** (136), doi:10.1098/rsif.2017.0367.

Hanson, J.W., S.G. Ortman, L.M.A. Bettencourt and L.C. Mazur (2019), 'Urban form, infrastructure, and spatial organization in the Roman Empire', *Antiquity*, **93** (369), 702–18.

Laurence, R., S. Esmonde Cleary and G. Sears (2011), *The City in the Roman West: c. 250 BC–c. AD 250*, Cambridge: Cambridge University Press.

Lo Cascio, E. (2009), 'Urbanization as a proxy of demographic and economic growth', in A.K. Bowman and A.I. Wilson (eds), *Quantifying the Roman Economy, Oxford Studies in the Roman Economy*, Oxford: Oxford University Press, pp. 87–106.

Pounds, N.J.G. (1969), 'The urbanization of the classical world', *Annals of the Association of American Geographers*, **59** (1), 135–57.

Rose, P. (2005), 'Spectators and spectator comfort in Roman entertainment buildings: a study in functional design', *Papers of the British School at Rome*, **73** (November), 99–130.

Scheidel, W. (2007), 'Demography', in W. Scheidel, I. Morris and R. Saller (eds), *The Cambridge Economic History of the Greco-Roman World*, Cambridge: Cambridge University Press, pp. 38–86.

Temin, P. (2012), *The Roman Market Economy*, Princeton, NJ: Princeton University Press.

Wallace-Hadrill, A. (1994), *Houses and Society in Pompeii and Herculaneum*, Princeton, NJ: Princeton University Press.

Wilson, A.I. (2011), 'City sizes and urbanization in the Roman Empire', in A.K. Bowman and A.I. Wilson (eds), *Settlement, Urbanization, and Population, Oxford Studies on the Roman Economy*, Oxford: Oxford University Press, pp. 161–95.

6. Associational life in the rounding out of dynamic cities: an in-depth methodology and application to Newcastle city-region in the nineteenth century

Michael Barke and Peter J. Taylor

INTRODUCTION

In her classic theory of economic development Jane Jacobs (1971) identifies explosive city growth as the key mechanism. Less well known is her description of a broader urban consequence of this rapid change that she terms the rounding out of the city, a new array of goods and services for the local population. Her list of these for her 'home town' Scranton, Pennsylvania, in the first half of the twentieth century are given as a zoo, a museum of natural history, a central public, reference library, several hospitals, 'several stuffy but imposing clubs', departmental stores, city departments for fire-fighting and public health services, and a trolley-car system (Jacobs 1970, p. 160). That is, although Jacobs defines a city as a settlement that experiences one or more momentous economic spurts, it only begins to more broadly function as we expect of a city during its period of rounding out. In this chapter we equate Jacobs's rounding out with development of an urban associational life, unplanned and voluntary creation of a thick layer of small social networks riding the rapid economic changes. They can be interpreted as buffers against the uncertainties and anonymity consequent upon urban growth (Neal 2013, p. 41). This process is explored for Newcastle upon Tyne, England, in the nineteenth century.

Our choice of Newcastle for this case study follows on from previous research on economic spurts through urban demographic growth that found this city to be one of the fastest growing cities in the world for two periods, 1800–49 and 1850–99 (Taylor et al. 2010). Subsequently we have described this economic success in detail, showing how Newcastle's division of labour grew in complexity as it was transformed from a centre for coal export at the beginning of the nineteenth century to become a large city-region and industrial powerhouse by century's end (Barke and Taylor 2014). This was also a period of massive rounding out in the form of a vast diversity of voluntary associations. Thus huge economic change was accompanied by equally momentous cultural, political and social additions to the activities of the city. In this contribution we outline the methods we have deployed to describe and analyse a growing associational life behind transformation into a multifaceted urban society.

The approach we take is to search out myriad associations of people within Newcastle city-region (roughly equivalent to today's county of Tyne and Wear) across the nineteenth century. Bringing people together involves a communicative process; each association can be thought of as a small network of people with a shared purpose (such as running a scientific society, celebrating an event or organizing a charity). We call these functional communicative associations (FCAs) and the rounding out of Newcastle is interpreted as the proliferation of these associations. However, the city is much more than an aggregation of little networks;

active citizens are to be found in multiple associations and their co-locations across FCAs thus creating a larger interlocking urban network, which is the representation of the associational life that we are attempting to measure. The outcome is a very large-scale data collection in which we have found 343 FCAs that provide 1329 lists of members (little networks) over the hundred years, and in which we have identified 7723 active participants. It is this data collection, illustrated with some preliminary results, that is the subject matter of this chapter. It is offered as a conceptualization and methodology that is portable for studying the associational life of cities as interlocking networks across different times and countries.

We proceed as follows. In the next section we introduce the literature on voluntary associations, focusing on debates about their meaning and importance, and how these relate to our case study of a rapidly changing city-region based on Newcastle. The data collection is then described in detail, and initial output in simple descriptive tables is provided. Further discussion of the data considers the credibility and utility of some early results. A new method needs to generate results that are expected to fulfil the credibility criteria, but also to produce novelty results since the unexpected constitutes new knowledge. In the concluding comments we look forward to more advanced analysis that uses co-presences of people in different FCAs to define unfolding interlocking networks as Newcastle's rounding out in the nineteenth century.

STUDIES OF ASSOCIATIONAL LIFE

There is a substantial academic literature on voluntary associations, and Smith and Shen (2002) provide a useful entrée based on the importance of the unit of analysis. They show that the most frequently studied subject of study is the individual volunteer or member. Second is the study of specific groups or organizations, or samples of them. Less frequently studied are voluntary associations in aggregate and their population rates, for example, in growth, size and decline, within a particular territory or comparison across several territories (McPherson 1983; Popielarz and McPherson 1995). In our study we combine the first (individuals) and third (multiple groups) of these approaches. Neal (2013, pp. 37–8) describes this duality between people and groups as a two-mode network structure from which two one-mode structures, person-to-person and group-to-group, can be derived. This enables the interlocking network analysis that is our investigative goal. Thus we search out lists of active participants in multiple groups, numerous voluntary associations, throughout the Newcastle city-region between 1800 and 1899. Our specific urban framing is of particular interest, given that most similar studies are based on single core cities (for example, Shapin 1974; Gorsky 1998a; Caglioti 2006), specific types of association (for example, Gorsky 1998b) or examples selected from a wide range of urban centres (for example, De Vries 2006; Morris 2006). In nineteenth-century Newcastle, economic growth was accompanied by a wider geographical expansion thereby creating a multi-nodal city-region (Barke and Taylor 2014). Hence our study includes an additional spatial agglomeration process.

Assessing the American Contribution

The literature on voluntary associations is dominated by American scholars and stems from Alexis de Tocqueville's famous quip, in the 1830s, that Americans were a 'nation of joiners' (Neal 2013, p. 41). This has led to what Skocpol et al. (2000) term the conventional wisdom that democracy and public life in the USA has long been embedded in voluntary membership groups. These organizations have been lauded as representing an alternative to centralized government actions and large-scale political and economic interests. This perspective accords with the findings of a large-scale historical study showing that smaller places and places outside the North East had greater numbers of voluntary groups per capita and that they were created and sustained most easily in 'slow growing communities that were relatively small and homogeneous' (Gamm and Putnam 1999, p. 549). This is at odds with the British literature and view of voluntary associations that argues that they were overwhelmingly urban in origin and more usually associated with rapidly growing cities (for example, Morris 1983, 1998). However, the 'localist' interpretation of voluntary association groups has been challenged also in the USA, and Skocpol et al. (2000, p. 528) quote earlier findings that 'American voluntary groups developed in close relationship to the representative and federal institutions of the U.S. state', and they conclude that the localist interpretation of the roots of American civic voluntarism is a weak interpretation. This equates with the provenance of some voluntary associations within our study region where the national government is directly involved in legislating for some local groupings.

A further possible contrast of US and British perspectives on voluntary associational groupings concerns their role as consensual agents. Associational culture in nineteenth-century Britain has been argued to be a means of building citizenship and promoting discipline and identity (Morris, 1983). That many organizations modelled themselves on each other is possibly supportive of this view. However, in the USA substantial evidence suggests that voluntary associations owed their origins to the desire of some groups to differentiate themselves, for example, to distinguish clearly between 'high culture' and 'low culture'. Dimaggio (1982) argues that the traditional elite in Boston built institutions in order to retain control of the city's social life and seal themselves off from the masses. Again, parallels can be found in our study. In nineteenth-century Newcastle, the Mechanics' Institute was founded in 1826 as a break away from the Literary and Philosophical Society, which was, in the view of the contemporary local historian and publisher Eneas Mackenzie, 'daily becoming more exclusive and aristocratic' (quoted in Middlebrook 1950, p. 228). This could lead to a reactive spiral of changes: Hall (2006) has argued that, while the origins of some associations could be caused by a desire to counteract conservative political elites that dominate public life, those elites then frequently responded by forming new associations that maintain and extend their public influence.

Furthermore, an important strand in the US literature explores the competing claims that voluntary associations are important instruments of societal integration (Babchuk and Edwards 1965) while an alternative group of researchers (McPherson 1983; Popielarz and McPherson 1995), using a large data-set of individuals similar to ours, albeit for a different century and country, argue that voluntary associations tend to be socially homogeneous, thus promoting relationships between similar people but potentially inhibiting contact between dissimilar ones. Our research strategy of examining the networks formed by active members of a wide variety of voluntary associations over a considerable time span has been devised precisely to explore these processes.

In summary, we argue that, to place voluntary associations more firmly within their societal context it is necessary to examine the networks formed by individuals, not just the associations themselves. How common was it for individuals to be members of several associations, for example? If they were, what type of organizations featured as key components in the entire superstructure of voluntary organizational culture? An important background feature in this approach is the persistence and volatility of voluntary associations. This, in turn, informs any discussion on the consensual (or otherwise) nature of voluntary associational culture. De Vries (2006) observes for the Netherlands that voluntary associations changed during the course of the nineteenth century, from being a unifying force to being representations of sectional interests, such as religious and political divisions. It is arguable that too few historical studies have sought to distinguish between different categories of organizations and have tended to focus on just a few leading examples in a particular place. General conclusions drawn are likely to be somewhat partial if there has been a failure to recognize the variety of societies including, for example, those that existed simply for pleasure, those whose main focus was to promote sociability, those that were symbolic in some way (for example, to evoke or reaffirm a particular value system), those that were campaigning for a specific objective, and those that produced goods or services or brought about change in relation to some objects or services (Smith and Freedman 1972). This categorization includes potential overlap, but the likelihood of differences of emphasis and operation are apparent. We argue that, in order to achieve robust conclusions on the persistence, volatility and characteristics of voluntary associations (and their role in the rounding out process), it is necessary to examine a large and varied sample of groups and their active members extending over a long time period.

Further Consideration of the British Context

Although theoretically important, the literature examined so far is predominantly American in origin. An examination of the literature specifically concerned with associational culture in the British historic context reveals three main groups. In one of the first detailed historical studies, the growth and change of the phenomenon in the early nineteenth century has been interpreted mainly through a historical materialist perspective (Morris 1983). Associations were 'vital to the distribution and mediation of power within British towns. They were part of the continuous recreation of urban elites in conditions of rapid social and economic change. They were the basis for the formation of middle-class identity' (Morris 1983, p. 96). In a later paper, however, the variety and density of associational culture is explained more in terms of the nature of urbanism itself, particularly as interpreted by the Chicago school of urban sociology: 'The potential for alienation and social disorganization, especially in periods of rapid change and migration, was countered by association' (Morris 1998, p. 292). In the same paper Morris concludes that 'Within the urban unit the societies sought to co-ordinate the activities of individuals who often knew each other but had imperfect knowledge of each other . . ..Central to these activities was the creation and exchange of information' (Morris 1998, p. 301). Thus the growth of voluntary associations was a product of the conditions of urbanization and the need to counter alienation and provide an arena for self-expression through the formation of networks. While the third group of literature makes no reference to this paradigm, it is related to this perspective as it is concerned with the nature of citizenship (Marshall 1950; Turner 1990, 2001). In conceptual and practical terms, the development of voluntary participative associations may be regarded as an exercise in citizenship. The latter is not a static and unchanging phenomenon, and it could be argued that one of the functions of voluntary associations was to broaden the experience of citizenship.

Citizenship is a form of entitlement and at least part of this is maintained and witnessed through voluntary associations and the networks within and between them. It is our view that the character of voluntary associations in north-east England in the nineteenth century can be best understood in relation to these networks, especially as indicated by active individuals. It is in the manifestation of these networks that we find indications of the broader functional role of voluntary associations. This is grounded firmly in the changing conceptualizations and practical role of citizenship. So, the study of networks within and between the voluntary associations of the Tyne and Wear area in the nineteenth century is a study of an important dimension of evolving citizenship. This examination throws light, for example, on the role of voluntary associations as controlling, consensual or divergent organizations, and provides a powerful justification for the examination of voluntary associations through a network approach. The methodology provides the means to tease out the role and functions of voluntary associations in the context of citizenship and to explore their potential contradictions.

This study follows the recommendation of Morris (1983, p. 96) that it is necessary to examine associations 'in particular provincial localities if their unity and fortunes as a coherent social development are to be demonstrated'. The use of the plural is also significant as studies of just one or a limited number of associational types (for example, Robinson 2014 on the Royal Society of St George), although of intrinsic interest within their own terms, serve only to provide a less than complete view of the phenomenon of English associational culture and thus offer a distorted perspective. The development of voluntary associations in a region is a complex phenomenon, if only because of the wide variety of these associations. Accordingly, we report on a data-set that is rich in depth and broad in temporal extent. We seek to examine the changing nature of voluntary associations within our study area in the nineteenth century in relation to networks. It is in the operation of these networks that we find clues relating to the broader operational role and character of voluntary associations. Inevitably, a number of specific issues arise.

For example, Morris suggests that many societies, in the early nineteenth century at least, were formed in response to specific urban challenges or crises. He contextualizes the growth of societies especially in late eighteenth and early nineteenth centuries in relation to the breakdown of consensus and stability (Morris 1983, p. 99) and the challenge to social authority represented by rapid rates of urbanization and its consequences. The growth of voluntary societies is interpreted as a means of 'gaining and asserting authority' by the middle-class elites and 'defending their social, economic and political power and privilege' (Morris 1983, p. 101). However, this is to place an interpretation of a long-lasting phenomenon – voluntary associations – within a very specific historical context and may not be a robust explanation over a longer time span, especially when associations of a very different character (non-crisis related) begin to emerge. While accepting much of Morris's argument, especially for the late eighteenth and early nineteenth centuries, we seek to place our analysis of voluntary associations in a longer historical context and, in particular, engage with Jacobs's broader hypothesis of rounding out as an urban phenomenon. Of particular concern in this perspective is the differentiation made by Morris (1983, p. 118) between the first and second halves of the nineteenth century, noting that, in the latter, 'The societies went on to increase and diversify, but worked within the shadow of state agencies and commercial competition'. This is undoubtedly the case (School Boards; Poor Law Guardians, and so on) but it does not necessarily follow that older associations disappeared or new associations, independent of the apparatus of the state, ceased to be created.

A further issue concerns the relationship of voluntary associations to class. Morris (1983, p. 113) is dismissive of the role of the gentry and aristocracy, observing that 'The aristocracy had a minimal involvement in the societies described here, and did not generate a network of voluntary societies of their own', and noting that very few were involved, except as patrons. Many of the associations included in this study did have members of the gentry as patrons, but it is also clear that in many the involvement was more than nominal. In the Newcastle Society of Antiquaries and the Natural History Society, for example, the gentry were actively involved, frequently attending meetings and, especially, engaged in activities of collecting, facilitating activities and contacts or communication outside the region, but even more in the publication of scientific papers. However, voluntary associations were also far from being the sole preserve of the middle and upper classes, and it has been argued that they played a fundamental role in the emergence and identity of, for example, the artisan elite (Crossick 1978). The significance of the networks formed by and through these associations was implied by Henry Mayhew in 1861 when writing of attending a meeting of the Woodcarvers' Society in a private house in Tottenham Street: 'Round the table are ranged the members of the Society – some forty or fifty were there on the night of my attendance – discussing the affairs of the trade' (quoted in Edwards 2011, p. 145). However, the suggestion that one group or class of individuals dominated societies and associations is susceptible to analysis, although rarely attempted, and our data enables some explorations on this issue.

Associational life in nineteenth-century England was largely, although certainly not solely, part of the growth of the middle class. However, the literature on the role of the middle class in voluntary associations demonstrates some contradictory observations. There is a general consensus that a prime outcome was that membership of societies enabled the middle classes to become cohesive and create an identity for themselves. However, more subtly, Morris (1983, p. 112) suggests that membership of voluntary associations was also favoured by the middle classes because they offered the opportunity for mild commitment in a situation of high mobility: 'The extreme flexibility of the voluntary society was attractive in the face of great uncertainty and rapid change.' These two views are at odds, with the former suggestive of stability and the latter volatility and high turnover. However, it is possible the latter perspective underestimates the significance of local status in being a member, especially a committee member, and indeed the opposite could be the case. Membership of a society offered status, stability and assurance; committee membership even more so. Nonetheless, there is substantial evidence of disputes between societies and between members of the same society, and these disputes would undoubtedly erode the coherence and identity of the membership. In addition, that voluntary associations are often in competition for members suggests a degree of fluidity. McPherson (1983, p. 520) has likened such organizations to 'a froth of bubbles, constantly sharing or exchanging members, growing and dying, and being absorbed and segmented in response to changing conditions'. Whatever the ultimate conclusion, these apparent contradictions are susceptible to consideration through our approach, demonstrating issues such as continuity, fragmentation or turnover.

Previous Newcastle Studies

Turning more specifically to research within our study area, a number of associations have been studied in our region. Foremost among these are the 'big three', the Literary and Philosophical Society of Newcastle upon Tyne, established 1793 (Watson 1897; Parish 1989), the Newcastle Antiquarian Society of 1813 (Breeze 2013) and the Natural History Society of Northumbria, 1829 (Goddard 1929; Hickling 1979; Alberti 2002). However, associational groupings as varied as the social and political affiliations of the 'unreformed' (that is, pre-Municipal Reform Act 1835) Corporation of Newcastle upon Tyne (Cook 1961), philanthropy (Burnett 2009), scientific societies (Campbell 1968), librarian associations (Duckett 1996), religious groupings (Mood 2005; Pickering 1981), migrant associations (MacRaild 2005), affiliations formed around newspaper production (Stokes 2006), trade unions (Allen, et al. 1971), patronage of the arts (Macleod 1989), women's sub-groups of otherwise male dominated associations (MacPherson and MacRaild 2006), arboreal culture and protection (Ito 2006) and several other, more informal, associations have also received attention. An example of the latter is the colony of artists that formed in the coastal village of Cullercoats (9 miles from Newcastle, but linked by rail) in the 1870s (Charleton 1886). Although the village was a well-known location for artists from the early nineteenth century (Larkins 1894; Newton and Gerdts 2003), it was in the last third of the century when it exhibited the classic artists' colony features of close personal relationships, mutual support that did not eschew friendly criticism, sharing of access to patronage, active socialization, and the choice of subject matter, but all in the absence of a formal agenda or committee.

The majority of this wealth of investigations have been studies of the organizations themselves with only limited recognition of the societal context and the nature of the networks they represented. Nevertheless, this variety and range of associations gives rise to interesting questions about their sheer numbers, their membership and turnover in comparison to other growing cities in the nineteenth century. The growth of an associational culture is well attested, but the detailed characteristics rewards quantitative comparative study. Did the Tyne and Wear area have a disproportionate share of these associations? Were the topical concerns of the associations replicated in other rapidly growing city regions? A plethora of voluntary associational groupings may be interpreted as indicative of a very specific civic culture, where the social environment (within the limitations of nineteenth-century class divisions) sought to encourage participation and involvement. In contrast with the general tone of most studies, one study of the civic community in Newcastle has demonstrated 'the local intelligentsia's heavy involvement in the making of populist meanings of the "civic", rather than simply dismissing them as an aloof elite cut off from the people' (Ito 2006, p. 9).

An important exception to the general tendency to deal only briefly with networks (frequently asserted but rarely measured) concerns Milne's (2007) study of 'business regionalism' in north-east England. What emerges from this study is not a sense of regional identity, but confirmation that the 'North East' consisted of a series of smaller localities: 'The industrial North East is best visualised as a polycentric space with shifting economic focal points and centres of gravity' (Milne 2007, p. 122). This is illustrated by the weak cross-ownership and investment from one part of the region to another. For example in 1895, 86 per cent of Sunderland shipowners' registered tonnage had been built on the

Wear, and owners based on the Tyne had 65 per cent of their tonnage built on that river. Swan Hunter fitted two out of every three ships they built with engines made by Tyneside firms, and the Wallsend Slipway Company supplied 24 different shipbuilders with at least one set of engines between 1870 and 1914, but three-quarters of these were supplied to the Tyneside firms of Swan Hunter (also in Wallsend) or Armstrongs in Scotswood, higher up the Tyne. Milne (2007, p. 127) traces a series of networks in the economic and business sphere that 'worked at two levels – the local and the national – while largely ignoring the regional level represented by the other towns and districts of the North East'. This inevitably raises the question, if business networks were so highly localized in the later nineteenth-century Tyne and Wear area, to what extent were other forms of organization and contact within the region similarly fragmented – including voluntary associations? Our study may confirm Milne's results across a wider ambit of activities but might also contradict with cooperation in city-regional associational life being much stronger than in economic work.

DATA TO STUDY ASSOCIATIONAL LIFE

The unit of analysis of our study are functional communicative associations in the specific context of rounding out a city. 'Functional' means the association pursues an agreed purpose. In practice this translates into our data collection including many more transient associations, often single instances of comings together, than in the literature. We see such temporary, even momentary, associations as particularly important in the rounding-out process during rapid change. 'Communicative' means the association encompasses a message, each providing something distinctive to urban life. In practice this translates as identifying the more committed members at the expense of those more passive, typically a focus on organizers, such as committee members, formally active members rather than ordinary members. Keeping to the spirit of the literature another criterion is that data collected must be available in public sources. This means the masonic lodges and private clubs are only included where information on them is freely available; in practice, this means in trade directories or newspapers. Finally, the FCAs are largely unplanned (bottom up) as a Jacobs process and are often the result of diffusion of cultural innovation, as in the spread of scientific societies in the first half of the nineteenth century and sports clubs in the second half across Britain, both very much including Newcastle. For instance, Newcastle's famed Lit and Phil openly followed the example of the Manchester Literary and Philosophical Society (Harbottle 1997, p. 54). In addition, we include local state organizations in the city-region, both elected and appointed (general-purpose councils, magistrates and grand juries, and two specific-function organizations – boards of Poor Law guardians and school boards – both required, from 1836 and 1871, respectively, by national legislation). All members of these state organizations were unpaid and therefore qualify as volunteers.

The method of data collecting was based upon early research of the world-city network (Taylor et al. 2002): the scavenger method – searching out any relevant information from myriad sources, in this instance publicly available archives and libraries (see Box 6.1). What we are searching for are lists of people organizing an FCA from which we can construct customized data. For each participation list members are scored in terms of

BOX 6.1 SOURCES FOR DATA COLLECTION

Tyne and Wear Archive:
Library
Special collections

The Literary and Philosophical Society of Newcastle upon Tyne:
Library
Society archive

Great North Museum (Hancock):
Library
Archive

The British Newspaper Archive:
Newcastle Courant
Newcastle Daily Journal
Newcastle Journal
Shields Daily Gazette and Shipping Telegraph
Shields Daily News
Sunderland Echo

The North of England Institute of Mining and Mechanical Engineers:
Electronic archive

Newcastle City Library:
Local Studies collection

Sunderland City Library:
Local Studies Centre

North Shields Library:
Local Studies section

South Shields Library:
Local Studies collection

Gateshead Central Library:
Local and Family History Service

their communicative potential ranging from 1 to 4. For instance, for a committee running a society or club, ordinary committee members score 2, functional officers (Secretary and Treasurer) score 3 and figurehead (Chair or President) scores 4, with others in less operational positions (for example, Patron or chaplain) scoring 1. We have been ravenous collectors of these lists, scoring each as appropriate. These are the 1329 participation lists from 343 FCAs during the nineteenth century that we mentioned in the introduction to this chapter. Thus the focus is as much on individuals as FCAs per se; these are the 7723 active persons found on the lists as further mentioned in the introduction. The lists are little networks of people communicating for a purpose; the key issue is that many people are found to be members of multiple lists and these indicate larger interlocking networks. This takes us to a network analysis that is beyond the scope of this chapter. In what follows we indicate some initial results that show the overall structure of the data and indications of possible connectivities.

Structure of the Data

There are many more participation lists than FCAs because we have data for many of the latter for several different years. Table 6.1 is an array of FCAs by number of participant lists. Well over half of the lists feature just once. These are largely momentary associations, specific to a particular event, such as a celebratory dinner or scientific meeting. However, many societies have annual elections to their organizing committee, and these are represented by FCAs with several participation lists over different years; for these FCAs this can mean many lists and in four cases there are more than 40 years' worth of participation lists. Table 6.2 shows these participation lists over decades and it can be seen that the number increases over time – reflecting both the growth and increasing complexity (that is, rounding out) of Newcastle and its city-region. It is worth noting that no decade falls below 50 lists, which suggests we will be able to generate robust results across the whole century.

We have divided the FCAs in two ways. First in relation to function, they are each identified as cultural, economic, political or social, and these are arrayed in Table 6.3. Again there are variations in frequencies but without the smallest (political/judicial) suggesting any robustness problems. We have more economic/commercial FCAs than

Table 6.1 Participation lists and functional communicative associations

Number of participant lists per FCA	Number of FCAs	% of total FCAs
1	202	58.89
2	40	11.66
3	23	6.71
4	11	3.21
5	11	3.21
6–10	32	9.33
11–20	13	3.79
21–30	5	1.46
31–40	2	0.58
More than 40	4	1.17

Table 6.2 Participation lists over time

Decade	Number of participation lists	% of total participant lists
1800–09	50	3.76
1810–19	62	4.67
1820–29	86	6.47
1830–39	130	9.78
1840–49	131	9.86
1850–59	137	10.31
1860–69	117	8.80
1870–79	160	12.04
1880–89	238	17.91
1890–99	218	16.40

Table 6.3 Types of functional communicative associations

Association types	Number of associations	% of total associations	Number of association lists	% of total association lists
Cultural/recreational	87	25.36	510	38.37
Economic/commercial	106	30.90	355	26.71
Political/judicial	61	17.78	192	14.45
Social (charity, public service)	89	25.95	272	20.47

others but, in relation to participations lists, cultural/recreational FCAs are the most common. This indicates the economic/commercial category has more FCAs with low or momentary lists, whereas the cultural FCAs are often organized over long time periods It should be mentioned that while carrying out the scavenger collecting we had no inkling that economic/commercial had the largest number of FCAs; we operated without targets and it just turned out that way. The same cannot be said for the political/judicial category, where we did exercise some control on numbers of FCAs. Owing to its formal nature, data for the political/judicial category is publicly abundant on a large scale and we chose to restrict our data collection to spaced years (for example, every five years for councils) in order not to overwhelm the data; we did not target this FCA type to be bottom of the list but it remains a reasonable representation of this FCA type.

The second division of the FCAs is more complicated and relates to the multi-nodal nature of the Newcastle city-region. Figure 6.1 shows a contemporary map of the city-region at the end of the century. It comprises very complex settlement patterns with numerous rail and road connections linking all main urban centres. For our analyses we have simplified this reality into a simple cartogram (Figure 6.2) from which we derive the spatial framing in Table 6.4. Each frame is separate and not summative. This means a frame combining place A with place B (FBAs covering both places) is distinct from place A (FBAs covering this one place) and place B (also FBAs covering one place). The framing is organized around five nodes: as well as Newcastle the nearest node is Gateshead just across the River Tyne, and the other three nodes are on the coast: North Shields (sometimes named Tynemouth) and South Shields either side of the mouth of the River Tyne, and Sunderland further south on the mouth of the River Wear. Our spatial framing features these five nodes, combinations thereof, and beyond. Six framings are Newcastle-centred. In order they are the city alone, with Gateshead, with the rest of the city-region (Metropolitan), with the county to the north (Northumberland), with the counties both north and south (County Durham) commonly referred to as the 'North East', and Tyneside combining the four nodes on the Tyne (that is, excluding Sunderland). There are also six spatial framings that do not include Newcastle, the other four nodes alone and two coastal links to South Shields, from the north with North Shields, and from the south with Sunderland. Again it should be emphasized that we had no spatial targets in our data collection, but we did attempt to ensure reasonable coverage across all nodes as the sources in Box 6.1 confirm. The results in Table 6.4 do show relatively weak coverage of Gateshead but this is partly owing to its close association with Newcastle. The difference between North Shields and South Shields was not expected and, although it is barely compensated for by South Shields bi-lateral links north and south, their

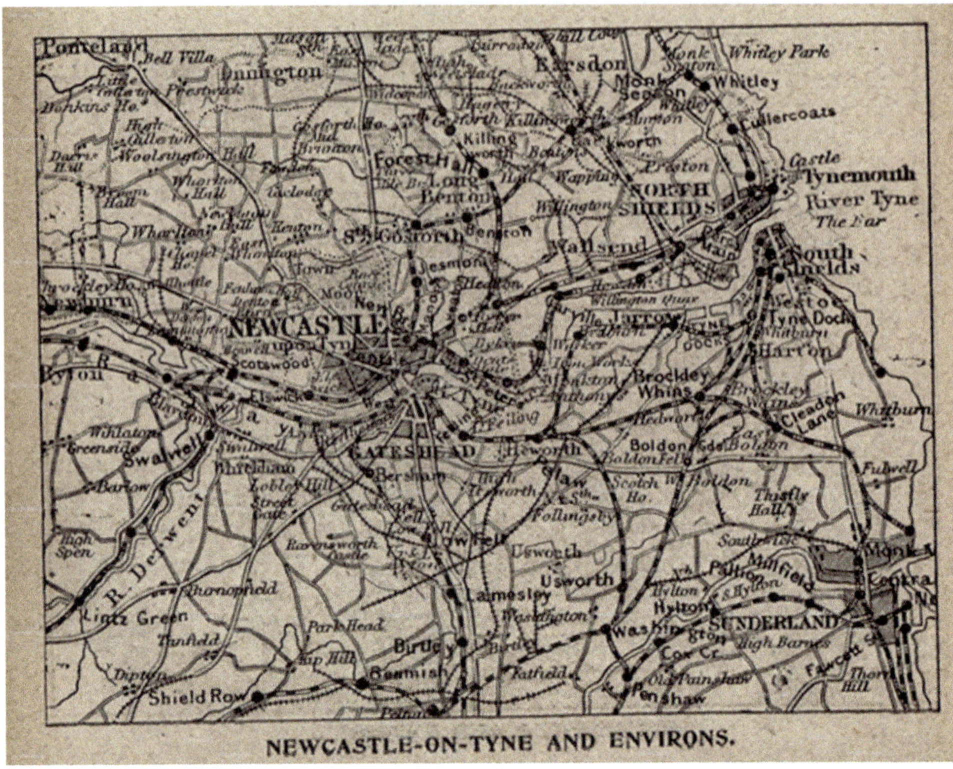

Source: *The British Almanac* (1902), London: The Stationers' Co.

Figure 6.1 Contemporary map of Newcastle city-region, c. 1900

number is reasonably robust at 24. However this is not the case for three spatial framings Newcastle/Northumberland and the two South Shields links previously referred to. For these we will not be able to make any evidential statements and, in addition, we will have to be circumspect when discussing the Tyneside framing. However at the other end of the scale the results are as expected, with Newcastle having most participation lists and Sunderland, the second city of the city-region, in second place for number of lists. Of the nodal combinations, it is the overall city-region, the 'Metropolitan scale', that ranks first, which is of particular interest to our study.

Finally there are types of individual participants that we can identify by their names and titles. In total these constitute a small proportion of all active persons but their numbers are large enough to make their identification worthwhile (that is, they can generate robust statements). These groups are listed in order of frequency in Table 6.5. Two stand out – women and clergy – but both only constitute about 5 per cent of the total. This is a very low proportion for women and reflects them being largely absent from the public realm in nineteenth-century English society; for clergy this is proportionally high, reflecting the religious nature of British society at the time. Other categories are much smaller but robust and potentially interesting for further analysis.

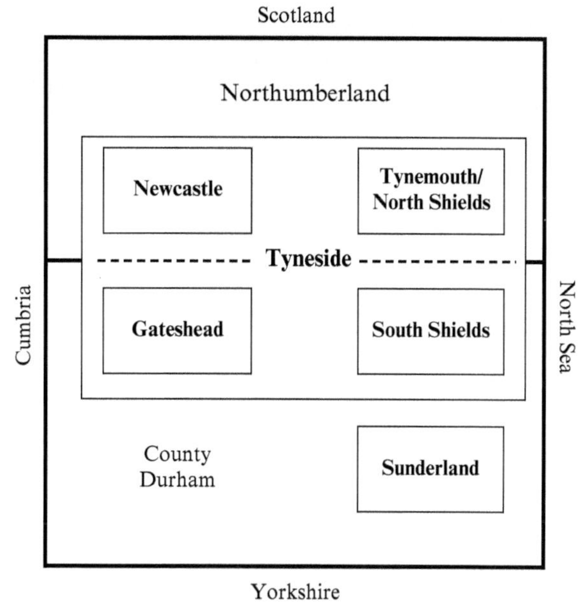

Figure 6.2 Schematic cartogram of Newcastle city-region

Table 6.4 Functional Communicative Associations within different spatial framings

Spatial framing	Number of associations	% of all associations
NEWCASTLE-CENTRED		
Newcastle	88	25.66
Newcastle & Gateshead	14	4.08
Newcastle City-Region/Metropolitan	44	12.83
Newcastle & Northumberland	5	1.46
Newcastle, Northumberland & Durham	26	7.58
Tyneside	14	4.08
TOTAL	191	55.69
NOT-NEWCASTLE		
Tynemouth/North Shields	46	13.41
Gateshead	18	5.25
Sunderland	61	17.78
South Shields	24	7.00
South Shields & Sunderland	1	0.29
South Shields & North Shields	2	0.58
TOTAL	152	44.31

Table 6.5 Specific groups of participants

Identifiable groups	Number of participants	% of total participants
Women	394	5.10
Clergy	384	4.97
Scientists	103	1.33
Aristocracy	92	1.19
MPs	91	1.18
Sirs	65	0.84
Military	52	0.67

Table 6.6 Latent connectivity over time

Decade	Latent connectivity	% total latent connectivity
1800–09	1832	2.59
1810–19	2471	3.49
1820–29	4637	6.55
1830–39	7461	10.53
1840–49	7749	10.94
1850–59	6457	9.12
1860–69	6955	9.82
1870–79	7803	11.02
1880–89	11949	16.87
1890–99	13521	19.09

Latent Connectivity

Although we are not reporting network analyses in this chapter, we can obtain a glimpse of what we might expect by focusing on the codes we have designated for each person in every FCA (that is, 1, 2, 3 and 4). As noted previously, codes are allocated to indicate enabling positions of persons in an FCA implying their importance for communication. For example, the chairman of a society represents his organization in the wider public realm (his communicative importance is coded 4), the secretary deals with the routine business of the society (coded 3), ordinary board members provide regular inputs to the organization (coded 2); and the patron allows his name to be used to provide the organization with status and respectability (minimal communicative importance scores just 1). Although not direct network measures of connectivity, these codes do evoke the latent communication within the data. Therefore, summing codes for both FCAs and individuals provide indications of their communicative respective importance, which we term latent connectivity. For instance, the sum of all the codes, 70835, represents the total latent connectivity in the data. This total can be divided in the same way as participations lists in previous tables. For instance, in Table 6.6 latent connectivity, also known as sum of codes, is distributed over time and can be compared with Table 6.2. Table 6.6 shows a similar increasing trend over the century but with slightly higher proportions for latent

Table 6.7 Latent connectivity of functional lists

Functions	Latent connectivity	% total latent connectivity
Cultural/recreational	30 825	43.52
Economic/commercial	14 731	20.80
Political/judicial	12 063	17.03
Social (charity, public service)	13 216	18.66

Table 6.8 Latent connectivity within different spatial framings

Spatial framing	Latent connectivity	% total latent connectivity
NEWCASTLE-CENTRED		
Newcastle	13 006	18.36
Newcastle & Gateshead	1125	1.59
Newcastle City Region/Metropolitan	23 832	33.64
Newcastle & Northumberland	430	0.61
Newcastle, Northumberland & Durham	5260	7.43
Tyneside	4115	5.81
TOTAL	47 768	67.44
NOT-NEWCASTLE		
Tynemouth/North Shields	4421	6.24
Gateshead	3469	4.90
Sunderland	10 710	15.12
South Shields	4226	5.97
South Shields & Sunderland	129	0.18
South Shields & North Shields	112	0.16
TOTAL	23 067	32.56

connectivity over time compared with simple participation. This suggests more complex associations later in the century, with more functional (communicative) roles.

In relation to FCA types, the four functions are shown on Table 6.7 and latent connectivity appears to accentuate the importance of cultural/recreational and political/judicial types compared with economic/commercial and social (Table 6.3). This shores up the robustness of the political/judicial. The large proportional fall for economic/commercial FCAs, of 6 per cent, is a consequence of this type having smaller participant lists and therefore less functional roles. The really big changes emerge with latent connectivity appear with the spatial framings (Table 6.8). Remarkably, latent connectivity proportions are lower for all five single-node framings, and combinations thereof, than for participation lists in Table 6.4. The main reason for this is the very large proportional increase for the city-region/metropolitan scale from 12.83 per cent to 33.64 per cent. The only other spatial framing to increase with connectivity is Tyneside (simply Metropolitan minus Sunderland). It seems two large functional spatial frames are inhabited by more complex FCAs with additional functional roles. This is an immensely interesting finding to be explored further in this research.

A more predictable outcome from focusing on latent connectivity is shown in Table 6.9. This features the specific groups of participants that are identifiable in the data, previously

Table 6.9 Latent connectivity of identifiable groups

Identifiable groups	Latent connectivity	% total latent connectivity
Clergy	5984	8.45
MPs	3476	4.91
Sirs	2655	3.75
Aristocracy	2436	3.44
Scientists	2427	3.43
Women	1317	1.86
Military	514	0.73

recorded in Table 6.5. Generally, the individuals in these groups are relatively important people. Thus in comparing these two tables, percentages of total latent connectivities for the groups are mainly larger than their percentages of persons in the data, and this is especially noticeable for the clergy. However, there is one major exception, topping the participation list percentages, women drop to next to bottom of the connectivity list, the only group to decline comparatively. Women's participation is both little and concentrated in more minor roles.

However, there is another important point to be made about Table 6.9. Although overall with more important roles compared with other participants, none of the identifiable groups are found to be particularly important within the wider pattern of associations; for instance, none reach even 10 per cent of total latent connectivity. Table 6.10 puts some names to the leading participants – the top 20 in latent connectivity is shown and all identifiable groups except women are well represented – and the proportional latent connectivities are individually all low. None comes even close to 1 per cent. This is by no means a set of data that lends itself to a 'who rules the city' narrative or a historical 'great men' narrative; it is truly illustrating a complex mix of associational activity, a rounding out in times of dynamic change that involves much common citizenship, 7723 people in all. This process is analysed in Table 6.11 where participants are classified by their levels of latent connectivity. Those scoring 1–2 are the large number of people who are members of just one FCA; this is nearly half of the participants, and between them they contribute just under 10 per cent of latent connectivity. Perhaps more interesting are those in the middle levels of latent connectivity. The two levels with the highest percentages of total latent connectivity are those scoring 10–19 and 20–39; here we are dealing with nearly 1300 people who are typically in five to 20 participation lists and contribute a third of all latent connectivity. This does not mean that those listed in Table 6.10 are not important, but that they are only the tip of a very big iceberg. It is by providing insights into the mass of this associational life that the data we have produced, and the methodology behind it, constitute a new, unique contribution to understanding cities and particularly their rounding out in periods of dynamic change.

CONCLUDING COMMENTS

We have two comments. First, the simple ordering of the data in the tables is just a prelude to much further analysis. Cross-tabulation of different variables is the next stage of

Table 6.10 Top 20 participants for latent connectivity

Rank	Participant	Latent connectivity	% total latent connectivity
1	Clayton, John	441	0.62
2	Bigge, Charles William	428	0.60
3	Swinburne, John Edward (Sir)	337	0.48
4	Embleton, Dennis (Dr)	325	0.46
5	Turner, William (Rev)	307	0.43
6	Bruce, John Collingwood (Rev, Dr)	292	0.41
7	Armstrong, William George (Sir, Lord)	258	0.36
8	Adamson, John (1)	240	0.34
9	Losh, James	236	0.33
10	Headlam, Thomas Emerson (Dr)	229	0.32
11	Trevelyan, Walter Caverley (Sir)	225	0.32
12	Dees, Robert Richardson	221	0.31
13	Hodgkin, Thomas (Dr)	219	0.31
14	Watson, Robert Spence (Dr)	213	0.30
15	Philipson, George Hare (2) (Prof, MD)	209	0.30
16	Liddell, Henry Thomas (MP, 2nd Baron Ravensworth)	202	0.29
17	Stevenson, James Cochran (MP)	201	0.28
18	Robson, Edward Capper (2)	199	0.28
19	Bell, Matthew (1) (MP)	198	0.28
20	Charlton, Edward (Dr)	193	0.27

Table 6.11 Participants in levels of latent connectivity

Level of latent connectivity	Number of participants	% of participants	Total latent connectivity	% of total latent connectivity
1–2	3539	45.82	6845	9.66
3–5	1526	19.76	5892	8.32
6–9	1038	13.44	7241	10.22
10–19	831	10.76	11 204	15.82
20–39	462	5.98	12 614	17.81
40–59	150	1.94	7275	10.27
60–99	100	1.29	7384	10.42
100–149	47	0.61	5594	7.90
150–199	13	0.17	2203	3.11
200–299	12	0.16	2745	3.88
300 +	5	0.06	1838	2.59
TOTAL	7723	100.00	70 835	100.00

analysis: we want to find out how, say, the FCA functional types relate to the spatial framing. Further along, the data has been specifically collected to facilitate network analyses using the co-memberships of participants, to move from two-mode data (persons/FCAs) to one-mode analyses (person to person networks and FCA to FCA networks). In this way we will describe the associational life of Newcastle city-region in the nineteenth century as a dense world of interlocking networks. This work is ongoing.

Second, we are submitting a methodology for consideration by other researchers on voluntary associations within cities. The unique feature, and its most important characteristic, is the sheer size of the data that is collected. In studying cities it is easy to focus on a few major associations and highlight a well-known coterie of important citizens, but for associational life to be fully understood as an integral part of complex city changes requires delving down to myriad associations and people far less appreciated. There is no easy way of doing this except through a large data collection exercise. In this chapter we have offered an example of how this might be done beyond our specific case study.

REFERENCES

Alberti, S.J.M.M. (2002), 'Placing nature: natural history collections and their owners in nineteenth century provincial England', *British Journal for the History of Science*, **35** (3), 291–311.

Allen, E., J.F. Clarke, N. McCord and D.J. Rowe (1971), *The North-East Engineers' Strikes of 1871*, Newcastle upon Tyne: Frank Graham.

Babchuk, N. and J.N. Edwards (1965), 'Voluntary associations and the integration hypothesis', *Sociological Inquiry*, **35** (2), 149–62.

Barke, M. and P.J. Taylor (2014), 'Newcastle's long nineteenth century: a world historical interpretation of making a multi-nodal city region', *Urban History*, **42** (1), 43–69.

Breeze, D. (ed.) (2013), *200 Years: the Society of Antiquaries of Newcastle upon Tyne, 1813-2013*, Newcastle upon Tyne: Society of Antiquaries of Newcastle upon Tyne.

Burnett, J.A. (2009), '"Department of Help for Skint Scotsmen!": Associationalism among Scots migrants in the North East of England, ca. 1859-1939', in T. Bueltmann, A. Hinson and G. Morton (eds), *Ties of Bluid, Kin and Countrie: Scottish Associational Culture in the Diaspora*, Ontario: University of Guelph, pp. 221–36.

Caglioti, D.L. (2006), 'Voluntary societies and urban elites in nineteenth century Naples', in G. Morton, R.J. Morris and B. De Vries (eds), *Civil Society, Associations, and Urban Places: Class, Nation and Culture in Nineteenth Century Europe*, Farnham: Ashgate, pp. 19–54.

Campbell, W.A. (1968), 'The Newcastle Chemical Society and its illustrious child', *Chemistry and Industry*, **43** (January), 1463.

Charleton, R.J. (1886), 'Cullercoats', *Magazine of Art*, **9**, 456–62.

Cook, M. (1961), 'The last days of the unreformed Corporation of Newcastle upon Tyne', *Archaeologia Aeliana*, 4th series, **39** (January), 207–28.

Crossick, G. (1978), *An Artisan Elite in Victorian Society*, London: Croom Helm.

De Vries, B. (2006), 'Voluntary societies in the Netherlands, 1750-1900', in G. Morton, R.J. Morris and B. De Vries (eds), *Civil Society, Associations, and Urban Places: Class, Nation and Culture in Nineteenth Century Europe*, Farnham: Ashgate, pp. 103–16.

Dimaggio, P. (1982), 'Cultural entrepreneurship in nineteenth century Boston: the creation of an organizational base for high culture in America', *Media, Culture and Society*, **4** (1), 33–50.

Duckett, B. (1996), 'Friend or foe? The Northern Counties Library Association, 1900-1922', *Library History*, **12** (1), 155–70.

Edwards, C. (2011), 'Tottenham Court Road: the changing fortunes of London's furniture street', *London Journal*, **36** (2), 140–60.

Gamm, G. and R.D. Putnam (1999), 'The growth of voluntary associations in America, 1840-1940', *Journal of Interdisciplinary History*, **29** (4), 511–57.

Goddard, T.R. (1929), *History of the Natural History Society of Northumberland, Durham and Newcastle upon Tyne, 1829-1929*, Newcastle upon Tyne: Andrew Reid & Co.

Gorsky, M. (1998a), 'Mutual aid and civil society: friendly societies in nineteenth century Bristol', *Urban History*, **25** (3), 302–22.

Gorsky, M. (1998b), 'The growth and distribution of English friendly societies in the early nineteenth century', *Economic History Review*, **51** (3), 489–511.

Hall, P.D. (2006), 'A historical overview of philanthropy, voluntary associations, and nonprofit organizations in the United States, 1600-2000', in W.W. Powell and R. Steinberg (eds), *The Non-Profit Sector: A Research Handbook*, New Haven, CT and London: Yale University Press, pp. 32–65.

Harbottle, S. (1997), *The Reverend William Turner: Dissent and Reform in Georgian Newcastle upon Tyne*, Leeds: Northern Universities Press.

Hickling, G. (1979), *The Natural History Society of Northumbria, 1829-1979*, Newcastle upon Tyne: Natural History Society of Northumbria.

Ito, K. (2006), 'The making of the civic community – Newcastle upon Tyne – 1850-1900', PhD thesis, University of Leicester.

Jacobs, J. (1970), *The Economy of Cities*, New York: Vintage Books.

Larkins, W.G. (1894), 'Reminiscences of artistic Cullercoats', *Tyneside*, **1** (1), 41.

Macleod, D.S. (1989), 'Private and public patronage in Victorian Newcastle', *Journal of the Warburg and Courtauld Industries*, **52** (January), 188–208.

MacPherson, D.A.J. and D.M. MacRaild (2006), 'Sisters of the brotherhood: female Orangeism on Tyneside in the late nineteenth and early twentieth centuries', *Irish Historical Studies*, **35** (137), 40–60.

MacRaild, D.M. (2005), 'Networks, communication and the Irish Protestant diaspora in Northern England, c.1860-1914', *Immigrants and Minorities*, **23** (2–3), 311–38.

Marshall, T.H. (1950), *Citizenship and Social Class and other Essays*, Cambridge: Cambridge University Press.

McPherson, J.M. (1983), 'An ecology of affiliation', *American Sociological Review*, **48** (August), 519–32.

Middlebrook, S. (1950), *Newcastle upon Tyne: It's Growth and Achievement*, Newcastle upon Tyne: Newcastle Journal.

Milne, G.J. (2007), 'Business regionalism: defining and owning the industrial North East, 1850-1914', in A. Green and A.J. Pollard (eds), *Regional Identities in North-East England, 1300-2000*, Woodbridge: Boydell, pp. 113–32.

Mood, J. (2005), 'Women in the Quaker community: the Richardson family of Newcastle, c. 1815-60', *Quaker Studies*, **9** (2), 204–17.

Morris, R.J. (1983), 'Voluntary societies and British urban elites, 1780-1850: an analysis', *Historical Journal*, **26** (1), 95–118.

Morris, R.J. (1998), 'Civil society and the nature of urbanism: Britain, 1750-1850', *Urban History*, **25** (3), 289–301.

Morris, R.J. (2006), 'Urban associations in England and Scotland, 1750-1914: the formation of the middle classes or the formation of a civil society?', in G. Morton, R.J. Morris and B. De Vries (eds), *Civil Society, Associations, and Urban Places: Class, Nation and Culture in Nineteenth Century Europe*, Farnham: Ashgate, pp. 139–57.

Neal, Z.P. (2013), *The Connected City: How Networks are Shaping the Modern Metropolis*, New York: Routledge.

Newton, L. and A.B. Gerdts (2003), *Cullercoats: A North-East Colony of Artists*, Bristol: Sansom.

Parish, C. (1989), *The History of the Literary and Philosophical Society of Newcastle upon Tyne, Vol 2 (1896-1989)*, Newcastle upon Tyne: Newcastle Lit & Phil.

Pickering, W.S.F. (1981), *A Social History of the Diocese of Newcastle*, Stocksfield: Oriel Press.

Popielarz, P.A. and J.M. McPherson (1995), 'On the edge or in between: niche position, niche overlap, and the duration of voluntary association membership', *American Journal of Sociology*, **101** (3), 698–72.

Robinson, L. (2014), 'English associational culture in Lancashire and Yorkshire, 1890s–c.1930s', *Northern History*, **51** (1), 131–52.

Shapin, S. (1974), 'Property, patronage, and the politics of science: the founding of the Royal Society of Edinburgh', *British Journal for the History of Science*, **7** (1), 1–41.

Skocpol, T, M. Ganz and Z. Munson (2000), 'A nation of organizers: the institutional origins of civic voluntarism in the United States', *American Political Science Review*, **94** (3), 527–46.

Smith, C.E. and A.E. Freedman (1972), *Voluntary Associations: Perspectives on the Literature*, Cambridge, MA: Harvard University Press.

Smith, D.H. and C. Shen (2002), 'The roots of civil society: a model of voluntary association prevalence applied to data on larger contemporary nations', *International Journal of Contemporary Sociology*, **43** (2), 93–133.

Stokes, W. (2006), 'The rise and fall of an enterprising provincial paper', *Journalism Studies*, **7** (3), 427–36.

Taylor, P.J., G. Catalano and D.R.F. Walker (2002), 'Measurement of the world city network', *Urban Studies*, **39** (13), 2367–76.

Taylor, P.J., A. Firth, M. Hoyler and D. Smith (2010), 'Explosive city growth in the modern world-system: an inventory derived from urban demographic changes', *Urban Geography*, **31** (7), 865–84.

Turner, B.S. (1990), 'Outline of a theory of citizenship, *Sociology*, **24** (2), 189–217.

Turner, B.S. (2001), 'The erosion of citizenship', *British Journal of Sociology*, **52** (2), 189–209.

Watson, R.S. (1897), *The History of the Literary and Philosophical Society of Newcastle upon Tyne, Vol 1 (1793–1896)*, London: W. Scott.

PART III

METHODS AND MODELS OF CITY–NETWORK INTERACTIONS

7. The structure of urban networks

Marc Barthelemy

THE IMPORTANCE OF URBAN NETWORKS

Cities are located at the crossroads of many different networks. These networks are physical entities such as infrastructures that enable movements of individuals, transportation of goods, and distribution of electricity, gas and water, but can also be virtual, such as the social and professional networks. In modern countries, these networks have evolved over a long time (sometimes centuries), have become very complex and display a large variety of elements and actors in their functioning. This complexity leads to the relatively unexpected result that their robustness is not very well understood, as demonstrated by the occurrence of large blackouts in 2003 in North America. In addition, many of these networks are coupled to each other, which changes their resilience behavior (Buldyrev et al. 2010), and indicates the importance of considering multilayer networks (Vivela et al. 2014) for understanding their fragility. Measuring the structure and evolution of these infrastructures is therefore fundamental for understanding and modeling the resilience of cities and for understanding the socio-economic processes that govern its functioning.

We focus here on physical networks, such as transportation systems where nodes are stations and links represent segments between consecutive stations. In particular, an important component of cities is the street network made of nodes that represent the intersections, and the links are segments of roads between consecutive intersections. These networks can be thought of as a simplified schematic view of cities, but which captures a large part of their structure and organization (Southworth and Ben-Joseph 2003), and contains a large amount of information about underlying and universal mechanisms at play in their formation and evolution. Identifying the main mechanisms in these systems is not a new task (Haggett and Chorley 1969; Xie and Levinson 2011), but the recent availability of digitalized maps, historical or contemporary, allows us to test ideas and models on large-scale cross-sectional and historical data. Understanding these networks is therefore a central task for constructing a quantitative science of cities (Barthelemy 2016; Batty 2013).

All these physical networks have in common that they are embedded in space: nodes have a position, and links have a certain length and hence a cost associated with their formation and maintenance. These 'spatial networks' (Barthelemy 2011, 2017) necessitate specific tools for their characterization and in this chapter we review some of the most important tools. Many studies on complex networks were about how to characterize them and what are the most relevant measures for understanding their structure (see, for example, Latora et al. 2001). In particular, the degree distribution and the existence of the second moment for an infinite network was shown to be critical when studying dynamical processes on networks (Barrat et al. 2008). These behaviors are therefore strongly connected to degree fluctuations and to the existence of hubs. In spatial networks, the physical constraints are usually large and prevent the appearance of these hubs (the most important and almost unique counterexample is the airline network; see, for example, Barrat et al. 2004). These constraints also impact other measures that are non-trivial for complex networks but . . . become irrelevant for spatial networks. Examples of irrelevant parameters include the clustering coefficient or the assortativity, and we see that the clustering coefficient is always large and conveys interesting information for spatial networks.

Finally these networks evolve on a substrate that is not neutral and we can expect a coupling between the structure of urban networks and socio-economic features of the city where they develop. In the final part of this chapter we investigate a simple theoretical framework that allows us to investigate this coupling and to relate important characteristics of the network (such as the number of nodes and the total length of the network) to socio-economic features, such as the gross metropolitan product and the population.

SPATIAL NETWORKS ARE USUALLY NEITHER SCALE-FREE NOR SMALL-WORLDS

Average Degree

Spatial networks (Barthelemy 2017) are graphs embedded in space. They are completely characterized by their adjacency matrix A and the list of position of nodes. We recall that a graph with N nodes and E edges can be described by its $N \times N$ adjacency matrix A which is defined as $A_{ij} = 1$ if i and j are connected and $A_{ij} = 0$ otherwise. If the graph is undirected then the matrix A is symmetric (for more details, see, for example, the graph textbook (Clark and Halton 1991). The degree of a node is by definition the number of its neighbors, and is given by $k_i = \sum_j A_{ij}$. The first simple indicator of a graph is then the average degree

$$\langle k \rangle = \frac{1}{N}\sum_i k_i = \frac{2E}{N} \tag{7.1}$$

where here and in the following the brackets $\langle \cdot \rangle$ denote the average over the nodes of the network. In particular the scaling of $\langle k \rangle$ with N indicates if the network is sparse (which is the case when $\langle k \rangle \to const.$ for $N \to \infty$).

Buhl et al. (2006) and Cardillo et al. (2006) report measurements for street networks in different cities in the world. Based on the data from these sources, Barthelemy and Flammini (2008) plotted (Figure 7.1) the number of roads E (edges) versus the number of intersections N. The plot is consistent with a linear fit with slope ≈ 1.44 which is consistent with the value $\langle k \rangle \approx 2.5$ measured in Buhl et al. (2006). The quantity $e = E/N = \langle k \rangle/2$ displays values in the range $1.05 < e < 1.69$, in between the values $e = 1$ and $e = 2$ that characterize tree-like structures and $2d$ regular lattices, respectively. Few exact values and bounds are available for the average degree of classical models of planar graphs. In general it is known that $e \leq 3$, while it has been shown in Gerke and McDiarmid (2004) that $e > 13/7$ for planar Erdös–Rényi graphs.

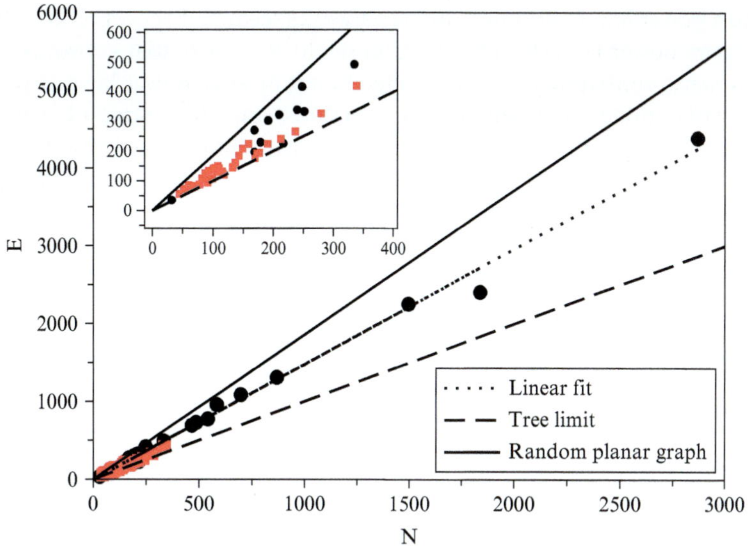

Source: Barthelemy and Flammini (2008).

Note: In the inset, we show a zoom for a small number of nodes.

Figure 7.1 Numbers of roads versus the number of nodes (intersections and centers) for data from Cardillo et al. (2006) (circles) and from Buhl et al. (2006) (squares)

Non-scale-free Degree Distributions

Quantities that depend very much on the spatial constraints turn out in general to be irrelevant for spatial networks: when physical constraints are strong or when the cost associated with the creation of new links is large, a cut-off appears in the degree distribution (Amaral et al. 2000) and the distribution is not broad. The prime example is the degree distribution which in many complex networks was found to be a broad law (see Albert and Barabasi 2002 and Latora et al. 2001) and in some cases was well fitted by a power law of the form

$$P(k) \sim k^{-\gamma} \tag{7.2}$$

with $1 < \gamma < 3$ in general (see Newman 2010, and for a discussion about power law fits see Broido and Clauset 2019). In the scale-free network case, degree fluctuations are very large which has a direct impact on many dynamical processes that take place on the network, such as epidemics (Barrat et al. 2008). Scale-free networks were also connected to preferential attachment (Barabasi and Albert 1999), as this provides a mechanism leading to a power law $P(k)$ with exponent $\gamma = 3$. However, it is in general incorrect to infer from a power law form for $P(k)$ to the existence of preferential attachment. Not only is the exponent measured usually different from $\gamma = 3$, but many other mechanisms could lead to a power law. In addition, many networks seem to display broad distributions that are not pure power laws (Broido and Clauset 2019). For spatial networks the degree has to satisfy steric constraints and we usually do not observe power laws. If we consider the road network, nodes represent the intersections, and the degree of a node is the number of streets starting from it and is therefore clearly limited as a result of space. As a consequence, the degree distribution for most spatial networks is not broad but displays a fast-decaying tail (such as an exponential)

$$P(k \gg 1) \ll k^{-\gamma} \tag{7.3}$$

for all finite values of γ. We note that the most important thing is not a power law degree distribution but to have a distribution such that the variance grows with the system size. Indeed, for many processes it is the second moment $\langle k^2 \rangle$ that governs the behavior. We have therefore to distinguish the two main classes of networks according to the limit of $\langle k^2 \rangle$ when $N \to \infty$. This in particular implies that any scale-free network with $\gamma > 3$ behaves as a network with a peaked distribution: a network with a fit giving $\gamma > 3$ will display the same behavior as an Erdös–Renyi graph for many processes, such as epidemic spread.

An important example that we thoroughly discuss in this chapter is the road network, but the conclusions are valid in general for planar networks. In a study of 20 German cities, Lämmer et al. (2006) showed that most nodes have four neighbors (the full degree distribution is shown in Figure 7.2) and that the degree rarely exceeds 5 for various world cities (Cardillo et al. 2006).

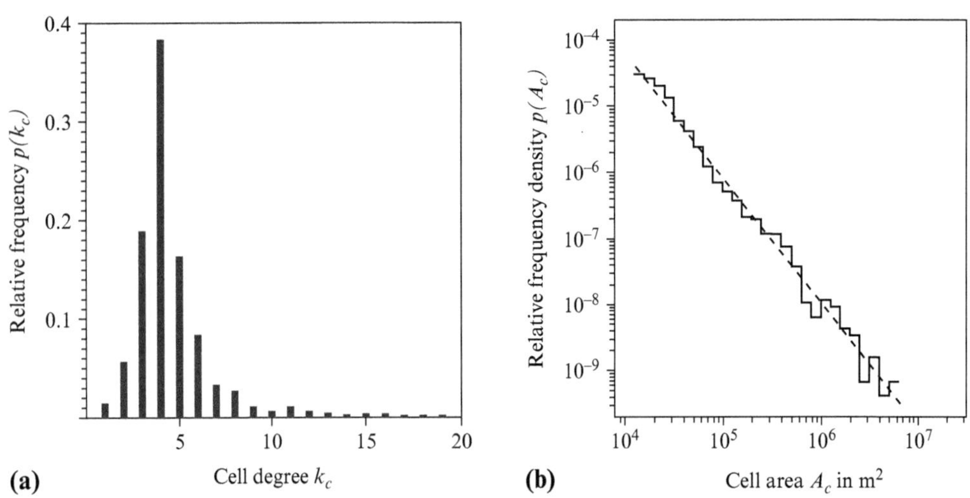

Source: Lämmer et al. (2006).

Note: The frequency distribution of the cell's surface areas A_c obeys a power law with exponent $\alpha \approx 1.9$.

Figure 7.2 *(a) Degree distribution of degrees for the road network of Dresden. (b) Frequency distribution of the cell's surface areas A_c for the road network of Dresden*

We note that in real-world cases such as the road network, it is natural to study the usual (or primal) representation where the nodes are the intersections and the links represent the road segment between the intersection. However, another representation, the dual graph can be of interest (see Porta et al. 2006) and is constructed in the following way: nodes represent roads and two nodes are connected if there is an intersection between the two corresponding roads. We can then measure the degree of a node which represents the number of roads which intersect a given road. Also, the shortest path length in this network represents the number of different roads that have to be taken to go from one point to another. Even if the road network has a peaked degree distribution, its dual representation can display broad distributions (Kalapala et al. 2006). Indeed, in Kalapala et al. (2006) measurements were made on the dual network for the road network in the US, England and Denmark, and showed large fluctuations with a power law distribution with exponent $2.0 < \gamma < 2.5$.

In 2005, Sienkiewicz and Hołyst studied a larger set made of public transportation networks of buses and tramways for 22 Polish cities and in 2009, von Ferber et al. studied the public transportation networks for 15 world cities. The number of nodes of these networks varies from $N = 152$ to 2811 in Sienkiewicz and Hołyst (2005) and in the range [1494,44629] in Von Ferber et al. (2009). Interestingly, Sienkiewicz and Hołyst (2005) observe a strong correlation between the number of stations and the population, which is not the case for the world cities studied in Von Ferber er al. (2009), where the number of stations seems to be independent from the population (see the find section of this chapter for a detailed discussion about the connection between socio-economic indicators and the properties of networks). For Polish cities the degree has an average in the range [2.48, 3.08] and in a similar range [2.18, 3.73] for Von Ferber er al. (2009). In both cases, the degree distribution is relatively peaked (the range of variation is usually of the order of one decade) consistently with the existence of physical constraints (Amaral et al. 2000).

Other empirical examples comprise power grids and water distribution networks that are important infrastructures in our societies. The topological structure of these networks was studied in different papers, such as Albert et al. (2004), Amaral et al. (2000) and Sole et al. (2008). In particular, Amaral et al. (2000) and Albert et al. (2004) consider the Southern Californian and the North American power grids. In these networks the nodes represent the power plants, distribution and transmission substations, and the edges correspond to transmission lines. These networks are typically planar (see, for example, the Italian case in Figure 7.3) and we expect a peaked degree distribution, decreasing typically as an exponential of the form $P(k) \sim \exp(-k/\langle k \rangle)$ where $\langle k \rangle$ of order 3 in Europe and 2 in the US. The other studies on US power grids confirm that the degree distribution is exponential (see Figure 7.4). In addition, we note that Albert et al. (2004) found a large redundance in the North American network with, however, 15 percent of cut-edges.

In addition to the distribution of electricity, our modern societies also rely on various other distribution networks. The resilience of these networks to perturbations is an important issue in the design and operating of these systems. Yazdani and Jeffrey (2010) studied the topological properties of the Colorado Springs Utilities and the Richmond (UK) water distribution networks (shown in Figure 7.5). Both these networks (of size $N = 1786$ and $N = 872$, respectively) are sparse planar graphs and display very peaked degree distributions – the maximum degree is 12, showing without any fitting that the distribution cannot be broad.

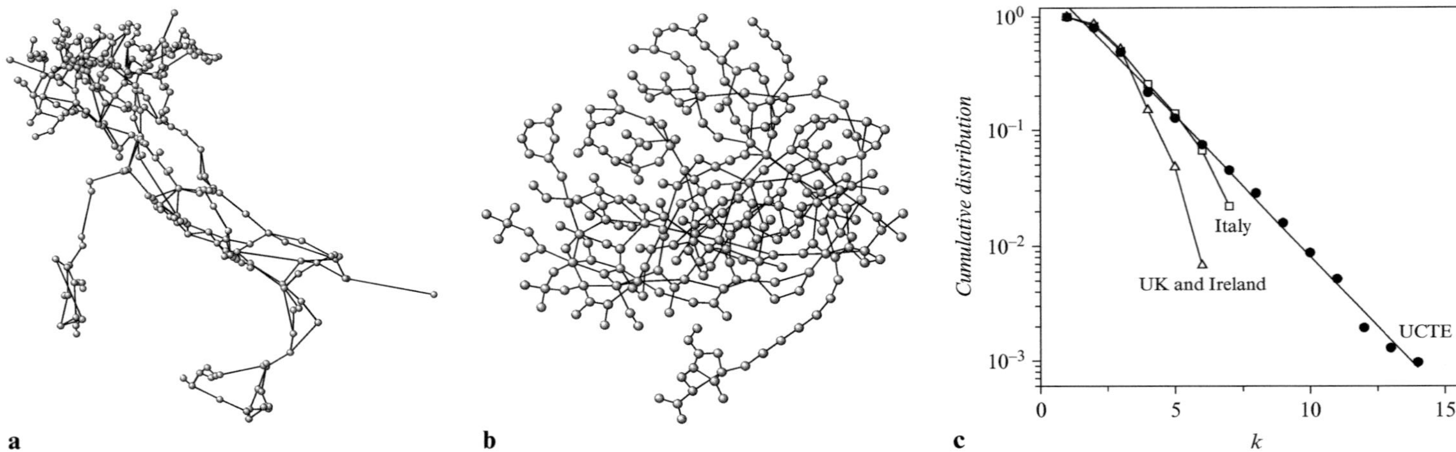

a b c

Source: Solé et al. (2008).

Note: In all cases the degree distribution is peaked and can be fitted by exponential.

Figure 7.3 (a) Map of the Italian power grid. (b) Topology of the Italian power grid. (c) Degree distribution for the European network (UCTE), Italy and the UK and Ireland

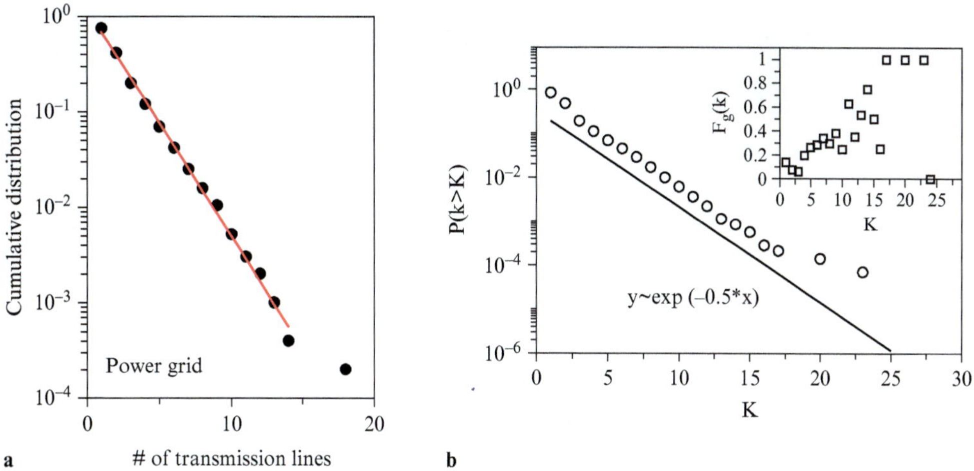

Source: Left panel (Amaral et al. 2000); right panel (Albert et al. 2004).

Note: The lines in both panels represent an exponential fit.

Figure 7.4 Degree distribution of substations in Southern California (left panel) and for the North American power grid (right panel)

Large Worlds: Average Shortest Path

There are in general many paths between two nodes in a connected network and if we keep the shortest one it defines a distance on the network

$$\ell(i,j) = \min_{paths\,(i \to j)} |path| \tag{7.4}$$

where the length $|path|$ of the path is defined as its number of edges (other distances can be defined using weights for each link). The *diameter* of the graph is defined as the maximum value of all $\ell(i,j)$ and the size of the network can also be estimated by averaging this distance over all pairs of nodes. Indeed, for a d-dimensional regular lattice with N nodes, this average shortest path $\langle \ell \rangle$ scales as

$$\langle \ell \rangle \sim N^{1/d}. \tag{7.5}$$

As most urban networks are embedded in $d = 2$ space, the good benchmark is the $2d$ dimensional lattice for which we have

$$\langle \ell \rangle_{d=2} \sim \sqrt{N}. \tag{7.6}$$

In a small-world network (see Watts and Strogatz 1998) constructed over a d–dimensional lattice, $\langle \ell \rangle$ has a very different behavior

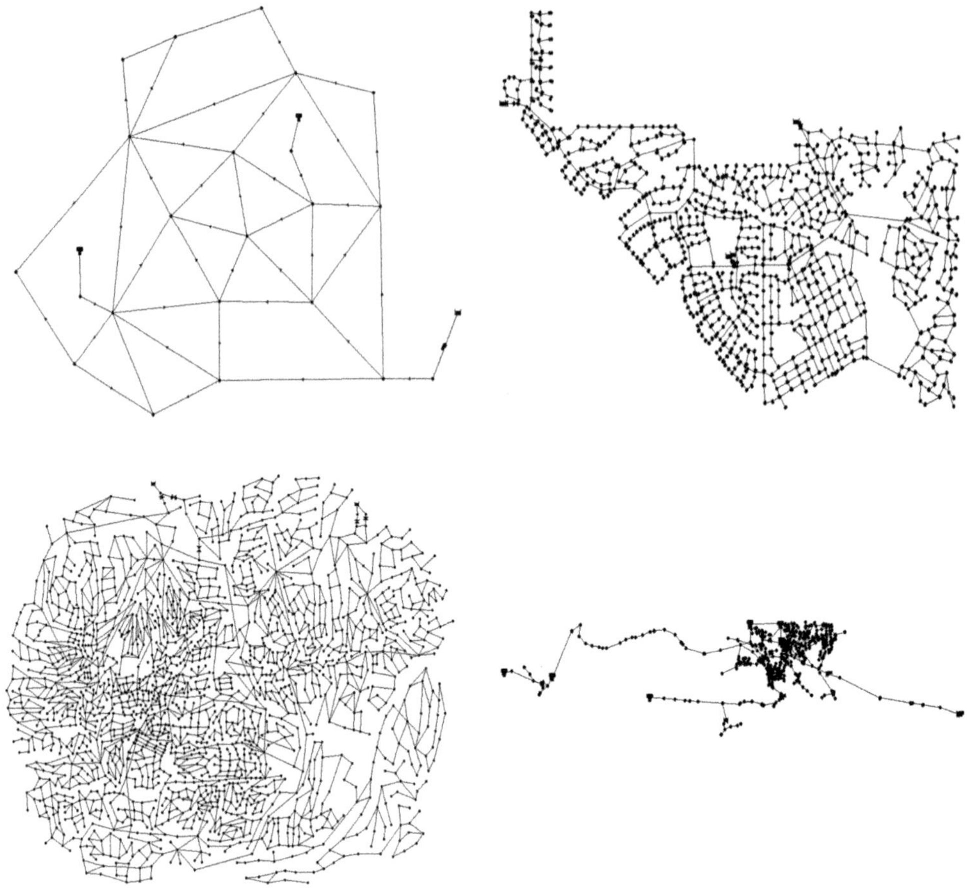

Source: Yazdani and Jeffrey (2010).

Note: Top left-hand side panel: synthetic networks ('Anytown', Walski et al. 1987). Bottom left-hand side panel: 'EXNET' (Farmani et al. 2004). Top right-hand side panel: Colorado Spring Utilities network. Bottom-right panel: Richmond (UK) water distribution network.

Figure 7.5 Representation of water distribution networks

$$\langle \ell \rangle \sim \log N \tag{7.7}$$

and is much smaller than for lattices $\log N \ll N^{1/d}$. The crossover from a large-world behavior $N^{1/d}$ to a small-world behavior with $\log N$ can be achieved for a density p of long links (or shortcuts) (Barthelemy and Amaral 1999), such that

$$pN \sim 1. \tag{7.8}$$

In most spatial networks, we do not have any shortcuts and the expected behavior is therefore given by equation 7.6. The effect of space could thus in principle be detected in the behavior of $\langle \ell \rangle(N)$. It should be noted that if the number of nodes is too small this can be a tricky task. In the case of brain networks, for example, a behavior typical of a three-dimensional network, is in $N^{1/3}$ and could easily be confused with a logarithmic behavior if N is not large enough. This small N effect can also be observed in earlier studies (Latora and Marchiori 2001; Sienkiewicz and Hołyst 2005; Viana et al. 2013), where the behavior of the average shortest path is not clear and could be fitted by a logarithm or a power law as well. We note however that in these studies the average shortest path is usually large compared to $\log N$ suggesting that the behavior of $\langle \ell \rangle$ might not be logarithmic with N but more likely scales as $N^{1/2}$: $\langle \ell \rangle \approx 16$ for $N = 124$ stations in Latore and Marchiori (2001) – in this case $\ln 124 \approx 5$ and 16 is closer to the two-dimensional result $\sqrt{124} \approx 11$; of order 10 in Sienkiewicz and Hoły et al. (2005) and in the range [6.4, 52.0] for $N \in [1500, 46\,000]$ in Von Ferber et al. (2009).

At the country level, Kurant and Thiran (2006) studied the railway system of Switzerland and major trains and stations in Europe (and the public transportation system of Warsaw, Poland). The Swiss railway network contains $N = 1613$ nodes and $E = 1680$ edges and all conclusions drawn in this case are consistent with a two-dimensional spatial network: the average degree is $\langle k \rangle \approx 2.1$, the average shortest path is ≈ 47, consistent with the $\sqrt{N} \approx 40$ result for a two-dimensional lattice (we also note that, as expected, the degree distribution is peaked and exponentially decreasing).

Other Properties: Segment Length and Clustering

Segment length and total length of the network

In Figure 7.6, we display the total length ℓ_T of the network versus N for the cities considered in Cardillo et al. (2006). Data are well fitted by a power function of the form

$$\ell_T = \mu N^{\beta} \tag{7.9}$$

with $\mu \approx 1.51$ and $\beta \approx 0.49$. In order to understand this result, one has to focus on the street segment length distribution $P(\ell_1)$. This quantity has been measured for London in Masucci et al. (2009) and its distribution can be fitted by Masucci et al. (2009)

$$P(\ell_1) \sim \ell_1^{-\gamma} \tag{7.10}$$

with $\gamma \simeq 3.4$, which implies that both the average and the dispersion are well defined and finite. If we assume that this result extends to other cities, it means that we have a typical distance ℓ_1 between nodes, which is meaningful (that is, fluctuations around this typical distance are relatively small). This typical distance between connected nodes then naturally scales as

$$\ell_1 \sim \frac{1}{\sqrt{\rho}} \tag{7.11}$$

where $\rho = N/L^2$ is the density of vertices and L the linear dimension of the ambient two-dimensional space. This implies that the total length scales are

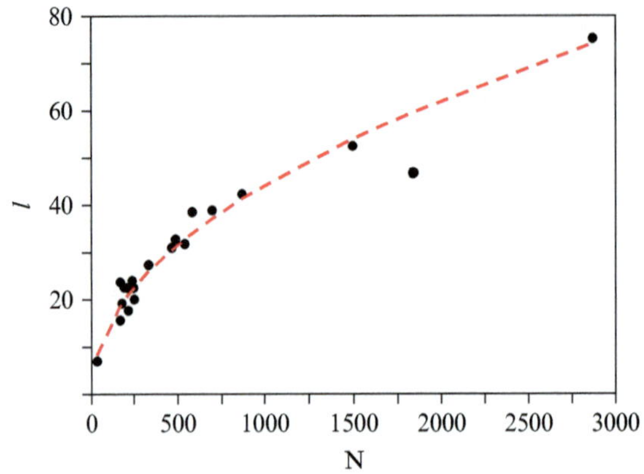

Source: Data from Cardillo et al. (2006) and figures from Barthelemy and Flammini (2008).

Note: The dashed line is a fit which predicts a growth as $\sqrt{N}$.

Figure 7.6 Total length versus the number of nodes

$$\ell_T \sim E\ell_1 \sim \frac{\langle k \rangle}{2} L\sqrt{N}. \tag{7.12}$$

This simple argument reproduces well the $\sqrt{N}$ behavior observed in Figure 7.6 and the value (given the error bars) of the prefactor $\mu \approx \langle k \rangle L/2$.

Clustering coefficient

The clustering coefficient of a node i of degree k_i is defined as

$$C(i) = \frac{E_i}{k_i(k_i - 1)/2} \tag{7.13}$$

where E_i is the number of edges among the neighbors of i. This quantity gives some information about local clustering and was the object of many studies in complex networks. For the Erdös–Rényi random graphs with finite average degree $\langle k \rangle$, the average clustering coefficient is simply given by $\langle C \rangle = p \sim \frac{\langle k \rangle}{N}$ where the brackets $\langle \cdot \rangle$ denote the average over the network. In contrast, for spatial networks, closer nodes have a larger probability of being connected, leading to a large clustering coefficient. Many studies use the clustering coefficient per degree classes which is given by

$$C(k) = \frac{1}{N(k)} \sum_{i/k_i = k} C(i) \tag{7.14}$$

where $N(k) = NP(k)$ is the number of nodes of degree k. The behavior of $C(k)$ versus k thus gives an indication how the clustering is organized when we explore different classes of degrees. However, in order to be useful, this quantity needs to be applied to networks with a large range of degree variations, which is usually not the case in spatial networks.

The Erdös–Rényi network is not planar and is not a spatial network. A better comparison

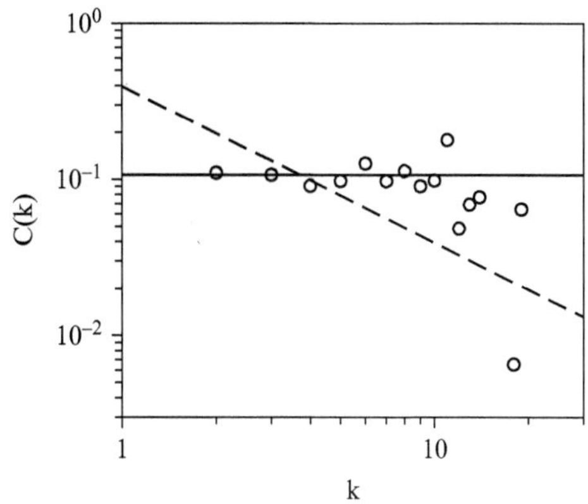

Source: Ravasz and Barabási (2003).

Note: The dashed line has a slope −1 and the solid line corresponds to the average clustering coefficient.

Figure 7.7 Scaling of the clustering C(k) for the power grid of the Western United States

point for spatial networks is then obtained with the random geometric graph (for a mathematical account see Penrose 2003). This network is obtained by connecting points in the two-dimensional plane if their distance is less than a given threshold (see also Barthelemy 2017 and references therein). Many quantities can be computed for this graph and in particular the clustering coefficient (Dall and Christensen 2002). For $d = 1$, we obtain Dall and Christensen (2002) $\langle C_1 \rangle = 3/4$ and for $d = 2$ $\langle C_2 \rangle = 1 - 3\sqrt{3}/4\pi \approx 0.58650$. The average clustering coefficient thus decreases from the value 3/4 for $d = 1$ to 0.58 for $d = 2$ and is independent from the number of nodes, in sharp contrast with Erdös–Rényi graphs for which $\langle C \rangle \sim 1/N$. Random geometric graphs are thus much more clustered than random Erdös–Rényi graphs. The main reason, which is valid for most spatial graphs, is that long links are prohibited or rare. This implies that if both i and j are connected to v, they lie in the same spatial neighborhood, and therefore have a large probability to be close enough and connected, leading to a large clustering coefficient.

We can illustrate this discussion on the empirical example of power grids, for which the clustering profile $C(k)$ is rather large and even independent of k as shown in the case of the network of the Western US (see Figure 7.7).

TYPOLOGY OF ROAD NETWORKS

Being able to compare different planar networks is of interest to many fields, from botanics to urban studies. If we can compare two road networks, we can compare two cities and very likely we can then construct families of cities with similar road network structures. The existence of families can then indicate different mechanisms governing the evolution of these systems, and can therefore help in uncovering how these systems grow and expand.

We first discuss a classification method based on the statistics of both the surface area and shape of faces. This is different from other classifications that we see in the next

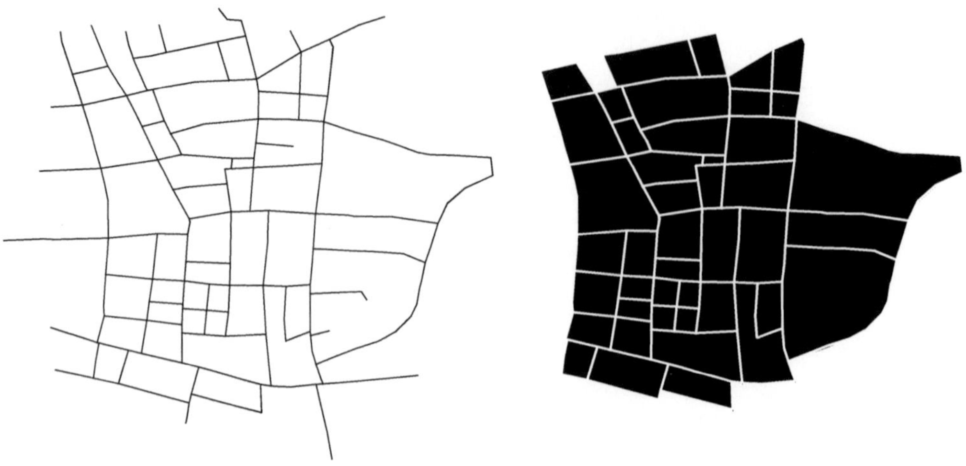

Source: Louf and Barthelemy (2014).

Note: The block representation does not take dead-ends into account.

Figure 7.8 From the street network to blocks: example of a street pattern taken in the neighbourhood of Shibuya in Tokyo (Japan) and the corresponding set of blocks

sections and which are based on the network itself and its topology alone. In the case of planar graph, where the geometry counts (as, for example, with city maps), the visual impression is mostly given by the faces and the pattern they form in the two-dimensional plane (see, for example, Figure 7.8).

The method that we discuss can in principle be applied to any random planar graph and we illustrate this approach using the example of street networks. Urban morphology and the structure of streets and roads are important components of cities, and can be thought as a simplified schematic view of cities, which captures a large part of their structure and organization (Southworth and Ben-Joseph 2003), and contain a large amount of information about underlying and universal mechanisms at play in their formation and evolution. Extracting common patterns between cities is a way towards the identification of these underlying mechanisms. At stake is the question of the processes behind the 'organic' patterns, which grow in response to local constraints, and whether they are preferable to the planned patterns which are designed under large-scale constraints. Also, comparing networks between them or with results of simulations implies the capacity to test the null hypothesis that two networks have been drawn from the same distribution.

We note that, strictly speaking, road networks are not always perfectly planar. An interesting study about this issue can be found in Boeing (2020), where the author examines street networks in 50 cities worldwide and variation of nonplanarity across single cities.

Statistics of Shape and Area of Blocks

The road network – usually planar (or at least to a good approximation, see Boeing 2020 – is made of faces that are termed *blocks* in this context: a city block is the smallest area surrounded by streets, and forms the basic unit of the urban fabric. We characterize these objects that are polygons and whose properties, such as the area or shape characterization, are easily measured.

First, the surface area A of a block is an important indicator and has been measured for various street networks (Fialkowski and Bitner 2008; Lämmer et al. 2006; Louf and Barthelemy 2014). Most street networks display a power law behavior for the surface area distribution $P(A)$

$$P(A) \sim \frac{1}{A^{\tau}} \tag{7.15}$$

with an exponent always of order $\tau \approx 2$ (Barthelemy 2011; Barthelemy et al. 2013; Lammer et al. 2006; Louf and Barthelemy 2014; Nicosia et al. 2018; Strano et al. 2012). This observation is in sharp contrast with the simple picture of an almost regular lattice which would predict a distribution $P(A)$ peaked around some value ℓ_1^2 where ℓ_1 is the typical linear size of a face. It is interesting to note that if we assume that $A \sim 1/\ell_1^2 \sim 1/\rho$ (where ρ is the density of nodes) and that ρ is distributed according to a law $f(\rho)$ with a finite $f(0)$, a simple calculation gives

$$P(A) \sim \frac{1}{A^2} f(1/A) \tag{7.16}$$

which behaves as $P(A) \sim 1/A^2$ for large A. This simple argument suggests that the observed value ≈ 2.0 of the exponent is 'universal' and reflects the random variation of the density. The downside of this result is that it does not allow distinguishing efficiently a graph from another.

A second characterization of a block is through its shape, with the form (or shape) factor Φ, defined in the geography literature in Haggett et al. (1977) as the ratio between the area of the block and the area of the circumscribed circle C

$$\Phi = \frac{A}{A_C}. \tag{7.17}$$

Note that another similar measure is $A/(\pi D^2/4)$ where D is the largest distance found in the polygon and is easier to compute. The quantity Φ is in $[0,1]$, and the smaller its value and the more anisotropic the block is. For example for a rectangle of sides a and b we easily find

$$\Phi_{rectangle} = \frac{4}{\pi} \frac{ab}{a^2 + b^2} \tag{7.18}$$

and varies from 0 for an infinitely elongated rod to $2/\pi \approx 0.64$ for a square. We note that there is not a unique correspondence between a particular shape and a value of Φ, but this measure gives a good indication about the face's shape (when it is a relatively simple polygon). Empirical observations on street networks for the distribution $P(\Phi)$ display important differences between cities. This leads to the first naive idea for classifying planar graphs according to the distribution of the shape of faces. The shape itself, however, is not enough to account for visual similarities and dissimilarities between planar graphs. Indeed, in the case of cities, we find, for example, that for cities such as New York and Tokyo, the distributions $P(\Phi)$ are similar (see Figure 7.9), but the visual similarity between both cities' layout is not obvious. One reason for this is that blocks can have a similar shape but very different surface areas: if two cities have blocks of the same shape in the same proportion but with totally different areas, they will look different. We thus need to combine the information about both the shape and the area of faces.

Towards a Typology of Planar Graphs

In order to construct a simple representation of planar graphs which integrates both area and shape, we rearrange the faces according to their area (on the y-axis) and display their Φ value on the x-axis (Figure 7.9). We divide the range of areas in (logarithmic) bins and the color of a block represents the area category to which it belongs. We describe quantitatively this pattern by plotting the conditional probability distribution $P(\Phi|A)$ of shapes, given an area bin (Figure 7.9, right). The colored curves represent the distribution of Φ in each area category, and the curve delimited by the gray area is the sum of all of these curves and is the distribution of Φ for all cells, which is simply the translation of the well-known formula for probability conditional distribution

$$P(\Phi) = \sum_A P(\Phi|A)P(A). \tag{7.19}$$

These figures give a 'fingerprint' of the planar graph which encodes information about both the shape and the area of faces. We then base a classification, not on $P(\Phi)$ but on the conditional probability $P(\Phi|A)$. We then first construct bins for the surface areas of faces. Typically for street networks, the block area in the range $A \in [10^3,10^5]$ (in square meters) dominate the total number of cells, and we can neglect very small blocks (of area $A < 10^3\text{m}^2$) and very large blocks (of area $A > 10^5\text{m}^2$). We thus sort, in this case, the blocks according to their area in two distinct bins

$$\alpha_1 = \{\text{cells}|A \in [10^3,10^4]\}$$

$$\alpha_2 = \{\text{cells}|A \in [10^4,10^5]\}$$

However, in general, we might have a very different area distribution and the binning should then be done accordingly. We denote by $f_\alpha(\Phi)$ the ratio of the number of faces with a form factor Φ that lie in the bin α over the total number of faces for that graph. We then define a distance d_α between two graphs G_a and G_b characterized by their respective $f_\alpha^{G_a}$ and $f_\alpha^{G_b}$ as

$$d_\alpha(G_a,G_b) = \int_0^1 |f_\alpha^{G_a}(\Phi) - f_\alpha^{G_b}(\Phi)|^n \, d\Phi \tag{7.20}$$

with $n = 1$ or $n = 2$ (other choices are also possible but this does not seem to be relevant in the typology construction). We then construct a global distance D between two cities by combining all area bins α

$$D(G_a,G_b) = \sum_\alpha d_\alpha(G_a,G_b)^2. \tag{7.21}$$

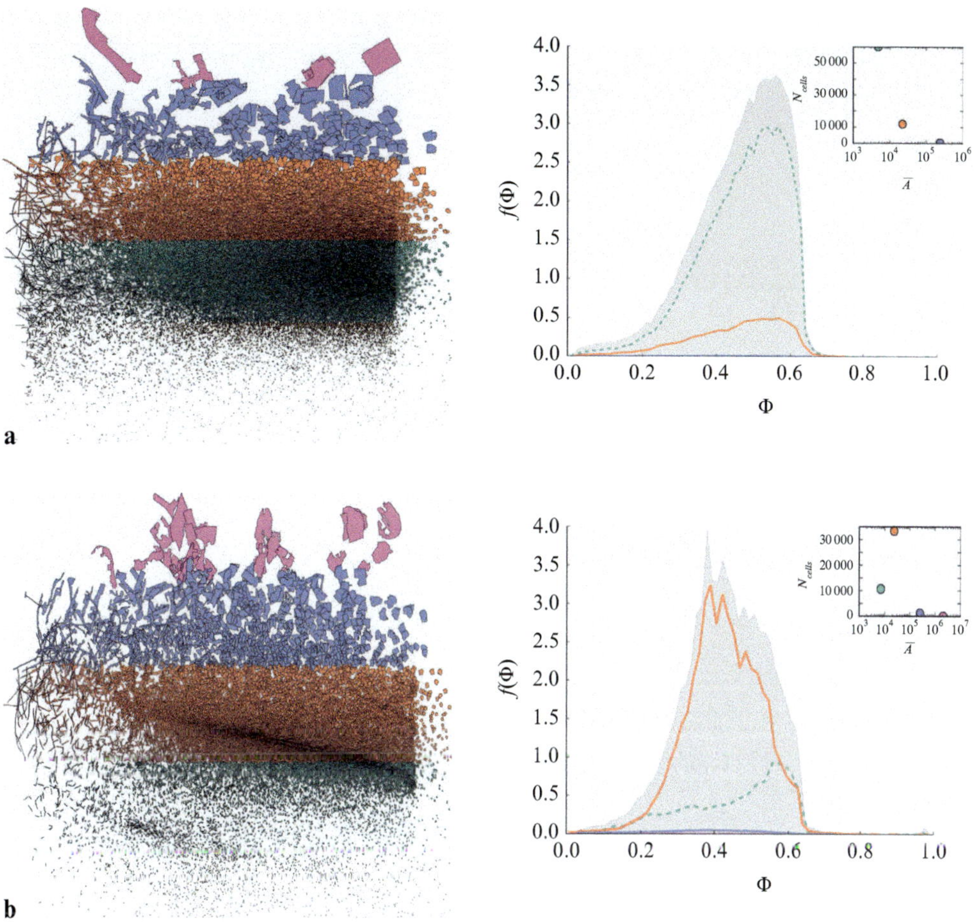

Source: Louf and Barthelemy (2014).

Note: Left-hand side column: we rearrange the blocks of a city according to their area (y-axis), and their Φ value (x-axis). The color of each block corresponds to the area category it falls into. Right-hand side column: we quantify this pattern by plotting the repartition of shapes per area category, as measured by Φ for each area category – represented by the solid and dashed curves. The contour of the gray shade is the sum of the distribution of Φ for all cells. As shown in the inset, intermediate area categories dominate the total number of cells, and are thus enough for the clustering procedure.

Figure 7.9 The fingerprints of Tokyo (top) and New York (bottom)

We note that from the mathematical perspective distances between graphs already exist, such as the graph edit distance (GED) (Gao et al. 2010) or the Gromov-Hausdorff distance (Mémoli 2007). These distances, however, are usually difficult to apply to large graphs and in general do not take into account the geometrical structure of the graph as encoded in the face area distribution.

At this point, we have a distance between two graphs and we can use a classical hierarchical clustering algorithm (Kaufman and Rousseeuw 1990). This is what was carried

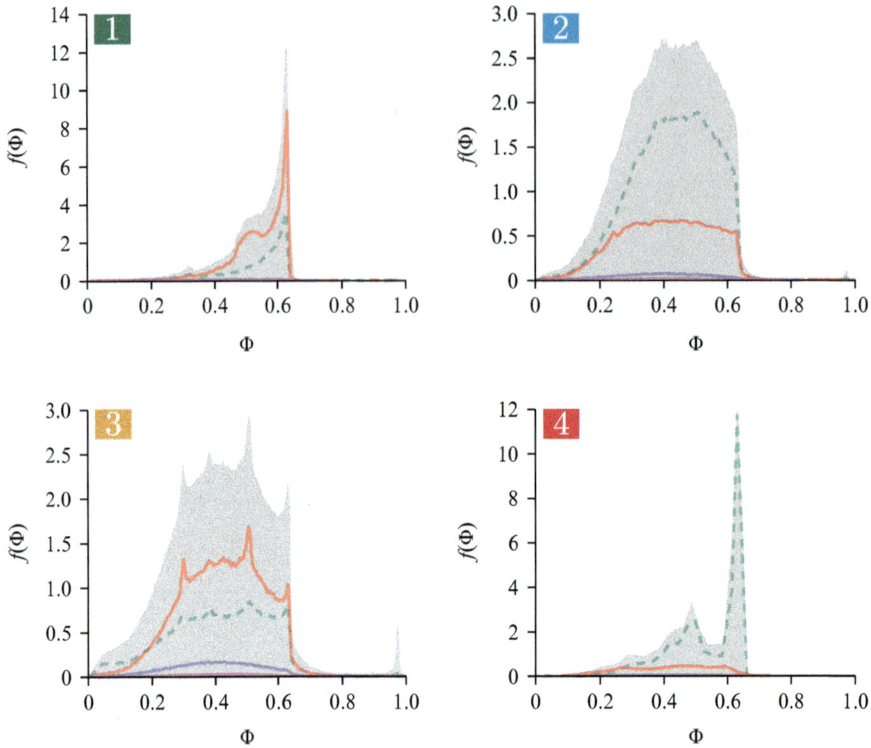

Source: Louf and Barthelemy (2014).

Note: Above: average distribution of the shape factor Φ for each group found by the clustering algorithm. Each curve corresponds to the area bin, and from small to large (color online): dashed green, orange and blue. We can observe on these plots here the main differences for the distributions of medium- and large-size blocks. Right: typical street pattern for each group (plotted at the same scale in order to observe differences both in shape and areas). Group 1 (top left), Buenos Aires; group 2, Athens; group 3, New Orleans; and group 4, Mogadishu.

Figure 7.10 Four groups of street networks

out for 131 cities resulting in a dendrogram (not shown here; Louf and Barthelemy 2014). At an intermediate level of this dendrogram we then identify four distinct categories of street networks, which are easily interpretable in the abundance of blocks with a given shape and with small or large area. In Figure 7.10 we show the average distribution of Φ for each category and the typical street patterns associated with each of these groups. The main features of each group are the following.

- In group 1 (comprising Buenos Aires, Argentina only) we essentially have blocks of medium size (in the bin α_2) with shapes that are dominated by the square shape and regular rectangles. Small areas (in bin α_1) are almost exclusively squares.
- Athens, Greece is a representative element of group 2, which comprises cities with a dominant fraction of small blocks with shapes broadly distributed.

Figure 7.10 (continued)

- Group 3 (illustrated by New Orleans, USA) is similar to group 2 in diversity of shapes but is more balanced in relation to areas, with a slight predominance of medium-size blocks.
- Group 4, which contains for this dataset the interesting example of Mogadishu, Somalia, displays small, square-shaped blocks, together with a small fraction of small rectangles.

The advantage of this method is that it is relatively simple and allows for an interpretation in shape and area of faces. However, a drawback is that we are completely neglecting correlations between the faces. This is particularly important in the city case where there is usually a particular level of homogeneity at a particular scale (what we usually call a 'neighborhood'). For New York City we can isolate the different boroughs (the Bronx, Brooklyn, Manhattan, Queens and Staten Island), and extract their fingerprint (not shown, see Louf and Barthelemy 2014). The fingerprint of New York (bottom, Figure 7.9) results therefore from the combination of different fingerprints for each of the boroughs. While Staten Island and the Bronx have very similar fingerprints, the others are different. Manhattan exhibits two sharp peaks at $\Phi \approx 0.3$ and $\Phi \approx 0.5$ which are the signature of a grid-like pattern with the predominance of two types of rectangles. Brooklyn and Queens exhibit a sharp peak at different values of Φ, also the signature of grid-like patterns with different rectangles for basic shapes.

A further step in this classification would thus be to find a method to extract regions where the faces features are strongly correlated and to include this information in the construction of classes. Despite the simplifications that this method entails, it allows for a quantitative and systematic comparison between different planar graphs where the geometrical information is of importance.

We end this section by noting that other approaches were proposed. In particular, Katifori and Magnasco (2012) and Mileyko et al. (2012) proposed to construct an approximate mapping from a planar graph to a tree. This method relies on the idea of deleting edges according to their weight and observing how the faces merge together. It has the advantage of being simple and allowing for a quantitative characterization of the loop organization in planar maps. It is certainly an interesting direction for future research in this field, and an important step would be achieved by generalizing this method to non-weighted networks.

Another approach was proposed in combinatorics studies. In 2004, Bouttier et al. (2004) extented a bijection between a planar graph (with a choice of an origin) and a tree, originally proposed by Schaeffer in his 1998 PhD thesis. This bijection is a process to encode a planar map in a simple way and, even if this method was originally developed mostly for combinatorial reasons, it has the advantage to be simple and could be useful for other, more practical purposes. The resulting tree depends on the choice of the origin and this is somehow a limitation of this method, but further work is needed in order to understand the impact of this choice on the statistical structure of the tree (we note that many quantities will not depend on this origin choice, which makes the method very useful for computing various quantities).

TIME EVOLUTION OF INFRASTRUCTURE NETWORKS

Road Networks Evolution: General Properties

So far, we have discussed properties on static networks, and in this section we characterize the time evolution of the road network. We first present results on the time evolving road network of the Groane region, which is a 125 square kilometer area located north of Milan (Strano et al. 2012) and for which seven snapshots for different times from 1833 to 2007 are available. This region evolved without central planning and is thus a good example of an organic evolution of urban systems.

At a smaller scale, we will discuss the evolution of central Paris between 1789 and 1999. This dataset provides an interesting case study, as Paris experienced large changes due to Haussmann in the middle of the ninetheenth century (see Barthelemy et al. 2013 for details and more references about this network). This is an opportunity to observe quantitatively the effect of top-down planning: until 1836, we are in the pre-Haussmann Paris, while from 1888 until now we are in the post-Haussmann period.

We first describe briefly the evolution of basic properties of the road network, such as the number of nodes, of links, the total length of the network, the area and the shape factor of blocks. The evolution of the network, characterized for example by the number of nodes N, has usually a non-constant growth rate. For Groane, the total number N grows by a factor of 20, from the original 255 nodes present at $t = 1833$ to more than 5000 nodes at $t = 2007$, with period of slow growth from 1833 to 1933, fast from 1933 to 1994 and slow again from 1994 to 2007. Interestingly enough, the number of nodes N in the network is a linear function of the number of people living in the area; that is, the number of people per road intersection remains constant over time (Figure 7.11a).

For Paris, the number of nodes increased from about 3000 in 1836 to about 6000 in 1888, and the total length increased from about 400 to almost 700 kilometers, all this in about 50 years. The same behavior of N versus the population is also found here: the number of nodes N is proportional to the population P and the corresponding increase rate is of order $dN/dP \approx 0.0021$. This seems to reflect that the density of the street network indeed represents a good indicator of the evolution of a region and its economic potential, but we do not have a clear theoretical argument for this.

In order to understand the evolution of the network and to compare different cases, it thus seems natural to consider the number of nodes N as the natural internal clock of the system, and to study the change of various network properties as a function of N. In this way we get rid of exogenous effects on the evolution of this system.

We first observe simple facts, such as the number of links K that grows almost linearly with N (Figure 7.11b), showing that the average degree is roughly constant despite massive historical changes, with a slight increase from $\langle k \rangle \simeq 2.57$ to 2.8 when going from 1914 to 1980 (Figure 7.11b, bottom). In the Paris case we also observe that K is proportional to N giving an average degree of order 3 (Barthelemy et al. 2013).

The total network length (Figure 7.11c) displays an increase of the form N^{β} where $\beta \simeq 0.54$. This result – also observed for the Paris case – is consistent with the evolution of two-dimensional lattices with a peaked link length distribution (see above and Barthelemy 2011; Barthelemy and Flammini 2008) which are described by a value $\beta = 1/2$. The quantity $r_N = [N(1) + (N(3)]/N$, where $N(k)$ is the number of nodes of degree k, gives additional information on the structure of the network. Indeed, the plot of r_N versus N (Figure 7.11d) shows that it steadily decreases from $r_N \simeq 0.87$ at $t = 1833$ to $r_N \simeq 0.835$ at $t = 2007$ signalling a transition from pre-urban to a urban phase.

In agreement with other results (Lämmer et al. 2006), the distribution of block areas in Groane at $t = 2007$ is a power law with reported in the inset, the exponent changes in time: it takes a value $\gamma \simeq 1.2$ at $t = 1833$ and converges towards $\gamma \simeq 1.9$ as the network grows. Since a larger exponent indicates a higher homogeneity of cell areas, we are thus witnessing here a process of homogenization of the size of cells. This appears to be a clear effect of increasing urbanization over time, with the fragmentation of larger cells of natural land into smaller urbanized cells. Accordingly, the relative dispersion of cell areas, shown in Figure 7.12b, decreases from 0.5 at $t = 1833$ to 0.26 at $t = 2007$, indicating that the variance of the distribution becomes smaller as N increases.

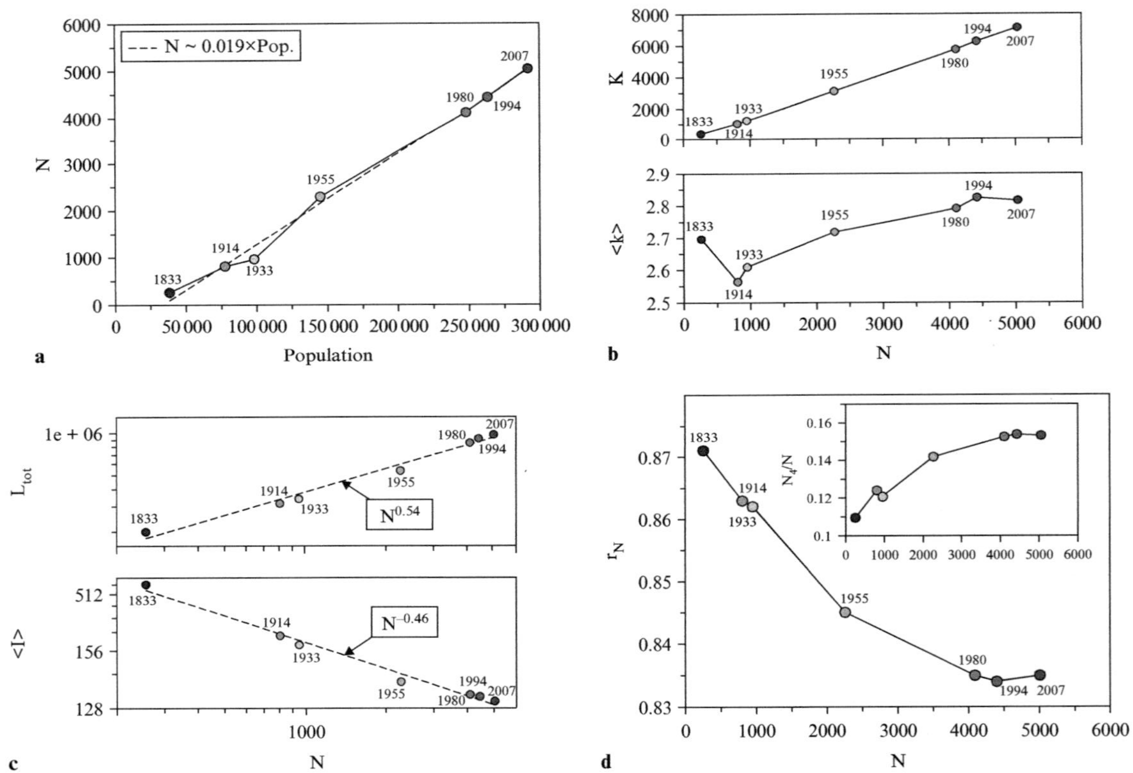

Source: Strano et al. (2012).

Note: Graph (a): number of nodes *N* versus total population (continuous line with circles) and its linear fit (dashed line). Graph (b): total number of edges *K*, and the average node degree á*k*ñ as a function of *N*. The number of edges increases slightly faster than linearly with the number of nodes, as shown by the slight increase in the average node degree. Graph (c): both the total network length L_{tot} (upper panel) and the average link length á*l*ñ (bottom panel) scale as power law functions of *N* (the corresponding fits are reported as dashed lines). Graph (d): as the network grows, the value of the ratio r_N between the number of nodes with degree $k = 1$ and $k = 3$, and the total number of nodes, decreases, indicating the presence of a higher number of four-ways crossings. In the inset we report the percentage of nodes having degree $k = 4$ as a function of *N*. Note that the relative abundance of four-ways crossings increases by 5 percent in two centuries.

Figure 7.11 Measures for the Groane region (Italy)

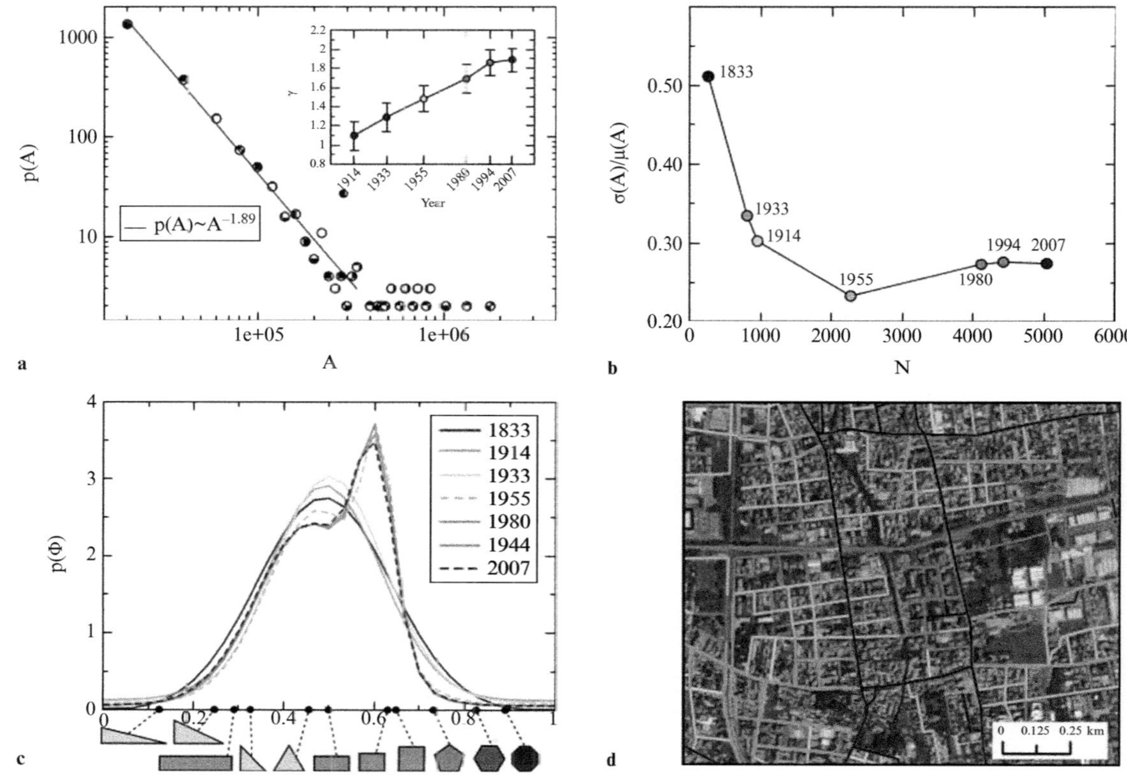

Source: Strano et al. (2012).

Note: Graph (a): the size distribution of cell areas at $t = 2007$ can be fitted with a power law $p(A) \sim A^{-\gamma}$, with an exponent $\gamma \simeq 1.9$. The values of γ increase over the years as shown in the inset. Graph (b): relative dispersion in the distribution of areas as a function of the network size N. Graph (c): distribution of cell shapes at each time, as quantified by the shape factor Φ. We also report, as a reference, the values of shape factors corresponding to various convex regular polygons. Graph (d): the map shows some typical cell shapes at different times.

Figure 7.12 Groane region (Italy)

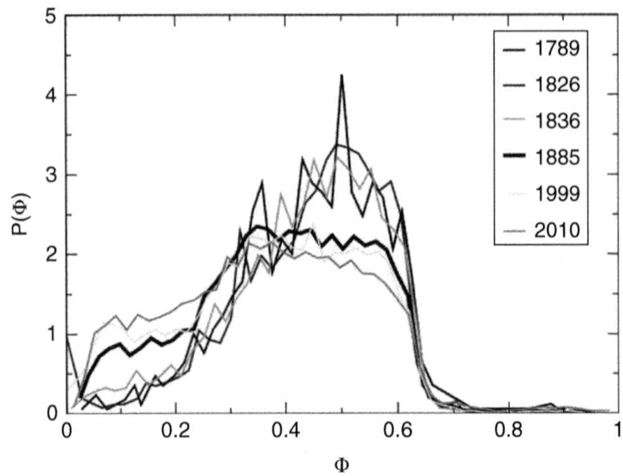

Source: Barthelemy et al. (2013).

Figure 7.13 Shape factor distribution for central Paris at different times

The distribution of the shape factor $P(\Phi)$ reported in Figure 7.12c clearly reveals the existence of two different regimes: for $t \leq 1933$ the distribution is well approximated by a single Gaussian function with an average of about 0.5 and a standard deviation of 0.25. Conversely, for $t \geq 1955$ the distribution of shape factors displays two peaks and can be fitted by the sum of two Gaussian functions. The first peak coincides roughly with the one obtained for $t \leq 1933$, while the second peak, centered at 0.62, signals the appearance for $t \geq 1955$ of an important fraction of regular shapes such as rectangles with sides of similar lengths. In Figure 7.12d some examples of the cell shapes at different times are shown. In the case of Paris we observe different distribution of $P(\Phi)$ according to the period considered (see Figure 7.13). Until 1836, the distribution is stable and we observe a dramatic change during the Haussmann period, with a larger abundance of blocks with small value of Φ. These small values correspond to elongated rectangles or triangles created by streets crossing the existing geometry at various angles, and highlights the effect of central planning that did not respect in this case the existing geometry to roads. This can also be seen in the distribution of the road orientation (Barthelemy et al. 2013), a quantity that was measured extensively for various cities (Kadish n.d.).

Road Networks: The BC Reveals Structural Changes

A road network grows by the addition of new streets (edges) and new junctions (nodes). We discuss the properties of these new links by looking at their length and centrality. Barthelemy et al. (2011) showed that the average length of new links steadily decreases over time. More interesting are the features revealed by the betweenness centrality (BC). The BC is a path-based measure of the importance of a node in relation to the amount of flow passing through it (Freeman 1977). More precisely, the BC for an edge e is defined as (and a similar definition for the BC of a node)

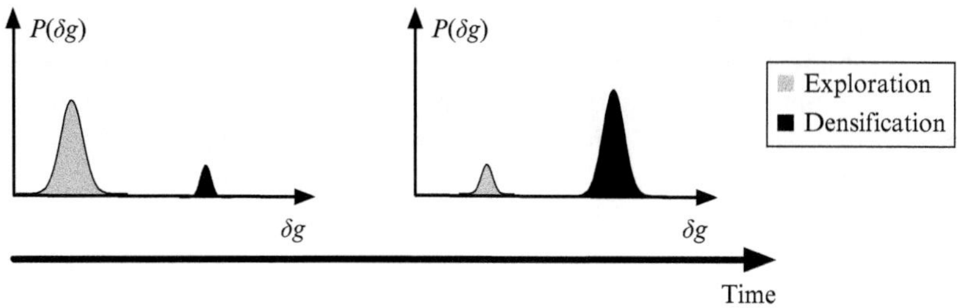

Note: At short times, exploration is the dominant process while at later time, densification is the most important one.

Figure 7.14 Schematic representation of the evolution of the BC impact distribution

$$g(e) = \frac{1}{\mathcal{N}} \sum_{s \neq t \in V} \frac{\sigma_{st}(e)}{\sigma_{st}}, \tag{7.22}$$

where σ_{st} is the number of shortest paths going from nodes s to t and $\sigma_{st}(e)$ is the number of these paths that go through e (Freeman 1977). Here $\mathcal{N}$ is a normalization constant, typically of order N^2 where N is the number of nodes, although the unnormalized version ($\mathcal{N} = 1$) can also be used if we do not compare graphs with different sizes.

In order to evaluate the impact of a new link on the overall distribution of BC in the graph (Figure 7.14) at time t, we first compute the average betweenness centrality of all the links of G_t as:

$$\bar{g}(G_t) = \frac{1}{(N(t)-1)(N(t)-2)} \sum_{e \in E_t} g(e) \tag{7.23}$$

where $g(e)$ is the betweenness centrality of the edge e in the graph G_t. Then, for each new link e^* added in the time window $[t - 1, t]$ we consider the new graph obtained by removing the link e^* from G_t (indicated by $G_t \backslash \{e^*\}$). We compute the average edge betweenness centrality for the graph $G_t \backslash \{e^*\}$, and the BC impact $\delta g(e^*)$ of edge e^* is then defined as

$$\delta g(e^*) = \frac{[\bar{g}(G_t) - \bar{g}(G_t \backslash \{e^*\})]}{\bar{g}(G_t)}. \tag{7.24}$$

That is, this quantity measures the relative variation of the graph average betweenness due to the removal of the link e^*.

Remarkably, the distribution of this quantity $\delta g(e^*)$ displays two well-separated peaks (for details and figures see Strano et al. 2012), with the first peak tending to increase in time while the second peak decreases, until they merge into only one peak in the final time section (1994–2007). These two peaks correspond to very different types of links. The first category consists in links that tend to bridge already existing links (first peak), while the second category concerns links that usually connect existing edges to new nodes (second peak). The distribution of BC impact thus suggests that the evolution of the road network is essentially characterized by two distinct, concurrent processes: densification, which corresponds to an increase of local density of the urban texture, and exploration, which corresponds to the expansion of the network towards non-urbanized areas. Since the amount of available land decreases over time, at earlier time sections (such as in 1833) the fraction of exploration is higher, while in the 1880s it becomes smaller until it completely disappears in 2007. It is important to stress that even if from the urbanism perspective the main conclusion with the existence of two main simple processes is expected, it is a non-ambiguous proof that can be reproduced and tested on other datasets. This quantitative, falsifiable aspect is crucial and distinguishes this approach from the more usual approaches in urbanism. For the sake of completeness, we note that other quantities such as the simplicity defined in Viana et al. (2013) also allows the observation of structural changes with results consistent with those discussed here.

Finally, we note that there is a strong correlation between the age of a road and its centrality. For the Groane case, we observe that highly central links are also usually the oldest links. In particular, the links constructed before $t = 1833$ have a much higher centrality than those added at later time sections. Strano et al. (2012) observe that the historical structure of oldest links mostly coincides with the highly central links at $t = 2007$: more than 90 percent of the 100 most central links in 2007 (and almost 60 percent of the top 1000) were already present in 1833.

In the case of Paris, most of the measures seem to show so far that Haussmann works did not change radically the structure of the city. However, we suspect that Haussmann's impact is very important on congestion and traffic, and should therefore be seen on the spatial distribution of centrality. In Figure 7.15, we show the maps of Paris at different times and we indicate the most central nodes (such that their centrality $g(i)$ is larger than $\max g/\alpha$ (with $\alpha = 10$). We can clearly see on this figure that the spatial distribution of the BC is not stable, displays large variations, and is not uniformly distributed over the Paris area (we represented here the node centrality, and similar results are obtained for the edge centrality). In particular, we observe that between 1836 and 1888, the Haussmann works had a dramatic impact on the spatial structure of the centrality, especially near the center of Paris. Central roads usually persist in time (Strano et al. 2012), but in this case, the Haussmann reorganization acted precisely at this level by redistributing the shortest paths which certainly had an impact on congestion inside the city. After Haussmann, stability of the network continued, including up to the present day.

It is interesting to note that these maps also provide details about the evolution of the road network in Paris during other periods, which seems to reflect what happened in reality and which we can relate to specific local interventions. For example, in the period 1789–1826 between the French Revolution and the Napoleonic empire, the maps shown in Figure 7.15 display large variations with a redistribution of central nodes which probably reflects that many religious and aristocratic domains and properties were sold and divided in order to create new houses and new roads, reducing congestion in Paris. During the period 1826–1836, which corresponds roughly to the beginning of the July Monarchy, the maps in Figure 7.15 suggest an important reorganization on the east side of Paris. This seems to correspond very well to the creation during that period of a new channel in this area (the channel 'Saint Martin') which triggered many transformations in the eastern part of the network.

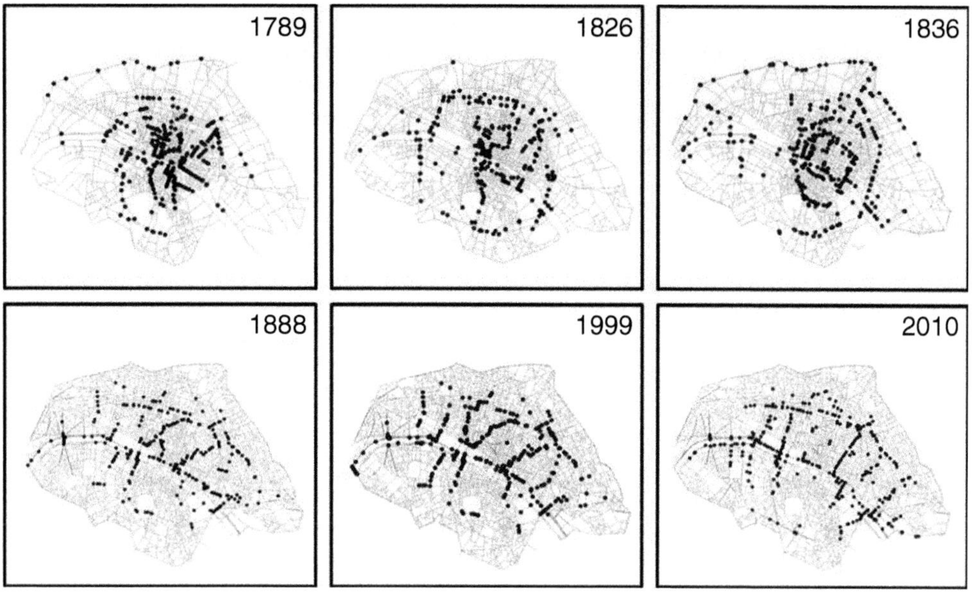

Source: Barthelemy et al. (2013).

Note: We observe for the different periods important reorganizations of the spatial distribution of centrality, corresponding to different specific interventions. In particular, we observe a very important redistribution of centrality during the Haussmann period with the appearance of a reticulated structure on the 1888 map.

Figure 7.15 Spatial distribution of the most central nodes in central Paris, France

Subways: Convergence to a Universal Structure

Subway systems have been developed to improve movement in urban areas, in particular from suburbs to the center, and almost 200 subway systems run through the largest agglomerations in the world and offer an efficient alternative to congested road networks in urban areas. The early history of subways is sometimes connected to large-scale planning, for instance, with the need to bring population from a growing periphery to the center where traditionally production and exchange usually take place. More broadly, it might seem that subway systems are engineered systems and intentionally structured in a core–periphery shape with their self-organization thus playing only a very minor role. This would be true if these subway systems were planned from their beginning in their current shape, but this is not the case for most networks. Their shape results from multiple actions, from planning within a time-limited horizon, set within the wider context of the evolution of the spatial distribution of population and related economic activities. Subway networks result from a superimposition of many actions, both at a central level with planning and at a smaller scale with the reorganization and regeneration of economic activity and the growth of residential populations. From this perspective, subway systems are self-organizing systems, driven by the same mechanisms and responding to various geographical constraints and historical paths. This self-organized view leads to the idea that – in addition to local peculiarities owing to the history and topography of the particular system – the topology of world subway networks should display general, universal features, within the limits of the physical geometry and cultural context in which their growth takes place.

Previous studies have explored the topological and geometrical static properties of these transit systems (Benguigui 1992, 1995; Derrible and Kennedy 2009; Levinson 2012; Sienkiewicz and Hołyst 2005), as well as their evolution in time (Leng et al. 2014; Roth et al. 2012; Sienkiewicz and Hołyst 2005), and we follow the results presented in Roth et al. (2012).

First, we note that large cities always have a subway system: approximately 25 percent of the cities of more than one million individuals have a subway system, 25 percent of those of more than 2 million, and all those above 10 million have a subway system (Roth et al. 2012). We focus on the largest networks in major world cities and ignore currently developing, smaller networks in many medium-sized cities. We thus consider most of the largest metropolitan networks (with at least 100 stations) which exist in major world cities. These are currently: Barcelona, Beijing, Berlin, Chicago, London, Madrid, Mexico, Moscow, New York City, Osaka, Paris, Seoul, Shanghai, and Tokyo. We focus on urban subway systems and we do not consider longer-distance heavy and light-rail commuting systems in urban areas, such as the Réseau Express Régional (RER) in Paris or the overground NetworkRail in London.

For most of these networks the average growth rate (measured by the number of stations built per year) is small, except for those in Seoul and Shanghai, which are more recently developed networks. More generally, we observe that for all networks, larger growth rates occur at earlier stages of the network, large fluctuations occur from one year to another, and the fraction of inactivity time (the time when no stations are built) is similar for all these networks, with an average of about 58 percent.

In order to compare the growth across all networks, we compute the average of the number of stations after a particular number of years since their creation. The result is shown in Figure 7.16 and exhibits a linear increase which indicates that as these networks become large, then for a few decades thereafter new stations represent an increasingly small percentage of existing stations. That is, the time evolution of all these networks is characterized by short additions and not by sudden, abrupt changes with a large number of stations added in a short time duration. Also, as in the case of road networks, in order to get rid of exogenous effects and to be able to compare different networks that experienced different historical developments, we choose the number of stations N as the 'clock' and we plot all the quantities versus N. The large subway networks considered here thus converge to a long time limit where there is always an increasingly smaller percentage of new stations added over time.

An important difficulty in the study of time-evolving (subway) networks is to characterize their evolution. This is typically so where the large number of potentially interesting measures might be a problem as it hides the most salient features of the system's evolution. We show that the introduction of a template greatly simplifies the study of these systems (Roth et al. 2012) and provides a guide to the relevant quantities to measure. By inspection, we observe that in most large urban areas, the network consists of a set of stations delimited by a ring that constitute the core. From this core, quasi-one-dimensional branches grow and reach out to areas of the city increasingly further from the core. We note that the ring, which is defined topologically as the set of core stations which are either

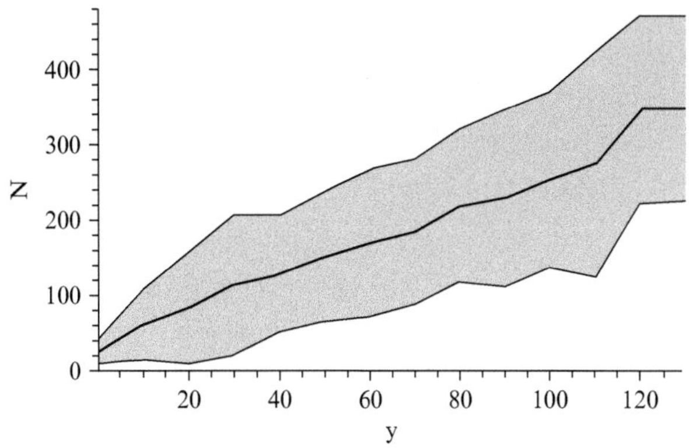

Source: Roth et al. (2012).

Note: The shaded area marks the standard deviation across all networks. The linear shape indicates that the growth of new stations from one decade to another goes to zero for all these networks, signaling the possible appearance of a stationary limit.

Figure 7.16 Evolution of the number of stations y years after creation, averaged over all networks

at the junction of branches or on the shortest geodesic path connecting these junction stations, exists or not as a subway line. For instance, in Tokyo, there is a such a circular line (the Yamanote line), while in Paris the topological ring does not correspond to a single line. It is interesting that the future Grand Paris project plans to build large circular rings around the existing Paris area.

The general structure can then be represented schematically as in Figure 7.17. This template suggests what the relevant measures are and in particular to distinguish the core from the branches. We then first characterize this structure with the parameter $\beta(t)$ defined as

$$\beta(t) = \frac{N_B(t)}{N_B(t) + N_C(t)} \tag{7.25}$$

where $N_B(t)$ and $N_C(t)$ 'respectively' represent the number of stations on branches and the number of stations in the core at time t. We plot in Figure 7.18 this parameter β averaged over networks as a function of N (together with the dispersion around the average). On this figure we may see that the average value of β seems to stabilize slowly to a value in $[0.35, 0.55]$.

The branches have a trivial topological structure (they are essentially one-dimensional lines) and their complexity resides in their spatial structure. We can then determine the average distance (in kilometers) from the geographic barycenter of the city to all core and branches stations, respectively: $\overline{D}_C(t)$ and $\overline{D}_B(t)$ (the barycenter is computed as the center of mass of all stations). This last distance provides information about the spatial extension of the branches and we can form the ratio $\eta(t)$

$$\eta(t) = \frac{\overline{D}_B(t)}{\overline{D}_C(t)} \tag{7.26}$$

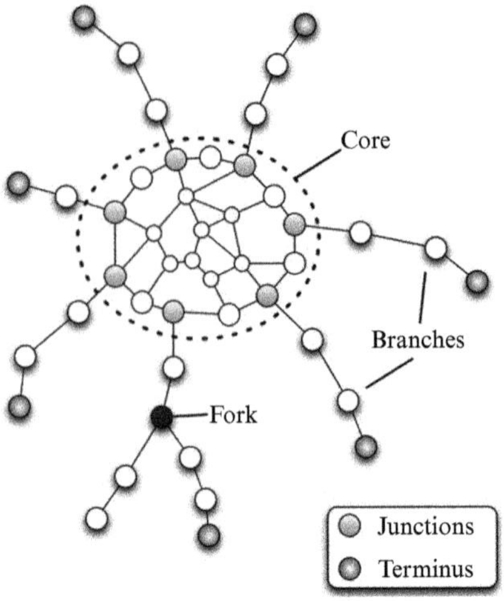

Source: Roth et al. (2012).

Note: A large ring encircles a core of stations. Branches radiate from the core and reach further areas of the urban system. The branches are essentially characterized by their size ($\beta(t)$), and their spatial extension (parameter $\eta(t)$). The core is characterized by its average degree ($\langle k_{core}\rangle(t)$), the fraction of nodes of degree 2 (f2), its number of stations $N_C(t)$ and its size $r_C(t)$.

Figure 7.17 Schematic structure of subway networks

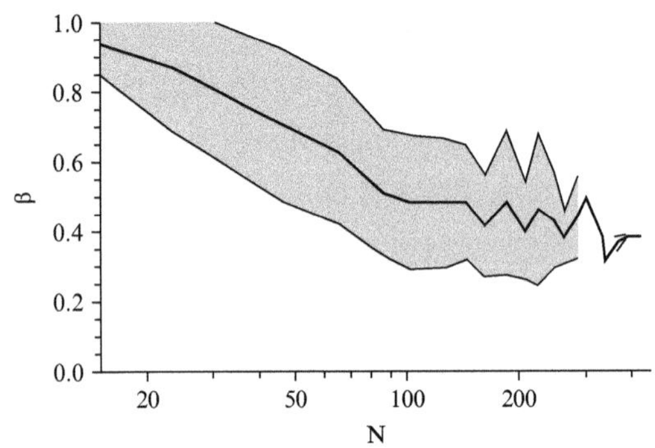

Source: Roth et al. (2012).

Figure 7.18 Parameter β as a function of the number of stations N averaged over 20 bins and showing the standard deviation

which gives a spatial measure of the amount of extension of the branches. The parameter η gives a precious indication about their extension and we show in Figure 7.19 the evolution of this parameter with N (the data is binned). Figure 7.19 shows that in the interval where we have the largest number of subways, the average value of η is around 2 with relatively large fluctuations which seem to decrease with N.

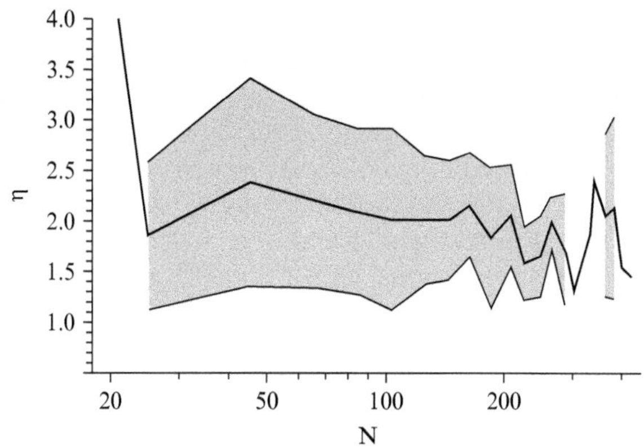

Source: Roth et al. (2012).

Figure 7.19 Evolution of the ratio η, which characterizes the spatial extension of branches relative to the core

The parameters β and η give an indication of the importance of the core but do not say anything about its structure. The core is in most cases an almost planar spatial network and can be characterized by many parameters (Barthelemy et al. 2011). It is important to choose those which are not simply related but ideally represent different aspects of the network. At each time step t, we characterize the core structure by its average degree

$$\langle k_{\text{core}}\rangle(t) = \frac{2E_C(t)}{N_C(t)} \tag{7.27}$$

where $N_C(t)$ is the number of core nodes and $E_C(t)$ the number of its edges (we note that this average degree is connected to the standard index $\gamma(t) = E_C(t)/(3N_C(t) - 6)$ where the denominator is the maximum number of links admissible for a planar network (see Haggett and Chorley 1969).

Another simple quantity which describes in more detail the level of interconnections in the core is given by the fraction f_2 of nodes in the core with $k = 2$. For subway systems, nodes with degree larger than 2 are acting as interconnecting stations, and the quantity f_2 thus reveals information about the functioning of the system. In the case of a well-interconnected system, this fraction will be small, while sparse cores with a few interconnections will have a larger fraction of $k = 2$ nodes.

These two quantities $\langle k_{\text{core}}\rangle$ and f_2 were measured in Roth et al. (2012) and showed that the average degree of the core display moderate variations around 2.5 (with a slow increase with N). This value is relatively small and indicates that the fraction of connecting stations is also small and means that most core stations belong to one single line with few allowing connections. More precisely, for subways with $N < 100$ the fraction of interconnecting stations is increasing with N – which probably corresponds to some organization of the subway – but for larger subways ($N > 100$), the percentage f_2 is increasing again, which probably corresponds to a densification process without the creation of new interconnections.

The branches constitute probably the most interesting part of these large networks. They comprise one-dimensional lines that explore space at a large distance from the center. We first consider the number N_B of these branches and propose a simple argument for estimating this quantity. If we assume that the distance between the starting points from the ring of two consecutive branches is constant (which probably correspond to an optimum in terms of distribution and cost), the number of branches is then proportional to the perimeter of the core structure. The core is a relatively dense planar graph and contains a number of nodes proportional to N and its perimeter thus scales as $\sqrt{N}$. These arguments thus lead to

$$N_B \sim \sqrt{N}. \tag{7.28}$$

We display the number of branches versus the number of stations N for the various networks considered here. A power law fit of the data presented in Figure 7.20 gives $N_B \sim N^b$ with $b \approx 0.6$ ($r^2 = 0.85$) consistent with this simple argument.

Finally, we note that the simple picture of a core and radial branches allows us to provide simple explanations to empirical observations that were difficult to understand using earlier frameworks such as fractals for example (Benguigui 1995) (see Roth et al. 2012 for a longer discussion on this issue).

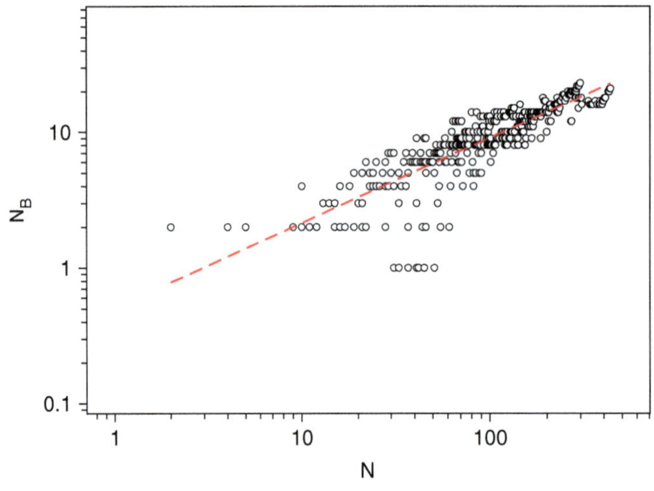

Source: Roth et al. (2012).

Note: The dashed line is a power law fit with exponent ≈ 0.6.

Figure 7.20 Loglog plot of the number of different branches versus the number of stations for the different subway networks considered here

CONNECTING THE STRUCTURE OF NETWORKS TO THE ECONOMICS OF THE SUBSTRATE

Transport systems such as subways are not mere geometrical structures growing in empty space: they are usually embedded in large, highly congested urban areas and it seems plausible that some properties of these systems find their origin in the interaction with the city they are in. Previous studies (Levinson and Krizek 2008; Xie and Levinson 2009) have shown that the growth and properties of transportation networks are tightly linked to the characteristics of urban environment. Levinson (2008) for instance, showed that rail development in London followed a logic of both induced supply and induced demand. That is, while the development of rail systems within cities answers a need for transportation between different areas, this development also has an impact on the organization of the city. Therefore, while the growth of transportation systems cannot be understood without considering the underlying city, the development of the city cannot be understood without considering the transportation networks that run through it. As a consequence, the subway system and the city can be thought of as two systems exhibiting a symbiotic behavior. Understanding this behavior is crucial if we want to gain deeper insights into the growth of cities and how the mobility patterns organize themselves in urban environments.

Many studies (Derrible and Kennedy 2009; Kansky 1963; Levinson 2012) explore the interplay between regional characteristics and the structure of transportation networks, but are usually very specific. In contrast, we discuss here a general coarse-grained approach, based on a cost–benefit analysis that accounts for the scaling properties of the main quantities characterizing transportation networks – the number of stations N_S, the total length L, the ridership R (number of passengers per unit time) – and connecting them with the main features of the substrate, which are the population P, the surface area A, the wealth characterized by the gross metropolitan product (GMP) G.

In the following section we discuss a simple framework that allow us to relate these different factors to each other and discuss in detail the application of these ideas to subways.

General Framework for the Scaling Theory

The cost–benefit analysis (Black 1971; Louf et al. 2013) seems to be the appropriate theoretical framework for understanding how network features and socio-economic indicators relate to one another. We therefore describe the growth of the network by iterative process where at each time step an edge e is built so that the cost function

$$Z_e = B_e - C_e \tag{7.29}$$

is maximum (for a general discussion about this type of models, see, for example, Barthelemy et al. 2017). The quantity B_e is the expected benefit and C_e the expected cost of edge e. We consider networks after they have been built, and we assume that they are in a steady-state for which we can write a cost function in the form

$$Z = \sum_e Z_e = B - C \tag{7.30}$$

where the summation is over all existing links. The quantity B is the total expected benefits and C the total expected costs, mainly owing to maintenance (in the steady state regime). We further assume that, during this steady-state, operating costs are balanced by benefits. In other words, we assume that

$$Z \approx 0. \tag{7.31}$$

Since lines and stations cost money to be maintained, we expect the network to adapt to the way it is being used. Therefore, we can reasonably expect that at first order the cost of operating the system is compensated by the benefits gained from its use.

Application to Subways

In the case of subways, the total benefits in the steady-state are simply connected to the total ridership R and the ticket price f over a given period of time. The costs, however, are due to the maintenance costs of the lines and stations, so that we can write (for a given period of time)

$$Z_{sub} = Rf - \varepsilon_L L - \varepsilon_S N_s \tag{7.32}$$

where L is the total length of the network, ε_L the maintenance cost of a line per unit of length, N_S the total number of stations and ε_S the maintenance cost of a station (for a given period time).

It is usually difficult to estimate the ridership of a system given its characteristics and those of the underlying city. Owing to its importance for planning purposes, the problem of estimating the number of boardings per station, given the properties of the area surrounding the stations, has been the subject of numerous studies (Kuby et al. 2004; Matsunakan et al. 2014). Here we are interested in the dependence of global, average behavior of the ridership on the network and the underlying city. Generally, we assume that the number R_i of people using the station i is a function of the surface area C_i serviced by this station – the coverage Derrible and Kennedy 2009 – and of the population density $\rho = \frac{P}{A}$ in the city.

$$R_i = \xi_i C_i \rho \tag{7.33}$$

where ξ_i is a random number of order one representing the fraction of people that are in the area serviced by the station and who use the subway. The main difficulty is in finding the expression of the coverage. It depends, a priori, on local particularities, such as the accessibility of the station, and should thus vary from one station to another. We take a simple approach and assume that on average

$$C_i \sim \pi d_0^2 \tag{7.34}$$

where d_0 is the typical size of the attraction basin of a given station. If we assume that it is constant, the total ridership can be written as

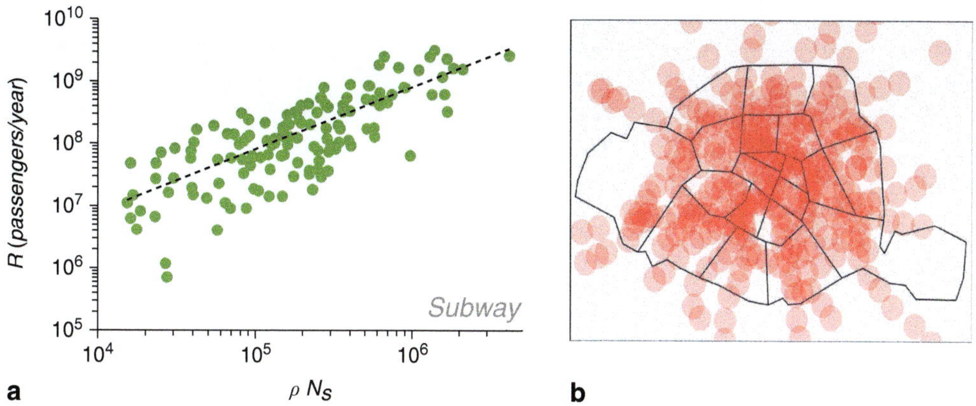

Source: Louf et al. (2014).

Note: Left: we plot the total yearly ridership R as a function of ρN_s. A linear fit on the 138 data points gives $R \approx 800\, \rho N_s$ ($R^2 = 0.76$) which leads to a typical effective length of attraction $d_0 \approx 500$ m per station. Right: map of Paris, France, with each subway station represented by a red circle of radius 500 meters.

Figure 7.21 Relationship between ridership and coverage

$$R = \sum_i R_i \sim \bar{\xi} \pi d_0^2 \rho N_s \tag{7.35}$$

where $\bar{\xi} = \frac{1}{N_s} \sum_i \xi_i$ is of order $O(1)$.

Using data gathered for 138 metro systems across the world (see Louf et al. 2014 for details), we plot the ridership R as a function of $N_s \rho$ on Figure 7.21 and observe that the data is consistent with a linear behavior.

We measured a slope of 800 square kilometers which gives an estimate for d_0

$$d_0 \approx 500 \text{ m}. \tag{7.36}$$

This result is illustrated in Figure 7.21 by representing each subway station of Paris with a circle of radius 500 meters. The distance d_0 appears here an intrinsic feature of user's behaviors: it is the maximal distance that an individual would walk to go to a subway station. The average interstation distance ℓ_1 is another distance characteristic of the subway system. Rigorously, this distance depends on the average degree $\langle k \rangle$ of the network so that $\ell_1 = \frac{2L}{N_s \langle k \rangle}$. However, Roth et al. (2012) found that for the 13 largest subway systems in the world, $< k > \in [2.1, 2.4]$, so that we can reasonably take $\langle k \rangle / 2 \approx 1$ and thus

$$\ell_1 \simeq \frac{L}{N_s}. \tag{7.37}$$

The interstation distance depends in general on many technological and economical parameters, but we expect that for a properly designed system it will match human constraints. Indeed, if $d_0 \ll \ell_1$, the network is not dense enough and in the opposite case, $d_0 \gg \ell_1$, the system is not economically interesting. We can thus reasonably expect that

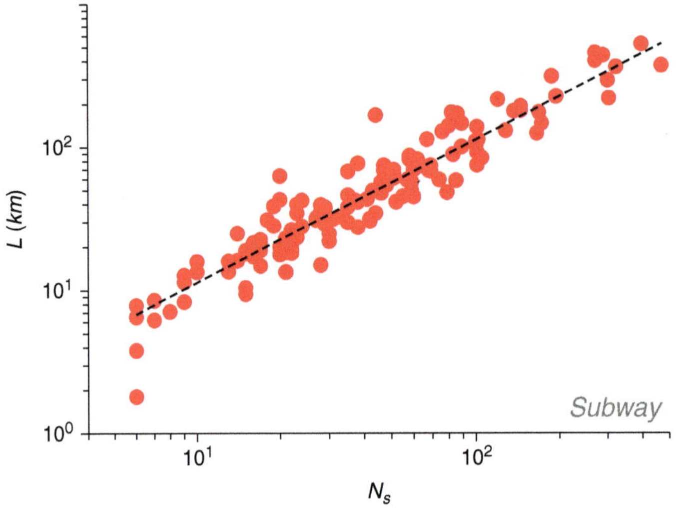

Source: Louf et al. (2014).

Note: Length of 138 subway networks in the world as a function of the number of stations. A linear fit gives $L \sim 1.13\, N_S$ ($R^2 = 0.93$).

Figure 7.22 Relationship between the length and the number of stations

the interstation distance fluctuates slightly around an average value given by twice the typical station attraction distance d_0

$$d_0 = \frac{\ell_1}{2} = \frac{L}{2N_s}. \tag{7.38}$$

It follows from this assumption that the interstation distance is constant and independent from the population size. In order to test our assumption, we plot on Figure 7.22 the total length of subway networks as a function of the number of stations. The data agrees well with a linear fit $L \sim 1.13N_S(r^2 = 0.93)$. The histogram of the inter-station length (not shown, see Louf et al. 2014) shows that it is narrowly distributed around an average value $\overline{\ell_1} \approx 1.2$ kilometers with a variance $\sigma \approx 400$ meters, consistently with the value found above for $d_0 \approx 500$ meters.

We can now express ℓ_1 in terms of the systems characteristics. Indeed, the total ridership now reads

$$R \sim \overline{\xi}\pi\rho\frac{L^2}{N_s}. \tag{7.39}$$

If we assume to be in the steady-state $Z_{sub} \approx 0$, using the results from equation 7.32 and 7.39, the total length of the network and the number of stations are connected at first order in $\varepsilon_s/\varepsilon_L$ by

$$L \sim \left(\frac{4\varepsilon_L}{\pi\xi f\rho} + \frac{\varepsilon_s}{\varepsilon_L}\right)N_s \tag{7.40}$$

and the interstation distance reads

$$\ell_1 = \frac{4\varepsilon_L}{\pi\xi f\rho} + \frac{\varepsilon_s}{\varepsilon_L}. \tag{7.41}$$

This relationship implies that the interstation distance increases with the station maintenance cost, and decreases with increasing line maintenance costs, density and fare. We thus see that the adjustment of ℓ_1 to match $2d_0$ can be made through the fare price (or subsidies by the local authorities or national government). We note that it would interesting to obtain reliable data about the maintenance costs and fares for subway systems in order to pursue in this direction and to test the accuracy of this prediction.

So far, we have a relationship between the total length L and the number of stations N_s, but we need another equation in order to compute their value. Intuitively, it is clear that the number of stations – or, equivalently, the total length – of a subway system is an increasing function of the wealth of the city. We thus assume a simple, linear relation of the form

$$N_s = \kappa\frac{G}{\varepsilon_s} \tag{7.42}$$

where G is the city's GMP, and κ the fraction of the city's wealth invested in public transportation. This relationship can equivalently be interpreted as the proportional relationship between the number of stations per person and the city's development, as measured by its GMP per capita. In Figure 7.23 (left) we plot the number of stations of different metro systems around the world as a function of the GMP of the corresponding city. A linear fit agrees relatively well with the data ($R^2 = 0.73$, dashed line), and gives $\frac{\varepsilon_s}{\kappa} \sim 10^{10}$ dollars/station. However, the dispersion around the linear average behavior is

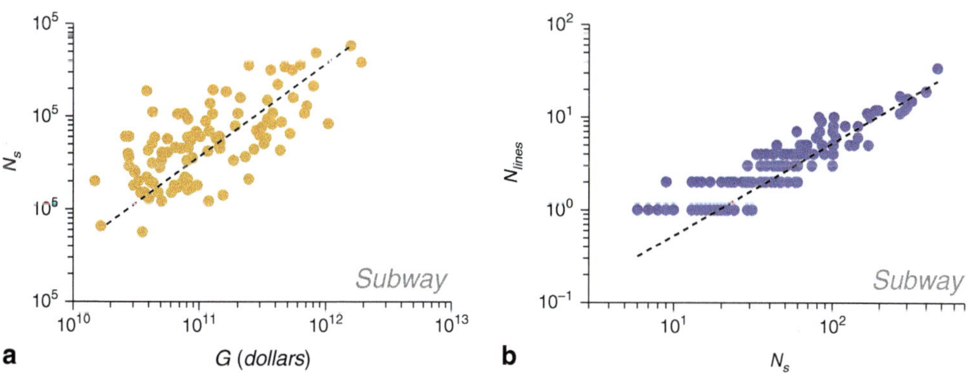

Source: Louf et al. (2014).

Note: Left: we plot the number of stations for the different subway systems in the dataset as a function of the Gross Metropolitan Product of the corresponding cities (obtained for 106 subway systems). A linear fit (dashed line) gives $N_s = 2.51\ 10{-}10\ G$ ($R^2 = 0.73$). Right: number of lines and number of stations. We plot the number of metro lines N_{lines} as a function of the number of stations N_s. A linear fit on the 138 data points gives $N_{lines} \approx 0.053\ N_s$ ($R^2 = 0.93$); that is, metro lines comprise on average 19 stations.

Figure 7.23 Left: size of the subway system and city's wealth; right: number of lines and number of stations

important: more specific data is needed in order to investigate whether differences in the construction costs and investments (or the age of the system) can explain the dispersion, or if other important parameters need to be taken into account.

Finally, we consider the number of different lines with distinct tracks. A natural question is how the number of lines N_{lines} scales with the number stations N_s, that is whether lines get proportionally smaller, larger or stay the same with the size of the whole system. We plot the number of lines as a number of stations on Figure 7.23 (right) and find that the data agree with a linear relationship between both quantities ($R^2 = 0.93$) showing that the number of stations per line is distributed around a typical value of 19, whatever the size of the system.

DISCUSSION

Despite many studies in graph theory and combinatorics, neurophysiology, botanic, geography and transport studies, among others, we do not have a full understanding of the structure and evolution of spatial networks. In addition, these networks are not just simple structures embedded in a substrate, but constitute the support for dynamical processes: it can be the traffic on transport networks, electric flow on power grids, or the spread of nerve impulses in neural networks. We therefore understand that there are potentially three different components that can influence the morphogenesis of these structures (Barthelemy 2017):

- the network itself;
- the dynamical process that takes place on it; and
- the substrate that can possibly react to the presence and the evolution of the network.

These considerations suggest that there are two important research directions about spatial networks, which can be summarized as follows.

First, the apparent universality of some of the statistical properties of these systems raises the important question of the existence of classes of spatial networks. Are these networks all similar, or can we construct a typology of these objects? What would be the main determinant(s) allowing us to construct this classification? We discussed a typology attempt based on the statistics of faces (and briefly approximate and exact mappings between a planar map and a tree), but further work is certainly needed on clarifying this question.

Second, the time evolution of these systems is not well understood and there is a crucial lack of theoretical models. So far, most existing models for spatial networks are relying on global optimal considerations and neglect mechanistic processes. It would probably be very helpful to construct a general framework for the formation and evolution of these systems which includes, for example, a cost–benefit type of growth together with feedback loops between the substrate and the network itself. This could allow us to discuss whether optimality is consistent with the dynamical evolution and to address more specific issues, such as the possible existence of different regimes characterized by different network structures, the impact of a finite volume on network properties, and the effect of the dimensionality of the embedding space (either two or three dimensional).

In order to make progress in these two directions, we probably first need to improve the tools for the characterization of spatial networks and their evolution. As we briefly saw in this chapter, most measures that are interesting for complex networks become irrelevant for spatial networks, triggering the question of relevant tools for their characterization. More advanced tools that integrate both topological and spatial properties are needed, not only for classifying spatial networks but also to understand the growth and the evolution of these systems.

It will also be important to introduce a new modeling framework. Most models and discussion on spatial networks usually rely on optimal models. In these models, the networks optimize a global quantity (the classic example being the minimum spanning tree which minimizes the total length) and most discussions concerning venation patterns in leaves or the network structure of the brain are in relation to optimal properties. However, most networks in the real world do not appear at once but grow by successive addition of one or a few links, and the relevance of optimal models is thus unclear. It will certainly be useful to propose a general conceptual framework for studying the co-evolution of spatial networks and their substrate, and to test whether realistic mechanisms can lead to optimal structures.

Finally, a more general discussion concerns the role of engineering, self-organization and evolution in the formation and the evolution of spatial networks. This discussion goes far beyond the problematic of spatial networks and concerns complex systems in general, but progress on this theoretical problem will certainly impact how we see and model networks embedded in space that grow in time.

REFERENCES

Albert, R. and A.-L. Barabasi (2002), 'Statistical mechanics of complex networks', *Rev Mod Phys*, **74** (1), 47–97.

Albert, R., I. Albert and G.L. Nakarado (2004), 'Structural vulnerability of the North American power grid', *Physical Review E*, **69** (2), 025103.

Amaral, L.A.N., A. Scala, M. Barthelemy and H.E. Stanley (2000), 'Classes of small-world networks', *Proceedings of the National Academy of Sciences*, **97** (21), 11149–52.

Barabási, A.-L. and R. Albert (1999), 'Emergence of scaling in random networks', *Science*, **286** (5439), 509–12.

Barrat, A., M. Barthelemy and A. Vespignani (2008), *Dynamical Processes in Complex Networks*, Cambridge: Cambridge University Press.

Barrat, A., M. Barthelemy, R. Pastor-Satorras and A. Vespignani (2004), 'The architecture of complex weighted networks', *Proceedings of the National Academy of Sciences*, **101** (11), 3747–52.

Barthelemy, M. (2011), 'Spatial networks', *Physics Reports*, **499** (1), 1–101.

Barthelemy, M. (2016), *The Structure and Dynamics of Cities*, Cambridge: Cambridge University Press.

Barthelemy, M. (2017), *Morphogenesis of Spatial Networks*, Cham: Springer.

Barthelemy, M. and L.A.N. Amaral (1999), 'Smallworld networks: evidence for a crossover', *Physical Review Letters*, **82** (15), 3180.

Barthelemy, M. and A. Flammini (2008), 'Modeling urban street patterns', *Physical Review Letters*, **100** (13), 138702.

Barthelemy, M., P. Bordin, H. Berestycki and M. Gribaudi (2013), 'Self-organization versus top-down planning in the evolution of a city', *Scientific Reports*, **3** (1), 1–8.

Batty, M. (2013), *The New Science of Cities*, Cambridge, MA: MIT Press.

Benguigui, L. (1992), 'The fractal dimension of some railway networks', *Journal de Physique I*, **2** (4), 385–88.

Benguigui, L. (1995), 'A fractal analysis of the public transportation system of Paris', *Environment and Planning A*, **27** (7), 1147–61.

Black, W.R. (1971), 'An iterative model for generating transportation networks', *Geographical Analysis*, **3** (3), 283–88.

Boeing, G. (2020), 'Planarity and street network representation in urban form analysis', *Environment and Planning B: Urban Analytics and City Science*, **47** (5), 855–69.
Bouttier, J., P. Di Francesco and E. Guitter (2004), 'Planar maps as labeled mobiles', *Electronic Journal of Combinatorics*, **11** (1), R69.
Broido, A.D. and A. Clauset (2019), 'Scale-free networks are rare', *Nature Communication*, **10** art. 1017, doi:10.1038/s41467-019-08746-5.
Buhl, J., J. Gautrais, N. Reeves, R. Solé, S. Valverde, P. Kuntz, et al. (2006), 'Topological patterns in street networks of self-organized urban settlements', *European Physical Journal B – Condensed Matter and Complex Systems*, **49** (4), 513–22.
Buldyrev, S.V., R. Parshani, G. Paul, H.E. Stanley and S. Havlin (2010), 'Catastrophic cascade of failures in interdependent networks', *Nature*, **464** (7291), 1025–8.
Cardillo, A., S. Scellato, V. Latora and S. Porta (2006), 'Structural properties of planar graphs of urban street patterns', *Physical Review E*, **73** (6), 066107.
Clark, J. and. D. Holton (1991), *A First Look at Graph Theory*, Singapore: World Scientific.
Dall, J. and M. Christensen (2002), 'Random geometric graphs', *Physical Review E*, **66** (1), 016121.
Derrible, S. and C. Kennedy (2009), 'Network analysis of world subway systems using updated graph theory', *Transportation Research Record: Journal of the Transportation Research Board*, **2112** (1), 17–25.
Farmani, R., D. Savic and G. Walters (2004), 'Exnet benchmark problem for multi-objective optimization of large water systems', in 'Modelling and control for participatory planning and managing water systems', IFAC Workshop, Venice, 29 September–1 October..
Fialkowski, M. and A. Bitner (2008), 'Universal rules for fragmentation of land by humans', *Landscape Ecology*, **23** (9), 1013–22.
Freeman, L.C. (1977), 'A set of measures of centrality based on betweenness', *Sociometry*, **40** (1), 35–41.
Gao, X., B. Xiao, D. Tao and X. Li (2010), 'A survey of graph edit distance', *Pattern Analysis and Applications*, **13** (1), 113–29.
Gerke, S. and C. McDiarmid (2004), 'On the number of edges in random planar graphs', *Combinatorics, Probability and Computing*, **13** (2), 165–83.
Haggett, P. and R.J. Chorley (1969), *Network Analysis in Geography*, London: Edward Arnold.
Haggett, P., A.D. Cliff and A. Frey (1977), *Locational Analysis in Human Geography*, 2nd edn, London: Edward Arnold.
Kadish, S. (n.d.), 'On and off the street grid: relative distributions of road orientations', accessed 6 March 2021 at https://vizual-statistix.tumblr.com/post/80468941142/unlike-like-emperor-kuzco-i-was-actually-born.
Kalapala, V., V. Sanwalani, A. Clauset and C. Moore (2006), 'Scale invariance in road networks', *Physical Review E*, **73** (2), 026130.
Kansky, K. J. (1963), 'Structure of transportation networks: relationships between network geometry and regional characteristics', PhD thesis, University of Chicago, Chicago, IL.
Katifori, E. and M.O. Magnasco (2012), 'Quantifying loopy network architectures', *PloS ONE*, **7** (6), e37994.
Kaufman, L. and P. Rousseeuw (1990), *Finding Groups in Data*, New York: Wiley.
Kivelä, M., A. Arenas, M. Barthelemy, J.P. Gleeson, Y. Moreno and M.A. Porter (2014), 'Multilayer networks', *Journal of Complex Networks*, **2** (3), 203–71.
Kuby, M., A. Barranda and C. Upchurch (2004), 'Factors influencing light-rail station boardings in the United States', *Transportation Research Part A: Policy and Practice*, **38** (3), 223–47.
Kurant, M. and P. Thiran (2006), 'Layered complex networks', *Physical Review Letters*, **96** (13), 138701.
Lämmer, S., B. Gehlsen and D. Helbing (2006), 'Scaling laws in the spatial structure of urban road networks', *Physica A: Statistical Mechanics and its Applications*, **363** (1), 89–95.
Latora, V. and M. Marchiori (2001), 'Is the Boston subway a small-world network?', *Physica A*, **314** (1–4), 109–113.
Latora, V., V. Nicosia and G. Russo (2017), *Complex Networks: Principles, Methods and Applications*, Cambridge: Cambridge University Press.
Leng, B., X. Zhao and Z. Xiong (2014), 'Evaluating the evolution of subway networks: evidence from Beijing subway network', *EPL (Europhysics Letters)*, **105** (5), 58004.
Levinson, D. (2012), 'Network structure and city size', *PLoS ONE*, **7** (1), e29721.
Levinson, D. and K. Krizek (eds) (2008), *Planning for Place and Plexus: Metropolitan Land Use and Transport*, New York: Routledge.
Louf, R. and M. Barthelemy (2014), 'A typology of street patterns', *Journal of The Royal Society Interface*, **11** (101), 20140924.
Louf, R., P. Jensen and M. Barthelemy (2013), 'Emergence of hierarchy in cost-driven growth of spatial networks', *Proceedings of the National Academy of Sciences*, **110** (22), 8824–9.
Louf, R., C. Roth and M. Barthelemy (2014), 'Scaling in transportation networks', *PloS ONE*, **9** (7), e102007.
Masucci, A., D. Smith, A. Crooks and M. Batty (2009), 'Random planar graphs and the London street network', *European Physical Journal B – Condensed Matter and Complex Systems*, **71** (2), 259–71.

Matsunaka, R., T. Oba, D. Nakagawa, M. Nagao and J. Nawrocki (2013), 'International comparison of the relationship between urban structure and the service level of urban public transportation – a comprehensive analysis in local cities in Japan, France and Germany', *Transport Policy*, **30** (November), 26–39.
Mémoli, F. (2007), 'On the use of Gromov-Hausdorff distances for shape comparison', in M. Botsch, R. Pajarola, B. Chen and M. Zwicker (eds), *Eurographics Symposium on Point-Based Graphics*, Geneva: Eurographics Association.
Mileyko, Y., H. Edelsbrunner, C.A. Price and J.S. Weitz (2012), 'Hierarchical ordering of reticular networks', *PloS ONE*, **7** (6), e36715.
Newman, M. (2010), *Networks: An Introduction*, Oxford: Oxford University Press.
Nicosia, V., R. Louf, V. Latora and M. Barthelemy (2018), 'Fragmentation models for road patterns', submitted.
Penrose, M. (2003), *Random Geometric Graphs: 5*, Oxford: Oxford University Press.
Porta, S., P. Crucitti and V. Latora (2006), 'The network analysis of urban streets: a primal approach', *Environmental Planning B: Planning and Design*, **33** (5), 705–25.
Ravasz, E. and A.-L. Barabasi (2003), 'Hierarchical organization in complex networks', *Physical Review E*, **67** (2), 026112.
Roth, C., S.M. Kang, M. Batty and M. Barthelemy (2012), 'A long-time limit for world subway networks', *Journal of The Royal Society Interface*, **9** (75), 2540–50.
Schaeffer, G. (1998), 'Conjugaison d'arbres et cartes com binatoires aléatoires', PhD thesis, University of Bordeaux 1.
Sienkiewicz, J. and J.A. Hołyst (2005), 'Statistical analysis of 22 public transport networks in Poland', *Physical Review E*, **72** (4), 046127.
Solé, R., M. Rosas-Casals, B. Corominas-Murtra and S. Valverde (2008), 'Robustness of the European power grids under intentional attack', *Physical Review E*, **77** (2), 026102.
Southworth, M. and E. Ben-Joseph (2003), *Streets and the Shaping of Towns and Cities*, Washington, DC: Island Press.
Strano, E., V. Nicosia, V. Latora, S. Porta and M. Barthelemy (2012), 'Elementary processes governing the evolution of road networks', *Scientific Reports*, **2** (1), 1–8.
Viana, M.P., E. Strano, P. Bordin and M. Barthelemy (2013), 'The simplicity of planar networks', *Scientific Reports*, **3** (1), 1–6.
Von Ferber, C., T. Holovatch, Y. Holovatch and V. Palchykov (2009), 'Modeling metropolis public transport', in C. Appert-Rolland, F. Chevoir, P. Gondret, S. Lassarre, J.-P. Lebacque and M. Schreckenberg (eds), *Traffic and Granular Flow'07*, Berlin: Springer, pp. 709–19.
Walski, T., E. Brill, J. Gessler, I. Goulter, R. Jeppson, K. Lansey, et al. (1987), 'Battle of the network models: epilogue', *Journal of Water Resource Planning and Management*, **113**, 191–203.
Watts, D.J. and D.H. Strogatz (1998), 'Collective dynamics of small-world networks', *Nature*, **393** (6684), 440–42.
Xie, F. and D. Levinson (2009), 'Modeling the growth of transportation networks: a comprehensive review', *Networks and Spatial Economics*, **9** (3), 291–307.
Xie, F. and D.M. Levinson (2011), *Evolving Transportation Networks*, New York: Springer.
Yazdani, A. and P. Jeffrey (2010), 'A complex network approach to robustness and vulnerability of spatially organized water distribution networks', paper presented at the Water Distribution System Analysis Conference, WDSA2010, Tucson, AZ, 19–23 September.

8. Modeling the co-evolution of cities and networks

Juste Raimbault

1. INTRODUCTION

1.1 Structuring Effects of Transportation Networks

The idea of possible causal relationships between territorial characteristics and transportation networks has fed a scientific debate that is still active. The underlying assumptions can be synthesized as more or less deterministic attributions of impacts of transportation infrastructures or of a new transportation mode on territorial transformations. Precursors of this reasoning can be tracked back to the 1920s: Burgess et al. (1925 p. 69) mention for example some 'modifications of forms of transportation and communication as determining factors of growth and decline cycles (of territories)' (p. 69). Methodologies to identify what is then termed *structuring effects* of transportation networks have been developed for planning in the 1970s. Bonnafous and Plassard (1974) situate the concept of structuring effects in the perspective of using the transportation offer as a planning tool (the alternatives are the development of an offer to answer to a congestion of the network, and the simultaneous development of associated offer and planning). They identify from an empirical viewpoint direct effects of a novel offer on the behavior of agents, on transportation flows and possible inflexions on socio-economic trajectories of concerned territories. Bonnafous et al. (1974) develop a method to identify these effects through the modification of the class of cities in a typology established a posteriori. More recently, Bonnafous (2017) recalls that the institution of permanent observatories for territories makes these analyses more robust, allowing a continuous monitoring of the territories that are the most concerned by the extent of a new infrastructure.

According to Offner (1993) who reformulates ideas already given by Plassard (1977) for example, a non-reasoned and out-of-context use of these methods has then been developed by planners and politicians who generally used them to justify transportation projects in a technocratic manner; through the argument of a direct effect of a new infrastructure on local development (for example, economic), politicians are able to ask for subsidies and to legitimate their action in front of the people. Offner (1993) insists on the necessity of a critical positioning on these issues, recalling that there exists no scientific demonstration of an effect that would be systematic. A special issue of the journal *L'Espace Géographique* (Offner et al. 2014) on that debate recalled that 'on the one hand' misconceptions and misuses were still greatly present in operational and planning communities, which can be explained 'for example' by the need to justify public actions, and, on the other hand, that a scientific understanding of relationship between networks and territories is still under construction.

1.2 Co-evolution of Cities and Networks

An alternative approach to relationship between transportation networks and territories is to consider them as co-evolving. The evolutive urban theory considers systems of cities as systems of systems at multiple scale, from the intra-urban microscopic level, to the macroscopic level of the whole system and through to the mesoscopic level of the city (Pumain 2008). These systems are complex, dynamic, and adaptive; their components co-evolve and the system answers to internal or external perturbations by modifying its structure and its dynamics. Interactions between cities are the main drivers of these evolutionary patterns. These interactions consist in material or informational exchanges, and the diffusion of innovation is therein a crucial component (Pumain 2010). These are necessarily carried by physical networks and 'more particularly' transportation networks. Thus, we expect from a theoretical view point strong interdependencies between cities and transportation networks at these scales, that is, a co-evolution.

From an empirical point of view, this has already been suggested by some studies; Bretagnolle (2003) reveals an increasing correlation in time between urban hierarchy and the hierarchy of temporal accessibility for the French railway network (which is a priori clearer for this measure than for integrated measures of accessibility that are prone to auto-correlation). This correlation is a clue to positive feedbacks between urban ranks and network centralities. According to Bretagnolle (2009), different regimes of interactions between cities and transportation networks have been identified: for the evolution of the French railway network, a first phase of adaptation of the network to the existing urban configuration was followed by a phase of co-evolution, in that causal relations became difficult to identify. The impact of the contraction of space-time by networks on patterns of growth potential had already been shown for Europe with an exploratory analysis in Bretagnolle et al. (1998).

Modeling results by Bretagnolle and Pumain (2010), and more particularly the different parametrizations of the Simpop2 model, unveil different regimes of interactions between networks and cities. The generic structure of the Simpop2 model is as follows (Pumain 2008): cities are characterized by their population and their wealth; they produce goods according to their economic profile; interactions between cities produce exchanges, determined by the offer and demand functions; populations evolve according to wealth after exchanges. The application of this model shows that the evolution of the railway network in the United States has followed a different dynamic, without hierarchical diffusion, shaping locally urban growth in some cases. This particular context of conquest of a space empty of infrastructures implies a specific regime for the territorial system. Other contexts reveal different impacts of the network in the short and long terms; Berger and Enflo (2017) study the impact of the construction of the Swedish railway network on the growth of urban populations, from 1800 to 2010, and find an immediate causal effect of the accessibility increase on population growth, followed on long term by strong inertia for population hierarchy. In each case, we observe the existence of structural dynamics in the long term, which correspond to the slow dynamics of the urban system structure, and witness structuring effects in the long term as Pumain (2014) puts it.

We 'however' differentiate the latest from the structuring effects previously mentioned which are subject to debates. At the level of the urban system, it is relevant to globally follow trajectories that were possible, and locally the effect has necessarily a probabilistic aspect. Moreover, we insist on the role of path-dependency for trajectories of urban systems; for example, the existence in France of a previous system of cities and network (postal roads) has strongly influenced the development of the railway network, or as Berger and Enflo (2017) showed for Sweden. Similarly, Chaudhuri and Clarke (2015) highlight the importance of historical events in coupled dynamics of the road network and territories, historical shocks that can be seen as exogenous and inducing bifurcations of the system that accentuate the effect of path-dependency. Therefore, for these structural dynamics in the long term, forecasting can be difficult.

1.3 Models of Co-evolution

Considering models as a fundamental component for the production of knowledge, as a knowledge domain in itself (Raimbault 2017a), this chapter focuses on simulation models concerned with the co-evolution of cities and networks. We propose to review models that integrate dynamically a strong coupling between territorial components and transportation networks, in that a clear conditioning of one by the other cannot be identified. We broadly designate by model of co-evolution simulation models that include a coupling of urban growth dynamics and transportation network growth dynamics. These are relatively rare, and most of them are still at the stage of stylized models. The efforts being relatively sparse and in very different domains, there is not much unity in these approaches, other than the abstraction of the assumption of an interdependency between networks and territorial characteristics over time. This sparsity may be due to a high compartmentalization of related disciplines (Raimbault 2017c). We review them through the prism of scales.

1.3.1 Microscopic and mesoscopic scales

Geometrical models Achibet et al. (2014) describe a co-evolution model on a very large scale (scale of a building), in which evolution of both network and buildings are ruled by a same agent, influenced differently by network topology and population density, and that can be understood as an agent of urban development. The model allows us to simulate an auto-organized urban extension and to produce district configurations. Even if it strongly couples territorial components (buildings) and the road network, described results do not imply any conclusion on the processes of co-evolution themselves.

A generalization of a geometrical local optimization model for network growth yields a co-evolution model for network topology and the density of its nodes (Barthelemy and Flammini 2009). The localization of new nodes is simultaneously influenced by density and centrality, yielding the looping of the strong coupling. More precisely, the global behavior of the model is the same as the network extension behavior. Centers then localize following a utility function that is a linear combination of nodes centrality averaged (in a local neighborhood) and of the opposite of density (dispersion owing to higher prices as a function of density). This utility is used to compute the probability of localization of new centers following a discrete choices model. The model allows us to show that the influence of centrality reinforces aggregation phenomena (in particular through an analytical resolution on a one-dimensional version of the model), and furthermore reproduces exponentially decreasing density profiles which are observed empirically (these exponential density profiles correspond to Clark's law (Chen 2010)).

Ding et al. (2017) introduce a model of co-evolution between different layers of the transportation network, and show the existence of an optimal coupling parameter in relation to inequalities for the centrality in network conception; if the road network is assimilated at a fine granularity to a population distribution, this model can be compared with the precedent model of co-evolution between the transportation network and the territory.

Economic models Levinson et al. (2007) take an economic approach, which is richer in relation to implied network development processes, similar to a four-step model (that is, including the generation of origin-destination flows and the assignment of traffic in the network) which takes into account travel cost and congestion, coupled with a road investment module simulating toll revenues for constructing agents, and a land-use evolution module updating activities and employment through discrete choice modeling. The exploration experiments show that co-evolving network and land uses lead to positive feedbacks reinforcing hierarchies. These are 'however' far from satisfying, since network topology does not evolve only as capacities, and flows change within the network, which implies that more complex mechanisms (such as the planning of new infrastructures) on longer timescales are not taken into account. Li et al. (2016) have extended this model by adding endogenous real estate prices and an optimization heuristic with a genetic algorithm for deciding agents.

Another viewpoint (Levinson and Chen 2005) is also presented as a model of co-evolution, but corresponds more to a predictive model based on Markov chains, and thus is closer to a statistical analysis than a simulation model based on these processes. Rui and Ban (2011) describe a model in which the coupling between land use and network topology is actioned with a weak paradigm, land use and accessibility having no feedback on network topology, and the land-use model being conditioned to the growth of the autonomous network.

Cellular automatons A simple hybrid model explored and applied to a stylized planning example of the functional distribution of a new district in Raimbault et al. (2014), relies on mechanisms of accessibility to urban activities for the growth of settlements, with a network adapting to the urban shape. The rules for network growth are too simple to capture more elaborate processes than a simple systematic connection (such as potential breakdown), but the model produces on a large scale a broad range of urban shapes reproducing typical patterns of human settlements. This model is inspired by Moreno et al. (2012) for its core mechanisms but yields a much broader generation of forms by taking into account urban functions.

On these relatively large scales, spanning from the urban to the metropolitan scale, mechanisms of population localization influenced by accessibility coupled with mechanisms of network growth optimizing particular functions seem to be the rule for this type of model. Similarly, Wu et al. (2017) couple a cellular automaton for population diffusion with a network optimizing local cost that depends on the geometry and on population distribution.

Models answering to more remote questions can also be linked to our problem. For example, in a conceptual way, a certain form of strong coupling is also used in Bigotte et al. (2010), who by using an operational research approach propose a network design algorithm to optimize the accessibility to amenities, taking into account both network hierarchy and the hierarchy of connected centers.

Thus, co-evolution models at the microscopic and mesoscopic scales globally have the following structure: (1) processes of localization or relocalization of activities (action, buildings) influenced by their own distribution and network characteristics; (2) network evolution, that can be topological or not, answering to very diverse rules – local optimization, fixed rules and planning by deciding agents. This diversity suggests the need to take into account the superposition of multiple processes ruling network evolution.

1.3.2 Urban systems modeling

On a macroscopic scale, co-evolution can be taken into account in models of urban systems. Baptiste (1999) proposes to couple an urban growth model based on migrations (introduced by the application of synergetics to systems of cities by Sanders 1992) with a mechanism of self-reinforcement of capacities for the road network, without topological modification. More precisely, the general principles of the model are the that (1) attraction and repulsion indicators enable each city to determine emigration and immigration rates and to make populations evolve; and (2) network topology is fixed in time, but capacities of links evolve. The rule is an increase in capacity when the flow becomes greater given a fixed parameter threshold during a given number of iterations. Flows are affected with a gravity model of interaction between cities. The final version of this model is presented by Baptiste (2010). General conclusions that can be obtained from this work are that this coupling yields a hierarchical configuration and that the addition of the network produces a less hierarchical space, allowing medium-sized cities to benefit from the feedback from the transportation network. These conclusions remain limited as simpler models without co-evolution such as that developed by Raimbault (2020a) also produce hierarchical urban systems.

The model proposed by Blumenfeld-Lieberthal and Portugali (2010) can be seen as a bridge between the mesoscopic scale and the approaches of urban systems, since it simulates migrations between cities and network growth induced by potential breakdown when detours are too large. In the continuity of Simpop models for systems of cities, Schmitt (2014) describes the SimpopNet model which aims at precisely integrating co-evolution processes in systems of cities on long time scales, typically via rules for hierarchical network development as a function of the dynamics of cities, coupled with those that depend on network topology. Unfortunately, this model remained a toy-model which was not explored nor further studied. Cottineau (2014) proposes an endogenous transportation network growth as the last building block of the Marius modeling framework, but it stays at a conceptual level since this brick has not yet been specified or implemented. To the best of our knowledge, there exists no model which is empirical or applied to a concrete case based on an approach of co-evolution by urban systems from the perspective of the evolutive urban theory.

We must note here the epistemological opposition of these geographical approaches inspired by the evolutive urban theory (Pumain 1997) to principles of economic geography. Fujita et al. (1999) introduce, for example, an evolutionary model able to reproduce and urban hierarchy an an organization typical of central place theory Banos et al. (2011), but that still relies on the notion of successive equilibriums, and moreover considers a Krugman-like model, that is a one-dimensional and isotropic space, in which agents are homogeneously distributed. This approach can be instructive on economic processes in themselves but more difficult on geographical processes, since these imply the embedding of economic processes in the geographical space for which spatial particularities not taken into account in this approach are crucial. Our work focuses on demonstrating to what extent this structure of space can be important and explicative, since networks, and even more physical networks, induce spatio-temporal processes that are path-dependent and thus are sensitive to local singularities and prone to bifurcations induced by the combination of these with processes at other scales (for example, the centrality inducing a flow).

We mention several other streams of research which are concerned with modeling co-evolution in systems of cities, such as economic geography (Schamp 2010). For example, Wal and Boschma (2011) model the co-evolution of firms and networks. Liu et al. (2013) empirically study the co-evolution of air transportation networks and company networks. Neal (2012) uses network models to investigate the concept of interlocking of firms and cities, of which dynamics can be understood as co-evolutive. Closer to the fields of planning and transportation, Land-use Transport Interaction (LUTI) models do not explicitly include co-evolution but are concerned with coupling aspects of both fields (Wegener and Furst 2004). This class of models can be generalized at the macroscopic scale to model larger systems of cities (Russo and Musolino 2012).

To synthesize the approaches at the macroscopic scale, existing models are mostly based on the evolution of agents (generally cities) as a consequence of their interactions, carried by the network, whereas the evolution of the network can follow different rules: self-reinforcement and potential breakdown. The general structure is globally the same as at larger scales, but ontologies are fundamentally different.

1.4 Proposed Approach

This literature review confirms the rarity of co-evolution models for cities and transportation networks at the macroscopic scale of a system of cities. As regards the existing examples, Baptiste (2010) is too specific on the processes included and the case study to which it is applied, whereas Schmitt (2014) does not include network growth processes other than topological breakdown and exogenously conditions link speeds. A generic family of models that can apply both to stylized and real systems of cities, and with a degree of flexibility in processes included, would therefore fill this gap in the literature.

In this chapter we introduce this family of co-evolution models. Our contribution is significant on several issues: (1) this type of generic model is introduced to model the co-evolution of cities and transportation networks; (2) we systematically explore a specification of the model on synthetic systems of cities and unveil its ability to produce co-evolutive regimes; and (3) we calibrate it on the French system of cities with population and railway network data.

The rest of this chapter is organized as follows. We first formalize the family of models and the specification studied. We then systematically explore its behavior on synthetic systems of cities, and describe its calibration for the French system of cities. Finally, we briefly present an alternative specification for network growth and discuss possible developments and applications of this work.

2. A FAMILY OF CO-EVOLUTION MODELS

We introduce in this section the family of models we propose for the co-evolution of cities and networks at a macroscopic scale.

2.1 Rationale

Our approach relies in a direct extension of the interaction model within a system of cities described by Raimbault (2020a), at a macroscopic scale with an ontology typical of systems of cities. For simplicity, we similarly retain a unidimensional description of cities by their population.

In relation to network growth, we propose to stay at a relatively aggregated and simplified level, allowing us to test growth heuristics at different levels of abstraction. In order to be flexible on model mechanisms, diverse processes can be taken into account, such as direct interactions between cities, intermediate interactions through the network, the feedback of network flows and a demand-induced network growth. Empirical characteristics emphasized by Thévenin et al. (2013) for the French railway network suggest the existence of feedback of network use, or of flows traversing it, on its persistence and its development, whose properties have evolved in time: a first phase of strong development would correspond to an answer to a high need of coverage, followed by a reinforcement of main links and the disappearance of weakest links.

The coupling between cities and the network is achieved the intermediate flows between cities in the network. These capture the interactions between cities and simultaneously have an influence on the network in which they flow.

2.2 Model Description

2.2.1 Generic formulation

The urban system is characterized by populations $\mu_i(t)$ and the network $G(t)$, with which can be associated a distance matrix $d_{ij}^G(t)$.

The parameter space of the model is composed of seven parameters, which are explained in detail in model equations. These are r_0 endogenous growth rate, w_G weight of spatial interactions, d_G spatial range of interactions, γ_G hierarchy of interactions, g_{max} maximal network growth, ϕ_0 network threshold and γ_S network growth hierarchy. We summarize in Table 8.1 these parameters with the corresponding processes and theoretical ranges.

Flows between cities V_{ij} are given by a gravity interaction potential, given a gravity span parameter d_G which gives the typical range of spatial interactions. The potential writes

$$V_{ij} = \left(\frac{\mu_i \mu_j}{\left(\sum_k \mu_k \right)^2} \right)^{\gamma_G} \cdot exp(-d_{ij}/d_G) \tag{8.1}$$

with d_{ij} network distance.

The evolution of populations then follows the specifications of Raimbault (2020a), i.e.

Table 8.1 Parameters of the model

Parameter	Notation	Process	Interpretation	Range
Growth rate	r_0	Endogenous growth	Urban growth	[0, 1]
Gravity weight	w_G	Direct interaction	Maximal growth	[0, 1]
Gravity gamma	γ_G	Direct interaction	Level of hierarchy	$[0, +\infty]$
Gravity decay	d_G	Direct interaction	Interaction range	$[0, +\infty]$
Maximal network growth	g_{max}	Network growth	Speed improvement	$[0, +\infty]$
Network threshold	ϕ_0	Network growth	Use threshold	$[0, +\infty]$
Network growth hierarchy	γ_S	Network growth	Reinforcement strength	$[0, +\infty]$

Note: We show the parameters of the model that vary in experiments, with the corresponding processes, interpretation and theoretical ranges.

$$\mu_i(t+1) = \Delta t \cdot \left[r_0 \cdot \mu_i + w_G \cdot \sum_j \frac{V_{ij}}{<V_{ij}>} \right]. \tag{8.2}$$

Here r_0 is the endogenous growth rate of the Gibrat model, giving the growth rates of cities without taking into account interactions. The parameter w_G fixes the balance between endogenous growth and growth induced by interactions, and $<\cdot>$ denotes the average.

As regards the network, we assume that it evolves following the equation

$$G(t+1) = F(G(t), \phi_{ij}(t)) \tag{8.3}$$

So that the assignment of flows within the network and a local variation of its elements can be taken into account through the function F. That is, equation (8.3) specifies that network links evolve according to their previous states and the flows traversing them.

This most generic form describes our family of models, which can yield different specifications of models, in particular by describing the network growth heuristic. Note that this model could be easily extended similar to Raimbault (2020a), by adding second-order effects of network centralities on population growth for example.

2.2.2 Distance-based network growth specification

We propose in a first time to consider patterns linked to distance only, and to specify a relationship on an abstract network as

$$d_{ij}^G(t+1) = F(d_{ij}^G(t), \phi_{ij}(t)), \tag{8.4}$$

that is, an evolution of the distance matrix only. In this spirit, we keep an interaction model strictly at a macroscopic scale, since a precise spatialization of the network would imply taking into account a finer scale that includes the local shape of the network which determines shortest paths.

Following a thresholded feedback heuristic, given a flow ϕ in a link, we assume its effective distance to be updated by:

$$d(t+1) = d(t) \cdot \left(1 + g_{max} \cdot \left[\frac{1 - \left(\frac{\phi}{\phi_0}\right)^{\gamma_s}}{1 + \left(\frac{\phi}{\phi_0}\right)^{\gamma_s}}\right]\right) \tag{8.5}$$

with γ_s a hierarchy parameter, ϕ_0 the threshold parameter and g_{max} the maximal growth rate at each step. This auto-reinforcement function can be interpreted in the following way: above a limit flow ϕ_0, the travel conditions improve, whereas they deteriorate below. The hierarchy of gain is given by γ_s, and since

$$\frac{1 - \left(\frac{\phi}{\phi_0}\right)^{\gamma_s}}{1 + \left(\frac{\phi}{\phi_0}\right)^{\gamma_s}} \rightarrow_{\phi\rightarrow\infty} - 1, g_{max} \tag{8.6}$$

is the maximal distance gain. This function is similar to that proposed by Tero et al. (2007), which uses

$$\Delta d = \Delta t \frac{\phi^{\gamma}}{1+\phi^{\gamma}} - d. \tag{8.7}$$

This function yields similarly a threshold effect, since the derivative vanishes at

$$\phi^* = \left(\frac{d}{1-d}\right)^{1/\gamma} \tag{8.8}$$

but this value depends on the distance and can more difficultly be adjusted to a value that can be interpreted.

Our specification is summarized in Figure 8.1 as a model workflow. We have a double feedback of populations and network on themselves, but also the interplay of one on the other, through the mediation of network flows. As interaction potentials depend on populations and network at a given time, the full history of both is crucial to determine it, and the model takes thus into account path-dependency of the system.

2.3 Indicators

To quantify the behavior of the model, we use different indicators, which give a grasp of typical properties of systems of cities produced by the model and on typical properties of temporal trajectories. These have been introduced by Raimbault (2020b) precisely to study the behavior of a macroscopic model of co-evolution for cities and transportation networks, in a very similar context.

The indicators are

- summary statistics of trajectories, taken as hierarchy $\alpha[\cdot]$, entropy $\varepsilon[\cdot]$ and average $\bar{\cdot}$ of distributions;

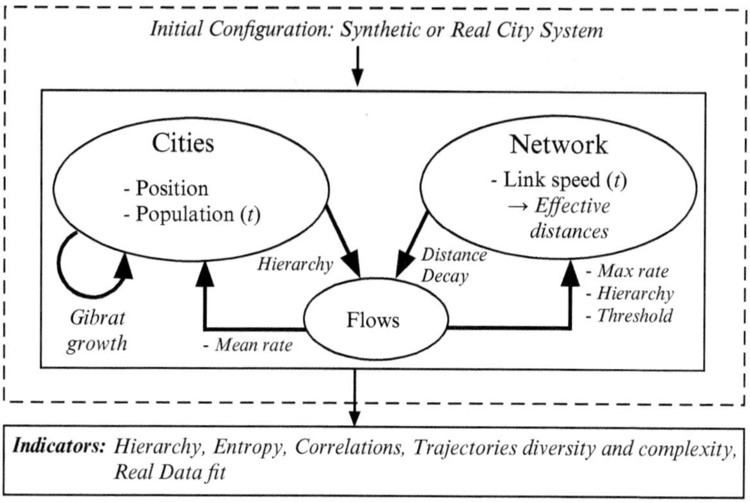

Note: Ellipses correspond to main ontological elements (cities, network and flows), whereas arrows translate processes for which associated parameters are given. The model is described in its broader ecosystem of initialization and output indicator.

Figure 8.1 Abstract representation of the model

- rank correlations $\rho[\cdot]$ between initial state and final state; and
- diversity $D[\cdot]$ and complexity $C[\cdot]$ of temporal trajectories.

These indicators are applied on variables representing both cities and the transportation network: populations $\mu_i(t)$, closeness centralities $c_i(t)$ and generalized accessibilities $Z_i(t)$.

2.4 Model Summary

Figure 8.1 shows the structure of the model as geographic objects and processes included in their relationships. The different components in the schema correspond to the different equations described previously. The self-loop on population of cities is Gibrat endogenous growth (parameter r_0, equation 8.2). The flows between cities are determined by equation 8.1 (parameters γ_G hierarchy and d_G distance decay). They influence city growth with a particular weight w_G (interpreted as mean growth rate owing to interactions, equation 8.2) and network evolution (parameters of maximal speed growth rate g_{max}, flow threshold ϕ_0 under which speeds decay and over which speeds increase, and hierarchy γ_s of this growth; equation 8.5).

3. RESULTS

The model is fully implemented in NetLogo, for the simplicity of coupling between heterogeneous components. Particular care is taken for the duality of network representation, both as a distance matrix and as a physical network, in order to facilitate the extension to physical network heuristics. Source code, data and results are openly available on the git repository of the project at https://github.com/JusteRaimbault/CityNetwork/tree/master/Models/MacroCoevol. Simulation results are available on the dataverse data repository at http://dx.doi.org/10.7910/DVN/TYBNFQ.

The model was explored and calibrated using the OpenMole workflow engine (Reuillon et al. 2013), which provides exploration methods and enables a transparent distribution of computation jobs on a computation grid. We recall that the purpose of the model and these explorations is (1) to investigate the properties of a macroscopic co-evolution model, and (2) to gain knowledge on real case processes by its application to real data.

3.1 Exploration on a Synthetic System of Cities

3.1.1 Synthetic setup

The model is first tested and explored on synthetic systems of cities as it is camed out by (Favaro and Pumain 2011), in order to understand some of its intrinsic properties. This approach is crucial for the study of these simulation models, to be able to distinguish the conclusions specific to the system studied and the conclusions that are general to processes studied.

In this case, we consider the model with an abstract network as specified above, that is without spatial description of the network and with evolution rules acting directly on d_{ij}^G given the previous specifications. A synthetic city system is generated following the heuristic used in the previous section: (1) N_S cities are randomly distributed in a homogeneous geographical space; and (2) populations are attributed to cities following an inverse power law, with a hierarchy parameter α_S and so that the largest city has a population equal to P_{max}, that is, following $P_i = P_{max} \cdot i^{-\alpha_S}$.

To simplify, several meta-parameters are fixed: the number of cities is fixed at $N_S = 30$, the maximal population at $P_{max} = 100\,000$ and the maximal network growth to $g_{max} = 0.005$. Final time is fixed at $t_f = 30$, which corresponds to distances divided approximatively by 5, in order to comply with an empirical constraint: this corresponds to the evolution of the travel time between Paris and Lyon from around 10 hours at the beginning of the century to 2 hours today, showed for example by Thévenin et al. (2013). We also neglect network effects at the second order by taking $w_N = 0$. We explore a grid in the extended parameter space α_S, ϕ_0, γ_s, w_G, d_G, γ_G. We describe the results for $\alpha_S = 1$, which is the closest to existing city systems (in comparison to 0.5 and 1.5, see the systematic review of the rank-size law estimations in Cottineau (2017).

3.1.2 Model behavior

The evolution of the average closeness centrality in time is shown in Figure 8.2 (top) for $w_G = 0.001$, and with variables (γ_G, ϕ_0) (the behavior is not sensitive to d_G). This evolution witnesses a transition as a function of the level of hierarchy: when the parameter for the level of hierarchy decreases, we observe the emergence of trajectories for which the average centrality increases in time, which corresponds to configurations in which all cities profit in average from accessibility gains. That is when the hierarchy of spatial interactions is very high (top-right panel of Figure 8.2), average centrality always decreases in time, which corresponds to a case where only the largest cities benefit from network improvement. On the contrary, low hierarchy (top-left panel of Figire 8.2) can lead to an increas-

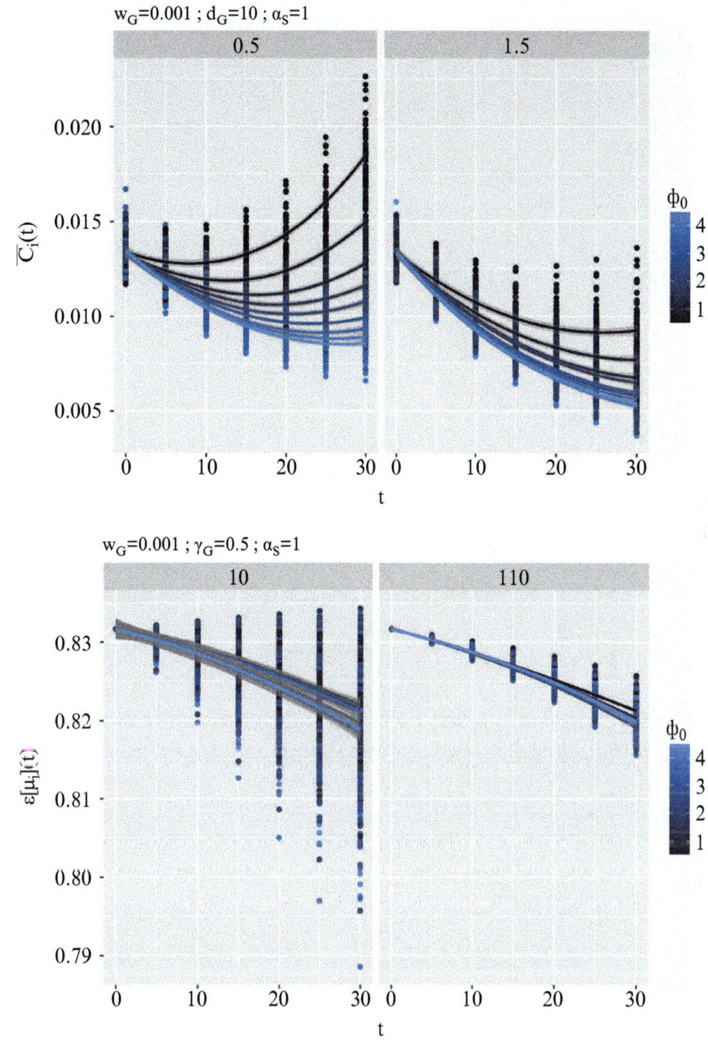

Note: Top panel: average closeness centralities $\bar{c}_i(t)$, which gives the average travel time to other cities, as a function of time, for varying hierarchy of gravity γ_G (rows) and varying network speed threshold φ_0 (color), at fixed weight of interactions $w_G = 0.001$ and gravity decay $d_G = 10$. Bottom panel: entropy of populations $\varepsilon\,[\mu_i]\,(t)$, which gives the concentration of population in cities, as a function of time, for varying d_G (columns) and φ_0 (color), at fixed $w_G = 0.001$ and $\gamma_G = 0.5$. See main text for interpretation.

Figure 8.2 Temporal behavior of the co-evolution model with abstract network on a synthetic system of cities

ing average centrality in time. This transition can be understood in relation to Sanders (2017), that is, a change of the dynamic regime of the urban system, as the qualititative behavior of indicators changes when parameters change.

As regards entropy of populations, for which the temporal trajectory is shown in Figure

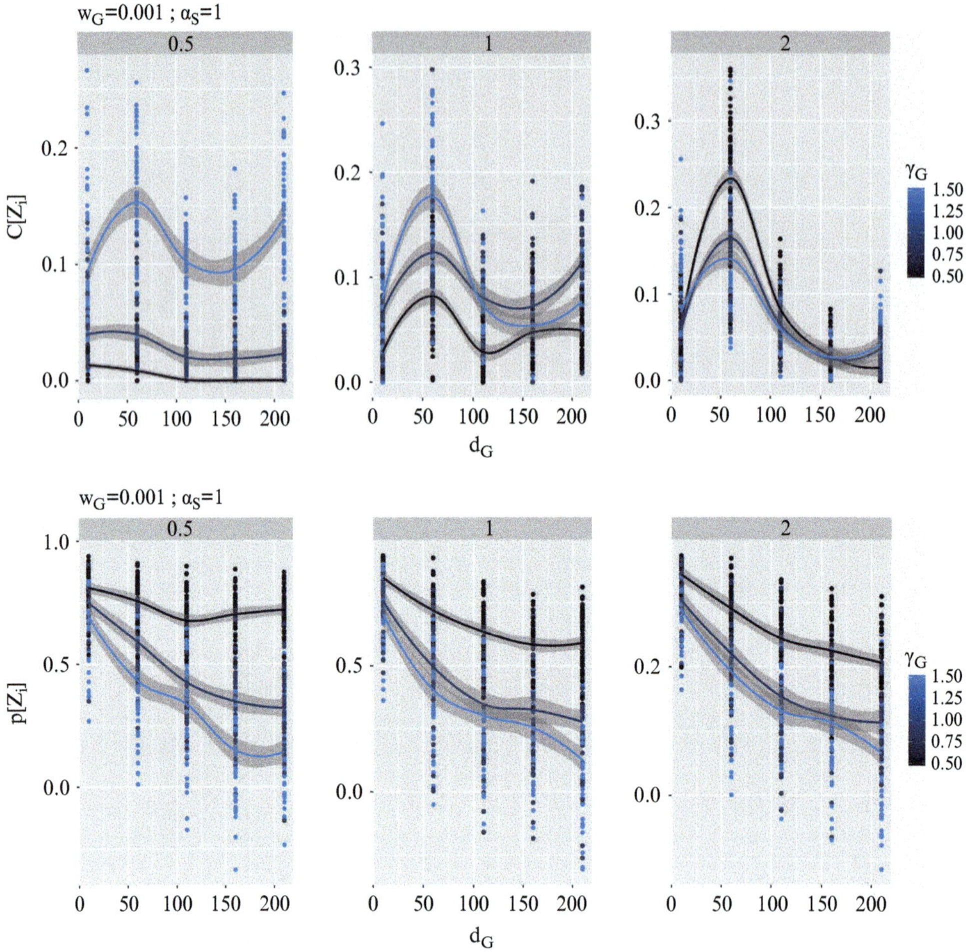

Note: Top panel: complexity of accessibilities $C[Z_i]$, which captures the non-monotony of accessibility trajectories, as a function of gravity decay d_G, for varying network speed growth threshold φ_0 (columns) and varying hierarchy of gravity γ_G (color), at fixed $w_G = 0.001$. Bottom panel: rank correlations of accessibilities $\rho[Z_i]$, which give the number of switches in the urban hierarchy between initial and final time, as a function of d_G, for the same parameters.

Figure 8.3 Aggregated behavior of the co-evolution model

8.2 (bottom panel), all parameters give a decreasing entropy. The entropy of population distributions is maximal when all cities have the same population, and we witness here therefore the progressive concentration of population into fewer cities.

Looking at the complexity of accessibility trajectories (top panel of Figure 8.3), we observe for values of $\phi_0 > 1.5$ a maximum of complexity as a function of interaction distance d_G, stable when w_G and γ_G vary (Figure 8.3, top panel, final column). Therefore, systems with local interaction or with system-wide interactions only produce much simpler trajectories of accessibility than the intermediate range producing the maximum at $d_G = 60$.

As the complexity is computed here as the number of changes in the variation of the temporal curve, the system-wide interactions seem to produce monotonous trajectories, as does the absence of interactions. This intermediate scale can thus be interpreted as producing regional subsystems, large enough for each to develop a particular level of complexity, and isolated enough to avoid the convergence of trajectories over the whole system. We reconstruct therein a spatial non-stationarity, and rejoin the concept of the ecological niche (an independent ecosystem in which there is co-evolution between the species, Holland 2012) localized in space: the emergent subsystems that are relatively independent, are good candidates to contain processes of co-evolution. The emergence of this intermediate scale can be compared to the modularity of the French urban system showed by Berroir et al. (2017). This is, to the best of our knowledge the first time that this niche analogy has been identified in a model of a system of cities, although similar conceptualizations in neighbor fields exist, such as in political science (Monstadt 2009) or in the study of technological change (Geels 2005) in which the concept of technological niche is central.

Finally, the behavior of rank correlations for accessibility (bottom panel of Figure 8.3) reveals that an increase in interaction distance d_G systematically leads to a decrease of the rank correlation $\rho[Z_i]$. This corresponds with an increase in the number of hierarchy inversions (in relation to a switch between two cities in their population ranks), since the lowest the correlation, the highest the rank switches are. This can be interpreted as an increase in overall system complexity. The hierarchy parameter diminishes this correlation, which means that a more hierarchical organization will impact a larger number of cities in the qualitative aspects of their trajectories. This effect is similar to the first mover advantage of (Levinson and Xie 2011), which unveils a path-dependency and an advantage to be rapidly connected to the network: in our case, the modifications in the hierarchy correspond to cities that benefit from their positioning in the network.

3.1.3 Co-evolutive behavior

We can now study the ability of the model to effectively produce co-evolutive dynamics. This property of the model is crucial for several reasons:

1. Being based on a strong coupling at the process level (or microscopic level) does not imply that effective statistical patterns of co-evolution (in the characterization proposed by Raimbault (2018a) emerge from model behavior at the macroscopic level. Indeed, this characterization of co-evolution follows the original definition from biology and considers statistical relationships estimated on a population of objects (in our case, cities). As shown by Raimbault (2017b), which applies the methods on a toy co-evolution model with circular relations at the microscopic level, some model regimes yield directed causal relationships between indicators at the macroscopic level, while others witness a circularity and thus a co-evolution.
2. There is no example in the literature in which the link between a co-evolution model (in relation to processes included, as Schmitt 2014 does) and its macroscopic causal patterns has been undertaken.
3. The spectrum of dynamical regimes the model can produce will inform on the existence of structuring effects (direct causality of city growth on transportation network growth or the other way around, without co-evolution) or of a more intricate relationship in the case of co-evolution.

To characterize co-evolutive dynamics, we apply a simple but generic method, particularly suited for the study of spatio-temporal causality patterns in territorial systems, which was introduced by Raimbault (2017b). This method uses a weak version of Granger causality, by classifying profiles of lagged correlations between variables. If an absolute maximum of correlations for a strictly positive or negative lag exists, then we have a direct causal link between the variables. A co-evolution between two variables X, Y will correspond to the existence of circular causal links $X \rightarrow Y$ and $Y \rightarrow X$. We denote here by $\rho_\tau[X,Y]$ the lagged correlation with delay τ. Examples for estimations of these lagged correlation profiles are shown in Figure 8.3, and we detail their interpretation in the following.

We can investigate the profiles for ρ_τ for varying parameter values. We generally observe (1) the systematic existence of a constant correlation at $\tau = 0$ and (2) the small variations of correlations that impose the need for a statistical test to ensure that we isolate a significant effect. Some samples of these curves are shown in Figure 8.4. These will be interpreted as causality regimes, but we first need to describe the method used to extract them.

We need to extend the original method, because of the zero-delay correlation and of the small variations (in the original development by Raimbault (2017b), most profiles exhibit significant variations and no correlation at zero delay). We impose here an additional statistical test: for $\tau_+ = \text{argmax}_{\tau>0} |\rho_\tau - \rho_0|$ and $\tau_- = \text{argmax}_{\tau<0} |\rho_\tau - \rho_0|$, a Kolmogorov-Smirnov test is used to compare the distributions of $\rho_{\tau_\pm}$ and of ρ_0. If they are declared different with a p-value smaller than 0.01, and if $|\rho_{\tau_\pm}| > |\rho_0|$, we accept the causality link between variables in the corresponding direction. A configuration is then coded by a representation of its graph between variables, given by the six discrete variables equal to 0 if there is no link between the variables (within all directed couples between population, accessibility and centrality) and 1 or –1 depending on the sign of the correlation if there exists a statistically significant link (in practice we observe only positive correlations).

We obtain overall 33 different configurations of links between variables, out of the 64 possible configurations (2^6 possible choices for positive correlations only), what confirms the existence of multiple causality regimes. The type of relations we obtain are particularly interesting regarding co-evolution. We indeed observe the following types of regimes:

- a configuration without any link between variables;
- 13 configurations of type 'structuring effect', that is for which the graph does not have any loop;
- a configuration of type 'indirect co-evolution', for which the graph has a loop of length three ($c_i \rightarrow X_i \rightarrow \mu_i \rightarrow c_i$);
- 18 configurations of type 'co-evolution', in which there exists at least a loop of length two (direct circular relation between two variables).

Among all these regimes, eight correspond to a graph with at least four links (which are then necessarily co-evolutive): we show these profiles in Figure 8.4. Two regimes witness a positive deviation of the correlation between population and accessibility for positive delays, increasing up to the maximal delay, what could be a clue of a reinforcement of population dynamics through centrality, stylized fact shown for the French system of cities by Bretagnolle (2009).

The regimes in which the centrality is co-evolving with population correspond to those

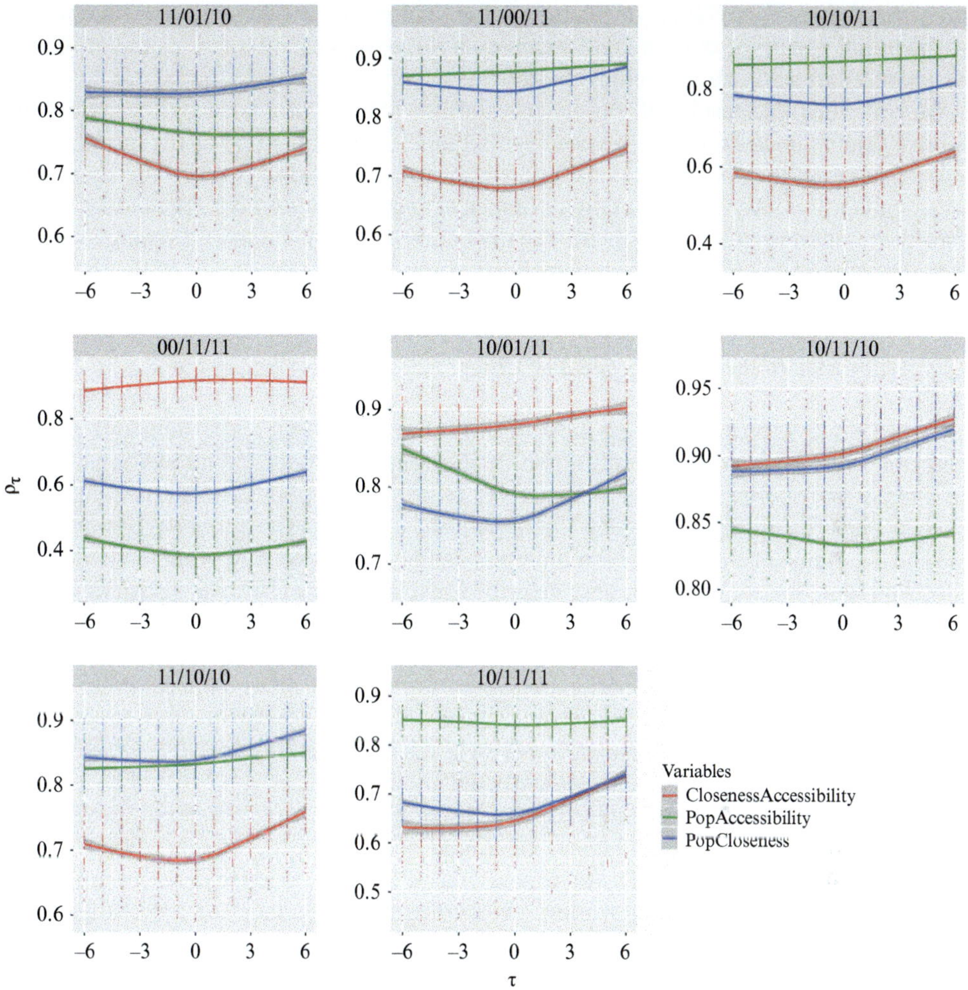

Note: For the eight configurations showing at least four links among variables (coded in the order of couples, by the existence or not of a link for $\tau+$ and for $\tau-$), these are the lagged correlation profiles $\rho\tau$ as a function of τ, for all couples of variables (color). The variation of these lagged correlations compared to no delays at $\tau = 0$, and more particularly the relative positioning of potential extrema for positive and negative delays, defines what we term a causality regime under particular assumptions of statistical significance (see main text).

Figure 8.4 Lagged correlations

where the co-evolution between the network and the territory is the strongest (since the accessibility depends on both), and are observed for large values of d_G (average $d_G = 183$ on 62 parameter points). In his way, this co-evolution is favored by long interaction ranges.

Finally, the regime with the largest number of links (which corresponds to the regime coded by '10/11/11', with co-evolution of population and centrality and of population and accessibility, and a causality of centrality on accessibility), is obtained for a long interaction range $d_G = 160$, a strong interaction hierarchy $\gamma_G = 1.5$, but a low hierarchy of the initial system of cities α_S. Far-reaching but hierarchical interactions in a uniform system of cities lead to a maximum entanglement among variables.

We confirm these results of variety in causality regimes produced by the model by applying the pattern space exploration algorithm (Chàel et al. 2015) to the model, with objectives being the six correlations studied above (evaluated as zero in the case of a non-significance). We mainly obtain a number of regimes produced by the model larger than those obtained previously (with negative correlations, 260 realized regimes out of $3^6 = 729$ possible). This short complementary study confirms the ability of the model to produce a large number of co-evolution regimes.

3.2 Application to the French System of Cities

In this section, we investigate how the model can be applied to a real system and knowledge can be gained from its calibration. The model is applied to the French system of cities on long-term dynamic data: the Pumain-INED database for populations, spanning from 1831 to 1999 (Pumain and Riandey 1986), with the evolving railway network from 1840 to 2000 (Thévenin et al. 2013). This a timespan can be associated with potential structural effects on long timespans, which were described by Bretagnolle (2003) as a positive correspondence between city growth and accessibility. This application aims, on the one hand, to test the ability of the model to reproduce a real dynamic of co-evolution and, on the other, to extract thematic information on processes through calibrated parameter values.

3.2.1 Data

We work on railway network data constructed by Thévenin et al. (2013). The French railway network is particularly interesting jointly with population data already presented, since the covered time span is relatively close, and this transportation mode has at times revealed the implication of public and private actors. It corresponds to different processes depending on the period, from a more decentralized management to a recently more centralized, and different technological materializations with, for example, the recent emergence of high-speed trains (Zembri 1997). For each date in the population database, we extract the simplified abstract network in which all stations and intersections with a degree larger than 2 are linked with abstract links which speed and length attributes correspond to real values, at a granularity of 1 kilometer. This yields the time–distance matrices between the cities included in the model.

3.2.2 Stylized facts

Before calibrating the model, we can observe the lagged correlation patterns in the dataset, by applying the causality regimes method (Raimbault 2017b). This empirical study should, on the one hand, allow us to verify well-known stylized facts and, on the other, produce a preliminary knowledge of empirical system behavior. We compute the closeness centrality through the network, given by $T_i = \sum_j exp{-d_{ij}/d_0}$, and we study the lagged correlation between its derivative ΔT_i and the derivative of the population ΔP_i, given by $\hat{\rho}_\tau = \hat{\rho}[\Delta P_i(t), \Delta T_i(t-\tau)]$ estimated on a moving window containing T_w successive dates. We show in Figure 8.5 the results obtained.

These results are important for at least two reasons. First, the behavior of the number

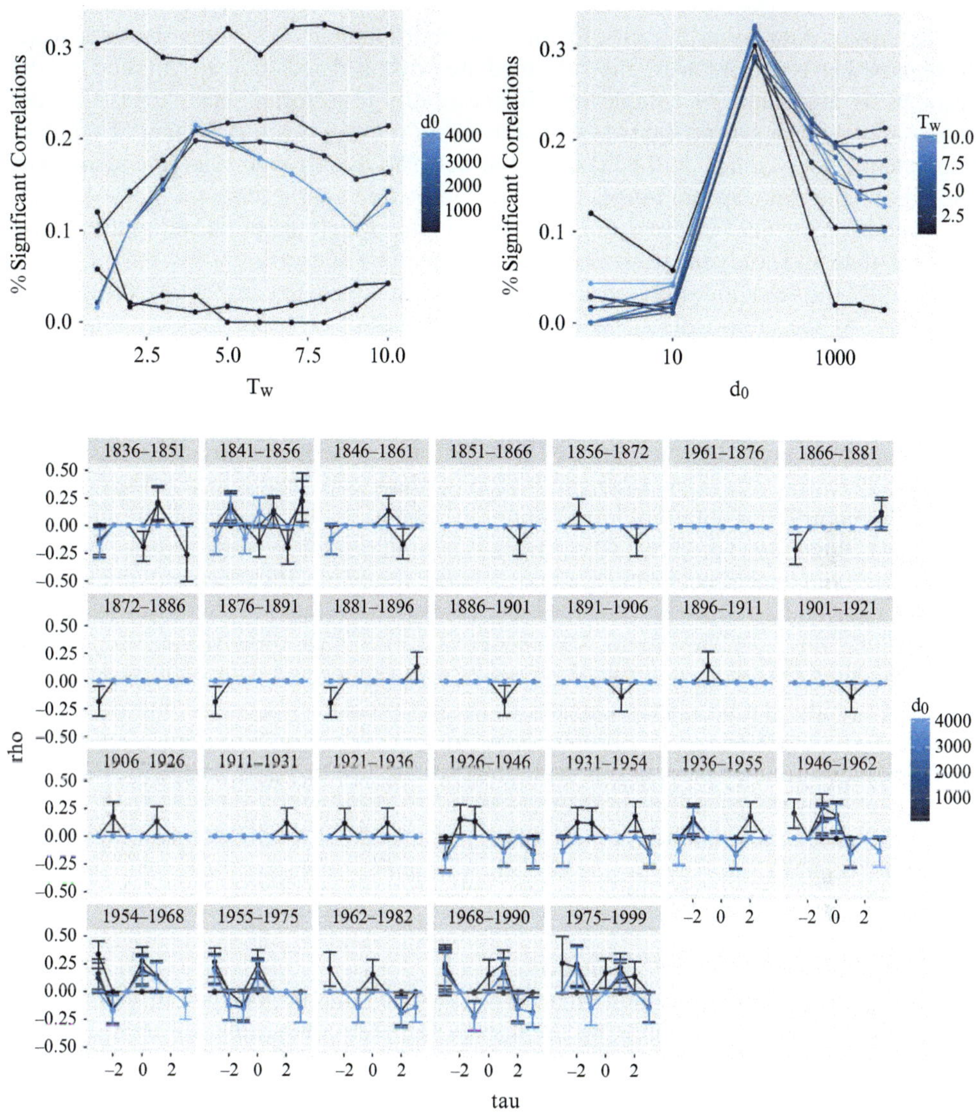

Note: Correlations are estimated on a window of duration 5 T_w, between population growth rates and the variations of closeness centrality with a decay parameter d_0 (see main text). Top-left panel: number of significant correlations (taken such that $p < 0.1$ at 95%) as a function of T_w for d_0 variable. Top-right panel: number of significant correlations as a function of d_0 for T_w variable. Bottom panel: for the optimal window $Tw = 4$, value of $\rho\tau$ as a function of τ, for all successive periods.

Figure 8.5 Empirical lagged correlations for the French system of cities

of significant correlations as a function of T_w and d_0 allows us to find stationarity scales in the system. We observe, on the one hand, a specific spatial scale that gives a maximum for all temporal windows, at $d_0 = 100$ *kilometers*, which suggests the existence of consistent regional subsystems, whose existence is stable in time. This value can thus be interpreted as the typical interaction distance for most intercity interactions at the national scale. It remarkably coincides with the intermediate scale isolated in the synthetic model through the maximum of complexity of accessibility trajectories. This means that the model applied on a synthetic city system already reproduces this stylized fact of regional subsystems observed here in the maximum of correlations, through emerging niches. On the other hand, long spatial ranges induce an optimal temporal scale, for $T_w = 4$ which corresponds to around 20 years; we identify it as the overall temporal stationarity scale of the system and study the lagged correlations for this value.

Secondly, the behavior of lagged correlations does not seem to comply with the existing literature. At the intermediate spatial scale, the values of ρ_+, ρ_- exhibit no regularity. On the whole system, until 1946 there is close to no significant effect, then no causality between 1946 and 1975 (maximum at $\tau = 0$, non-significant minimum), and a five-year shift of accessibility causing population after 1968 (the effect, however, staying doubtful). We do not reproduce the correlation effect between network centrality and place in the urban hierarchy advocated by Bretagnolle (2003), which leads us to question the existence of the 'structural co-evolution' in the long term described by Bretagnolle in Offner et al. (2014). What Bretagnolle (2003) obtains is a simultaneous correspondence between growth rate and level of connectivity to the network (and not with network dynamic), but not in a co-evolution, since no statistical relationship is exhibited.

We rejoin the recent results of Mimeur et al. (2017) that show the statistical non-significance of the correlation between growth rate and evolution of network coverage and accessibility, at a zero delay. Our results are less precise on the class of cities studied (they differentiate large and small cities, and work on a larger panel), but are more general as they study variable delays and accessibility ranges, and are thus complementary. We show that determining if city growth causes network growth, or the contrary is not possible through those methods as results are not significant. For this reason, we cannot expect to calibrate consistently the model on regimes themselves, and will therefore consider first-order objectives .

3.2.3 Calibration of the abstract model

Expected results of the calibration on real data concern both the generally accurate reproduction of real city population growth dynamics, that is, to what extent the inclusion of a dynamical network can increase the explanatory power for trajectories, and how realistic the evolution of network distance is. We continue to work with the abstract model.

Model evaluation The population trajectories are evaluated similarly to Raimbault (2020a), with the logarithm of the mean square error on cities and on time for populations. We add a calibration indicator for distance, given by

$$\varepsilon_D = log\left[\sum_t \sum_{i,j} (d_{ij}(t) - \tilde{d_{ij}}(t))^2\right] \tag{8.6}$$

where $d_{ij}(t)$ are observed distances and $\tilde{d}_{ij}(t)$ the simulated distances. It is simply a cumulated squared-error, as used for the comparison of origin – destination matrices in a similar case of simulation of a transportation network in Jacobs-Crisioni and Koopmans (2016).

Results We proceed to a non-stationary calibration, on the (ε_P, ε_D) objectives, that is, the squared-error on populations and on distances. The estimation is calculated with a moving window on periods with a duration of 20 years. The calibration is undertaken with a standard genetic algorithm (NSGA2) provided by the model exploration platform OpenMole (Reuillon et al. 2013). Figure 8.6 shows the obtained Pareto fronts, and Figure 8.7 the evolution in time of parameter values for the optimal solutions.

We observe a large variability of the shape of Pareto fronts for the bi-objective calibration on population and distance, which witnesses more or less difficulty in simultaneously adjusting population and distance. Some periods, such as 1891–1911 and 1921–1936, are close to having a simultaneous objective point for the two objectives, which would equate to a good correspondence of the model with both trajectories of cities and the trajectory of the network during these periods.

In comparison with calibration results of the model with the static network of Raimbault (2020a), when comparing the performances for the objective ε_G, we find periods where the static is clearly better (1831 and 1841, for example) and others where the co-evolutive model is better (1946 and 1962). Thus, taking into account the co-evolution helps in some instances to have a better reproduction of population trajectories.

The values of optimal parameters in time, shown in Figure 8.7, seem to contain some signal. The evolution of w_G and γ_G are coherent with the evolutions observed for the static model. For d_G, the model mostly reaches the maximal distance, and the evolution is difficult to interpret. However, the evolution of ϕ_0 could be a sign of a TGV effect in recent periods, through the secondary peak for population after 1960. Indeed, the construction of high-speed lines has shortened distances between cities at the top of the hierarchy, and an increase of the threshold ϕ_0 corresponds to an increase of the selectivity for a potential diminution of distances.

The calibrated g_{max} can finally be interpreted according to the history of the railway network (at least for all points in the Pareto front): a significant secondary peak in the first years, a minimum in the years corresponding to the stabilization of the network (1900), and an increase until today linked to the increase of train speeds and the opening of high-speed lines.

We have here to an extent indirectly quantified interaction processes through the network and the processes of network adaptation to flows, using the of example of a real system.

3.3 Model with a Physical Network

We finally sketch the outline of a specification of the model with a physical network, which would correspond to a hybrid model combining different scales. The objective of this specification would be, on the one hand, to study the difference in trajectories compared with the abstract network, that is to quantify the importance of economies of scale (owing to common links), congestion and the possible compromises to make in order to spatialize the network. On the other hand, it would help to understand to what extent it is possible to produce realistic networks in comparison to autonomous network growth models (see Xie and Levinson 2009), for example. This specification follows the frame of Li et al. (2014), who model the co-evolution between transportation corridors and the growth of main poles at a regional scale. Note that possible ontologies for fully

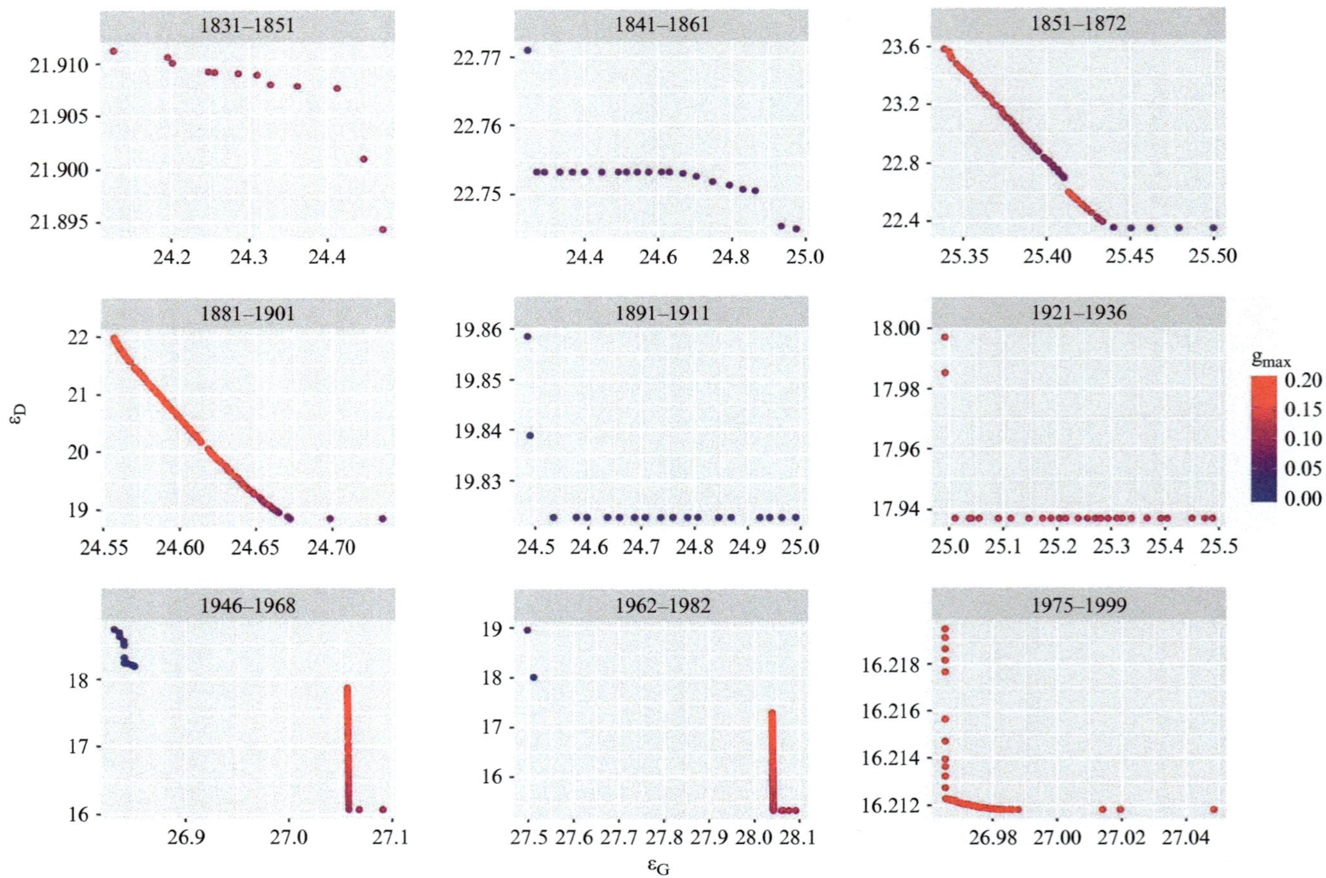

Note: Fronts are given for each calibration period and are colored according to network speed maximal growth g_{max}.

Figure 8.6 Pareto fronts for the bi-objective calibration between population ε_P and distance ε_D objectives

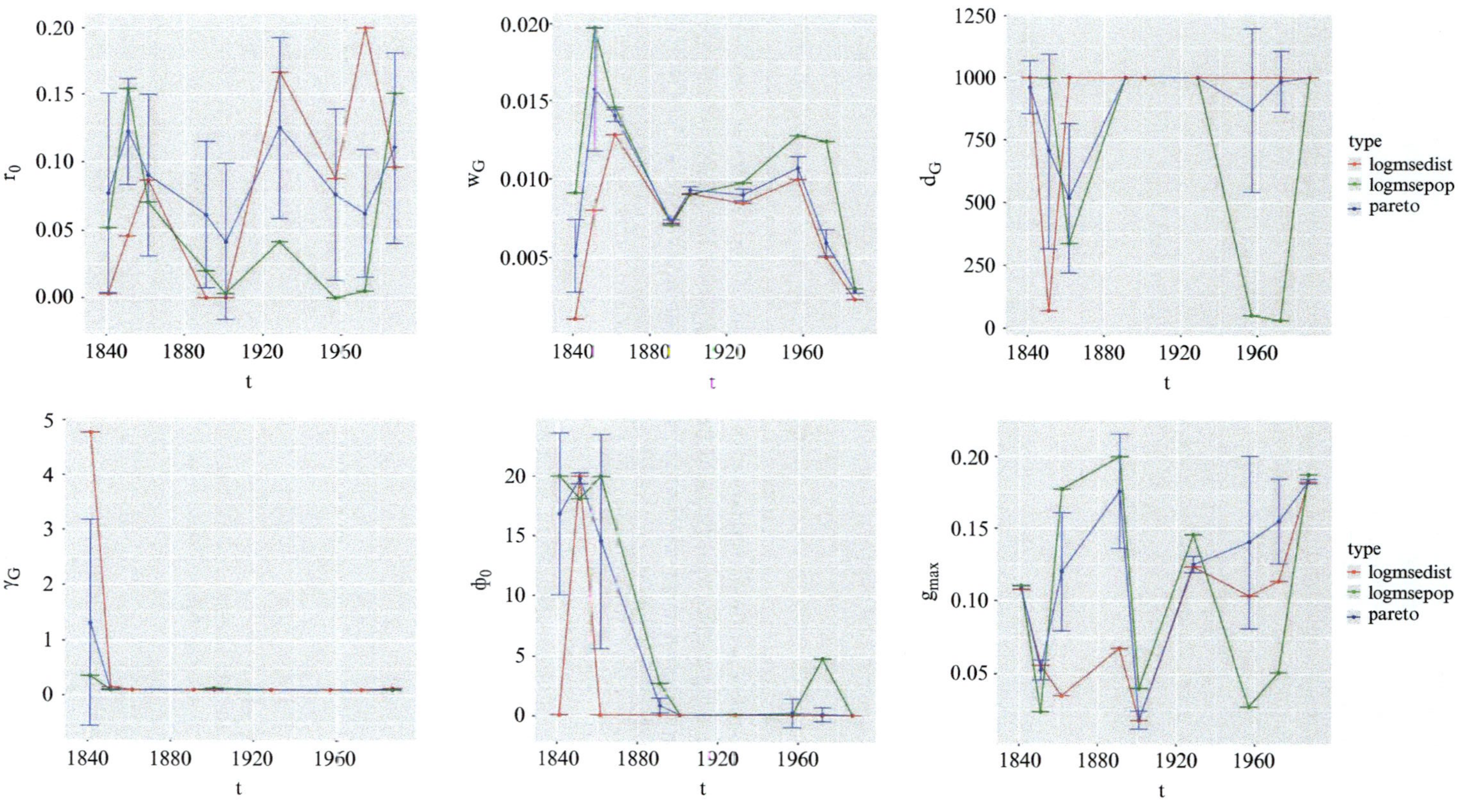

Note: From left to right and top to bottom, values of calibrated parameters r_0 Gibrat growth rate, w_G weight of gravity interactions, d_G distance decay of gravity interactions, γ_G hierarchy of gravity interactions, φ_0 threshold of network speed growth, and g_{max} maximal network speed growth. These are given, respectively, for the full Pareto front (blue), the optimal point in the sense of the distance (red) and the optimal point in the sense of the population (green).

Figure 8.7 Temporal evolution of optimal parameters

mesoscopic co-evolution models seem to be very different from that which we used here, closer to cellular automata urban morphogenesis models (Raimbault 2019a). They can in particular include multiple processes for the growth of transportation infrastructure networks (Raimbault 2018b).

The physical network we implement aims to satisfy a greedy criteria of local time gain. More precisely, we assume a self-reinforcement similar to that of Tero et al. (2010). This network growth heuristic captures the improved link performance (speed, in our case) for links which have a high usage, whereas links with a low usage have a relatively low improvement, or even a decrease in performance.

A specification analog to the one used previously assumes a growth for each link, given also in a logic of self-reinforcement by:

$$d(t+1) = d(t) \cdot \left(1 + g_{max} \cdot \left[\frac{\phi}{\max \phi}\right]^{\gamma_s}\right) \tag{8.7}$$

if ϕ is the flow in the link and $d(t)$ its effective distance. The threshold specification used previously does not allow a good convergence in time, in particular with the emergence of local oscillation phenomena, and we therefore use a specification where all links improve and the hierarchy γ_s gives the relative evolution between links with low and high flows.

In order to test the model on a synthetic system, we need to generate a random initial network. Therefore, we start from a grid network, remove a fixed proportion of links (40 percent), and perturb the position of vertices with a small noise. This produces a network more realistic than a simple grid, which could be resulting of a topology for example. We finally connect cities to the network through the geographically shortest path. Links initially have all the same impedance. The resulting initial network is shown in Figure 8.8 (left-hand panel).

We then evolve this network according to the model with equation 8.7 for network growth. An example of a configuration obtained with this specification is shown in Figure 8.8 (right-hand panel). The good convergence properties (visual stabilization of network structure during restricted experiments) suggest the potentialities offered by this specification, but systematic exploration is beyond the scope of this chapter.

4. DISCUSSION

The study of particular trajectories within a system of cities enables us to answer specific thematic questions. For example, the influence of medium-sized cities on the global trajectory of the system can be assessed through this model. More precisely, depending on parameters, a given city can influence the trajectories of all cities and the network evolution. Very high hierarchy parameters will favor only the biggest cities, but lower values will give more importance to the rest of the distribution, in particular to medium-sized cities. The quantification of their contribution in indicator variance could, for example, inform their role in the system. The drivers of a successful trajectory for those medium-sized cities can also be investigated, that is, whether if a higher population growth or connectivity is obtained owing to a higher centrality or to the proximity of a large city, for example.

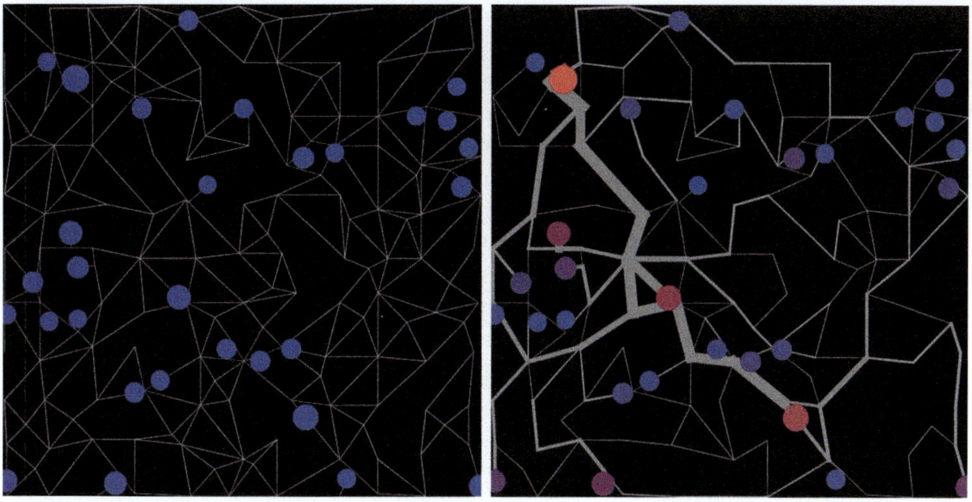

Note: Left-hand panel: initial random configuration, with uniform impedances. Right-hand panel: final configuration obtained after 100 iterations. Circles are cities distributed in space, the relative size of which gives the population and color level the total population growth. Physical transportation links are in grey and their thickness relates to the flow traversing them.

Figure 8.8 Example of configuration obtained with a self-reinforcing network

For the application to a real system, the mapping of deviation to the model in time can suggest regional particularities, that is local factors exogenous to the model that have a high impact on population growth patterns.

We also expect to be able, through the model, to compare urban systems in different geographical and political contexts, and at different scales. This should foster understanding of the implications of planning actions on the interactions between networks and territories. For example, the French railway network has emerged through multiple operators, on the contrary to the Chinese high-speed railway network, for which a more precise development could be considered.

Another development would be to superimpose different layers of the transportation network as different multi-modeling components, on, for example, a dynamic highway database for the case study we took. This would allow us to make the model more complex and compare which transportation mode fits better within co-evolutionary dynamics. In a broader perspective, the family of models we introduced could be used for a more general benchmark of models of growth for systems of cities based on interactions, which would also include, for example, the Favaro–Pumain model focusing on innovation (Favaro and Pumain 2011), or the Marius model family focusing on economic dynamics (Cottineau 2014). A systematic comparison of these models on various systems of cities would shed light on concurrent explanations for city growth and possible superimposition of processes.

5. CONCLUSION

We have given a first insight into models of co-evolution for cities and networks at macroscopic scales, by introducing a family of models and studying the properties of a specification on synthetic data and on the French system of cities. This work paves the way for a finer understanding of entangled interactions in systems of cities.

Coming back to the question of structuring effects of transportation networks, that is how to model and identify these effects (interpreted as directed causal relationships between cities and transportation networks) or, more generally, how to understand them in a co-evolution framework, we have confirmed from the modeling and empirical perspective that the issue is highly complex. First, our empirical analysis for the French system of cities does not reveal any possible effect, in contradiction to previous results in the literature, which means that these structuring effects are either difficult to identify empirically or do not exist. Secondly, our modeling experiments show that there exist theoretically some dynamical regimes in which these effects really exist, and others in which interactions are more intricate, in particular when there is co-evolution, confirming that these can be modeled in the more general framework of co-evolution. In that context, model calibration produces indirect knowledge on underlying processes.

The empirical identification of these on real systems, for example through a correspondence between model regimes and calibrated parameters, however, remains an open issue beyond the scope of this chapter, for example, owing to issues of equifinality for which an efficient inverse problem heuristic for simulation models must be established.

ACKNOWLEDGEMENTS

Results obtained in this paper were computed on the vo.complex-system.eu virtual organization of the European Grid Infrastructure (http://www.egi.eu). We thank the European Grid Infrastructure and its supporting National Grid Initiatives (France-Grilles in particular) for providing the technical support and infrastructure.

REFERENCES

Achibet, M., S. Balev, A. Dutot and D. Olivier (2014), 'A model of road network and buildings extension co-evolution', *Procedia Computer Science*, **32** (June), 828–33.

Banos, A., D. Moreno, C. Pivano and P. Taillandier (2011), 'Christaller, toujours vivant!', *Cybergeo: European Journal of Geography*, accessed 23 March 2021 at https://journals.openedition.org/cybergeo/24877.

Baptiste, H. (1999), 'Interactions entre le système de transport et les systèmes de villes: perspective historique pour une modélisation dynamique spatialisée', PhD thesis, Centre d'études supérieures de l'aménagement, Tours.

Baptiste, H. (2010), 'Modeling the evolution of a transport system and its impacts on a French urban system', in P. Mathis (ed.), *Graphs and Networks: Multilevel Modeling*, 2nd edn, London: Wiley-ISTE, pp. 67–89.

Barthelemy, M. and A. Flammini (2009), 'Co-evolution of density and topology in a simple model of city formation', *Networks and Spatial Economics*, **9** (3), 401–25.

Berger, T. and K. Enflo (2017), 'Locomotives of local growth: the short-and long-term impact of railroads in Sweden', *Journal of Urban Economics*, **98** (March), 124–38.

Berroir, S., N. Cattan, F. Dobruszkes, M. Guérois, F. Paulus and C. Vacchiani-Marcuzzo (2017), 'Les systèmes urbains français: une approche relationnelle', *Cybergeo: European Journal of Geography*, doi.org/10.4000/cybergeo.27945.

Bigotte, J.F., D. Krass, A.P. Antunes and O. Berman (2010), 'Integrated modeling of urban hierarchy and transportation network planning', *Transportation Research Part A: Policy and Practice*, **44** (7), 506–22.

Blumenfeld-Lieberthal, E. and J. Portugali (2010), 'Network cities: a complexity-network approach to urban dynamics and development', in B. Jiang and X. Yao (eds), *Geospatial Analysis and Modelling of Urban Structure and Dynamics*, Dordrecht: Springer, pp. 77–90.

Bonnafous, A. (2017), 'Les observatoires permanents comme instruments d'évaluation ex-post: le cas français', in International Transport Forum, *L'évaluation ex-post des investissements et interventions publiques dans les transports*, Paris: OECD Publishing, doi:10.1787/9789282108277-3-fr.

Bonnafous, A. and F. Plassard (1974), 'Les méthodologies usuelles de l'étude des effets structurants de l'offre de transport', *Revue économique*, **5** (2), 208–32.

Bonnafous, A., F. Plassard and D. Soum (1974), 'La detection des effets structurants d'autoroute: Application à la vallée du Rhône', *Revue économique*, **25** (2), 233–56.

Bretagnolle, A. (2003), 'Vitesse et processus de sélection hiérarchique dans le système des villes françaises', in D. Pumain and M.-F. Mattei (eds), *Données urbaines*, vol. 4, Paris: Anthropos-Economica, pp. 309–22.

Bretagnolle, A. (2009), 'Villes et réseaux de transport: des interactions dans la longue durée, France, Europe, États-Unis', PhD thesis, Université Panthéon-Sorbonne – Paris I, Paris.

Bretagnolle, A. and D. Pumain (2010), 'Comparer deux types de systèmes de villes par la modélisation multi-agents', in G. Weisbuch and A. Zwirn (eds), *Qu'appelle t-on aujourd'hui les sciences de la complexité? Langages, réseaux, marchés, territoires*, Paris: Eyrolles, pp. 271–99.

Bretagnolle, A., D. Pumain and C. Rozenblat (1998), 'Space-time contraction and the dynamics of urban systems', *Cybergeo: European Journal of Geography*, doi.org/10.4000/cybergeo.373.

Burgess, E.W., R.D. McKenzie and L. Wirth (1925), *The City*, Chicago, IL: University of Chicago Press.

Chaudhuri, G. and C. Clarke, Keith (2015), 'On the spatiotemporal dynamics of the coupling between land use and road networks: does political history matter?', *Environment and Planning B: Planning and Design*, **42** (1), 133–56.

Chen, Y. (2010), 'A new model of urban population density indicating latent fractal structure', *International Journal of Urban Sustainable Development*, **1** (1–2), 89–110.

Chérel, G., C. Cottineau and R. Reuillon (2015), 'Beyond corroboration: strengthening model validation by looking for unexpected patterns', *PLoS ONE*, **10** (9), e0138212.

Cottineau, C. (2014), 'L'évolution des villes dans l'espace post-soviétique. Observation et modélisations', PhD thesis, Université Paris 1 Panthéon-Sorbonne.

Cottineau, C. (2017), 'Metazipf: a dynamic meta-analysis of city size distributions', *PLoS ONE*, **12** (8), 1–22.

Ding, R., N. Ujang, H. bin Hamid, M.S.A. Manan, R. Li and J. Wu (2017), 'Heuristic urban transportation network design method, a multilayer coevolution approach', *Physica A: Statistical Mechanics and its Applications*, **479** (August), 71–83.

Favaro, J.-M. and D. Pumain (2011), 'Gibrat revisited: an urban growth model incorporating spatial interaction and innovation cycles', *Geographical Analysis*, **43** (3), 261–86.

Fujita, M., P. Krugman and T. Mori (1999), 'On the evolution of hierarchical urban systems, *European Economic Review*, **43** (2), 209–51.

Geels, F.W. (2005), 'Processes and patterns in transitions and system innovations: refining the co- evolutionary multi-level perspective', *Technological Forecasting and Social Change*, **72** (6), 681–96.

Holland, J.H. (2012), *Signals and Boundaries: Building Blocks for Complex Adaptive Systems*, Cambridge, MA: MIT Press.

Jacobs-Crisioni, C. and C.C. Koopmans (2016), 'Transport link scanner: simulating geographic transport network expansion through individual investments', *Journal of Geographical Systems*, **18** (3), 265–301.

Levinson, D. and W. Chen (2005), 'Paving new ground: a Markov chain model of the change in transportation networks and land use', *Access to Destinations*, Bingley: Emerald, pp. 243–66.

Levinson, D. and F. Xie (2011), 'Does first last? The existence and extent of first mover advantages on spatial networks', *Journal of Transport and Land Use*, **4** (2), 47–69.

Levinson, D.M., F. Xie and S. Zhu (2007), 'The co-evolution of land use and road networks', in R. Allsop, M. Bell and B. Heydecker (eds), *Transportation and Traffic Theory*, Bingley: Emerald, pp. 839–59.

Li, T., J. Wu, H. Sun and Z. Gao (2016), 'Integrated co-evolution model of land use and traffic network design', *Networks and Spatial Economics*, **16** (2), 579–603.

Li, Y., D. Lu and Y. Tian (2014), 'Modeling corridor and growth pole coevolution in regional transportation network', *Transportation Research Record: Journal of the Transportation Research Board*, **2466** (1), 144–52.

Liu, X., B. Derudder and C.G. García (2013), 'Exploring the co-evolution of the geographies of air transport aviation and corporate networks', *Journal of Transport Geography*, **30** (June), 26–36.

Mimeur, C., F. Queyroi, A. Banos and T. Thévenin (2017), 'Revisiting the structuring effect of transportation

infrastructure: an empirical approach with the French Railway Network from 1860 to 1910', *Historical Methods: A Journal of Quantitative and Interdisciplinary History*, **51** (2), 65–81.
Monstadt, J. (2009), 'Conceptualizing the political ecology of urban infrastructures: insights from technology and urban studies', *Environment and Planning A*, **41** (8), 1924–42.
Moreno, D., D. Badariotti and A. Banos (2012), 'Un automate cellulaire pour expérimenter les effets de la proximité dans le processus d'étalement urbain: le modèle Raumulus', *Cybergeo: European Journal of Geography*, doi:10.4000/cybergeo.25353.
Neal, Z. (2012), 'Structural determinism in the interlocking world city network', *Geographical Analysis*, **44** (2), 162–70.
Offner, J.-M. (1993), 'Les "effets structurants" du transport: mythe politique, mystification scientifique', *Espace géographique*, **22** (3), 233–42.
Offner, J.-M., F. Beaucire, M. Delaplace, A. Frémont, O. Ninot, A. Bretagnolle, et al. (2014), 'Les effets structurants des infrastructures de transport', *Espace Geographique*, **43** (1), 51–67.
Plassard, F. (1977), *Les autoroutes et le développement regional*, Lyons: Presses Universitaires de Lyon.
Pumain, D. (1997), 'Pour une théorie évolutive des villes', *L'Espace géographique*, **26** (2), 119–34.
Pumain, D. (2008), 'The socio-spatial dynamics of systems of cities and innovation processes: a multi-level model', in S. Albeverio, D. Andrey, P. Giordano and A. Vancheri (eds), *The Dynamics of Complex Urban Systems*, Mendrisio: Springer/Verlag-Physics, pp. 373–89.
Pumain, D. (2010), 'Une théorie géographique des villes', *Bulletin de la Société géographie de Liège*, **55** (2), 5–15.
Pumain, D. (2014), 'Les effets structurants ou les raccourcis de l'explication géographique', *Espace géographique*, **43** (1), 65–7.
Pumain, D. and B. Riandey (1986), 'Le fichier de l'ined', *Espace, populations, sociétés*, **4** (2), 269–77.
Raimbault, J. (2017a), 'An applied knowledge framework to study complex systems', in A. Chapoutout, D. Krob, A. Roussel and F. Stephan (eds), *Proceedings of the 8th International Conference on Complex Systems Design & Management*, Paris: CESAMES, pp. 31–45.
Raimbault, J. (2017b), 'Identification de causalités dans des données spatio-temporelles', in G. Del Mondo (ed.), *Spatial Analysis and GEOmatics 2017*, Rouen: Université de Rouen, pp. 244–59.
Raimbault, J. (2017c), 'Models coupling urban growth and transportation network growth: an algorithmic systematic review approach', *Plurimondi*, **17**.
Raimbault, J. (2018a), 'Caractérisation et modélisation de la co-évolution des réseaux de transport et des territoires', PhD thesis, Univeristé Paris 7 Denis Diderot.
Raimbault, J. (2018b), 'Multi-modeling the morphogenesis of transportation networks', in T. Ikegami, N. Virgo, O. Witkowski, M. Oka, R. Suzuki and H. Iizuka (eds), *Artificial Life Conference Proceedings*, Cambridge, MA: MIT Press, pp. 382–3.
Raimbault, J. (2019), 'An urban morphogenesis model capturing interactions between networks and territories', in L. D'acci (ed.), *The Mathematics of Urban Morphology*, Cham: Birkhäuser, pp. 383–409.
Raimbault, J. (2020a), 'Indirect evidence of network effects in a system of cities', *Environment and Planning B: Urban Analytics and City Science*, **47** (1), 138–55.
Raimbault, J. (2020b), 'Unveiling co-evolutionary patterns in systems of cities: a systematic exploration of the SimpopNet model, in D. Pumain (ed.), *Theories and Models of Urbanization*, Cham: Springer, pp. 261–78.
Raimbault, J., A. Banos and R. Doursat (2014), 'A hybrid network/grid model of urban morphogenesis and optimization', in M.A. Aziz-Alaoui, C. Bertelle, X.Z. Liu and D. Olivier (eds), *4th International Conference on Complex Systems and Applications*, Le Havre: Université de Normandie, pp. 51–60.
Reuillon, R., M. Leclaire and S. Rey-Coyrehourcq (2013), 'Openmole, a workflow engine specifically tailored for the distributed exploration of simulation models', *Future Generation Computer Systems*, **29** (8), 1981–90.
Rui, Y. and Y. Ban (2011), 'Urban growth modeling with road network expansion and land use development', in A. Ruas (ed.), *Advances in Cartography and GIScience*, vol. 2, Berlin: Springer, pp. 399–412.
Russo, F. and G. Musolino (2012), 'A unifying modelling framework to simulate the spatial economic transport interaction process at urban and national scales', *Journal of Transport Geography*, **24** (September), 189–97.
Sanders, L. (1992), Système de villes et synergétique, Paris: Anthropos-Economica.
Sanders, L. (2017), '*Peupler la terre – De la préhistoire à l'ère des métropoles*, Tours: Presses Universitaires Francois Rabelais.
Schamp, E.W. (2010), 'On the notion of co-evolution in economic geography', in R. Boschma and R. Martin (eds), *The Handbook of Evolutionary Economic Geography*, Cheltenham, UK and Northampton, MA, USA: Edward Elgar, pp. 432–50.
Schmitt, C. (2014), 'Modélisation de la dynamique des systèmes de peuplement: de SimpopLocal à SimpopNet', PhD thesis, Université Paris 1.
Tero, A., R. Kobayashi and T. Nakagaki (2007), 'A mathematical model for adaptive transport network in path finding by true slime mold', *Journal of Theoretical Biology*, **244** (4), 553–64.
Tero, A., S. Takagi, T. Saigusa, K. Ito, D.P. Bebber, M.D. Fricker, et al. (2010), 'Rules for biologically inspired adaptive network design', *Science*, **327** (5964), 439–42.

Thévenin, T., R. Schwartz and L. Sapet (2013), 'Mapping the distortions in time and space: the French railway network 1830–1930', *Historical Methods: A Journal of Quantitative and Interdisciplinary History*, **46** (3), 134–43.
Wal, A.L.J.T. and R. Boschma (2011), 'Co-evolution of firms, industries and networks in space', *Regional Studies*, **45** (7), 919–33.
Wegener, M. and F. Fürst (2004), 'Land-use transport interaction: state of the art', SSRN Paper No. 1434678, accessed May 2019 at https://papers.ssrn.com/sol3/papers.cfm?abstract_id=1434678.
Wu, J., R. Li, R. Ding, T. Li and H. Sun (2017), 'City expansion model based on population diffusion and road growth', *Applied Mathematical Modelling*, **43** (March), 1–14.
Xie, F. and D. Levinson (2009), 'Modeling the growth of transportation networks: a comprehensive review', *Networks and Spatial Economics*, **9** (3), 291–307.
Zembri, P. (1997), 'Les fondements de la remise en cause du schéma directeur des liaisons ferroviaires à grande vitesse: des faiblesses avant tout structurelles', *Annales de géographie*, **593** (January), 183–94.

9. Ties through place: socio-material network analyses in urban studies

Meg Bartholomew and Alasdair Jones

1. INTRODUCTION

In the interdisciplinary field of urban studies, the relevance of space and geography to social bonds is often assumed; our physical presence in the world informing a presumption of material sensitivity. However, the importance of urban space and morphology to social interactions has come into question with the rising ubiquity of communications technologies (Woods 2000). Concomitantly, rapid urbanisation, media consumption and rising mobility have helped to foster urban conditions in which hitherto clear identity markers, such as neighbourhood or ethnicity, have begun to have less bearing on how and where any individual self-identifies as belonging (Morley 2000). This chapter investigates these interactions between physical materiality and social connection in the formation of community in the contemporary 'urban age' (Burdett and Rode 2018). The chapter reports the socio-material subset of the findings of the Measuring Community in an Urban Age (MCiUA) study funded by LSE Cities.

In that study we wanted to identify recent social scientific studies from the fields of sociology, urban studies, geography, psychology and computational analysis that have employed an understanding of social and or infrastructural networks to understand issues related to community and belonging. To do this, we undertook a quasi-systematic literature review at a heuristic level, methodically assessing relevant texts published from 2000 onwards. In this review, relevance of review materials took precedence over an exhaustively comprehensive assembly (Booth et al. 2016). Two databases were used for the search, LSE Library Database and Google Scholar. These databases are considered to have extensive collections as well as effective algorithms for returning relevant responses, and use of both is recommended for broad coverage (Brophy and Bawden 2005). In addition to the database searches, key references identified manually from previous research in related fields were used as material for the study, for generators of key words and for further citation links. The review used key search terms drawn from both the urban (neighbourhood, community, urban, city, spatial*, embed* and place) and network studies (social network*, 'social network', 'social network analysis' and ties) literatures in a variety of combinations.[1] Owing to funding limitations and practical research constraints, only texts returned from these searches written in English were included in the study.

The titles and abstracts of the top 100 results of each search were reviewed and relevant titles were inputted into a database. These search engines rank results by key words in the title, publication, author and number of citations (Google 2018). This may have limited the results to mainstream journals and clearly named articles, evidenced by a number of relevant writings found subsequently through citations from these sources; however, these were not included in the review owing to the constraints of the study. Titles included in the database were then read in more detail and codified by network focus and methodology. Seventy-one texts were added to our database and, of these, 60 were deemed relevant to the research goals of the study. Of these remaining studies, 30 were explicitly socio-material in methodology and conceptualisation, and it is these studies (and a synthesis of their premises, methods and findings) that form the basis of this chapter.

The chapter is divided into four sections: definitions and interpretations of socio-materiality and of community from a historical and a network perspective; a thematic synthesis of the findings; a quantitative analysis investigating theoretical groupings of this literature; and a discussion on methodologies used by the studies reviewed. The purpose of the chapter is to: synthesise extant approaches for understanding community from a socio-material perspective; interrogate the relevance of linking social to spatial investigation; and identify how social and spatial data, concepts and measures are practically linked in these studies. As an externality, we also report gaps and blind spots in the literature that we identified as indicating further areas for investigation or potentially fruitful collaborations in socio-material studies of community.

2. DEFINITIONS

Socio-materiality is a term coined to describe the interdependencies between human action or behaviour and physical objects or environments, where either is seen as equally shaping or as shaped by the other (Jones 2013). The term is often used in reference to technology (Orlikowski 2007), however, in this chapter it is used in a broader sense to encompass all relationships between the social and material world. This social-environmental entanglement, in spatially orientated studies, has also been termed double-embeddedness, where a relationship between two people 'may be embedded in a local structure of other relationships, in turn embedded in geographic space' (Habinek et al. 2015, p.27). The term double-embeddedness was originally used to contrast the structural and cultural determinants of human action (Baker and Faulkner 2009) without discounting other relational forms of influence. To avoid confusion, socio-material embeddedness is used in this chapter to encompass both understandings.

Community is a heavily critiqued (Day 2006) yet resolutely enduring concept. It is a term that conveys a sense of solidarity with others without being specific about the characteristics of that implied closeness, and it is used loosely in common language and much debated in academic use. Recognised in the social sciences as referring 'to those things which people have in common, which bind them together, and give them a sense of belonging with one another' (Day 2006, p. 1), it is the substance of those ties that is the contested component. As Blokland-Potters (2017, p. 41) puts it, '[t]o understand how we imagine communities, we should ask what sort of relations can convey symbols that we may identify with and that help us place ourselves socially'; we need to interrogate the basis under which an individual can be said to belong to or identify with a particular community. Using a network approach enables an investigation of belonging based on the form of the relationship between two people, where a tie might represent a social relationship or a spatial relationship – two people who are linked by friendship or by living in the same neighbourhood, for example. A socio-material network approach overlays the social and the spatial ties to consider their interaction(s) and combined effect(s).

Increased attention to the concept of community evolved out of the rapid transition from rural to urban society during the Industrial Revolution (Delanty 2010). Tönnies (1940) separated the idea of community from the concept of society where community became associated with strong ties of the family and close social relations and society was seen to be governed by more transactional, impersonal interactions. These have been loosely interpreted as strong and weak ties in social network analysis (Granovetter 1983). Subsequent interpretations of Tönnies's work, including writing by Simmel, applied a negative connotation to weak societal ties (Delanty 2010), idolising rural life as the quintessential form of community instead of the more loose connections of urban society (Simmel 1903 [1971]).

The Chicago School continued in this vein of seeking to transpose the allegedly superior communication and obligation from rural communities as a means to foster a virtuous 'moral order' in urban environments (Ballis Lal 2007). These urban–rural, society–community, weak tie–strong tie dichotomies were challenged by other theorists who saw the city as 'a community composed of an aggregation of smaller communities, such as families, professional groups, etc.' (Maunier 1910, p. 543). More recently, the ways these aggregations of strong and weak ties manifest particularly in cities and the role they play, including information exchange, job opportunities, social cohesion and civic participation (Granovetter 1973; Hipp and Perrin 2006; Kavanaugh et al. 2005), have been revisited.

This broad perspective on the role of both strong and weak ties in the definition and formation of community was the approach taken for the wider MCiUA literature review, and eight broad but overlapping typologies of relations were discerned. Four of these were social typologies (which we term eponymous relationships, familiar strangers, institutional relationships and avocational relationships) and four were spatial typologies (neighbourhoods, third places, activity spaces and online spaces). These typologies are set out in Figure 9.1.

Turning first to the social typologies, eponymous ties are those held with other people known by name, for example, the 'alters' (other people named by the ego), who would be listed in a name-generator survey of acquaintances. These ties correspond to what Delanty (2010) characterises as the post-Industrial Revolution definition of community. Familiar strangers are people who are encountered regularly in daily life and who are not known by name but are recognisable faces; those who provide important interactions that have been linked to concepts of cognitive neighbourhood (Blokland-Potters 2017) and

Social dimension

Eponymous relationships – relationships on a first name basis
Familiar strangers – regularly encountered strangers in everyday life
Institutional ties – people known through institutions, formally and informally
Avocational associations – others with shared interests

Spatial dimension

Neighbourhood – ties based on living in the same place
Third places – people known from regular favourite locations
Activity spaces – people encountered through routine everyday mobility
Online relationship – ties built with others through online interactions

Figure 9.1 Typologies of ties that form community

personal identity formation (Fingerman 2009). Examples are the local bus driver or shop attendant and those with a shared sense of ethnic identity or other distinct characteristics of homophily. Institutional relationships are formed in institutions, such as schools, local councils, universities, health centres or other service providers that form important links in everyday life. Finally, avocational relationships are ties made through hobbies and recreational pursuits. These common interests are themselves one of the two prominent defining characteristics (alongside common place of residence) of 'community' in dictionary definitions of the term (Oxford 2012).

In studies that take a more spatial approach to community, relational typologies span neighbourhood, third place, activity space or online ties. Neighbourhood is based on relationships established with those living around the same area. Third places is a concept popularised by Ray Oldenburg (1999) identifying the places other than home and work that are important in people's lives. These include places such as parks, cafés and community centres, but may be as seemingly benign as a bench on a street. They are generally places that involve social interaction but can also be places to go to be publicly alone. Activity space is a concept used widely in planning and economics as mapping the geographic area covered through daily activity (Chapin 1968). It is an area that has grown with the availability of large data sets that track everyday movements, such as mobile phone data, but can also be elicited from surveys or diaries (Patterson and Farber 2015). The last spatially derived typology is the idea of online communities, where the space of interaction is the Internet rather than a physical space, although even in these studies some assortativity was found along cultural and geographic lines (Leskovec and Horvitz 2008; Spiro et al. 2016).

These typologies are not mutually exclusive and a given tie may be of more than one type, such as where a neighbour becomes a close friend, or a third place is the site of an avocational interest. Familiar strangers and activity spaces are the most similar typologies and distinguished only through research approach; namely, whether the focus of a given study is individual mobility (activity space) or frequent interactions (familiar strangers). Some theories suggest that each of these types of relationships informs a multi-faceted sense of identity that underpins our perception of belonging (Fingerman 2009; Young 1990). This is especially the case in studies where '[t]he city is not only viewed as an aggregation of multiple networks, but also as an interconnection node between networks' (Pflieger and Rozenblat 2010, p. 2725).

In our review, overlaps between relational typologies were frequently across the social/spatial distinction (see Figure 9.2), highlighting the relevance of purposively socio-material investigations, and reinforcing argument that 'places and networks should no longer be considered as separate interactive processes, but rather as forms of spatial organisation that are intrinsically co-produced' (Pflieger and Rozenblat 2010, p. 2725). Eponymous relationships are often associated with activity spaces, the friends and family seen at home, work, after work and on the weekends, where 33 per cent of all individual travel can be related to leisure (Frei and Axhausen 2011). Institutional ties can often, but not necessarily, be found in neighbourhoods, schools, local health-care facilities, local councils, and so on. Avocational ties are often formed in third places, through recreational activities at a local park or socialising at a gig, and even online hobbies may have physical meeting places, as when political movements coalesce online and then form physical demonstrations (McCaughey and Ayers 2003). The familiar stranger concept is closely

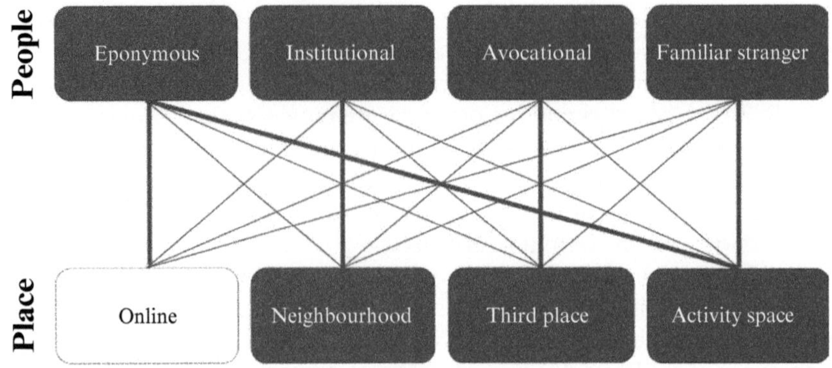

Figure 9.2 Diagram of socio-material overlaps in community formation

related to activity spaces. In practice, any one of these categories can overlay with any other, as demonstrated by the lighter lines in Figure 9.2, however, some are more closely related than others, as discussed previously and as shown by the thicker lines in Figure 9.2.

3. THEMATIC SYNTHESIS OF FINDINGS FROM THE REVIEW

Through our research we identified three types of study that seek an overtly socio-spatial understanding of community. The first are studies using big-data sets that interrogated combined social and spatial effects, the second used methods that gathered information about both place and people, and the third is the literature on neighbourhoods – by definition socio-spatial.

3.1 Studies Using Spatially Embedded Large Data Sets

A topical place to start is with studies that use large data sets. These data sets lend themselves to the analysis of socio-spatial information as a great deal of big data includes geolocation information (Tsou 2015). Big data refers to large volumes of frequently produced variable data in text, audio, video, location and social media format (Gandomi and Haider 2015). Some believe that gathering big data, such as mobile phone data, will render other forms of data collection, for example, census approaches, obsolete (ONS 2019). In our literature review there were six big-data studies that contributed empirically to a relational understanding of community.

A very interesting application of big-data analysis is in the study of activity spaces and familiar strangers, one of the hardest phenomena to capture using other methodological techniques. Recent studies using mobile phone data have found that 'friends are more likely to share urban space than random user pairs, and they also share more locations' (Xu et al. 2017, p. 468), implying that subsets of the population with probably similar attributes (McPherson et al. 2001) co-locate in unique parts of the city, potentially linked to enclaves and other geographic segregations. The regularity of repeated encounters with others was further investigated by Sun et al. (2013), who found a strong tendency for repeated spatial co-presence in their study, with three-quarters of people coming upon the same other people on a regular basis. For most people in their study sample the probability of co-presence was above random, particularly for a small group of other people encountered with high regularity. These repeated encounters were most likely in the morning and are related to the regularity of routine of any individual. Moreover, the network of encounters generated have strong small-world properties with high clustering and low path length (Sun et al. 2013) suggesting a person has multiple presences in clusters of synchronised people, spatially and temporally linking co-present unconscious groups of familiar strangers across the day across the city.

This more than random grouping of individuals in parts of a city corresponding to their daily activity spaces may explain why another study found that it is just as likely that your friends will know each other, and collectively form a friendship group, in a small town as it is in a city (Schlapfer et al. 2014). These authors found that as city size increases, a single person is more likely to know more people, however, the probability that the people they know also know each other remains the same, independent of overall population size. These studies imply that people from towns and cities form similarly composed friendship clusters and interact with a particular group of familiar strangers linked to the places where they spend the most time. It may be that this gives them a sense of belonging, however, these large data sets have no qualitative component to explore this potential mechanism.

From a places perspective, studies combining mobility from mobile phone data with interests from accessed websites or social media show that the frequency of meeting others with similar interests increases with the density of locations serving that interest in a given area (Trestian et al. 2009). This density of points of interest also correlates with higher visitation to those locations (Xu et al. 2017); in addition research has indicated that the physical connectivity of a place (how many other places or streets connect to it) could be related to its popularity (Agryzkov et al. 2017). These studies point to the importance of an agglomeration of activity in central locations with connected street permeability to building a sense of community. An interesting finding generated through the analysis of big data is that the places chosen to meet up on weekdays differ depending on whether the meeting is between close friends or weak ties, but on the weekend there is no differentiation, and that certain places, such as shopping malls, have higher bonding and bridging capacity as a consequence (Xu et al. 2017). This has important implications for urban transportation programming. The final big-data study added an online spatial dimension, finding that the online presence of businesses and notice boards (or online networks of third places in an urban neighbourhood) contributed to a sense of place and community for residents of that neighbourhood (Bingham-Hall and Law 2015).

3.2 Studies Combining Discrete Sets of Social and Spatial Data

The studies that collected data both on people and place used a variety of techniques, including sensors, surveys, observations, interviews, public open space (POS) surveys, trackers, mapping and place- or name-generator surveys. They mixed these data collection techniques in order to analyse associations between people and place. These studies can be grouped by a focus on the importance of distance in eponymous relationships (including the role of face-to-face and other modes of communication) and on the types of place that are important to community and well-being.

Moore et al. (2013, p. 774) used a survey that collected information on demographics, social networks and time use to explore 'the intertwined relationship between the individuals' social networks, and the temporal and spatial characteristics of their daily social activities'. They found that duration of activities is affected by the distance of travel to the activity as well as the number of ties participating in the event, with access to the central business district (CBD) and network density having positive relationships with the number of people involved. If community is a type of social pattern characterised by repeated patterns of interaction (Neal 2013), this indicates the societal benefits of creating cities with accessible transport and for place-based planning that provides localised services and activities, and reduces the need for long-distance travel. Half of an individual's strong ties generally live within 25 kilometres of that individual (probably the same city). Similarly, socialising frequency is distance dependent, with local strong ties socialising with one another every other day or so, those living up to 30 kilometres away every fourth day or so and those living 50 kilometres away almost once a week. It is worth noting that ties living 400 kilometres or further away only meet as often as with those who live 125 kilometres away (Larsen et al. 2006). One study using a mixture of sensors and surveys capturing details on relationships demonstrated that closeness in relationships is directly linked to length of time spent in face-to-face social interactions (Oloritun et al. 2013), and even with advanced telecommunications, most telephone calls and text messages are brief, local and concerned with coordinating meetings, while email is used more for ties further away (Larsen et al. 2006). So removing blockages, such as time and distance, for face-to-face interaction would appear to be critical for strong community formation.

The studies combining social and spatial data that we reviewed, focused on places considered to be the types of spaces that are important for community formation. Among these studies, one of the most extensive was a qualitative study of an area starting with discussion groups with residents to gather a sense of important people and places in their local area, followed by site observations and in-depth interviews, including place- and name-generation techniques (Cattell et al. 2008). The authors of this study found that 'public spaces are more than just simply containers of human activity, they possess subjective meanings that accumulate over time' (Catell et al. 2008, p. 556) and that it is a diverse range of, often mundane, places that are important to people and impact their well-being. Mundanity appeared to be an important factor as all places evoked a sense of comfort where the unique attractors were activity focused rather than aesthetically attractive, an important lesson for urban design. The shared and collective use of public spaces was important as the 'routine encounters with familiar others in familiar places can bestow "feel good" effects, and help maintain a sense of the reliability of persons and things, but unexpected encounters – in places which attract strangers as well as locals – are often experienced as especially cheering' (Cattell et al. 2008, p. 557).

Also, having spaces where people can connect with the ethnicity-based networks with which they identify was important for a sense of pride and cultural identity, as well as providing places for people to be reflective though not necessarily in solitude. Regularity of use, longevity of establishment and potential to dwell without reason are all constitutive factors in the importance of place to individuals in these authors' findings. Specific places were identified as especially important in this respect, including parks and green spaces, shopping areas (including markets) and the forecourts of residential spaces.

Francis et al. (2012) also find that shops are important meeting and socialising spaces for building a sense of community, as they provide 'opportunities for proximity to others, passive social contact, and casual interactions' (Francis et al. 2012, p. 406). Using a mix of survey, POS audit and geographical information system (GIS) data Francis et al. found that the frequency of use of open spaces, shops, community centres and schools does not necessarily correlate with their reported importance, and many places that are only visited occasionally remain important for community identity. In addition, 'involvement in more neighbouring activities (comprised of Residents' Associations, street parties, community action groups, welcome evenings and mothers' groups) and school or work activities was significantly and positively associated with sense of community' (Francis et al. 2012, p. 407). Hickman (2013) also found that shops, in particular, are important places of social interaction and symbolise the health of a neighbourhood together with other third places, including libraries, cafés, pubs, libraries, local parks, play areas, churches and mosques.

The mundanity of spaces for fostering connection is also identified by Simões Aelbrecht (2016) who observed interactions in a popular public space and conducted ethnographic interviews to understand more about the social interactions between strangers. It was found that it was the intermediate and leftover spaces and 'in-between activities such as waiting and queuing, which established favourable conditions for 'time out' and created the spatial opportunities for social interaction with strangers to occur' (Simões Aelbrecht 2016, p. 142). The more intersections, between desire lines across spaces and between the different activities, the more likely it was that unplanned encounters would occur. These encounters have been suggested to have important implications for both identity formation (Fingerman 2009) and social cohesion (Hipp and Perrin 2006).

Finally, Nast and Blokland (2014) suggest that instead of relying on random occurrences to build the connections necessary for a sense of community, it is the interactions that occur in institutional settings, such as schools, that lead to greater tie generation. They specifically look at mixing across class and ethnic divides in a school in Germany and find that setting specific similarities, such as parent's mutual engagement with the school, can lead to the breaking down of social boundaries that exist outside the school, and potentially to a more inclusive community.

3.3 Neighbourhood Studies

Neighbouring is one of the most heavily studied areas of community, with over a quarter of the texts reviewed in MCiUA addressing this phenomenon. The association of a physical cognitive neighbourhood within a walkable boundary around the home (Talen and Koschinsky 2013) as a source of community, and the question of the loss of a sense of community as a function of changing spatial scales and physical form of neighbourhoods, goes back to the modern interpretation of the rural village as the default definition of community (Wirth 1938), with people tied together by place. Whether villages really provided the supportive, cohesive environment that is implied by community, or instead enforced restrictive conformity, is debatable (Young 1990); nevertheless a sense of belonging to the place where we live, and the other people that live there, continues to be a salient social force (for example, Small 2012).

Discussions of neighbourhood in the literature can be broadly grouped into those concerned with propinquity and those concerned with demographic features. The former deal with whether distance is still a factor in the formation of direct relationships, an area touched upon already in our discussion of eponymous relationships. Studies concerned with demographics, by contrast, are interested in who people make relationships with, and deal with questions of homogeneity and social segregation.

In the literature we reviewed, moving or living close to others was found to positively impact relationships in at least seven studies (Butts et al. 2011; Daraganova et al. 2012; Drouhot 2017; Habinek et al. 2015; Hipp and Perrin 2009; Kebede and Butterfield 2009; Völker and Flap 2007). Most of these studies were undertaken in developed economies, where results were consistent across a variety of settings (Drouhot 2017). A single study was conducted in an emerging economy (Ethiopia), and it likewise found propinquity to be relevant to interpersonal relationship strength (Kebede and Butterfield 2009). Two studies found that moving in to a neighbourhood affects the probability of initiating a relationship with any other individual in that neighbourhood (Daraganova et al. 2012; Habinek et al. 2015) and moving away negatively affects the likelihood of that relationship continuing to an even greater extent (Habinek et al. 2015). Network structure in a neighbourhood, aggregate mean degree, edge length and local clustering, as well as the probability of a tie between two random individuals, can be predicted well by properties such as the mean nearest neighbour distance (Butts et al. 2011).

Among these studies, Völker and Flap (2007) found that having one neighbour in a personal network increases the chance of having another; that is, neighbours come in packages. The same authors also found that half of their respondents did not have any neighbourhood ties, impacted most by either having more friends outside the area or living within a large family household (Völker and Flap 2007). Ryley and Zanni (2013) also found that over 40 per cent of their sample had no neighbours in their social network. One study on neighbourhood attachment found it is dependent on having both strong and weak ties close by, and also found wider social cohesion is a positive affect associated more with weak ties over strong. As the authors put it:

> Strong ties have a particularly strong positive effect on perceived cohesion with the local neighbourhood, but have virtually no effect on cohesion with the larger community. On the other hand . . . weak ties not only increased perceived cohesion with the local neighbourhood, they also increased perceived cohesion with the larger community. This finding is consistent with structural network arguments and bridging social capital theories that the minimal time investment of such ties allows more time for creating numerous linkages. And these numerous linkages then seem important for fostering cohesion across various geographical contexts. (Hipp and Perrin, 2006, p. 2518)

Other research has concentrated on the demographics of neighbour–neighbour relationships on the basis that communities need a 'spatial or demographic anchor around which relationships and social capital can coalesce' (Neal 2015, p. 376). The findings of these studies are broadly that longevity of both the physical neighbourhood environment (Guest et al. 2006) and length of residence in it (Guest et al. 2006; Völker et al. 2007) has positive effects in creating neighbour relationships. Older people (Völker and Flap 2007; Wissink and Hazelzet 2012), people with children (Frei and Axhausen 2007; Hipp and Perrin 2009; Wissink and Hazelzet 2012), and those with fewer resources (Völker et al. 2007) tend to have more neighbours in their networks than others in the sample. Socio-economic homogeneity is linked to denser neighbour networks (Völker et al. 2007) and even within neighbourhoods that are socio-economically mixed, networks tend to be stratified along these lines (Blokland and Van Eijk 2010; Hipp and Perrin 2009; Wissink and Hazelzet 2012).

Varying impacts have been found for ethnic and cultural diversity or urbanisation in the strength of neighbourhood ties from negligible effect, with no significant correlations found (Hipp and Perrin 2009; Völker et al. 2007) to tending to discourage ties (Blokland and Van Eijk 2010), although this may depend on spatial segregation effects (Laurence 2017; Neal 2015). One study found that people with higher levels of education had more neighbours in their network (Völker and Flap 2007), although this contradicts findings that higher education leads to more dispersed networks (Völker and Flap 2007) and less local involvement (Drouhot 2017). These demographic effects have important implications for urban planning policy, especially as regards policy responses to processes of gentrification (Paton 2014). The bridging situations under which increased social capital and associated benefits of access to opportunity, trust and cooperation is transferred within neighbourhoods was investigated by a study in Los Angeles which showed that regular encounters in multiple locations was a key factor in generating collective efficacy, connection between generations and social exchange fostering generalised trust in the community (Browning et al. 2017). Creating places for regular encounters, planned or spontaneous, that bridge demographic divides appears to have potential in helping to foster a strong sense of community.

The findings from the socio-material studies reviewed show that 'spatial arrangements of neighbourhoods (and at a larger scale also cities) may still play a significant role in the formation of local communities and neighbourly interactions' (Mahmoudi Farahani 2016, p. 362). This implies important issues for consideration in urban planning and design practice when it comes to developing approaches orientated towards the creation of neighbourhoods and cities that foster a sense of community. The studies show that a sense of community can exist in a variety of personal interactions, from close relationships to familiar strangers. These different interactions appear to play different roles in the formation of personal identity, sense of well-being, inclusion and social cohesion. Limited definitions of community that restrict analysis to only eponymous ties may miss out on important weaker links, an issue discussed further in the final section of this chapter. The next section, which traces academic disciplines and practices, aims to identify other potential blind spots in the selected literature, which we argue ought to be addressed in order to strengthen the application of socio-material methods and perspectives across different fields.

4. QUANTITATIVE ANALYSIS OF CITATION SOURCING

The thematic contours of the literature review set out in the preceding section were corroborated with an analysis using R of all the links to common citations between the authors (see Figure 9.3). The analysis also sought to identify referencing insularity between different schools of thought in the same vein as Peris et al.'s (2018) study of urban systems literature. A directed edge list (Kolaczyk and Csárdi 2014) data set was created tying the main author to every author in every citation in the list of references. Authors who were only referred to by one main author (they had an in-degree of only one and no out degree) were deleted to obtain a core group of inter-referring authors. A community

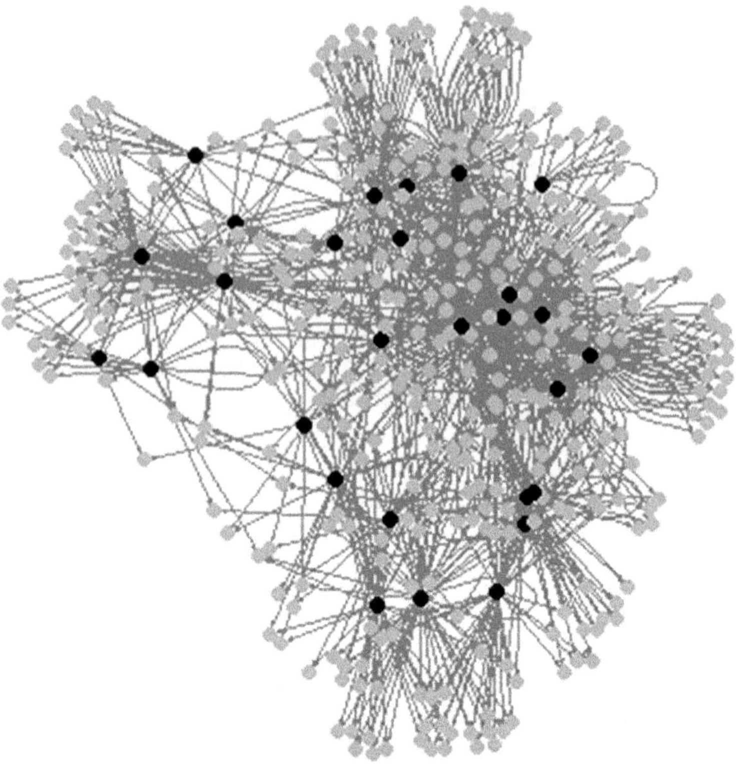

Note: Authors from the set of reviewed texts are in black.

Figure 9.3 Author citation network

detection algorithm[2] was used on this core data set to determine whether there were any heavily connected groups of authors that cross-reference both each other and the same source material. A diagram of these communities is provided in Figure 9.4.

The dendogram (Figure 9.5) shows the same information as the community detection network diagram (Figure 9.4) visually grouping the communities under branches. As can be seen in the dendogram, the algorithm grouped the citation links into three main groups, with subgroups based on cross-referencing within the publications. The first group (1) is of scholars using computational analysis of large data sets, corresponding almost completely to the big-data group identified above. The second group (2) is made up of two subgroups with a built environment (2A) or urban sociology (2B) focus. This general common focus for each subgroup was determined by finding synergies between the most commonly cited authors within the subgroup, from investigating their university profile to see how they self-defined their own area of interest. This second group contained most of the people-place mixed-methods papers. The third grouping (3) has a community development subgroup (3A) – small and slightly removed, and the core subgroups of behavioural science (3B), mathematical sociology (3C) and social network analysis (3D), again as defined by commonly cited author's own, or their university department's,

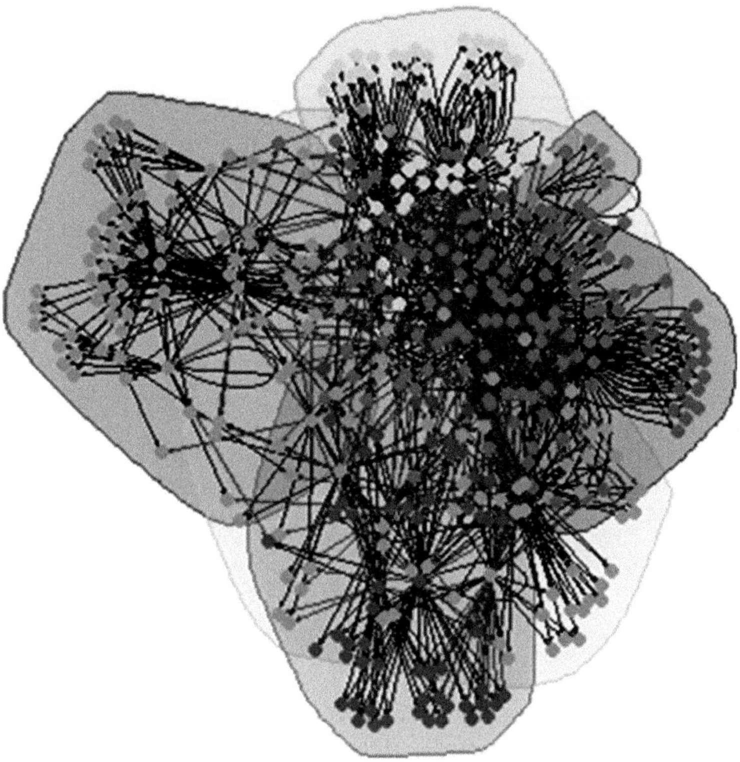

Figure 9.4 Plot of community detection of citation links between authors

profile, which dominated the neighbourhood studies. The community detection indicates that these subgroups generally draw on the same subsets of literature and cross-reference work within the same sub-field which is likely to limit the breadth of perspective that can be reached.

Methodological distinctions can also be made between the groups of studies identified in Figures 9.4 and 9.5. Specifically, groups 1 and 3 use quantitative methods and group 2 uses predominantly qualitative analysis. Arguably, all of the group's studies could be improved through the use of more mixed-methods research approaches (Domínguez and Hollstein 2014). Combining qualitative understandings of, for example, the intentions behind forming and maintaining particular kinds of relationships (as per methods deployed by founding network analysts, Jones 2018, and as advocated by Small 2017) with more numerical methods, could lead to greater perspective on applicability of the qualitative findings. In addition, qualitative interrogation can be used as a means to unravel the biases that may be contained within mathematical models or inductive reasoning (Fischer 2003). In particular, the data analytics group (1) are dependent on new types of data that have become available, and collaborating with qualitative researchers may allow for a productive invention of new techniques to analyse information and imagine what can be learned from it (Mohr and Ghaziani 2014). This mixed-methods approach may also result in a weakening of the discipline-specific divisions evident in the citation network analysis.

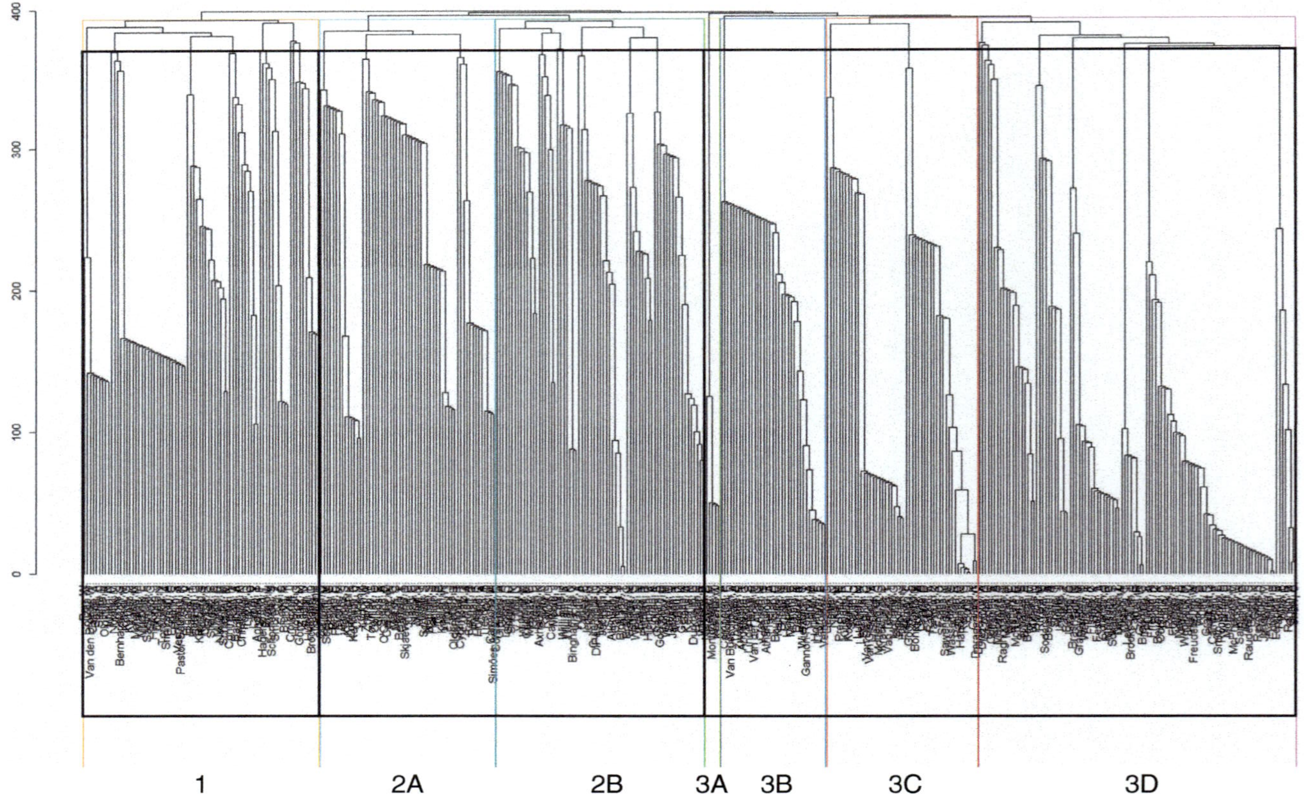

Figure 9.5 Dendogram of citation analysis community detection

In the qualitative papers we reviewed, the most cited authors can be grouped according to three substantive themes; Talja Blokland, George Galster, Ash Amin and Tim Butler working on urban inequality, discrimination and diversity; Pierre Bourdieu, Alejandro Portes, James Coleman and Robert Putnam working on social capital; and Barry Wellman working on community, cyberspace and conceptions of urban networks. By contrast, the quantitative studies draw more heavily from social network theory, in particular, with Barry Wellman as the most influential author but also Mark Granovetter, Claude Fischer, Katherine Faust, Ronald Burt, John Hipp and Tom Snijders featuring prominently. Robert Putnam is by far the most cited author on social capital for the quantitative group. Two texts, 'Birds of feather: homophily in social networks' (McPherson et al., 2001) and 'Neighborhoods and violent crime: a multilevel study of collective efficacy' (Sampson et al. 1997) are among the most frequently cited papers for the quantitative-based studies. It appears both qualitative and quantitative approaches have the joint aim of understanding urban forces with the potential to shape policy, but despite this shared objective, cross-referencing of works produced in the contrasting methodological paradigm is notably limited.

5. METHODOLOGIES OF SOCIO-MATERIAL DATA COLLECTION AND ANALYSIS

Investigating how social and spatial phenomena overlap often implies complex, multi-level data collection and analysis methods. Data capturing personal mobility through space has been typically more difficult to capture and has only recently become more accessible with new data sources and methodologies leading to a revival of interest in spatial analysis (Baur et al. 2014). As the findings from the literature attest, the resulting insights can be powerful. This section offers an overview of the data collection techniques and analysis methods undertaken by the socio-material studies reviewed in the preceding sections of this chapter.

5.1 Data Collection and Sources

In the studies concerned with community as a socio-spatial concept reviewed for MCiUA, the most popular data collection instruments were interviews (Bingham-Hall and Law 2015; Blokland and Van Eijk 2010; Cattell et al. 2008; Daraganova et al. 2012; Habinek et al. 2015; Hickman 2013; Hipp et al. 2011; Kebede and Butterfield 2009; Larsen et al. 2006; Moore et al. 2013; Nast and Blokland 2014; Völker and Flap 2007; Simões Aelbrecht 2016) and surveys (Browning et al. 2017; Daraganova et al. 2012; Drouhot 2017; Entwisle et al. 2007; Francis et al. 2012; Guest et al. 2006; Hipp and Perrin 2006, 2009; Moore et al. 2013; Oloritun et al. 2013; Wissink and Hazelzet 2012; Völker et al. 2007). Slightly less than half of the interview-based studies involved sample sizes of fewer than 100 people (Bingham-Hall and Law 2015; Cattell et al. 2008; Kebede and Butterfield 2009; Larsen et al. 2006; Nast and Blokland 2014), while studies employing larger samples, most within a range of 180 to 900 respondents (Blokland and Van Eijk 2010; Daraganova et al. 2012; Habinek et al. 2015; Hickman 2013; Moore et al. 2013; Völker and Flap 2007), often supplemented ego-generators with interviews. Two studies using interviews had an unreported sample size (Hipp et al. 2011; Simões Aelbrecht 2016). Surveys were also mostly in a range from approximately 100 up to 1000 respondents. Eight studies used name generators that asked respondents to list the names of people in their family or group of friends, sometimes where they performed a specific role, such as someone they turn to when they need advice or places that they spent time in or were important to them. Neal and Neal (2017) provide a detailed discussion on the complexities of composing name-generator questions in an editorial and show that the results can be severely biased by the structure of the question. As with any tools relying on recollection, interviews and surveys, with or without name-generator questions, can suffer from framing and response bias (Wilkinson 2018).

Addressing this bias is part of the appeal of using big-data sources as they theoretically capture naturalistic everyday behaviour recorded with digital devices, in particular mobile phone use and location data, but also social media, such as Twitter or other geo-tagged content, and smart cards (Liu et al. 2015). The data sets in this review have very high numbers (up to 10 million) of unique user data points (Agryzkov et al. 2017; Schlapfer et al. 2014; Sun et al. 2013; Trestian et al. 2009; Xu et al. 2017) and, given the scale of these data sets, the results are often interpreted as generalisable to wider populations. As well as complex privacy issues to manage when using big data (de Montjoye et al. 2013), large data sets require cleaning and skill in interpretation (Sivarajah et al. 2017), however, and an awareness of gaps and biases in the data (be they systematic or particular to a given data set) before results can be extrapolated as more general trends. For example, social media such as Twitter may exclude significant parts of the population (Bingham-Hall and Law 2015) or location triangulation from mobile phone towers only records activity when the phone is on or in use (Jiang et al. 2017). A key example of this, relating to the methodological concerns of ego name generators, is the difference in reported degree (people known by an individual) between studies using mobile, name generator and social media data. For instance, in one of the mobile phone network studies we reviewed, the degree value was found to be 8.5 people (Xu et al. 2017), in studies using ego name generators average degree values of 15 (Wicki et al. 2018) and 24 (Mok et al. 2010) were reported, while for an online social media network study (Hofstra et al. 2017) an average degree value of 336 was reported. To mitigate this measurement problem, one study, using fewer data points than those we referred to previously, used a mixed-method approach, which verified results from the data analytics by interview, which the researchers reported gave them increased certainty in their findings and provided additional insights (Bingham-Hall and Law 2015).

The final group of methodological tools employed in the studies reviewed for this chapter are place-based collection instruments, including field observations, mapping (including open space audits) and place-name generators. These all collect information in and about specific areas. Four studies used field observations (Agryzkov et al. 2017; Cattell et al. 2008; Nast and Blokland 2014; Simoes Aelbrecht 2016), a methodology championed by Gehl and Svarre (2013) and used widely in urban design fields. Five used some form of mapping including the use of GIS (Agryzkov et al. 2017; Cattell et al. 2008; Francis et al. 2012; Larsen et al. 2006; Simões Aelbrecht 2016), and one used the ego name-generator technique to ask respondents to name places that were important to them (Cattell et al. 2008).

Bridging these methodological paradigms, the studies that combined techniques across people-sourced, big-data-sourced and place-sourced collection tools were particularly insightful. Agryzkov et al. (2017) used big data and supplemented it with observations, Cattell et al. (2008) used observations and place-name generators with general interviews on the meaning of places to people, Larsen et al. (2006) combined interviews with mapping, and Nast and Blokland (2014) and Simoes Aelbrecht (2016) overlaid interviews with observation. Only two collected and overlaid both social networks and spatial networks rather than analysing them discretely. One study adopted a novel model that used sensors embedded in mobile phones with supplementary online surveys (Oloritun et al. 2013) and the other used POS audits, pedometers, surveys and GIS data to understand the role of public space in creating community (Francis et al. 2012).

Finally, only two studies used census data which may relate to the amount of detail and type of information required to undertake network studies. It may, however, also indicate other issues with the census, including mismatch between census areas and cognitive neighbourhoods or meaningful areas of data enquiry for definitions of community. As the evidence for an ecological approach to individual outcomes, such as the impact of social or infrastructural networks on sense of community and belonging, gains more traction, inclusion of some network information in public census-type collection instruments may warrant further consideration.

5.2 Comparison of Data Collection Instruments to Type of Analysis

A comparison of the type of data collected in the studies reviewed for this chapter with the methodology used to analyse that data in these studies was undertaken. The studies were grouped by those that used people-based data collection instruments (interview, survey and ego generator), big-data sets (mobile phone, social media, smart cards and other big-data sources), or place-based instruments (observation, mapping tools including GIS, public open space audits and sensors).

The comparison showed that qualitative analysis was lacking in the studies using people-based data collection instruments (these being studies primarily concerned with neighbourhood) and in big-data-based studies; as mentioned previously, only one study included interview verification of the findings from mobile phone data. With the consideration that sense of belonging and community is something more than merely the sum of social relationships, this additional layer of information and analysis can be valuable for drawing inferences and contextualising any patterns observed. Also, regression analysis is used virtually exclusively to analyse people-based data, perhaps because the structure of interviews and surveys is typically well-suited to this type of analysis where independent variables are compared with dependent variables. Place data, by contrast, often lacks clear independence of variables and is therefore more reliant on network analysis techniques. Statistical inference techniques for network data including exponential random graph modelling for static networks (Snijders et al. 2006) or stochastic actor-orientated models for dynamic networks (Snijders et al. 2010) are being further developed. A surprising result is the lack of use of GIS for place-based data in urban network studies and there is scope for research that integrates GIS and network-based understandings of urban community.

6. CONCLUSION

Over the course of human history, society has been intricately linked to space, culture was exchanged through physical movement and trade (Clark 2016), and community formed in close proximity (Delanty 2010). With increasing urbanisation during the nineteenth century, as well as the rise of mass migration and of online communication (Morley 2000), the spatial features of community have morphed. These processes have in turn underpinned calls to defend the relevance of material conditions to social interaction (Storper and Venables 2004). With a resurgent concern for spatiality during the 1990s (Baur et al. 2014), and access to new intrinsically geographic data sources in the twenty-first century (Tsou 2015), space (albeit with an added technological emphasis; Castells 2005) has once again become of interest. Despite these developments, it is still the case that 'almost nothing is known about the joint effects of network structure and geographic position' (Habinek et al. 2015, p. 28) on our social worlds. In response to this observation, this review has identified recent urban studies research that specifically combines social and spatial inquiry. Our preceding synthesis of this literature strongly indicates that space is profoundly instrumental in how we form, enact and maintain social relations in cities.

When it comes to studying community as a socio-spatial phenomenon, however, our citation analysis shows that studies are divided along disciplinary lines and, while they may draw from a few mutual sources, most literature referenced by each group is indicative of intellectual silos. Bridging these could undoubtedly lead to novel research perspectives. This is the same conclusion for methods of data collection and analysis where particular methods are preferred by particular groups or topics. This may be in part owing to the suitability of matching data and method, but there is scope for a greater emphasis on mixed-methods studies in this domain (both in relation to the collection and analysis of data; Small 2011).

Combined socio-spatial approaches to researching community allow for a clearer understanding of the combined effects of geography, the built environment and social networks on urban social life. As new data sets and methods become more accessible, socio-material analysis can become more widespread and interdisciplinary teams more common, entwining social with spatial analysis. To create impact from this cross-dimensional field of research it will be important to refer to a wide range of sources and perspectives, to seek out interdisciplinary teams and to encourage urban social network scholars to ask how social relations are formed in, enacted through and mediated by space and place.

NOTES

1. The asterisks indicate that these were truncated search terms in our literature review. Additional key words of 'well-being', 'belonging' and 'exclusion' were also added, which allowed us to explore a further interest in the application of socio-spatial approaches to researching community in relation to these particular substantive areas.
2. Community detection algorithms 'identify groups of nodes that are heavily connected among themselves, but sparsely connected to the rest of the network' (Yang et al. 2016, p. 1).

REFERENCES

Agryzkov, T., P. Martí, L. Tortosa and J. Vicent (2017), 'Measuring urban activities using Foursquare data and network analysis: a case study of Murcia (Spain)', *International Journal of Geographical Information Science*, **31** (1), 100–121.

Baker, W. and R.R Faulkner. (2009), 'Social capital, double embeddedness, and mechanisms of stability and change', *American Behavioral Scientist*, **52** (11), 1531–55.

Ballis Lal, B. (2007), 'Robert E. Park, Ernest W. Burgess, and urban social research', in G. Ritzer (ed.), *The Blackwell Encyclopedia of Sociology*, Oxford: Blackwell, doi:10.1002/9781405165518.wbeosr075.

Baur, N.L., L. Hering, A. Raschke and C. Thierbach (2014), 'Theory and methods in spatial analysis, towards integrating qualitative, quantitative and cartographic approaches in the social sciences and humanities', *Historical Social Research*, **39** (2), 7–50.

Bingham-Hall, J. and S. Law (2015), 'Connected or informed? Local Twitter networking in a London neighbourhood', *Big Data and Society*, **2** (2), 1–17.

Blokland, T. and G. Van Eijk (2010), 'Do people who like diversity practice diversity in neighbourhood life? Neighbourhood use and the social networks of "diversity-seekers" in a mixed neighbourhood in the Netherlands', *Journal of Ethnic and Migration Studies*, **36** (2), 313–32.

Blokland-Potters, T. (2017), *Community as Urban Practice*, Oxford: Polity Press.

Booth, A., A. Sutton and D. Papaioannou (2016), *Systematic Approaches to a Successful Literature Review*, 2nd edn, Los Angeles, CA: Sage.

Brophy, J. and D. Bawden (2005) 'Is Google enough? Comparison of an internet search engine with academic library resources', *Aslib Proceedings*, **57** (6), 498–512.

Browning, C., C. Calder, B. Soller, A. Jackson and J. Dirlam (2017), 'Ecological networks and neighborhood social organization', *American Journal of Sociology*, **122** (6), 1939–88.

Burdett, R. and P. Rode (eds) (2018), *Shaping Cities in an Urban Age*, London: Phaidon Press.

Butts, C.T., R.M. Acton, J.R. Hipp and N.N. Nagle (2011), 'Geographical variability and network structure', *Social Networks*, **34** (1), 82–100.

Castells, M. (2005), 'Spaces of flows, spaces of places: materials for a theory of urbanism in the information age', in B. Sanyal (ed.), *Comparative Planning Cultures*, New York: Routledge, pp. 45–64.

Cattell, V., N. Dines, W. Gesler and S. Curtis (2008), 'Mingling, observing, and lingering: everyday public spaces and their implications for well-being and social relations', *Health and Place*, **14** (3), 544–61.

Chapin, F. (1968), 'Activity systems and urban structure: a working schema', *Journal of the American Institute of Planners*, **34** (1), 11–18.

Clark, G. (2016), *Global Cities: A Short History*, Washington, DC: Brookings Institution Press.

Daraganova, G., P. Pattison, J. Koskinen, B. Mitchell, A. Bill, M. Watts, et al. (2012), 'Networks and geography: modelling community network structures as the out of both spatial and network processes', *Social Networks*, **34** (1), 6–17.

Day, G. (2006), *Community and Everyday Life*, London: Routledge.

De Montjoye, Y, C. Hidalgo, M. Verleysen and V. Blondel (2013), 'Unique in the crowd: the privacy bounds of human mobility', *Scientific Reports*, **3**, art. 1376, doi:10.1038/srep01376.

Delanty, G. (2010), *Community*, 2nd edn, New York: Routledge.

Domínguez, S. and B. Hollstein (eds) (2014), *Mixed Methods Social Networks Research: Design and Applications (Structural Analysis in the Social Sciences)*, Cambridge: Cambridge University Press.

Drouhot, L. (2017), 'Reconsidering "community liberated": how class and the national context shape personal support networks', *Social Networks*, **48** (January), 57–77.

Entwisle, B., K. Faust, R. Rindfuss and T. Kaneda (2007), 'Networks and contexts: variation in the structure of social ties', *American Journal of Sociology*, **112** (5), 1495–533.

Fingerman, K. (2009), 'Consequential strangers and peripheral ties: the importance of unimportant relationships', *Journal of Family Theory & Review*, **1** (2), 69–86.

Fischer, F. (2003) 'Public policy as discursive construct: social meaning and multiple realities', in F. Fischer (eds), *Reframing Public Policy Discursive Politics and Deliberative Practices*, Oxford and New York: Oxford University Press

Francis, J., B. Giles-Corti, L. Wood, and M. Knuiman (2012), 'Creating sense of community: the role of public space', *Journal of Environmental Psychology*, **32** (4), 401–9.

Frei, A. and K. Axhausen (2007), *Size and Structure of Social Network Geographies*, Zurich: ETH Zurich.

Frei, A. and K. Axhausen (2011), 'Modeling spatial embedded social networks', Working Paper Transport and Spatial Planning No. 685, Institute for Transport Planning and Systems (IVT), ETH Zurich, Zurich.

Gandomi, A. and M. Haider (2015), 'Beyond the hype: big data concepts, methods, and analytics', *International Journal of Information Management*, **35** (2), 137–44.

Gehl, J. and B. Svarre (2013), *How to Study Public Life*, Washington, DC: Island Press.

Google (2018), 'Google Scholar Metrics', accessed 20 October 2018 at https://scholar.google.co.uk/intl/en/scholar/metrics.html#metrics.
Granovetter, M. (1973), 'The strength of weak ties', *American Journal of Sociology*, **78** (6), 1360–80.
Granovetter, M. (1983), 'The strength of weak ties: a network theory revisited', *Sociological Theory*, **1** (1), 201–33.
Guest, A., J. Cover, R. Matsueda and C. Kubrin (2006), 'Neighborhood context and neighboring ties', *City & Community*, **5** (4), 363–85.
Habinek, J., J.L. Martin and B.D. Zablocki (2015), 'Double-embeddedness: spatial and relational contexts of tie persistence and re-formation', *Social Networks*, **42** (C), 27–41.
Hickman, P. (2013), '"Third places" and social interaction in deprived neighbourhoods in Great Britain', *Journal of Housing and the Built Environment*, **28** (2), 221–36.
Hipp, J. and A. Perrin. (2006), 'Nested loyalties: local networks' effects on neighbourhood and community cohesion', *Urban Studies*, **43** (13), 2503–24.
Hipp, J.R. and A.J. Perrin (2009), 'The simultaneous effect of social distance and physical distance on the formation of neighborhood ties', *City & Community*, **8** (1), 5–25.
Hipp, J.R., R.W. Faris and A. Boessen (2011), 'Measuring "neighborhood": constructing network neighborhoods', *Social Networks*, **34** (1), 128–40.
Hofstra, B., R. Corten, F. Van Tubergen and N. Ellison (2017), 'Sources of segregation in social networks: a novel approach using Facebook', *American Sociological Review*, **82** (3), 625–56.
Jiang, S., J. Ferreira and M.C. Gonzalez (2017), 'Activity-based human mobility patterns inferred from mobile phone data: a case study of Singapore', *Big Data, IEEE Transactions*, **3** (2), 208–19.
Jones, A. (2018), 'Revisiting Bott to connect the dots: an exploration of the methodological origins of social network analysis', *Forum Qualitative Sozialforschung/Forum: Qualitative Social Research*, **19** (2), art. 5.
Jones, M. (2013), 'Untangling sociomateriality', in P. Carlile, D. Nicolini, A. Langley and H. Tsoukas (eds), *How Matter Matters: Objects, Artifacts, and Materiality in Organization Studies*, Oxford Scholarship Online, ch. 9, doi:10.1093/acprof:oso/9780199671533.001.0001.
Kavanaugh, A., D. Reese, J. Carroll and M. Rosson (2005), 'Weak ties in networked communities', *The Information Society*, **21** (2).
Kebede, W. and A. Butterfield (2009), 'Social networks among poor women in Ethiopia', *International Social Work*, **52** (3), 357–73.
Kolaczyk, E. and G. Csárdi (2014), *Statistical Analysis of Network Data with R (Use R!)*, New York: Springer.
Larsen, J., K. Axhausen and J. Urry (2006), 'Geographies of social networks: meetings, travel and communications', *Mobilities*, **1** (2), 261–83.
Laurence, J. (2017), 'Wider-community segregation and the effect of neighbourhood ethnic diversity on social capital: an investigation into intra-neighbourhood trust in Great Britain and London', *Sociology*, **51** (5), 1011–33.
Leskovec, J. and E. Horvitz. (2008), 'Planetary-scale views on a large instant-messaging network', *Proceedings of the 17th International Conference on World Wide Web*, New York: Association for Computing Machinery, pp. 915–24.
Liu, Y., X. Liu, S. Gao, L. Gong, C. Kang, Y. Zhi, et al. (2015), 'Social sensing: a new approach to understanding our socioeconomic environments', *Annals of the Association of American Geographers*, **105** (3), 1–19.
Mahmoudi Farahani, L. (2016), 'The value of the sense of community and neighbouring', *Housing, Theory and Society*, **33** (3), 357–76.
Maunier, R. (1910), 'The definition of a city', *American Journal of Sociology*, **15** (4), 536–48.
McCaughey, M. and M.D. Ayers (2003), *Cyberactivism: Online Activism in Theory and Practice*, New York: Routledge.
McPherson, M., L. Smith-Lovin, and J.M Cook. 'Birds of a feather: homophily in social networks', *Annual Review of Sociology*, **27** (1), 415–44.
Mohr, J. and W. Ghaziani. (2014), 'Problems and prospects of measurement in the study of culture', *Theory and Society*, **43** (3), 225–46.
Mok, D., B. Wellman, J. Carrasco, G. Pflieger and C. Rozenblat (2010), 'Does Distance Matter in the Age of the Internet?', *Urban Studies*, **47** (13), 2747–83.
Moore, J., J. Carrasco and A. Tudela (2013), 'Exploring the links between personal networks, time use, and the spatial distribution of social contacts', *Transportation*, **40** (4), 773–88.
Morley, D. (2000), *Home Territories: Media, Mobility, and Identity*, London: Routledge.
Nast, J. and T. Blokland (2014), 'Social mix revisited: neighbourhood institutions as setting for boundary work and social capital', *Sociology*, **48** (3), 482–99.
Neal, Z. (2013), *The Connected City: How Networks are Shaping the Modern Metropolis*, New York and London: Routledge.
Neal, Z. (2015), 'Making Big communities small: using network science to understand the ecological and

behavioral requirements for community social capital', *American Journal of Community Psychology*, **55** (3–4), 369–80.

Neal, Z. and J. Neal (2017), 'Network analysis in community psychology: looking back, looking forward', *American Journal of Community Psychology*, **60** (1–2), 279–95.

Office for National Statistics (ONS) (2019), 'Statistical uses for mobile phone data: literature review', ONS Methodology Working Paper Series No. 8, accessed 10 February 2019 at https://www.ons.gov.uk/methodology/methodologicalpublications/generalmethodology/onsworkingpaperseries/onsmethodologyworkingpaperseriesno8statisticalusesformobilephonedataliteraturereview#appendix-references.

Oldenburg, R. (1999), *The Great Good Place: Cafes, Coffee Shops, Bookstores, Bars, Hair Salons, and Other Hangouts at the Heart of a Community*, New York: Marlowe.

Oloritun, R., A. Pentland and I. Khayal (2013), 'Dynamics of human social networks: people, time, relationships, and places', paper presented at the COINs13 Conference, Santiago, 11–13 August.

Orlikowski, W. (2007), 'Sociomaterial practices; exploring technology at work', *Organization Studies*, **28** (9), 1435–48.

Oxford (2012), *Oxford Latin Dictionary*, 2nd edn, Oxford: Oxford University Press.

Paton, K. (2014), *Gentrification: A Working-Class Perspective*, Aldershot: Ashgate.

Patterson, Z. and S. Farber (2015), 'Potential path areas and activity spaces in application: a review', *Transport Reviews*, **35** (6), 1–22.

Peris, A., E. Meijers and M. van Ham (2018), 'The evolution of the systems of cities literature since 1995: schools of thought and their interaction', *Networks and Spatial Economics*, **18** (3), 533–54.

Pflieger, G. and C. Rozenblat (2010), 'Introduction: urban networks and network theory: the city as the connector of multiple networks', *Urban Studies*, **47** (13), 2723–35.

Ryley, T.J. and A.M. Zanni (2013), 'An examination of the relationship between social interactions and travel uncertainty', *Journal of Transport Geography*, **31** (C), 249–57.

Sampson, R.J., S.W. Raudenbush and F. Earls (1997), 'Neighborhoods and violent crime: a multilevel study of collective efficacy', *Science*, **277** (5328), 918–24.

Schlapfer, M., L.M.A. Bettencourt, S. Grauwin, M. Raschke, R. Claxton, Z. Smoreda et al. (2014), 'The scaling of human interactions with city size', *Journal of the Royal Society Interface*, **11** (98), 20130789, doi:10.1098/rsif.2013.0789.

Simmel, G. (1903) *The Metropolis and Mental Life*, repr. 1971 in D. Levine (ed.), *On Individuality and Social Forms: Selected Writings*, Chicago, IL and London: University of Chicago Press, pp. 30–35.

Simões Aelbrecht, P. (2016), '"Fourth places": the contemporary public settings for informal social interaction among strangers', *Journal of Urban Design*, **21** (1), 124–52.

Sivarajah, U., M.M. Kamal, Z. Irani and V. Weerakkody (2017), 'Critical analysis of big data challenges and analytical methods', *Journal of Business Research*, **70** (C), 263–86.

Small, M.L. (2011), 'How to conduct a mixed method study: recent trends in a rapidly growing literature', *Annual Review of Sociology*, **37**, 55–84.

Small, M.L. (2012) *Villa Victoria: The Transformation of Social Capital in a Boston Barrio*, Chicago, IL: University of Chicago Press.

Small, M.L. (2017) *Someone to Talk to*, Oxford: Oxford University Press.

Snijders, T., P. Pattison, G. Robins and M. Handcock (2006), 'New specifications for exponential random graph models', *Sociological Methodology*, **36** (1), 99–153.

Snijders, T.A.B., G.G. Van de Bunt and C.E.G. Steglich (2010), 'Introduction to stochastic actor-based models for network dynamics', *Social Networks*, **32** (1), 44–60.

Spiro, E.S., Z.W Almquist and C.T Butts (2016), 'The persistence of division: geography, institutions, and online friendship ties', *Socius*, **2** (2), 1–15.

Storper, M. and A. Venables (2004), 'Buzz: face-to-face contact and the urban economy', *Journal of Economic Geography*, **4** (4), 351–70.

Sun, L., K. Axhausen, D. Lee and X. Huang (2013), 'Understanding metropolitan patterns of daily encounters', *Proceedings of the National Academy of Sciences of the United States of America*, **110** (34), 13774–9.

Talen, E. and J. Koschinsky (2013), 'The walkable neighbourhood: a literature review', *International Journal of Sustainable Land Use and Urban Planning*, **1** (1), 42–63.

Tönnies, F. (1940), *Fundamental Concepts of Sociology: (Gemeinschaft Und Gesellschaft)*, New York: American Book Company.

Trestian, I., A. Kuzmanovic, S. Ranjan and A. Nucci (2009), 'Measuring serendipity: connecting people, locations and interests in a mobile 3G network', *IMC 2009 – Proceedings of the 2009 ACM SIGCOMM Internet Measurement Conference*, New York: Association for Computing Machinery, pp. 267–79.

Tsou, M. (2015), 'Research challenges and opportunities in mapping social media and big data', *Cartography and Geographic Information Science*, **42** (Suppl. 1), 70–74.

Völker, B. and H. Flap (2007), 'Sixteen million neighbors: a multilevel study of the role of neighbors in the personal networks of the Dutch', *Urban Affairs Review*, **43** (2), 256–84.

Völker, B., H. Flap and S. Lindenberg (2007), 'When are neighbourhoods communities? Community in Dutch neighbourhoods', *European Sociological Review*, **23** (1), 99–114.
Wicki, M., Guidon, S., Axhausen, K. and Bernauer, T. (2018), 'Social networks, mobility behaviour and societal impacts. Field report: survey methods and response behaviour', ISTP Paper, April, ETH Zurich.
Wilkinson, N. (2018), *An Introduction to Behavioural Economics*, 3rd edn, London: Red Globe Press.
Wirth, L. (1938), 'Urbanism as a way of life', *American Journal of Sociology*, **44** (1), 1–24.
Wissink, B. and A. Hazelzet, A. (2012), 'Social networks in "Neighbourhood Tokyo"', *Urban Studies*, **49** (7), 1527–48.
Woods, N. (2000), 'The political economy of globalization', in N. Woods (eds), *The Political Economy of Globalization*, London: Palgrave, pp. 1–19.
Xu, Y., A. Belyi, I. Bojic and C. Ratti (2017), 'How friends share urban space: an exploratory spatiotemporal analysis using mobile phone data', *Transactions in GIS*, **21** (3), 468–87.
Young, I. (1990), *Justice and the Politics of Difference*, Princeton, NJ: Princeton University Press.

10. Theory and method in macro-comparative social network analysis

Matthew C. Mahutga and Robert Clark

INTRODUCTION

Scholars have long suspected the importance of social networks (Emerson 1972; Simmel 1908 [1971]). Networks facilitate the flow of information across large populations (Milgram 1967; Watts 2003). They confer positional advantages to central actors (Freeman 1979) or to those whose network partners may be centrally located (Bonacich 1987). In other instances, power derives from those whose partners are relatively isolated (Bonacich 1987; Cook et al. 1983) or who are resource dependent (Cook 1977; Emerson 1962; Molm 1990). Networks are even thought to benefit those who have access to non-redundant information via structural holes (Burt 1997, 2004) and weak ties (Granovetter 1973).

Over time, as network analysis evolved, studies moved beyond dyadic exchange (for example, Cook and Emerson 1978) and interpersonal relations (for example, Cook 1977) to tackle larger and more complex networks at higher levels of abstraction. In particular, network actors were not only individuals operating within small groups, but were also organizations, cities (Taylor 2004) or countries (Galtung 1971) that were tied to one another through affiliations or the exchange of valued resources. Moreover, analysts developed utilized methods to identify multiple actors within the same network that occupy structurally similar roles or positions (Snyder and Kick 1979).

Political economy scholars have used some of these network principles to reveal a core/periphery hierarchy among nation-states in the capitalist world economy (Galtung 1971), featuring an integrated core, isolated periphery and an intermediate group of countries in the semiperiphery (Wallerstein 1974). Network scholars from this tradition empirically locate countries along this hierarchy and sometimes link their structural positions in the world economy to developmental outcomes (for example, Snyder and Kick 1979).

In this chapter, we review foundational and contemporary network studies that span several domains and theoretical traditions, focusing on states and cities of the world economy and the exchange relationships they form. We discuss both theoretical and methodological innovations that have advanced this area of the network literature during the twentieth and early twenty-first centuries. In doing so, we characterize macro-level networks as more than metaphorical constructs, that is, as a web of relations that are highly structured and ultimately consequential for constituent actors that occupy particular positions within this framework.

SOCIAL NETWORKS AS METAPHOR AND SOCIAL STRUCTURE

Few sociological concepts have gained greater intellectual cache in the past 30 years than social networks. However, the term is used in very different ways in both scholarly and popular discourses. Social networking sites, such as Facebook, Twitter, Snapchat, and so on, allow users to send and receive inputs (messages, photographs, updates, and so on) from individuals with whom they are linked. Migrant networks are conceptualized as the 'interpersonal [and other kinds of social ties] ties linking kin, friends, and community members in their places of origin and destination' (Poros 2011). Organizational scholars also conceptualize a network as a distinct mode of economic organization distinct from both hierarchies (vertically integrated firms) and markets (arm's-length transactions between otherwise independent firms) (Granovetter 1985; Powell 1990). Some link the increasing ubiquity of social networks to processes of globalization that connect place-bound actors across space and thereby create a new 'network society' (Castells 1996). We refer to this variegated sociological concept as the network metaphor. It implies the sometimes vague intuition that actors (often individuals) are linked to each other, and this linking phenomena matters as something we care about.

A parallel conceptualization of social networks has also grown over this period into a unique branch of social science focused upon 'relationships among social entities, and on the patterns and implications of these relationships' (Wasserman and Faust 1994, p. 3). This branch of social science, social network analysis (SNA), shares an affinity with the network metaphor in that SNA analysts understand social actors as 'interdependent, rather than independent, autonomous units' linked together with relational ties that are 'channels for transfer or "flow" of resources' (Wasserman and Faust 1994, p. 4). Moreover, these relational ties are patterned into real social structures that provide 'opportunities for or constraints on individual action' (Wasserman and Faust 1994, p. 4). What makes SNA distinct, however, is the formal treatment of social networks. Social network analysts collect systematic data on the relationships between a set of predetermined actors, and utilize graph-theoretic and statistical methods to characterize the pattern of these relationships. They also try to understand how global characteristics of the networks matter to the actors who occupy them.

MACRO-COMPARATIVE SOCIAL NETWORK ANALYSIS

The network metaphor played a role in early criticisms of macro-comparative social science. In the post-war period, modernization theory was an influential explanation for the persistent economic stagnation of less-developed countries. What these countries lacked was the proper types of motivation or economic incentives at the individual level, or economic policy at the national level. Dependency theory provided a counter-narrative rooted in the network metaphor. Here, modernization theory's critical error was treating less-developed countries in isolation from the larger world economy in which they were embedded. Economic stagnation had less to do with factors internal to these countries, and more to do with the relationships between these countries and others (Frank 1969; Portes 1973). More powerful countries in the core of the world economy exploited less powerful countries in the periphery.

Over time, macro-comparative social science treated this nascent network explanation for underdevelopment more formally. One of the earliest formal treatments was that of Galtung (1971), who theorized a 'feudal interaction structure' comprised of two positions: center and periphery. Core countries inculcate dense trade links, spreading their ties among central and peripheral countries. Peripheral countries, however, only forge relationships with countries in the center and are isolated from one another. These different structural positions allowed core states to exercise bargaining power in their trade relationships with peripheral states. For example, core countries have more possible trading partners in the periphery than do peripheral countries in the core. Thus, import prices in the center are subject to greater competition than are import prices in the periphery. Moreover, peripheral nations are thought to be more vulnerable to demand fluctuations among the few export partners in the center (Dahi and Demir 2008; Greenaway and Milner 1990).

By the late 1970s, kindred social scientists began to employ formal methods of social network analysis. These studies measured social networks among countries using various types of relationships. Snyder and Kick (1979) analyzed cross-national relationships of international trade, diplomatic exchanges, military interventions and joint treaty memberships to characterize the overall structure of these exchanges and to compare rates of economic growth across network positions. Snyder and Kick found that these relations resembled a 'core–periphery' network, characterized by one large component to which individual network members are more or less attached. Core actors reside at the center of the network and forge relations with both core and peripheral countries. Conversely, peripheral countries reside at the margins of the network and forge relationships primarily with core countries. The identity of countries in each position coincided with this new structural intuition: most of the advanced industrial democracies resided in the core, while extremely poor, developing countries populated the periphery. Moreover, Snyder and Kick (1979) found that countries in the core experienced more rapid economic growth than those in the periphery. Additional work in this tradition measured the world economy with similar types of relational data, but found a less clear link between core positions in the world economy and economic growth (Van Rossem 1996).

Subsequent researchers analyzed multi-relational commodity trade data to characterize the structure of the world economy (Breiger 1981; Nemeth and Smith 1985). They disaggregated trade by level of industrial sophistication. The lowest end consisted of raw materials and animal products. The middle-range commodities were lightweight/low-wage manufactured goods. The most sophisticated commodities were high-technology/heavy manufacturing. In addition to characterizing the structure formed by these trade relationships as a core/periphery structure, these analysts also compared the pattern of trade across the type of trade relationship. They found that commodities with low levels of industrial sophistication flow from the periphery to the core, while commodities with high levels of industrial sophistication flow from the core to the periphery (Smith and White 1992). This pattern exemplified another key mechanism of underdevelopment in this tradition – the unequal exchange between the core and periphery.

Studies in this lineage incorporated a longitudinal component. This enabled them to examine dynamic change in the structure of world trade, as well as the mobility of individual countries within it. For example, Smith and White (1992) observed that the pattern of unequal exchange discussed above became more extreme between 1965 and

1980, and that few countries experienced significant upward mobility. Mahutga (2006) found that a core–periphery structure provided a better fit to the trade data over time and that only three countries (Singapore, South Korea and Turkey) experienced more upward mobility from 1965 to 2000 than would be expected by chance. The longitudinal component is increasingly prevalent, as evinced by its adoption in macro-comparative network analyses with different intellectual lineages. For example, Kim and Shin (2002) examine the degree to which globalization leads to more or less centralized and more regionalized networks over time. They employ measures of graph centralization, which characterize the departure of a graph from the maximum centralized star graph (see Figure 10.1 in the next section).

Another, more recent branch of macro-comparative social network analysis examines social networks among cities. In contrast to the work reviewed above, this research draws theoretical inspiration from urban geographers and sociologists (for example, Friedmann 1986; Sassen 2001; Taylor 2004). These studies are most interested in the relationship between economic globalization and urbanization and congeals under the heading of 'global cities' research. It attempts to quantify the structure of the world-city system by analyzing city-to-city relationships. Research on global cities takes as given the argument that (1) cities vary in how they are linked to each other and to the global economy, (2) this variation effects both the type of economic activity contained within cities and the developmental consequences of that activity, and (3) political conflict within cities should follow from the relationship of (2) to (1). Prototypical global cities are the command and control centers of an expanding global economy, from which transnational corporate (TNC) headquarters manage a spatially diffused production system.

A key task in this literature is the quantification of prominence or status in the world-city system. Early network metaphors hypothesized that New York, Paris, London and Tokyo were the global cities. Empirical research utilizing city-to-city airline passenger flows (Smith and Timberlake 2001), TNC headquarter–subsidiary relations (Alderson and Beckfield 2004), and interlocking producer service firms (Derudder et al. 2010) tends to support this proposition (cf. Friedmann 1995; Sassen 2001). New York, Paris, London and Tokyo stand out as exceptionally prominent world cities because resident headquarters tend to send subsidiaries to many other cities, contain a disproportionate share of important offices in the producer–service firm network, and receive large flows of airline passengers to/from many other cities. Similar to the country-centered macro-comparative network analyses reviewed above, world-city analysts focus on structural hierarchy or inequality where a handful of global cities operate as command and control centers of the whole world economy. Other cities possess descending levels of status and prominence in the various networks analyzed. While no network analyses of world cities have directly assessed the extent to which world cities conform to a core-periphery network using formal core–periphery models as described below, the structural intuition is especially comparable (cf. Derruder and Taylor 2005; Neal 2013).

In summary, macro-comparative social network analyses examine the network relationships between macro units of analysis (for example, countries or cities). The relationships examined vary from trade, diplomats, military bases, airline travel, TNC headquarter–subsidiary relationships, and so on. These studies share in common a concern with structural hierarchy or inequality. They employ social network tools to characterize the degree of hierarchy in a particular structure and to characterize individual actors as more or less central in these hierarchical structures. They also relate the structures, and/or the individual positions within these structures, to other phenomena, including patterns of economic organization, and economic development. In the following section, we provide a pedantic discussion of the common social network tools employed by this literature. We then discuss some of the more recent developments in the field before concluding.

MACRO-COMPARATIVE SOCIAL NETWORK METHODS

Centrality

Various notions of power, prominence and prestige lie at the substantive heart of the macro-comparative literature engaged with SNA. For that reason, network concepts of centrality feature prominently in this literature. Some studies thus proceed directly from observed relational data to measurements of the centrality of each node. Others employ a family of techniques termed role and position analysis. We address each in turn, beginning with a family of centrality measures. Figure 10.1 displays a simple star graph with which to illustrate these various measures.

Since notation conventions vary, we define ours here. Observed matrices are denoted by A. Vertices (or nodes) in these matrices are denoted by *ij*. The count of all vertices in a matrix is denoted by *N*. We differentiate between individual relationships (*k*) within a set of matrices (*R*). Thus, A_{ij}^{k} would denote the directed (or sending) tie from *i* to *j* in matrix A^k, which belongs in a multi-relational set of *R* matrices such that the full count of the strength (or number) of ties from *i* to *j* across all matrices in *R* could be written as

$$\sum\nolimits_{k=1}^{R} A_{ij}^{k}. \tag{10.1}$$

To measure the centrality of node *i* in matrix *A*, many macro-comparative researchers employ one or more of six definitions of centrality (see Freeman 1979 for a review of three). The most basic measure of centrality is degree centrality (C_d). Degree centrality measures the number of ties sent from and received by a vertex as in

$$C_d(i) = \sum\nolimits_{j=1}^{N} ij, \tag{10.2}$$

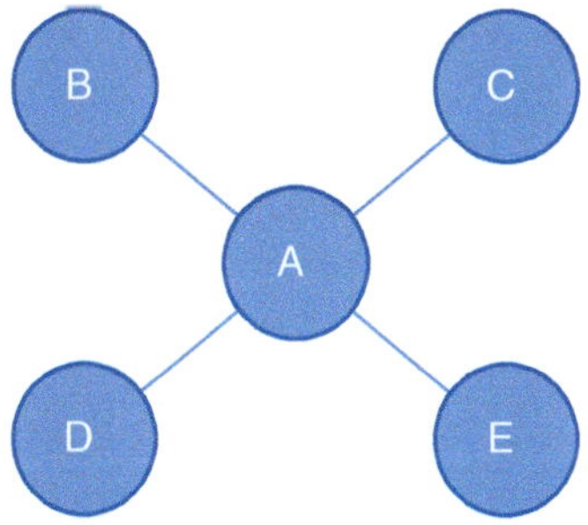

Figure 10.1 *Star graph*

where j indexes the maximum of N nodes to which i is connected. In a symmetrical matrix (where every tie from i to j is reciprocated by a tie from j to i), it is the number of vertices adjacent to each vertex. In an asymmetric matrix, degree centrality is calculated separately for the outgoing and incoming ties between i and it alters. Outgoing ties are sometimes referred to as 'out degree', while incoming ties are referred to as 'in degree'. In the symmetric matrix in Figure 10.1, A has a degree centrality of 4, and B–E have a degree centrality of 1. The maximum degree centrality in a symmetric matrix is given by N–1, so we could normalize these as 1 and .25.

Another measure of centrality is closeness centrality, which captures the degree to which a vertex is close to all other vertices. Formally, the closeness of vertex i can be defined by

$$C_c(i) = \left[\frac{\sum_{j=1}^{N} d(i,j)}{N-1}\right]^{-1}, \tag{10.3}$$

where d is the length of the shortest path connecting i to all of its alters and $^{-1}$ is the inverse (Freeman 1979). In Figure 10.1, for example, there is only one path to get from B to C, which goes through A. The same is true for all paths connecting B–E. Each of these paths has a length of 2. Similarly, there is one path to get from A to B, or from A to C, or from A to D, or from A to E. Each of these paths has a length of 1. Thus, the sum of d for B–E is 7, and the sum of d for A is 4. The closeness centrality score for A is $4/4^{-1} = 1$. The closeness centrality for B–E is $7/4^{-1} = .571$.

Betweenness centrality captures the extent to which a given vertex is connected to other vertices that are not connected to each other. We can define betweenness centrality with

$$C_B(i) = \frac{\sum_{j<k} \frac{g_{jk}(i)}{g_{jk}}}{\frac{(N-1)(N-2)}{2}}, \tag{10.4}$$

where g refers to a geodesic (shortest path). g_{jk} is the number of geodesics between each pair of vertices (jk), and $g_{jk}(i)$ is the number of these on which i sits. $C_B(i)$ is thus the sum of the proportion of the geodesics between each pair of vertices on which i sits across all pairs of vertices. The normalization (N–1)(N–2)/2 is for undirected graphs as in Figure 10.1 ((N–1)(N–2) for directed graphs). In Figure 10.1, there is one geodesic connecting each pair of vertices not including A (there are six pairs), and A sits on all of them while B–E sit on none of them. Thus, A has a betweenness centrality score of 6/6 = 1, and B–E have betweenness centrality scores of 0/6 = 0.

Degree, closeness and betweenness centrality are local measures of centrality insofar as the centrality of the focal vertex (i) depends primarily on its own location in the network. The last three measures of centrality are more global in that the centrality of the focal vertex (i) depends on its own location in a network, and the location of the other vertices to whom i is connected. To illustrate the utility of these centrality measures, we introduce a new exemplar network taken from Clark and Mahutga (2013). In Figure 10.2, B, C and D have equal degree (2), as do E and F (1). A has degree of 4. However, the relative

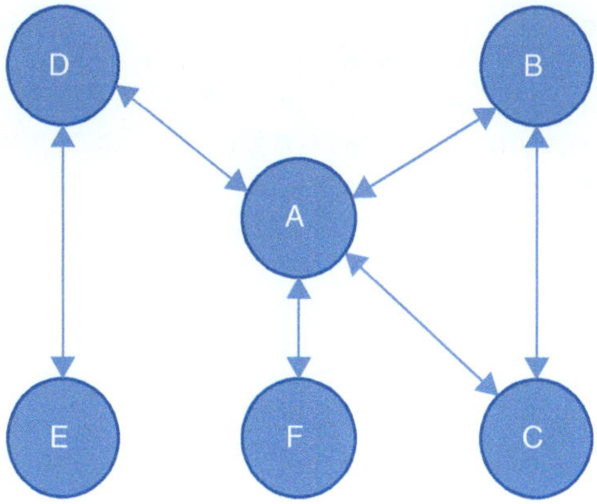

Figure 10.2 *Graph*

ranking of centrality will also depend on the centrality of the partners to whom these vertices are connected.

Eigenvector centrality is a measure for which the centrality of the focal vertex (i) depends on both its own centrality and the centrality of its neighborhood (Bonacich 1972; see also Boyd et al. 2013; Neal 2011). Formally, eigenvector centrality ($C_e(i)$) is given by

$$C_e(i) = \sum_{j=1}^{n} A_{ij}\lambda_j, \tag{10.5}$$

where A_{ij} is the tie between vertex i and j in A, and λ_j is the centrality (eigenvalue) of vertex j. It thus weights the tie between i and j by the centrality of i. i is awarded greater centrality for being connected to central actors than less central actors, all else equal. A similar measure is beta centrality (Bonacich 1987). As with eigenvector centrality, the centrality of the focal vertex (i) depends on the centrality of its neighbors. Unlike eigenvector centrality, though, the centrality of i's neighborhood can either increase or decrease the centrality of i. Formally, beta centrality ($C_\beta(i)$) is given by

$$C_\beta(i) = \sum_{j=1}^{n} A_{ij}(\alpha + \beta_{cj}), \tag{10.6}$$

where A_{ij} is the tie between vertex i and j in A, α is a normalizing constant and β is an attenuation factor that affects the size and direction of j's centrality on the centrality if i.

Coreness is the final measure of centrality we cover, and it is frequently invoked in the macro-comparative SNA literature. Originally defined by Borgatti and Everett (1999), the intuition begins with the overall structure of a network and works backward to defining centrality by the location of a vertex in that structure. The intuition 'entails a dense, cohesive core and a sparse, unconnected periphery', or a network with 'one group to which all actors belong to a greater or lesser extent' (Borgatti and Everett 1999, pp. 375, 376). Core vertices belong in the dense, cohesive group and peripheral vertices belong to a group that is unconnected except by virtue of their connection to core vertices. Figure 10.3 illustrates two versions of the concept originally outlined by Borgatti and Everett

Discrete

	A	B	C	D	E	F	G	H	I	J	K	L	M	N	O	P
A	0	1	1	1	1	1	0	1	0	0	0	1	0	0	1	1
B	1	0	1	1	1	0	1	0	0	1	0	0	0	1	0	0
C	1	1	0	1	1	1	0	0	1	0	0	0	1	0	1	1
D	1	1	1	0	1	0	0	1	0	0	0	1	0	0	0	0
E	1	1	1	1	0	0	1	0	0	0	0	0	1	0	0	1
F	1	0	1	0	0	0	0	0	0	0	0	0	0	0	0	0
G	0	1	0	0	1	0	0	0	0	0	0	0	0	0	0	0
H	1	0	0	1	0	0	0	0	0	0	0	0	0	0	0	0
I	0	0	1	0	0	0	0	0	0	0	0	0	0	0	0	0
J	0	1	0	0	0	0	0	0	0	0	0	0	0	0	0	0
K	0	0	0	0	0	0	0	0	0	0	0	0	0	0	0	0
L	1	0	0	1	0	0	0	0	0	0	0	0	0	0	0	0
M	0	0	1	0	1	0	0	0	0	0	0	0	0	0	0	0
N	0	1	0	0	0	0	0	0	0	0	0	0	0	0	0	0
O	1	0	1	0	0	0	0	0	0	0	0	0	0	0	0	0
P	1	0	1	0	1	0	0	0	0	0	0	0	0	0	0	0

Continuous

	A	B	C	D	E	F	G	H	I	J	K	L	M	N	O	P
A	0	1	1	1	1	1	0	1	0	0	0	1	0	0	1	1
B	1	0	0	1	0	0	1	0	0	1	0	0	0	1	0	0
C	1	0	0	1	0	0	0	0	1	0	0	0	1	0	1	1
D	1	1	1	0	0	0	0	1	0	0	0	1	0	0	0	0
E	1	0	0	0	0	0	1	0	0	0	0	0	1	0	0	1
F	1	0	0	0	0	0	0	0	0	0	0	0	0	0	0	0
G	0	1	0	0	1	0	0	0	0	0	0	0	0	0	0	0
H	1	0	0	1	0	0	0	0	0	0	0	0	0	0	0	0
I	0	0	1	0	0	0	0	0	0	0	0	0	0	0	0	0
J	0	1	0	0	0	0	0	0	0	0	0	0	0	0	0	0
K	0	0	0	0	0	0	0	0	0	0	0	0	0	0	0	0
L	1	0	0	1	0	0	0	0	0	0	0	0	0	0	0	0
M	0	0	1	0	1	0	0	0	0	0	0	0	0	0	0	0
N	0	1	0	0	0	0	0	0	0	0	0	0	0	0	0	0
O	1	0	1	0	0	0	0	0	0	0	0	0	0	0	0	0
P	1	0	1	0	1	0	0	0	0	0	0	0	0	0	0	0

Figure 10.3 Discrete and continuous core/periphery structures

(1999). An ideal-typical illustration of a discrete concept is shown on the left-hand side. Each vertex is either inside or outside the core. Core vertices are maximally connected to each other, and peripheral vertices are linked only to core vertices. The right-hand image depicts an ideal-typical continuous model. Here, individual vertices are more or less core-like in their pattern of exchanges with the rest of the network, where core-like vertices interact with most other vertices, and peripheral vertices interact primarily with the core. Both concepts are deployed in the literature, though only the latter can be compared with the continuous centrality metrics reviewed above.

Formally, we can define continuous coreness with $C_C(i)$ as the vector **C** of non-negative numbers for which CC^T maximizes $r(A, \delta)$, where T refers to transpose, r is a measure of fit (say, correlation), A is an observed matrix and δ is the matrix of expected values formed by CC^T. Multiple formalizations of C, δ and r have been proposed (for example, Borgatti and Everett 1999; Boyd et al. 2010), but we calculate the continuous coreness score for each vertex in Figure 10.2 using the procedure developed by Borgatti and Everett (1999) as implemented in UCInet. Table 10.1 reports these results for Figure 10.2, together with eigenvector, and two versions of beta centrality. We also include results for degree, closeness and betweenness for comparison.

The relative ranking of vertices with eigenvector centrality is close to that of degree centrality, but the deviations are interesting. Vertex D has less eigenvector centrality than vertices B and C despite equal degree. This is because D is penalized for spending one of its degree on the relatively isolated vertex E. Similarly, vertex E has less eigenvector centrality than vertex F because F is tied to the most central vertex A, while E is tied to D. This relative ranking is identical to the coreness results in the final column of Table 10.1.[1] Unsurprisingly, the relative ranking of eigenvector centrality and coreness is also identical to that of beta centrality when $\beta = .010$ because all three centrality definitions

Table 10.1 *Centrality of vertices in Figure 10.2 under varying definitions*

	Degree	Closeness	Betweenness	Eigenvector	Beta		Coreness
					β = 0.01	β = −0.01	
A	0.800	0.833	0.800	0.628	1.000	1.000	0.903
B	0.400	0.625	0.400	0.455	0.506	0.494	0.237
C	0.400	0.556	0.000	0.455	0.506	0.494	0.237
D	0.400	0.556	0.000	0.321	0.504	0.496	0.192
E	0.200	0.500	0.000	0.135	0.251	0.249	0.038
F	0.200	0.417	0.000	0.264	0.256	0.244	0.183

Note: All scores normalized according to the literature except for Beta centrality, which is normalized by dividing each score by the maximum observed.

give greater weight to vertices that are more often tied to other central vertices. The scores when β = −.010 reverse this relative ranking in the two interesting cases discussed above. Here, D has more beta centrality than B and C, since it is rewarded (rather than penalized) for spending one of its degree on vertex E. Similarly, F has more beta centrality than E as it is rewarded for linking with D rather than with the most central A.

A casual review of the macro-comparative literature suggests that coreness, eigenvector centrality and beta centrality are most frequently deployed (in that order), and for obvious reasons. They capture the intuition that central actors are more powerful than isolated actors, and that the centrality of the neighborhood matters. Beta centrality with a − β is a special case for contexts in which it might be advantageous to be connected to relatively isolated others (see discussion of Clark and Mahutga 2013 in the final section of this chapter).

Roles and Positions

Another commonly employed family of tools in the macro-comparative networks literature is role and position analyses. Succinctly, 'there are two key aspects to the positional and role analysis of social networks: identifying social positions as collections of actors who are similar in their ties with others, and modeling social roles as systems of ties between actors or between positions' (Wasserman and Faust 1994, p. 351).

The intuition is straightforward. If it is true that macro units are interdependent so that each plays a role in a hierarchically organized system, then it should be possible to discover sets of actors with more or less interchangeable patterns of relationships in a network. With these sets in hand, it should be possible to test hypotheses about how the sets interact with each other, given theories at the system level. As has been noted elsewhere, the practical process

> involves (1) measuring the similarity between actors with a formal definition of equivalence (i.e., structural, regular, etc.), (2) grouping similar actors into mutually exclusive positions in such a way that intergroup similarity is minimized, and (3) understanding (modeling) the ways in which the various positions interact with each other to understand their various roles in the network. (Lloyd et al. 2009, p. 58)

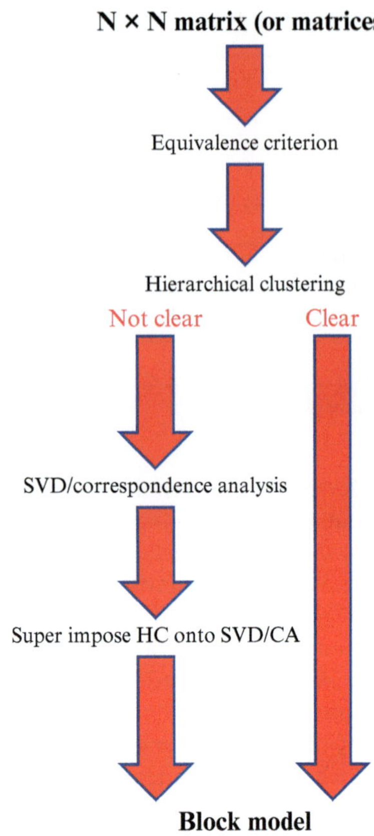

Figure 10.4 Role and position analysis

Figure 10.4 provides a schematic depiction of a role and position analysis.

In Figure 10.4, the process begins with a matrix (or set of matrices). The matrix is then evaluated by an equivalence criterion, which produces an N × N matrix in which A_{ij} is the relational equivalence of vertex *i* and *j*. What is key is that the equivalence of vertex *i* and *j* is a function of the similarity of pattern of their relationships to others in the network, the similarity of the other vertices to which they are connected and the formal definitions of similarity (the equivalence criterion). If multiple relationships are used, this equivalence sums over all relationships. There are many equivalence criteria in the literature, and we review two below.

The second step attempts to separate all of the vertices into equivalent groups (or blocks). Hierarchical clustering is one approach. Here, the process begins with 1 equivalent group (each vertex in one group) and then iteratively partitions the matrix until each vertex is in its own group (n groups). The clustering results can sometimes be quite clear – there are clearly *k* number of groups to which each vertex in A belongs. With complicated networks, they are often less than perfectly clear. In these instances, it can be helpful to analyze the N × N equivalence matrix with a scaling technique (for example, an SVD

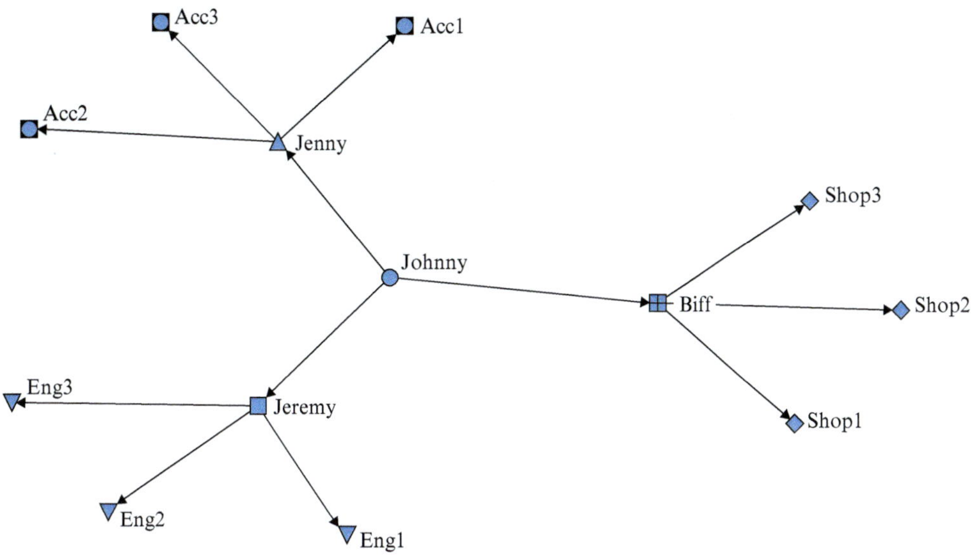

Note: Equivalent positions denoted by node symbol.

Figure 10.5 Structural equivalence at a widget factory

or correspondence analysis) and superimpose these results onto multiple hierarchical partitions.

In what follows, we discuss two common equivalence criteria, and then illustrate the procedure using a simple data set. Two pieces of the macro-comparative SNA literature provide exemplars of the two most common equivalence criteria. The first is Snyder and Kick (1979). In their analysis, they used structural equivalence as their criterion. Conceptually, structural equivalence identifies sets of actors who relate in the same way with identical others. For example, Figure 10.5 imagines a widget factory in which the relationship is 'gives orders to'. We know only the individual or department names of the employees. Johnny only gives orders. Jeremy, Jenny and Biff give orders to different sets of employees, but receive orders from the same employee. Finally, the departmental workers are similar in that the only receive orders, but different in that they receive orders from different vertices. With structural equivalence, there would be seven unique positions – Johnny, Jeremy, Jenny, Biff, engineers, accountants and shop workers. Perhaps the most straightforward measure of structural equivalence is simply the correlation coefficient between the row, column or both profiles for each pair in a network.

The other classic piece is Smith and White (1992). Smith and White used the same block modeling procedure as Snyder and Kick except that they employed the concept of regular equivalence. Regular equivalence is motivated by the notion that equivalent vertices should be defined by the similarity of their pattern of relations and the equivalence (rather than identity) of their alters. To illustrate, Figure 10.6 displays the same graph as in Figure 10.5. A regular equivalent partition would identify three positions rather than seven. Johnny, Jeremy–Jenny–Biff and workers. Johnny sends to three equivalent others, and does not receive (i.e., he has no in-degree, but three out-degree to others

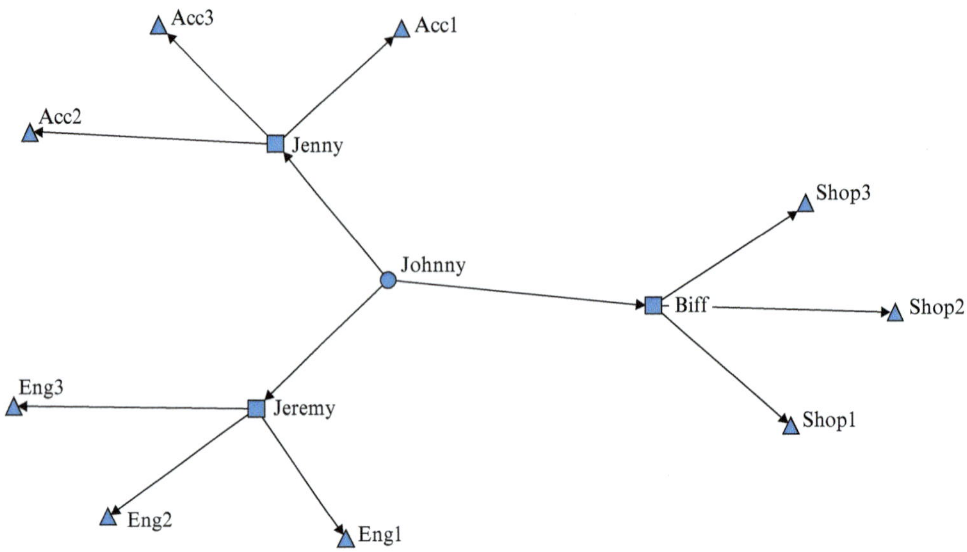

Note: Equivalent positions denoted by node symbol.

Figure 10.6 Regular equivalence at the widget factory

who have the same in-degree and out-degree). Jeremy-Jenny-Biff receive from the same other and send to equivalent others (they have the same in-degree and out-degree, and their alters have the same in-degree and out-degree). The workers receive from equivalent others (they each have the same in-degree and no out-degree, and their alters have the same in-degree and out-degree). Also, the equivalence of the vertices in any position is a function of the similarity of their ties and the similarity of the ties of those with whom they are connected. The identity of those with whom they are connected is irrelevant. Regular equivalence is thus a recursive concept – algorithms to measure it must start with relational patterns to initially define equivalencies, then update these equivalencies in a second step by considering the equivalence of the alters in the first step, and so on.

There is no evidently superior criterion. The suitability of each does and should depend entirely on the theoretical question at hand. While structural equivalence has a simple formal definition (or several simple formal definitions), the same is less clear for regular equivalence (see Boyd and Jonas 2001; White and Reitz 1983). Recall the analysis of a socio-matrix (or matrices) with regular or structural equivalence results in a matrix of equivalencies, where the entries are the equivalence between each pair. These numbers can range from 0 to 1 (or 100, depending on how they are scaled). One option would be to define, a priori, a threshold (α) by which to declare any given pair of vertices equivalent (E), so that i is equivalent to j if $E \geq \alpha$. However, no theory exists to define these thresholds and complicated, real-world data sets rarely yield perfectly equivalent sets. Thus, the next step in the diagram in Figure 10.4 seeks to identify a plurality of cut-points with which to divide vertices into equivalent sets based on their measured equivalence. In Figure 10.7, we see a hierarchical clustering (HC) result of a structural equivalence matrix from the widget factory data described above. At the low end,

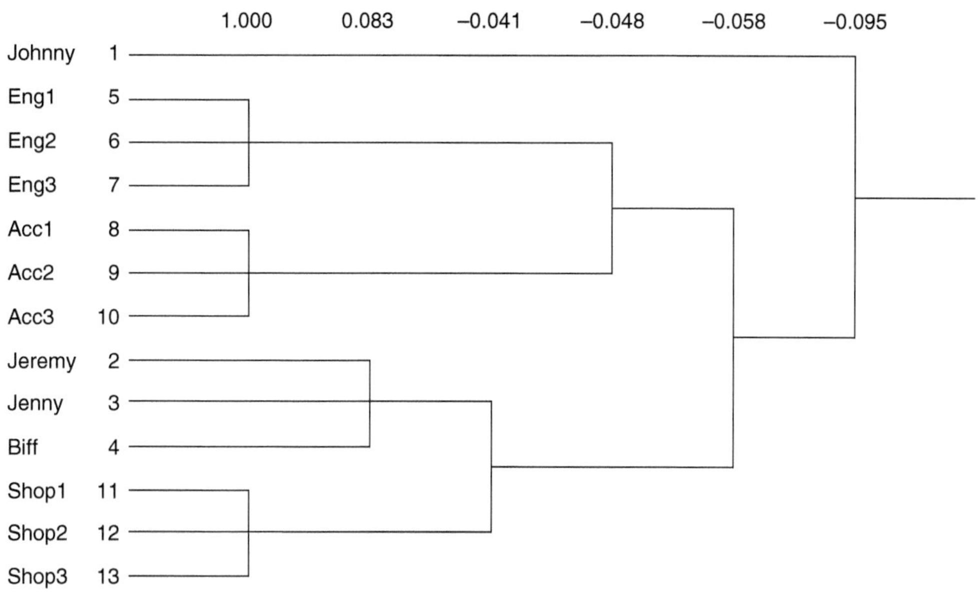

Note: Based on single link hierarchical clustering routine in UCInet.

Figure 10.7 Hierarchical clustering of structural equivalence at a widget factory

the results split Johnny from everyone else at –.095. At the high end (equivalence of 1.00), there are the perfectly structural equivalent groups identified with the intuition above: Johnny, Jeremy, Jenny and Biff, and then the accountants, engineers and shop workers.

Figure 10.8 shows the same results of a regular equivalence matrix of the widgets factory.[2] At the lowest measured level of equivalence (10.971), the procedure separated Johnny, Jeremy, Jenny and Biff from the others. The next partition separated Johnny into his own group. The procedure concludes with the three perfectly regularly equivalent (for example, RE = 100) positions discussed above. In either of these cases, there is no ambiguity; a researcher would simply select the perfectly equivalent groups. However, real-world cases are less obvious. Figure 10.9 shows a hierarchical clustering of a structural equivalence matrix based on the countries trade data held in UCInet (Smith and White 1992; Wasserman and Faust 1994), which only includes 24 countries. The smallest observed equivalence is .267. The largest is .696 (Argentina and Brazil). In between those two there are no clear partitions; many splits include one or two countries, a phenomenon known as chaining.

In these instances, it is useful to compare a continuous scaling of the equivalencies with various partitions. To proceed, we analyze the equivalence matrix with reduced ranked decompositions such as SVD or correspondence analysis. Singular value decomposition decomposes a matrix into three matrices: U, d and V. U summarizes the information in the rows, V summarizes the information in the columns and d is a diagonal matrix of singular values that weights the dimensions in U and V. Singular value decomposition can completely reproduce a matrix with

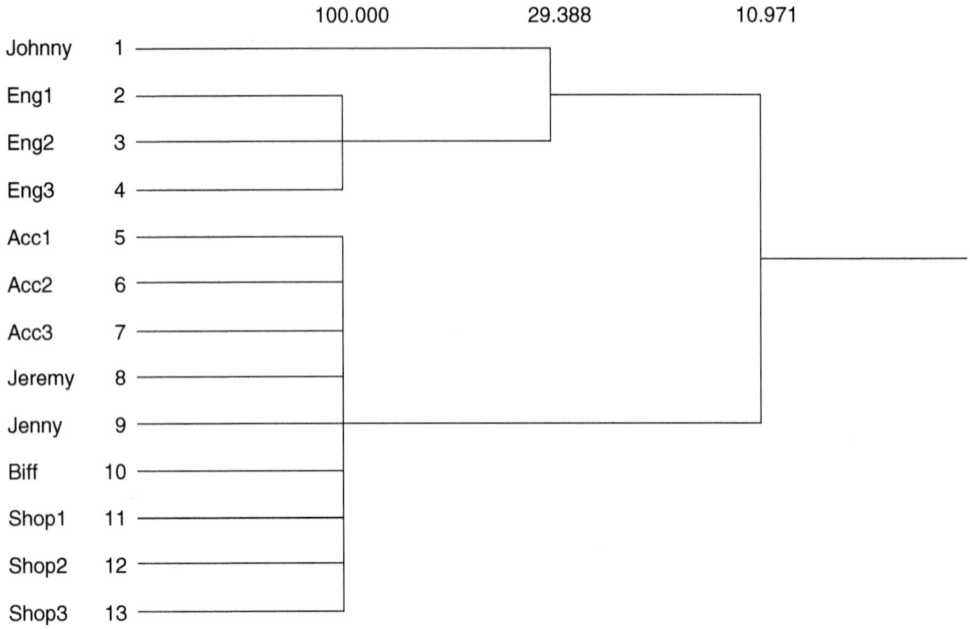

Note: Based on single link hierarchical clustering routine in UCInet.

Figure 10.8 Hierarchical clustering of regular equivalence at a widget factory

$$\sum_{k=1}^{n} U_k \, \mathrm{d}_k V_k^T, \tag{10.7}$$

where T refers to the transpose of V. Correspondence analysis (CA) is just an SVD of a matrix transformed with $A_{ij} / \sqrt{A_{i\cdot}.A_{j\cdot}}$, where A is the matrix of equivalencies and . refers to the row/column marginal, respectively. This transforms the matrix so that the marginals equal approximately 1 for each vertex.[3] There are as many as n dimensions in an SVD (or CA), where n corresponds to the number of vertices in the matrix. But the *UdV* dimensions are ordered hierarchically in terms of the amount of variance they can explain in the data, and each dimension is orthogonal. It is often the case that 1, 2 or 3 dimensions can explain all of the meaningful variation in an equivalence matrix. We can assess the fit of different solutions with

$$100 \times \frac{\sum_{m=1}^{M} \lambda_m^2}{\sum_{k=1}^{d} \lambda_k^2}, \tag{10.8}$$

where k indexes the full set of singular values in d, and m indexes the subset of singular values in the M singular values a user proposes. The result thus gives you the proportion of variance in the full set of d explained by the subset in M.

Figures 10.10 and 10.11 display a correspondence analysis of the structural and regular equivalences of the widget factory in Figures 10.5 and 10.6. CA (and SVD) places vertices

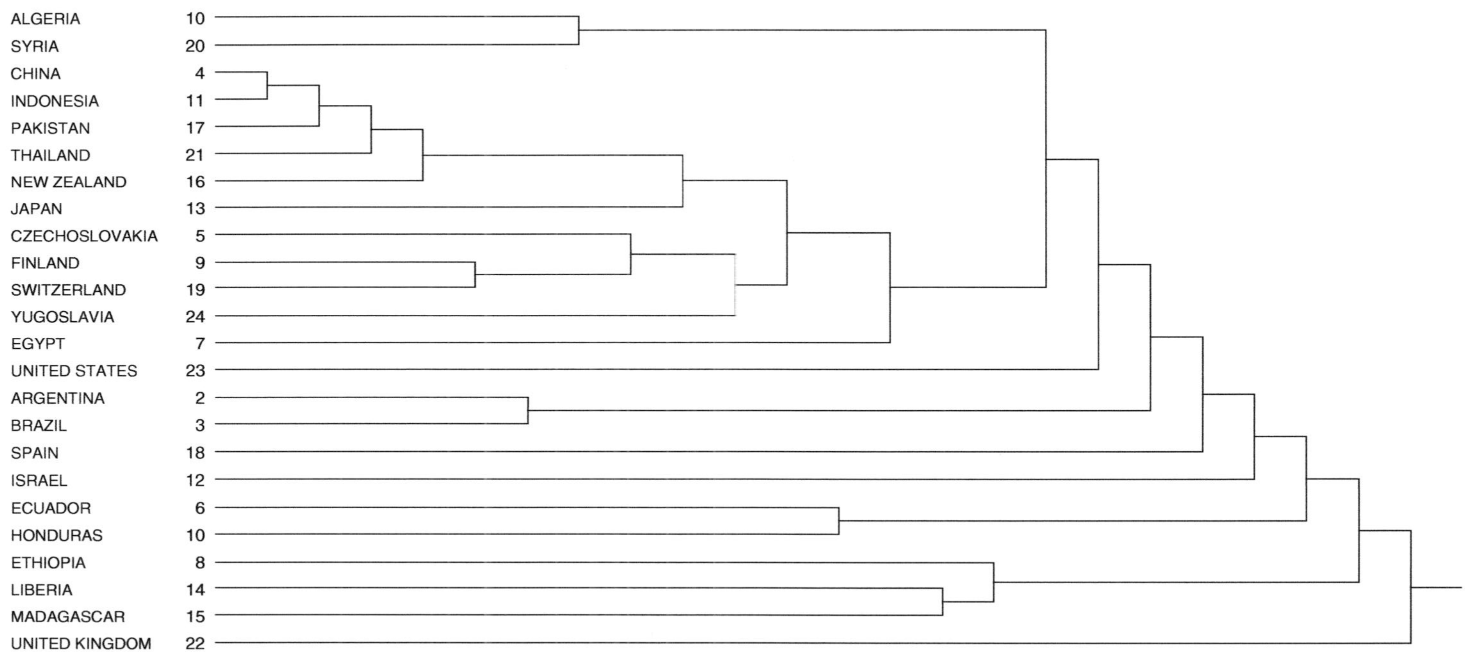

Note: Based on single link hierarchical clustering routine in UCInet.

Figure 10.9 Hierarchical clustering of structural equivalence of countries trade data

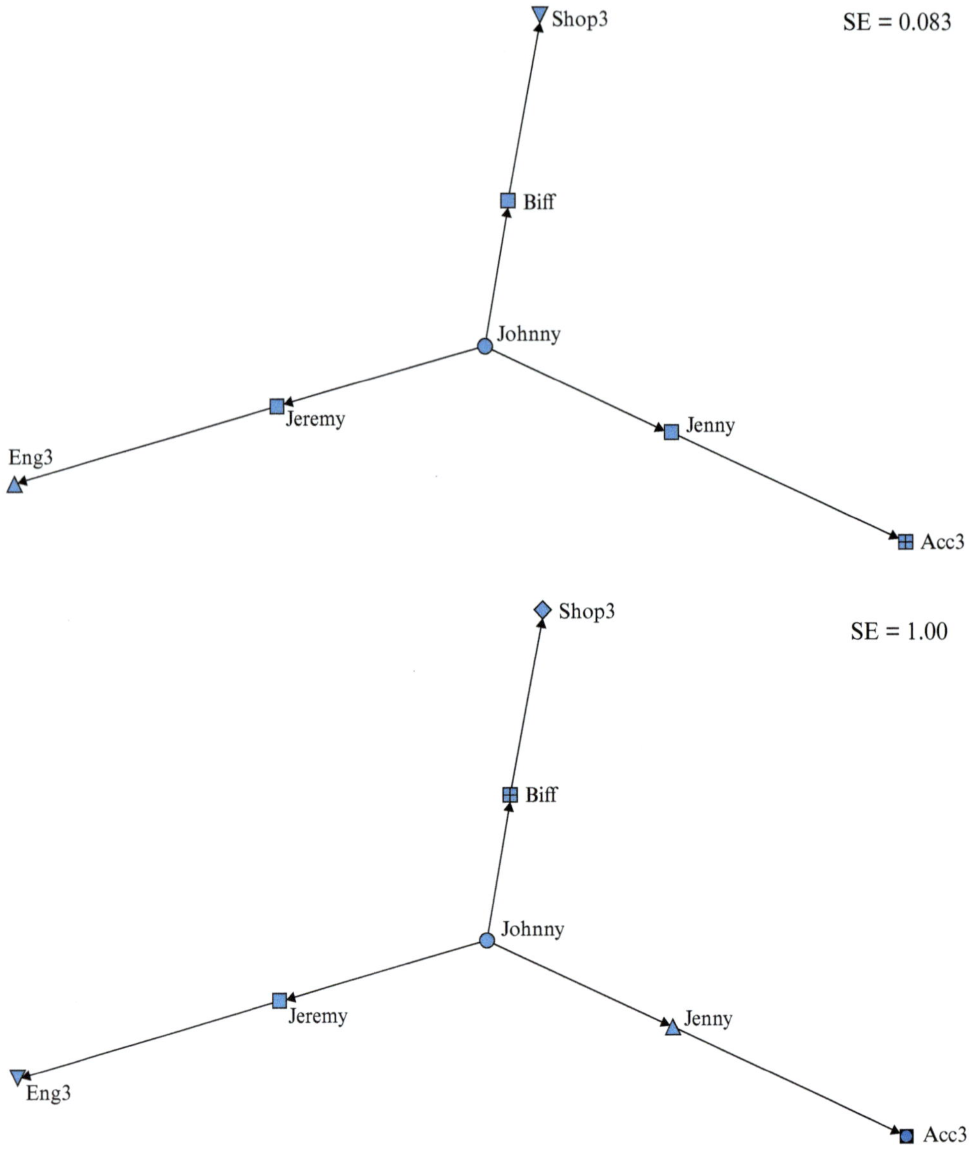

Figure 10.10 Superimposition of hierarchical clustering on correspondence analysis of structural equivalence of widgets factory

close to others with whom they have similar patterns of relations, and far from those with whom they have dissimilar relations. Since the underlying matrices are equivalences, the distance between vertices is readily interpretable as the degree to which they are structurally or regularly equivalent. In each case, we use two dimensions of the CA results to model the degree of equivalent between vertices. The first dimension is on the X axis, and the second dimension is on the Y axis. With the structural equivalence results, we

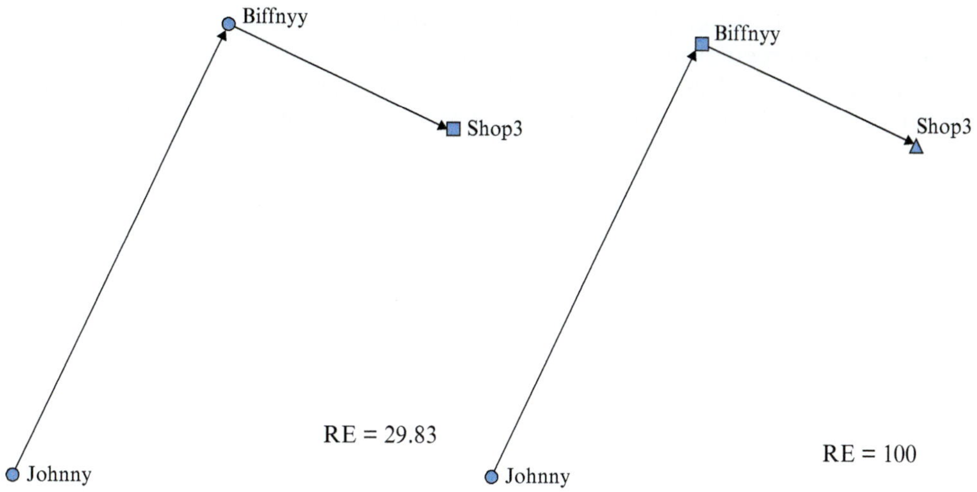

Figure 10.11 Superimposition of hierarchical clustering on correspondence analysis of regular equivalence of widgets factory

can explain about 46.1 percent of the variation with the first two dimensions. With the regular equivalence results, we can explain 100 percent of the variation with the first two dimensions.

In the top panel of Figure 10.10, the three workers in the shop, engineering and accounting departments collapse into one position, respectively. This is because these workers are perfectly structurally equivalent in their various departments, so the CA results plot them in the same space. The node shapes in the top panel are based on the HC cluster results when structural equivalence (SE) = .083. Recall from Figure 10.7 that .083 is the level of structural equivalence at which Jeremy, Jenny and Biff remained in the same HC group. Thus, these three vertices have the same symbol in the top panel to signify their joint membership in a structurally equivalent group of .083. The bottom panel codes the node shapes according to a structural equivalence criterion of 1.00. Jeremy, Jenny and Biff now each have a different shape, which corresponds to the final partition in the HC of SE results in Figure 10.7. Were there any doubt that Jeremy, Jenny and Biff occupy unique positions upon a visual inspection of the HC results in Figure 10.7, this doubt would be eliminated upon quick examination of Figure 10.10, since CA places these three vertices relatively far from each other. That is, the seven unique positions identified in Figure 10.10 map perfectly onto the final partition in Figure 10.7.

Figure 10.11 conducts the same exercise with the regular equivalence results. It is immediately obvious that there are three regularly equivalent positions – the workers, the middle managers and the boss – as the equivalent vertices are placed on top of each other. Thus, the two-group solution at regular equivalence 29.388 (where Johnny and the middle managers share the same shape) is inferior to the three-group solution at regular equivalence 100.

The final step in Figure 10.4 is the construction of the block model. To recap, we began with a matrix of the relationship 'gives orders to' at a widget factory. We knew only the names of the employees (or their departments) and who they gave orders to and received orders from. We then measured the degree of structural and regular equivalence of the vertices in this matrix. We used the equivalence matrices to partition the actors into equivalent groups, and employed correspondence analysis as a guide to choosing between multiple partitions. The resulting block models appear in Tables 10.2 and 10.3.

Tables 10.2 and 10.3 are block models based on structural and regular equivalence, respectively. The cells are the average number of ties sent or received by the vertices in the corresponding blocks. Note that we have successfully reduced the complexity in the matrix from 13 vertices to seven in Table 10.2, since the engineers, accountants and shop floor workers are collapsed into structurally equivalent groups. From this block model, it would be relatively clear that one position only gives orders, and must be the boss. Three others only send orders to one other group, and therefore must be managers of that group. The latter three only receive orders, and must be workers under the manager to which they are tied. Table 10.3 yields the block model of regular equivalence. Here we have reduced the complexity of the original matrix even more, from 13 vertices to three. The intuition is the same, though perhaps more immediately obvious: bosses only give orders, managers give and receive orders, and workers only receive orders. Both approaches would provide us with a clear intuition of the authority structure at the widget factory, but one is more parsimonious than the other (for better or worse).

Table 10.2 Widget block model from structural equivalence

	Boss	Acc. manager	Eng. manager	Shop manager	Accounting	Engineering	Shop
Boss	0	**1**	**1**	**1**	0	0	0
Acc. manager	0	0	0	0	**1**	0	0
Eng. manager	0	0	0	0	0	**1**	0
Shop Manager	0	0	0	0	0	0	**1**
Accounting	0	0	0	0	0	0	0
Engineering	0	0	0	0	0	0	0
Shop	0	0	0	0	0	0	0

Note: Cells are the average number of ties between members of each block.

Table 10.3 Widget block model from regular equivalence

	Boss	Middle management	Workers
Boss	0	**1**	0
Middle management	0	0	**0.333**
Workers	0	0	0

Note: Cells are the average number of ties between members of each block.

THE STATE OF THE FIELD

We conclude with a review of the contemporary literature, focusing on issues surrounding the classification of nation-states in the world economy, the extent to which states exhibit mobility, and recent extensions into political, social, urban and organizational networks. These areas highlight empirical contributions that confront and advance theoretical foundations regarding the existence of three distinct zones in the world economy, the stability of the core/periphery hierarchy and the primacy of economic relations in the modern world system.

Orthodox conceptions of the world economy identify three distinct zones (core, semiperiphery and periphery) comprising a single division of labor across multiple polities. Several studies find that world income is distributed trichotomously (Arrighi and Drangel 1986; Babones 2005), and one early network analysis suggests the presence of an intermediate stratum in the world economy lying between a core and periphery (Steiber 1979). Nevertheless, most subsequent network studies involving international trade group countries into more than three categories. For example, Snyder and Kick (1979) identified 10 country clusters. Kick (1987) and Kick and Davis (2001) identified 11 country clusters. Nemeth and Smith (1985) identified eight country clusters. These studies were based on the concept of structural equivalence discussed above. Smith and White (1992) and Mahutga (2006) identified five country clusters. Mahutga and Smith (2011) observed six country clusters. Van Rossem (1996) observed four country clusters. These studies were based on the regular equivalence criterion discussed above. Thus, on one hand, this variance reflects different levels of generality in the equivalence criterion employed. On the other, it is also consistent with arguments made elsewhere that the world-system hierarchy may best be conceptualized as continuous rather than categorical (see Chase-Dunn 1998; Mahutga 2006). Recent work proceeds deductively and forces a three-tier solution to the network partition of trade (Clark 2012; Clark and Beckfield 2009). Many studies find significant associations between a country's structural position in world trade and developmental outcomes (Clark and Beckfield 2009; Mahutga and Smith 2011; Nemeth and Smith 1985; Snyder and Kick 1979). Nevertheless, the pioneering classification by Snyder and Kick (1979) remains the industry standard (for example, Alderson and Nielsen 1999; Beckfield 2003; Clark 2017; Wejnert 2005), despite several critiques (Bollen 1983; Clark and Beckfield 2009).

Several studies also consider the extent to which countries in the international trade network experience mobility along the core/periphery hierarchy (Clark 2010; Mahutga 2006; Mahutga and Smith 2011; Smith and White 1992). That is, to what extent are countries able to become more (or less) core-like over time? Most studies in this genre adopt sample periods that span several decades, and they offer a range of interpretations for their results. Ultimately, though, most agree that the international trade network features some instances of mobility, particularly among middle-tier countries and the emerging economies in East Asia. Related studies link changes in the trade position to economic growth (Clark 2010; Mahutga and Smith 2011) and improved trade terms (Clark and Cason 2015). At first glance, this suggests that countries should adopt new trade partners in order to become more core-like. However, recent work shows that the largest gains from trade are achieved by forming relationships with isolated partners, rather than with central partners (Clark and Mahutga 2013).

Other recent network analyses of international trade depart from the classic core–periphery lineage in important ways. Mahutga (2014a, 2014b, 2014c) uses commodity specific trade to operationalize positional power in global commodity chains/production networks (GCC/GPN) (for example, Coe et al. 2008; Gereffi and Korzeniewicz 1994; Gereffi et al. 2005). Here, power is an exchange-theoretic conceptualization in that the power of i is a function of both its resources (which make i a more or less attractive exchange partner) and its ability to inculcate dependent ties (which create higher switching costs for i's partners than for i). Since industries globalize in different ways, positional power accrues to different types of positions in different industries. Mahutga (2014a) thus deduces two types of industry-specific network centrality positions that capture the average positional a country's resident firms. They borrow from the distinction of in-degree and out-degree centrality, and from eigenvector and beta centrality. The two measures are

$$P_j^B = \sum\nolimits_{i=1}^{n} \log\left(y_{ij}/x_i \cdot +1\right) \tag{10.9}$$

and

$$P_j^P = \sum\nolimits_{i=1}^{n} \log(x_{ji}/y_i \cdot +1), \tag{10.10}$$

where y is imports, x is exports and | is the row or column marginal. The superscripts ^{P} and ^{B} refer to industries for which the powerful firms are buyers and producers, respectively. Consistent with intuitions from the GCC/GPN literatures, Mahutga shows that these correlate with wages (Mahutga 2014a), the location of manufacturing (Mahutga 2014c) in two archetypically buyer and producer-driven industries (garments and transport equipment). He also shows that wage growth in these positions is greater than that in more dependent positions (Mahutga 2014b).

One issue that has garnered some growing attention is how and if world cities and the world system co-constitute. That is, does a city's position in the world-city system map on to the world-system position of the country in which it is located? Early social science theory suggested different answers to this question. On one hand, Hymer (1972) believed that capitalist development at the city level would reproduce international inequality at the country level. On the other hand, world-city scholars including Friedmann (1986) and Sassen (2001) believed globalization would create a world-city system that decoupled in significant ways from the inter-state world system. The first empirical treatment of this question was Alderson and Beckfield (2004). Using inter-city headquarter–subsidiary relationships to quantify the structure of the world-city system, they find a close mapping of a city's position in the world-city hierarchy, and the world-system position of the country in which it was located. Others find similar results (for example, Alderson et al. 2010), but some note a significant amount of upward mobility among some semiperipheral cities (Mahutga et al. 2010). A particularly innovative set of studies provide an entirely new empirical context to examine this question by examining the spatiality of the network formed by ownership between multinational firms (Rozenblat 2018; Rozenblat et al. 2017).

The vibrant world-cities literature deserves special attention, both because of its prominence in the general field of urban studies and the theme of this volume. A key task in this literature has been the quantification of power and position in the world-city system.

This emphasis has led to a proliferation of methods to quantify power and position. We believe this proliferation has both positives and negatives. On the positive side, much of this methodological innovation is interesting in its own right (for example, Neal 2011; Taylor 2004). On a potentially negative side, however, the multiplicity of measurement strategies must by their very nature produce variegated findings as regards which cities have power and position in the world-city system. This is not necessarily bad insofar as these variegated findings may produce a more triangulated consensus. However, we view one of the more important tasks facing the world-cities research as that of theory articulation and hypothesis testing in respect of outcomes of interest to urban political economists, such as urban and regional development, economic structure, and inequality and stratification (for example, Alderson 2018; Neal et al. 2019). The empirical side of these efforts would benefit from some consensus around the best operationalization of the right-hand side of these equations, a problem also true of the older research on the world system (see above).

Beyond international trade and world cities, other cross-national network studies focus on treaty relationships (Bandelj and Mahutga 2010; Bandelj et al. 2013) and telecommunications (Clark and Hall 2011; Koster 2007; Sandholtz and Gray 2003; Sun and Barnett 1994). More interest, however, has shifted towards international organizations. This work is notable not only for its burgeoning popularity, but also for the distinctive manner in which network ties are calculated. Organizational ties are conceived of as affiliation networks, in which organizations serve as hubs that attract multiple actors. Thus, instead of an exchange or flow occurring across nodes, relationships are formed through mutual participation in an organizational setting. Recent applications consider network hierarchies in intergovernmental organizations (Beckfield 2008, 2010) and international nongovernmental organizations (INGOs) (Hughes et al. 2009; Paxton et al. 2015). Intergovernmental organization and INGO networks have the capacity to foster trade relationships (Ingram et al. 2005), while diffusing economic policy (Cao 2009, 2010), democratic principles (Torfason and Ingram 2010) and human rights standards (Greenhill 2010). Centrality in these networks is also positively associated with economic growth, tertiary school enrollment and voting power at the World Bank through a range of relational mechanisms (Clark 2008, 2017). We anticipate future work will further explore how core/periphery relationships might manifest in a wide range of organizational network structures.

NOTES

1. The correlation $r(A, \delta)$ for the scores in the last column is .718.
2. In Figure 10.7, we define structural equivalence based on the correlation of row profiles. In Figure 10.8, we define regular equivalence with the REGE algorithm in UCInet.
3. There are variants. See Weller and Romney (1990).

REFERENCES

Alderson, A. (2018), 'Urban development and the world city system: inter-city relations and the fate of U.S. cities', paper presented at the UCR Political Economy Seminar, 15 February, Riverside, CA.

Alderson, A. and J. Beckfield (2004), 'Power and position in the world-city system', *American Journal of Sociology*, **109** (4), 811–51.
Alderson, A. and F. Nielsen (1999), 'Income inequality, development, and dependence: a reconsideration', *American Sociological Review*, **64** (4), 606–31.
Alderson, A., J. Beckfield and J. Sprague-Jones (2010), 'Inter-city relations and globalization: the evolution of the global urban hierarchy, 1981–2007', *Urban Studies*, **47** (9), 1899–923.
Arrighi, G. and J. Drangel (1986), 'The stratification of the world-economy: an exploration of the semiperipheral zone', *Review*, **10** (1), 9–74.
Babones, S. (2005), 'The country-level income structure of the world-economy', *Journal of World-Systems Research*, **11** (1), 29–55.
Bandelj, N. and M. Mahutga (2010), 'Structures of globalization: evidence from the evolution of the world-wide network of bilateral investment treaties (1959–2009)', *International Journal of Comparative Sociology*, **54** (2), 95–123.
Bandelj, N., M. Mahutga and K. Shorette (2013), 'Signaling demand for foreign investment: the rise of post-socialist countries in the global bilateral investment treaties network', *Europe-Asia Studies*, **67** (6), 870–92.
Beckfield, J. (2003), 'Inequality in the world polity: the structure of international organization', *American Sociological Review*, **68** (3), 401–24.
Beckfield, J. (2008), 'The dual world polity: fragmentation and integration in the network of intergovernmental organizations', *Social Problems*, **55** (3), 419–42.
Beckfield, J. (2010), 'The social structure of the world polity', *American Journal of Sociology*, **115** (4), 1018–68.
Bollen, K. (1983), 'World system position, dependency, and democracy: the cross-national evidence', *American Sociological Review*, **48** (4), 468–79.
Bonacich, P. (1972), 'Factoring and weighting approaches to status scores and clique identification', *Journal of Mathematical Sociology*, **2** (1), 113–20.
Bonacich, P. (1987), 'Power and centrality: a family of measures', *American Journal of Sociology*, **92** (5), 1170–82.
Borgatti, S. and M.P. Everett (1999), 'Models of core/periphery structures', *Social Networks*, **21** (4), 375–95.
Boyd, J. and K. Jonas (2001), 'Are social equivalences ever regular? Permutation and exact tests', *Social Networks*, **23** (2), 87–123.
Boyd, J., W. Fitzgerald, M. Mahutga and D. Smith (2010), 'Computing continuous core/periphery structures for social relations data using MINRES SVD', *Social Networks*, **32** (2), 125–37.
Boyd, J., M. Mahutga and D. Smith (2013), 'Measuring centrality and power recursively in the world city network: a reply to Neal', *Urban Studies*, **50** (8), 1641–7.
Breiger, R. (1981), 'Structures of economic interdependence among nations', in P. Blau and R. Merton (eds), *Continuities in Structural Inquiry*, London: Sage, pp. 353–79.
Burt, R. (1997), 'The contingent value of social capital', *Administrative Science Quarterly*, **42** (2), 339–65.
Burt, R. (2004), 'Structural holes and good ideas', *American Journal of Sociology*, **110** (2), 349–99.
Castells, M. (1996), *The Rise of the Network Society*, Oxford: Blackwell.
Cao, X. (2009), 'Networks of intergovernmental organizations and convergence in domestic economic policies', *International Studies Quarterly*, **53** (4), 1095–130.
Cao, X. (2010), 'Networks as channels of policy diffusion: explaining worldwide changes in capital taxation, 1998–2006', *International Studies Quarterly*, **54** (3), 823–54.
Chase-Dunn, C. (1998), *Global Formations: Structures of the World Economy*, New York: Rowman and Littlefield.
Clark, R. (2008), 'Dependency, network integration, and development', *Sociological Perspectives*, **51** (3), 629–48.
Clark, R. (2010), 'World-system mobility and economic growth, 1980–2000', *Social* Forces, **88** (3), 1123–51.
Clark, R. (2012), 'World-system position and democracy, 1972–2008', *International Journal of Comparative Sociology*, **53** (5–6), 367–99.
Clark, R. (2017), 'Quotas operandi: examining the distribution of voting power at the IMF and World Bank', *Sociological Quarterly*, **58** (4), 595–621.
Clark, R. and J. Beckfield (2009), 'A new trichotomous measure of world-system position using the international trade network', *International Journal of Comparative Sociology*, **50** (1), 5–38.
Clark, R. and T. Cason (2015), 'Getting on good terms: examining cross-national variation in the terms of trade, 1990–2012', *Sociology of Development*, **1** (3), 374–99.
Clark, R. and J. Hall (2011), 'Migration, international telecommunications, and human rights', *Sociological Forum*, **26** (4), 870–96.
Clark, R. and M. Mahutga (2013), 'Explaining the trade-growth link: assessing diffusion-based and structure-based models of exchange', *Social Science Research*, **42** (2), 401–17.
Coe, N., P. Dicken and M. Hess (2008), 'Global production networks: realizing the potential', *Economic Geography*, **8** (3), 271–95.
Cook, K. (1977), 'Exchange and power in networks of interorganizational relations', *Sociological Quarterly*, **18** (1), 62–82.

Cook, K. and R. Emerson (1978), 'Power, equity, and commitment in exchange networks', *American Sociological Review*, **43** (5), 721–39.

Cook, K., R. Emerson, M. Gillmore and T. Yamagishi (1983), 'The distribution of power in exchange networks: theory and experimental results', *American Journal of Sociology*, **89** (2), 275–305.

Dahi, O.S. and F. Demir (2008), 'South–South trade in manufactures: current performance and obstacles for growth', *Review of Radical Political Economy*, **40** (3), 266–75.

Derudder, B. and P. Taylor (2005), 'The cliquishness of world cities', *Global Networks*, **5** (1), 71–91.

Derudder, B., P.J. Taylor, P. Ni, A. De Vos, M. Hoyler, H. Hanssens, et al. (2010), 'Pathways of change: shifting connectivities in the world city network, 2000-08', *Urban Studies*, **47** (9), 1861–77.

Emerson, R. (1962), 'Power-dependence relations', *American Sociological Review*, **27** (1), 31–41.

Emerson, R. (1972), 'Exchange theory, part I: a psychological basis for social exchange' and 'Exchange theory, part ii: exchange relations and networks', in J. Berger, M. Zelditch and B. Anderson (eds), *Sociological Theories in Progress*, Boston, MA: Houghton-Mifflin, pp. 38–87.

Frank, A.G. (1969), *Latin American: Underdevelopment or Revolution*, New York: Monthly Review.

Freeman, L. (1979), 'Centrality in social networks: conceptual clarification', *Social Networks*, **1** (3), 215–39.

Friedmann, J. (1986), 'The world city hypothesis', *Development and Change*, **17** (1), 69–84.

Friedmann, J. (1995), 'Where we stand: a decade of world city research', in P. Knox and P. Taylor (eds), *World Cities in a World System*, Cambridge: Cambridge University Press, pp. 21–47.

Galtung, J. (1971), 'A structural theory of imperialism,' *Journal of Peace Research*, **8** (2), 81–117.

Gereffi, G. and M. Korzeniewicz (1994), *Commodity Chains and Global Capitalism*, Westport, CT: Praeger Press.

Gereffi, G., J. Humphrey and T. Sturgeon (2005), 'The governance of global value chains', *Review of International Political Economy*, **12** (1), 78–104.

Granovetter, M. (1973), 'The strength of weak ties', *American Journal of Sociology*, **78** (6), 1360–80.

Granovetter, M. (1985), 'Economic action and social structure: the problem of embeddedness', *American Journal of Sociology*, **91** (3), 481–510.

Greenaway, D. and C. Milner (1990), 'South-South trade theory, evidence and policy', *World Bank Research Observer*, **5** (1), 47–68.

Greenhill, B. (2010), 'The company you keep: international socialization and the diffusion of human rights norms', *International Studies Quarterly*, **54** (1), 127–45.

Hughes, M., L. Peterson, J. Harrison and P. Paxton (2009), 'Power and relation in the world polity: the INGO network country score, 1978–1998', *Social Forces*, **87** (4), 1711–42.

Hymer, S. (1972), 'The multinational corporation and the law of uneven development', in J. Bhagwati (ed.), *Economics and World Order*, New York: Macmillan, pp. 113–40.

Ingram, P., J. Robinson and M. Busch (2005), 'The intergovernmental network of world trade: IGO connectedness, governance, and embeddedness', *American Journal of Sociology*, **111** (3), 824–58.

Kick, E. (1987), 'World-system structure, national development, and the prospects for a socialist world order', in T. Boswell and A. Bergesen (eds), *America's Changing Role in the World-System*, New York: Praeger, pp. 127–55.

Kick, E. and B. Davis (2001), 'World-system structure and change: an analysis of global networks and economic growth across two time periods', *American Behavioral Scientist*, **44** 10), 1561–78.

Kim, S. and E.-H. Shin (2002), 'A longitudinal analysis of globalization and regionalization in international trade: a social network approach', *Social Forces*, **81** (2), 445–71.

Koster, F. (2007), 'Globalization, social structure, and the willingness to help others: a multilevel analysis across 26 countries', *European Sociological Review*, **23** (4), 537–51.

Lloyd, P., M. Mahutga and J. de Leeuw (2009), 'Looking back and forging ahead: thirty years of social network research on the world-system', *Journal of World-System Research*, **15** (1), 48–85.

Mahutga, M. (2006), 'The persistence of structural inequality? A network analysis of international trade, 1965–2000', *Social Forces*, **84** (4), 1863–89.

Mahutga, M. (2014a), 'Global models of networked organization, the positional power of nations and economic development', *Review of International Political Economy*, **21** (1), 157–94.

Mahutga, M. (2014b), 'Global production networks and international inequality: making a case for a meso-level turn in macro-comparative sociology', *Journal of World-Systems Research*, **20** (1), 11–37.

Mahutga, M. (2014c), 'Production networks and the organization of the global manufacturing economy', *Sociological Perspectives*, **57** (2), 229–55.

Mahutga, M. and D. Smith (2011), 'Globalization, the structure of the world economy, and economic development', *Social Science Research*, **40** (1), 257–72.

Mahutga, M., X. Ma, D. Smith and M. Timberlake (2010), 'Economic globalization and the structure of the world-city system: the case of airline passenger data', *Urban Studies*, **47** (9), 1925–47.

Milgram, S. (1967), 'The small-world problem', *Psychology Today*, **1** (1), 61–7.

Molm, L. (1990), 'Structure, action, and outcomes: the dynamics of power in social exchange', *American Sociological Review*, **55** (3), 427–47.

Neal, Z. (2011), 'Differentiating centrality and power in the world city network', *Urban Studies*, **48** (13), 2733–48.
Neal, Z. (2013), 'Brute force and sorting processes: two perspectives on world city network formation', *Urban Studies*, **50** (6), 1277–91.
Neal, Z.P., B. Derudder and P.J. Taylor (2019), 'Should I stay or should I go: predicting advanced producer services firm expansion and contraction', *International Regional Science Review*, **42** (2), 207–29.
Nemeth, R. and D. Smith (1985), 'International trade and world-system structure: a multiple network analysis', *Review*, **8** (4), 517–60.
Paxton, P., M. Hughes and N. Reith (2015), 'Extending the INGO network country score, 1950–2008', *Sociological Science*, **2** (May), 287–307.
Poros, M. (2011), 'Migrant social networks: vehicles for migration, integration and development', *Migrant Information Source*, accessed 19 December 2017 at https://www.migrationpolicy.org/article/migrant-social-networks-vehicles-migration-integration-and-development.
Portes, A. (1973), 'Modernity and development: a critique', *Studies in Comparative International Development*, **8** (3), 247–79.
Powell, W.W. (1990), 'Neither market nor hierarchy: network forms of organization', in B.M. Staw and L.L. Cummints (eds), *Research in Organizational Behavior*, vol. 12, Greenwich, CT: JAI Press, pp. 295–336.
Rozenblat, C. (2018), 'Urban systems between national and global: recent reconfiguration through transnational networks', in C. Rozenblat, D. Pumain and E. Velasquez (eds), *International and Transnational Perspectives on Urban Systems*, Singapore: Springer, pp. 19–49.
Rozenblat, C., F. Zaidi and A. Bellwald (2017), 'The multipolar regionalization of cities in multinational firms' networks', *Global Networks*, **17** (2), 171–94.
Sandholtz, W. and M. Gray (2003), 'International integration and national corruption', *International Organization*, **57** (4), 761–800.
Sassen, S. (2001), *The Global City: New York, London, Tokyo*, Princeton, NJ: Princeton University Press.
Simmel, G. (1908), *On Individuality and Social Forms: Selected Writings*, repr. 1971, Chicago, IL: University of Chicago Press.
Smith, D. and M. Timberlake (2001), 'World city networks and hierarchies, 1977–1997: an empirical analysis of global air travel links', *American Behavioral Scientist*, **44** (10), 1656–78.
Smith, D. and D. White (1992), 'Structure and dynamics of the global economy: network analysis of international trade 1965–1980', *Social Forces*, **70** (4), 857–93.
Snyder, D. and E. Kick (1979), 'Structural position in the world system and economic growth, 1955–1970: a multiple-network analysis of transnational interactions', *American Journal of Sociology*, **84** (5), 1096–126.
Steiber, S. (1979), 'The world system and world trade: an empirical exploration of conceptual conflicts', *Sociological Quarterly*, **20** (1), 23–36.
Sun, S.-L. and G. Barnett (1994), 'The international telephone network and democratization', *Journal of the American Society for Information Science*, **45** (6), 411–21.
Taylor, P. (2004), *World City Network: a Global Urban Analysis*, London: Routledge.
Torfason, M. and P. Ingram (2010), 'The global rise of democracy: a network account', *American Sociological Review*, **75** (3), 355–77.
Van Rossem, R. (1996), 'The world system paradigm as general theory of development: a cross-national test', *American Sociological Review*, **61** (3), 508–27.
Wasserman, S. and K. Faust (1994), *Social Network Analysis: Methods and Applications*, Cambridge: Cambridge University Press.
Wallerstein, I. (1974), 'The rise and future demise of the world capitalist system: concepts for comparative analysis', *Comparative Studies in Society and History*, **16** (4), 387–415.
Watts, D. (2003), *Six Degrees: The Science of a Connected Age*, New York: Norton.
Wejnert, B. (2005), 'Diffusion, development, and democracy, 1800–1999', *American Sociological Review*, **70** (1), 53–81.
Weller, S. and A.K. Romney (1990), *Metric Scaling: Correspondence Analysis*, New York: Sage.
White, D.R. and K.P. Reitz (1983), 'Graphs and semigroup homomorphisms on networks of relations', *Social Networks*, **5** (2), 193–234.

11. The role of proximity and distance in inter-urban networks

Thomas Sigler, Kirsten Martinus and Petr Matous

INTRODUCTION

As the chapters in this book reveal, cities are embedded within myriad network types at multiple scales linking people and places both within and between cities across the globe. These links are uneven, with some people and places more connected than others, depending on their proximity, which is the focus of this chapter. Within and between cities, proximity can be related to geography as well as other dimensions reflecting closeness. For example, consider a close childhood friend you grew up with. Your spatial or geographic proximity would have been fairly close when you were young, becoming more distant if you decided to move away to pursue a career or tertiary studies. However, you may still regularly see each other and communicate via social media, even if you or your friend then get a job in another city. In this instance, your geographic proximity to this friend has decreased, but your social proximity has remained constant, or even increased. In the following sections, we explain proximity as a concept and how different types collectively shape the connectedness of different cities.

GLOBAL URBAN NETWORKS IN DAILY LIFE

The above scenario is a simple example of a complex issue that network scientists have tried to examine through the concept of proximity. Understanding how people and places are linked in a multitude of social, economic, political and organisational relationships, across time and space provides insight into interesting research and policy questions. Complex global networks can be segmented to smaller spatial scales, such as the individual, or aggregated to the neighbourhood, city, provincial, national or regional scales. Likewise, networks can comprise relationships within and/or between firms, organisations, institutions and governments; these serve as foundational to the diverse empirical studies approached from a network perspective. Although these can often reveal vast and complex networks, ego networks radiating from individual (that is, node-level) daily social and economic interactions are the building blocks of globally connected systems.

Proximity has increasing salience in the digital era as many interactions between networked actors have shifted from physical to virtual. In the past, interpersonal networks were almost exclusively formed through face-to-face interactions at a specific location. Although telecommunications have enabled humans to interact across large distances for approximately a century and half, the past 25 years have witnessed rapid expansion of digital communication in various forms through increasingly affordable and accessible communications devices, platforms, network access and related technologies. As sociologist Manuel Castells (1996 [2010]) observed in *The Rise of the Network Society*, humans are now connected through instantaneous flows of knowledge, information and capital. He argued that networks are a fundamental concept for understanding contemporary human organisation. While this and previous waves of innovation have led commentators to lament the 'death of geography' (Bates 1996; Kolko 2000), research has repeatedly proven that 'distance matters' across a variety of networked phenomena (Han et al. 2018; Mok et al. 2010).

Furthermore, the complexity of network relationships makes analysis difficult, given the various perspectives, contexts and interests of individuals at different scales. For example, the chief executive officer of a multinational oil company may make corporate decisions which lead to environmental destruction at the global level, but voice concern over the disappearance of parkland in their own residential neighbourhood. It is not surprising, then, that different disciplines in the social sciences have increasingly looked to networks and network science to explain complex social and economic relations. Global production networks theory (cf. Henderson et al. 2002) examines, via interviews with corporations, places as international nodes and linkages of commodity value chains in the production of goods and services, such as electronics, and more recently in the recycling or reuse of unwanted goods, such as e-waste. World city network (cf. Taylor 2004) theory explores the international nodes and linkages of cities by mapping trans-national corporate relations (via office or subsidiary locations) or other human flows (via labour migration data or passenger flight data). This provides insight into the role and position of different places in the global network of cities, as well as the complex urban relationships playing out across the globe. While global production networks and world city networks approach international economic relationships in slightly different ways, they highlight that geographic distance is not the only factor determining why cities interact with each other, with farther away cities sometimes having stronger ties and relationships than closer cities.

DISTANCE AND PROXIMITY

Distance and proximity are two terms we use in everyday communication. Distance generally refers to how far one thing is from another. In absolute terms, we might say something is 50 kilometres away. That might be precise enough if we are discussing the driving distance to the next town along a highway, but we can be more specific by adding digits and make it 50.259 kilometres, and so on. Distance can also be relative; 50 kilometres can be close, if we are referring to an approaching asteroid's distance from Earth, or far if it is the nearest service station and our car is running short of fuel. Alternatively, we might simply say we will be there in about an hour. Thus what some consider near, others may consider distant, and vice-versa, and our interpretation depends on the circumstances. Kevin Lynch's (1960) work in *The Image of the City* demonstrated how social class, among other human factors, can have a strong impact on residents' perceptions of how the various elements of the city are arranged. He found car owners had a broader perspective on urban space than pedestrians, but were less acquainted with spatial relationships on a finer scale.

To further contextualise the issue of operationalising geographical distance in both absolute and relative terms, its measurement can be subject to multiple interpretations. For example, we can measure the cargo shipment between New York and Los Angeles via absolute straight-line distance, or 'as the crow flies', as 3944035 metres, or 3944 kilometres. This is known as geodesic, or Euclidian, distance (after the Greek mathematician Euclid). However, unless the cargo is transported by drone or helicopter, this figure needs to be contextualised as it can take anywhere between a few hours by air to several weeks by non-motorised transport. Another way of measuring distance is by road distance, or Minowski distance. As roads conform to natural terrain features and orthogonal street grids, 3944 Euclidian kilometres turns into 4503 kilometres using the interstate highway network to get from New York to Los Angeles by truck.

The distance between New York and Los Angeles is also contingent upon the cities' relative connectivity. As the two most populous cities in the United States, there are several direct flights between them each day. Thus, by air the two could be said to be well connected, perhaps even more so than smaller cities closer to one another. Most commonly, however, routine inter-continental freight shipments are deployed by a combination of rail and truck, and do not take a direct route. Instead, cargo is routed through a series of warehouses and depots, minimising cost for the carrier and maximising efficiency. Thus, a more useful measure would be Chebyshev distance, which is the minimum sum of vectors between A and B or, in this example, between the transport segments connecting New York and Los Angeles. This is the basis of Global Navigation Satellite Systems (GNSS) based applications indicating optimum route choice. While this could be measured in kilometres or miles, a more useful measure would be of time in hours and minutes. This calculation would factor in impedance, which is defined as the resistance to traversing a network. Impedance can be enumerated in time – accounting for traffic delays, downtime in warehouses, and so on – or cost. When we add these segments together, a cargo shipment between the two cities may travel more than double the Euclidian distance between them (or more in our example, if it has to go through the Panama Canal), but may arrive more quickly than cargo to smaller cities in nearby states. Why? Owing to the sheer volume of daily shipments between them, New York and Los Angeles are well connected, translating into high network proximity despite relatively low spatial proximity.

TYPES OF PROXIMITY

One framework to conceptualise non-spatial proximity derives from economic geography (Balland et al. 2016). Primarily concerned with the exchange of information between firms and institutions, these proximity types have emerged to explain different varieties of closeness. Balland (2012) and others (Boschma 2005; Lagendijk and Oinas 2005) have articulated a number of these types, each of which adds a new dimension to proximity that can be applied to relationships between people, firms, institutions, organisations or other actors, which translate into relationships between places. These proximity types are listed next.

- Cognitive proximity refers to a common knowledge or technical base (Nooteboom 2000; Knobens and Oerlemans 2006), indicating that actors within a networked system share a common educational or professional background or understanding. While this facilitates strong collaborations to solve issues, it may stymie creative thinking with actors having similar ideas and knowledge on a topic. Between cities, cognitive proximity may be created when there is a common industry base or other similarities in socio-economic profile. This may assist in the transfer of policy or strategies from one place to another to address common issues or future directions. For instance, cities associated with the energy sector (for example, producers of oil, gas and coal) will have many of the same multinational corporations, such as Chevron, Shell, Hess and BHP, in each city. These cities may also form governance and strategic links, such as through the World Cities Energy Partnership, as well as through the small-business and start-up worlds to solve common problems through innovation and entrepreneurship.

- Institutional proximity is defined through common sets of norms, practices, incentives and other conventions (Martinus and Sigler 2018). In corporations, these are associated with common legal structures. A simple example is that a research team distributed across various global universities and research laboratories might find it easier to work together if they have the same goals and objectives. Individuals in disparate academic contexts are often brought together through common goals (for example, advancing science), and motivated by similar factors, such as research output benchmarks or global research indicators. This contrasts with joint projects between corporations and universities as organisations whose respective interests in profits and public knowledge production may be competing. In urban networks, institutional proximity will manifest through common governance frameworks (such as memorandums of understanding, MOUs) or political collaboration. The latter is illustrated by the way governments operate together when trade blocs are put into place, such as the Association of Southeast Asian Nations (ASEAN), the North American Free Trade Agreement (NAFTA) and the Southern Common Market (MERCOSUR) (Duina 2006), or when cities are part of twin-city relationships (Jayne et al. 2013).
- Organisational proximity is defined by the way actors come together as part of a large-scale system. An example would be the various city-based offices of a large multinational firm. This may involve complex organisational structures of company branches, as well as joint ventures and subsidiaries (Boschma 2005). In inter-urban networks, this plays out in the networks of particular industry sectors, which display a strong hierarchical organisation in how corporations locate branches in cities across the globe. This may be owing to the way knowledge in particular industries is accumulated in some places over others, and the need for corporations to access this knowledge leads them to locate in particular cities (Martinus and Sigler 2018). For example, the strong global presence in the advanced producer services sectors of financial, business, advertising and legal services in New York, London, Tokyo, Hong Kong and Sydney accounts for a high level of interaction between these 'world cities' (Beaverstock et al. 1999). It is also apparent in the way some cities play mediating, or 'relational' roles in the global economy (Sigler 2013), such as the way that Hong Kong is a gateway to China, and Dubai to the Middle East.
- Social proximity refers to closeness in relation to friendships, family and common acquaintances. This is particularly relevant when trust is crucial, with ties forged

through interpersonal linkages. Social proximity can be articulated on a personal scale (for example, person X knows person Y) or on a collective scale (for example, person X is of the same ancestry as person Y). In the latter, cultural or religious ties may serve as a proxy for proximity where a common language, race or ethnicity connect unrelated parties. A community is articulated as this, regardless of existing relationships. In urban network terms, we see connections between and within cities along social dimensions. Cities sharing a common history, connected to each other through human interactions ranging from migration, to commercial, to sporting networks. For instance, this might create respective networks between cities in which major baseball, cricket, ice hockey or soccer teams compete. Shaw (2010), for example, found that cities within the British colonial Commonwealth retain strong links in contemporary development.

These four proximity types are not alone in being cited in the academic literature on networks. Geographical (spatial) proximity, for example, still matters in social networks. This is reminiscent of Tobler's (1970) first law of geography, which postulates that while all things are related to all others, near things are more related than distant things. To this end, the literature suggests that geographical proximity stimulates social proximity and, to a lesser extent, institutional and cognitive proximity (Hansen 2015). However, as our examples have demonstrated, this is not always true – particularly in instances in which another proximity type exerts more influence. For example, a migrant labourer in a foreign country is more likely to have strong ties to their city of origin than within the city in which they work. Researchers, such as Hansen (2015), have suggested that non-spatial proximity can substitute geographical proximity in some instances.

In theory, the distinctions between various proximity types are straightforward. In practice, however, there is a high degree of overlap between them. Connectivity between cities may be reinforced between multiple parallel systems, including transportation and infrastructure links, common institutions and organisations, and varying degrees of cognitive and social proximity. Torre and Rallet (2005) suggest there is often overlap between geographical proximity and organisational proximity, for example, when two cities within the same nation-state are constrained or governed by the same set of standards. A final example of this overlap is presented through cities with similar economic functions; for instance, two cities whose dominant industry is tourism. Both are likely to have similar ways of doing business (for example, cognitive proximity through trained hotel and theme park operators), similar institutions (for example, tour agencies, hotel chains and car rental agencies) and social practices, as tourists often demand the same set of standards from one resort to the next). It is thus perhaps no surprise that Seine-et-Marne, France, and Urayasu, Japan, are formal 'sister cities' of Orlando, Florida, all of which are home to Disney theme parks.

PROXIMITY AND URBAN NETWORK STRUCTURES

Networks are systems of entities termed nodes that are connected together through some type of tie or link. Many systems can be represented as networks. In social networks, the nodes are people and the links are their relationships. A pair of nodes connected through a tie is a dyad. In airline networks, nodes are airports and ties are direct flights between them. Real-world networks are embedded in physical and social spaces, which can be characterised by the various types of proximities described in the previous section. Institutional proximity and administrative boundaries are known to shape the inter-urban commuting and telecommunication networks (Blondel et al. 2010), and physical distance influences the cost of the creation, operation and maintenance of links in infrastructure networks.

Imagine a global road network in which all roads lead directly and only to Rome, without crossing one another anywhere else. You could get from anywhere in the world to any other place with only one set of traffic lights on the way. The problem is that for most origin–destination pairs, you would need to travel for a long time and the junction in Rome would probably be busy. Moreover, any disruption in Rome would paralyse the entire global network. This type of road network structure would not be optimal.

Let us consider another example. From the viewpoint of administrative costs, it might appear efficient to centralise reporting structures of large organisations. If every unit reports directly to the same central unit, the entire organisation can be connected through a small number of chains of command. However, after considering the distance in relation to tasks and responsibilities between different units, as well as the capacity of the central unit, again, this organisational structure may not always be optimal.

In order to better understand non-physical proximity, it is necessary to understand how the various types of proximities between nodes in networks are relevant for the probability of link (tie) creation and for the flows of people, resources and information within the links. It is also important to understand how diverse costs of long links influence optimal topologies of network and structures of network processes.

One of the most common means of understanding these concepts within a network structure is social network analysis (SNA), a method of analysis associated with understanding social structures through networks and graph theory. Network graphs in SNA are created using matrices which tell us how nodes are connected. Depending upon the type of data, network graphs can also include the direction of flows between nodes.

If we are exploring trade networks of Singapore as a port city, we can either examine the absolute number of other port cities it connects to or we can divide connections into ports receiving exports from Singapore and ports it imports goods from. This gives greater understanding of how information, resources and people flow through a network.

Social network analysis measures the network distance of nodes from each other in different ways, identifying the most important nodes in a network through centrality. This also helps us measure network proximity in different ways. There are many measures of centrality, which – depending on the phenomenon under analysis – can either be interpreted to understand the local importance of a node, or the importance of a node within the overall network. These are some of the most common in urban analysis:

1. Betweenness centrality of a node is the number of shortest network pathways between all pairs of nodes that go through the node. In inter-urban networks, this indicates cities that act as bridges, brokers or mediators between other cities. In street networks, these are intersections or road segments that are vital to the overall network (Crucitti et al. 2006). Betweenness centrality has been associated with intermediary roles in

various networks, and is thus sometimes considered as an anomaly in that it often identifies a set of nodes that other measures do not. In relation to this, Guimera et al. (2005) identify the importance of cities such as Port Moresby and Anchorage in air hubs based on their betweenness centrality. In these examples, neither is particularly important in the global network, but many smaller airports would be cut off from the system entirely without them. Similarly, Krätke (2014) finds that Tokyo and Detroit have high betweenness centrality but relatively less degree centrality (discussed in item 3) in the global automotive industry, meaning that they connect to specific subnetworks as intermediaries.

2. Closeness centrality of a node is a measure of its average nearness to other nodes. This enumerates proximity as the inverse sum of the shortest path (see the network distance discussion in the 'The length of links and the distance on a network' section) between the node and all others. We might think of a city with high closeness centrality as one with access to other cities. Closeness centrality is perhaps the most global measure as it takes into account the role of each node in the entire network. For example, if we were to construct a network from social media data, such as Instagram or Twitter, we would find that global cities, such as Sydney, Toronto and Dubai, would have high closeness centrality. This is owing to their multi-cultural populations, which are likely to include residents from nearly every country in the world, meaning the number of steps to reach someone in a small village in Moldova, Ethiopia or Papua New Guinea would be relatively lower than an equally large city with a culturally homogenous population. Closeness has been used in analyses at the intra-urban scale to look at street grid integration (Jiang and Claramunt 2004), but finds most utility in studies incorporating large communities of actors (Newman 2001).
3. Degree centrality of a node is the number of its connections to other nodes, with in-flows (for example, immigration) measured by in-degree and out-flows (for example, emigration) measured by out-degree. This is the simplest measure of centrality, and the most local, accounting only for direct connections. Thus degree centrality gives us little sense of positionality within the overall network. High-degree centrality is associated with a large number of direct connections, and studies using both in-degree and out-degree centrality are able to distinguish between how cities either receive flows, contribute flows, or both. Thus unless only local and first-order ties are of interest, degree centrality is perhaps best combined with other centrality measures, as numerous studies do (Ducruet and Notteboom 2012).
4. Eigenvector centrality of a node is its level of connectedness to other highly connected nodes, taking into consideration the entire network. Internet search engines rely on a form of eigenvector centrality for ranking the importance of websites. Thus, we might think of eigenvector centrality as high among actors that are highly connected to the inner circle of the network. For this reason, it has been one of the most often used centrality measures in urban research (Neal 2013). If we were studying the fashion industry, it is likely that cities such as Milan and Paris would have high eigenvector centrality, as their designers would be connected to counterparts in other fashion capitals, such in New York, Tokyo and London. Thus, it could be argued that the influence of designers in Milan or Paris is greater than in other cities as they are more likely to be connected to important and influential people. In urban research, eigenvector centrality can be associated with power, particularly where the strength of

a city is reflected through the connectivity of its firms, institutions or organisations (Sigler and Martinus 2017).

Just as various proximity types are related, there is strong correlation between particular measures of centrality. However, the levels of correlation between different centrality indices depend on the structure of the network. For example, in a network with one major core and no significant divides, nodes with high closeness centrality (which are near the centre of the network as regards the number of steps in the network) may also have many links (degree) and be on many shortest pathways between other nodes (betweenness). In this type of network, major centrality indices will rank nodes in a similar fashion.

However if, for example, a network is fragmented into cliques, centrality indices may diverge. A node with many links within a clique on the periphery of a network will have a high degree but relatively low closeness centrality and node with a couple of links spanning different cliques will have relatively low degree but high betweenness centrality.

BIRDS OF A FEATHER FLOCK TOGETHER

Social scientists long ago realised that people are more likely to connect to similar and proximate people. Rich people are more likely to befriend other rich people and popular individuals are more likely to interact with other popular individuals. This tendency is known as homophily (McPherson et al. 2001). The common tendency in social networks for high-degree nodes to be connected to other high-degree nodes is termed assortativity.

However, similarity does not necessarily stimulate connectivity and, as we noted at the beginning of the chapter, neither does physical (spatial) proximity. Connecting to dissimilar and distant people or organisations across administrative, institutional or socio-economic boundaries, or bridging gaps in social networks can be advantageous, primarily because others may not be doing that enough (Todo et al. 2016). Connecting across distance can provide access to precious diverse resources and opportunities for brokerage (Burt 2000). In technological networks, nodes with many connections are often connected to nodes with only a few connections, such as the World Wide Web where typically many obscure sites link to a popular site (Barthélemy 2011). Firms with partnerships in distant regions can tap into diverse local knowledge that is spatially embedded within localised interpersonal interactions (Mackinnon et al. 2004; Fitjar and Rodríguez-Pose 2011). Inter-regional links help organisations become more creative, and to recover faster after regional disasters since they can receive support from their partners in unaffected areas (Todo et al. 2015, 2016).

THE LENGTH OF LINKS AND THE DISTANCE ON A NETWORK

Let us now imagine a non-urban community in a remote region with very limited access to transportation and communication infrastructure (such as the region described in Matous et al. 2013). The friction of distance is high; people interact mainly with their neighbours (short links) and rarely with people who live beyond the walking distance from their house (long links). This illustrates the concept of network distance, which is defined by the number of links on the shortest paths between pairs of nodes within the network. The network distances in these networks are large, as it takes many hops over many short links from neighbour to neighbour for a common social norm to spread across physical space.

These types of social networks emerge when transportation is expensive, communication is possible only locally, or if friendships are common only with similar people in the socio-economic space. The constraint on the length of links leads to high clustering within such networks. Social networks are considered clustered if friends of friends are also likely to be friends with one another. If long links in the physical or social space are impossible, this is often the consequence. In a context where everyone connects only to proximate peers in physical or social space, if you are connected to two people, they must be proximate to you and therefore likely to also be proximate and connected to one another.

The limit on long links prevents the creation of network hubs, which are extremely high-degree nodes. Hubs are an important feature of many types of networks as they significantly increase connectivity and decrease network distances in many real-world networks. These are networks with several very high-degree nodes and a vast majority of very low-degree nodes, where the degree distribution of the nodes may resemble the power law. A power-law distribution is a statistical distribution describing many natural and human-made phenomena in which most common events are of a very small magnitude, with some events of a disproportionally larger magnitude also occurring. For example, visitor frequency has been approximated by the power law. While many, if not most, tourists to Italy would visit Rome, Venice or Florence, it is considerably less likely that they would visit a small and remote village without hotels, restaurants or major attractions. These networks are referred to as scale-free, and are a popular subject of study in physics (Barabási and Albert 1999; Barabási 2009).

Not all distributions in nature follow the power law. For example, there are no people who are many times taller than the average human. The term scale-free has become commonly used for power-law distributions simply because these distributions do not have a typical scale. Although it might not be appropriate, it makes more sense to talk about a woman of a typical weight than a woman of a typical wealth. Most people's weight is relatively close to the mean of the distribution, but most people earn less than the mean income and some people earn many times more than that (although wealth is not perfectly approximated by power law either). In relation to a city, we might think of urban populations as being close to scale-free, as most large cities will have a population of between 1 million and 5 million, with a handful of very large megacities and extended metropolitan regions having populations of more than 10 million. Similarly, most airports will be connected to a small number of other airports, while a few major global hubs (for example, Tokyo's Narita airport and London's Heathrow airport) are connected to hundreds of hubs.

When geographical proximity matters, scale-free networks are unlikely to emerge (see the analysis of rail networks and inter-urban commuting traffic of Kurant and Thiran 2006 and Montis et al. 2007). The influence of various types of proximities on the formation and maintenance of links supports the existence of smaller regional and local hubs as opposed to larger global hubs in global networks. Among different types of transportation networks, global maritime cargo networks tend to be closest to scale-free networks as there are only weak spatial proximity constraints on the creation of links and high-degree nodes. Some types of sea cargo can be transported directly to the opposite side of the world, and hub ports are disproportionally larger than other ports (Hu and Zhu 2009). As a consequence of long links and the large size of some seaports, 50 percent of all possible origin pair destinations in the world are connected within a network distance of two or less. That is, cargo can be transported between half of all pairs of ports either directly or with only one interchange on the way (Kaluza et al. 2010).

Long links make hubs possible by enabling them to connect to more nodes than are available in their vicinity. In some instances, long links can connect only to hubs. An intercontinental flight is unlikely to connect two small-town airports. The payoff of connecting to a hub is typically what compensates for the increased cost of the long link. We may be more willing to connect to a person who is distant in social space if the person is highly influential.

In social networks, short links are more common and, therefore, social networks tend to be clustered. Real social networks also tend to include some long-distance connections. Even a small number of long links can significantly decrease the average distances within a network. Most readers have probably heard that, as a global society, we are connected within six degrees of separation. Research suggests that it could be even less (Watts 2004). Although clustered interactions are prevalent, the short network distances around the world are the result of relatively less common, long social connections. In network science, highly clustered networks combined with short average pathways are termed small-world networks (Watts and Strogatz 1998). If you go to a party in a city where you have never been before and find that you have a friend in common with someone you just met, this is thanks to the small-world phenomenon enabled by a small number of long links.

WHICH PART OF A NETWORK IS IMPORTANT?

While important hubs can emerge in any part of a small-world network, in large-world networks composed only of short links, the most important nodes are those that are centrally located, even if these nodes do not have more connections than the nodes on the spatial periphery.

Just as we can measure the importance of a node in a network by centrality, we can also measure the centrality of links. For example, streets and crossings in city centres have the highest betweenness centrality as they lie on the shortest pathways between many parts of the city, and therefore the central areas tend to suffer from congestion. To combat this, urban planners develop long links by flyovers, bypasses and tunnels that create shortcuts in the urban road networks around the city centre. As regards inter-urban connections, Wójcik (2013) documented the importance of 'NY-LON' in financial networks characterised by the intense interactions between firms in New York and London. This could be measured in numbers of daily calls, emails, flights, messages, reciprocal visitors, and so on.

Compared with road networks, inter-urban airline networks are less constrained by geographical proximity and instead are designed to reduce network distances. For example, it may be preferable to change planes only once, albeit on a circuitous flight path, than to change two or three times taking the most direct geodesic route. When the cost of a longer link is relatively low in relation to the cost of an additional node, network centrality diverges from central locational measures derived from geographical information systems (GIS). Alice Springs (population 28 000) is the largest settlement near the geographical centre of Australia, for example, but fewer flights and passengers pass through its airport than through airports on the geographical fringes of Australia in large cities such as Sydney or Melbourne.

A final example is the Japanese energy companies, which often supply electricity to firms headquartered on the other side of the country (Matous and Todo 2016). Owing to their abundance of long links, the companies that are topologically most central to the Japanese network are, unlike other major Japanese firms, geographically distributed in rural regions outside main Japanese industrial clusters and population centres. The implication for the resilience of the system is that any region-specific disaster (for example, earthquakes) can hit a topologically central node representing an energy company and cause temporary disruptions. Power outages will be transferred through long links around the entire country. However, no region-specific shock would destroy the spatially distributed topological core of the Japanese production network.

CONCLUSION

The purpose of this chapter has been to demonstrate how proximity in urban networks can refer to multiple relationships, some geographic and others not. Conventionally, proximity refers to closeness. This can be absolute if measured in metres, or more commonly relative closeness in reference to something else such as spatial, or geographic, proximity.

Our main focus, however, has been on non-physical proximity or when two objects are close in a way other than physical. Proximity in this instance means well connected, something that can be articulated and elaborated in a number of ways. Geographers interested in inter-urban networks have identified several proximity types outside of spatial, including cognitive proximity, institutional proximity, organisational proximity and social proximity. Each of these highlights a different way urban actors are connected, such as how firms are connected through shared knowledge, researchers are connected through shared objectives, or people are globally connected through their respective linguistic, cultural, ethnic or religious networks.

Social network analysis offers a number of tools for understanding proximity. At the node level, centrality is a good indicator of position within a network. Nodes with higher closeness centrality tend to be closer to others within the system, and various types of centrality measure either system-wide (eigenvector) or local (degree) characteristics. The tendency for similar actors to be well connected is captured by homophily and assortativity.

Network distance is perhaps the most straightforward way to capture proximity using network analysis. It tells us how far two nodes are from one another, that is, how many steps it takes to move between nodes. Research in network science advises us that many phenomena follow a scale-free distribution, meaning that few nodes are much more connected than others. Although similar patterns can be often observed in inter-urban flows, whether in reference to transport routing, tourism or corporate firm linkages, strictly scale-free behaviour is rare in networks embedded in geographical space; there is a physical limit to how large network hubs and flows can become in physical space.

As we now live in a digital era, conducting research on network connectivity across geographical distances is perhaps more accessible than ever. Voluminous amounts of user-generated information on flows can be found online, and in a few basic steps we can collect spatial data on a number of network relationships. There are therefore numerous avenues for future research on proximity in urban networks. Understanding the mechanisms by which systemic relationships are formed is critical in better unpacking the role of cities within inter-urban networks.

REFERENCES

Balland, P.A. (2012), 'Proximity and the evolution of collaboration networks: evidence from research and development projects within the global navigation satellite system (GNSS) industry', *Regional Studies*, **46** (6), 741–56.

Balland, P.A., J.A. Belso-Martínez and A. Morrison (2016), 'The dynamics of technical and business knowledge networks in industrial clusters: embeddedness, status, or proximity?', *Economic Geography*, **92** (1), 35–60.

Barabási, A.L. (2009), 'Scale-free networks: a decade and beyond', *Science*, **325** (5939), 412–13.

Barabási, A.L. and R. Albert (1999), 'Emergence of scaling in random networks', *Science*, **286** (5439), 509–12.

Barthélemy, M. (2011), 'Spatial networks', *Physics Reports*, **499** (1–3), 1–101.

Bates, S. (1996), 'The death of geography, the rise of anonymity, and the Internet', *American Enterprise*, **7** (2), 50–52.

Beaverstock, J.V., R.G. Smith and P.J. Taylor (1999), 'A roster of world cities', *Cities*, **16** (6), 445–58.

Blondel, V., G. Krings and I. Thomas. (2010), 'Régions et frontières de téléphonie mobile en Belgique et dans l'aire métropolitaine bruxelloise', *Brussels Studies. La revue scientifique pour les recherches sur Bruxelles/Het wetenschappelijk tijdschrift voor onderzoek over Brussel/The Journal of Research on Brussels*, **42**, accessed 22 February 2021 at http://journals.openedition.org/brussels/799.

Boschma, R. (2005), 'Proximity and innovation: a critical assessment', *Regional Studies*, **39** (1), 61–74.

Burt, R.S. (2000), 'The network structure of social capital', *Research in Organizational Behavior*, **22**, 345–423.

Castells, M. (1996), *The Rise of the Network Society: The Information Age: Economy, Society, and Culture*, vol. 1, 2nd edn 2010, Chichester: John Wiley & Sons.

Crucitti, P., V. Latora and S. Porta (2006), 'Centrality in networks of urban streets', *Chaos: an Interdisciplinary Journal of Nonlinear Science*, **16** (1), 015113.

Ducruet, C. and T. Notteboom (2012), 'The worldwide maritime network of container shipping: spatial structure and regional dynamics', *Global Networks*, **12** (3), 395–423.

Duina, F.G. (2006), *The Social Construction of Free Trade: The European Union, NAFTA, and Mercosur*, Princeton, NJ: Princeton University Press.

Fitjar, R.D. and A. Rodríguez-Pose (2011), 'Innovating in the periphery: firms, values and innovation in Southwest Norway', *European Planning Studies*, **19** (4), 555–74.

Guimera, R., S. Mossa, A. Turtschi and L.N. Amaral (2005), 'The worldwide air transportation network: anomalous centrality, community structure, and cities' global roles', *Proceedings of the National Academy of Sciences*, **102** (22), 7794–9.

Han, S.Y., M.H. Tsou and K.C. Clarke (2018), 'Revisiting the death of geography in the era of big data: the friction of distance in cyberspace and real space', *International Journal of Digital Earth*, **11** (5), 451–69.

Hansen, T. (2015), 'Substitution or overlap? The relations between geographical and non-spatial proximity dimensions in collaborative innovation projects', *Regional Studies*, **49** (10), 1672–84.

Henderson, J., P. Dicken, M. Hess, N. Coe and H.W.C. Yeung (2002), 'Global production networks and the analysis of economic development', *Review of International Political Economy*, **9** (3), 436–64.

Hu, Y. and D. Zhu (2009), 'Empirical analysis of the worldwide maritime transportation network', *Physica A: Statistical Mechanics and its Applications*, **388** (10), 2061–71.

Jayne, M., P. Hubbard and D. Bell (2013), 'Twin cities: territorial and relational geographies of "worldly" Manchester', *Urban Studies*, **50** (2), 239–54.

Jiang, B. and C. Claramunt (2004), 'A structural approach to the model generalization of an urban street network', *GeoInformatica*, **8** (2), 157–71.

Kaluza, P., A. Kölzsch, M.T. Gastner and B. Blasius (2010), 'The complex network of global cargo ship movements', *Journal of the Royal Society Interface*, **7** (48), 1093–103.

Knoben, J. and L.A. Oerlemans (2006), 'Proximity and inter-organizational collaboration: a literature review', *International Journal of Management Reviews*, **8** (2), 71–89.

Kolko, J. (2000), 'The death of cities? The death of distance? Evidence from the geography of commercial Internet usage', in I. Vogelsang and B. Compaine (eds), *The Internet Upheaval: Raising Questions, Seeking Answers in Communications Policy*, Cambridge, MA: MIT Press, pp. 73–98.

Krätke, S. (2014), 'How manufacturing industries connect cities across the world: extending research on "multiple globalizations"', *Global Networks*, **14** (2), 121–47.

Kurant, M. and P. Thiran (2006), 'Extraction and analysis of traffic and topologies of transportation networks', *Physical Review E*, **74** (3), 036114.

Lagendijk, A. and P. Oinas (2005), 'Proximity, external relations, and local economic development', in A. Lagendijk and P. Oinas (eds) *Proximity, Distance and Diversity, Issues on Economic Interaction and Local Development*, Farnham: Ashgate, pp. 3–22,

Lynch, K. (1960), *The Image of the City*, Boston, MA: MIT Press.

Mackinnon, D., K. Chapman and A. Cumbers. (2004), 'Networking, trust and embeddedness amongst SMEs in the Aberdeen oil complex', *Entrepreneurship and Regional Development*, **16** (2), 87–106.

Martinus, K. and T.J. Sigler (2018), 'Global city clusters: theorizing spatial and non-spatial proximity in inter-urban firm networks', *Regional Studies*, **52** (8), 1041–52.

Matous, P. and Todo, Y. (2016), 'Energy and resilience: the effects of endogenous interdependencies on trade network formation across space among major Japanese firms', *Network Science*, **4** (2), 141–63.

Matous, P., Y. Todo and D. Mojo, D. (2013), 'Boots are made for walking: interactions across physical and social space in infrastructure-poor regions', *Journal of Transport Geography*, **31** (C), 226–35.

McPherson, M., L. Smith-Lovin and J.M. Cook (2001), 'Birds of a feather: homophily in social networks', *Annual Review of Sociology*, **27** (1), 415–44.

Mok, D., B. Wellman and J. Carrasco. (2010), 'Does distance matter in the age of the internet?', *Urban Studies*, **47** (13), 2747–83.

Montis, A.D., M. Barthélemy, A. Chessa and A. Vespignani (2007), 'The structure of interurban traffic: a weighted network analysis', *Environment and Planning B: Planning and Design*, **34** (5), 905–24.

Neal, Z. (2013), 'Does world city network research need eigenvectors?', *Urban Studies*, **50** (8), 1648–59.

Newman, M.E. (2001), 'Scientific collaboration networks. II. Shortest paths, weighted networks, and centrality', *Physical Review E*, **64** (1), 016132.

Nooteboom, B. (2000), *Learning and Innovation in Organizations and Economies*, Oxford Scholarship Online, accessed 22 February 2021 at https://oxford.universitypressscholarship.com/view/10.1093/acprof:oso/9780199241002.001.0001/acprof-9780199241002.

Shaw, T.M. (2010), 'Comparative commonwealths: an overlooked feature of global governance?', *Third World Quarterly*, **31** (2), 333–46.

Sigler, T.J. (2013), 'Relational cities: Doha, Panama City, and Dubai as 21st century entrepôts', *Urban Geography*, **34** (5), 612–33.

Sigler, T.J. and K. Martinus (2017), 'Extending beyond "world cities" in World City Network (WCN) research: urban positionality and economic linkages through the Australia-based corporate network', *Environment and Planning A*, **49** (12), 2916–37.

Taylor, P. J. (2004), *World City Network: A Global Urban Analysis*, New York: Psychology Press.

Tobler, W.R. (1970), 'A computer movie simulating urban growth in the Detroit region', *Economic Geography*, **46** (June), 234–40.

Todo, Y., P. Matous and H. Inoue (2016), 'The strength of long ties and the weakness of strong ties: knowledge diffusion through supply chain networks', *Research Policy*, **45** (9), 1890–906.

Todo, Y., K. Nakajima and P. Matous (2015), 'How do supply chain networks affect the resilience of firms to natural disasters? Evidence from the Great East Japan Earthquake', *Journal of Regional Science*, **55** (2), 209–29.

Torre, A. and A. Rallet (2005), 'Proximity and localization', *Regional Studies*, **39** (1), 47–59.

Watts, D.J. (2004), *Six Degrees: The Science of a Connected Age*, London: W.W. Norton.

Watts, D.J. and S.H. Strogatz (1998), 'Collective dynamics of "small-world" networks', *Nature*, **393** (6684), 440–42.

Wójcik, D. (2013), 'The dark side of NY–LON: financial centers and the global financial crisis', *Urban Studies*, **50** (13), 2736–52.

12. About being in the middle: conceptions, models and theories of centrality in urban studies

Michiel van Meeteren

INTRODUCTION

A commonly encountered perspective in textbooks explains the existence of cities as function of being in the middle of something. Cities are centres of political, cultural, economic and/or religious life, explaining the large religious buildings, stock exchanges or pantheons in the city. According to many (for instance, Bird 1977, p. 1; Friedmann 1968, p. 236; Lefebvre 1974 [1991], pp. 331–34), the associated notion of centrality amounts to a key building block of what the city is about. This chapter surveys the urban theories that try to understand the benefits, the side effects and the desires associated with being in the middle.

Urban centrality's core idea is that being in the middle exerts spatial effects. The middle can induce a desire to be there, a centripetal tendency. Authors (for example, Bobek 1927) emphasize the magnetic attraction of urban places, invoking images of being pulled to bright lights in the big city. Alternatively, being in the middle can be something we are pulled away from, a centrifugal tendency, for instance when the middle is too expensive, too crowded or too dirty. These centripetal and centrifugal tendencies were first theorized as urban phenomena by Schlüter (1899, cited in Müller-Wille 1978, p. 50) in Germany and by H.G. Wells (1902, cited in Bird 1977, p. 104) in the anglophone world, and were further elaborated in the subsequent century. Colby (1933) provides an overview of what these tendencies entail in early twentieth-century cities. The balance between countervailing centripetal and centrifugal tendencies generates distinctive divisions of labour and zonal patterns of urban fabric within a particular technological conjuncture. Colby (1933) notes how new technological possibilities, transportation modalities and changing economic fortunes continuously upset these equilibriums and change centralities (Van Meeteren et al. 2016a).

From this abstract perspective of centrifugal and centripetal tendencies, all location theory is theory about being in the middle. However, it is easy to complicate the matter by unpacking the question, what is the theory about? Being in the middle of a landfill will surely exert a different balance of centrifugal and centripetal tendencies than being in the middle of a sun-bathed park. How do we conceive of the middle? As a point in the middle of a plane (that is, an imaginary flat surface through or joining material objects), region, network or (social) structure? There might be several middles to be considered. Who or what is desiring to be in the middle and how strongly is it desired? A car demolition company might offer less to be in the middle than the headquarters of a multinational corporation, generating sorting effects. Sorting effects explain how the aggregated small differences in preferences of individual actors lead to distinctive zones, neighbourhoods and regions within cities (Clark 2008). When we want to account for the differences in who and what wants to be in which middle, we need not one single theory of centrality, but several, emphasizing different dynamics and trade-offs about being in the middle. Moreover, centrality theories work on different scales. Thus far, real intra-city examples have been used but centrality theories equally discuss the dynamic between cities within regions or, even, between regions. This chapter puts most emphasis on the latter.

The avid reader will notice that materials discussed in this chapter might appear old-fashioned. The theories were developed in the early twentieth century when scholars were coming to terms with modern urban change: skyscrapers, telephones and the automobile were new features in the urban landscape. Subsequently, in the 1950s and 1960s Zeitgeist, formalizing these notions in abstract theories of centrality really took root. This was a period when scholars were increasingly interested in formulating and mathematically testing abstract theory, often using newly available computational tools (Barnes 2003; Johnston et al. 2019; Van Meeteren 2019a). Particularly in geography, many defined the substance of the discipline as being specialized in explaining abstract spatial patterns that were valid across contexts. Consequently, scholars prioritized the extraction of context-independent causal mechanisms and abstracting them to a point where they could explain many different instances (Bunge 1966; Haggett 1965; Ullman 1980). Another characteristic of this 'spatial science' era is exploration of the possibilities of 'relative space' (Harvey 2006, pp. 121–3). According to Harvey (2006, p. 122) 'on the mundane level of geographical work', exploring relative space means that 'The uniqueness of location and individuation defined by bounded territories in absolute space gives way' to a perspective where the places and locations are studies relative to each other in terms of distance, cost, relations, [and] time'. Theorizing with relative space meant that geographers were less concerned with specific locations and, instead, wanted to understand the spatially articulated processes that mediate the relationship between different places. Spatial science experimented with computational methods and different geometrical representations (for example, Bunge 1966; Tobler 1961) to calibrate these spatial relationships. Regarding centrality, it entailed thinking abstractly how being in the middle affects spaces, organizations and individuals. Since the 1960s, geography has expanded its spatial-theoretical repertoires even further (Harvey 2006) and therefore this chapter chiefly relies on mid-twentieth century classic references to emphasize the relative space perspective. However, for each theory, contemporary applications will be indicated when sketching the evolution and current (ir)relevance of theoretical ideas.

As the notion of abstraction plays an important role in theories of centrality, the chapter first elaborates different aspects of abstraction. Thereafter, sections follow on centrality and land-use theory, central place theory, centre–periphery theory and urban network theory. The chapter concludes with reflecting on the relevance of these notions for the contemporary spatial disciplines.

THE POWER OF ABSTRACTION

Abstraction comes from the Latin verb *abstrahere*, which means to draw away. To abstract is to identify those properties of an object relevant or even essential to the question at hand. We ignore those characteristics that do not interest us and focus on those that do. In the social sciences, abstraction means isolating the crucial elements of a causal mechanism while bracketing the inessential elements (Sayer 1992, p. 138).

A first relevant aspect of abstraction is about modelling, as centrality theories tend to be based on models of centrality. Modelling is an abstraction technique where idealized representations of reality are created in order to demonstrate specific isolated properties. Models are heuristic devices that communicate particular causal hypotheses but do not pretend to provide a comprehensive account of reality. Instead, they inform research questions and empirical inquiry to test the fit between a model and a segment of the world outside (Haggett 1965, pp. 19–23; Haggett and Chorley 1967; Van Meeteren 2019b). It is only when models find empirical justification that they start to become part of established theory: statements that have the pretention to explain (parts of) reality (Barnbrock 1974; Saey 1968). More complex models might provide more comprehensive explanations for specific cases as several causal mechanisms and their interactions are grasped simultaneously. However, this comes at the expense of validity across cases as combinations of causal mechanisms tend to vary from place to place (Van Meeteren et al. 2016b). This is why model-builders are inclined to emphasize Ockham's razor, the problem-solving principle whereby we strive to provide answers that make the fewest assumptions. Unless there is a clear reason to do otherwise, Ockham's razor suggests preferring simple models (and theories) over those that are more complicated.

A second aspect of abstraction relates to the question of what is the middle. When geographically articulated phenomena are isolated in a model, its properties are projected onto a hypothetical space. That projection requires choice of a geometrical system. Different projections accentuate different properties in a model. When we think of projection on a surface to emphasize the role of physical nearness and distance, abstraction in Euclidian geometry associated with topographic maps is appropriate. When connections between objects are emphasized, abstraction in topological geometry associated with representations of networks is apt (Van Meeteren et al. 2016c). These abstractions are not mutually exclusive, and transforming between geometrical systems can be insightful to explore spatial relations (Tobler 1961, cited and elaborated in Bunge 1966; Getis 1963). However, different geometrical abstractions might define another middle: the middle of the network might not be in the middle of the plane (Chisholm 1975, p. 105).

The third aspect of abstraction regards what is being in the middle. The theories discussed in this chapter were first formulated based on substantive research questions. It is, for instance, farms, firms, shops, consumers or social relations that are in the middle. Many have tried to generalize these theories by abstracting them to more general theories of centrality in relation to accessibility or communication. This suggests a two-step abstraction process: after making a model that accounts for a specific mechanism, this mechanism is further reduced to an even more abstract concept that accounts for other phenomena as well. For instance, in central place theory, a theory that explains a centrality in shopping becomes a theory of communication where shopping is regarded a special case of communication (Claval 1986). Especially in the 1960s, formulating general principles that could account abstractly for many different phenomena were considered a highly desirable outcome of geographic research (Berry and Pred 1961 [1965]; Bunge 1966). Although geographers currently tend to be more sceptical on the usefulness of grasping the world in a single formula, formulating these theories is still aspired to by many in science. Consequently, other disciplines such as physics have filled this abandoned geographical niche (for example, Bettencourt and West 2010; see also Derudder and Van Meeteren 2019). Regardless of this, when debating the usefulness of general formulations on high levels of abstraction it is important to understand the methodological pitfalls. Although it is possible to see the city as an abstract networked or hierarchical system and infer meaningful properties of it, it does not mean cities can be reduced to this geometrical representation (Harary and Rockey 1976; Sack 1972).

A final caution as regards abstraction posits the question of being in the middle of what? Geographies of centrality are often what Warntz (1957) termed 'macroscopic geographies'. The aggregated behaviour of many individuals generates an opportunity landscape that is emergent on the scale of the macrogeographic system. Consequently, individual behaviour cannot be deduced from this geography because of the ecological fallacy (Van Meeteren and Poorthuis 2018). At best, we can make assumptions what a rational human being would do in that opportunity landscape but those economic-man assumptions have proven to be poor predictors of individual geographic behaviour (Wolpert 1964; Massey 1973; Barnes 1988). The rationality assumption's pitfalls beg important considerations since centrality is ultimately made through the actions of individuals, whether entrepreneurs, governments or consumers (Barton 1978; Vance 1970). Therefore, although abstract models of centrality might provide an adequate explanation of urban structure in the past, and describe potentialities in the present, they are not infallible when predicting urban evolution in the future.

CENTRALITY AND LAND-USE THEORY: PUTTING A PRICE ON BEING IN THE MIDDLE

Classic firm location theories are also centrality theories. They provide models that explain the optimum central location if the transport costs of inputs and outputs for the firm or the flow of customers is to be optimized (Alonso 1975 provides a thorough introduction). However, as firms' needs differ enormously, these are principally microscopic geographies that do not easily scale to the city level. This is not so with Von Thünen's (1826 [1966]) model of agricultural land use (see Garrison and Marble 1957 for an early model-based elaboration). The Von Thünen model posits a single marketplace on an isotropic plane. Based on the differential transportation costs and prices of agricultural products, the probable agrarian production profile of the surrounding area of the isolated town is estimated. The model results in an image of concentric rings of different agricultural products. To simplify the model's conclusion: heavy products that command a high price at the market are grown close to the market, while products with lower margins that exceed lower transport costs are produced further away.

William Alonso (1960) translated Von Thünen's ideas into a theory of urban land rent, effectively formulating a model that puts a market price on being in the middle of a city. Different land-use functions (agricultural, residential, industrial and commercial) bid different rents for use of the most central space, since different functions gain differentially from being in the middle. Higher land prices encourage more intense land use, resulting in taller and denser built environments. For Alonso, differential transportation costs are still the causal mechanism. The result (Figure 12.1) is a model that explains differentiations of functions and/or density in the city. Friedmann (1968) replaces the transport accessibility function for an information-based function, arguing that information- and innovation-rich environments are more valuable to some actors or land uses than others.

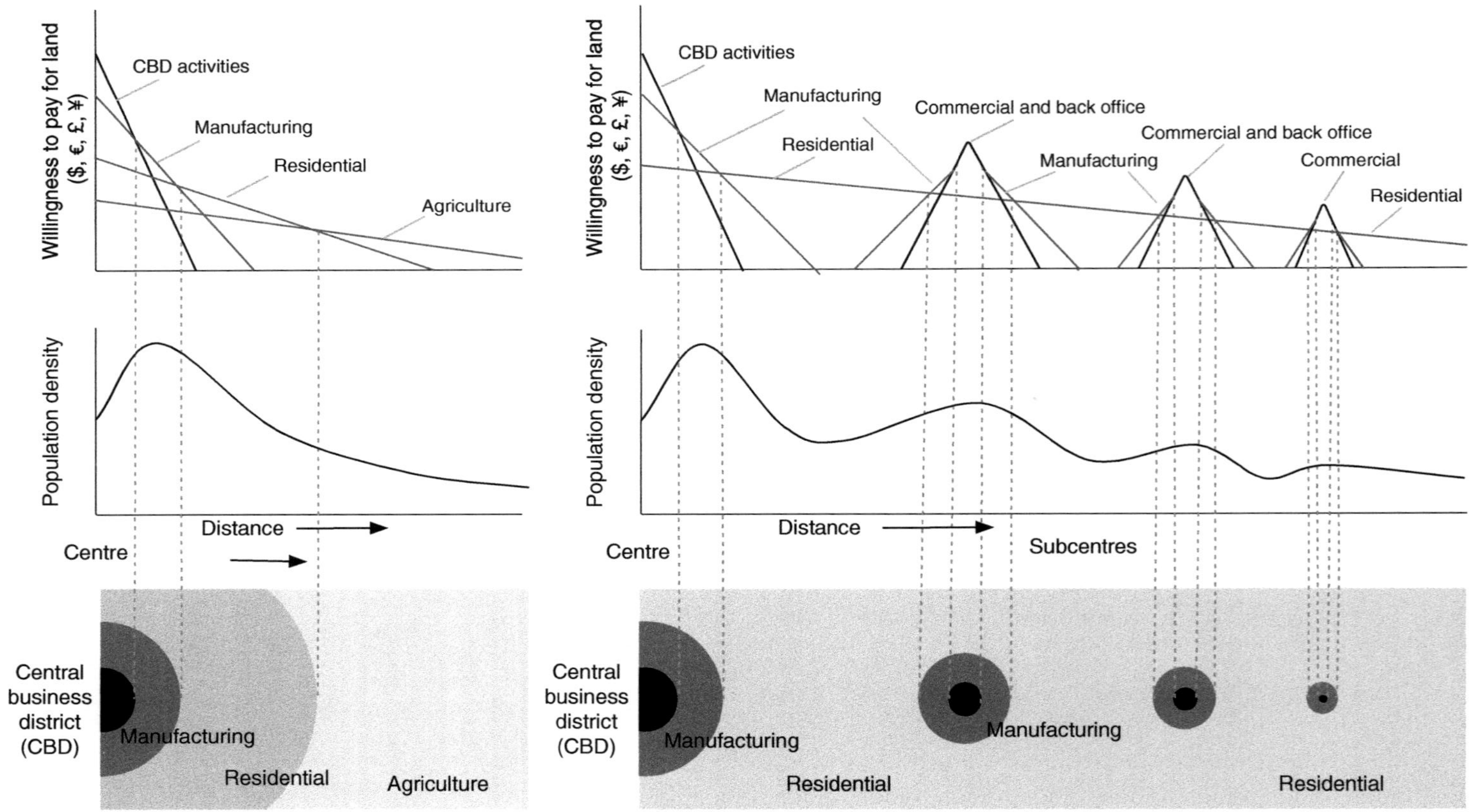

Source: Redrawn from Clark (2000, p. 144), based on Bourne (1981) and Cadwallader (1996); reproduced with permission from William Clark and Wiley-Blackwell.

Figure 12.1 Putting a price on centrality: the Von Thünen–Alonso model of land use in its monocentric and polycentric varieties

Later formulations (for example, Baker 1982) attempt to abstract individual preferences into a generic category of 'utility' that could account for the demand surface in bid rents. This last abstraction in particular is subject to criticism (Barnes 1988, 1996, pp. 63–8) as it obscures the notion of who wants to be in the middle and why. Alonso's theory became a central target of criticism on location theory in the 1970s (Harvey 1973) as it seemed to naturalize given patterns of land use in a city and to obscure housing inequalities, institutionalized racism and real-estate politics. Ultimately, it is entrepreneurs and politics that capitalize on land values and might even pump them artificially. Krijnen (2018), provides a recent overview of these issues. Moreover, she provides navigation in the debates around gentrification, as Alonso's assumption that all rational residents prefer spacious new car-orientated suburbs is outdated.

Regardless of valid critique, Alonso's model has stood the test of time relatively well for its original purpose: providing a model that puts a price on being in the middle (Van Nuffel and Saey 2006; Verhetsel et al. 2010). Perhaps the most important augmentation of the theory for its original purpose is the necessity of taking urban structures into account that have competing centres, that is, a polycentric model (Clark 2000).

CENTRAL PLACE THEORY

Historical settlement geography finds that the genesis of cities is associated with the relationship of cities with their *umland*[1] (Gradmann 1916, cited in Müller-Wille 1978). Central place theory models this city–*umland* connection. As the city–*umland* relationship can take many forms, there is a large variety of central place theories. According to this criterium, Von Thünen's model is also a (usually unrecognized) central place theory (Ullman 1941). The two most well-known archetypes of central place theory (CPT) are those of Christaller (1933 [1966]) and Lösch (1940 [1954]), which consider the supply of services and goods from the city to the *umland*. These two archetypical formulations are not comprehensive theories of urban location and only explain some functions of urban relationships (Preston 1975). The main difference between the two is that Löschian theories allocate *umlands* according to principles of monopolistic competition between central function providers, while Christallerian theories are built on the propensity of consumer travel. The models look visually very similar but are built on incompatible axioms, and only converge under special circumstances (Saey 1973). Moreover, many of the derived theories built from these two archetypes are incompatible with the foundational statements. This necessitates careful specification when discussing CPT (Van Meeteren and Poorthuis 2018). This section first discusses the Christallerian original before addressing attempts at further abstraction.

The empirical starting point of Christaller's (1933 [1966]) CPT is that the spacing of settlement sizes, in his example in Southern Germany, exhibits interesting regularities. If you travel between two cities of the same size, you might notice that the smaller towns in between tend to show a regular size pattern of alternating smaller and larger towns. It is this regularity that Christaller's central place model theorizes. The causal mechanism underpinning Christaller's (1933 [1966]) CPT is that every central place provides a limited selection of central goods to its complementary region or catchment area (the *umland* of a central place function). *Umland* regions are simultaneously under the influence of several central places with different levels of service provision. If needs cannot be provided by the nearest central place because its offerings are too limited, people tend to go to a better equipped central place – in Christallerian terms a place of a higher level – further away. This is self-evident: in a village you might be able to find a bakery or a hairdresser in your local shopping street, but if you want to buy expensive jewellery or frequent a shop that specializes in trading rare and vintage musical instruments you need to go to a bigger town.

The microfoundations of Christaller's CPT (Van Meeteren and Poorthuis 2018) theorize that every central good has a specific range with a lower and upper limit. The lower limit of the range (often termed the threshold, Berry and Garrison 1958a) is defined by the minimum number of goods or services that need to be sold so a central function can exist. The upper limit of the range (hereafter range, Berry and Garrison 1958a) is the maximum distance an average consumer is willing to travel before the consumption of central good in this central place is forgone. The following simple calculations intend to clarify these basic concepts:

- Hypothesize that a hairdresser needs to do 12 haircuts a day on average if s(he) is to stay in business. That means $365 \times 12 = 4380$ haircuts on a yearly basis.
- Hypothesize that the average person will go to the hairdresser every three months. That means every person procures $12 \times (1/3) = 4$ haircuts per year.

This means that the minimum population in the complementary area for a full-time hairdresser to exist is $4380/4 = 1095$ inhabitants, this is the threshold.

Hypothesize that the range for a haircut is 5 kilometres. If people have to travel more than 5 kilometres, everybody will start cutting their own hair instead. This means that the average population density of a complementary region (5 kilometre radius, translating in a surface area of $\pi \times 5^2 = 78.5$ km^2) needs to be $1095/78.5 = 13.9$ inhabitants per square kilometre for a full-time hairdresser to stay in business. In reality there is often leeway between the minimum complementary area (*umland*) of the threshold and the maximum complementary area (*umland*) of the range. This leeway is indicated by the difference between the 'Real range' and 'Ideal range' in Figure 12.2. A smart entrepreneur can make use of this room as there is potential for an extra supplier or expansion of existing suppliers (Getis and Getis 1966)

The model simplifies. Specialization and complementarity (having several musical instruments shops divided by genre or instrument family) can emerge in higher-density areas or between two interacting places. Moreover, complementary regions are intersected by roads, overlap and are plagued by congestion, which change real ranges (Van Meeteren and Poorthuis 2018). However, we have to remember Ockham's razor and bracket these complications, although many can these days be taken into account with global information system (GIS) software (Van Meeteren and Poorthuis 2018).

In Christaller's CPT, every central function has a specific combination of range and threshold but they nevertheless group in distinct baskets of central functions. We can think of this as typical central place environments: small-town shopping streets, ribbon shops, high streets or malls (see Borchert 1998; Carol 1962). In idealized isotropic circumstances where transportation costs are equal across the plane, population is evenly distributed, people always frequent the nearest central place, and there is no

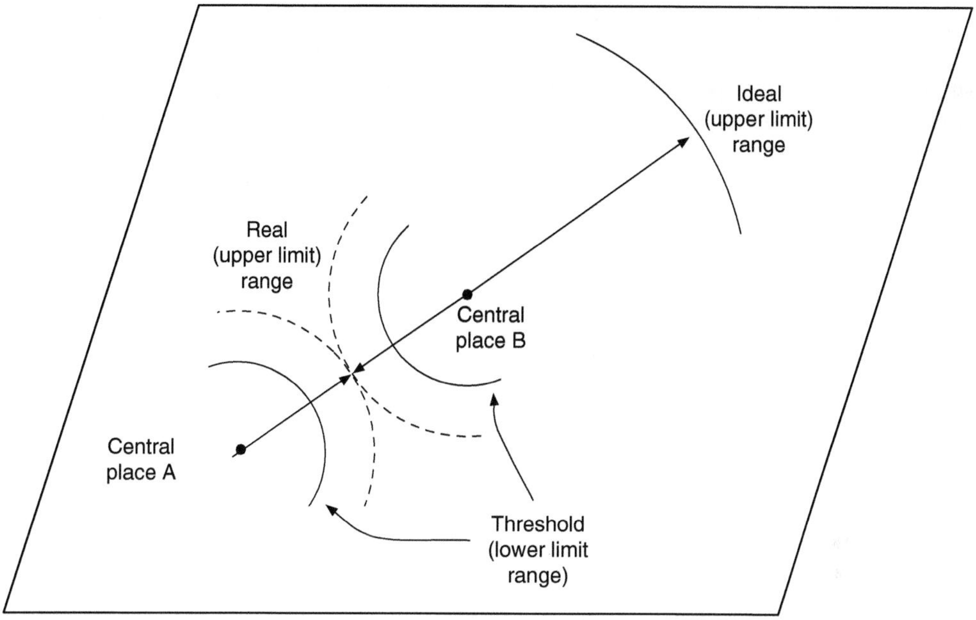

Source: After Getis and Getis (1966, p. 221), reproduced with permission from Arthur Getis.

Figure 12.2 The interplay between the ideal range, the real range and the threshold when defining complementary regions

multipurpose shopping; this interaction of central goods in Christaller's CPT generates a model landscape of perfectly nested hexagons between places (Figure 12.3). The hexagons emerge from equally dividing up the leftover areas of circular central places (Getis and Getis 1966 provide a step-by-step elaboration). Apart from this *versorgungsprinzip*, roughly translated into English as marketing principle, Christaller (1933 [1966]) also formulated idealized models of urban structure based on traffic and administrative principles.

Although idealized models, Christaller-type figures can be discerned in the urban landscape when the conditions of urban genesis show some conformity to Christaller's assumptions (Van Nuffel and Saey 2005). Nevertheless, the contemporary urban landscape hardly ever resembles the idealized geometric models that in the 1960s caught the imagination. However, more interesting when understanding contemporary centrality is that the interplay between range and threshold seems to be currently still valid (Boussauw et al. 2014; Van Meeteren and Poorthuis 2018), although the most central places in cities have often moved to more peripheral locations (Clark 2000).

Christaller had made already make a leap of abstraction when empirically testing his model. In 1933, the excess telephone connections over what was to be expected of a city of a certain size were considered a sufficient indicator to calculate the influence on the *umland*. As telephones became ubiquitous midway through the twentieth century, research resorted to other proxies to measure centrality, such as wholesale ratios, newspaper circulation or counting shops (Davies 1967; Preston 1975; Siddall 1961). In the

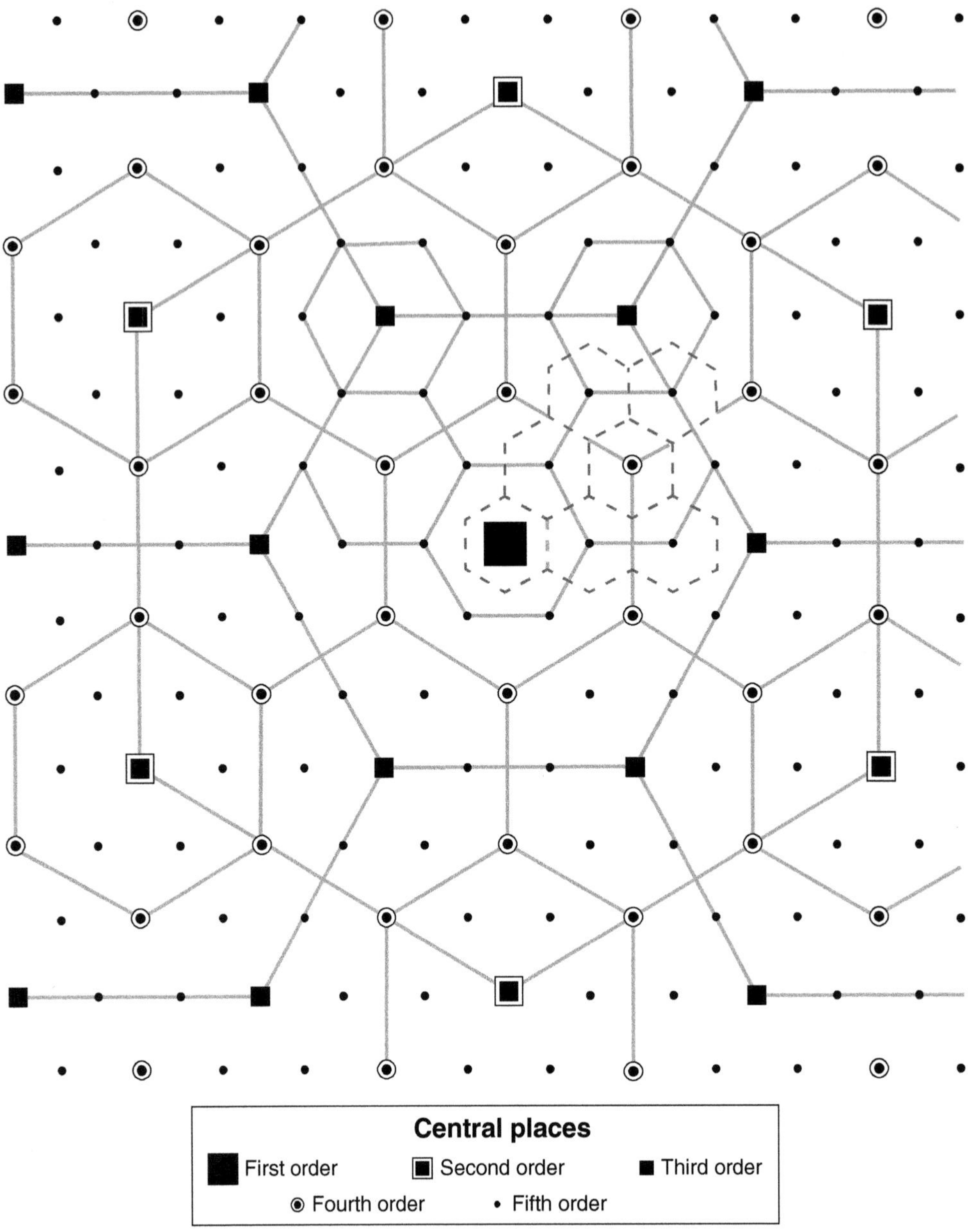

Source: Redrawn by Mark Szegner from Saey (1973, p. 183), with permission from the Royal Dutch Geographical Society and Pieter Saey.

Figure 12.3 A Christaller-type central place system according to the marketing principle with five orders twenty-first century, machine learning and big data can come to the operationalization rescue (Van Meeteren and Poorthuis 2018) but scholars also find interesting results using indicators of accessibility and central functions (Zhong et al. 2017). The notion of hierarchical relationships between the larger and smaller central places received only very limited attention in Christaller's (1933 [1966]) original treatise (Buursink 1975). Later, however, abstracting a theoretical hierarchy between settlements was defined as the core characteristic of CPT (Berry and Garrison 1958b; Parr 1978). In the 1960s, theorizing the *umland* as a catchment area analogous to rivers became increasingly popular. The rationale behind comparing central properties of rivers and cities is the promise of an abstract notion of centrality being universally valid for different analogically structured phenomena (Woldenberg and Berry 1967 take this to the next level). Even if research had little to do with the procurement of central goods and services, a wide variety of research questions were routinely squashed in the mould of the central place model (Bird 1977; Vance 1970). For instance, innovations were thought to trickle down through the urban hierarchy (Berry 1972; see Pred 1977 for a critique) although the causal mechanism for these processes is far from Christaller's microfoundations. Central place theory became synonymous with the idea of *the* urban hierarchy that exerted causal effects on the level of a system, probably aggravated by the theory's widespread use in spatial planning (Blotevogel 1996) and the adoption of the abstracted variety in mainstream economics (Fujita et al. 1999). In any event, it is important to pay close attention to what scholars actually mean when they invoke CPT, regardless whether they praise or criticize it.

CENTRE–PERIPHERY THEORY

In Christaller's CPT, centre and *umland* are in a symbiotic relationship. The centre needs the clients from the *umland*, while the *umland* receives services from the centre. Centre–periphery theory, however, emphasizes unequal relationships between centre and periphery and has a connotation of winners and losers.[2] Centre–periphery theory again originates in the 1920s and 1930s when two University of Chicago scholars, sociologist Roderick McKenzie (1927) and geographer Charles Colby (1933), tried to make sense of the Windy City. Their divergent interpretations of centre–periphery relationships have coloured the literature ever since. McKenzie (1927) describes centre–periphery relationships as instances of dominance. The centre is the locus of hierarchy that commands the economy in the periphery, while the benefits disproportionally accrue to the centre. McKenzie (1927, p. 31) noted that innovations in transportation technology were strengthening this centre–periphery relationship. For him, with normative undertone, 'the city has ever been the symbol of civilization; which, even in its crudest form, represents a spatial pattern of a fixed center of dominance with tributary subordinate districts' (McKenzie 1927, p. 32). Colby (1933), distilled from the same empirical case the aforementioned shifting spatial equilibrium account. While the centre becomes stronger (centripetal forces), the periphery benefits through sorting processes (centrifugal forces). Although the centre might gain corporate headquarters and dominating functions, Colby clearly saw that with suburbanization some the city's riches were moving in the opposite direction.

Centre–periphery theory became prominent in the 1950s and 1960s. The immediate post-1945 environment and the advent of macroeconomic development policy put regional economic inequalities centre stage. Colby's students, Harris (1954) and Ullman (1958) conducted empirical studies on centre–periphery relationships in the United States, showing how the most urbanized regions there (the central manufacturing belt, now sometimes called the rustbelt) were growing to the detriment of the periphery. Both theorize how improved connections between core and periphery could induce centrifugal forces. Moreover, a United Nations commission argued that economic discrepancies between centre and periphery were not only detrimental at the national scale, but also internationally. The developing countries were being dominated by the centre through unequal terms of trade (Prebisch 1950). Note that the mechanisms are similar to those of McKenzie (1927), but that the moral evaluation of this situation had made a 180-degree turn after World War II: civilization was about spreading wealth to counterbalance the centralizing metropolis.

Apart from the terms of trade argument, it was Swedish economist Gunnar Myrdal (1957)[3] who cogently described a causal mechanism and spatial model for core–periphery relationships. The driving force of accumulation in the centre is a process of circular and cumulative causation. Growth begs more growth and decline begs decline. If a region[4] is doing well economically, it will attract inward migration of talented people, inward capital flows, infrastructure investment and other centripetal tendencies. These have to come from somewhere, and tend to be from less well-off areas. These are backwash effects, where the centre's gain is the periphery's loss. The centre's relative advantage is strengthened, making the relative unevenness with the periphery larger. Myrdal also identified centrifugal forces in his model, termed spread effects. Owing to pressure on the centre, sorting effects not unlike the Alonso model occur that lead to effective diffusion of growth. Crucially, however, spread and backwash effects work on different scales: spread effects in the immediate vicinity of the centre and backwash effects in the further periphery (see Chapman et al. 2003; Gaile 1980; Figure 12.4).

Initially, there was a strong confidence about governments intervening to help overcome centre–periphery polarizations. Uneven development was argued to be an intermediary symptom until all regions in a national economy would reach a mature developed state (Friedmann 1966). Until the 1970s, some success with spatial Keynesianism (Brenner 2004) policy was achieved, for instance, in Belgium (Van Meeteren et al. 2016a) and France (Veltz 1996). The long economic crisis of the 1970s largely crushed belief in spatial-economic planning. In some instances, tectonic economic shifts from areas known as 'core' to areas known as 'periphery' (from rustbelt to sunbelt, see Smith and Dennis 1987) took place. This incited scepticism of spatial Keynesianism and the theories underpinning it (Brenner 2004; Veltz 1996). Although the, often recently decolonialized, third world was a welcome playground for the development-orientated core–periphery model (Friedmann 1966), Prebisch's 1950 report spawned a critical literature. *Dependencia* and world-systems theories (for example, Amin 1974; Frank 1966; Hopkins and Wallerstein 1977; Santos 1974) emphasize that the relationship between core and periphery cannot be symbiotic or developmental as long as the core benefits disproportionally from the periphery. Drawing on Marx-inspired theories of uneven development and imperialism, these theories use the core–periphery model to emphasize that the development of the centre is due to underdevelopment of the periphery. These critical strands get the upper hand in the 1970s as global economic inequality exacerbates. The most prominent

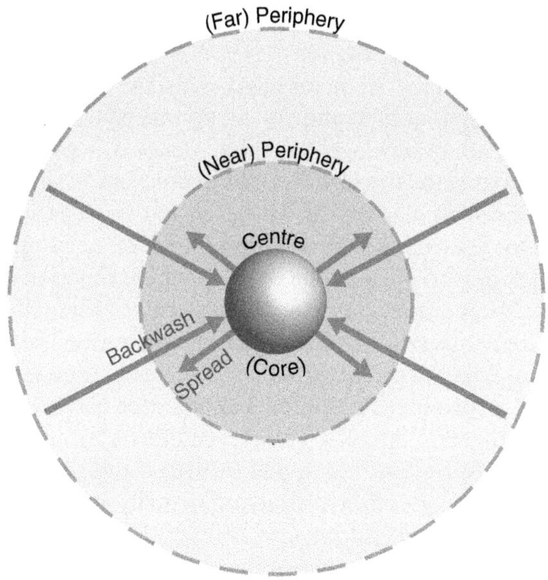

Source: Figure by Mark Szegner, inspired by Chapman et al. (2003).

Figure 12.4 Myrdal's (1957) core–periphery model

convert is John Friedmann (Friedmann and Weaver 1977, pp. 186–216), who originally helped synthesize the development version of the core–periphery model (Friedmann 1966). Friedmann subsequently recasts a new critical version of his theory as the world-city hypothesis (Friedmann and Wolff 1982) where powerful corporations located in cities in the core, in the middle of deindustrialization, accumulate and dominate, to the detriment of cities in the periphery.

Centre–periphery theory was from its inception formulated on a high level of abstraction. The notion of urban dominance, central to McKenzie's account (1927), is ambiguous and operationalized in many, often incompatible, ways. Colby's (1933) version is more concrete and indicates that economic variables, such as rent and transportation costs, do the majority of the work in the centrifugal and centripetal relationships. This focus on tangible transportation costs is prominent in the geographers' accounts and in Friedmann's (1966) treatise. These are centre–periphery relationships that are easily and meaningfully mapped in Euclidian geometry. However, as Copus (2001) notices, it is distance costs that have been increasingly irrelevant in the late twentieth and early twenty-first centuries. He therefore proposes a distinction between spatial and aspatial peripherality, where the causal mechanisms that are not explicitly articulated in relative space can be subsumed under the latter. Following that argument, there is a case to be made that the centre–periphery relationship is more meaningfully depicted as a network. Spatial or not, core–periphery relationships between cities and associated uneven development endure as an important empirical phenomenon in our contemporary world (Brown et al. 2010; Van Meeteren and Bassens 2016).

URBAN NETWORK THEORY

Technological change has made it increasingly sensible to discuss centre–periphery relationships through the spatial metaphor of an archipelago instead of than assuming contiguous territories. The archipelago metaphor denotes a geographical form where territorial 'islands' are connected through flows of people, information, capital and goods interspersed by less-connected places (Van Meeteren and Bassens 2016; Veltz 1996). The notion was first used to describe the speckled geography of early capitalist relationships (Braudel 1984, p. 30). A central place is one of the possible functions of a city, and some other types of cities may have little relation with their *umland* (Siddall 1961). For instance, port cities, with their forelands and hinterlands, are also gateways (Bird 1977; Vance 1970) that connect islands in a networked archipelago. The same argument is easily made for airport hubs or, indeed, McKenzie's corporate headquarter places as recast in the world-city model. Here, using the term 'hinterworld' is even more appropriate than a hinterland (Derudder and Taylor 2018; Taylor and Derudder 2016). Archipelago structures, owing to their networked character, suggest that abstracting them in a network is useful (although not exclusively, see Van Meeteren and Bassens 2016).

When relationships between cities are seen as a network, these relationships are expressed in the topological language of 'nodes' and 'edges' (Neal 2013, p. 15). Nodes represent the objects that the relationships connect, and edges represent the relationships between nodes. Abstraction of an empirical phenomenon in network language requires nodalization (Van Meeteren et al. 2016c). For inter-urban network analysis, this usually means that cities are considered nodes and relationships between cities are considered edges.

An attraction of network thinking is that it provides different conceptions and measurements of centrality. There are two general notions of centrality in network analysis: structural centrality of the network as a whole, and the centrality of individual nodes within a network (Freeman 1979). The notion of structural centrality enables gauging the degree to which a structure is centralized. It also enables calculating and comparing whether an urban network is functionally dominated by a single city or whether it is more polycentric (Burger and Meijers 2012; see Pereira et al. 2012 for a Euclidian geometrical measure). Moreover, structural centrality allows comparing different urban (sub)systems. For instance, Taylor and Derudder (2016, ch. 5) develop the notion of 'global network connectivity' to gauge the structural composition of contributing countries to the world-city network.

Individual centrality measures provide different indications of how central a specific node is in the network. These can provide convincing explanations for geohistorical phenomena. For instance, in a classic paper using network-analytical methods, Pitts (1965) shows how the emergence of Moscow as the dominant city in Russia can be explained by the city's centrality in Russia's medieval transport networks. There are several indictors of centrality in a network and all depict a different property of a node's role (Freeman 1979; Neal 2013 provide more thorough introductions). The three most common are:

- Degree centrality expresses the number of edges a node is connected to. It is commonly standardized to account for the total amount of nodes in the network.
- Closeness centrality is a measure that expresses how close a node is, on average, to

all the other nodes in the network by looking at the shortest paths to get from a node to all the other nodes, again often standardized. We could say that the lower the closeness centrality, the higher the peripherality in the network.

- Betweenness centrality is a measure of how often you have to pass through the focal node to travel the shortest pathways between two other nodes. This is important for making arguments whether nodes have a type of middleperson function.

All these measures are mathematical constructs which have no a priori social causal mechanism attached to them. Hence when these centrality measures are used, a hypothesis about the causal mechanisms that these mathematical constructs represent needs to be formulated first. It is for this reason that it is necessary to caution against using overly complex centrality measures as these can be difficult to interpret in relation to causal mechanisms (Neal 2013).

Figure 12.5 clarifies some of the categories. The Belgian urban system of cities and roads is first nodalized: the highways and most important cities are extracted in the simplified diagram on the left-hand side. The table compares the three centrality measures where the top five scores are in bold. It is clear that Brussels is the most central node in the Belgian urban network, and Ghent and Antwerp are important hubs. The high betweenness centrality of Brussels suggests a very centralized character in the Belgian urban system. Note, for instance, that Leuven and Mechelen are in the core (high closeness) but relatively unimportant for the other measures. Bruges scores high on betweenness as it completely controls the gates to Ostend. If you think I went too fast and should have included road distance, railroads, or your favourite town that got dropped off the map, you have just discovered why nodalization is difficult. Slicing up an empirical object in a language of nodes and edges has an arbitrariness to it where many slippery judgement calls about borderline cases have to be made (Garrison 1960, p. 137). For instance, if the analyses had included cities across the country border, such as Lille, Aachen, Luxembourg or Maatstricht, centralities would look different already. For example, if we include the foreign cities connected with the dotted lines in the analyses, Antwerp's betweenness centrality more than doubles (from 0.11 to 0.28), while the standardized centrality values of Brussels go down with 0.08 (betweenness), 0.07 (closeness) and 0.13 (degree). Therefore, depending on the topical and geographical framing, there is not a single statement about Belgium made on the basis of this graph that cannot be disputed. Yet I would invite the reader to immerse himself or herself in the history of the Belgian settlement geography (Van Meeteren et al. 2016a) and gauge to what extent these findings 'make sense'.

Thus far, urban network analysis has been depicted as merely a different way of operationalizing and abstracting substantive arguments. For instance, we can analyse CPT through network analytical methods (Nystuen and Dacey 1961; Neal 2013, p. 116) or centre–periphery theories (Smith and Timberlake 1995; Taylor and Derudder 2016), and Neal (2011) demonstrates how to operationalize McKenzie's (1927) dominance categories in network terms. However, many authors have argued that urban network theory can also be a substantive theory of its own (Derudder and Neal 2019). Tangible networks, whether social or physical, work as a transmission mechanism that make people, places and information accessible. The city is then considered as the connector of multiple networks (Pflieger and Rozenblat 2010; Pred 1977; Rozenblat 2010). The empirical and theoretical question here is whether multiple different networks (such as airlines, goods,

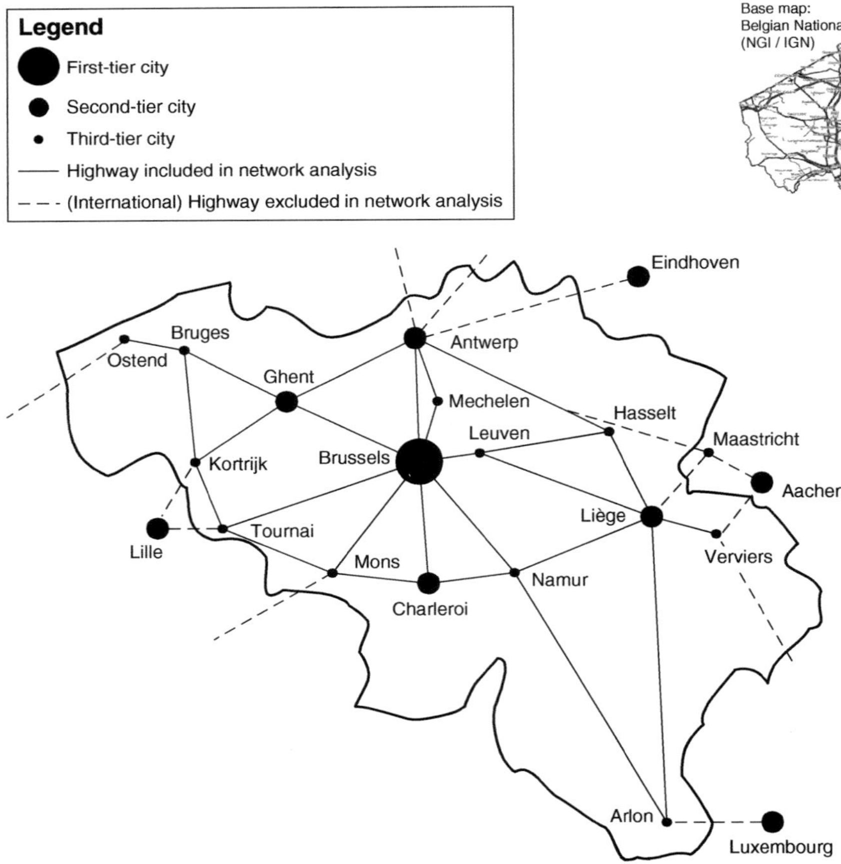

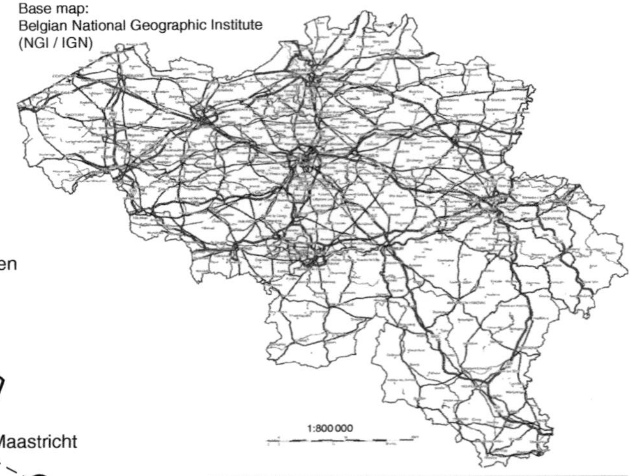

City	Standardized degree centrality	Standardized closeness centrality	Standardized betweenness centrality
Antwerp	**0.27**	**0.52**	**0.11**
Arlon	0.13	0.38	0
Bruges	0.20	0.38	**0.13**
Brussels	**0.53**	**0.63**	**0.48**
Charleroi	0.20	0.45	0.02
Ghent	**0.27**	**0.52**	**0.26**
Hasselt	0.20	0.44	0.09
Kortrijk	0.20	0.39	0.03
Leuven	0.20	**0.48**	0.08
Liège	**0.33**	0.43	**0.18**
Mechelen	0.13	0.42	0
Mons	0.20	0.43	0.01
Namur	**0.27**	**0.50**	0.12
Ostend	0.07	0.28	0
Tournai	0.20	0.45	0.06
Verviers	0.07	0.31	0

Figure 12.5 Belgium crudely nodalized

organizations, Internet backbone) together shape an emergent inter-urban network structure on which abstract theories such as central flow theory can be formulated (Derudder and Taylor 2018; see also Ducruet et al. 2011; Pflieger and Rozenblat 2010; Smith and Timberlake 1995). The result is an abstract notion of urban networks that is often contrasted with the hierarchies that are abstracted from CPT. Some (for example, Camagni and Salone 1993) argue that the former will supersede the latter. Others tend to emphasize the complementarity of perspectives and caution against a network fever (Shearmur and Doloreux 2015) where older perspectives are too easily jettisoned in favour of casting everything as a network (Parnreiter 2014; Shearmur and Doloreux 2015; Van Meeteren and Bassens 2016).

CONCLUSION

Centrality is such a central concept when thinking about cities and their relationships that it is unsurprising that a large amount of theorizing on the topic has already been undertaken. It is a crowded intellectual field of ideas, and rightly so. Since so many authors engage with a topic, that topic is more likely to cause controversy. As different authors internalize a concept and make it work for themselves, different interpretations of theories and many misunderstandings in theoretical discussions appear, not because of a lack of engagement, but because an idea is so popular (Van Meeteren et al. 2016b). This is definitely the case with the old-fashioned theories of centrality discussed in this chapter, which were very influential in the heyday of relative space thinking. Moreover, as these theories are also generational peers, some of the tenets of different centrality theories got tangled up and hybridized in some accounts, compounding conceptual confusion. The different meaning of 'hierarchy' in concrete versions of centre–periphery theory and CPT is the exemplary example. Similarly it can be argued, as I have done here, that network theories of centrality are just another way of (how we understand) being in the middle. To others, the notion of urban network has been considered to be the antithesis to the notions of urban centrality of the past (Camagni and Salone 1993). These are issues subject to continuing debate (Derudder and Neal 2019). This chapter, apart from offering a basic introduction, has chiefly been trying to clear up some of the confusions between these positions so that debates can be more than semantic. By untangling the different theories of centrality, their genealogy, and the main causal mechanisms that they describe, both in their more abstract and their more concrete forms, a set clean of conceptual tools becomes visible that can be used to grasp the contemporary city. As technology is once again changing our cities, perhaps with automatic vehicles, the breakthrough of the platform economy and the ubiquitous connectivity of the digital, what being in the middle is about might change. However, that does not mean that we have to completely replace our toolbox. Augmenting our existing tools might also suffice. The advantage of using the old tools is that we get a clearer view of what changes and what stays the same. This is important, because only the boldest or the brashest scholars would dare to claim that the future city will no longer be about being in the middle.

ACKNOWLEDGEMENTS

Ben Derudder and Alois Humer provided constructive feedback on earlier drafts. Mark Szegner assisted with redrawing Figures 12.3 and 12.4.

NOTES

1. The German word *umland* (surrounding lands) is often mistakenly translated into English as hinterland (the lands behind the city) which suggests an eccentric location (Bird 1977).
2. This chapter considers core-periphery and centre–periphery nomenclature as synonymous.
3. At the same time, others such as Albert Hirschman, François Perroux and Bert Hoselitz formulated similar theoretical ideas. Friedmann (1966), Friedmann and Weaver (1979), and Gaile (1980) provide introductions.
4. Myrdal explicitly referred to national economies in his analysis; however, Friedmann (1966) makes the case that these centre–periphery relationships can be located at different scales from the urban to the global.

REFERENCES

Alonso, W. (1960), 'A theory of the urban land market', *Papers and Proceedings of the Regional Science Association*, **6** (1), 149–57.
Alonso, W. (1975), 'Location theory', in J. Friedmann and W. Alonso (eds), *Regional Policy: Readings in Theory and Applications*, Cambridge, MA: MIT Press, pp. 35–63.
Amin, S. (1974), 'Accumulation and development: a theoretical model', *Review of African Political Economy*, **1** (1), 9–26.
Baker, R.G.V. (1982), 'Place utility fields', *Geographical Analysis*, **14** (1), 10–28.
Barnbrock, J. (1974), 'Prologomenon to a methodological debate on location theory: the case of von Thünen', *Antipode*, **6** (1), 59–66.
Barnes, T.J. (1988), 'Rationality and relativism in economic geography. An interpretative essay of the homo economicus assumption', *Progress in Human Geography*, **12** (4), 473–96.
Barnes, T.J. (1996), *Logics of Dislocation*, New York and London: Guilford Press.
Barnes, T.J. (2003), 'The place of locational analysis: a selective and interpretive history', *Progress in Human Geography*, **27** (1), 69–95.
Barton, B. (1978), 'The creation of centrality', *Annals of the Association of American Geographers*, **68** (1), 34–44.
Berry, B.J.L. (1972), 'Hierarchical diffusion: the basis of developmental filtering and spread in a system of growth centers', in N.M. Hansen (ed.), *Growth Centers in Regional Economic Development*, New York and London: Free Press and Collier-Macmillan, pp. 108–38.
Berry, B.J.L. and W.L. Garrison (1958a), 'A note on central place theory and the range of a good, *Economic Geography*, **34** (4), 304–11.
Berry, B.J.L. and W.L. Garrison (1958b), 'The functional bases of the central place hierarchy', *Economic Geography*, **34** (2), 145–54.
Berry, B.J.L. and A.R. Pred (1961), *Central Place Studies: A Bibliography of Theory and Applications*, repr. 1965, Philadelphia, PA: Regional Science Research Institute.
Bettencourt, L. and G. West (2010), 'A unified theory of urban living', *Nature*, **467** (7318), 912–13.
Bird, J.H. (1977), *Centrality and Cities*, London: Routledge & Kegan Paul.
Blotevogel, H.H. (1996), 'Zentrale orte: zur karriere und krise eines konzepts in geographie und raumplanung', *Erdkunde*, **50** (1), 9–25.
Bobek, H. (1927), 'Grundfragen der stadtgeographie', *Geographischer Anzeiger*, **28** (7), 213–24.
Borchert, J.G. (1998), 'Spatial dynamics of retail structure and the venerable retail hierarchy', *GeoJournal*, **45** (4), 327–36.
Bourne, L. (1981), *The Geography of Housing*, New York: John Wiley.
Boussauw, K., M. van Meeteren and F. Witlox (2014), 'Short trips and central places: the home-school distances in the Flemish primary education system (Belgium)', *Applied Geography*, **53**, 311–22.
Braudel, F. (1984), *The Perspective of the World*, London: William Collins Sons.
Brenner, N. (2004), *New State Spaces*, Oxford and New York: Oxford University Press.
Brown, E., B. Derudder, C. Parnreiter, W. Pelupessy, P.J. Taylor and F. Witlox (2010), 'World city networks and

global commodity chains: toward a world-systems' integration', *Global Networks: a Journal of Transnational Affairs*, **10** (1), 12–34.

Bunge, W.W. (1966), *Theoretical Geography*, 2nd edn, Lund: Gleerup.

Burger, M.J. and E.J. Meijers (2012), 'Form follows function? Linking morphological and functional polycentricity', *Urban Studies*, **49** (5), 1127–49.

Buursink, J. (1975), 'Hierarchy. A concept between theoretical and applied geography', *Tijdschrift voor Economische en Sociale Geografie*, **66** (4), 194–203.

Cadwallader, M. (1996), *Urban Geography: An Analytical Approach*, Upper Saddle River, NJ: Prentice Hall.

Camagni, R. and C. Salone (1993), 'Network urban structures in Northern Italy: elements for a theoretical framework', *Urban Studies*, **30** (6), 1053–64.

Carol, H. (1962), 'The hierarchy of central functions within the city. Principles developed in a study of Zurich, Switzerland', in K. Norborg (ed.), *Proceedings of the IGU Symposium in Urban Geography Lund 1960*, Lund: Gleerup, pp. 555–76.

Chapman, D., D. Pratt, P. Larkham and I. Dickins (2003), 'Concepts and definitions of corridors: evidence from England's Midlands', *Journal of Transport Geography*, **11** (3), 179–91.

Chisholm, M. (1975), *Human Geography: Evolution or Revolution*? Hardingsworth: Penguin Books.

Christaller, W. (1966), *Central Places in Southern Germany*, repr. 1966, Englewood Cliffs, NJ: Prentice Hall.

Clark, W.A.V. (2000), 'Monocentric to policentric: New urban forms and old paradigms', in G. Bridge and S. Watson (eds), *A Companion to the City*, Oxford: Blackwell, pp. 141–55.

Clark, W.A.V. (2008), 'Geography, space, and science: perspectives from studies of migration and geographical sorting', *Geographical Analysis*, **40** (3), 258–75.

Claval, P. (1986), 'Communication theory as a basis for urban theory', in C.S. Yadav (ed.), *New Directions in Urban Geography*, New Delhi: Concept, pp. 137–48.

Colby, C.C. (1933), 'Centrifugal and centripetal forces in urban geography', *Annals of the Association of American Geographers*, **23** (1), 1–20.

Copus, A.K. (2001), 'From core-periphery to polycentric development: concepts of spatial and aspatial peripherality', **9** (4), 539–52.

Davies, W.K.D. (1967), 'Centrality and the central place hierarchy', *Urban Studies*, **4** (1), 61–79.

Derudder, B. and Z.P. Neal (2019), 'Uncovering links between urban studies and network science', *Networks and Spatial Economics*, **18** (3), 441–6, doi:10.1007/s11067-019-09453-w.

Derudder, B. and P.J. Taylor (2018), 'Central flow theory: comparative connectivities in the world-city network', *Regional Studies*, **52** (8), 1029–40.

Derudder, B. and M. van Meeteren (2019), 'Engaging with "urban science"', *Urban Geography*, **40** (4), 555–64, doi:10.1080/02723638.2019.1585138.

Ducruet, C., D. Ietri and C. Rozenblat (2011), 'Cities in worldwide air and sea flows: a multiple networks analysis', *Cybergeo: European Journal of Geography*, art. 528, doi:10.4000/cybergeo.23603.

Frank, A.G. (1966), 'The development of underdevelopment', *Monthly Review*, **18** (4), 17–31.

Freeman, L.C. (1979), 'Centrality in social networks: conceptual clarification', *Social Networks*, **1** (3), 215–39.

Friedmann, J. (1966), *Regional Development Policy: A Case Study of Venezuela*, Cambridge, MA and London: MIT Press.

Friedmann, J. (1968), 'An information model of urbanization', *Urban Affairs Quarterly*, **4** (2), 235–44.

Friedmann, J. and C. Weaver (1979), *Territory and Function: The Evolution of Regional Planning*, London: Edward Arnold.

Friedmann, J. and G. Wolff (1982), 'World city formation: an agenda for research and action', *International Journal of Urban and Regional Research*, **6** (3), 309–44.

Fujita, M., P. Krugman and T. Mori (1999), 'On the evolution of hierarchical urban systems', *European Economic Review*, **43** (2), 209–51.

Gaile, G.L. (1980), 'The spread-backwash concept', *Regional Studies*, **14** (1), 15–25.

Garrison, W.L. (1960), 'Connectivity of the interstate highway system', *Papers and Proceedings of the Regional Science Association*, **6** (1), 121–37.

Garrison, W.L. and Marble, D.F. (1957), 'Spatial structure of agricultural activities', *Annals of the Association of American Geographers*, **47** (2), 137–44.

Getis, A. (1963), 'The determination of the location of retail activities with the use of a map transformation', *Economic Geography*, **39** (1), 14–22.

Getis, A.J. and Getis (1966), 'Christaller's central place theory', *Journal of Geography*, **65** (5), 220–26.

Gradmann, R. (1916), 'Schwäbische stadte', *Die Erde – Zeitschrift der Gesellschaft für Erdkunde zu Berlin*, pp. 425–57.

Haggett, P. (1965), *Locational Analysis in Human Geography*, London: Edward Arnold.

Haggett, P. and R.J. Chorley (1967), 'Models, paradigms and the new geography', in R.J. Chorley and P. Haggett (eds), *Models in Geography*, London: Methuen, pp. 19–42.

Harary, F. and J. Rockey (1976), 'A city is not a semilattice either', *Environment and Planning A*, **8** (4), 375–84.

Harris, C.D. (1954), 'The market as a factor in the localization of industry in the United States', *Annals of the Association of American Geographers*, **44** (4), 315–48.
Harvey, D. (1973), *Social Justice and the City*, London: Edward Arnold.
Harvey, D. (2006), 'Space as a keyword', in D. Harvey, *Spaces of Global Development*, London and New York: Verso, pp. 117–48.
Hopkins, T.K. and I. Wallerstein, I. (1977), 'Patterns of development of the modern world-system', *Review (Fernand Braudel Center)*, **39** (1–4), 11–45.
Johnston, R.J., R. Harris, K. Jones, D. Manley, W.W. Wang and L. Wolf (2019), 'Quantitative methods I', *Progress in Human Geography*, **43** (6), 1133–42, doi:10.1177/0309132518774967.
Krijnen, M. (2018), 'Gentrification and the creation and formation of rent gap', *City: Analysis of Urban Trends, Culture, Theory, Policy, Action*, **22** (3), 437–46.
Lefebvre, H. (1974), *The Production of Space*, repr. 1991, Oxford and Cambridge, MA: Blackwell.
Lösch, A. (1940), *The Economics of Location*, 1954 trans. W.H. Woglom, New Haven, CT and London: Yale University Press.
Massey, D. (1973), 'Towards a critique of industrial location theory', *Antipode*, **5** (3), 33–9.
McKenzie, R.D. (1927), 'The concept of dominance and world-organization', *American Journal of Sociology*, **33** (1), 28–42.
Müller-Wille, C.F. (1978), 'The forgotten heritage: Christaller's antecedents', in B.J.L. Berry (ed.), *The Nature of Change in Geographical Ideas*, DeKalb, IL: Northern Illinois University Press, pp. 38–64.
Myrdal, G. (1957), *Rich Lands and Poor, the Road to World Prosperity*, New York and Evanston: Harper & Row.
Neal, Z.P. (2011), 'Differentiating centrality and power in the world city network', *Urban Studies*, **48** (13), 2733–48.
Neal, Z.P. (2013), *The Connected City, How Networks are Shaping the Modern Metropolis*, Abingdon and New York: Routledge.
Nystuen, J.D. and M.F. Dacey (1961), 'A graph theory interpretation of nodal regions', *Papers and Proceedings of the Regional Science Association*, **7**, 29–42.
Parnreiter, C. (2014), 'Network or hierarchical relations? A plea for redirecting attention to the control functions of global cities', *Tijdschrift voor Economische en Sociale Geografie*, **105** (4), 398–411.
Parr, J.B. (1978), 'Models of the central place system: a more general approach', *Urban Studies*, **15** (1), 35–49.
Pereira, R.H.M., V. Nadalin, L. Monasterio and P.H.M. Albuquerque (2012), 'Urban centrality: a simple index', *Geographical Analysis*, **45** (1), 77–89.
Pflieger, G. and C. Rozenblat (2010), 'Introduction. Urban networks and network theory: the city as the connector of multiple networks', *Urban Studies*, **47** (13), 2723–35.
Pitts, F.R. (1965), 'A graph theoretic approach to historical geography', *Professional Geographer*, **17** (5), 15–20.
Prebisch, R. (1950), 'The economic development of Latin America and its principal problems', No. 50.II.G.2, United Nations Department of Economic Affairs, Lake Success, NY.
Pred, A.R. (1977), *City-Systems in Advanced Economies*, London: Hutchinson.
Preston, R.E. (1975), 'A comparison of five measures of central place importance and of settlement size', *Tijdschrift voor Economische en Sociale Geografie*, **66** (3), 178–87.
Rozenblat, C. (2010), 'Opening the black box of agglomeration economies for measuring cities' competitiveness through international firm networks', *Urban Studies*, **47** (13), 2841–65.
Sack, R.D. (1972), 'Geography, geometry, and explanation', *Annals of the Association of American Geographers*, **62** (1), 61–78.
Saey, P. (1968), 'A review article: towards a new paradigm? Methodological appraisal of integrated models in geography', *Bulletin of the Ghana Geographical Association*, **12**, 51–60.
Saey, P. (1973), 'Three fallacies in the literature on central place theory', *Tijdschrift voor Economische en Sociale Geografie*, **64** (3), 181–94.
Santos, M. (1974), 'Geography, Marxism and underdevelopment', *Antipode*, **6** (3), 1–9.
Sayer, A. (1992), *Method in Social Science, a Realist Approach*, 2nd edn, London and New York: Routledge.
Schlüter, O. (1899), 'Bemerkungen der siedelungsgeographie', *Geographische Zeitschrift*, **5**, 65–84.
Shearmur, R.G. and D. Doloreux (2015), 'Central places or networks? Paradigms, metaphors, and spatial configurations of innovation-related service use', *Environment and Planning A*, **47** (7), 1521–39.
Siddall, W.R. (1961), 'Wholesale-retail trade ratios as indices of urban centrality', *Economic Geography*, **37** (2), 124–32.
Smith, N. and W. Dennis (1987), 'The restructuring of geographical scale: coalescence and fragmentation of the northern core region', *Economic Geography*, **63** (2), 160–82.
Smith, D.A. and M.F. Timberlake (1995), 'Conceptualising and mapping the structure of the world system's city system', *Urban Studies*, **32** (2), 287–302.
Taylor, P.J. and B. Derudder (2016), *World City Network*, 2nd edn, London and New York: Routledge.
Tobler, W.R. (1961), 'Map transformations of geographic space', PhD thesis, University of Washington, Seattle, WA.

Ullman, E.L. (1941), 'A theory of location for cities', *American Journal of Sociology*, **46** (6), 853–64.
Ullman, E.L. (1958), 'Regional development and the geography of concentration', *Papers and Proceedings of the Regional Science Association*, **4**, 179–98.
Ullman, E.L. (1980), 'Introduction', in R.R. Boyce (ed.), *Geography as Spatial Interaction*. Seattle, WA: University of Washington Press, pp. 3–27.
Van Meeteren, M. (2019a), 'Statistics do sweat: Situated messiness and spatial science', *Transactions of the Institute of British Geographers*, **44** (3), 454–7, doi:10.1111/tran.12298.
Van Meeteren M. (2019b), 'Urban system', in A.M. Orum (ed.), *The Wiley-Blackwell Encyclopedia of Urban and Regional Studies*, Chichester: Wiley.
Van Meeteren, M. and D. Bassens (2016), 'World cities and the uneven geographies of financialization: unveiling stratification and hierarchy in the world city archipelago', *International Journal of Urban and Regional Research*, **40** (1), 62–81.
Van Meeteren, M. and A. Poorthuis (2018), 'Christaller and "big data": recalibrating central place theory via the geoweb', *Urban Geography*, **39** (1), 122–48.
Van Meeteren, M., K. Boussauw, B. Derudder and F. Witlox (2016a), 'Flemish diamond or ABC-axis? The spatial structure of the Belgian metropolitan area', **24** (5), 974–95.
Van Meeteren, M., Z.P. Neal and B. Derudder (2016c), 'Disentangling agglomeration and network externalities: a conceptual typology', *Papers in Regional Science*, **95** (1), 61–80.
Van Meeteren, M., A. Poorthuis, B. Derudder and F. Witlox (2016b), 'Pacifying Babel's tower: a scientometric analysis of polycentricity in urban research', *Urban Studies*, **53** (6), 1278–98.
Van Nuffel, N. and P. Saey (2005), 'Commuting, hierarchy and networking: the case of Flanders', *Tijdschrift voor Economische en Sociale Geografie*, **96** (3), 313–27.
Van Nuffel, N. and P. Saey (2006), 'Beyond Christaller: regionalisation of residential markets – the example of North Belgium', *Die Erde – Journal of the Geographical Society of Berlin*, **137** (1–2), 75–99.
Vance, J.E. (1970), *The Merchant's World: The Geography of Wholesaling*, Englewood Cliffs, NJ: Prentice Hall.
Veltz, P. (1996), *Mondialisation, villes et territoires*, Paris: Presses Universitaire de France.
Verhetsel, A., I. Thomas and M. Beelen (2010), 'Commuting in Belgian metropolitan areas: the power of the Alonso-Muth model', *Journal of Transport and Land Use*, **2** (3), 109–31.
Von Thünen, J.H. (1826), *Von Thünen's Isolated State*, repr. 1966, Oxford: Pergamon Press.
Warntz, W. (1957), 'Geography of prices and spatial interaction', *Papers and Proceedings of the Regional Science Association*, **3** (1), 118–29.
Wells, H.G. (1902), 'The probable diffusion of great cities', in H.G. Wells, *Anticipations*, London: Chapman & Hall, pp. 33–65.
Woldenberg, M.J. and B.J.L. Berry (1967), 'Rivers and central places: analogous systems?', *Journal of Regional Science*, **7** (2), 129–39.
Wolpert, J. (1964), 'The decision process in spatial context', *Annals of the Association of American Geographers*, **54** (4), 537–58.
Zhong, C., M. Schlapfer, S. Muller Arisona, M. Batty, C. Ratti and G. Schmitt (2017), 'Revealing centrality in the spatial structure of cities from human activity patterns', *Urban Studies*, **54** (2), 437–55.

PART IV

NETWORK PROCESSES WITHIN CITIES

13. The city of opportunity: designing Cities4People

Karima Kourtit, Peter Nijkamp and Tigran Haas

It takes *hands* to *build* a *house*, but only *hearts* can *build* a home.
(Unattributed)

1. THE *HOMO URBANUS*

Since the early history of mankind, human beings have tried to build settlements that were fit for purpose. In a nomadic or low-technology society human settlements had to be by necessity flexible and small. Large cities hardly existed and were only built to demonstrate or strengthen political and military power. However, even these cities were – compared with current standards – relatively small and their management demanded enormous efforts (for example, import of agricultural produce, fresh water, sewerage systems and jobs for citizens). Cities of more than 1 million inhabitants were, until a few centuries ago, a miraculous urbanistic exception (see Tellier 2009). This situation lasted in most countries until the age of the Industrial Revolution (mid-nineteenth century) when cities started to grow. With the worldwide uninterrupted population growth since two centuries ago (the first demographic revolution), cities went through an accelerated growth path. This has led over the past half a century to a double urbanization process: most cities were growing and became large cities, while several large cities reached the size of megacities (more than 10 million inhabitants). This development is clearly demonstrated in China which currently has more than 60 cities with a population exceeding 1 million inhabitants, while the number of megacities in this country is also showing rapid growth. Worldwide, we witness a megatrend towards more and bigger cities. This revolutionary phenomenon in the history of urbanism is sometimes termed the New Urban World (Kourtit 2019).

In the past decade, our world has reached a stage where more people are residing and living in urban areas than in rural areas. This new species of mankind is sometimes named the *homo urbanus*. The city has become the most attractive settlement place for the majority of people on our planet. The United Nations has, therefore, named this new epoch in the geographic-demographic history of our world the urban century.

It should be noted that the massive concentration of millions of people in urban agglomerations prompts intriguing policy, management and research questions (see, for instance, Acs 2006; Geyer 2009; Haas and Westlund 2017). There is an uncontested need for a thorough reflection on the sustainability and manageability of large population concentrations, in relation to fit-for-purpose hardware (for example, built environment, material infrastructure), up-to-date software (for example, communication channels for people, spatial accessibility for all and educational facilities) and appropriate humanware (for example, medical and social healthcare, safe living environment and desired social capital). It is one of the greatest contemporaneous policy challenges to create and maintain healthy, happy and liveable cities in the decades to come.

A new form of urbanism seems to be needed, in which a proper management of public or common spaces – in combination with effective guidelines for private spaces in the city – may lay the foundation stones for sustainable and liveable urban agglomerations in the future (see, for example, Farrell and Haas 2017). The urban challenges are formidable and vast: how do we reconcile the diverse preferences and varying spatial behaviours of the *homo urbanus* with the existing – sometimes rigid and strictly controlled – urban morphology? How do we accommodate millions of new urban inhabitants in cities that should provide an attractive and sustainable living environment to its clients? How do we create and maintain a balance between, on the one hand, public or common spaces that are open to everyone and can be shared with others and, on the other, safe and private urban spaces that have a restricted access? How do we position and manage individual urban agglomerations as part of a globally connected network society (Castells 1998). Perhaps the most intriguing question of all is how can we create and maintain an urban agglomeration that offers the best possible opportunities for everyone, so that cities and urban areas become attractive and sustainable places to live, to work and to enjoy, on the basis of their measurable maximum urban quality (the XXQ-principle advocated by Nijkamp 2008)?

Against the background of the need to care for the quality of life and life satisfaction of *homo urbanus*, this chapter highlights the city (or urban agglomeration) as a great place of human opportunities to reach a satisfactory level of well-being, rather than to conceive of the city as an arena of social conflicts and crime. We use the modern enabling theory as a frame of reference for analysing the great potential of cities to satisfy human needs. We focus specifically on public or common spaces in cities as place-based and human-centred pockets that may generate new ideas on well-being, social capital, tolerance and creativity in urban areas. In this context, the recently developed concept of the microcosmic city is highlighted. This chapter offers a new and promising perspective on cities of the future, based on the urban opportunity principle, sometimes articulated as 'chances4all'.

The chapter is organized as follows. After this general introduction into the rise of the *homo urbanus* in the urban century, section 2 is devoted to the relevance of smart cities and to the cognitive principle of a microcosmic city based on Simon's decomposition principle which can be unfolded into the happiness–health–home (3H-city) triangle. The concept of the city of opportunity is introduced in section 3, in which the city is seen as an enabling agent. This concept is based on the possibilism approach as articulated *inter alia* in Sen's capability theory. This approach is then translated, in section 4, into the action domain of (mainly) public spaces in cities which act as platforms for interaction, vibrant life and urban buzz, using the urban piazza as a frame of reference. Section 5 focuses on a postmodern cultural-urbanism orientation on city life. Section 6 is devoted to urban data analysis, while the systematic use of complex urban data for city planning – based on decomposed urban dashboards – is discussed in section 7. The chapter also sketches the research agenda associated with the above conceptualization. This conceptual contribution concludes with some retrospective and prospective remarks.

2. SETTING THE SCENE: THE MICROCOSMIC CITY

Many cities are in full swing. Urban agglomerations such as Shanghai, Dhaka, Lagos or São Paulo show amazing dynamics in relation to population growth, land use developments, urban economic disparities and environmental quality conditions. Global developments (for example, climate change, shifts in economic heartlands, spatial mobility and international migration, and pervasive worldwide use of digital technology) appear to lead in the urban century to unprecedented pressures on cities to provide effective solutions to many serious challenges in our New Urban World. This concern has also prompted the rise of the current fashionable concept of a smart city. This new development in knowledge-based or cognitive city policies calls for a critical judgement in our review.

In the vein of the general sustainable city movement, we currently witness an increasing popularity of smart cities ideas. The profile and success conditions of a smart city has in become an interesting research topic in urban research and policy (see, for instance, Winters 2010; Caragliu et al. 2011). Smart city research appears to originate from various disciplines, ranging from economics to political science, from ecology to urban architecture and planning, and from digital technology (mainly information and communication technology, ICT) to network analysis. Modern cities have turned into multi-faced organisms, acting in complex power fields that are also co-determined by other, often competing, urban agglomerations – both neighbouring and remote – and by generic regional, national and global driving forces. This challenge calls for new policy and management concepts which may be difficult to pursue (Nam and Pardo 2011). There is no doubt that the appealing notion of a smart city has gained much popularity among policy-makers, private firms and the research community generally. A smart city is often seen as a shiny beacon for a successful urban performance strategy.

A range of definitions of a smart city can be found in the literature, sometimes complementary, sometimes contrasting (see also Torres et al. 2005; Deakin 2010). It is often assumed that cities may be termed smart, if they exploit the potential of innovative digital network technology applications, for example, for e-government services (see, for example, Nijkamp and Cohen-Blankshtain 2013; Norris and Reddick 2013; Zhuhadar et al. 2017). It has become evident in the meantime that the assessment of the extent to which digital technology provides cities with a competitive sustainable advantage is overloaded with many empirical and methodological problems and uncertainties. There is a need for quantitative indicators that allow us to assess evidence-based differences among smart cities in respect of a baseline case. However, defining a baseline needs not only a transparent, measurable and large-range urban database, but also the identification of a fit-for-purpose strategic institutional setting which may be subject to change owing to the changing preferences of citizens (often referred to as institutional endogeneity). A smart city is a multi-faceted interconnected concept that needs the involvement of many disciplines. In this context, Kourtit et al. (2015, p. 210) note the 'tendency to assess smart city performance not only from an economic perspective but also from an evolutionary, partly non-economic valuation perspective, in order to suggest solutions for cities to have a competitive edge in a global network economy'.

It is increasingly accepted that cities are spatially integrated agents or network constellations of individual and collective interest, in which information collection, information use and information sharing offer an unprecedented value added for dedicated strategies on quality of life, ecological sustainability and economic vitality (TNO 2012). Smart cities may be supported by using digital network information, for instance, in domains such as human health, mobility, energy use, education, knowledge transfer and urban governance. The basis of this idea on cities as multi-faceted spatial networks originates from Castells's

(1998) seminal work on networks, making a distinction between space of places and space of flows, next to a space of information. All these spaces may be integrated in a smart city concept aimed at maximizing XXQ as an overall policy objective.

Despite the appeal of the smart city concept, it is also increasingly recognized that this popular perspective on city policy is generic and may not be of direct effective help to urban management at the level of daily living spaces of citizens (see Vanolo 2013). If the city is a human-made connectivity system (Neal 2013), bottom-up and decentralized initiatives ought to be respected as well. In most instances – especially in the developed world – the manoeuvring space for drastic urban changes (land use, built environment, transformation of economic functions, and so on) is limited. Therefore, it is an important question how smart city ideas – as a set of generic, cognitive policy objectives – can be geared towards improvement or better utilization of the existing urban space (for example, through mobility policy, green space initiatives, brownfield development, adjusted public amenities, public entertainment space and better waste management). This would also call for new plans on retro-fitting the existing urban space or remodelling specific urban districts or transport arteries from a micro-based, human-centred perspective. This may require a drastic de-construction of the urban space.

In a recent paper (see Kourtit et al. 2019) the notion of a microcosmic city was introduced. A microcosm is a small place or community that contains all (or most) essential elements of a much larger place or community. It is thus a typical miniature model of a larger unit that functions as a small, characteristic ecosystem in a much larger environment. From this perspective, an urban district or neighbourhood or even a street may be regarded as a representation or typical subset (microcosm) of a larger urban constellation. Thus, using Simon's (1955) decomposition principle based on bounded rationality (see Gigerenzar and Selten 2001), we may argue that a complex urban system can be deconstructed into microcosmic network units which are the critical agents for the creation of urban welfare, satisfaction and well-being. This means that user appreciation of a city and of its intra-urban microcosmic constituents is determined by a range of decisive factors for urban clients, in particular:

- The happiness of people; this factor refers to the psychic feelings of well-being and life satisfaction of individuals or groups as a consequence of spatially disaggregated opportunities, to be found in particular districts or neighbourhoods of the city concerned. There is an extant recent literature on the geography of happiness which highlights the importance of small-scale urban life satisfaction (see, for an overview, Wahlström et al. 2020a).
- The health conditions of people living in a city; these factors refer to the external factors that determine the mental and physical health feelings of residents of visitors in a city. The factors comprise, *inter alia*, quality of life (QoL) and social cohesion (including social capital) in the city. The factors have also been investigated extensively in the recent literature (see, for example, Hawe and Shiell 2000; Ma et al. 2012; Murayama et al. 2012; Westlund 2016).
- The home characteristics of urban neighbourhoods; these features refer to the quality and attractiveness of the built environment, the quality of the housing stock, the accessibility through appropriate and convenient transport infrastructure, the safety situation in specific city neighbourhoods, the local historical-cultural

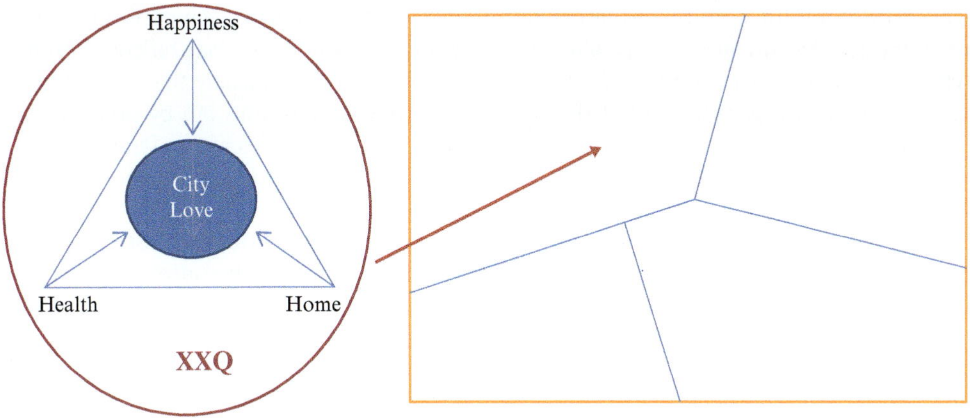

Figure 13.1 Decomposition of an XXQ-city into microcosmic areas in an urban space whose 3H features impact on city love

heritage, the public and private amenities in the area concerned, and so on. These features and aspects of the microcosmically decomposed urban structure into districts, communities and neighbourhoods shape the self-identity (home) feelings of residents, visitors, businessmen or artists. The emphasis here is mainly on public spaces or shared goods in cities.

This list describes three critical factors for an appreciative city (in particular, appreciative urban neighbourhoods), namely, happiness, health and home, here named as the 3H-city triangle. These factors represent the conditions that shape the user appreciation for city life, referred to as city love in a recent publication (see Kourtit et al. 2020; Wahlström et al. 2020b). Thus, the above arguments can be synthesized in the conceptual and stylized model of urban space in Figure 13.1.

The critical characteristics incorporated in the 3H-city triangle do not automatically and deterministically determine a high appreciation for city life (a high city love). They may be necessary conditions, but by no means sufficient conditions. This is articulated in the next section where the city of opportunity, based on enabling characteristics, is introduced.

3. CITIES AS SOURCES OF OPPORTUNITIES

Cities are connected platforms for human action and activities. Cities with a high XXQ provide the desired conditions for living, working, entertaining, socializing, and so on. Cities offer user shared spaces for a multiplicity of functions: economic, social, technological, cultural and administrative. The 3H-city triangle offers a multivariate performance criterion for well-functioning cities and their neighbourhoods, while the microcosmic perspective takes for granted that a decomposed urban space is the action platform for users or clients. As indicated in section 2, the use of and user conditions for urban shared space – in particular public space – shape and induce human activities that contribute to urban welfare and appreciation of the city (city love). There is, however, no guarantee that favourable internal and external living conditions – as reflected in the 3H-city concept – lead automatically to a higher city love. We argue here that the 3H-city triangle comprises, at best, a set of desirable or necessary conditions, but by no means a set of sufficient conditions. The city is a basket full of opportunities whose fruits have to be actively acquired. The city is not a passive manna from heaven for opportunity seekers; it is an active space (see Geenhuizen and Ratti 1998).

In the recent literature on urbanism we encounter increasingly the new notion of the city of opportunity (see PwC 2016). This concept refers to the agglomeration advantages of a city that allow residents and visitors to exploit the city's benefits, for example, for business, culture, innovation, education, employment or entertainment. It is worth noting that some cities have decided to go for a smart specialization (for instance, Venice for cultural heritage and Singapore for business and science), while others (for example, Paris, London, San Francisco and Amsterdam) offer a broad package of services to their clients, for example, residents and visitors.

The question whether the soul and body of a city lead to a predictable or deterministic outcome, in relation to its attractiveness for and appreciation by its clients, brings us back to an old discussion in geography, namely, the concept of determinism. This intellectual debate was initiated by the French geographer Vidal de la Blache (1903), who contradicted the, by then popular, traditional way of geographical thinking, coined environmental determinism. According to Cloke et al. (1991, p. 64), the latter concept 'regarded all the facets of human activity (from farming practices to political systems) as ultimately determined in character by the natural-environmental context', whereas Vidal de la Blache's conceptualisation of the relationship between people and their surroundings means that 'the natural environment offers possible avenues for human development, the precise one chosen being very much a human decision' (Cloke et al. 1991, p. 65). The latter viewpoint is known as possibilism. This concept reflects the idea that the complex interactions between milieux and civilizations may lead to a variety of *genres de vie* (manifestations of lifestyles). Thus, the environment provides a range of spatial socio-economic, cultural or technological opportunities in space from which people can choose, but the outcome is not deterministic.

The theory of possibilism was further developed by the French historian Lefebvre (1971), who argued: 'there are no necessities, but everywhere possibilities; and man, as a master of the possibilities, is the judge of their use' (see Johnston et al. 2000, p. 609). Thus, regions, and consequently also cities, act as the dynamic, mutually interacting meeting place between human societies and their environment.

The dilemma of determinism and possibilism has prompted a wealth of publications. For example, Taylor (1951) has made an attempt to bridge this dichotomy between these two fundamentally different approaches by introducing the notion of neo-determinism, which takes for granted that in reality various hard-manoeuvring constraints do exist or may be accepted by inhabitants, which hardly harm the free decision space of individuals or groups, even though these constraints may be binding. Another strand of intermediate approaches to the determinism–possibilism dilemma can be found in the cultural ecology school (Steward 1955, 1977; Milton 1996), which studies human geography through the lenses of the processes by which a society adapts to its environment. Consequently, human–environment interactions in an adaptive learning context are critical in understanding spatial evolution. A recent informed review on the paradox in environmental determinism and possibilism can be found in Fekadu (2014).

In an article by Nijkamp (2016), the possibilism approach was revisited by introducing the concept of a resourceful region. A resourceful region is

> an area that is pro-actively driven by a smart combination of *economic potential* (e.g., capital provision), *spatial networks in terms of accessibility and connectivity* (e.g., locational conditions, cyberspace access), *historico-cultural support mechanism* (e.g., entrepreneurial spirit), *ecologically sustainable quality conditions, and educational and creativeness facilities* (e.g., institutions of higher education). (Nijkamp 2016, p. 205, original emphasis)

Regional development policy would then have the task of optimizing and re-combining all promising opportunities in a given region. The region is in his view an action platform that enables the choice among various developmental options.

In the same vein, we may now interpret the notion of a *city of opportunity*. This is an enabling urban agglomeration that offers a portfolio of choice options which have to be combined in a smart manner so as to achieve the highest urban performance (XXQ) from a set of available choice opportunities.

The city as a flexible enabling agency for creating new opportunities for people is increasingly recognized in spatial and urban planning. This view on the city as a form of active space ties in with the interpretation of an urban society as a learning and adaptable social mechanism, which in interaction with its users creates many desired options for developing their social, cultural and economic life (see also Bailly 2010).

This enabling interpretation of cities as sources of manifold opportunities is inspired by the capability theory introduced by Sen (1985, 1992, 1993, 2005, 2008). He positions inequality and disparity conditions in the context of access to possibilities and functionings, which may impact human well-being by a voluntary choice for a specific bundle of functionings. Consequently, choosing, employing and optimizing in a smart manner a portfolio of capabilities may increase someone's well-being. This approach has also played a critical role in the debate on justice and feminism (Nussbaum 2000). Several publications have in the past decades been devoted to the capabilities approach (see, for example, Crocker 1992; Qizilbash 1996; Dowding 2006; Kaufman 2006; Comim et al. 2008).

The capability approach has great relevance for urbanism, as it addresses conditions underlying the well-being effects of less privileged groups or of deprived areas in a city. This holds in particular for urban segregation, depressed neighbourhoods or polluted streets (see also, Edwards and Pellé 2011).

The capability approach has a specific importance in the context of our microcosmic approach to a decomposed city, as – despite socio-economic or cultural-ethnic similarities in terms of intra-cohesion in city neighbourhoods – significant disparities continue to exist between such areas. This may substantially affect the levels of well-being, happiness or satisfaction among individuals and groups in different urban areas (see, for example, Bruni and Porta 2007). We will therefore, now focus on disaggregate shared user spaces in a city.

4. SHARED SPACES AND NETWORKS IN THE CITY OF OPPORTUNITY

Urban agglomerations show a high degree of internal heterogeneity from a socio-economic and historical-cultural perspective. We find cities with particular districts for creative classes (see, for example, Evans 2009; Comunian 2011), or cultural districts (see, for example, Scott 2000; Storper and Scott 2009; Sacco et al. 2013). These specific districts were not a result of centralized and deterministic blueprint planning, but were rooted in spontaneous or evolutionary urban dynamics which enabled these attractive areas to emerge. The modern city appears to be an effective seedbed for spatially segmented mushroom behaviour of people and activities. The city of opportunity enables the rise of unplanned initiatives which contribute to a better performance of urban areas through a 3H-capability perspective or approach. Through the supply of attractive urban amenities, spatial accessibility, the protection of historical-cultural heritage, a favourable social climate or the provision of good health conditions, an appealing city climate may emerge or can be provided. Thus, city governance can contribute to the success of a city of opportunity, but cannot guarantee in advance its success in relation to a high XXQ of the city concerned.

A particular role in the city of opportunity is played by public or common space (areas in a city that, without any exclusion, are freely accessible to clients, for example, streets, squares, parks, waters and public amenities such as libraries). These public spaces are not only functional for daily activities (for example, mobility from home to work or from home to school), but are also social interaction platforms where people can meet or entertain themselves (see Lefebvre 1971; Gehl 1987; Madanipour 2003). Public spaces add to the liveability of cities and hence to city life (based on the 3H-city triangle model). These public arenas for human interplay are an essential and connecting thread in city life. Public spaces may be red (built environment, streets and infrastructure), green (gardens and green spaces such as parks) or blue (lakes, ponds, rivers and canals). The balanced integration of these functions in one city is a challenge (see Thompson 2002; Cormona 2010; Scott 2017). Public spaces favour human health, social contacts and appreciation for the city, and therefore contribute significantly to urban happiness (see Waldinger and Schulz 2010). In general, shared spaces in a city may thus be regarded as the enabling infrastructure for the city of opportunity. This city may favour higher welfare levels, more tolerance and social innovation (see also, Engelen et al. 2014), provided accessibility to and connectivity with public spaces in a city is guaranteed. The emerging microcosmic city is based on decentralized, interconnected urban areas in which network linkages and new information technology form the backbone for the city of opportunity (see also, Ratti and Claudel 2016).

The above-advocated concept of the city of opportunity is in agreement with earlier developed concepts on the architecture, functionalities and interactions of ancient cities, inspired by the ancient Roman piazza concept for city design (see Kourtit et al. 2014). The piazza architecture sketches a functional, spatially decomposed and systematically structured map of an ideal-typical multi-tasking city (see Figure 13.2). The piazza model can be interpreted as follows.

The urban piazza is an appropriately metaphorical strategy in its own right and embodies strategically well-connected multi-levels for maximizing XXQ as an overall policy

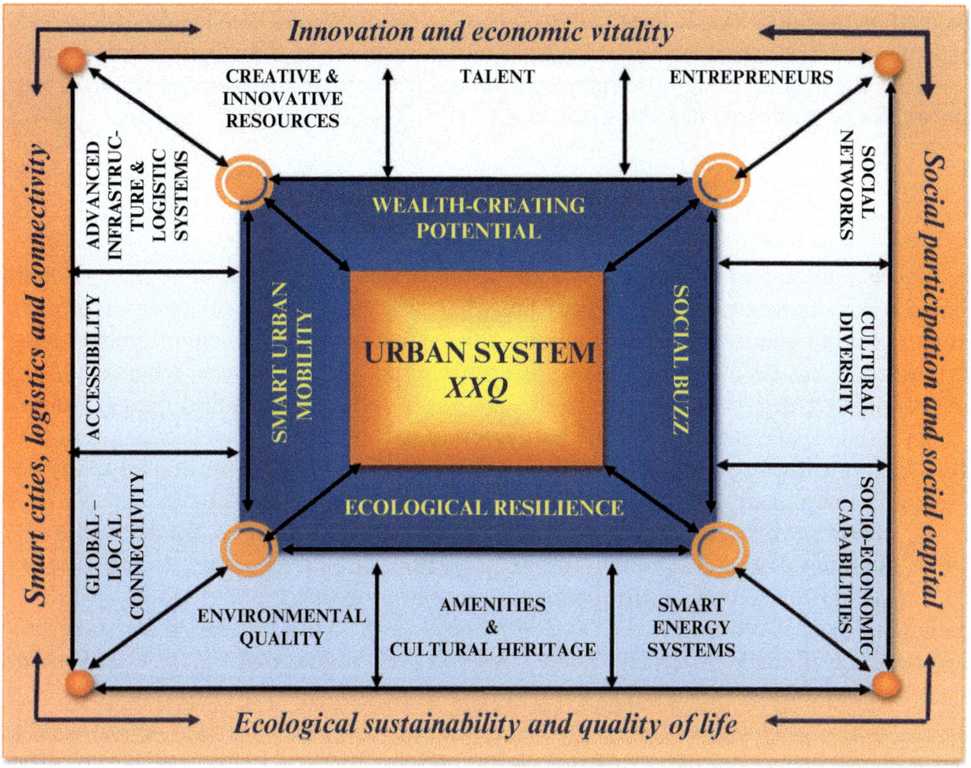

Source: Kourtit et al. (2014).

Figure 13.2 The New Urban World urban piazza network

objective function and strategic performance of the network city concerned (Mahaffy et al. 2009; Kourtit et al. 2014). This holistic approach not only helps in the decomposition of an XXQ-city into microcosmic areas, but also gives a very clear understanding and balanced view on the complex and adaptive city system from multiple ways or insights into opportunities and challenges. Also, this approach supports the ability to learn and innovate from past and future encountered performance conditions by exploring, monitoring and evaluating the intersections of the multidimensional set of urban dynamic network patterns (Kourtit and Nijkamp 2018) in order to enhance efficiency with a view to maximizing the XXQ of the city concerned.

The spatially decomposed structure of a piazza architecture of a city has a strong internal logic in relation to interrelated segmented urban spaces, which in a network combination adds to the presence of a rich portfolio of opportunities for urban agents.

The piazza architecture maps out a stylized, decomposed constellation of a functionally segmented urban agglomeration. It captures the relevance of public spaces for the economic functioning and logical structuring of an urban system, based on smart specialization in functionally demarcated, but mutually connected, urban network spaces. Thus, the piazza model may be seen as an illustrative architecture for a city of opportunities, inspired by the microcosmic approach to the urban economy and its related decomposed structure of intra-urban spaces. An important focal point in the recent literature has been the introduction of cultural-urbanism perspectives on urban architecture and urban spaces. This is addressed in section 5.

5. PUBLIC SPACE AND CULTURAL URBANISM

Public space is a space4all. Nevertheless, the access to and use of public or common urban spaces is not shared equally in everyday life. There are often significant socio-economic, ethnic, age and gender differences among citizens using the urban environment. This phenomenon has been articulated in particular in (post-)feminist new geography (see, for example, Ardener 1993; McDowell and Sharp 1999; Nelson and Seager 2005). Urban public space is an important battleground for social and spatial equality in the city. Consequently, in pursuing the goal of an inclusive city, due attention paid to public space – as a prominent pillar of local equity – is warranted.

The difference between place and *space* is not always evident in the literature on urban planning, social geography, urban geography, urban sociology, urban design, environmental psychology, anthropology and urban economics. For example, in urban geography, place is often viewed as a geographically and historically specific instance of the social use of space. Alternatively, de Certeau (2011) argues that space is a practised place, while according to Massey (1994) both space and place are the projections of social and economic inequality. The main challenge of inclusive and culturally connected shared places in a city is to create urban environments that are convivial to all groups.

The above observations prompt the question: what is a good public space? There are numerous spatial and social qualities that can make a good public space. Characteristics related to size, scale, degree of physical enclosure, amenities, aesthetics and other variables evidently matter; public spaces at different times and in dissimilar contexts might change in their role of accommodating various heterogeneous groups of people in the city (Carmona et al. 2010). These changing roles also mean the emergence of changing conditions for various social and economic groups, for those inhabiting the adjacent urban areas and for those people visiting or passing by (Amin 2008). This complexity in the city of opportunity problematizes the notion of public space and begins to redefine the concept and systems of public spaces, where no universal vocabulary exists (Haas and Olsson 2014). Public space means a common and shared space, with free access and no fear of entry, as has been emphasized frequently in feminist geography (see, for example, Valentine 1989; Duncan 1996; Ruddick 1996; Binnie 1997; Kostela and Pain 2000; Pain 2001; Fenster 2007; Crimmion 2013).

In this context, the issue of culture and contextualized urbanity is relevant. The city is not just a geographical and static entity, but is the outcome of social processes, in which static and kinetic elements are coming together. This new, cultural urbanism perspective becomes increasingly pivotal in understanding and shaping all-inclusive spaces in the city. For example, Meyer (2014) looks at cultural urbanism as a conceptual approach that has seven major constituents: (1) paying attention to the context and history of place as well as its narratives, (2) understanding local preferences of all inhabitants, (3) paying attention to a diversity of users and uses, (4) providing a variety of products, (5) changes to be different in space and place, (6) establishing high-quality open space and public realms, and (7) creation of higher real-estate values owing to all above factors (see also Hayden 1981; Greed 1994).

It is worth noting that cultural urbanism addresses in particular the everyday, the temporal, the occasional, but also the timeless dimensions of city life, something which flagship buildings are almost never able to do. It is an approach that explores the spatial and social fabric of the city and the kineticism of relationships on the tangible ground that defines the use and management of the built environment. It opens the differences in building environments, which foster community interaction, enabling us to go deeper in order to see and understand what makes cities so unique and interesting, as well as what is the foundation for creating genuine and authentic places. Such a paradigm also allows the dwellers and citizens to preserve the fine-grain urban distinctions, celebrating their tasks, narratives and histories, as well as differences that occur in space and time and between all different groups of users. In this context, Massey (1994) emphasizes the key in the issues of space, where space must be conceptualized integrally with time and where a new thinking of space – not as some absolute independent dimension, but as constructed out of social relations – must be brought in. This is most evident in common spaces in the city.

In the light of the previous observations, it is important to refer here to Whyte (1980), who studied, in the vein of observational urbanism, a series of urban spaces in New York City and commented on why some were successful, while others were not. Among Whyte's main and bearing ideas and findings are (1) the social life in public spaces contributes fundamentally to the quality of life of individuals and society, (2) designers have a moral responsibility to create physical places that facilitate civic engagement and community interaction, (3) public spaces should be designed from the bottom up, not top down, and (4) design should start with a thorough understanding of the way people use spaces and the way they would like to use spaces.

These processes and practices are dynamic and fluid; they are constantly re-created and re-formed, even though the spaces they construct are changed and transformed (Staeheli and Martin 2000). In line with that, Wirth (1938) and Tonkiss (2014) argue that cities are essentially social forms, not necessarily built forms. City-making is a social process and the intricate and close relationship between the social and physical shaping of urban environments is crucial for creating social identity and inclusive public spaces. Just as space, the network of processes and relationships that connect places with each other (Massey 1994) may be coded with a social identity. Consequently, social, economic, planning and technological ideologies are central to the ways in which urban spaces are constructed (Bjelland et al. 2013).

Finally, if we are going to reach a city of hope or a city of opportunity, we need a thorough exploration of key elements of cultural urbanism, where a nuanced understanding of public space is brought in. 'Culture is the software of cities just as the built environment is its hardware', says Calthorpe (2013, pp. 1–6). According to him software and hardware co-evolve with each coming time and generation, 'culture informing and transforming the hardware of a city while technological change and infrastructure redirect the city culture' (Calthorpe 2013, pp. 77–90). The important thing to keep in mind, as Calthorpe (2013, pp. 35–48) observes is that 'while each place is unique, there are universal human traits that set the fundamental DNA of great cities: human scale, diversity of action, and social interaction'. This view is reflected in our 3H-city conceptualization advocated in this chapter.

6. CITIES4PEOPLE: SOCIAL INTERACTIONS AND NETWORKS

According to the founding father of urban sociology, Max Weber (1922 [1978]), the city is a human construct, designed and governed by man-made rules, institutions and conventions. The city is meant to accommodate people. In a recent study, Wahlström et al. (2020a) argue that a city comprises two constituents, viz. a '*body*' and a '*soul*'. The body refers to the hardware of a city (for example, built environment, infrastructure, facilities etc.), while the soul refers to the human appreciation and perception of tangible and intangible city life (including social contacts).

City life is not only characterized by individualistic or atomistic behaviour of residents or visitors. It is based on an urban social organism of interacting human beings, living and operating not only for social or altruistic reasons, but also for egoistic or economic reasons (Castells 1998). The urban piazza was historically designed as a smart and functional way for facilitating human and social interactions for a wide variety of purposes and motives of residents. Consequently, the expression 'cities4people' does not only refer to the challenge to create a liveable urban environment for separate individuals in an urban area, but also to the challenge to create tangible infrastructures and intangible suprastructures that facilitate human and social interactions in a city. This calls for both physical networks (roads, public spaces and so on) and immaterial networks (community centres, digital communication systems, and so on). The modern city is clearly a connected city (Neal 2013), but also a communication city, in which both material and virtual interactions – in numerous forms – play a prominent role. Consequently, the traditional urban agglomeration advantages (scale, localization and urbanization benefits) are increasingly supplemented and extended with an unprecedented rise of ICT facilities, in particular in the form of digital technology opportunities (Duranton and Puga 2004). These interactive cities benefit from an abundance of sharing, matching and learning mechanisms.

Interactive or networked cities are, in general, knowledge-based (both codified and tacit) and, more recently, data orientated (see Roberts and Setterfield 2010). Cities offer the great advantage of combining both physical proximity and virtual proximity (Torre and Wallet 2014). Kourtit (2016b) has even introduced the notion of hyper-proximity in the modern ICT age, in which non-geographic proximity becomes an increasingly dominant feature in our network society.

Despite the rapidly rising importance of virtual social interactions, it ought to be emphasized that face-to-face contacts are still an indispensable core feature of a modern urbanized society (Glaeser et al. 1996; Buechel and Von Ehrlich 2016). In particular, public space is a *sine qua non* for human and social interactions. The current pedestrianization of many city centres demonstrates citizens' and visitors' need for human-interaction orientated development of cities. Social media may often even become a catalyst for intensifying social interactions in a city, as location-based devices make it so much easier to identify and find the locations of people in the city (Rosenblat and Mobius 2004). It therefore seems plausible that the digital revolution reflected in social media is not leading to a direct substitution for virtual contacts, but offers reinforcing complementary support systems for urban communications and interactions (Gentzkow et al. 2011). Networked (or connected) cities may thus offer a new impetus for the essence of a city, namely, social communication and physical interaction (Neal 2013). Face-to-face and virtual interaction may thus become two mutually reinforcing mechanisms in an urbanized world.

There is more to say about the benefits of a digital city. A digital urban society appears to have a great potential for unprecedented volumes of social interactions, as is witnessed by social media and use of smartphones. These digital technologies do not only offer many new opportunities for information and data handling, but may as a downside also reduce the quality of face-to-face interactions and hence have an impact on feelings or social well-being (Rotondi et al. 2017). Thus, the virtual network character of modern urban technology has great social and economic implications for city life, as is also exemplified by social communication studies undertaken by Arribas-Bel et al. (2016) and Kourtit et al. (2019). Indeed, digital technology networks in and between cities may not only create 'magic in the air' (Katz 2006), but also shadows on the 3H profile of city life.

7. URBAN ANALYTICS IN THE CITY OF OPPORTUNITY: AN EXPLORATION

The New Urban World does not only result in great scientific challenges at the global level of rapid urbanization, but also prompts many meso- and micro-scale research questions at the intra-city level. First, it ought to be recognized that cities have turned into complex spatial and virtual, planar and non-planar networks, with a strong physical and digital interaction among people, activities and information. Thus, interdependency and connectivity are a first feature of a microcosmic city in which opportunities are to be optimized. The analysis of this complex interdependency is an important research task.

Secondly, a city has an organized, dynamically adaptable socio-economic, demographic-cultural network structure which is based on the decomposed architecture of a microcosmic settlement. In this conceptual framing, the city has a multi-layered, evolutionary structure. The successful management of this complex non-linear system is fraught with many hurdles and difficulties, so that the achievement of a high performance is far from easy (see also the previously mentioned XXQ principle).

The data analytics involved in modern complex urban dynamics may use various steps and ingredients.

1. The operational analysis of an opportunity-rich, microcosmic city needs to take into account the morphology, networks, transition trends, resilience capacity and adaptive potential of decomposed network cities, both physical and invisible. This calls for research on urban networks of different scale and nature.
2. A complex, multi-layered urban system needs, for a proper understanding of its fabric and evolution, an integrated space–time activity-based research approach, in which individual and group choices on opportunities and options related to the microcosmic constellation of the city are investigated and modelled.

3. Since a complex, multi-layered urban system with a mix of private and public spaces often exhibits a fluctuating behaviour in population size, cultural-ethnic composition, employment, mobility patterns, social capital and socio-economic disparities, it is necessary to identify tipping points which reflect bifurcations in urban dynamics, in order to simulate alternative futures for urban neighbourhood spaces to better project and protect the future(s) of a microcosmic city (see also Batty 2005).
4. Urban agglomerations are based on multi-scalar spatial networks, characterized by complex, dynamic, non-linear and emergent behaviour which are hard to model by means of traditional statistical and modelling approaches, and hence it may be necessary to resort to spatial complexity models (see Reggiani and Nijkamp 2009) in order to trace critical transitions for the identification of early warnings in urban management and design.
5. Cities have turned into big-data machines (Batty 2005) and various tools are needed to monitor, experiment and analyse this avalanche of data in a digital urban environment, through the use of sophisticated data-handling methods, such as:
   (a) computational neural networks;
   (b) spatial cellular automata techniques;
   (c) spatial agent-based models (cf. Hommes 2013);
   (d) spatial epidemic models for social contagion issues;
   (e) evolutionary network and spatial interaction models (cf. Frenken 2006);
   (f) multi-scalar geoscience-based methods;
   (g) three-dimensional simulation models for dynamic microcosmic urban design;
   (h) social percolation and spatial filter models (cf. Salomon et al. 2000); and
   (i) heuristic data-mining techniques.
6. The 3H-city dimensions of a microcosmic city need to be studied from a micro-behavioural, opportunity-orientated and bounded rationality perspective at local and neighbourhood level; this approach may resort to spatial and dynamic micro-simulation models.
7. The impact of social media on opinion formation, living habits and spatial behaviour in cities is another and increasingly important issue (see, for example, Bond et al. 2012), which calls for advanced social network diffusion models on circulation of information, for instance, on the impact of small changes in social-spatial networks on opinion dynamics, spread of local (dis)agreement or social manipulation (see, for example, Flache and Macy 2011).
8. The information monitoring, handling and management by public agencies in a city is based on a flexibility in governance structures with a view to the interest of citizens. This interactive policy perspective may be supported by the design and use of urban management dashboards, outlined in section 8.

The analytics of a microcosmic city is not only necessary for mapping out current structures and spatial movements, but is also needed for future projections, in particular with a view to the identification of critical transitions and vulnerabilities at the local and district level of urban spaces. Monitoring all relevant information is necessary for ensuring the city love feelings of agents in a 3H-city triangle constellation of urban networks.

8. URBAN MANAGEMENT DASHBOARDS IN THE CITY OF OPPORTUNITY

In recent years an extant literature has emerged in administrative and political sciences and in business and spatial planning, which has focused attention on the design and use of new operational and user-orientated navigation tools for strategic management and policy-making. These tools have to be able to cope with daily needs of corporate organizations, in the form of interactive, learning and flexible decision support systems that are able to capture a wide range of relevant information in a big-data environment. In this section we offer a sketch of an urban management dashboard system that is able to digest and present strategic information on decomposed urban spaces for a proper and interactive handling of business, sustainability, appreciation and land use information in a microcosmic city.

A city dashboard is a prototype tool for a smart combined virtual–physical place, even at street level, that displays a complete and stylized picture of all opportunities related to important or prioritized socio-economic activities and actions of multi-stakeholders regarding the fulfilment of strategic goals for the future of places in complex city life. It also provides a clear snapshot of public venues' performance in which the local community is strongly involved and becomes part of the ongoing process by sharing their own interests, priorities and ideas on different dimensions of sustainable future well-being indicators in urban areas that align and motivate everyone towards the common goals of maximizing XXQ (Kaplan and Norton 1996, 2001; Brignall 2002; Ho and Chan 2002; Miller 2005; De Waal 2007; De Waal et al. 2009; Kitchin et al. 2015; Kitchin and Perng 2016; Kourtit 2016a).

The architecture of an operational dashboard provides the possibility for two-way physical and digital communication and interaction, and enhances the voice of multiple stakeholders regarding their ideas and concerns about all aspects and conditions that shape their city love.

This strategic, integrated and interactive navigation tool provides an intelligent roadmap that helps to translate a common vision of the city into an operational strategy, in order to set clear strategic goals and objectives along the way with critical conditions and appropriate and smart key measures (removing obstacles through an ongoing planning process) that enable – with a high degree of efficiency and flexibility – them to delegate responsibility for operational activities and to focus on the desired outcomes to achieve the shared vision towards a place 4 all regarding future sustainable development in terms of a maximum XXQ (Nijkamp and Kourtit 2011; Kourtit 2016a).

The core element of this approach is that it helps in managing the city through an effective translation and strategic alignment, with an understanding of the established vision for the future of appropriate places of city life and related strategies. On the one hand, this takes place through a balanced and prioritized set of smart critical success factors (CSFs) and key performance indicators (KPIs) (predictors from both an economic and non-economic perspective) in order to be able to support or take balanced decisions and to develop smart solutions to keep the city, with its multiplicity of functions and its multi-actors, on track (see Kourtit and Nijkamp 2017). On the other hand, it helps to collect a wide range of relevant (real-time and digital) data and information sources in a massive unstructured big-data environment (see Kourtit and Nijkamp 2018).

This ongoing complex and multi-faceted strategic decision process may generate a high added value (see also, Morecroft 1984; Foil and Huff 1992; Mintzberg 1994; Mintzberg et al 1998; Tan and Platts 2003) for both private and public multi-actors in terms of an efficient and effective collection and transformation of (traditional and new) 'big data' and user-friendly data visualizations into 'actionable intelligence'. This may be instrumental to refurbish and reprofile the performance of a city as an 'intelligent community' (Kourtit and Nijkamp 2018) in shaping a promising future.

9. CONCLUDING REMARKS

The conceptual framework outlined in the previous sections has emphasized the importance of a micro- and meso-based urban system's perspective, in particular, regarding the spatial network organization of shared urban spaces, interdependencies in human spatial behaviour, the exploration of tipping and bifurcation points at the interface of neighbourhoods, infrastructure and the city as whole (see also, OECD 2012), the constituents of city love captured in the 3H-city triangle, and the policy handles needed to govern a microcosmic city. To understand these complex mechanisms at a disaggregate scale of urban public space, especially in respect of big-data, human–machine cognitive interactions and artificial intelligence may be needed to reduce complexity to simplicity (see, for example, Reggiani and Nijkamp 2009; West 2016). The city of opportunity is certainly work in progress.

REFERENCES

Acs, Z. (ed.) (2006), *The Growth of Cities*, Cheltenham, UK and Northampton, MA, USA: Edward Elgar.
Amin, A. (2008), 'Collective culture and urban public space', *City*, **12** (1), 5–24.
Ardener, S. (1993), *Women and Space: Ground Rules and Social Maps*, Oxford: Berg.
Arribas-Bel, D., K. Kourtit and P. Nijkamp (2016), 'The socio-cultural sources of urban buzz', *Environment and Planning C*, **34** (1), 188–204.
Bailly, A. (2010), 'Regional geography', *Encyclopaedia of Life Support Systems* (EOLSS), vol. 2, *Geography*, accessed 14 March 2010 at www.eolss.net/sample-chapters/c01/e6-14-03-10.pdf.
Batty, M. (2005), *Cities and Complexity*, Cambridge, MA: MIT Press.
Binnie, J. (1997), 'Coming out of geography: towards a queer epistemology?', *Environment and Planning D: Society and Space*, **15** (1), 223–37.
Bjelland, M.D., D.R. Montello, J.D. Fellman, A. Getis and J. Getis (2013), *Human Geography: Landscapes of Human Activities*, New York: McGraw-Hill.
Bond, R.M., C.J. Fariss, J.J. Jones, A.A. Kramer, C. Marlow, J.E. Settle, et al. (2012), 'A 61-million person experiment in social influence and political mobilisation', *Nature*, **489** (7415), 295–8.
Brignall, S. (2002), 'The unbalanced scorecard: a social and environmental critique', in A. Neely, A. Walters and R. Austin (eds), *Performance Measurement and Management 2002: Research and Action: Papers from the Third International Conference on Performance Measurement and Management*, Cranfield: Cranfield School of Management, pp. 85–92.
Bruni, L. and P.L. Porta (2007), *Handbook on the Economics of Happiness*, Cheltenham, UK and Northampton, MA, USA: Edward Elgar.
Buechel, K. and M. Von Ehrlich (2016), 'Cities and the structure of social interactions: evidence from mobile phone data', Diskussionsschriftencredresearchpaper13, Universitaet Bern, Departement Volkswirtschaft – CRED.
Calthorpe, P. (2013), *Urbanism in the Age of Climate Change*, Washington, DC: Island Press.
Caragliu, A., C. Del Bo and P. Nijkamp (2011), 'Smart cities in Europe', *Journal of Urban Technology*, **18** (2), 65–82.

Carmona, M., T. Heath, T. Oe and S. Tiesdell (2010), *Public Places, Urban Spaces*, Boston, MA: Elsevier/Architectural Press.

Castells, M. (1998), *End of Millennium: The Information Age: Economy, Society and Culture*, New York: Wiley-Blackwell.

Certeau, M. de (2011), *The Practice of Everyday Life*, Berkeley, Los Angeles, CA and London: University of California Press.

Cloke, P., C. Philo and D. Sadler (1991), *Approaching Human Geography*, London: Chapman.

Comim, F., M. Qizilbash and S. Alkire (2008), *The Capability Approach*, Cambridge: Cambridge University Press.

Comunian, R. (2011), 'Rethinking the creative city', *Urban Studies*, **48** (6), 1157–79.

Cormona, M. (2010), *Public Places, Urban Spaces*, Boston, MA: Architectural Press/Elsevier.

Crimmion, A. (2013), 'The slutwalks: reappropriation through demonstration', *Manchester Metropolitan University Gender and Urban Space Gender Forum*, **42** (6), 34–54.

Crocker, D. (1992), 'Functioning and capability, the foundations of Sen's and Nussbaum's development ethic (part 1)', *Political Theory*, **20** (4), 584–612.

Deakin, M. (2010), 'SCRAN: the Smart Cities (inter) Regional Academic Network supporting the development of a trans-national comparator for the standardisation of eGovernment services', in C. Reddick (ed.), *Comparative E-Government: An Examination of E-Government across Countries*, Berlin: Springer, pp. 425–47.

De Waal, A. (2007), *Strategic Performance Management: A Managerial and Behavioural Approach*, Basingstoke: Palgrave.

De Waal, A.A., K. Kourtit and P. Nijkamp (2009), 'The relationship between the level of completeness of a strategic performance management system and perceived advantages and disadvantages', *International Journal of Operations and Production Management*, **29** (12), 1242–65.

Dowding, K. (2006), 'Can capabilities reconcile freedom and equality?', *Journal of Political Philosophy*, **14** (3), 323–36.

Duncan, N. (1996), 'Renegotiating gender and sexuality in public and private spaces', in N. Duncan (ed.), *Destabilizing Geographies of Gender and Sexuality*, London: Routledge, pp. 127–89.

Duranton, G. and D. Puga (2004), 'Micro-foundations of urban agglomeration economies', in J.V. Henderson and J.F. Thisse (eds), *Handbook of Regional and Urban Economics*, Amsterdam: Elsevier, pp. 2063–117.

Edwards, J.M. and S. Pellé (2011), 'Capabilities for the miserable; happiness for the satisfied', *Journal of the History of Economic Thought*, **33** (3), 335–55.

Engelen, E., S. Johal, A. Salento and K. Williams (2014), 'How to build a fairer city', *Guardian*, 24 September, accessed 24 September 2014 at http://www.theguardian.com/cities/2014/sep/24/manifesto-fairer-grounded-city-sustainable-transport-broadband-housing.

Evans, G. (2009), 'Creative cities, creative space and urban policy', *Urban Studies*, **48** (5–6), 1003–40.

Farrell, K. and T. Haas (2017), 'Rapid urban growth and the future of public spaces: shifting development agendas and a new roadmap', *International Journal of Arts and Humanities*, **1** (1), 57–73.

Fekadu, K. (2014), 'The paradox in environmental determinism and possibilism: a literature review', *Journal of Geography and Regional Planning*, **7** (7), 132–9.

Fenster, T. (2005), 'Gender and the city: the different formations of belonging', in L. Nelson and J. Seager (eds), *A Companion to Feminist Geography*, Oxford: Blackwell, pp. 242–56.

Flache, A. and M.W. Macy (2011), 'Small worlds and cultural polarization', *Journal of Mathematical Sociology*, **35** (1), 146–76.

Frenken, U. (2006), *Innovation, Evolution and Complexity Theory*, Cheltenham, UK and Northampton, MA, USA: Edward Elgar.

Foil, C.M. and A.S. Huff (1992), 'Maps for managers: where are we? Where do we go from here', *Journal of Management Studies*, **29** (3), 267–85.

Geenhuizen, M.S. van and R. Ratti (1998), 'Managing openness in transport and regional development: an active space approach', in K. Button, P. Nijkamp and H. Priemus (eds), *Transport Networks in Europe*, Cheltenham, UK and Northampton, MA, USA: Edward Elgar, pp. 84–102.

Gehl, J. (1987), *Life between Buildings: Using Public Space*, Washington, DC: Island Press.

Gentzkow, M.A., J.M. Shapiro and M. Sinkinson (2011), 'The effect of newspaper entry and exit on electoral politics', *American Economic Review*, **101** (7), 2980–3018.

Geyer, M. (2009), *International Handbook of Urban Policy*, Cheltenham, UK and Northampton, MA, USA: Edward Elgar.

Gigerenzar, G. and R. Selten (2001), *Bounded Rationality*, Cambridge: MIT Press.

Glaeser, E.L., B. Sacerdote and J.A. Scheinkman (1996), 'Crime and social interactions', *Quarterly Journal of Economics*, **111** (2), 507–48.

Greed, C. (1994), *Women and Planning, Creating Gendered Realities*, London: Routledge.

Haas, T. and K. Olsson (2014), 'Transmutation and reinvention of public spaces through ideals of urban planning and design', *Space and Culture*, **17** (1), 59–68.
Haas, T. and H. Westlund (eds) (2017), *In the Post-Urban World – Emergent Transformations of Cities and Regions in the Innovative Global Economy*, London: Routledge.
Hawe, P. and A. Shiell (2000), 'Social capital and health promotion: a review', *Social Science and Medicine*, **51** (6), 871–85.
Hayden, D. (1981), 'What would a non-sexist city be like', C.R. Stimpson, E. Dixler, M.J. Nelson and K. Yatrakis (eds), *Women and the American City*, Chicago, IL: University of Chicago Press, pp. 167–84.
Ho, S.K. and Y.L. Chan (2002), Performance Measurement and the Implementation of Balanced Scorecards in Municipal Governments, *Journal of Government Financial Management*, 51, 8–19.
Hommes, C.H. (2013), *Behavioural Rationality and Heterogeneous Agents in Complex Economic Systems*, Cambridge: Cambridge University Press.
Johnston, R., D. Gregory, G. Pratt and M. Watts (2000), *The Dictionary of Human Geography*, Cambridge: Blackwell.
Kaplan, R.S. and D.P. Norton (1996), *The Balanced Scorecard – Translating Strategy into Action*, Boston, MA: Harvard Business School Press.
Kaplan R.S. and D.P. Norton (2001), 'Transforming the balanced scorecard from performance measurement to strategic management: part 1', *Account Horizons*, **15** (1), 87–104.
Katz, J. (2006), *Magic in the Air: Mobile Communication and the Transformation of Social Life*, New York: Routledge.
Kaufman, A. (2006), 'Capabilities and freedom', *Journal of Political Philosophy*, **14** (3), 289–300.
Kitchin, R. and S.-Y. Perng (2016), *Code and the City*, London: Routledge.
Kitchin, R., T.P. Lauriault and G. McArdle (2015), 'Knowing and governing cities through urban indicators, city benchmarking and real-time dashboards', *Regional Studies, Regional Science*, **2** (1), 6–28.
Koskela, H. and R. Pain (2000), 'Revisiting fear and places women's fear of attack and the built environment', *Geoforum*, **31** (2), 269–80.
Kourtit, K. (2016a), 'Towards a sustainable *i*-city: intelligent transition management of digital places', *Quality Innovation Prosperity*, **21** (1), 151–64.
Kourtit, K. (2016b), 'Super-proximity and spatial development', *Journal of Regional Research*, **36**, 215–31
Kourtit, K. (2019), *The New Urban World: Economic-Geographical Studies on the Performance of Urban Systems*, Aachen: Shaker.
Kourtit, K. and P. Nijkamp (2017), 'Critical performance factors for large world cities – in search of qualitative causal patterns by means of rough set analysis', *REGION*, **4** (1), 51–70.
Kourtit, K. and P. Nijkamp (2018), 'Big data dashboards as smart decision support tools for i-cities – an experiment on Stockholm', *Land Use Policy*, **71** (February), 24–35.
Kourtit, K., P. Elmlund and P. Nijkamp (2019), 'Neighbourhood proximity: a microcosmic complex systems perspective on cities', in A. Torre and D. Gallaud (eds), *Handbook of Proximity Relations*, Cheltenham, UK and Northampton, MA, USA: Edward Elgar (forthcoming).
Kourtit, K., P. Nijkamp and J. Östh (2020), 'My home is my castle assessment of city love in Sweden', *International Journal of Information Management*, 23 October, doi:10.1016/j.ijinfomgt.2020.102213.
Kourtit, K., P. Nijkamp and M. Partridge (2015), 'Challenges of the new urban world', *Applied Spatial Analysis and Policy*, **8** (3), 199–215.
Kourtit, K., P. Nijkamp, R.S. Franklin and A. Rodríguez-Pose (2014), 'A blueprint for strategic urban research: the "urban piazza"', *Town Planning Review*, **85** (1), 97–126.
Lefebvre, H. (1971), *Everyday Life in the Modern World*, London: Harper Torchbooks.
Ma Y., X. Qin, R. Chen, N. Li, R. Chen and Z. Hu (2012), 'Impact of individual-level social capital on quality of life among AIDS patients in China', *PLoS ONE*, **7** (11), e48888.
Madanipour, A. (2003), *Public and Private Spaces of the City*, New York: Routledge.
Mahaffy, M., S. Cowan and D. Urge-Vorsatz (2009), 'The factor of urban morphology in greenhouse gas emission: a research overview', *IOP Conference Series: Earth and Environmental Science*, **6** (202001), 1–12, doi:10.1088/1755-1307/6/20/202001.
Massey, D. (1994), *Space, Place, and Gender*, Minneapolis, MN: University of Minnesota Press.
McDowell, L. and P.J. Sharp (1999), *A Feminist Glossary of Human Geography*, New York: Arnold.
Meyer, T. (2014), 'Cultural urbanism: distinct urban places that promote social interaction and create value', *Planning Magazine*, April, 14–19.
Miller, C.A. (2005), 'New civic epistemologies of quantification: making sense of indicators of local and global sustainability', *Science, Technology and Human Values*, **30** (3), 403–32.
Milton, K. (1996), *Environmentalism and Culture Theory: Exploring the Role of Anthropology in Environmental Discourse*, New York: Routledge.
Mintzberg, H. (1994), *The Rise and Fall of Strategic Planning*, New York: Prentice Hall.
Mintzberg, H., B. Alstrand and J. Lampel (1998), *Strategy Safari*, New York: Free Press.

Morecroft, J.D. (1984), 'Strategy support models', *Strategic Management Journal*, **5** (3), 215–29.

Murayama, H., Y. Fujiwara and I. Kawachi (2012), 'Social capital and health: a review of prospective multilevel studies', *Journal of Epidemiology*, **22** (3), 179–87.

Nam, T. and T.A. Pardo (2011), 'Conceptualization of smart city with dimensions of technology, people and institutions', in *Proceedings of the 12th Annual International Conference on Digital Government Research*, New York: Association for Computing Machinery, pp. 282–91.

Neal, Z. (2013), *The Connected City*, New York: Routledge.

Nelson, L. and J. Seager (2005), *A Companion to Feminist Geography*, London: Blackwell.

Nijkamp, P. (2008), 'XXQ factors for sustainable urban development: a systems economics view', *Romanian Journal of Regional Science*, **2** (1), 1–34.

Nijkamp, P. (2016), 'The "resourceful region"', *Journal of Regional Research*, **36**, 191–214.

Nijkamp, P. and G. Cohen-Blankshtain (2013), 'The importance of ICT for cities: e-governance and cyber perceptions', in J. Klaasson, B. Johansson and C. Karlsson (eds), *Metropolitan Regions*, Berlin: Springer, pp. 295–308.

Nijkamp, P. and K. Kourtit (2011), 'New urban Europe: a place 4 all', *Urban Governance in the EU Current Challenges and Future Prospects*, Brussels: Committee of the Regions, DG RTD, European Commission, pp. 61–76.

Norris, D.F. and C.G. Reddick (2013), 'Local e-government in the United States', *Public Administration Review*, **73** (1), 165–75.

Nussbaum, M. (2000), *Woman and Human Development: The Capability Approach*, Cambridge: Cambridge University Press.

Organisation for Economic Co-operation and Development (OECD) (2012), *Redefining 'Urban': A New Way to Measure Metropolitan Areas*, Paris: OECD, doi:10.1787/9789264174108-en.

Pain, R. (2001), 'Gender, race, age and fear in the city', *Urban Studies*, **38** (5–6), 899–913.

PwC (2016), 'Cities of opportunity', New York, accessed April 2016 at https://www.pwc.com/us/en/cities-of-opportunity/. . ./cities-of-opportunity-7-report.pdf.

Qizilbash, M. (1996), 'Capabilities, well-being and human development: a survey', *Journal of Development Studies*, **33** (2), 143–62.

Ratti, C. and M. Claudel (2016), *The City of Tomorrow: Sensors, Networks, Hackers, and the Future of Urban Life*, New Haven, CT: Yale University Press.

Reggiani, A. and P. Nijkamp (eds) (2009), *Complexity and Spatial Networks*, Berlin: Springer.

Roberts, M. and M. Setterfield (2010), 'Endogenous regional growth: a critical survey', in M. Setterfield (ed.), *Handbook of Alternative Theories of Economic Growth*, Cheltenham, UK and Northampton, MA, USA: Edward Elgar, pp. 431–50.

Rosenblat, T. and M. Mobius (2004), 'Getting Closer or Drifting Apart?', *Quarterly Journal of Economics*, **119** (3), 971–1009.

Rotondi, V., L. Stanca and M. Tomasuolo (2017), 'Connecting alone: smartphone use, quality of social interactions and well-being', *Journal of Economic Psychology*, **63** (C), 17–26.

Ruddick, S. (1996), 'Constructing difference in public spaces: race, class, and gender as interlocking systems', *Urban Geography*, **17** (2), 132–51.

Sacco, P.L., G. Ferilli, T.B. Tavano and M. Nuccio (2013), 'Culture as an engine of local development processes', *Growth and Change*, **44** (4), 571–88.

Salomon, S., G. Weisbruck, L. de Arcangelis, N. Jan and D. Stauffer (2000), 'Social percolation models', *Physika A*, **277** (1), 239–47.

Scott, A. (2000), *The Cultural Economy of Cities*, London: Sage.

Scott, A.J. (2017), 'Urbanization, work and community: the logic of city life in the contemporary world', *Quality Innovation Prosperity*, **21** (1), 9–30.

Sen, A. (2008), 'The economics of happiness and capability', in L. Bruni, F. Comim and M. Pugno (eds), *Capabilities and Happiness*, Oxford: Oxford University Press, pp. 16–27.

Sen A.K. (1985), *Commodities and Capabilities*, Amsterdam: Elsevier.

Sen A.K. (1992), *Inequality Reexamined*, Oxford: Oxford University Press.

Sen, A.K. (1993), 'Capability and well-being', in M. Nussbaum and A.K. Sen (eds), *The Quality of Life*, Oxford: Clarendon Press, pp. 30–53.

Sen, A.K. (2005), 'Human rights and capabilities', *Journal of Human Development*, **6** (2), 151–66.

Simon, H.A. (1955), 'On a class of skew distribution functions', *Biometrika*, **42** (3–4), 425–40.

Staeheli, L.A. and M.P. Martin (2000), 'Spaces for feminism in geography', *Annals of the American Academy of Political and Social Science*, **571** (1), 135–50.

Steward, J. (ed.) (1955), *Theory of Culture Change: The Methodology of Multilinear Evolution*, Urbana, IL: University of Illinois Press.

Steward, J. (1977), 'Evolution and ecology', in J.C. Steward and R.F. Murphy (eds), *Essays on Social Transformations*, Urbana, IL: University of Illinois Press.

Storper, M. and A. Scott (2009), 'Rethinking human capital, creativity and urban growth', *Journal of Economic Geography*, **9** (2), 147–67.
Tan, K.H. and K. Platts (2003), 'Linking objectives to actions: a decision support approach based on cause–effect linkages', *Decision Sciences*, **34** (3), 569–93.
Taylor, T.G. (ed.) (1951), *Geography in the Twentieth Century*, London: Methuen.
Tellier, L.-N. (2009), *Urban World History; An Economic and Geographical Perspective*, Quebec: Presses de l'Université de Quebec.
Thompson, C.W. (2002), 'Urban open space in the 21st century', *Landscape and Urban Planning*, **60** (2), 59–72.
TNO (2012), 'A framework for sustainable urbanisation', TNO, Delft, accessed 3 October 2012 at http://www.tno.nl/downloads/iB_Urban_Strategy_en.pdf.
Tonkiss, F. (2014), *Cities by Design: The Social Life of Urban Form*, London: Polity Press.
Torre, A. and F. Wallet (2014), 'Introduction to the role of proximity in relations in regional and territorial development processes', in A. Torre and F. Wallet (eds), *Regional Development and Proximity Relations*, Cheltenham, UK and Northampton, MA, USA: Edward Elgar, pp. 94–134.
Torres, L., V. Pina and B. Acrete (2005) 'E-government developments on delivering public services among EU cities', *Government Information Quarterly*, **22** (2), 217–38.
Valentine, G. (1989), 'The geography of women's fear', *Area*, **21** (4), 385–90.
Vanolo, A. (2013), 'Smart mentality: the smart city as disciplinary strategy', *Urban Studies*, **51** (3), 883–98.
Vidal de la Blache, P. (1903), *Tableau de la Géographie de la France*, Paris: Machette,
Waldinger, R.J. and M.S. Schulz (2010), 'What's love got to do with it? Social functioning, perceived health, and daily happiness in married octogenarians', *Psychology and Aging*, **25** (2), 422–31.
Wahlström, M., K. Kourtit and P. Nijkamp (2020a), 'How to make cities the home of people – a "soul and body" analysis of urban attractiveness', *Land Use Policy*, 2 June, 104734, doi:10.1016/j.landusepol.2020.104734.
Wahlström, M., K. Kourtit and P. Nijkamp (2020b), 'Planning cities 4 people–a body and soul analysis of cities and their urban neighbourhoods', *Public Management Review*, **22** (5), 687–700, doi:10.1080/14719037.2020.1718190.
Weber, M. (1922), *The City*, *Wirtschaft und Gesellschaft*, repr. 1978, Tübingen: J.C.B. Mohr (Paul Siebeck).
West, G. (2016), *Scale: The Universal Laws of Growth, Innovation, Sustainability, and the Pace of Life in Organisms, Cities, Economies, and Companies*, New York: Penguin Press.
Westlund, H. (2006), *Social Capital in the Knowledge Economy: Theory and Empirics*, Cheltenham, UK and Northampton, MA, USA: Edward Elgar.
Whyte, H.W. (1980), *Social Life of Small Urban Spaces*, Washington, DC: Conservation Foundation.
Winters, J.V. (2010). 'Why are smart cities growing? Who moves and who stays', *Journal of Regional Science*, **20** (10), 1–18.
Wirth, L. (1938), 'Urbanism as a way of life', *American Journal of Sociology*, **44** (1), 1–24.
Zhuhader, L., E. Thrasku, S. Marklin and P.O. de Pablos (2007), 'The next wave of innovation', *Computers in Human Behavior*, **66** (C), 273–81.

14. Community organizing and interorganizational network changes in a justice system reform coalition in Chicago

Brian D. Christens and Daniel G. Cooper

INTRODUCTION

Community organizing occurs when residents collaboratively investigate and take collective action on issues that concern them. These issues of concern often include safety, educational quality, community development, improvements to neighborhoods and the physical environment, and access to health care, affordable housing, transportation and employment opportunities. Often, these concerns are local manifestations of macro-level forces, such as neoliberal globalization and structural racism. In order to achieve changes in local policies and systems in spite of these forces, residents often seek to build long-lasting local organizing initiatives that are capable of sustaining social action, holding decision-makers and institutions accountable to address concerns of local residents across a variety of issues. Organizing is therefore a way for local residents to enhance their quality of life in tangible ways while simultaneously resisting and challenging oppressive sociopolitical forces and working to advance social justice.

Many local organizing initiatives are built as coalitions or federations of smaller geographically- or institutionally situated efforts. Just as labor-organizing federations support efforts across many different workplaces, community organizing federations (often termed initiatives) support efforts within and among community centers, faith-based institutions, schools, neighborhood organizations, and other nonprofit and voluntary organizations. For example, the Southwest Organizing Project (SWOP) in Chicago currently lists 37 member institutions including faith-based institutions (for example, a Lutheran church, an Ethiopian Hebrew congregation, and the Inner-City Muslim Action Network, itself a separate federation), a number of elementary, middle, and high schools, a community development corporation, and several social service organizations. Many times, this type of federated structure is used for two purposes: (1) to equip local organizing committees or efforts within each member institution (sometimes termed affiliates), which may organize around issues of their own choosing, and (2) to create a broader federated structure that can carry out campaigns on issues of concern across the initiative or federation.

Organizing federations such as these often have hundreds or even thousands of members who participate on a voluntary basis, across a variety of different types of meetings. Unlike most forms of civic participation and voluntarism, which primarily seek to engage people for sporadic volunteer labor, donations, or expressive participation such as rallies and protests, participants in community organizing initiatives are encouraged to become leaders, and to progress toward greater levels of responsibility and prominence within the initiative. A primary role of the (typically small) paid staff of a local organizing initiative is to identify, engage, train and support the development of these volunteer leaders.

This process of leadership development is grounded in principles and/or models for community organizing that have become widespread in the field of community organizing. Volunteer leaders and staff organizers become familiar with one or more of these models and adaptively apply organizing principles to their local context. For example, one principle that is fundamental to many grassroots organizing initiatives is that the people affected by a problem are the people to address that problem. This principle helps organizing groups to identify potential leaders and to distinguish what they do from advocacy efforts, in which groups or individuals act on behalf of others (Speer and Christens 2014). Leaders and staff organizers conduct trainings on principles such as these, and the philosophies that undergird their practice. They encourage fellow residents to reflect on organizing principles as they undertake local action. In this way, they build on the accumulated wisdom that has enabled the formation of shared identity and progress on social issues in other local organizing initiatives.

Many of these practice-based models conceptualize community organizing as a cyclical process involving four phases: (1) assessment or listening, (2) research, (3) mobilization or action and (4) reflection (Speer et al. 1995). The first phase, listening or assessment, often involves a series of one-to-one or small group meetings in which residents invite one another to share their concerns and develop a deeper understanding of others' hopes and concerns. Distilling and prioritizing shared concerns then leads into research processes in which groups of resident leaders conduct meetings with decision-makers and stakeholders in the systems that create or sustain problematic community conditions. As this research identifies potential policy or systemic levers to address community problems, the groups will mobilize large numbers of participants in carefully orchestrated actions, in which decision-makers (for example, elected officials and corporate leaders) are publicly pressured to take specific steps toward resolving problems. This is followed by collective reflection processes among participants that help the initiative to adapt and learn from its practices and the responses of relevant institutions and community systems.

A single community organizing initiative or federation may be simultaneously moving through iterations of this cycle on multiple issues at the same time. A cycle of organizing on a particular community issue may progress rapidly or occur over the course of years. In addition, local federations are often participants in regional, statewide, national, and trans-national campaigns and coalitions organizing around issues at extra-local scales. Community organizing, as a phenomenon, therefore, can alter (1) interpersonal networks, (2) interorganizational networks, and (3) inter-city networks. Next, we examine some of the ways that community organizing tends to affect networks at each of these scales.

Interpersonal Networks

For many community organizing initiatives, a central principle is that power is built on the strength of interpersonal relationships. For example, one prominent organizer (Chambers 2003, p. 28) writes simply that 'power takes place in relationships'. This view is especially deeply rooted in congregation-based organizing initiatives, which are sometimes said to employ relational models for organizing. In a study of this type of community organizing, Robinson and Hanna (1994, p. 80) found that 'relational, or one-to-one, meetings are the source of the organization's power and its basic method for finding and acquiring energy, talent, and ideas'. One-to-one meetings – brief semi-structured conversations between participants – are a prominent feature of organizing initiatives that are seeking to employ a relational model, or that have been influenced by relational models of community organizing.

Participants are trained in the practice of one-to-one meetings, which involve 'careful listening and the judicious use of probing questions' (Robinson and Hanna 1994, p. 85) to elicit another person's history and formative experiences, their present situation, their hopes for the future and their concerns about what may prevent those hopes from being realized. These conversations serve many purposes within community organizing initiatives, in addition to identifying the issues of greatest concern to local residents. By urging many participants to carry out one-to-one meetings with their neighbors and fellow members, this process broadens participants' networks of interpersonal relationships. In a qualitative study specifically focused on one-to-one meetings, Christens (2010) found that those engaged in organizing interacted with many people that they probably would not have encountered under other circumstances, that this process helped them to develop new understandings of social, political and economic systems, and that this combination of new social ties and new sociopolitical perspectives helped to strengthen participants' commitment to civic involvement and action. The relationships that are developed through organizing may become friendships, but the primary purposes of relationship development in many organizing initiatives is mutual understanding, trust and the identification of, and shared commitment to, goals for collective action.

This widespread relational model places particular emphasis on building relationships through numerous meetings between dyads of participants. Although other models of community organizing tend to be less focused on interpersonal relationship development, they nevertheless often emphasize the importance of relationships and collective identity. In many cases, organizing initiatives seek to tap into dense interpersonal networks that exist, for instance, in urban neighborhoods. Through issue identification, framing and mobilization processes, these initiatives also alter relationships and collective identity (Stoecker 1995). Although each initiative is unique and complex, a broad-brush distinction between more relational models and other models of community organizing is often a difference in preferred sequence of events. Relational models prefer to build interpersonal relationships and then have issue selection and mobilization occur subsequently, whereas other models (for example, many neighborhood-based efforts or more single-issue efforts) identify issues and seek to forge interpersonal relationships in the process of mobilization and collective action.

Finally, many long-standing organizing initiatives have historically primarily sought adult participants, but this has changed in the new millennium, as the power and potential of youth organizing has been repeatedly demonstrated. By 2010, a scan of the field identified 160 youth organizing initiatives in the US (Torres-Fleming et al. 2010), and this number has undoubtedly continued to increase. Moreover, many initiatives that were exclusively adult organizing have broadened to include young people. A spectrum exists from nearly entirely youth-led organizing efforts to intergenerational efforts with both youth and adult leaders. In all of these initiatives, however, collaborative relationships are being forged across generations in ways that disrupt age segregation and encourage youth civic engagement (Camino and Zeldin 2002).

Interorganizational Networks

Organizing also creates and alters relational ties between organizations. Many organizing initiatives exist as federated structures of other local organizations, such as faith-based institutions, schools and neighborhood organizations. The organizing initiative creates a superstructure for members of these institutions to collaborate on social action, and thereby creates new ties and strengthens existing ties between local organizations. These interorganizational ties include, for instance, flows of information and resources, and collaborations on issue-based campaigns. In addition, as organizing initiatives undertake research on community issues, participants form new ties with officials, decision-makers and stakeholders in numerous community institutions. For example, in Speer et al.'s (2003) description of a community organizing initiative working to reduce vacant housing and violent crime in Camden, New Jersey, members of the organizing initiative held meetings with local officials (for example, the mayor, the city council, the Camden Redevelopment Agency, the city attorney, fire department, police department, public works and building inspectors), state officials (for example, the governor's office, state legislators, the New Jersey Department of Environmental Affairs and the New Jersey Housing and Mortgage Finance Agency) as well as with local banks and community development corporations.

Holding meetings with local decision-makers enables organizing participants to gain insights into multiple facets of community issues, and into the network of relationships between various prominent actors in the community's power structure. Some organizing initiatives use the term research actions for these meetings, others the term power one-to-ones. Christens et al. (2014) identify these types of meetings as one of the ways that organizing initiatives broker relationships between participants and decision-makers, in the process building ties that can be mobilized and a deep understanding of the local power map as it relates to specific community issues.

Power that is built through relationships – between individuals and between organizations – is often exercised through actions that are rooted in conflict. Achieving a policy goal might involve pressuring, or even embarrassing, a lawmaker in public and in front of a large group of constituents. However, not all forms of organizing use a conflict-prone approach, which is often associated with Saul Alinsky who rose to prominence organizing Chicago stockyard laborers in the middle of the twentieth century (Alinsky 1972). Consensus organizing seeks to share power among stakeholders, who are often place based, and reach a consensus about goals and strategies for improving neighborhoods. In a similar vein, community-building approaches seek to capitalize on local assets and resources, and work collaboratively with stakeholders to build local capacity to improve places from within (Eichler 2007). Both of these approaches involve building relationships across geographically defined areas, but they do not necessarily challenge broader power structures that are responsible for neighborhood conditions. DeFilippis et al. (2010) argue that organizing efforts that de-emphasize conflict limit the possibilities of building power that can contest the root causes of most community problems. However, within interorganizational networks, philosophies may diverge on the efficacy of conflict versus consensus-building approaches to achieving objectives.

Interorganizational network analysis has been employed in the study of coalitions, but more rarely in studies of grassroots community organizing. For example, Feinberg et al. (2005) conducted a cross-sectional study of 23 coalitions, and found that the average closeness in the interorganizational networks was positively related to coalition functioning, and the average level of betweenness was negatively related to functioning. Thus, coalitions in this study benefitted from more dispersed relationships between organizations rather than a network that was brokered by a few key organizations.

Longitudinal studies of coalition networks have tended to focus on single coalitions, and several have provided key insights into coalition dynamics and the conditions necessary for coalitions to build capacity and achieve their goals. For example, in a nine-year study of an environmental health coalition including subsets of eight Native American tribes surrounding a superfund site in Oklahoma, Kegler et al. (2010) found that although communications and information-exchange capacity increased over time, a key to the coalition's functioning was the development of relationships with extra-local entities, such as state and federal agencies. Studies by Bess et al. (Bess 2015; Bess et al. 2012) of a youth violence prevention coalition in Tennessee found that increases in collaborative capacity among the formal members in the coalition may have been developed to the detriment of broader networks of collaboration between organizations in the city. These studies suggest some of the complexities of coalition network dynamics and provide conceptual and methodological tools that could be used to generate empirical insights into community organizing network dynamics.

Inter-city Networks

Although community organizing initiatives are typically built locally, they are often connected to efforts in other cities through regional, statewide, national and trans-national networks. For example, an organizing initiative in San Diego, California, the San Diego Organizing Project (SDOP), is currently organizing within 31 local member faith-based institutions (http://www.sdop.net/about). The SDOP works on a variety of local issues, including affordable housing, compassionate policing and reforming the criminal justice system. PICO California (http://www.picocalifornia.org/) is a statewide federation of 14 local organizing federations (including SDOP) that is leading campaigns on issues including environmental justice, funding for education, access to health care, and civic engagement and voter development. This federation is part of the Faith in Action (FIA) National Network (https://faithinaction.org), which currently operates in 17 US states and has affiliates in Central America and Rwanda through FIA International (https://faithinactioninternational.org). This network provides training and support to local organizing initiatives, and builds relationships across cities and regions within the US and in other countries where it has been invited to assist with building local power to address injustices. It also facilitates organizing and mobilization at the national level in the US on a number of issues, including immigration, living wages, and health care and health equity.

The FIA National Network is one of the major national federations supporting local organizing based primarily in faith-based institutions. Other networks and federations, such as the Gamaliel Foundation (http://www.gamaliel.org) and the Industrial Areas Foundation (IAF) (http://www.industrialareasfoundation.org) exist to support similar local federations around the US (and, in the IAF's case, in Canada, the UK and Germany). These networks have grown to the extent that there is now at least one of these federations in nearly every large or mid-sized city in the US (Wood et al. 2013). Moreover, neighborhood-based organizing that does not structure its membership primarily around congregations or other community institutions has continued in many places, despite the demise in 2010 of the largest national network that supported this type of organizing, that is, the Association of Community Organizations for Reform Now (ACORN). For example, the Texas Organizing Project (TOP) (http://organizetexas.org), a statewide neighborhood-based organizing federation with offices in Houston, Dallas and San Antonio, is currently engaged in campaigns on issues, including immigration, education, justice system reform and health care. In turn, the TOP is an affiliate of the Center for Popular Democracy (http://populardemocracy.org/), which currently supports organizing efforts in 33 states.

Local, statewide, national and international networks that support grassroots organizing also interface and collaborate with an array of other organizations, campaigns and social movements. Labor unions and health-care advocacy groups, for instance, have worked in coalitions with grassroots organizing federations, at local, state and federal levels. Longer-standing multi-issue organizing groups also often collaborate with more issue-defined organizing groups and social movements, such as Occupy Wall Street and Black Lives Matter (Taylor 2016). Finally, professional organizations in fields such as social work, education and public health have collaborated and built coalitions with organizing networks, and philanthropic foundations have sometimes been supportive or directly involved in these initiatives.

In US cities, where long-standing patterns of segregation and disadvantage exist, there is potentially a unique overlap between geographically situated or place-based organizing efforts, and social movements pushing for greater racial equity. Many place-based organizing efforts in disadvantaged neighborhoods of color aim to improve local conditions by attracting more resources and programs. At the same time, there is a growing recognition that unjust systems, such as the harsh punishment of Black and Brown students in schools, and the disproportionately harsh punishment of Black and Brown people by the criminal justice system, are some of the biggest drivers of problems that place-based organizing efforts are trying to solve (that is, poverty, unemployment and youth violence). Pastor et al. (2009) identified movements for regional equity as an emerging trend in cities across the US, where local groups are increasingly shifting their targets of change to broader systems and policies that are responsible for creating and maintaining inequitable community conditions. Regional coalitions, such as the Boston-based Action for Equity and the Chicago-based Grassroots Collaborative have pushed for city and regional policies that could address broader racial equity concerns, such as gentrification. However, there are fewer examples of citywide coalitions that attempt to change systems and policies in targeted ways that can also bring more resources to low-income and highly segregated neighborhoods.

Summary: Community Organizing and Multiscalar Networks

Community organizing is therefore constantly building and altering social and institutional networks at multiple scales within and across contemporary cities. Organizing initiatives operate as cross-level networks, in which changes in interorganizational networks, for example, may affect interpersonal or inter-city network dynamics (Brass et al. 2004). Accounts of specific organizing initiatives have emphasized the importance of these network dynamics to the development and exercising of social power to bring about policy and systems changes (for example, Christens et al. 2014). Yet, systematic network data collection, and the use of social network analysis, have been rare in studies of community organizing. Also, studies of urban networks have not often emphasized the roles that organizing may be playing in creating or altering relational structures or processes. In the next section, were therefore examine an organizing network in Chicago, Illinois, that has been studied using interorganizational network methods. Interorganizational collaborations are particularly critical for understanding alignment, capacity-building, and the development of broad-based power. Within cities, the more that various organizations' priorities, philosophies and methods for collaboration are aligned, the greater the likelihood that their efforts to change policies and systems will be effective (Speer et al. 2003; Wolff et al. 2016).

CHICAGO'S RIGHT ON JUSTICE NETWORK

Background

The Right on Justice (ROJ) Network was a coalition of grassroots community organizers and restorative justice practitioners in Chicago that formed in 2014.[1] It provides an example of a coalition that operates at the nexus of place-based community organizing for local capacity, and citywide organizing for greater racial equity policies and practices. The network consisted of 25 groups that were engaged in educational and criminal justice reform efforts since before the ROJ network was formed. Roughly half of the organizations that comprised the ROJ were primarily engaged in grassroots organizing, at both geographic-specific and citywide levels. For instance, the SWOP, described in the previous section, was a constituent member of the ROJ. Some of the organizing groups that participated in the ROJ network have a local neighborhood focus, while others operate citywide. Some, for example the SWOP, are engaged in organizing around a variety of issues, while others have been more specifically focused on educational and/or criminal justice system reforms. Much of the SWOP's interpersonal relationship-building occurs on the southwest side of Chicago, and many of their institutional relationships are with area lawmakers and organizations. Yet they are also engaged with citywide policy reform efforts. Many policy-change efforts require obtaining enough support from lawmakers across cities and regions in order to enact legislation that might have equitable benefits for disadvantaged neighborhoods across an urban area.

The other, approximately, half of the organizations that formed the ROJ network were focused less on grassroots organizing, and more on building the capacity of nonprofit organizations to provide resources and services to young people of color. More specifically, they have engaged in advocacy for and implementation of restorative justice practices as alternatives to school suspension and expulsion, as well as youth incarceration in Chicago. Restorative justice is an approach to criminal justice that seeks to create settings in which those involved in offenses – as victims, offenders and/or members of the community in which the offense took place – can take part in facilitated discussions. The goals are to have those who have harmed others take responsibility for their actions, to give victims active roles that can be part of a healing process, and to restore social bonds in the community. Restorative justice practices, which are often used as alternatives to more punitive or exclusionary approaches to justice, such as incarceration or suspension from school, can reduce recidivism, promote healing and help address some of the inequities in the criminal justice system, such as racial disparities in incarceration (Wadhwa 2016).

A commonality among ROJ members was the growing recognition that restorative justice practices hold promise as a tool for addressing the school-to-prison pipeline and mass incarceration that disproportionately affects Black and Brown people. Although the use of these practices had been growing in places such as Chicago, there were few policies that explicitly mandated its use as an alternative to practices such as school expulsion or youth incarceration. The groups in the ROJ network that were more focused on restorative justice represent advocates for this approach, as well as practitioners; and as these practices have spread, there has been a need for some degree of standardization of these practices (Rossner and Bruce 2016).

This distinction between restorative justice and community organizing is not a mutually exclusive dichotomy into which each group fits neatly. Most organizations have had some level of prior engagement or interface with both organizing and advocacy, as well as direct practice with individuals, families, and/or schools. Furthermore, some groups provide technical assistance or research that influences restorative justice practices and/or organizing efforts. The coalition did, however, bring together many organizations that had not worked directly with one another prior to the initiative. Further, each organization came into the coalition with its own history and practices, often its own geographic target areas, and its own framework for changing systems. More broadly, these types of issue-based coalition efforts have become more common among community organizing initiatives (Christens and Speer 2015).

Prior to the formation of the ROJ network, there were many different organizations interested in pursuing alternatives to the status quo in school discipline and criminal justice in the city. An impetus for the formation of the ROJ network was the recognition that restorative justice practitioners and advocates shared many of the same school and criminal justice system reform goals as did grassroots community organizing initiatives in the city. In order to provide effective alternatives at a scale large enough to meaningfully alter community trends, practitioners would benefit from standardized practices and a coordinated campaign to expand the diversion of young people into these programs. The thought was that if a coalition was formed in which representatives and leaders of a number of different organizations involved in (1) restorative justice efforts and (2) grassroots organizing could learn about one another's work and build trusting relationships, that more coordinated and effective efforts to change institutional and civic policies and practices could result. Specifically, the goals of the coalition involved aligning approaches and targets of action, deeper collaboration and consistent messaging. These goals were ultimately intended to further the objective of systems and policy changes that would mandate the decreased reliance on punishment in schools and communities.

Methods

Surveys were collected from lead representatives of each organization at two time-points: during the formation of the ROJ coalition, and approximately 18 months into its existence. The surveys included questions about each group's prioritization of various issues and each group's involvement in different forms of practice. Surveys also included several questions on interorganizational networks, assessing levels of collaboration and shared philosophies or approaches between groups (along five-point Likert-type scales). This approach to network measurement in a coalition context drew on previous studies, such as Nowell's (2009) study of interorganizational collaboratives. As part of an action research partnership, these survey data have been used alongside other data sources for practical insights into the coalition's progress toward its goals (see Cooper and Christens 2019). For the purposes of this chapter, however, we are most interested in examining the interorganizational network changes that occurred as a consequence of the coalition's activities, and what can be learned from these data about the roles that community organizing and coalitions play in shaping interorganizational networks within cities.

Results

Coalition member organizations were coded as primarily using either community organizing and advocacy (N = 14) or restorative justice practices (N = 11), based on the extent to which they reported using restorative practices and community organizing in their organizational efforts. Figures 14.1 and 14.2 illustrate the geographic distribution of these organizations across the city of Chicago. Figure 14.1 shows where organizations are located relative to Chicago's long-standing pattern of racial and economic segregation. The south and west sides of the city are largely disadvantaged communities of color, with this pattern having remained remarkably durable since the 1960s (Cooper 2019; Delmelle 2015). A major aim of the coalition was to ensure that participating organizations represented a broad geographic swathe of the city, and in particular, communities on the south and west sides of the city.

Figure 14.2 displays coalition members' locations in relation to the density of nonprofit organizations is in each of Chicago's community areas.[2] The density of nonprofit organizations and civic leaders in a geographic area is thought to be related to social capital and informal social control. For instance, nonprofit density has been shown to be inversely related to violent crime (Sampson 2012). These figures provide additional context about the coalition and how its members are situated within urban space. This enhances understanding of how cities are organized but offers very little on how social and institutional networks interact across this urban space. Although there may be latent network potential in a given community based on the high number of nonprofit organizations, methods such as social network analysis are necessary to examine how effectively organizations are building relationships.

One of the main objectives of the ROJ coalition was to build strong relationships between organizational members – representatives who held positions of leadership within their respective organization – to strengthen the alignment on approaches to changing policy and targets of action. It was thought that representation across geographic space would ensure that the unique concerns and policy-change desires of diverse stakeholders were represented in the coalition's goals. Survey response data, together with social network data, provide two different means of assessing how successful the coalition was in aligning its goals and approaches to change. The survey asked respondents to rank what they viewed as the most important approaches to change for the coalition. These included public education and messaging, restorative justice practices in school and community settings, organizing with youth and families most affected by the issues,

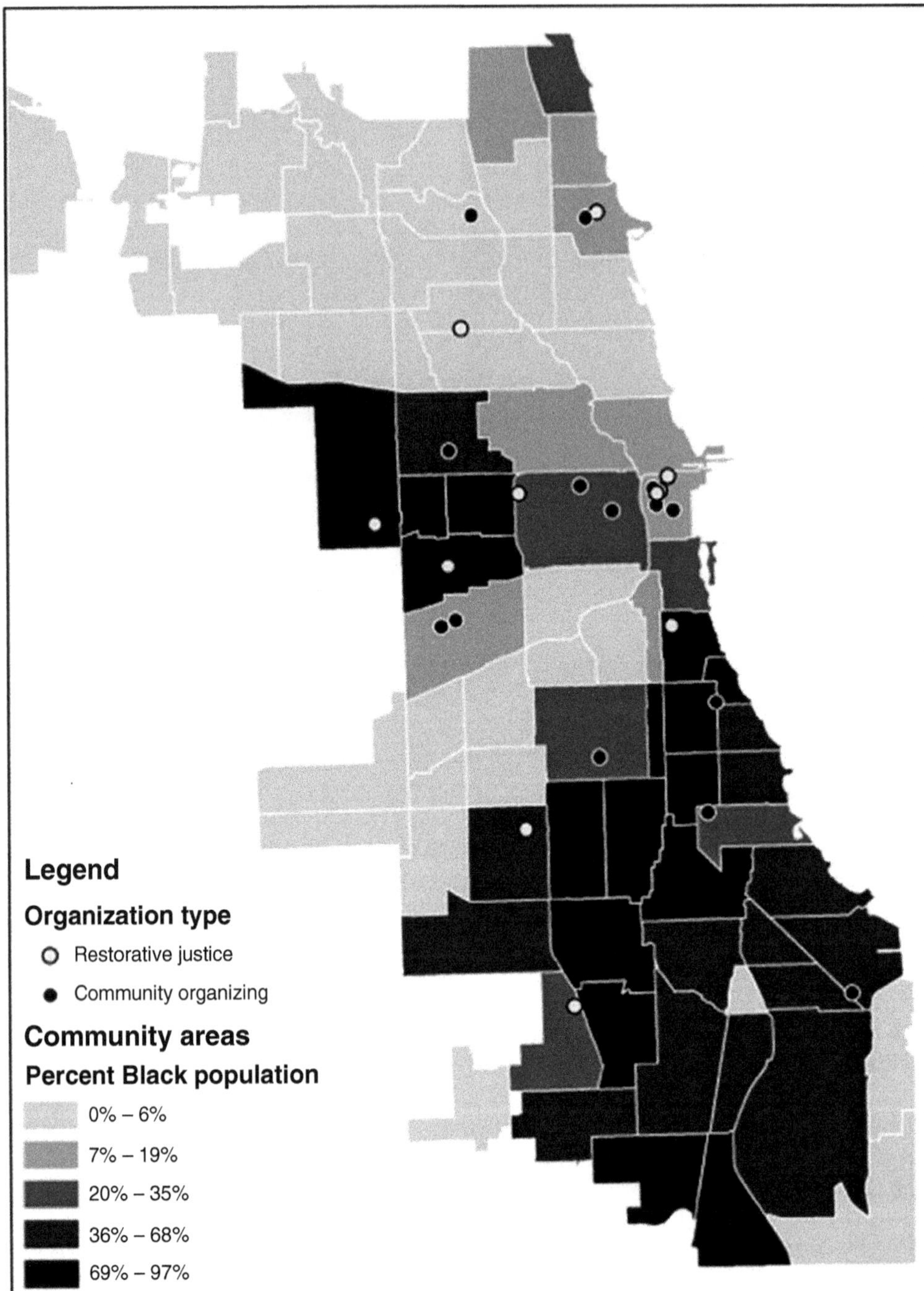

Figure 14.1 Map of ROJ coalition organizations in Chicago's community areas shaded according to percentage Black population

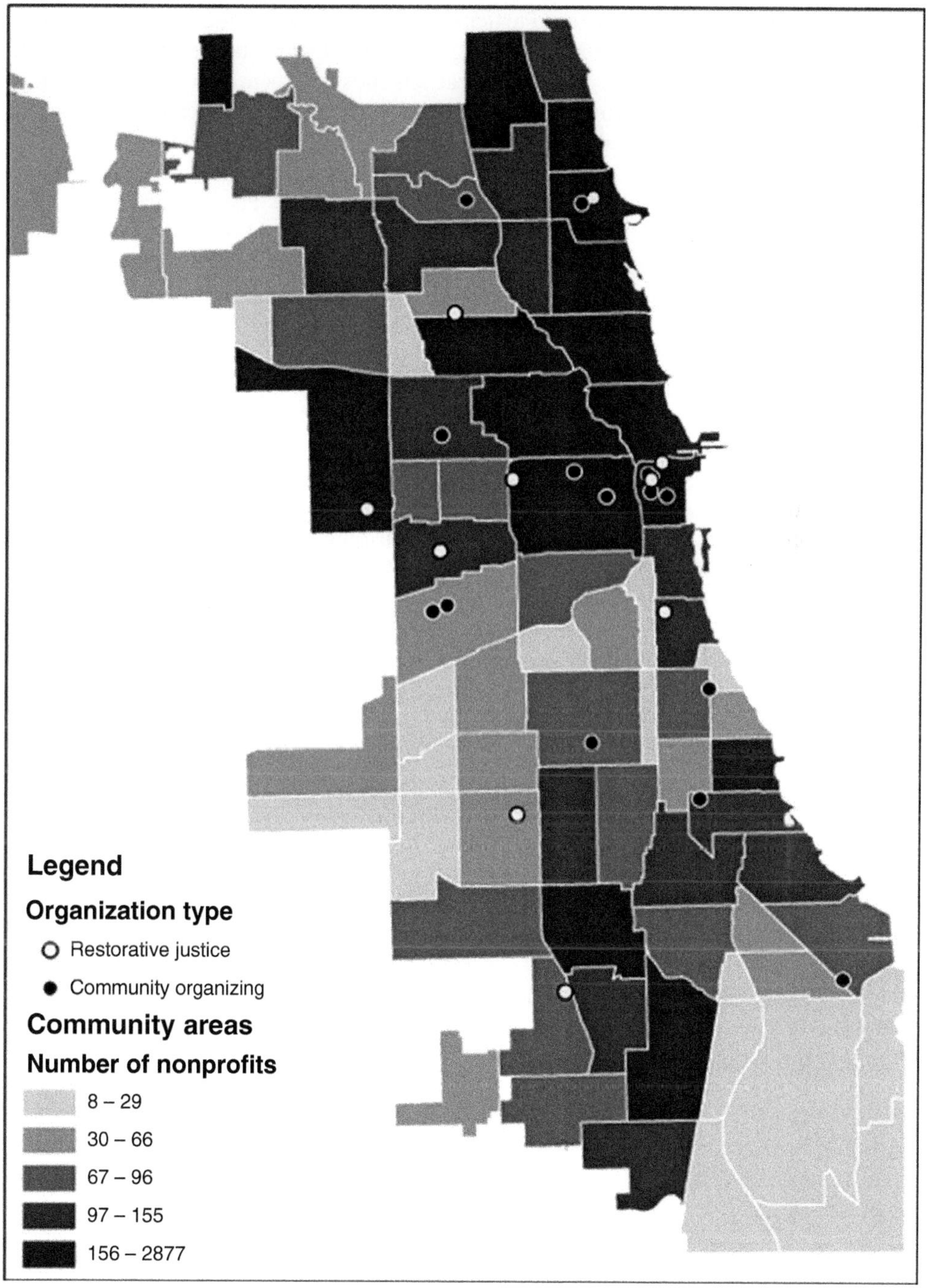

Figure 14.2 *Map of ROJ coalition organizations in Chicago's community areas shaded according to number of nonprofit organizations*

and other programs, to name a few. Participants were asked to rank these approaches from most important for the coalition to utilize, to least important (see Figure 14.3). At baseline, the two approaches that organizations most frequently ranked in their top three were 'Organizing with youth and families', identified by 36 percent of members, and 'Community-based restorative justice diversion programs', identified by 32 percent.

During the second wave of survey data collection, there was some evidence of change and re-alignment on the most important approaches to change. For instance, more than twice the number of organizations ranked work to change neighborhood-level determinants in their top five strategies during wave 2 (see Figure 14.3). A large increase was also observed on the importance that groups placed on expanding restorative justice practices in schools. Several of the strategies that involved community and youth organizing were ranked among the top five strategies by fewer groups at wave 2, probably owing to some of the complexities and tensions observed in the coalition (see Cooper and Christens 2019). Although the coalition members still displayed variance in their preferred approaches to change by the time of the second survey, there was nonetheless some convergence, and particularly around the restorative justice and neighborhood-level social determinants approaches that were central to the goals of the coalition.

The ROJ coalition rallied around a core goal of pushing for policies that would reduce the use of harsh punishment and, most critically, shift money from criminal justice systems and into community-based efforts (Lugalia-Hollon and Cooper 2018). More members identifying the importance of addressing neighborhood social determinants, rather than specific policies or programs, can be seen as reflecting the coalition's broader goal and theory of change. That is, even though many organizations may have had particular place-based goals and agendas, or typically advocated expansion of programs in their service or organizing areas, there was convergence around the importance of addressing social determinants through policy and systems changes. These data are indicative of some relationship-building and strategic alignment that took place among coalition members. That is, we would expect there to be little if any change in rankings if organizations were not aligning around the collective goals of the coalition and, instead, preferred to stick with their traditional organizational approaches. Shifts toward shared goals, in contrast, are indicative of aligned agendas for community change.

Social network data add more nuance to these findings, and allow for a deeper understanding of the degree to which relationships were built or altered across this citywide network. A dimension of the network survey measured whether organizations reported sharing a similar philosophy or approach to change with other organizations in the network. These social network data provide a relational and bi-directional assessment of the data presented in Figure 14.3. This is useful for descriptive purposes but, more importantly, has been used to provide coalition members with empirical insights into coalition progress toward greater alignment. Do organizations that primarily engage in community organizing report sharing a similar philosophy only to others who traditionally organize?

Figures 14.4 and 14.5 illustrate the self-reported changes over time in whether organizations reported a similar philosophy and approach with others in the network. Organizations that primarily used community organizing strategies appear in black, and those that primarily use restorative justice practices appear in white. Several hybrid organizations that were engaged in both organizing and restorative justice appear in grey. Lines between organizations represent only 'strong' or 'very strong' (for example, 4 or 5

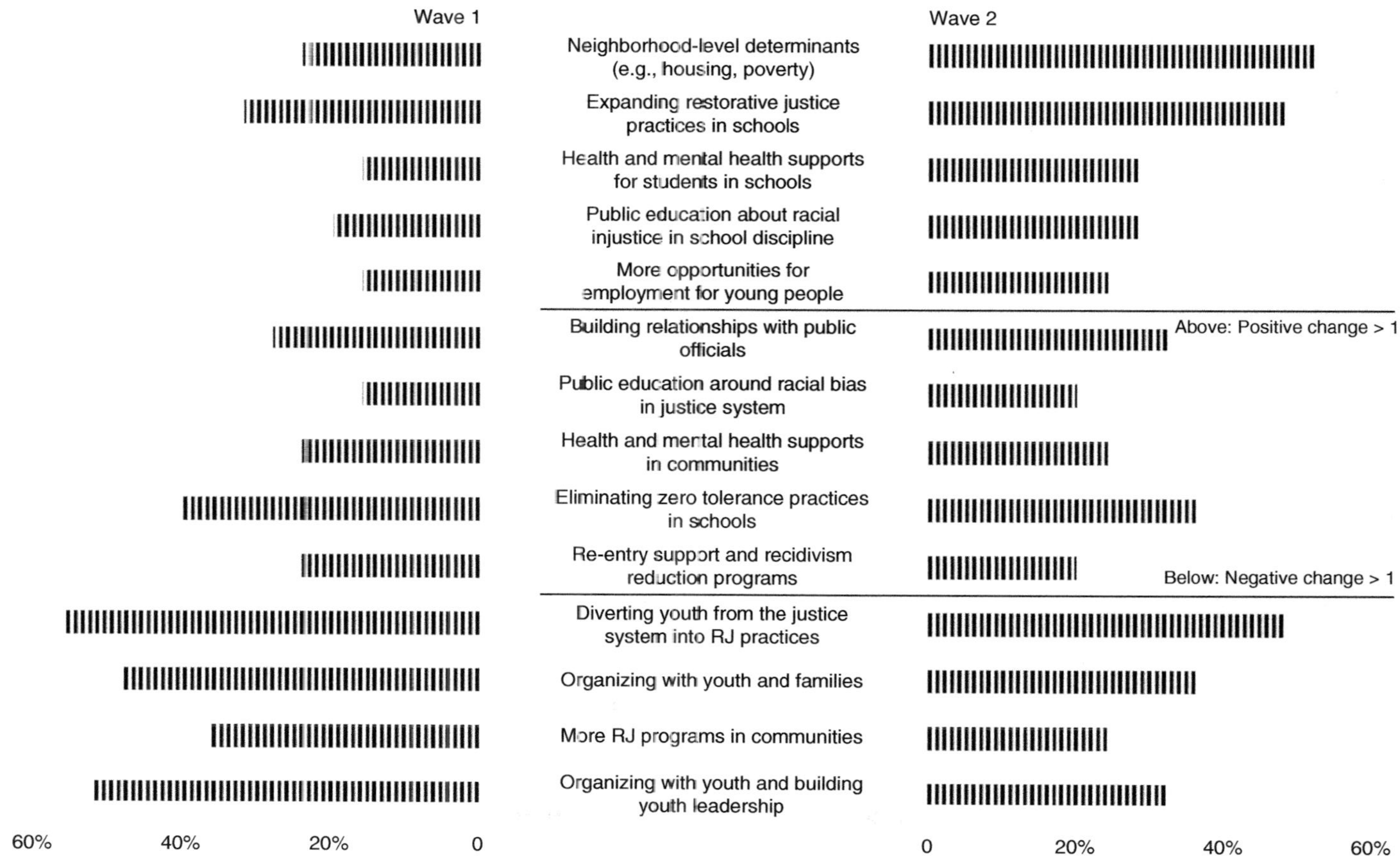

Figure 14.3 Changes in respondents' rankings (top five) of most important strategies for the coalition at baseline (wave 1) and 18 months into coalition formation (wave 2)

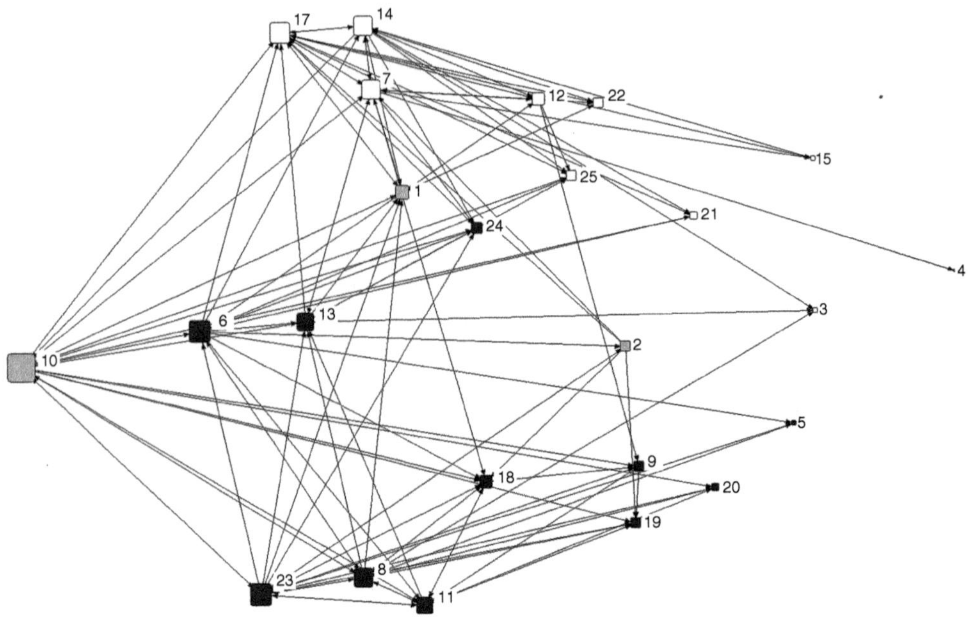

Figure 14.4 Perceptions of shared philosophy between organizations at time 1, with restorative justice-focused organizations in white, community organizing initiatives in black and hybrids in grey

on the five-point Likert scale) agreement about shared philosophy and approach. The size of each organizational node indicates degree, a measure of centrality indicating level of connectedness (undirected) to other nodes in the network. The larger the organization appears in the network diagram, therefore, the more central it is within the network by this metric. The nodes are arranged spatially in these diagrams using the principal components analysis layout in NetDraw, which places nodes in proximity based on their relational similarities (Borgatti 2002).

Network maps indicate convergence around a shared philosophy and approach to change. Across the entire network, for both restorative justice-focused organizations or organizing entities, there was an increase in reporting strong agreement of sharing a similar philosophy and approach to the coalition's work. For the whole network, the mean degree at time 1 was 13.84. This increased to 19.2 at time 2. These changes in perceptions of shared philosophy represent an important condition for successfully sustaining a coalition, more specifically, for furthering the goals of greater racial equity (Wolff et al. 2016). In community organizing efforts, the importance of aligning around a shared philosophy and approach cannot be understated. Organizations in cities have long and complicated histories of territorialism, competition for scarce resources, and varying levels of past success working in coordinated ways with one another. Thus, relationship-building is of critical importance to the success and longevity of organizing networks, and for building power. It is not necessarily a given that alignment can be achieved easily.

Some organizing entities are steeped in traditions of confrontation in their approach to changing policy; for example, holding public officials accountable in a public forum.

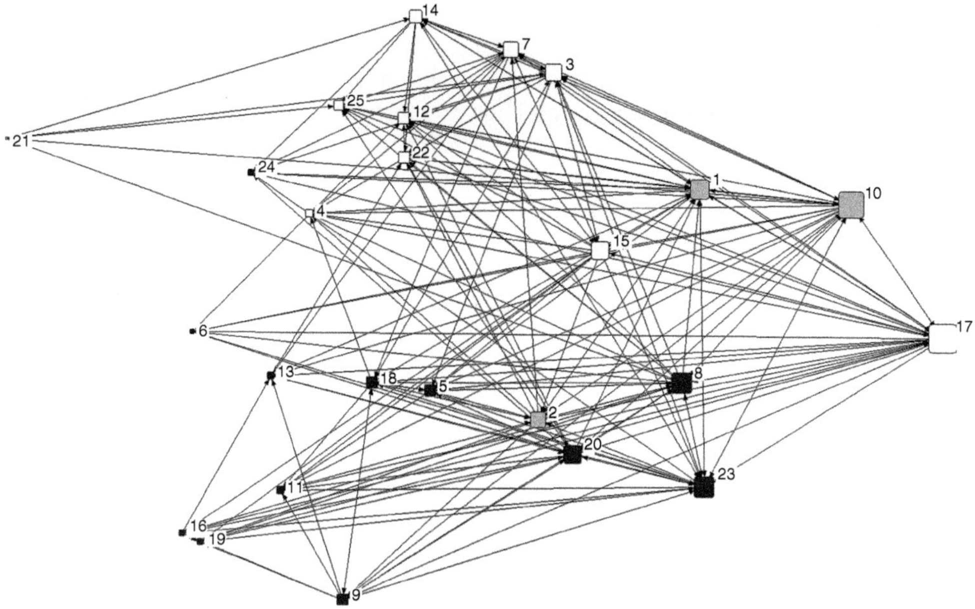

Figure 14.5 Perceptions of shared philosophy between organizations at time 2, with restorative justice-focused organizations in white, community organizing initiatives in black and hybrids in grey

Others may seek to build consensus and avoid conflict with these same public officials. The ROJ coalition had to navigate this tension and seeming contradiction in seeking a unified approach to systems change. Organizations that primarily deliver service rather than push for policy and systems changes are often less inclined to confront and challenge public officials, typically preferring to build relationships and engender support for their work. These potentially contradictory modes of organizing and action present a real barrier to achieving unified power as a citywide network or organizations, all of which represent their own constituents and networks in different geographic locations across the city. As the network diagrams in Figures 14.4 and 14.5 suggest, although more organizations came to agreement about sharing a philosophy or approach, these ties remained strongest among the two sub-networks which focused, respectively, on community organizing or restorative justice practice.

The coalition spent much of the first 18 months of its existence negotiating these differences and building relationships that would enable unified approaches to emerge. Yet this was not without tension and disagreement, nor was comprehensive unification ever fully achieved on every decision or action. As an example, several organizations attended an event and publicly voiced support for a public official who was the target of a confrontational demonstration by other coalition members during the same week. In this example, the difference in philosophy and approach was more than a mere abstraction. Whereas one organization preferred to hold an action to demand more accountability from a public official, another preferred to invite the same official to a public education event honoring allies of justice system reform. Some members viewed showing explicit support for a target of another's action as undermining that action, which inevitably strained organizational relationships and ultimately presented a challenge for building unified power. This led to prolonged discussions about which goals and actions could be fully supported by the entire coalition, and how to reconcile their differences in approach. Ultimately, compromises have to be reached if a coalition is going to succeed in sustaining itself to the point where members still feel a sense of alignment on philosophy and approach to change. Strong relationship-building and endorsement of an overarching strategy, such as pushing for greater racial equity across the city, is critical for being able to navigate these differences that arise in networks.

Even though there were some examples of divergence, there were also numerous examples of how the network converged in concrete ways. For example, three organizations that had previously only engaged in community organizing and advocacy ultimately started delivering restorative justice practices within their geographic target areas. This occurred because other coalition members helped build the capacity and infrastructure for these groups to start new restorative justice efforts in their communities. At the same time, numerous coalitions that had previously not been engaged in any organizing or advocacy work helped organize local constituents to take part in demonstrations, organized public education symposia, and met with lawmakers to discuss how the coalition's justice system reform agenda could benefit local communities.

A final example of how this interorganizational network aligned is evident in how it targeted different systems for change. Prior to the formation of the ROJ coalition, grassroots organizing entities and restorative justice organizations typically worked around either school discipline alternatives and policies, or community-based alternatives to incarceration and broader justice system reform. The coalition's theory of change linked both of these as part of the broader school-to-prison-pipeline that removes youth of color from school early on, initiating a lifelong cycle of contact with the criminal justice system. Removing organizations from a siloed approach that treated schools and community settings as separate was one aim of the coalition.

Figures 14.6 and 14.7 provide network maps displaying changes in the degree to which coalition members reported engaging in impactful work with other members (as opposed to Figures 14.4 and 14.5, which displayed relationships based on perceptions of shared philosophy). Degree is again used as the measure of centrality (again indicated by the size of the node), and lines only display connections that were rated as either 'strong' or 'very strong'. As with Figures 14.4 and 14.5, nodes are arranged using principal components analysis. The descriptive data presented by these diagrams provide further evidence of convergence in the network, and whole-network metrics of cohesion also indicate that convergence took place over time – mean degree for impactful collaboration at time 1 was 16.04, and this increased to 19.04 at time 2. Inspecting the network diagrams in Figures 14.6 and 14.7, the data suggest that a larger number of organizations began to bridge the initial divide between restorative justice practice and grassroots community organizing practice by forging new impactful collaborations.

This citywide, interorganizational coalition has increased the alignment and capacity of its network in many ways over the longitudinal survey period, across geography, organizational type and approach to systemic change. This alignment has probably played a role in in the coalition's ability to exercise power in Chicago, and in its achievement of a number of substantive victories. For example, the ROJ was successful in pushing the

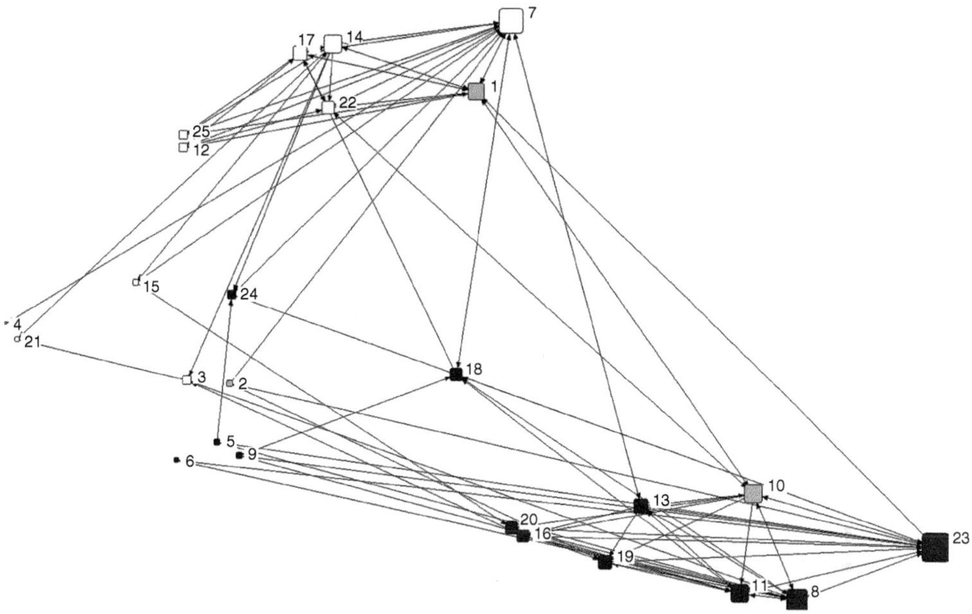

Figure 14.6 *Engaging in impactful collaborations at time 1, with organizations focused on community strategies in white and organizations focused on school strategies in black*

Cook County government to double its investment in restorative justice alternatives in Chicago communities. Much of this funding seeded coalition members to scale up their program efforts with young people and young adults. This met a major aim of the coalition, to expand investment in restorative justice alternatives. Moreover, the coalition's main framing of the issue – that justice systems are inequitable and costly to schools and communities of color – has radiated throughout the city. Coalition members released a report about how costly the justice system is (Communities United et al. 2016), together with a website mapping incarceration costs by Chicago community (https://chicagosmilliondollarblocks.com). This message about costliness, and the necessity of reinvesting into communities most affected by harsh discipline, reached numerous constituents, stakeholders and public officials, many of whom now use the figures provided by the coalition as public talking points.

In addition, the coalition helped secure commitment from the county to create the nation's first restorative justice court, which has just begun a pilot phase on the south side of Chicago. Coalition members met with local judges and collaborated on the design of a court that is beginning to divert many cases to restorative justice processes, with many coalition members serving as key collaborators and service providers. The court has already received national recognition as a promising alternative to incarceration (Kunichoff 2017). This also met a major aim of the coalition, to enact policies that mandate the use of restorative justice alternatives instead of harsh punishment. The coalition also has a campaign to pass a bill in the state legislature that would mandate the reclassification of specific felony offenses, and repurpose any cost savings associated

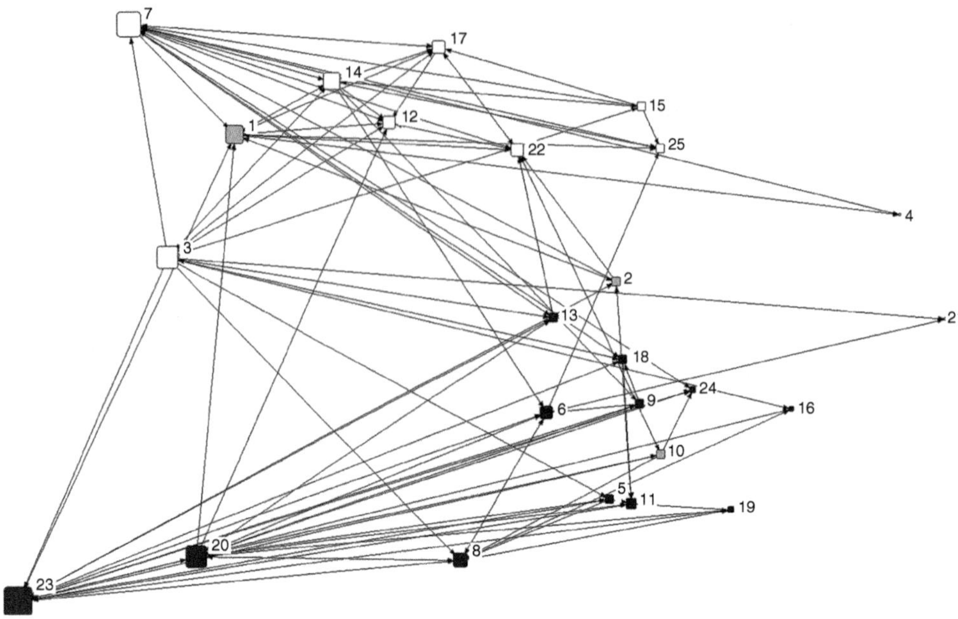

Figure 14.7 *Engaging in impactful collaborations at time 2, with organizations focused on community strategies in white and organizations focused on school strategies in black*

with lower incarceration rates into community-based efforts. The bill has the support of several key legislators and is currently moving through a committee.

CONCLUSIONS

Community organizing initiatives, in the course of their efforts to build power and achieve systemic changes, alter networks at multiple scales within cities and between cities. This chapter has described how community organizing develops and changes interpersonal, interorganizational and inter-city networks, and then descriptively examined the interorganizational dynamics of a coalition of local organizing and restorative justice organizations in Chicago. There are several features of this citywide coalition that have implications for the field of community organizing as well as the study of citywide, interorganizational networks. First, although community organizing has historically paid a great deal of attention to intentional learning processes, with frequent cycles data collection and reflection, social network analysis can add a powerful tool to this process. It can describe and display some of the mechanisms through which organizations are aligning and where there might be divergence that needs to be addressed through further relationship-building.

Second, although coalition approaches are very common in community organizing, working intentionally to build a network of organizers and service providers, across urban geographies and across sectors, holds a great deal of potential for building more successful citywide movements. The ROJ coalition had an ambitious goal of merging disparate approaches and organizations under one broad umbrella with a racial equity lens: the overarching goal of which was to dismantle the school-to-prison-pipeline and shift public investment into restorative justice school- and community-based alternatives. Successfully building relationships across urban space and navigating differences in organizational goals and philosophies is challenging in and of itself. Doing so can prove useful in building sustainable campaigns and movements where a common language and framework is adopted across an entire city. Strong and aligned interorganizational networks can serve as building blocks for successful systems and policy change efforts.

Finally, social network analysis holds particular promise as a tool in mixed-methods attempts to understand community organizing and related systems changes. Interorganizational network analysis has been applied to a far greater extent in the study of coalitions of neighborhood organizations, public and nonprofit service providers, and issue-focused advocates (for example, Bess 2015; Feinberg et al. 2005; Kegler et al. 2010; Nowell 2009) than it has specifically in the study of grassroots community organizing networks. This chapter provides descriptive insights into the functioning of a coalition partially composed of community organizing initiatives. Action researchers should continue to look for opportunities to employ interorganizational network analyses in studies of organizing efforts both to generate practical insights for grassroots leaders, and to contribute to the still incipient understanding of multi- and cross-level network dynamics in community organizing.

NOTES

1. The coalition was formally active from 2014 to 2017, although many of the groups involved continue to collaborate, for instance, on the Reimagine Justice campaign.
2. Chicago has 77 community areas that are based on US Census tracts and defined by University of Chicago sociologists dating back to the 1930s. These areas have remained consistent since that time.

REFERENCES

Alinsky, S.D. (1972), *Rules for Radicals: A Pragmatic Primer for Realistic Radicals*, New York: Vintage Books.

Bess, K.D. (2015), 'Reframing coalitions as systems interventions: a network study exploring the contribution of a youth violence prevention coalition to broader system capacity', *American Journal of Community Psychology*, **55** (3–4), 381–95.

Bess, K.D., P.W. Speer and D.D. Perkins (2012), 'Ecological contexts in the development of coalitions for youth violence prevention: an organizational network analysis', *Health Education & Behavior*, **39** (5), 526–37.

Borgatti, S. (2002), *NetDraw: Graph Visualization Software*, computer program, Cambridge, MA: Analytic Technologies.

Brass, D.J., J. Galaskiewicz, H.R. Greve and W. Tsai (2004), 'Taking stock of networks and organizations: a multilevel perspective', *Academy of Management Journal*, **47** (6), 795–817.

Camino, L. and S. Zeldin (2002), 'From periphery to center: pathways for youth civic engagement in the day-to-day life of communities', *Applied Developmental Science*, **6** (4), 213–20.

Chambers, E.T.D. (2003), *Roots for Radicals: Organizing for Power, Action, and Justice*, New York: Continuum.

Christens, B.D. (2010), 'Public relationship building in grassroots community organizing: relational intervention for individual and systems change', *Journal of Community Psychology*, **38** (7), 886–900.

Christens, B.D. and P.W. Speer (2015), 'Community organizing: practice, research, and policy implications', *Social Issues and Policy Review*, **9** (1), 193–222.

Christens, B.D., P.T. Inzeo and V. Faust (2014), 'Channeling power across ecological systems: social regularities in community organizing', *American Journal of Community Psychology*, **53** (3–4), 419–31.

Communities United, Right on Justice, Make the Road New York, Padres and Jóvenes Unidos (2016), 'The $3.4 million mistake: the cost of mass incarceration and criminalization, and how justice reinvestment can build a better future for everyone', accessed 10 May 2019 at https://communitiesunited.org/sites/apncorganizing.org/files/PROOF%20Communities%20United%20JR%20report%20%28for%20print%29%20%281%29.pdf.

Cooper, D.G. (2019), 'Now is the time to get serious about equitable development in south and west side communities', 4 April, Metropolitan Planning Council, Chicago, IL, accessed 1 May 2019 at https://www.metroplanning.org/news/8715/Now-is-the-Time-to-Get-Serious-About-Equitable-Development-in-South-and-West-Side-Communities.

Cooper, D.G. and B.D. Christens (2019), 'Justice system reform for health equity: a mixed methods examination of collaborating for equity and justice principles in a grassroots organizing coalition', *Health Education & Behavior*, **46** (1S), 62S–70S.

DeFilippis, J., R. Fisher and E. Shragge (2010), *Contesting Community: The Limits and Potential of Local Organizing*, New Brunswick, NJ: Rutgers University Press.

Delmelle, E. (2015), 'Five decades of neighborhood classifications and their transitions: A comparison of four US cities, 1970–2010', *Applied Geography*, **57** (February), 1–11.

Eichler, M. (2006), *Consensus Organizing: Building Communities of Mutual Self Interest*, Thousand Oaks, CA: Sage.

Feinberg, M.E., N.R. Riggs and M.T. Greenberg (2005), 'Social networks and community prevention coalitions', *Journal of Primary Prevention*, **26** (4), 279–98.

Kegler, M.C., J. Rigler and M.K. Ravani (2010), 'Using network analysis to assess the evolution of organizational collaboration in response to a major environmental health threat', *Health Education Research*, **25** (3), 413–24.

Kunichoff, Y. (2017), 'Should communities have a say in how residents are punished for crime? A new "restorative justice" court in Chicago will test this idea, by soliciting broader input on how offenders can make amends and stay out of jail', *The Atlantic*, May, accessed 10 May 2019 at https://www.theatlantic.com/politics/archive/2017/05/chicago-restorative-justice-court/524238/.

Lugalia-Hollon, R. and D. Cooper (2018), *The War on Neighborhoods: Policing, Prison, and Punishment in a Divided City*, Boston, MA: Beacon Press.

Nowell, B. (2009), 'Profiling capacity for coordination and systems change: the relative contribution of stakeholder relationships in interorganizational collaboratives', *American Journal of Community Psychology*, **44** (3–4), 196–212.

Pastor, M., C. Benner and M. Matsuoka (2009), *This Could be the Start of Something Big: How Social Movements for Regional Equity are Reshaping Metropolitan America*, Ithaca, NY: Cornell University Press.

Robinson, B. and M.G. Hanna (1994), 'Lessons for academics from grassroots community organizing: a case study – the Industrial Areas Foundation', *Journal of Community Practice*, **1** (4), 63–94.

Rossner, M. and J. Bruce (2016), 'Community participation in restorative justice: rituals, reintegration, and quasi-professionalization', *Victims & Offenders*, **11** (1), 107–25.

Sampson. R.J. (2012), *Great American City: Chicago and the Enduring Neighborhood Effect*, Chicago, IL: University of Chicago Press.

Speer, P.W. and B.D. Christens (2014), 'Community organizing', in V. Chien and S. Wolfe (eds), *Foundations of Community Psychology Practice*, Thousand Oaks, CA: Sage, pp. 220–36.

Speer, P.W., J. Hughey, L.K. Gensheimer and W. Adams-Leavitt (1995), 'Organizing for power: a comparative case study', *Journal of Community Psychology*, **23** (1), 57–73.

Speer, P.W., M. Ontkush, B. Schmitt, P. Raman, C. Jackson, K.M. Rengert, et al. (2003), 'The intentional exercise of power: community organizing in Camden, New Jersey', *Journal of Community & Applied Social Psychology*, **13** (5), 399–408.

Stoecker, R. (1995), 'Community, movement, organization: the problem of identity convergence in collective action', *Sociological Quarterly*, **36** (1), 111–30.

Taylor, K.-Y. (2016), *From #BlackLivesMatter to Black Liberation*, Chicago, IL: Haymarket Books.

Torres-Fleming, A., P. Valdes and S. Pillai (2010), *2010 Field Scan*, Brooklyn, NY: Funders Collaborative on Youth Organizing.

Wadhwa, A. (2016), *Restorative Justice in Urban Schools: Disrupting the School-to-Prison Pipeline*, New York: Routledge.

Wolff, T., M. Minkler, S.M. Wolfe, B. Berkowitz, L. Bowen, F.D. Butterfoss, et al. (2016), 'Collaborating for equity and justice: moving beyond collective impact', *Nonprofit Quarterly*, **23** (4), 42–53.

Wood, R.L., K. Partridge and B. Fulton (2013), *Building Bridges, Building Power: Developments in Institution-Based Community Organizing*, Longmont, CO: Interfaith Funders.

15. Racial/ethnic residential segregation and urban spatial networks in the United States

Kathryn Freeman Anderson and Joseph Galaskiewicz

1. INTRODUCTION

An old assumption in geography is that everything is related to everything else, but near things are more related than distant things (Tobler 1970). For this reason, the focus of urban studies has been on the neighborhood as the primary unit of analysis (Craven and Wellman 1973). However, urban geographers and planners have long known that people circulate in much larger circles than their immediate neighborhoods, and more recent work using big-data approaches have confirmed this on a large scale (Wang et al. 2018). Individuals and families move about the metropolitan area throughout the day in order to meet a wide variety of daily needs, including work, education and leisure.

It is also evident that these broad movement patterns vary by the socio-demographic characteristics of residents and households. However, there is little research on the ways in which involuntary racial/ethnic/class-based residential segregation impacts the manner in which people move about the city and the extent of their spatial networks. Our central question is, does exclusionary residential segregation affect the travel distances, trip frequency and mode of travel of households in segregated areas independent of households' financial and spatial resources? That is, do households that are in segregated neighborhoods tend to have more restricted or broader spatial networks than comparable households in integrated neighborhoods?

We conceptualize two ways of understanding this problem. On the one hand, segregated areas of exclusion are often bereft of organizational resources, thus, residents need to travel farther to access services compared to residents in more integrated neighborhoods. On the other hand, physical and social constraints may prevent this. The situation becomes more or less costly depending upon the available transportation networks in and out of a community. Also, residents of these neighborhoods often are socially stigmatized either by their ascriptive characteristics or the neighborhood they live in. Thus, they are forced to limit their travels to their segregated neighborhood. The first adds undue travel costs; the other two limit access to organizational resources outside of neighborhood boundaries. Either way, segregation based on spatial exclusion exacerbates urban inequality by limiting access to opportunities for some sub-populations. We begin this chapter with a review of the literature on travel behavior, focusing on individual and household factors that are associated with distances traveled, the mode of transportation and trip frequency. Then we discuss exclusionary residential segregation, why it should affect travel times, and how the transportation infrastructure and social stigma can restrict residents' mobility.

2. SOCIAL EXCLUSION[1]

There is considerable research on the distances traveled, the mode of transportation and trip frequency among various socio-demographic groups. In the planning literature, the problem is framed in relation to social exclusion, where different sub-populations are at a disadvantage travel-wise to others. Studying the movement of people in cities has evolved from simply looking at the road system and the accessibility of different points to work, recreational, service or shopping opportunities to activity modeling (or activity spaces), which focuses on the activities that people engage in (work, shopping, and so on), the chaining or interdependencies of activities within a trip or tour, household resources and household members' trips, travel to these activities (time, mode, and so on), and the spatial and time constraints they face (Jones 1990). The latter is the most interesting, because the poor, for example, face more serious space–time constraints than the general population (McCray and Brais 2007). They may reside just outside the central business district (CBD) and thus be proximate to jobs, recreation, services and retail, but their jobs may start and stop at odd times, they may have to walk some distance to bus or train stops (or the first/last mile problem), buses and trains travel only at certain times, fares can be expensive, train stations are often elevated or underground making it difficult for the physically disabled, their neighborhoods may be crime-ridden, streets and intersections may be dangerous, residents may not know schedules or about travel delays, and so on. Moreover, traveling with children often compounds the problems, for example, strollers have to be folded up when traveling on buses (McCray and Brais 2007). From a methodological standpoint, various measures for road system and activity modeling approaches are described in detail in Kwan (1998).

The research on who is more likely to be excluded is conclusive. The dependent variables are typically travel distances, travel mode and trip frequency. As regards independent variables, four socio-demographic variables stand out: gender, socio-economic status, immigration status and age. Key household characteristics include household income, number of adults and children, and number of automobiles. Some factors cut across socio-demographics and household characteristics; for example, possession of a motor vehicle license, the location of the household in the central city, inner ring suburbs or the outlying suburbs, and so on. Except for a few studies, race and ethnicity are not typical correlates of travel patterns, and we found only one recent study that explicitly looked at the effects of residential segregation using Twitter data (Wang et al. 2018). In general, Wang et al. (2018) found that people from more racially segregated areas had similar mobility patterns to their integrated counterparts, but they tended to travel within other segregated spaces. However, the extant literature has considered other variables that can impact mobility. We first review these considerations, grouped by the individual-level and household-level characteristics mentioned previously, and then we discuss how we may carry over these ideas to the study of residential segregation in the next section.

2.1 Socio-demographic Variables

There is considerable research that shows that women tend to travel shorter distances on a daily basis than do men (Kwan 1999a, 1999b; Stead 1999; Manaugh et al. 2010; Chen 2017b; Feng et al. 2017). Women are also more likely to use mass transit than are men

(Silva et al. 2006; Manaugh et al. 2010). However, women are more likely than men to take more trips, especially home-based trips (Lu and Pas 1999; Kuppam and Pendyala 2001; Manaugh et al. 2010). Partially, this is owing to part-time employment, women working in service jobs located near to the home, women's responsibility for child-rearing and women being less likely to own a motor vehicle (McCray and Brais 2007). However, some studies have found that these differences persist when comparing full-time employed men and women, and women with and without children. That is, gender truly seems to constrain women's travel more than it does men's.

Research results regarding age are also definitive. People who are in their early middle years tend to travel greater distances than those who are younger and those who are older (Stead 1999; Akar et al. 2016; Chen et al. 2017; Feng et al. 2017). They also take more trips (Lu and Pas 1999; Silva et al. 2012). Moreover, Silva et al. (2012) found that older residents, compared with younger residents, were more likely to take driving trips than transit trips. The reasons offered simply note that, in this age group, people are more likely to be employed, have resources and have the physical ability to travel more. Also, it could be that at this stage in the life course, people have young children or elderly parents who need to be cared for.

A less-studied individual-level characteristic is socio-economic status, measured typically by educational attainment and occupational status. It seems that residents with a higher education and/or more prestigious occupations tend to travel farther distances (Giuliano and Small 1993; Akar et al. 2016; Feng et al. 2017). They also take more trips (Jang 2005). Speculation on the causes focuses on where opportunities are located. On the one hand, some scholars have noted that professional and managerial jobs tend to be more highly concentrated spatially, while lower-status jobs tend to be more evenly distributed across the metropolitan area (hUallachain and Reid 1991; Stoll et al. 2000; Desmet and Fafchamps 2006). If higher-status individuals are living in the suburbs and commuting to downtown centers, it seems reasonable that their travel distances would be farther. On the other hand, others have argued that those with more education and higher-status jobs may have more cultural capital, which means having particular tastes in dining, entertainment and personal services (Murdie 1965; Friedland et al. 2007; Yaish and Katz-Gerro 2012). Thus, they may travel farther to obtain these.

Immigrant status has also received less attention. Research shows that newly arrived immigrants are more likely to spend more time traveling and to use mass transit and carpooling, but, as duration of residence lengthens, they typically do not differ in their use of transit compared with the rest of the population (Blumenberg and Smart 2010, 2011; Jang and Yao 2014). Travel distances also tend to equalize with time spent in the United States. However, in contrast to this work, Newbold et al. (2017), in a study of Ontario, found that the native-born population typically traveled longer distances. Specifically, they found that among immigrants, Blacks (mean travel = 12.4 kilometers), Whites (12.2 kilometers), Filipinos (11.5 kilometers) and South Asians (12.2 kilometers), travel distances were longer than for other groups (10.2–11.3 kilometers). Further, these mean distances were all slightly higher for native-born residents compared with those of their foreign-born counterparts. While the lengthening of travel distances may simply mean that immigrants are being assimilated into a larger labor market the longer they stay, the long commutes of these minority groups may be owing to residential segregation.

2.2 Household Characteristics

The most common household characteristic studied is income. Many studies have found that those with higher household incomes travel farther distances (Sermons and Koppelman 2001; Lucas 2004; Akar et al. 2016; Chen 2017b; Feng et al. 2017), take more frequent trips (Lu and Pas 1999; Wallace et al. 1999; Kuppam and Pendyala 2001; Jang 2005) and take fewer non-motorized trips (Silva et al. 2012). Another variable associated with travel distances and trip frequency is the number of cars in the household (Wallace et al. 1999; Jang 2005; Akar et al. 2016; Chen et al. 2017). The number of adults and children is also positively correlated with trip frequency (Goulias et al. 1990; Lu and Pas 1999; Wallace et al. 1999; Kuppam and Pendyala 2001; Jang 2005), but see Stead (1999) for different results, which found no significant association for household size. Goulias et al. (1990) also found that the number of mid-age children (5 to 15) increases the number of school and social tours, while the number of female adults is correlated with the number of shopping and school trips. Many of these household effects could be conflated with individual effects, for example, education, socio-economic status, race or ethnicity, and age and gender.

3. SPATIAL EXCLUSION

In addition to being excluded owing to their social characteristics, residents can be excluded by being physically segregated from others. With few exceptions (see Wang et al. 2018), these two literatures have largely not intersected. Broadly, urban scholars have long noted the impact that residential segregation can have on the life chances of individuals and families embedded within these contexts. Although the legal enforcement of racial residential segregation in the United States in particular was outlawed in 1968 with the passage of the Fair Housing Act, it remains an important and persistent part of the American urban landscape. As a product of institutionalized racism, segregation has and continues to shape the experiences of these residents, leading some to refer to residential segregation as the structural lynchpin of American racial inequality (Charles 2003). From a scholarly perspective, key works in the 1990s brought renewed attention to the plight of segregated neighborhoods and, since then, many studies have documented the difficulties of segregation in urban life (Massey and Denton 1993; Charles 2003).

While residential segregation can be based on various criteria and is often voluntary, we focus on residential segregation which is involuntary, based on residents' racial or ethnic and socio-economic characteristics, and excluded from the economic, political and social life of the city.[2] Marcuse (1997) refers to these neighborhoods as outcast ghettos to distinguish them from enclaves and what he terms citadels. In the US context, ascriptive characteristics, such as race or ethnicity, often intersect with class. For instance, Massey and Denton (1993) in their seminal work *American Apartheid*, note how segregation serves to concentrate poverty in space. Their argument is that as poverty is disproportionately concentrated in the Black community, when the group is also spatially concentrated through segregation, so too are the effects of poverty. Empirical work has supported this assertion (Krivo et al. 1998; Massey and Fischer 2000; Iceland and Wilkes 2006). This work has also shown that White poverty is much more evenly distributed throughout White society and White residential areas (Massey and Denton 1993; Massey et al. 1994).

The empirical work on the consequences of exclusionary or outcast segregation are definitive. With poverty, especially concentrated poverty, follows a variety of social problems that then also become disproportionately concentrated in the Black community. Furthermore, when examining the confluence of race and class, other work has shown that middle-class Black segregated neighborhoods tend to be adjacent to these urban problems and face greater exposure to them (Pattillo 1999). Empirically, a robust body of literature has demonstrated these effects. For example, this work has shown that segregated areas experience limited employment sites, fewer opportunities for economic advancement, and lower-quality education (Wilson 1987, 1996; Massey and Denton 1993; Krivo et al. 1998; Charles 2003). Similarly, a number of indicators of social disorder are more prevalent in these communities, such as family dissolution, crime and substance use (Wilson 1987; Greenberg and Schneider 1994; Shihadeh and Flynn 1996). Social disorder is often also accompanied by physical disorder, so that these neighborhoods often experience poorer housing quality, decreased access to services, housing code violations, vacant lots, broken windows, litter, graffiti and abandoned buildings (Shihadeh and Flynn 1996; Williams 1999; Chang et al. 2009), as well as greater exposure to a wide variety of environmental pollutants and hazards (Bullard 2005). Further, segregation has also been linked to poor health outcomes, poor health behaviors, and limited access to high-quality healthcare (LaVeist and Wallace 2000; Williams and Collins 2001; Subramanian et al. 2005; Larson et al. 2009; Anderson 2016).

While the bulk of this literature, especially as it relates to Black segregation, largely paints segregation in a negative light as part of a process of residential discrimination, the concentration of a minority group in space alone may not necessarily imply negative outcomes for those communities. An alternative to this can be found in the literature on immigrant enclaves, which in the post-1965 era of immigration mainly entails immigrant communities that are non-White, Latinos and Asians in particular (Alba et al. 2000; Logan et al. 2002). In contrast to the work on exclusionary segregation, these areas are often robust communities, which are self-sustaining through the enclave economy and provide a basis for social support and social networks for newly arrived migrants (Portes and Bach 1985; Logan et al. 2002). Indeed, research has found that these areas typically have a greater density of organizations and institutions (Anderson 2017b), which may indicate that these groups would have smaller spatial networks or travel patterns as many of their needs can be met within the enclave, including employment, food, socializing and opportunities for civic engagement. However, this residential mode is distinct from exclusionary segregation or Marcuse's outcast segregation, which is typically enforced from outside the community, and represents a method for social exclusion (Massey and Denton 1993). Further, immigrant enclaves are relatively rare, and would not reflect the modal residential type of most racial or ethnic minorities in the Unites States, particularly those who are not recent voluntary migrants to the Unites States (Portes and Bach 1985; Massey and Denton 1993).

Despite the large and growing body of literature that has documented the myriad ways that segregation can impact the life chances of the residents of these areas, relatively few studies have undertaken an examination of the spatial elements of the segregated community and the ways that space itself can serve as a form of inequality, which can exacerbate and further perpetuate the effects of segregation (Logan 2012). Logan (2012) called for more attention to spatial thinking within sociology as a discipline as a whole, as well as within our thinking about the effects of segregation. The central argument of this work is that spatial position within an urban system can confer specific advantages or disadvantages of access to valued community resources that Logan (2012) referred to as spatial inequality.

One example of this inequality is the spatial mismatch hypothesis, which is a longstanding example of attention to segregation and spatial thinking (Kain 1968). This work has demonstrated that macroeconomic changes in the economy and rapid suburbanization has led to a spatial mismatch between segregated neighborhoods and where low-skilled jobs are located (Kain 1968, 1992; Wilson 1987; Holzer 1992; Ihlanfeldt and Sjoquist 1998; Preston and McLafferty 1999). These jobs have moved out into the suburbs and are no longer in close proximity to where low-skilled workers are located in poor and minority communities, further exacerbating racial inequality. The empirical work on the spatial mismatch hypothesis, however, is mixed. Some studies found much longer travel times for Black men and women than for Whites (Gabriel and Rosenthal 1996; Preston and McLafferty 2016; McLafferty and Preston 2019) and that Black adolescents have longer commuting times than White adolescents (Holloway 1996). However, other studies have found shorter or no difference in travel times for Blacks and Whites (Taylor and Ong 1995), transportation access not spatial proximity matters more for employment (Ong and Miller 2005; Grengs 2010), and the relationship between trip length and wages remains unclear (McLafferty and Preston 2019). Job location (Weinberg 2004), job growth (Raphael 1998) and the likelihood that employers employ Blacks also matter (Hellerstein et al. 2008). Distance does not seem determinative, and race rather than place may matter more. For example, an increase in managerial and professional jobs in the center of the city are not accessible to residents of the outcast ghetto, because the latter lack the necessary human, cultural and social capital that employers seek (Wilson 1996).

Here, we take on this challenge to consider more broadly how segregation may be related to urban spatial networks, and we consider multiple possible outcomes. On the one hand, if organizational resources are lacking within the segregated community, this may prompt residents of these areas to branch out further to meet their daily needs and would then have broader travel patterns. On the other hand, if the urban design, transportation networks, and social stigma attached to the neighborhood and its residents make these travels more difficult, then the residents may be forced to forgo particular needs or seek out substandard options for their needs, and would thus have more constrained spatial networks. We consider each of these perspectives in turn.

4. RESIDENTIAL SEGREGATION AND THE DISTRIBUTION OF SERVICES

A large and growing body of literature has documented a number of spatial inequalities in the distribution of a wide variety of service providers and community organizations. This work has shown that, in general, poor and minority communities are less likely to have a number of different kinds of establishments. For instance, this work has shown that segregated Black neighborhoods are less likely to have business retail establishments, such as childcare centers, banks and credit unions, grocery, convenience and hardware stores, pharmacies, laundries and grooming facilities (Small and McDermott 2006). Beyond business establishments, a complementary body of work has shown that segregated areas have fewer social services, such as welfare agencies, non-profit social services, and mental health and substance abuse centers (Small and Stark 2005; Small and McDermott 2006; Allard 2009; Kissane 2010; Galaskiewicz et al. 2011, 2013; Sampson 2012; Marwell and Gullickson 2013). Some of these studies have examined the issues in an in-depth fashion to discuss the various ways that individuals navigate these spatial constraints and the importance of these resources as anchors in the community (Allard 2009; Small 2009; Galaskiewicz et al. 2012; Abramson 2015).

This line of research has exploded in the area of health-related services in particular. This is most notable in the food-deserts literature, which has shown that poor and minority communities lack access to a wide variety of food resources, such a supermarkets and restaurants (Beaulac et al. 2009; Walker et al. 2010). Furthermore, this work shows that, even when available, these food sources tend to be of lower quality, are less likely to have fresh fruits and vegetables, more likely to stock junk foods, and are typically higher in price (Morland et al. 2002b; Moore and Diez Roux 2006; Powell et al. 2007; Smiley et al. 2010; Bower et al. 2014). Similarly, a complementary literature has shown a similar pattern with fitness and recreational facilities, in that poor and minority communities are less likely to have these facilities in close proximity (Estabrooks et al. 2003; Wilson et al. 2004; Popkin et al. 2005; Moore et al. 2008), and that they tend to be of lower quality where available (Powell et al. 2007; Vaughan et al. 2013). Some work has also examined healthcare facilities (Hayanga et al. 2009; Dai 2010; Gaskin et al. 2012; Dinwiddie et al. 2013; Ko et al. 2014; Anderson 2017a, 2017b). These studies consistently demonstrate that minority segregated communities lack a wide variety of healthcare establishments compared with their White counterparts, though some of this effect can be accounted for with measures for the socio-economic status of such areas (Hayanga et al. 2009; Gaskin et al. 2012; Anderson 2017a, 2017b).

Moreover, several recent studies have linked the unequal distribution of these resources to important diet and health outcomes. For example, research has connected the distribution of food stores and supermarkets to personal food choices, with more of these resources leading to higher-quality diet and greater fruit and vegetable consumption (Morland et al. 2002a; Laraia et al. 2004; Jetter and Cassady 2006; Blitstein et al. 2012). Related studies have correlated this constrained diet quality to greater body mass index (BMI) and obesity in those communities, which are indicators closely linked to overall health outcomes (Inagami et al. 2006; Morland et al. 2006; Zick et al. 2009; Gibson 2011; Bader et al. 2013; Ohri-Vachaspati et al. 2013). In addition to diet, other studies have demonstrated that the lack of recreational facilities in neighborhoods is also related to greater BMI and lower rates of physical fitness (Black et al. 2010, 2014; Gordon-Larsen et al. 2006). Thus, lacking these resources can have consequences for those communities and may be an important factor in trying to understand neighborhood-level inequities.

This work provides some compelling evidence for why residents of segregated communities might be forced to travel further to meet the needs of daily life, as these communities are less likely to have a variety of important service providers in close proximity. This places a greater burden on disadvantaged communities in the ability of individuals to accomplish daily life tasks, which can be exacerbated for those with limited transportation mobility (Abramson 2015; Widener 2017). Further, there are temporal aspects of accessibility that can further restrict the use of service (for example, bus schedules and hours of operation) for residents reliant on public transportation (Kwan 2013).

However, most work in this area has not empirically examined where individuals are going for their daily needs in light of what is physically proximate to them. One exception is a study that found that Black and Latino segregated areas of Houston have fewer physicians' offices, and that this unequal distribution of physicians led to increased drive times to their physicians' office for residents of Black and Latino segregated neighborhoods (Anderson 2018). Similarly, a few small-scale studies have shown that low-income shoppers must routinely travel outside their immediate food environment to buy groceries at supermarkets rather than local convenience stores, as reported in a study on Detroit (LeDoux and Vojnovic 2013). Food aid recipients, as part of the Women, Children, and Infants (WIC) program, in another study traveled over a mile on average to buy food (Hillier et al. 2011), and research on Minneapolis recorded an average travel distance of over 3.4 kilometers from home to access more affordable, higher-quality options than were readily available nearby (Shannon 2016).

In contrast, other work on the consequences of the spatial provision of these resources would suggest that people are perhaps not traveling further to meet their needs and are instead opting for sub-optimal services and service providers that are in close proximity. For example, as regards the food-deserts literature, previous work demonstrates that people without food sources in close proximity are less likely to eat fresh fruits and vegetables (Morland et al. 2002b; Jetter and Cassady 2006; Blitstein et al. 2012). As this work has shown that small grocers and convenience stores, which are more common in food deserts compared with large supermarkets, are less likely to carry fresh fruits and vegetables, this would suggest that people living in food deserts are opting for the inferior food choices within their communities instead of traveling further to the large supermarkets outside their communities (Morland et al. 2002a; Jetter and Cassady 2006). Thus, another alternative is that people living in segregated communities may be less mobile and more likely to stay within their segregated communities despite the lack of local provision. This may be a consequence of the transportation options, both public and private, that are available to residents of these areas. We discuss this possibility next.

5. RESIDENTIAL SEGREGATION AND THE TRANSPORTATION INFRASTRUCTURE

The urban design and the position of segregated neighborhoods within the urban design may also have consequences for the spatial inequality that segregated communities face. Underlying a great deal of the current work on spatial inequality in access to resources, employment or otherwise, is that 'everything is related to everything else, but near things are more related than distant things' (Tobler 1970, p. 234). That is, people are more likely to access some job or amenity if it is located nearby, which would place a burden on individuals living within segregated communities who have few jobs and amenities in their neighborhoods. Inherent to any discussion of segregation is a physical separation, which is often created or reinforced by the urban infrastructure, which can create physical barriers between parts of the city (Grannis 1998; Sharkey 2013). Further, if the infrastructure is limited for efficiently and cheaply moving people about the city (such as with public transportation options), it may be difficult for these residents to overcome the limitations of their position within the network (Sharkey 2013).

Previous work in transportation geography has addressed the myriad ways that individuals experience travel burden and access to the places that they need to get to (Nicholls 2001). For example, Kwan (1998) has used both point integral measures, in addition to space–time indices of accessibility. Point integral, or gravity, measures focus on a single actor or location and the availability of opportunities around that individual. Conversely, space–time, or cumulative opportunity, measures focus on travel patterns and what is available to an individual on those travel lines. Both of these measures, though, emphasize the importance of location within the transportation system and how the urban design can facilitate or make it more difficult to access needed resources within a particular space.

As it relates to racial or ethnic residential segregation, Census figures consistently demonstrate that poor and minority individuals are less likely to own a private vehicle and are more likely to rely on public and shared forms of transportation, which may be limiting for the mobility networks of such individuals (US Census Bureau 2010, tables S0801 and B08006). As a public good, the transportation network should benefit residents throughout an urban area. Public transportation, if designed correctly, could have the potential to alleviate many spatial disparities in access if the system can provide quick access to needed goods and services. However, this perspective assumes that transportation networks, both driving and mass transit, are designed in ways to provide access to segregated neighborhoods.

Related to this issue, research has shown that transportation resources, similar to other public goods, are not equally distributed throughout urban areas. For example, many of the hub-and-spoke highway systems developed and built in the middle of the twentieth century were intended to quickly move White suburban dwellers to and from the urban core for work opportunities (Duany et al. 2000; Leinberger 2008; Sharkey 2013). These systems were not built to benefit minority neighborhoods, and often physically carved up historically minority neighborhoods. We often see a similar story with public transportation options. In an era of urban renewal and gentrification, inner-city neighborhoods are now becoming spatially advantageous zones again, but as a consequence, many poor and minority residents are being priced out of these spaces and displaced into the inner-ring suburbs (Kawabata and Shen 2007; Goodling et al. 2015). Similarly, new light-rail transit options, walkways and bicycle paths, are often being built to advantage young middle-class professionals instead of the inner-city poor (Grengs 2001; McKenzie 2013; Lubitow et al. 2017). Thus, from this perspective, the mobility of residents living within minority segregated communities may be limited.

6. STIGMATIZED RESIDENTS

Discrimination can also affect the travel patterns of those located in outcast ghettos. Trait-based discrimination takes place when others react negatively to people with particular traits, for example, racial characteristics, because they are in the wrong place at the wrong time. Place-based discrimination occurs when others react negatively to people because they are associated with a particular place, for example, an outcast ghetto. We argue that both types of spatially situated discrimination can prevent residents of segregated areas from venturing outside their neighborhoods.

Feagin's (1991) description of the continuing significance of race illustrates how trait-based discrimination can be spatially situated. Feagin (1991) notes that Blacks are more likely to face trait-based discrimination when they move into public accommodations, such as large stores and city restaurants. When Blacks encounter strangers, they are coded by those around them on their ascriptive traits. If they are in a space which is not typically frequented by people with those traits, the sojourner is likely to be avoided, rejected, or attacked physically or verbally (Feagin 1991). Centner (2008) made a similar point. He argued that space is imbued with particular social meanings that let in some and exclude others. He defined a group's spatial capital as 'the capacity to stake exclusionary claims, perceived by others as socially legitimate, on urban space that could reasonably be open to others' (Centner 2008, p. 194). That is, a certain group, in his example dot.comers in San Francisco, could lay exclusive claim to specific places for their own consumption. Others would then be excluded on the basis of their not having the right traits. In many ways this is similar to Lofland's (1971) findings of how in-groups appropriate public space and then make a legitimate claim to exclude others who are members of the out-group.

Research has also found evidence of place-based discrimination. While it is well known that place-based discrimination is important in the mortgage market (Massey and Denton 1993), Wilson (1996) described how employers in Chicago used stereotypes about place to reject young Black workers seeking employment. The rationale given is that if one is from the inner city, one has suffered irreparable harm from the culture of that neighborhood. As Wilson (1996, p. 115) states, 'The most common belief among the employers was that the social dislocations in the inner city are mainly a function of the environment in which blacks live'. He then quotes one employer, 'I necessarily can't tell from looking at an address whether someone's from Cabrini Green or not, but if I could tell, I don't think I'd want to hire them. Because it reflects on (their) credibility' (Wilson 1996, p. 116). Another simply said, 'If somebody gave me their address, uh, Cabrini Green, I might unavoidably have some concerns' (Wilson 1996, p. 114). Place-based discrimination at the interpersonal level implies that the dominant group has knowledge of where someone is from, and it is often unclear. However, if a person has a disparaged racial or ethnic identity and is known to be from a disparaged neighborhood, it is less likely that they would venture outside their neighborhood for jobs or necessities.

Under what conditions might both forms of discrimination prevent residents of outcast ghettos to travel outside their neighborhoods? We suspect when racial or ethnic exclusion is the greatest. Thus, the greater the residential segregation of the outcast group from the rest of the metropolitan population, the greater the concentration of poverty in these neighborhoods, and the more the mass media portray these neighborhoods and residents as disruptive or criminally inclined, the more the residents of these area will experience hostility when venturing outside their neighborhoods, and thus the less likely they will do so.

7. FUTURE DIRECTIONS

Although we present these two theoretical outcomes of segregation in relation to daily mobility, this largely remains an empirical question that needs to be answered. Transportation and urban geographers have considered how people move about their day to fulfill their daily needs, but on the whole, these patterns have not been tied to segregation, at least in terms of understanding the consequences of segregation (Farber et al. 2015). However, geography has recently made great strides in relation to thinking about and measuring segregation as a spatial–temporal process, and that work has introduced a number of new methods and data sources, which could be employed to answer this question (Farber et al. 2013, 2015; Park and Kwan 2018). The basis of this argument is that segregation is an ever-evolving construct, that as people move about an urban area throughout their daily lives, they may be more or less likely to encounter other types of people and have different exposures (Park and Kwan 2018). Also, we should expect that these exposures will change throughout the day as people continue to move around.

Turning this idea on its head, this notion could be gainfully employed in understanding the ways in which exclusionary residential segregation in particular may pattern daily life. This could be readily accomplished through the data and analysis advances found in the activity spaces literature. This chapter, grounded in transportation geography, has sought to measure individuals' spatial networks by accounting for the spatial–temporal dimensions of their daily routes or activities (Wong and Shaw 2011; Park and Kwan 2018). What these studies have traditionally sought to do is to better understand people's movement and utilization of space in a more intuitive fashion, instead of being confined to predefined administrative units (Farber et al. 2015), and new tools for data collection and advances in spatial analysis could facilitate this in studies on the link between residential segregation and travel patterns.

Some of this new research on activity spaces has considered the role of segregation. In particular, recent methodological advances in data science and wearable technology have opened up numerous possibilities for examining these outcomes in sophisticated ways. However, the focus is usually on measuring segregation, or the extent to which people are segregated throughout the day, and using activity spaces to better understand how and when segregation occurs. For example, several scholars have used the activity-spaces approach in an analysis of big data to better measure segregation by examining segregation that occurs beyond the residence, such as in school, work or other various daily activities (Wong and Shaw 2011; Farber et al. 2015; van Ham and Tammaru 2016; Park and Kwan 2018). Similarly, several recent studies have employed GPS data from mobile phones to track the daily movement of individuals throughout the day. Unlike previous studies, which rely primarily on self-report and recall from daily travel diaries, these technologies can provide a comprehensive view of exactly where and how individuals are moving throughout their day (Calabrese et al. 2015; Yip et al. 2016; Xiao et al. 2019). Other work has also employed big data from geo-tagged social media in order to examine spatial and temporal mobility throughout the day (Östh et al. 2018; Wang et al. 2018).

In a limited way, this research has been applied to the question of how segregation can impact daily mobility, as a consequence of residential segregation, as opposed to as a way to measure or understand segregation throughout the day. For example, in the Wang et al. (2018) study, the authors employ geo-tagged Twitter data across the 50 largest metropolitan areas in the United States, and they found that in terms of the average radius and spread of travel, individuals' travel circles did not differ substantially by racial/ethnic residential segregation. However, they found that residents of minority neighborhoods mainly did not travel to non-poor and White neighborhoods (Wang et al. 2018). Essentially, these residents were just as likely to move around the city, but their movement was largely confined to other segregated communities (Wang et al. 2018). In response to this study, though, some have pointed to the problems of the representativeness and spatial distribution of who constitutes the population of Twitter users (Vallée 2018).

Similarly, examining economic segregation in non-US contexts, others have found a connection between economic segregation and daily mobility using these advances in data collection techniques. For example, in a study of economic segregation (concentration of poverty and affluence) in metropolitan areas in Sweden, Östh et al. (2018) found that daily activity mobility results in residents frequenting less economically segregated spaces, but that a large portion of the population experiences a degree of economic segregation throughout the day. Although average Gini coefficients for the Swedish population are low compared with some other places (though these figures are higher for residents of higher poverty areas), these figures are notably reduced when taking into account time-geographical daily mobility measures (Östh et al. 2018). In another example, in a study of activity spaces between gated, exclusive communities versus non-exclusive communities across Beijing, the authors find substantial differences across these communities in relation to the extensiveness of their spatial networks (Wang et al. 2012). Specifically, they find that residents of institutionally privileged neighborhoods have more constrained activity spaces as more of their daily needs can be met locally (Wang et al. 2012). Thus, from this initial evidence using these developing analytical tools, we could easily export this approach to the study of the consequences of racial/ethnic residential segregation as well.

Despite these advances and recent studies, a number of questions remain, and the relationship between racial or ethnic segregation and the extent or patterns of mobility remains a mainly unanswered question, especially for racial segregation and a US context, on which the bulk of the literature on racial or ethnic segregation is currently focused. Moreover, this chapter, especially from urban and transportation geography, has mostly considered segregation as the outcome in question, rather than thinking about mobility and patterns of mobility as a consequence of segregation. Thus, in order to address the question of how segregation shapes mobility patterns, future work should more carefully address these considerations.

NOTES

1. Our literature review benefited greatly from three dissertations. We want to acknowledge and thank Misra (1999), Cooperman (2008) and Chen (2017a) for their excellent literature reviews.
2. We focus on spatial exclusion based on the intersectionality of race/ethnicity and class, however, other status characteristics could be the basis of exclusion as well, for example, sexual orientation, religion, criminality, age and immigrant status.

REFERENCES

Abramson, C.M. (2015), *The End Game: How Inequality Shapes our Final Years*, Cambridge, MA: Harvard University Press.

Akar, G., N. Chen and S.I. Gordon (2016), 'Influence of neighborhood types on trip distances: spatial error models for Central Ohio', *International Journal of Sustainable Transportation*, **10** (3), 284–93.

Alba, R.D., J.R. Logan and B.J. Stults (2000), 'The changing neighborhood contexts of the immigrant metropolis', *Social Forces*, **79** (2), 587–621.

Allard, S.W. (2009), *Out of Reach: Place, Poverty, and the New American Welfare State*, New Haven, CT: Yale University Press.

Anderson, K.F. (2016), 'Residential segregation and health outcomes in the United States: moving beyond Black and White', *Population Review*, **55** (2) 45–74.

Anderson, K.F. (2017a), 'Racial residential segregation and the distribution of auxiliary health care practitioners across urban space', *Research in the Sociology of Health Care*, **35**, 145–67.

Anderson, K.F. (2017b), 'Racial residential segregation and the distribution of health-related organizations in urban neighborhoods', *Social Problems*, **64** (2), 256–76.

Anderson, K.F. (2018), 'Racial/ethnic residential segregation, the distribution of physician's offices and access to health care: the case of Houston, Texas', *Social Sciences*, **7** (8), 119–37.

Bader, M.D.M., O. Schwartz-Soicher, D. Jack, C.C. Weiss, C.A. Richards, J.W. Quinn, et al. (2013), 'More neighborhood retail associated with lower obesity among New York City public high school students', *Health & Place*, **23** (September), 104–10.

Beaulac, J., E. Kristjansson and S. Cummins (2009), 'A systematic review of food deserts, 1966-2007', *Preventing Chronic Disease*, **6** (3), A105.

Black, C., G. Moon and J. Baird (2014), 'Dietary inequalities: what is the evidence for the effect of the neighbourhood food environment?', *Health & Place*, **27** (May), 229–42.

Black, J.L., J. Macinko, L.B. Dixon and G.E. Fryer Jr (2010), 'Neighborhoods and obesity in New York City', *Health & Place*, **16** (3), 489–99.

Blitstein, J.L., J. Snider and W.D. Evans (2012), 'Perceptions of the food shopping environment are associated with greater consumption of fruits and vegetables', *Public Health Nutrition*, **15** (6), 1124–9.

Blumenberg, E. and M. Smart (2010), 'Getting by with a little help from my friends . . . and family: immigrants and carpooling', *Transportation*, **37** (3), 429–46.

Blumenberg, E. and M.J. Smart (2011), 'Migrating to driving: exploring the multiple dimensions of immigrants' automobile use', in K. Lucas, E. Blumenberg and R. Weinberger (eds), *Auto Motives*, Bingley: Emerald, pp. 225–51.

Bower, K.M., R.J. Thorpe Jr, C. Rohde and D.J. Gaskin (2014), 'The intersection of neighborhood racial segregation, poverty, and urbanicity and its impact on food store availability in the United States', *Preventive Medicine*, **58** (January), 33–9.

Bullard, R.D. (2005), *The Quest for Environmental Justice: Human Rights and the Politics of Pollution*, San Francisco, CA: Sierra Club Books.

Calabrese, F., L. Ferrari and V.D. Blondel (2015), 'Urban sensing using mobile phone network data: a survey of research', *ACM Computing Surveys*, **47** (2), 25:21–25:20.

Centner, R. (2008), 'Places of privileged consumption practices: spatial capital, the dot-com habitus, and San Francisco's Internet boom', *City & Community*, **7** (3), 193–223.

Chang, V.W., A.E. Hillier and N.K. Mehta (2009), 'Neighborhood racial isolation, disorder and obesity', *Social Forces*, **87** (4), 2063–92.

Charles, C.Z. (2003), 'The dynamics of racial residential segregation', *Annual Review of Sociology*, **29**, 167–207.

Chen, N., C.-H. Wang and G. Akar (2017), 'Geographically weighted regression approach to investigate spatial variations in activity space', *Transportation Research Record*, **2671** (1), 40–50.

Chen, Y.-J. (2017a), 'Structural analysis on activity-travel patterns, travel demand, socio-demographics, and urban form: evidence from Cleveland metropolitan area', PhD dissertation, Ohio State University, Columbus, OH.

Chen, Y.-J. (2017b), 'Structural analysis on activity-travel patterns, travel demand, socio-demographics, and urban form: evidence from Cleveland metropolitan area', *City and Regional Planning*, Columbus, OH: Ohio State University.

Copperman, R.B.A. (2008), 'A comprehensive assessment of children's activity-travel patterns with implications for activity-based travel demand modeling', PhD dissertation, University of Texas at Austin, Austin, TX.

Craven, P. and B. Wellman (1973), 'The network city', *Sociological Inquiry*, **43** (3–4), 57–88.

Dai, D. (2010), 'Black residential segregation, disparities in spatial access to health care facilities, and late-stage breast cancer diagnosis in metropolitan Detroit', *Health & Place*, **16** (5), 1038–52.

Desmet, K and M. Fafchamps (2006), 'Employment concentration across U.S. counties', *Regional Science and Urban Economics*, **36** (4), 482–509.

Dinwiddie, G.Y., D.J. Gaskin, K.S. Chan, J. Norrington and R. McCleary (2013), 'Residential segregation, geographic proximity and type of services used: evidence for racial/ethnic disparities in mental health', *Social Science & Medicine*, **80** (March), 67–75.

Duany, A., E. Plater-Zyberk and J. Speck (2000), *Suburban Nation: The Rise of Sprawl and the Decline of the American Dream*, New York: North Point Press.

Estabrooks, P.A., R.E. Lee and N.C. Gyurcsik (2003), 'Resources for physical activity participation: Does availability and accessibility differ by neighborhood socioeconomic status?', *Annals of Behavioral Medicine*, **25** (2), 100–104.

Farber, S., T. Neutens, H.J. Miller and X. Li (2013), 'The social interaction potential of metropolitan regions: a time-geographic measurement approach using joint accessibility', *Annals of the Association of American Geographers*, **103** (3), 483–504.
Farber. S, M. O'Kelly, H.J. Miller and T. Neutens (2015), 'Measuring segregation using patterns of daily travel behavior: a social interaction based model of exposure', *Journal of Transport Geography*, **49** (December), 26–38.
Feagin, J.R. (1991), 'The continuing significance of race: antiblack discrimination in public places', *American Sociological Review*, **56** (1), 101–16.
Feng, J., M. Dijst, B. Wissink and J. Prillwitz (2017), 'Changing travel behaviour in urban China: evidence from Nanjing 2008-2011', *Transport Policy*, **53** (January), 1–10.
Friedland, L., D.V. Shah, N.-J. Lee, M.A. Rademacher, L. Atkinson and T. Hove (2007), 'Capital, consumption, communication, and citizenship: the social positioning of taste and civic culture in the United States', *Annals of the American Academy of Political and Social Science*, **611**, 31–50.
Gabriel, S.A. and S.S. Rosenthal (1996), 'Commutes, neighborhood effects, and earnings: an analysis of racial discrimination and compensating differentials', *Journal of Urban Economics*, **40** (1), 61–83.
Galaskiewicz, J., G. Hobor, B.M. Duckles and A.V. Mayorova (2012), 'Caregivers' social capital and satisfaction with their children's service providers', *Sociological Perspectives*, **55** (2), 241–66.
Galaskiewicz, J., O.V. Mayorova and B.M. Duckles (2013), 'Studying the roles of nonprofits, government, and business in providing activities and services to youth in the Phoenix metropolitan area', *Annals of the American Academy of Political and Social Science*, **647**, 50–82.
Galaskiewicz, J., S. Savage and D. Duerr (2011), 'Exploring the organizational dimensions of a "healthy community"', paper presented at the Thirteenth Annual Healthcare Organizational Research Association Conference, Seattle, WA, 10 June.
Gaskin, D.J., G.Y. Dinwiddie, K.S. Chan and R.R. McCleary (2012), 'Residential segregation and the availability of primary care physicians', *Health Services Research*, **47** (6), 2353–76.
Gibson, D.M. (2011), 'The neighborhood food environment and adult weight status: estimates from longitudinal data', *American Journal of Public Health*, **101** (1), 71–8.
Giuliano, G. and K.A. Small (1993), 'Is the journey to work explained by urban structure?', *Urban Studies*, **30** (9), 1485–500.
Goodling, E., J. Green and N. McClintock (2015), 'Uneven development of the sustainable city: shifting capital in Portland, Oregon', *Urban Geography*, **36** (4), 504–27.
Gordon-Larsen, P., M.C. Nelson, P. Page and B.M. Popkin (2006), 'Inequality in the built environment underlies key health disparities in physical activity and obesity', *Pediatrics*, **117** (2), 417–24.
Goulias, K., R. Pendyala and R. Kitamura (1990), 'Practical method for the estimation of trip generation and trip chaining', *Transportation Research Record*, **1285** (1), 47–56.
Grannis, R. (1998), 'The importance of trivial streets: residential streets and residential segregation', *American Journal of Sociology*, **103** (6), 1530–64.
Greenberg, M. and D. Schneider (1994), 'Violence in American cities – young black-males is the answer, but what was the question', *Social Science & Medicine*, **39** (2), 179–87.
Grengs, J. (2001), 'Does public transit counteract the segregation of carless households? Measuring spatial patterns of accessibility', *Transit Research Record*, **1753** (1), 3–10.
Grengs, J. (2010), 'Job accessibility and the modal mismatch in Detroit', *Journal of Transport Geography*, **18** (1), 42–54.
Hayanga, A.J., H.E. Kaiser, R. Sinha, S.M. Berenholtz, M. Makary and D. Chang (2009), 'Residential segregation and access to surgical care by minority populations in US counties', *Journal of the American College of Surgeons*, **208** (6), 1017–22.
Hellerstein, J.K., D. Neumark and M. McInerney (2008), 'Spatial mismatch or racial mismatch?', *Journal of Urban Economics*, **64** (2), 464–79.
Hillier, A., C.C. Cannuscio, A. Karpyn, J. McLaughlin, M. Chilton and K. Glanz (2011), 'How far do low-income parents travel to shop for food? Empirical evidence from two urban neighborhoods', *Urban Geography*, **32** (5), 712–29.
Holloway, S.R. (1996), 'Job accessibility and male teenage employment, 1980–1990: the declining significance of space?', *Professional Geographer*, **48** (4), 445–58.
Holzer, H.J. (1991), 'The spatial mismatch hypothesis: what has the evidence shown?', *Urban Studies*, **28** (1), 105–22.
hUallachain, B.O. and N. Reid (1991), 'The Location and growth of business and professional services in American metropolitan areas, 1976-1986', *Annals of the Association of American Geographers*, **81** (2), 254–70.
Iceland, J. and R. Wilkes (2006), 'Does socioeconomic status matter? Race, class, and residential segregation', *Social Problems*, **53** (2), 248–73.
Ihlanfeldt, K.R. and D.L. Sjoquist (1998), 'The spatial mismatch hypothesis: a review of recent studies and their implications for welfare reform', *Housing Policy Debate*, **9** (4), 849–92.

Inagami, S., D.A. Cohen, B.K. Finch and S.M. Asch (2006), 'You are where you shop: grocery store locations, weight, and neighborhoods', *American Journal of Preventive Medicine*, **31** (1), 10–17.
Jang, T.Y. (2005), 'Count data models for trip generation', *Journal of Transportation Engineering*, **131** (6), 444–50.
Jang, W. and X. Yao (2014), 'Tracking ethnically divided commuting patterns over time: a case study of Atlanta', *Professional Geographer*, **66** (2), 274–83.
Jetter, K.M. and D.L. Cassady (2006), 'The availability and cost of healthier food alternatives', *American Journal of Preventive Medicine*, **30** (1), 38–44.
Jones, P. (1990), *Developments in Dynamic and Activity-based Approaches to Travel Analysis*, Aldershot: Avebury.
Kain, J.F. (1968), 'Housing segregation, negro employment, and metropolitan decentralization', *Quarterly Journal of Economics*, **82** (2), 175–97.
Kain, J.F. (1992), 'The spatial mismatch hypothesis: three decades later', *Housing Policy Debate*, **3** (2), 371–460.
Kawabata, M. and Q. Shen (2007), 'Commuting inequality between cars and public transit: the case of the San Francisco Bay Area, 1990-2000', *Urban Studies (Routledge)*, **44** (9), 1759–80.
Kissane, R.J. (2010), '"We call it the badlands": how social-spatial geographies influence social service use', *Social Service Review*, **84** (1), 3–28.
Ko, M., J. Needleman, K.P. Derose, M.J. Laugesen and N.A. Ponce ((2014), 'Residential segregation and the survival of U.S. urban public hospitals', *Medical Care Research and Review*, **71** (3), 243–60.
Krivo, L.J., R.D. Peterson, H. Rizzo and J.R. Reynolds (1998), 'Race, segregation, and the concentration of disadvantage: 1980-1990', *Social Problems*, **45** (1), 61–80.
Kuppam, A.R. and R.M. Pendyala (2001), 'A structural equations analysis of commuters' activity and travel patterns', *Transportation*, **28** (1), 33–54.
Kwan, M.-P. (1998), 'Space-time and integral measures of individual accessibility: a comparative analysis using a point-based framework', *Geographical Analysis*, **30** (3), 191–216.
Kwan, M.-P. (1999a), 'Gender and individual access to urban opportunities: A study using space-time measures', *Professional Geographer*, **51** (2), 210.
Kwan, M.-P. (1999b), 'Gender, the home-work link, and space-time patterns of nonemployment activities', *Economic Geography*, **75** (4), 370–94.
Kwan, M.-P. (2013), 'Beyond space (as we knew it): toward temporally integrated geographies of segregation, health, and accessibility', *Annals of the Association of American Geographers*, **103** (5), 1078–86.
Laraia, B.A., A.M. Siega-Riz, J.S. Kaufman and S.J. Jones (2004), 'Proximity of supermarkets is positively associated with diet quality index for pregnancy', *Preventive Medicine*, **39** (5), 869–75.
Larson, N.I., M.T. Story and M.C. Nelson (2009), 'Neighborhood environments disparities in access to healthy foods in the US', *American Journal of Preventive Medicine*, **36** (1), 74–81.
LaVeist, T.A. and J.M. Wallace (2000), 'Health risk and inequitable distribution of liquor stores in African American neighborhood', *Social Science & Medicine*, **51** (4), 613–17.
LeDoux, T.F. and I. Vojnovic (2013), 'Going outside the neighborhood: The shopping patterns and adaptations of disadvantaged consumers living in the lower eastside neighborhoods of Detroit, Michigan', *Health & Place*, **19** (1), 1–14.
Leinberger, C.B. (2008), *The Option of Urbanism: Investing in a New American Dream*, Washington, DC: Island Press.
Lofland, L.H. (1971), *A World of Strangers: Order and Action in Urban Public Space*, San Francisco, CA: University of California Press.
Logan, J.R. (2012), 'Making a place for space: spatial thinking in social science', *Annual Review of Sociology*, **38** (August), 507–24.
Logan, J.R., R.D. Alba and W.Q. Zhang (2002), 'Immigrant enclaves and ethnic communities in New York and Los Angeles', *American Sociological Review*, **67** (2), 299–322.
Lu, X. and E.I. Pas (1999), 'Socio-demographics, activity participation and travel behavior', *Transportation Research Part A: Policy and Practice*, **33** (1), 1–18.
Lubitow, A., J. Rainer and S. Bassett (2017), 'Exclusion and vulnerability on public transit: experiences of transit dependent riders in Portland, Oregon', *Mobilities*, **12** (6), 924–37.
Lucas, K. (2004), *Running on Empty: Transport, Social Exclusion and Environmental Justice*, Bristol: Policy Press.
Manaugh, K., L.F. Miranda-Moreno and A.M. El-Geneidy (2010), 'The effect of neighbourhood characteristics, accessibility, home–work location, and demographics on commuting distances', *Transportation*, **37** (4), 627–46.
Marcuse, P. (1997), 'The enclave, the citadel, and the ghetto – what has changed in the post-Fordist US city', *Urban Affairs Review*, **33** (2), 228–64.
Marwell, N.P. and A. Gullickson A. (2013), 'Inequality in the spatial allocation of social services: government contracts to nonprofit organizations in New York City', *Social Service Review*, **87** (2), 319–53.

Massey, D.S. and N.A. Denton (1993), *American Apartheid: Segregation and the Making of the Underclass*, Cambridge, MA: Harvard University Press.
Massey, D.S. and M.J. Fischer (2000), 'How segregation concentrates poverty', *Ethnic and Racial Studies*, **23** (4), 670–91.
Massey, D.S., A.B. Gross and K. Shibuya (1994), 'Migration, segregation, and the geographic concentration of poverty', *American Sociological Review*, **59** (3), 425–45.
McCray, T. and N. Brais (2007), 'Exploring the role of transportation in fostering social exclusion: the use of GIS to support qualitative data', *Networks and Spatial Economics*, **7** (4), 397–412.
McKenzie, B.S. (2013), 'Neighborhood access to transit by race, ethnicity, and poverty in Portland, OR', *City & Community*, **12** (2) 134–55.
McLafferty, S. and V. Preston (2019), 'Who has long commutes to low wage jobs? Gender, race, and access to work in the New York Region', *Urban Geography*, **40** (9), 1270–90.
Misra, R. (1999), 'Toward a comprehensive representation and analysis framework for non-worker activity – travel pattern modeling', PhD dissertation, University of Texas at Austin, Austin, TX.
Moore, L.V. and A.V. Diez Roux (2006), 'Associations of neighborhood characteristics with the location and type of food stores', *American Journal of Public Health*, **96** (2), 325–31.
Moore, L.V., A.V.D. Roux, K.R. Evenson, A.P. McGinn and S.J. Brines (2008), 'Availability of recreational resources in minority and low socioeconomic status areas', *American Journal of Preventive Medicine*, **34** (1), 16–22.
Morland, K., A.V. Diez Roux and S. Wing (2006), 'Supermarkets, other food stores, and obesity: the atherosclerosis risk in communities study', *American Journal of Preventive Medicine*, **30** (4), 333–9.
Morland, K., S. Wing and A.D. Roux (2002a), 'The contextual effect of the local food environment on residents' diets: the atherosclerosis risk in communities study', *American Journal of Public Health*, **92** (11), 1761–7.
Morland, K., S. Wing and A.D. Roux and C. Poole (2002b), 'Neighborhood characteristics associated with the location of food stores and food service places', *American Journal of Preventive Medicine*, **22** (1), 23–9.
Murdie, R.A. (1965), 'Cultural differences in consumer travel', *Economic Geography*, **41** (3), 211–33.
Newbold, B.K., D.M. Scott and C. Burke (2017), 'Immigrant status and commute distance: an exploratory study based on the greater Golden Horseshoe', *Transportation*, **44** (1), 181–98.
Nicholls, S. (2001), 'Measuring the accessibility and equity of public parks: a case study using GIS', *Managing Leisure*, **6** (4), 201–19.
Ohri-Vachaspati, P., K. Lloyd, D. DeLia, D. Tulloch and M.J. Yedidia (2013), 'A closer examination of the relationship between children's weight status and the food and physical activity environment', *Preventive Medicine*, **57** (3), 162–7.
Ong, P.M. and D. Miller (2005), 'Spatial and transportation mismatch in Los Angeles', *Journal of Planning Education and Research*, **25** (1), 43–56.
Östh, J., I. Shuttleworth and T. Niedomysl (2018), 'Spatial and temporal patterns of economic segregation in Sweden's metropolitan areas: a mobility approach', *Environment and Planning A: Economy and Space*, **50** (4), 809–25.
Park, Y.M. and M.-P. Kwan (2018), 'Beyond residential segregation: a spatiotemporal approach to examining multi-contextual segregation', *Computers Environment and Urban Systems*, **71** (September), 98–108.
Pattillo, M.E. (1999), *Black Picket Fences: Privilege and Peril among the Black Middle Class*, Chicago, IL: University of Chicago Press.
Popkin, B.M., K. Duffey and P. Gordon-Larsen (2005), 'Environmental influences on food choice, physical activity and energy balance', *Physiology & Behavior*, **86** (5), 603–13.
Portes, A. and R.L. Bach (1985), *Latin Journey: Cuban and Mexican Immigrants in the United States*, Berkeley, CA: University of California Press.
Powell, L.M., S. Slater, D. Mirtcheva, D. Bao and F.J. Chaloupka (2007), 'Food store availability and neighborhood characteristics in the United States', *Preventive Medicine*, **44** (3), 189–95.
Preston, V. and S. McLafferty (1999), 'Spatial mismatch research in the 1990s: progress and Potential', *Papers in Regional Science*, **78** (4), 387–402.
Preston, V. and S. McLafferty (2016), 'Revisiting gender, race, and commuting in New York', *Annals of the American Association of Geographers*, **106** (2), 300–310.
Raphael, S. (1998), 'The spatial mismatch hypothesis and black youth joblessness: evidence from the San Francisco Bay Area', *Journal of Urban Economics*, **43** (1), 79–111.
Sampson, R.J. (2012), *Great American City: Chicago and the Enduring Neighborhood Effect*, Chicago, IL and London: University of Chicago Press.
Sermons, M.W. and F.S Koppelman (2001), 'Representing the differences between female and male commute behavior in residential location choice models', *Journal of Transport Geography*, **9** (2), 101–10.
Shannon, J. (2016),' Beyond the supermarket solution: linking food deserts, neighborhood context, and everyday mobility', *Annals of the American Association of Geographers*, **106** (1), 186–202.

Sharkey, P. (2013), *Stuck in Place: Urban Neighborhoods and the End of Progress toward Racial Equality*, Chicago, IL: University of Chicago Press.

Shihadeh, E.S. and N. Flynn (1996), 'Segregation and crime: the effect of black social isolation on the rates of black urban violence', *Social Forces*, **74** (4), 1325–52.

Silva, J. de A. e, T.F. Golob and K.G. Goulias (2006), 'Effects of land use characteristics on residence and employment location and travel behavior of urban adult workers', *Transportation Research Record*, **1977** (1), 121–31.

Silva, J. de A. e, C. Morency and K.G. Goulias (2012), 'Using structural equations modeling to unravel the influence of land use patterns on travel behavior of workers in Montreal', *Transportation Research Part A: Policy and Practice*, **46** (8), 1252–64.

Small, M.L. (2009), *Unanticipated Gains: Origins of Network Inequality in Everyday Life*, New York: Oxford University Press.

Small, M.L. and M. McDermott (2006), 'The presence of organizational resources in poor urban neighborhoods: an analysis of average and contextual effects', *Social Forces*, **84** (3), 1697–724.

Small, M.L. and L. Stark (2005), 'Are poor neighborhoods resource deprived? A case study of childcare centers in New York', *Social Science Quarterly*, **86** (5), 1013–36.

Smiley, M.J., A.V. Diez Roux, S.J. Brines, D.G. Brown, K.R. Evenson and D.A. Rodriguez (2010), 'A spatial analysis of health-related resources in three diverse metropolitan areas', *Health & Place*, **16** (5), 885–92.

Stead, D. (1999), 'Planning for less travel: identifying land use characteristics associated with more sustainable travel patterns', PhD thesis, Barlett School of Planning and Architecture, University College London.

Stoll, M.A., H.J. Holzer and K.R. Ihlanfeldt (2000), 'Within cities and suburbs: racial residential concentration and the spatial distribution of employment opportunities across sub-metropolitan areas', *Journal of Policy Analysis and Management*, **19** (2), 207–31.

Subramanian, S.V., D. Acevedo-Garcia and T.L. Osypuk (2005), 'Racial residential segregation and geographic heterogeneity in black/white disparity in poor self-rated health in the US: a multilevel statistical analysis', *Social Science & Medicine*, **60** (8), 1667–79.

Taylor, D.B. and P.M. Ong (1995), 'Spatial mismatch or automobile mismatch? An examination of race, residence and commuting in U.S. metropolitan areas', *Urban Studies*, **32** (9), 1453–73.

Tobler, W.R. (1970), 'A computer movie simulating urban growth in the Detroit region', *Economic Geography*, **46** (June), 234–40.

US Census Bureau (2010), *American Community Survey 2008–2012*, Washington, DC: US Census Bureau.

Vallée, J. (2018), 'Urban isolation and daytime neighborhood social composition from Twitter data', *Proceedings of the National Academy of Sciences*, **115** (51), E11886–E11887.

Van Ham, M. and T. Tammaru (2016), 'New perspectives on ethnic segregation over time and space. A domains approach', *Urban Geography*, **37** (7), 953–62.

Vaughan, K., A. Kaczynski, S. Wilhelm Stanis, G. Besenyi, R. Bergstrom and K. Heinrich (2013), 'Exploring the distribution of park availability, features, and quality across Kansas City, Missouri by income and race/ethnicity: an environmental justice investigation', *Annals of Behavioral Medicine*, **45** (1), 28–38.

Walker, R.E, C.R. Keane and J.G. Burke (2010), 'Disparities and access to healthy food in the United States: a review of food deserts literature', *Health & Place*, **16** (5), 876–84.

Wallace, B., F. Mannering and G.S. Rutherford (1999), 'Evaluating effects of transportation demand management strategies on trip generation by using poisson and negative binomial regression', *Transportation Research Record*, **1682** (1), 70–77.

Wang, D., F. Li and Y. Chai (2012), 'Activity spaces and sociospatial segregation in Beijing', *Urban Geography*, **33** (2), 256–77.

Wang, Q., N.E. Phillips, M.L. Small and R.J. Sampson (2018), 'Urban mobility and neighborhood isolation in America's 50 largest cities', *Proceedings of the National Academy of Sciences*, **115** (30), 7735–40.

Weinberg, B.A. (2004), 'Testing the spatial mismatch hypothesis using inter-city variations in industrial composition', *Regional Science & Urban Economics*, **34** (5), 505–32.

Widener, M.J. (2017), 'Comparing measures of accessibility to urban supermarkets for transit and auto users', *Professional Geographer*, **69** (3), 362–71.

Williams, D.R. (1999), 'Race, socioeconomic status, and health: the added effects of racism and discrimination', *Annals of the New York Academy of Sciences*, **896**, 173–88.

Williams, D.R and C. Collins (2001), 'Racial residential segregation: a fundamental cause of racial disparities in health', *Public Health Reports*, **116** (5), 404–16.

Wilson, D.K, K.A. Kirtland, B.E. Ainsworth and C.L. Addy (2004), 'Socioeconomic status and perceptions of access and safety for physical activity', *Annals of Behavioral Medicine*, **28** (1), 20–28.

Wilson, W.J. (1987), *The Truly Disadvantaged: The Inner City, the Underclass, and Public Policy*, Chicago, IL: University of Chicago Press.

Wilson, W.J. (1996), *When Work Disappears: The World of the New Urban Poor*, New York: Knopf.

Wong, D. and S.-L. Shaw (2011), 'Measuring segregation: an activity space approach', *Journal of Geographical Systems*, **13** (2), 127–45.
Xiao, Y., D. Wang and J. Fang (2019), 'Exploring the disparities in park access through mobile phone data: evidence from Shanghai, China', *Landscape and Urban Planning*, **181** (1), 80–91.
Yaish, M. and T. Katz-Gerro (2012), 'Disentangling "cultural capital": the consequences of cultural and economic resources for taste and participation', *European Sociological Review*, **28** (2), 169–85.
Yip, N.M., R. Forrest and S. Xian (2016), 'Exploring segregation and mobilities: application of an activity tracking app on mobile phone', *Cities*, **59** (November), 156–63.
Zick, C.D., K.R. Smith, J.X. Fan, B.B. Brown, L. Yamada and L. Kowaleski-Jones (2009), 'Running to the store? The relationship between neighborhood environments and the risk of obesity', *Social Science & Medicine*, **69** (10), 1493–500.

16. Urban mobility and segregation examined through networked travel activity

Susan A. Burtner and Alan T. Murray

1. INTRODUCTION

Individuals make travel decisions every day that feed into a vast network of spatial interactions. Unfortunately for some, ongoing economic, social and psychological effects related to the United States' history of racism, segregation, and discrimination have led to significant disparities in urban mobility across racial groups (Ray and Preston 2009). The effect of segregation in residential neighborhoods has been extensively studied, and more attention is being brought to other spatial contexts in which people interact. However, comparatively little research has taken a comprehensive approach to both segregation and urban mobility by analyzing residential, work and social environments together. The utility of this approach lies in the ability to characterize physical spaces by their potential for interaction between racial groups, and whether or not group networks of activity reveal the different accessibility levels of neighborhoods in an area.

This chapter seeks to address how networks and associated metrics can help to explain broad urban mobility patterns and the differences that exist between racial groups. In order to model urban mobility, we first construct networks of travel activity based on race. We then derive various network-based measures and a neighborhood-interaction potential matrix that compares travel between neighborhoods in our study area. We present results for the Los Angeles and Orange County area as a case study because it has a large, diverse population and continues to be a focus of segregation research (Johnston et al. 2006).

2. BACKGROUND

Numerous fields, including political science, economics, sociology, history and geography, have put considerable effort into studying why segregation exists and how it has affected modern society (Alexander 2012; Johnston et al. 2006; Massey and Denton 1989; Ray and Preston 2009; Redford 2017; Reilly et al. 2002; Schelling 1971). The United States has an unfortunate history of institutionalized segregation: before the Civil Rights Movement, White and non-White Americans could not live in the same neighborhoods, use the same shops, restrooms or transportation, nor eat at the same restaurants (or were separated into different sections). This was the reality of life until major federal legislation (including the Civil Rights Act 1964, the Voting Rights Act 1965 and the Fair Housing Act in 1968) and an abundance of state-level legislation deemed segregation unconstitutional (Richardson and Luker 2014). Segregation persisted for decades through Jim Crow laws, when racially restrictive zoning, targeted advertising and discriminatory lending heavily pushed White individuals to move into White neighborhoods and non-Whites into non-White neighborhoods (Redford 2017). Research continues to examine the extent to which people are racially separating based on residential and other contexts.

2.1 Residential and Work Segregation

The most enduring form of segregation has persisted through residential choices. A lack of racial mixing in the United States has contributed to ethnically homogeneous households and, by extension, neighborhoods. According to the 2010 Census, over 97 percent of the United States population identified as only one race and about 3 percent of individuals indicated two or more races (Humes et al. 2011). Of the 97 percent of individuals who identified as one race, about 72 percent identified as White, 13 percent as Black or African American, 5 percent as Asian, 0.9 percent as American Indian and Alaska Native, 0.5 percent as Native Hawaiian or Pacific Islander and 6 percent as one other race (Hume et al. 2011).

To analyze the extent to which race characterizes neighborhoods, the United States Census Bureau has been reporting on neighborhood-level segregation following each decennial census since 1890 (Glaeser and Vigdor 2012). Residential segregation has declined in most major metropolitan areas (Glaeser and Vigdor 2012). Immigration, gentrification and a change in racial attitudes have made most neighborhoods more diverse, but equality remains a major issue. Unequal access to jobs, health-care providers and affordable housing has proven to be a complex subject relative to segregation (Glaeser and Vigdor 2012).

While residential segregation has remained a vibrant topic of interest in the United States, only within the past decade have new perspectives on segregation been explored. The push for increased scholarly attention on segregation outside of residential contexts was largely motivated by people spending a significant amount of time outside their homes (Hall et al. 2019). Research has shown that working environments can be segregated across different socio-demographic dimensions, but often are far more diverse than residential areas (Hall et al. 2019).

2.2 Activity Spaces

Critics of research on segregation in confined spaces, whether residential, work, social, recreational or otherwise, have optimistically viewed the concept of activity spaces as an improvement in measuring the exposure of human interaction in peoples' daily lives (Kwan 2013). Work in activity spaces looks at the amount of racial exposure individuals experience when they visit different spaces (Jones and Pebely 2014). The size of the activity space (usually determined by associated administrative boundaries) is spatially joined with demographic variables and typically used in regression analyses. Other approaches create completely new spatial areas and analyze environmental variables within a daily path or derived ellipse, or incorporate time as a weighting factor in how significantly a place contributes to an activity space (Sharp et al. 2015; Zenk et al. 2011). In all of these cases, only a specific subset of a population is in the sample (such as families with young children), and the types and number of activities are usually predetermined. Scholars have pressed for analysis of the dynamic flows of travel across the demographic spectrum, but within the context of studying segregation, little current research exists that spatially and temporally addresses travel activity for different racial groups (Kwan 2013).

2.3 Network Approaches

Network approaches have long been used in transportation studies, travel behavior research and beyond; decades of methodological developments from sociology and computer science in particular have contributed profoundly to the growth of network science (Barabási 2003; Newman 2010). Networks offer a structure for examining both the relational attributes and spatial proximity of their represented entities. Several terms are used to describe networks: networks (also termed graphs) contain nodes (or vertices) and links (or edges) that connect the nodes. Links can represent connections or flows and nodes can represent entities or events, and together they represent the dynamics of relationships in a system. Networks serve as an abstraction of the real world, which can then be used in simulation, summary description or planning.

Researchers using network techniques from a geographic perspective have begun to explore questions regarding social issues in spatial contexts (Carrasco et al. 2008; Johnson and Gilles 2008; Ratti et al. 2010). Less research exists at the intersection of social and travel activity, largely because the needed data is difficult to come by. One group of researchers used surveys to collect data on travel activity with friends and family, and claim that social networks are not merely an attribute of travel networks, but that they cause travel (Carrasco et al. 2008). Social anchor points define the spatial locations of social ties and activities, and through the creation of a network, can reveal the number of people within a social network that engage in co-located travel (Carrasco et al. 2008). Still others have used an abstraction of social transactions (such as telephone calls or geo-located Twitter tweets) to create maps that reflect human interactions (Ratti et al. 2010; Wang et al. 2018). Networks help to bridge geographic and relational space, but most of the current research deals with mapping slices of social communication or slices of the population. Existing research does not address how a representative subset of the population actually navigates spaces across a region.

3. METHODS

This section illustrates our approach to deriving large networks of travel activity by racial groups, and characterizing them through different network statistics. Measuring urban mobility and segregation inherently involves estimating the extent to which people move and cluster in space (Cheba and Saniuk 2016; Massey and Denton 1989). Urban mobility is often measured through people's urban transport choices and constraints, but can be incredibly complex (Cheba and Saniuk 2016). For measuring segregation, the dissimilarity index is a popular measure, but despite its widespread use, researchers have argued that one measure alone cannot reflect extreme segregation (hypersegregation) that particular groups experience (Massey and Denton 1989). Other dimensions of segregation have been introduced, including evenness, exposure, clustering, centralization and concentration (Massey and Denton 1989). However, all of these measures are based on comparing population counts for one group in an area to population counts for a

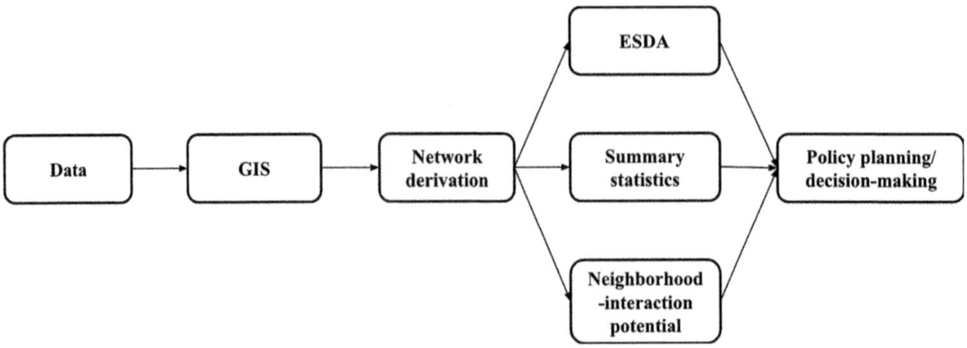

Figure 16.1 Overview of methodology

broader area. This means that individuals are only defined by their presence within a single geographic context, ignoring the dynamic movement of individuals in and out of a variety of geographic spaces. Our methodology addresses the movement of individuals between neighborhoods by integrating all trips made by individuals in a day. A summary of the entire methodology is given in Figure 16.1. Details for each step, starting with our data and study area, are outlined in sections 3.1–3.6.

3.1 Data and Study Area

Our study area encompasses the counties of Los Angeles and Orange, located in Southern California in the United States of America (key statistics for these regions are given in Tables 16.1 and 16.2). An important distinction to make is that Los Angeles County geographically extends far beyond the city of Los Angeles and, together with Orange County, make up the Los Angeles Metropolitan Statistical Area as defined by the United States Census Bureau. Los Angeles and Orange County together reflect a study area that is rich in ethnic and racial diversity. Los Angeles in particular has been touted as the quintessential postmodern city, and unlike other major metropolitan areas, has significant Hispanic and Asian populations (Johnston et al. 2006). Despite its reputation as a vision for modernity and diversity, research shows that as regards residential segregation, Los Angeles is no different than most other major metropolitan areas (Johnston et al. 2006). Not surprisingly, it continues to be a major focus of race-based research (Redford 2017).

The California Household Travel Survey (CHTS) is one of the only openly available data sources for studying travel activity for the counties of Los Angeles and Orange, and contains travel destinations for 8219 households in Los Angeles County and 2401 households in Orange County (Kunzmann and Daigler 2013). The survey was conducted between 2010 and 2012, with households selected based on stratified sampling of the population. Several modes of data collection were used, including telephone, mail and online interviews. Some individuals were given a Global Positioning System device to track their locations (Kunzmann and Daigler 2013).[1] In addition, several different demographic variables were collected for individuals, such as their race, age, gender, and home, school or work information. The variable of race included Hispanic,[2] White, Asian (Asian Indian, Japanese, Chinese, Korean, Filipino and Vietnamese), Black or African

Table 16.1 *Los Angeles (LA) and Orange County population attributes*

	US Census, 2010		CHTS, 2013	
	LA County	Orange County	LA County	Orange County
Population or sample size	9.818.605	3.010.232	20.457	6.163
Age and sex				
Persons under 5 years	6.1%	5.9%	3.9%	3.3%
Persons under 18 years	21.7%	21.9%	20.4%	20.0%
Persons 65 years and over	13.6%	14.8%	15.6%	18.6%
Female persons	50.7%	50.6%	50.8%	50.7%
Race				
White alone, not Hispanic or Latino	26.2%	40.5%	49.4%	68.9%
Black or African American alone	9.0%	2.1%	6.5%	1.1%
American Indian and Alaska Native alone	1.4%	1.0%	0.4 %	0.5%
Asian alone	15.3%	21.0%	8.4%	9.0%
Native Hawaiian and Other Pacific Islander alone	0.4%	0.4%	0.4%	0.2%
Hispanic or Latino	48.6%	34.2%	35.0%	20.1%

Source: Kunzmann and Daigler (2013); United States Census Bureau (2019).

Table 16.2 *Los Angeles (LA) and Orange County geographic attributes*

	LA County	Orange County
Average population per square mile	2419.6	3807.7
Land area in square miles	4057.9	790.6
Number of neighborhoods	270	48

Source: Humes et al. (2011); Los Angeles Times (2013).

American, Native Hawaiian or Pacific Islander (Guamanian, Samoan and Fijian), and American Indian or Alaska Native. With location activities and the race of each survey respondent, we were able to construct complete logs of travel for individuals within a 24-hour period.

A key aspect of our methodology attempts to address the potential for groups of different races to interact. Thus, while individual trip destinations can be mapped to single points, we must aggregate group travel activity locations at some meaningful spatial unit. One possible unit for aggregation is the neighborhood, which is often a source of social capital, community engagement, and identity (Chaskin 1997). However, the neighborhood as a spatial unit can be difficult to define as many people enact relationships that transcend the defined boundaries of a geographic space, and neighborhood delineations must be cognitively meaningful to residents. For the United States Census, a neighborhood is synonymous with a Census tract. However, few would agree that Census tracts are truly representative of neighborhood boundaries. The regional newspaper the Los Angeles Times attempts to better represent the neighborhoods of Los Angeles County and Orange County with the developed Mapping L.A. (Los Angeles Times 2013, p. 1) product.[3] The curators of Mapping L.A. argue that 'the shapes of the [Census] tracts are frequently out of synchronization with the geographical, historic and socioeconomic associations that define communities' (Los Angeles Times 2013, p. 1). The Mapping L.A. boundaries of neighborhoods incorporate the perception from Los Angeles County and Orange County residents about where their neighborhoods actually start and end. Each neighborhood definition is based on a Census tract but 'follows a set of principles intended to make it visually and statistically coherent' (Los Angeles Times 2013, p. 1). Some of the modifications involve including unincorporated areas, eliciting user revisions to boundaries, and updating boundaries according to maps by the Los Angeles County Department of Regional Planning. With confidence in the alignment of the cartographic boundaries and resident perceptions, we analyzed travel activity at the resolution of these neighborhoods for which there is a potential for spatial interaction to occur.

3.2 Geographic Information Systems

Geographic information systems (GIS) are systems for handling geographic or spatial data. There are four basic functions of GIS: input, storage, analysis and output (Anselin and Getis 1992). Geographical information systems are particularly useful for examining network data as they have the ability to store spatial attributes for both nodes and edges and provide a platform for visualizing spatial interactions (Murray 2000). A GIS was necessary for deriving our networks as it systematically helped turn neighborhoods defined by polygons into nodes and travel movements defined by points into edges.

Section 3.1 enumerated our choice to infer spatial interaction at the resolution of neighborhoods. To do this, the topology of the activity locations were considered in respect of perceived neighborhood boundaries. In GIS, this can be operationalized by the process of containment and involves finding the points (location activities) that fall completely within a polygon (a neighborhood).[4] Each record in the CHTS travel logs had a latitude longitude and an index for the order in which the location was visited. For example, an individual may have gone from home (Place = 1), to work (Place = 2), to the grocery store (Place = 3), and then back home (Place = 4), with coordinates for each. If that place fell into one of the Mapping L.A. neighborhood boundaries, it assumed the attributes of the neighborhood polygon. Other GIS functionality involved spatial summary, where the neighborhood centroid was derived. This facilitated geovisualization, but also allowed for spatially connecting neighborhoods once the networks were derived.

3.3 Network Derivation

After data underwent GIS processing, we were able to express individual movements from one neighborhood to another by chaining sequential visits. Extending the previous example, for the individual that went to four destinations in 24 hours, three new records were created that listed the neighborhood names of Place 1 and Place 2, Place 2 and Place 3, and Place 3 and Place 4. After completing this for all individuals, the new records were then combined into separate edge lists according to race. An edge list is a data structure that contains columns for the start and end nodes of a network or, in our case, the origin and destination neighborhoods in which people traveled (Butts et al. 2008). Grouping

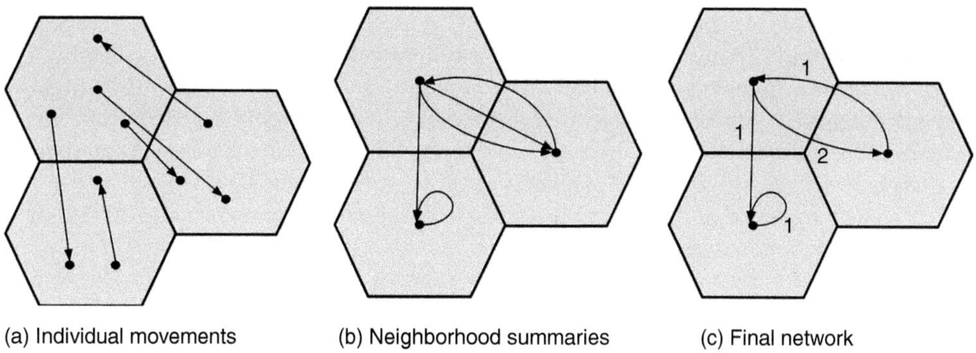

Source: The set E in (c) contains four edges, with three edge weights equal to 1 and one edge weight equal to 2.

Figure 16.2 A summary of network derivation: (a) individual movements, (b) neighborhood summaries and (c) final network

by unique neighborhood pairings, we derived edge weights equal to the total number of individuals who traveled between those two areas. This was repeated for each race, resulting in six directed networks.

Mathematically, a network graph is represented by $G = (V, E)$ where, $V = \{v_1, v_2, \ldots v_i, v_j\}$ is the set of point abstractions of the neighborhood centroids, and $E = \{e_1, e_2, \ldots. e_k\}$ is the set of edges representing travel between two neighborhoods. Because our networks are directed, an edge can be defined between any two neighborhoods i and j as $e_k = \overrightarrow{v_i v_j}$. The edge weight, equal to the sum of trips taken between two neighborhoods for a group, can be represented by $w(e_k)$. Figure 16.2 is a visual summary of the network derivation process.

3.4 Exploratory Spatial Data Analysis

Given a spatial network(s), exploratory spatial data analysis is essential for developing insights. Exploratory spatial data analysis is a framework for summarizing, assessing and interacting with spatial data, often in a graphical environment (Anselin 1999). It is based on and an extension of GIS, allowing a user to instantaneously perform operations such as selecting, deleting or rotating aspects of spatial objects (Anselin 1999).

More specifically, ESDA allowed us to: analyze simultaneously (1) trip derived information, such as the average number and total distance of daily trips for individuals in a group, and (2) visualize the derived networks of group activity. Within each network visualization, the Mapping L.A. neighborhoods provide geographic context. The results of ESDA are in section 4.1.

3.5 Summary Statistics

Network-based statistics were derived to quickly assess the characteristics and properties of the group travel activity. These included the degree, size, diameter, and clustering coefficient of a network, all of which provide some notion of connectivity (Butts et al. 2008;

Dehmer et al. 2014; Newman 2010). Measures of connectivity map well onto the concepts of segregation and urban mobility in that we can imagine that less connected neighborhoods are more segregated or may contain a population with low mobility, while highly connected neighborhoods may be less segregated or contain a highly mobile population. Extending the mathematical notation used in section 3.3, we define the following statistics for each network (Butts et al. 2008; Dehmer et al. 2014; Newman 2010):

Size – The number of nodes $|V|$ in the network, though in our analysis we give both the number of nodes and edges $|E|$.

Average degree – In a directed network, the ratio of the number of edges $|E|$ to the numbers of nodes $|V|$:

$$deg_{avg} = \frac{|E|}{|V|} \tag{16.1}$$

Diameter – The longest of all possible shortest path distances (denoted by d) between any two nodes v_i and v_j in the network:

$$diameter = max_{v_i, v_j}\, d(v_i, v_j) \tag{16.2}$$

Density – The ratio of the number of observed edges to the number of all possible edges:

$$density = \frac{|E|}{|V| * (|V| - 1)} \tag{16.3}$$

Clustering coefficient (global) – The probability that adjacent nodes of a node are connected. The local clustering coefficient is calculated first:

$$C_{v_i} = \frac{\text{number of completely connected triples centered at node } i}{\text{number of any connected triples centered at node } i} \tag{16.4}$$

Then, the global clustering coefficient C is given by:

$$C = \frac{1}{|V|} \cdot \sum_{v_i \in V} C_{v_i} \tag{16.5}$$

3.6 Neighborhood-interaction Potential

The measures outlined in section 3.5 offer some insight on the internal dynamics and connectivity of travel activity by race in Los Angeles County and Orange County on an individual neighborhood basis, but are limited in various ways. In this section we propose a measure of neighborhood-interaction potential, designed to better assess urban mobility and segregation simultaneously. Since the nodes in associated networks are point abstractions of neighborhoods, they contain geographic meaning that can be leveraged across neighborhoods. To incorporate this geographic information, we propose the neighborhood-interaction potential. The measure can be thought of as the amount of overlap between two derived travel networks, as it gives the proportion of trips that occur between the same neighborhoods in respect of all other trips taken by the two groups. The idea of neighborhood-interaction potential is related to the network science concept of assortativity and subgraph isomorphism in graph theory. Assortativity describes similarity between nodes that share a common value and is often expressed as some form of the Pearson correlation coefficient (Newman 2010). While a Pearson correlation coefficient would only measure similarities between nodes with similar degree or categorical label within the same network, the measure for neighborhood-interaction potential allows for comparison of edges between the same nodes (neighborhoods) across different network configurations.

Similarly, subgraph isomorphism is agnostic to how nodes exhibit relationships across different networks. Subgraph isomorphism characterizes components within a network that are structurally similar. For relational data, this is the process of graph-matching within a network to find unconnected subgraphs that have functionally similar roles (Dehmer et al. 2014). Subgraph isomorphism was not created to compare networks that have components of structural similarity at the same locations in geographic space. There has been research in geography, public health, and so on that conceptually deals with spatially explicit networks, point overlap or the congruence of activity spaces (Koblin et al. 2017; Murray 2000; Wang et al. 2015). However, as mentioned in section 2, most of this work only covers very broad or approximated spatial contexts, such as only looking at neighborhoods in which individuals live or pursue sexual relations or using cell towers to map out activity spaces. Even with a variety of applications, data and metrics for analyzing spatial social networks are limiting (Sarkar et al. 2019). The proposed neighborhood-interaction potential has both spatial and relational elements by uniquely comparing differences between networks at the same locations. Through this network overlap, we can both summarize and give a proxy for how spatial interactions may differ between neighborhoods across racial groups.

Building off of the mathematical notation in sections 3.3 and 3.5, the networks of two groups A and B are given by:

$$\begin{aligned} G_A &= (V_A, E_A) \\ G_B &= (V_A, E_B) \end{aligned} \tag{16.6}$$

A new graph, G_Z, can be derived such that:

$$\begin{gathered} G_Z = G_A \cap G_B \mid w_A(e_k) > 0 \text{ and } w_B(e_k) > 0 \\ \forall e_k \in E_Z \end{gathered} \tag{16.7}$$

That is, we want to find the intersection of two networks by finding the set of edges from both networks that have edge weights greater than zero on the same edges. Then, the overlap between network A with network B is derived by dividing the sum of edge weights in network A by the sum of weights in the network intersection:

$$\frac{\sum_{e_k \in E_A} w_A(e_k)}{\sum_{e_k \in E_Z} w_Z(e_k)} \tag{16.8}$$

Conversely, the overlap between network *B* with network *A* is:

$$\frac{\sum_{e_k \in E_B} w_B(e_k)}{\sum_{e_k \in E_Z} w_Z(e_k)} \tag{16.9}$$

The results of the differences in neighborhood-interaction potential are given in section 4.3.

4. RESULTS

4.1 Exploratory Spatial Data Analysis

Exploratory spatial data analysis helped us assess the representativeness of the survey populations against census information and draw some initial conclusions on our network derivations. The largest population in the sample was of individuals who identified as White, followed by those who identified as Hispanic, Asian, Black or African American, American Indian or Alaska Native, and Native Hawaiian or Pacific Islander. There was a fairly even spread between women and men. Similar to census findings, Hispanic children and young adults made up a disproportionate number of all individuals under the age of 5, between 5 and 17, and between 18 and 24. Of individuals over the age of 65, the number of those who identified as White was much higher than the number of those who identified as other races. The respondents who reported being American Indian or Alaska Native or Native Hawaiian or Pacific Islander were under 200 (85 and 119 respectively), while the other groups numbered well over 1000 (1400 for Black or Africans Americans, 2326 for Asians, 8340 for Hispanics and 14350 for Whites).

In comparison with the state of California averages, citizens of Los Angeles and Orange Counties took more trips and traveled longer distances (Kunzmann and Daigler 2013). As seen in Table 16.3, most residents from this survey in a day traveled between 27 and 67 miles over four to seven trips. When the networks for American Indians or Alaska Native and Native Hawaiians or Pacific Islanders were visualized, it became apparent that these groups were clear outliers with extremely sparse networks (27 and 44 neighborhoods with travel activity for the Native Hawaiian or Pacific Islander and

Table 16.3 LA and Orange County CHTS data by race and average daily trips and distance traveled

Race	Average number of trips	Average total distance traveled (miles)
Asian	5.5654	32.7697
Black or African American	6.1367	29.6173
Hispanic	6.1100	27.0676
White	5.9983	41.3075

American Indian or Alaska Native groups, versus 167, 214, 286 and 292 for the Black or African American, Asian, Hispanic, and White groups, respectively). In particular, individuals who identified as American Indian or Alaska Native traveled much farther than any other group, on average 66.1911 miles. Owing to this finding and the low sample sizes, we decided that comparisons with the American Indian or Alaska Native and Native Hawaiian or Pacific Islander populations would be difficult, and removed these groups from further analyses.

Visualizations of the travel activity networks for the remaining populations are given in Figure 16.3. An initial observation is of the wide geographic coverage of the Asian, Hispanic and White networks. The Hispanic network in particular appears to be the most dense, though the White network seems to have more edges with higher travel volume. Hardly any travel by Black or African Americans occurs in Orange County but seems to tightly cluster in the urban center of Los Angeles County, an area that is noticeably sparse in the Asian and White networks.

4.2 Summary Statistics

After deriving and visualizing the networks, we then calculated summary statistics. Table 16.4 shows the size (vertices and edges), average degree, density, diameter and global clustering coefficient for each network. Despite differences in size, there are some common characteristics across all networks. One shared aspect is their sparseness, with each network falling far below the number of possible edge connections (as seen by the low density values of Table 16.4). The low clustering coefficient for all networks may also be indicative of individuals engaging in a lot of within-neighborhood activities. As connectivity between neighborhoods increases, the diameter of the network should decrease, since there should be several options between neighborhoods to access other neighborhoods. Our results show that this is the case: the diameter of the networks are inversely proportional to the population sizes and network sizes.

4.3 Neighborhood-interaction Potential

Table 16.5 gives the matrix of pairwise neighborhood-interaction potentials. The matrix can be read as 'what is the likelihood that group A (a group in the first column) took trips to and from the same neighborhoods as group B (a group in the first row)?' Values in the bottom-left triangle of the matrix correspond to the formula specified in equation 16.8, and values in the upper-right triangle of the matrix correspond to the formula specified in equation 16.9. An important note is that the diagonal of the matrix is 1, validating that the measure correctly describes the potential for neighborhood interaction when looking at the same group. The highest neighborhood-interaction potential occurs between the Asian network with the White network, with Asians conducting a nearly 89 percent overlap in the amount of travel between the same neighborhoods as Whites. The next highest neighborhood interaction value occurs between Asians and Hispanics, with the Asian network experiencing a nearly 83 percent overlap with the Hispanic network. The lowest potential for neighborhood interaction occurs between Hispanics and Black or African Americans and Whites and Black or African Americans, both at about 48 percent.

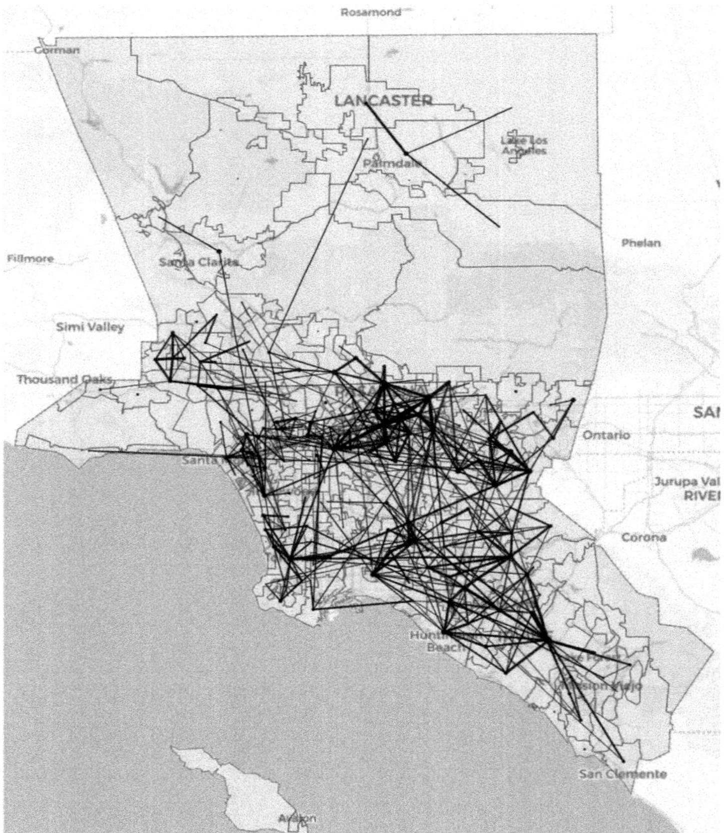

(a) Travel activity for Asian individuals

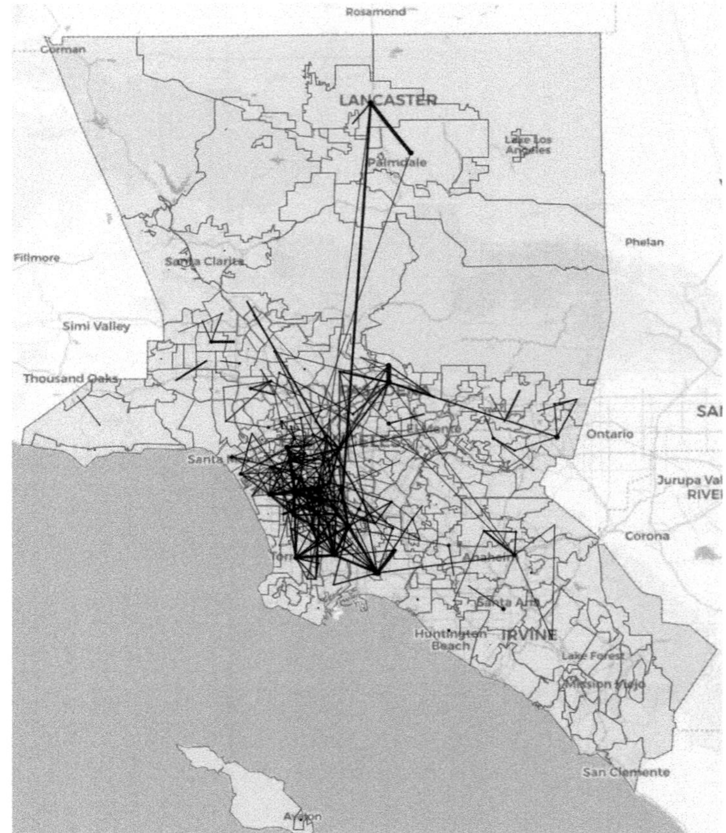

(b) Travel activity for Black or African American individuals

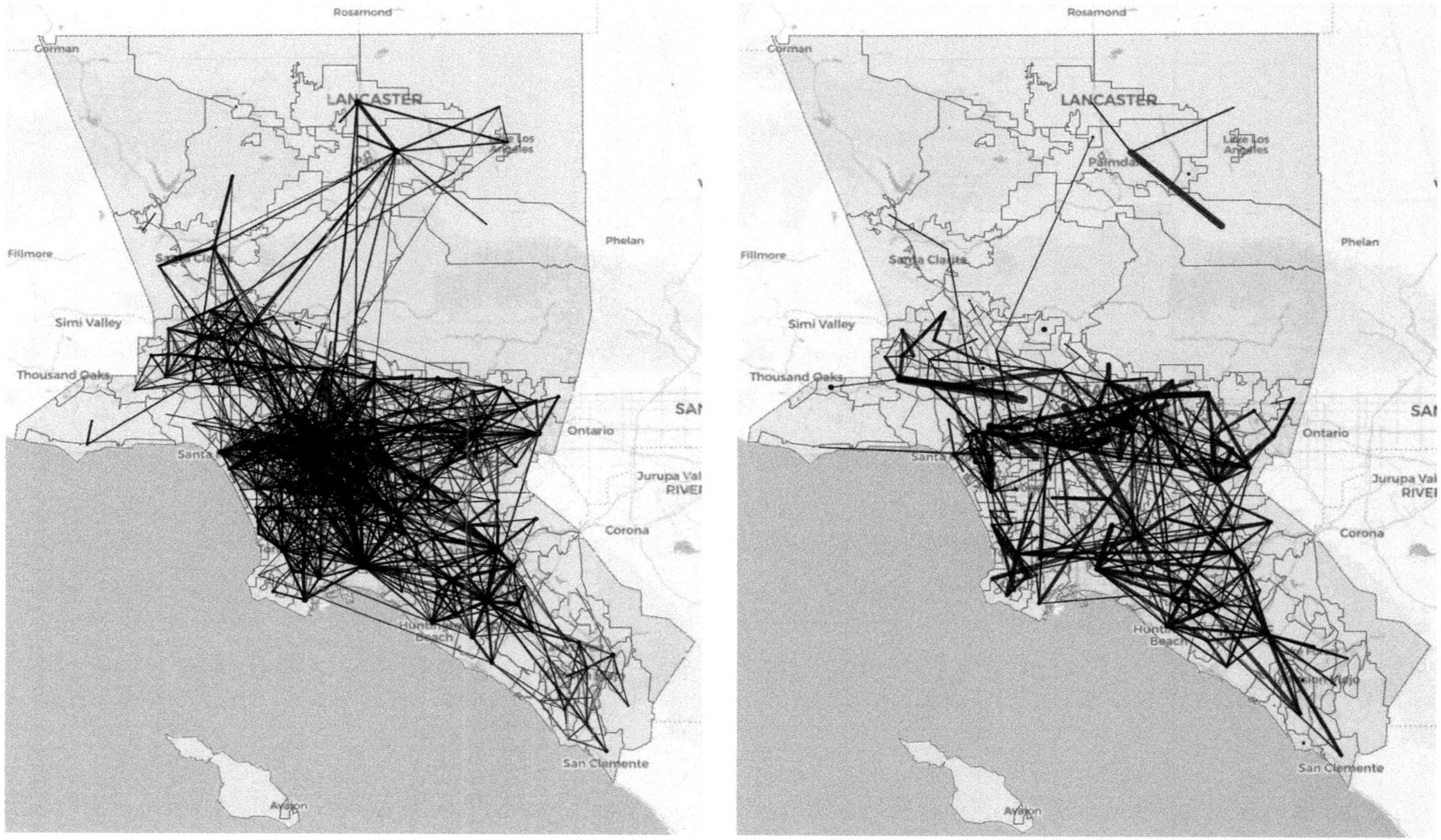

(c) Travel activity for Hispanic individuals

(d) Travel activity for White individuals

Note: The minimum edge weight is 2 and the maximum edge weight is 1671.

Figure 16.3 Travel networks by race, with the edge width proportional to the logarithm of the edge weight

Table 16.4 Summary statistics

Race	Vertices	Edges	Average degree	Density	Diameter	Clustering coefficient
Asian	214	908	4.2430	0.0198	13	0.2429
Black or African American	167	615	3.6826	0.0221	16	0.2058
Hispanic	286	2587	9.0455	0.0316	8	0.2398
White	292	3511	12.0240	0.0412	6	0.3110

Table 16.5 Neighborhood-interaction potential matrix

	Asian	Black or African American	Hispanic	White
Asian	1.0000	0.4918	0.8266	0.8825
Black or African American	0.5587	1.0000	0.8321	0.7319
Hispanic	0.5715	0.4798	1.0000	0.7822
White	0.6965	0.4770	0.8134	1.0000

5. DISCUSSION

For the network representing travel by White individuals, the ESDA and summary statistics tell a story of high connectivity. After removing the American Indian or Alaska Native and Native Hawaiian or Pacific Islander populations, the White group on average traveled to the most neighborhoods and longer distances than any other group. The average degree for the network is about 12, nearly three times larger than the Asian and Black or African American networks. High connectivity appears in the network visualizations as well: the White network is the most geographically expansive, spanning 292 different neighborhoods.

The breadth of the travel activity by Hispanic individuals is not far behind that of the White network. The Hispanic group on average traveled the shortest of distances of all racial groups, and yet the expanse of their travel covers 286 different neighborhoods. This finding is particularly worth noting given the differences in the White and Hispanic populations sizes and looking at the network visualizations. There are almost 15 percent and 49 percent more Whites than Hispanics in Los Angeles County and Orange County, respectively, but their travel networks by the number of neighborhoods is quite close. Still, despite a difference of only six neighborhoods from the Hispanic network, there are 924 more edge connections in the White network. While this may not seem terribly surprising given the large population size, it may partly explain how Whites are afforded relatively high urban mobility owing to residential and work or social dispersion. Another reason for this difference may be that Hispanic individuals also have a high amount of residential dispersion throughout the study area, but they may only need to travel within their neighborhood or to nearby neighborhoods for work or social activities.

While White and Hispanic individuals experience the largest geographic expanse and connectivity among the neighborhoods they visit, this is not the case for the network representing Black or African American individuals, the smallest group in the sample. Although Blacks and African Americans on average traveled a daily total distance slightly higher than Hispanics, the diameter of their network was double. The network visualization confirms this image of extremely low mobility, and provides further support for the previous work characterizing the hypersegregation of Blacks and African Americans. This can also be seen in the low average degree, which may imply that most activity takes place within the neighborhood, or that very few adjacent neighborhoods in which Blacks or African Americans visit are connected. The lack of connection may be because of poor transportation infrastructure, lack of accessible economic opportunities, social exclusion or any number of reasons.

The Asian population, the second smallest population in the sample, seems to experience some of the lowest urban mobility of all the groups. Individuals who identified as Asian took on average the fewest number of trips daily but had the second highest average for total daily distance traveled. The density of the Asian network was the lowest of all networks but the clustering coefficient was the second highest. It seems that while individuals in this network frequent the same clusters of neighborhoods, these neighborhoods are more geographically dispersed. The increase in travel costs because of distance may be preventing Asian individuals from engaging in more activities outside of the home.

Adding the values of the neighborhood-interaction potential allows us to look at how two given racial groups are exposed to one another throughout their travels. An interesting dynamic appears when comparing the White, Hispanic, and Asian travel networks. While the proportion of the White network that overlaps with the Hispanic network is strong (about 81 percent), the inverse is slightly weaker (about 78 percent). The same effect takes place for the Asian-White network overlaps: while the proportion of the Asian network that overlaps with the White network is very strong (again about 89 percent), the proportion of the White network that overlaps with the Asian network is much weaker (about 70 percent). The difference in these travel inconsistencies can partly be explained by the varying population sizes, but may also be indicative of the need for particular racial groups (such as Hispanics and Asians) to visit co-ethnic or co-racial neighborhoods in order to access cultural venues to which fewer White individuals go. However, even with starkly contrasting population sizes, the Asian and White networks generally share high neighborhood-interaction potential.

Unfortunately, the results for the neighborhood-interaction potential also show that travel by African American or Black individuals is unequivocally segregated. The neighborhood-interaction potential values for every race are lowest with the Black or African American network. While this group does have the smallest population, the CHTS data shows that, even though the Asian and Black or African American populations only differ by about 2 percent, there is over a 10 percent difference in the neighborhood-interaction potentials between Hispanics and Blacks or African Americans and Whites and Black or African Americans. It is interesting to note that the neighborhood-interaction potential of Blacks with Hispanics is higher (83 percent) than with Whites (73 percent). Overall the neighborhood-interaction potential values show that of all the places Black or African Americans go, they are most likely to see Hispanics, but for every other race and everywhere they go, they are least likely to see Black or African Americans.

We can dig further into the networks to show which neighborhoods are the sites of racial diversity.[5] Our results show that of any race, Whites seem to be the most self-segregating. There are many neighborhoods which only individuals who identify as White visit, but no neighborhoods which only one non-White race had visited, based on the survey data. The neighborhoods recording only visits by White individuals include Agua Dulce, Angeles Crest, Cleveland National Forest, Elizabeth Lake, Hasley Canyon, Hidden Hills, Laguna Coast Wilderness Park, Northwest Antelope Valley, Unincorporated Catalina Island and Westlake Village. It is notable that several of these neighborhoods (Angeles Crest, Cleveland National Forest, Elizabeth Lake, Hasley Canyon, Laguna Coast Wilderness Park, Antelope Valley and Unincorporated Catalina Island) have large areas devoted to recreational spaces while also being areas of low residential population. It could be that White individuals have the strongest capacity to self-segregate because they have the largest population and the most mobility, travel or otherwise.

5.1 Policy Planning and Decision-making

The results of our analyses have a number of implications for policy planning and decision-making. The first is that the continued hypersegegreation of Blacks and African Americans cannot be ignored. With such a geographically constrained travel network, Blacks and African Americans may not have exposure or access to vital health and economic resources, and counties like Los Angeles and Orange County must remain cognizant of this. Given how reliant this part of Southern California is on automobiles, the distance by which different demographic groups travel may also require thoughtful consideration. For example, if Asians are one of the smallest groups in Los Angeles County and Orange County yet travel farther on average than other groups, the counties might consider how automobile policy changes (for example, laws on carpooling or highway improvement projects) might affect these populations together with the neighborhoods in which they tend to cluster. Finally, if it is indeed the case that recreational areas tend to be spaces of exclusion for certain groups, then the cities, counties and organizations involved in providing services encompassing open spaces and recreational areas might encourage policies to promote broader usage and interaction with these areas.

5.2 Limitations and Future Work

There are a few limitations to our methodology that may be improved in future work and which are highlighted here. The first concerns the issue of time: the movements between neighborhoods that provided the foundation for deriving our networks were only considered in a two time-period framework. Furthermore, we treated all neighborhood transitions as occurring at the same time, and the variability of time periods is not being considered in this analysis. Future work should try to further anatomize travel activity by time. In addition, it is unfortunate that there were so few data points for the American Indian or Alaska Native and Native Hawaiian or Pacific Islander populations, especially when exploratory spatial data analysis showed that America Indians and Alaska Natives were on average traveling nearly three times farther daily than any other racial group. The public policy, health and economic implications mentioned in section 5.1 are equally relevant here, and future work should be devoted to these underrepresented groups.

6. CONCLUSIONS

The combination of exploratory spatial data analysis, summary statistics and our neighborhood-interaction potential give depth and variety to the examination of urban mobility and segregation in Los Angeles and Orange County. The travel networks that were derived have elements of both relational and geometric networks, and the strength of this construction is that we could apply existing network techniques (our summary statistics) while embedding geographic meaning (particularly using ESDA) and explicitly determine values for neighborhood-interaction potential. Our results exposed notable disparities in urban mobility and interaction potential between different races, such as that the White and Hispanic populations experienced similar levels of geographic spread but that Whites traveled to far more neighborhoods than Hispanics, perhaps implying similar amounts of residential dispersal but higher within-neighborhood travel for Hispanics. However, Asians and Black or African Americans had varying levels of clustered traveling, but while Asians traveled to more clustered neighborhoods that were geographically dispersed, Blacks and African Americans traveled to and within a very confined set of neighborhoods. It was interesting to note that despite population differences of only around 2 percent between Asians and Blacks or African Americans in parts of the study area, there was nearly a 10 percent difference in potential for interaction with Blacks or African Americans than Asians for nearly all other racial groups. The finding that many of the neighborhoods in which only Whites visit are recreational areas also has important policy implications for how these sites act as spaces of exclusion for particular races, and cities and counties should try to assess how these public spaces can be used more equitably. Finally, as mentioned in section 5.2, the urban mobility of American Indians or Alaska Natives and Native Hawaiians or Pacific Islanders should be more strongly considered in the future, as there are undoubtedly important mobility disparities in these underrepresented groups that merit attention. Regardless, the objective of the presented methodology was to better incorporate analyses of urban mobility and segregation together through a network approach, and measures such as the neighborhood-interaction potential can help decision-makers more succinctly realize the extent to which urban mobility and segregation changes throughout an entire area between different racial groups.

ACKNOWLEDGEMENTS

The authors would like to thank Dr Adam Davis and the GeoTrans Laboratory at the University of California, Santa Barbara, for providing a cleaned and collated version of the California Household Travel Survey data. We would also like to thank the National Science Foundation for their financial support and training through the Network Science and Big Data program.

NOTES

1. The respondents who were given the GPS device produced between three and seven days worth of information, however, the final survey publishes only a representative day for each individual. More on how this

was determined can be found in the California Household Survey final report: http://www.dot.ca.gov/hq/tpp/offices/omsp/statewide_travel_analysis/Files/CHTS_Final_Report_June_2013.pdf (accessed 1 August 2019).

2. In the CHTS, Hispanic is given as a racial designation in an other category even though Hispanic is not a race and was intended to encompass a broader group of individuals who speak Spanish. To preserve the integrity of the data collection, we retain this designation in the hope that it appropriately captures the social construct of race, ethnicity and culture together.
3. As mentioned in section 3.1, the United States Census Bureau recognizes both counties as comprising the Los Angeles Metropolitan Area (https://factfinder.census.gov/faces/tableservices/jsf/pages/productview.xhtml?src=bkmk, accessed 1 August 2019) and the Mapping L.A. boundaries match this geographic extent.
4. More specifically, this was operationalized in R using the package sp (Pebesma et al. 2018).
5. To determine the most and least diverse neighborhoods based on the proportion of travel activity of each race, we simply summed the total number of visits in each neighborhood and calculated the percentage of visits by each race. In the case of integration, we wanted to minimize the distance between all of these values. If each column in a neighborhood table represents the percentage of visits by each race, we used the Euclidean (or L2) norm of all column differences to find the neighborhoods that are the most integrated, or conversely, most segregated. Mathematically, if we represent the percentage of visits in a neighborhood by each of four racial groups *A*, *B*, *C*, and *D*, then the Euclidean norm $|D|$ of differences of visits by race for neighborhood *i* is equal to:

$$|D_i| = \sqrt{(A-B)^2 + (A-C)^2 + (A-D)^2 + (B-C)^2 + (B-D)^2 + (C-D)^2} \tag{16.10}$$

The neighborhoods with the lowest values of $|D_i|$ would be considered the most diverse, and the neighborhoods with the highest values of $|D_i|$ would be the least diverse.

REFERENCES

Alexander, M. (2012), *The new Jim Crow: Mass Incarceration in the Age of Colorblindness*, New York: New Press.

Anselin, L. (1999), 'Interactive techniques and exploratory spatial data analysis', *Geographical Information Systems: Principles, Techniques, Management and Applications*, **1** (1), 251–64.

Anselin, L. and A. Getis (1992), 'Spatial statistical analysis and geographic information systems', *Annals of Regional Science*, **26** (1), 19–33.

Barabási, A.-L. (2003), *Linked: The New Science of Networks*, New York: Perseus.

Butts, C., D. Hunter, M. Handcock, S. Bender-deMoll, J. Horner and L. Wang (2008), 'network: a package for managing relational data in R', *Journal of Statistical Software*, **24** (2), 1–36.

Carrasco, J.A., B. Hogan, B. Wellman and E.J Miller (2008), 'Collecting social network data to study social activity-travel behavior: an egocentric approach', *Environment and Planning B: Planning and Design*, **35** (6), 961–80.

Chaskin, R.J. (1997), 'Perspectives on neighborhood and community: a review of the literature', *Social Service Review*, **71** (4), 521–47.

Cheba, K. and S. Saniuk (2016), 'Urban mobility–identification, measurement and evaluation', *Transportation Research Procedia*, **14**, 1230–39.

Dehmer, M., V. Kraus, F. Emmert-Streib and S. Pickl (2014), *Quantitative Graph Theory*, Baton Rouge, FL: CRC Press.

Glaeser, E. and J. Vigdor (2012), 'The end of the segregated century: racial separation in America's Neighborhoods, 1890–2010', Technical Report 66, January, Manhattan Institute, New York.

Hall, M., J. Iceland and Y. Yi (2019), 'Racial separation at home and work: segregation in residential and workplace settings', *Population Research and Policy Review*, **38** (5), 1–24.

Humes, K.R., N.A. Jones and R.R. Ramirez (2011), 'Overview of Race and Hispanic Origin: 2010', technical report, March, United States Census, Suitland-Silver Hill, MD.

Johnson, C. and R.P. Gilles (2003), 'Spatial social networks', in B. Dutta and M.O. Jackson (eds), *Networks and Groups*, Heidelberg: Springer, pp. 51–77.

Johnston, R., M. Poulsen and J. Forrest (2006), 'Modern and post-modern cities and ethnic residential segregation: is Los Angeles different?', *Geoforum*, **37** (3), 318–30.

Jones, M. and A.R. Pebley (2014), 'Redefining neighborhoods using common destinations: social characteristics of activity spaces and home census tracts compared', *Demography*, **51** (3), 727–52.

Koblin, B.A., J.E. Egan, V. Nandi, J.M. Sang, M. Cerdá, H.-V. Tieu, et al. (2017), 'Congruence of home, social

and sex neighborhoods among men who have sex with men, NYCM2M study', *Journal of Urban Health*, **94** (3), 364–74.
Kunzmann, M. and V. Daigler (2013), '2010-2012, California Household Travel Survey final report', technical report, June, California Department of Transportation, Austin, TX.
Kwan, M.-P. (2013), 'Beyond space (as we knew it). Toward temporally integrated geographies of segregation, health, and accessibility: space–time integration in geography and GIScience', *Annals of the Association of American Geographers*, **103** (5), 1078–86.
Los Angeles Times (2013), 'Mapping L.A.', *Los Angeles Times* project, September, accessed 1 August 2019 at http://maps.latimes.com/about/.
Massey, D.S. and N.A. Denton (1989), 'Hypersegregation in US metropolitan areas: Black and Hispanic segregation along five dimensions', *Demography*, **26** (3), 373–91.
Murray, A.T. (2000), 'Spatial characteristics and comparisons of interaction and median clustering models', *Geographical Analysis*, **32** (1), 1–18.
Newman, M. (2010), *Networks: An introduction*, Oxford: Oxford University Press.
Pebesma, E., R. Bivand, B. Rowlingson, V. Gomez-Rubio, R. Hijmans, M. Sumner, et al. (2018), 'sp: classes and methods for spatial data', technical report, June, Comprehensive R Archive Network (CRAN).
Ratti, C., S. Sobolevsky, F. Calabrese, C. Andris, J. Reades, M. Martino, et al. (2010), 'Redrawing the map of Great Britain from a network of human interactions', *PloS One*, **5** (12), e14248.
Ray, B. and V. Preston (2009), 'Geographies of discrimination: variations in perceived discomfort and discrimination in Canada's gateway cities', *Journal of Immigrant & Refugee Studies*, **7** (3), 228–49.
Redford, L. (2017), 'The intertwined history of class and race segregation in Los Angeles', *Journal of Planning History*, **16** (4), 305–22.
Reilly, K., S. Kaufman and A. Bodino (2002), *Racism: A Global Reader*, New York: M.E. Sharpe.
Richardson, C.M. and R.E. Luker (2014), *Historical Dictionary of the Civil Rights Movement*, Lanham, MD: Rowman & Littlefield.
Sarkar, D., C. Andris, C.A. Chapman and R. Sengupta (2019), 'Metrics for characterizing network structure and node importance in spatial social networks', *International Journal of Geographical Information Science*, **33** (5), 1–23.
Schelling, T.C. (1971), 'Dynamic models of segregation', *Journal of Mathematical Sociology*, **1** (2), 143–86.
Sharp, G., J.T. Denney and R.T. Kimbro (2015), 'Multiple contexts of exposure: activity spaces, residential neighborhoods, and self-rated health', *Social Science & Medicine*, **146** (December), 204–13.
United States Census Bureau (2019), 'QuickFacts: Los Angeles County, California; Orange County, California', June, technical report, United States Department of Commerce, Washington, DC.
Wang, Q., N.E. Phillips, M.L. Small and R.J. Sampson (2018), 'Urban mobility and neighborhood isolation in America's 50 largest cities', *Proceedings of the National Academy of Sciences*, **115** (30), 7735–40.
Wang, Y., C. Kang, L. Bettencourt, Y. Liu and C. Andris (2015), 'Linked activity spaces: embedding social networks in urban space', in M. Helbich, J.J. Arsanjani and M. Leitner (eds), *Computational Approaches for Urban Environments*, Cham: Springer, pp. 313–36
Zenk, S.N., A.J. Schulz, S.A. Matthews, A. Odoms-Young, J. Wilbur, L. Wegrzyn, et al. (2011), 'Activity space environment and dietary and physical activity behaviors: a pilot study', *Health & Place*, **17** (5), 1150–61.

17. The impact of urban social life on sexual networks and health

Patrick Janulis and Michelle Birkett

INTRODUCTION

It probably comes as no surprise that sexual partnerships are not random; physical attraction, personality and shared values are just some of the factors that drive an individual's selection of sexual partners. Although we are conscious of many of the factors that drive selection, sexual relationships – as with all social relationships – are shaped also by social norms and environmental constraints, some of which we might be entirely unaware of (Bearman et al. 2004). Furthermore, individual partnership decisions, driven by these preferences, norms and constraints, build upon each other to form complex patterns. When viewed at the population level, these complex patterns have substantial consequences for health and the spread of disease.

This chapter details the norms, preferences and constraints that shape sexual partnership selection, how each of these interact with both physical place and population, and how individual decisions about partnership selection can create patterns that have substantial consequences for the spread of disease, particularly in urban populations. In discussing this process, we focus particularly on social network analysis, as it allows us to describe the specific relational mechanisms that lead to disease spread. We organize our discussion of these specific mechanisms into two distinct sections: network structural mechanisms and network compositional mechanisms. Following the discussion of these network mechanisms, we explore how social life, venues and online/mobile applications may impact sexual networks in urban regions, and then end on a brief discussion of limitations and future directions in this work. Each section surveys a wide range of literature on the formation and dissolution of sexual partnerships, and describes the impact of sexual networks on the spread of sexually transmitted infections (STIs) – in particular human immunodeficiency virus (HIV). We end with a discussion of how these network mechanisms are influenced by the real world and explore why, in order to understand disease dynamics, it is important to first have in-depth knowledge of the population of interest and the relevant social contextual forces impacting that population.

SEXUAL NETWORKS, POPULATION DISPARITIES AND HIV

The study of sexual contact networks to understand the flow of disease through a population has grown rapidly since the early days of the HIV epidemic. The foundation of this work can be seen in both the contact tracing widely used in epidemiology and famously applied to track the first cluster of acquired immune deficiency syndrome (AIDS) patients

(Auerbach et al. 1984) and in early use of mathematical modeling to better understand the spread of HIV/AIDS (Anderson et al. 1986; Hyman and Stanley 1988). In particular, modelers quickly recognized the challenge of using traditional epidemic modeling, which relied on random mixing, when complex contact patterns were identified as being instrumental in the spread of HIV (Hethcote and Yorke 1984; Hyman and Stanley 1988; Klovdahl 1985). For a comprehensive look at the early work on social and sexual networks and the spread of disease, see Morris (1993). Research examining sexual networks now provides substantial evidence documenting their enormous impact on the severity and longevity of STI epidemics as well as on disparities within diseases (Friedman et al. 2009; Keeling and Eames 2005). We emphasize that although this chapter attempts to provide a thorough overview on the topic of urban sexual networks and their impact on the spread of STIs, the breadth and diversity of the literature on sexual networks precludes us from providing a comprehensive review of this work.

Many of the topics covered in this chapter are relevant to the epidemiology of multiple STIs (such as chlamydia or syphilis). However, owing to its particular salience, examples will be largely drawn from the literature on the US HIV epidemic, with an emphasis on the sexual networks of men who have sex with men (MSM) in urban samples. We hope that these examples will highlight areas of research in which network analysis provides particularly powerful insight into topics which are not easily explained by traditional individual-level approaches. For example, there are substantial disparities in HIV incidence by race and ethnicity, both within the general population (Aral et al. 2008) and within specific subpopulations, such as MSM (Prejean et al. 2011). In particular, Black/African American MSM are at substantially greater risk of acquiring HIV than their White or Latino MSM peers (Centers for Disease Control 2016). However, meta-analyses suggest that Black MSM report equal or lower levels of the primary risk behaviors for HIV acquisition (such as receptive anal sex) relative to White MSM peers (Millett et al. 2007, 2012). Accordingly, the structure of sexual networks may help explain the magnitude and persistence of these disparities even without observing differences at the level of individual risk behavior. This example will be a persistent focus of our discussion given the enormous challenge of racial disparities across STIs and the importance of networks to understanding and addressing this phenomenon.

NETWORK MECHANISMS

This section describes relational patterns across two distinct network features: network compositional features and network structural features. The former refers to who tends to connect to whom and the latter describes structural patterns that effect the connectivity across sexual networks. Both of these have the ability to impact the spread of infectious disease.

Compositional

The study of sexual networks largely emerged from the recognition that random mixing inadequately described the formation of sexual partnerships. Consequently, a large body of work has examined the tendencies for sexual partnership to vary by partner characteristics. We examine next how partner similarity and dissimilarity across a variety of attributes drive infectious disease dynamics within sexual networks.

Homophily

A primary influence on sexual mixing patterns is a homophily, or a tendency for individuals with similar demographic characteristics to partner with each other (Adimora and Schoenbach 2013). These patterns in partner selection may develop owing to conscious preferences, or to unconscious opportunity structures that make it easier for like individuals to partner with each other, a feature witnessed across a wide variety of social networks (McPherson et al. 2001). Many studies have found substantial racial/ethnic homophily within sexual networks (Amirkhanian 2014; Hamilton and Morris 2015; Laumann and Youm 1999). For example, extremely high rates of racial homophily have been found among US urban samples of MSM (Birkett et al. 2015; Clerkin et al. 2011; Fujimoto and Williams 2015; Janulis et al. 2018b; Raymond and McFarland 2009). Similar results have even been found for studies outside the US and in non-MSM populations, such as in five different cities in sub-Saharan Africa (Kenyon and Colebunders 2013).

However, this does not mean that all populations report similar levels of racial/ethnic homophily. White MSM consistently report lower levels of racial homophily than do Black MSM (Amirkhanian 2014; Tieu et al. 2016). Higher levels of homophily within Black MSM may attenuate existing disparities in HIV prevalence, given that this network segregation may place Black MSM closer to other individuals who are HIV positive within these sexual networks (Amirkhanian 2014; Janulis et al. 2018b). However, it is important to note that studies have shown that homophily alone cannot sustain racial disparities indefinitely (Goodreau et al. 2017), and that other mechanisms are likely to contribute and should be explored.

Despite the large number of studies documenting racial homophily in sexual networks and their importance to sustaining STI disparities, far less attention has been paid to understanding the emergence and causes of this homophily. A few studies have attempted to document these declared preferences that may partially drive homophily in sexual networks. For example, studies indicate that Black MSM are viewed as the least preferred potential sexual partners by MSM of other races (Phillips et al. 2016; Raymond and McFarland 2009). These preferences are also frequently documented among individuals seeking sexual partners online (Paul et al. 2010; Phua and Kaufman 2003; Sweeney and Borden 2009). Consequently, racial hierarchies, racism and sexual stereotypes in preferred sexual partners may constrain sexual partner choices among racial minority MSM and promote network structures, such as racial homophily, that exacerbate the STI disparities witnessed in these groups (Phillips et al. 2016; Wilson et al. 2009).

Heterophily

While homophily may be a strong force in the selection of sexual partners, heterophily (that is, disassortative mixing) has also been documented along a number of sexual partner characteristics. Most notably, several studies have found disassortativity by age in both urban and non-urban heterosexual (Bearman et al. 2004; Doherty et al. 2009) and MSM (Bingham et al. 2003; Mustanski and Newcomb 2013) samples. For example, studies of adolescents have indicated a preference for older partners, particularly among females (Bearman et al. 2004; Lee et al. 2003; Morrison-Beedy et al. 2013), while other studies indicate age-discordant mixing may be associated with STI infections (Catania et al. 1996; Kraut-Becher and Aral 2006). Among MSM, studies have found conflicting results regarding whether age differences vary across race/ethnicity, with some finding no differences (Grey et al. 2015; Tieu et al. 2016; Whiteside et al. 2015) and others finding larger age disparities between partners of Black MSM (Berry et al. 2007; Mustanski and Newcomb 2013). On the whole, more age discordant mixing has been reported among MSM compared with their heterosexual peers (Glick et al. 2012).

Perhaps the most salient form of disassortative mixing for the spread of STIs is discordance by disease status, given that this type of mixing is required for disease to propagate through a network. While many studies have cataloged correlates of disassortative mixing by HIV status, we focus here on intentional assortative mixing by disease status, known as serosorting, used as a risk-reduction strategy (Eaton et al. 2009). Serosorting has been documented in a wide variety of studies of MSM (Purcell et al. 2017), mostly from urban samples, and may have increased in prevalence over time (Khosropour et al. 2016; Snowden et al. 2014). However, serosorting has been met with substantial controversy given that this technique relies on accurately knowing one's own and one's partner's serostatus, both of which are unlikely to be wholly accurate (Niccolai et al. 2002; Zablotska et al. 2009). Nonetheless, a recent review and meta-analysis (Purcell et al. 2017) found that serosorting increased risk relative to no condomless sex but decreased risk relative to condomless serodiscordant sex, suggesting serosorting is only an effective strategy among MSM who inconsistently use condoms.

Similarly, recent research into biomed sorting among urban and suburban MSM suggests that individuals may be engaging in intentional risk reductions strategies by only engaging in condomless sex with individuals either on pre-exposure prophylaxis (PrEP) or with an undetectable viral load (UVL) (Newcomb et al. 2016; Storholm et al. 2017). The study of biomed sorting may increase in importance as HIV preventions strategies increasingly rely on biomedical strategies (that is, PrEP use decreasing the probability of HIV negative individuals and reducing the likelihood of onward transmission by achieving a UVL for HIV-positive individuals). Yet, as in serosorting, biomed sorting also relies on accurate knowledge of both the individual's partner's and the individual's own viral load and or PrEP adherence, which is unlikely to always be the case (Mustanski et al. 2018; Sewell et al. 2017). Therefore, future research is required to comprehensively evaluate the relative benefit of biomed sorting as a successful HIV prevention strategy and its impact on the spread of other STIs (Nguyen et al. 2018).

Importantly, the aggregate effect of increasing use of PrEP and proportion of HIV individuals with a UVL (that is, treatment as prevention) implicitly relies on disassortative mixing by biomed status to obtain maximum effect. For treatment as prevention, this relies on increasing the number of virally suppressed individuals that would otherwise transmit HIV to an HIV-negative partner. For PrEP, this relies on increasing the number of HIV-negative individuals on PrEP that would otherwise be more likely to contract HIV from HIV-positive partners. However, population level impact of PrEP and treatment as prevention are likely to be reduced if network typologies are structured so that PrEP or antiretroviral therapy (ART) use are clustered within particular subpopulations. Indeed, recent modeling work examining this phenomenon found that ART clustering had a substantial impact on HIV incidence and differed from the effects of serosorting (Delva and Helleringer 2016). This may be especially problematic as this effect has the potential to exacerbate racial disparities given the substantial variation in ART and PrEP uptake and adherence among racial minority MSM (Calabrese et al. 2017; Rebeiro et al. 2017). However, the extent to which ART and PrEP clustering exist in sexual networks must be more thoroughly documented to better understand this phenomenon.

Structural

While the attributes of partners and compositions of sexual networks contribute to the spread of disease, the structure of sexual networks also has a substantial impact. In a seminal study, Bearman et al. (2004) examined the interplay of network structure with partner selection to examine the sexual network of a single high school. This study documented how norms against partner swapping (that is, dating a former partner's current partner's former partner) were essential to understanding the overall structure of long chain paths with few cycles that made this network conducive for the spread of STIs. This section explores network structure such as these and how they impact the spread of disease through sexual networks.

Degree disassortativity

We have already highlighted the importance of individual attributes, such as age and race, in shaping the likelihood for two individuals to form a sexual relationship, but another important structural feature which drives partnership selection, and which has been shown to do so across numerous studies, is dissortativity by degree. Disassortativity by degree describes a structural feature in which individuals with very few sexual partners are more likely to be sexually partnered with individuals with many sexual partners (Haraldsdottir et al. 1992; Laumann and Youm 1999). Although this property is a commonly documented feature of many networks, its consequences for the study of sexual networks may be especially important for disease transmission (Gupta et al. 1989; Jones and Handcock 2003). For example, Laumann and Youm (1999) documented how variation in the level of disassortative mixing by degree may contribute to racial disparities in HIV. They found that, among individuals with only a single sexual partner in the past year, African American participants were five times more likely to have a partner with four or more partners in the past year. Similarly, Haraldsdottir et al. (1992) found that MSM in Iceland reported disassortative mixing by degree, with MSM tending to partner with those who are dissimilar in their number of sexual partners, and that this mixing pattern was also found to increase the magnitude of simulated epidemics.

Monogamy, serial partnership and concurrency

Thus far we have largely ignored how the number and timing of sexual partners may impact the spread of STIs in sexual networks. Perhaps unsurprisingly, this area has received substantial attention in sexual network research. Monogamy is a common feature of sexual networks, with individuals showing a strong tendency to form a sexual connection with only one person at a time (Adimora and Schoenbach 2013). However, this tendency has probably decreased in recent decades with sizable minorities (for example, 17 per cent of women and 23 per cent of men among sexually active US adults) now reporting non-monogamy in the past year (Aral and Leichliter 2010). Monogamy is inherently protective against the spread of STIs given that only future partners are exposed to any infection that is contracted by either partner. However, even within serial monogamous partnerships (that is, multiple partners but only one partner at a time), the spacing between partners may have considerable influence on the spread of disease through these networks. For example, a study of clinic patients in England (Mercer et al. 2013) found that the median gap between sexual partners was 14 days for men and 24 days for women, well within the infectious periods for bacterial and viral STIs and within periods of highest viral load for individuals who recently acquired HIV. Similar results have been found in US urban samples (Foxman et al. 2006).

Monogamous partnerships can also be contrasted with concurrent partnerships, or the overlap of sexual partnerships in time. Concurrency has been shown to dramatically increase the spread of disease since it reduces the time between acquiring and passing the infection and eliminates the sequential protection of time, thereby potentially dramatically increasing the size of connected components in the network. Early simulation work by Morris and Kretzschmar (1997) established the important potential for concurrency to shape disease transmission. This work was subsequently verified in a wide variety of simulation and observational studies (Morris et al. 2010). This includes several studies that have found associations between concurrency and STI/HIV infection (Ellen et al. 2005; Hamilton and Morris 2015; Neaigus et al. 2013; Pines et al. 2016; Rosenberg et al. 1999). A study in five US cities (Glick et al. 2012), found that higher rates of concurrency among MSM may also partially explain the higher rates of STIs compared with those of heterosexuals, even with higher rates of condom use being reported among MSM. However, it is important to note that there is still a degree of uncertainty among researchers regarding the extent to which concurrency drives epidemics, particularly HIV (Aral 2010; Mah and Halperin 2010; Sawers and Stillwaggon 2010). Nonetheless, variation in concurrency across racial/ethnic groups has been implicated as a driver of racial disparities in HIV. This may be owing to differences in the timing of sexual partners explaining STI disparities even without differences in the total number of partners (Hamilton and Morris 2015; Krivitsky and Morris 2017). For example, Krivitsky and Morris (2017) found higher levels of concurrency among African Americans in a national sample of US adults. Networks simulated from these estimates also suggested African Americans were more likely to be in large connected components within sexual networks, thereby placing them at higher risk of contracting STIs. However, other studies have found mixed results comparing concurrency rates across racially diverse samples of urban MSM, with several studies finding no significant variation across race/ethnicity (Janulis et al. 2018b; Mustanski et al. 2015; Rosenberg et al. 2012) while others have found these differences (Bohl et al. 2009).

Although future work is necessary to determine if and when concurrency may contribute to racial disparities among MSM, some contradictions in this literature are likely owing to the complex number of ways concurrency can be measured (Helleringer et al. 2011; Kretzschmar and Morris 1996; Nelson et al. 2007). Also, the manner in which concurrency influences risk has been overlooked by many studies. That is, concurrency places an individual's partners at higher risk for STI infection rather than the individual reporting concurrency themselves (Neaigus et al. 2013; Rosenberg et al. 2012). Future research into concurrency must account for the complex relationship between concurrency and the spread of STIs, as well as be extremely careful to utilize measures of concurrency that fit the given research question.

SOCIAL LIFE, VENUES, GEOGRAPHY AND ONLINE/MOBILE APPLICATIONS

Thus far we have discussed structures that may drive the spread of STIs through sexual networks and, to a lesser extent, preferences or norms that may drive these network formations. However, constraints on urban social life have also been substantially implicated in the network formations that we have discussed. In particular, research has demonstrated how daily interactions in social spaces such as work, school and neighborhoods impact sexual partner selection. For example, Kalmijn and Flap (2001) document how partners who met at work or school may increase assortative mixing by constraining available partners choices to partners similar in various characteristics, such as education and socioeconomic status. In recent years, this work has been extended to documenting the impact of social venues, neighborhood and mobile phone applications on the spread of STIs within sexual networks.

In particular, HIV research has increasingly turned to the study of venues used to meet sexual partners as an important influence on sexual partner formation and location for potential intervention (Frost 2007; Weir et al. 2003). For example, studies in San Francisco (Stephens et al. 2014) and Jackson (Oster et al. 2013) have found that a large number of MSM meet their partner in a relatively small number of venues. Some venues may also increase racial homophily by catering specifically to certain racial groups such as White MSM (Choi et al. 2011). As for other forms of homophily, research has been divided regarding how venues may impact partnership homophily on sexual risk taking (for example, condomless sex). One study found that individuals with similar levels of sexual risk behavior were more likely to attend the same venues types (Young et al. 2017). However, others studies (Fujimoto et al. 2013) found the opposite effect, indicating heterophily by risk potentially owing to risk compensation where individuals attending risky venues take extra precautions to account for the higher level of risk. Other have used data to identify high-risk venues by aggregating viral loads of individuals who attend these venues (Brantley et al. 2017). Similarly, high-risk venues have identified key central venues within sexual and venue networks that were key to the spread of gonorrhea (De et al. 2004).

Similar to venues, geography of sexual partnerships within cities has also been examined in an attempt to understand the structure of sexual networks. For example, several studies have found spatial assortative mixing of sexual partners in cities (Rothenberg et al. 2005; Zenilman et al. 1999). Research has also found that long-distance sexual partnerships across regions and jurisdictions may be particularly impactful in later stages of STI epidemics (Hippe and Jolly 2012). Furthermore, neighborhood characteristics may influence the association between geography and sexual networks. For example, one study found that individuals residing in neighborhoods with high HIV-prevalence tended to have partners who were further away geographically, suggesting a potential dispreference for partners within these neighborhoods (Gindi et al. 2011). Similarly, higher levels of poverty have been associated with a greater likelihood to be in the connected component (Fichtenberg et al. 2010) and have concurrent partners (Adimora et al. 2013). As an alternative approach, others have examined the effect that individuals relocating neighborhoods has on sexual networks. Relocating to areas with greater economic advantage was shown to decrease concurrency (Cooper et al. 2015) and decrease the number of new partners within participants' sexual networks (Linton et al. 2017). Similar to the effect of relocation, incarceration may also impact sexual networks. For example, female-dominated adult sex ratios within African American communities, owing to incarceration and higher mortality among African American males, may increase disassortative mixing by degree that contributes to STI disparities in this population (Oser et al. 2017; Pouget 2017). Similarly, involvement within the criminal justice system has also been found to increase the turnover in an individual's social network that may negatively affect sexual health outcomes among MSM (Schneider et al. 2017).

However, there is evidence to suggest these geographically defined spaces may play a decreasing role in understanding sexual networks, given the rapid increase in use of the Internet and mobile applications (apps) to meet sexual partners. A study of MSM in Baltimore found that, between 2009 and 2014, the Internet became a more frequent location to find partners while bars/clubs became less frequent (Jennings et al. 2015). Another study of young MSM, in Chicago, found that a majority of partners in the past six months were met online or via mobile phone (Janulis et al. 2018a). Given the prevalence of meeting partners online, several studies have begun to incorporate these digital spaces into their network analysis, finding high levels of connectivity around a relatively small number of venues (Bradshaw et al. 2016; Brantley et al. 2017; Niekamp et al. 2013). However, methodologies used to understand physical spaces may not be compatible with mobile apps, given that the assumptions used for physical venues may not apply to digital spaces (for example, the probability of meeting a partner if an individual uses the same app). Accordingly, methods must grow and mature to account for these changing assumptions, particularly given the high level of adoption and importance these spaces will probably play in the spread of STIs.

CHALLENGES AND FUTURE DIRECTIONS

Despite the large accumulation of research on sexual networks that has greatly improved our understanding on the spread of STIs, substantial challenges remain. In this section, we discuss two particular areas of focus that remain essential to improving our understanding of sexual networks and STIs: sampling and measurement error.

Sampling error is a major concern when studying sexual networks. For studies examining whole networks with a roster of individuals, this includes traditional non-response bias among individuals who do not provide information regarding their sexual partners. Similarly, many whole network studies of sexual networks suffer from arbitrary bounding of these networks in space and time (Eames et al. 2015), the impact of which has largely gone unexplored in this work. For traditional approaches, this limits nomination of sexual partners to individuals who are part of the same arbitrary social organization, such as a school (Bearman et al. 2004). For non-traditional approaches, full information regarding sexual partnerships may be limited to individuals who appear in the same study (Friedman et al. 1997) or attend the same venues (Remple et al. 2007). Since individuals are unlikely to form sexual partners exclusively among these individuals, each of these designs probably introduces error and bias into the observed network structure. Accordingly, the influence of these design decisions should be carefully considered at all stages of study design and analysis. Sampling error may be especially acute when participants within the whole network are referred to participate through network contacts, a feature common in this work (Boodram and Williams 2013), thereby introducing greater selection effects on participation. For studies of personal networks, research suggests that personal network inventories may provide biased estimates of within and between mixing patterns (Ghani et al. 1998). However, sophisticated approaches have been developed to accurately capture mixing patterns with alternative techniques such as snowball sampling (Boily et al. 2000) or using exponential random graph models to estimate whole-network mixing parameters (Krivitsky and Morris 2017; Smith 2012).

In addition to sampling error, measurement remains an incredibly challenging aspect in the study of sexual networks. Socially desirable responding (for example, reporting fewer partners for impression management) and errors in recall have been two main areas of focus in the work. Reports obtained from both sexual partners generally find relatively high levels of agreement (Bell et al. 2000). This is particularly true if participants remember to include the partner in their network elicitation (Phillips II et al. 2017), probably because forgetting appears to be the main cause of recall error (Bell et al. 2007; Brewer et al. 1999). Fortunately, techniques can be used to reduce forgetting, such as explicitly probing for partners with unknown names (Brewer et al. 1999) and cueing people about partners met at locations where they frequently interact or meet partners (Brewer et al. 2005). However, measurement error will exist even in the most well-designed studies, given the complex nature of sexual partnerships. In particular, the measurement of the timing of sexual partnerships may be particularly prone to error (Helleringer et al. 2011), although others have found this error to be relatively minor (Brewer et al. 2006; Morris and O'Gorman 2000). Yet, just as methodological studies should continue to promote more accurate measurement, analytic approaches should continue increase in sophistication to model measurement error in sexual networks (Goyal and De Gruttola 2018).

While sampling and measurement error for sensitive behavior is not a novel problem (Catania et al. 1990), these issues may be especially acute for the study of sexual networks, particularly when gathering highly specific and intrusive information, such as the full names of recent sexual partners among people recently infected with STIs. This issue is especially problematic when error may confound important associations. For example, a qualitative investigation into the acceptability of sexual network research among a diverse sample of MSM found racial differences in how participants preferred to recruit sexual partners into research studies (Biello et al. 2017). Failure to account for these differences may create differential rates of participation or elicitation of sexual networks across racial/ethnic subgroups.

Potential error in sampling and self-reporting of sexual partnerships has fed excitement regarding alternative approaches, such as using phylogenetic data from genetic sequences of HIV, to infer contact network information (Fichtenberg et al. 2010). However, the association between phylogenetic transmission trees and sexual networks remains highly uncertain (Frost and Volz 2013; Giardina et al. 2017; Grabowski and Redd 2014; Leventhal et al. 2012; McCloskey et al. 2016; Robinson et al. 2013; Stadler and Bonhoeffer 2013; Villandre et al. 2016). In addition, many of the techniques used to find clusters of closely related infections may be substantially impacted by relative sampling rates (Poon 2016) and this approach may be subject to many of the same biases in traditional epidemiological studies (Grabowski and Redd 2014). Given this uncertainty, triangulation of data among multiple approaches may be particularly insightful. For example, Wertheim et al. (2017) found only 53 per cent of genetically linked individuals were named during partner service interviews in data from the New York City Department of Public Health, suggesting each approach may be important for characterizing local HIV epidemics. Research in this area should continue to clarify the limitations and strengths of each approach (Delva et al. 2016). This work is especially important because identifying and seeking out sexual partners of recently infected individuals (that is, partner services) remains a major focus of STI prevention activities in urban public health departments (Hogben et al. 2016). Yet, this research has traditionally been distinct from research on sexual networks (Delva and Helleringer 2016) and network-informed approaches for targeting sexual partners (Nikolopoulos et al. 2016). Increased collaboration across these fields would probably benefit both areas of research. Furthermore, given the finite available resources to public health departments, the development of tools to increase the ease of capturing and leveraging this data (Davidson et al. 2018; Plant et al. 2012) will probably become increasingly important as more sophisticated approaches to using this data become available.

CONCLUSION

This chapter has documented the preferences and constraints that shape sexual partnership selection, how these selection processes manifest into important network structures shaping the spread of HIV and other STIs, and how each of these network forces interact with both physical and social spaces in urban regions. This research highlights the importance of both in-depth knowledge of the impacted population as well as comprehensive information of the relevant social context that is required to fully understand disease dynamics. As methods to understand sexual networks continue to mature, the study of these networks will undoubtedly continue to enhance our understanding of the spread of disease and inform preventive interventions.

REFERENCES

Adimora, A.A. and V.J. Schoenbach (2013), 'Social determinants of sexual networks, partnership formation, and sexually transmitted infections', in *The New Public Health and STD/HIV Prevention*, New York: Springer, pp. 13–31.

Adimora, A.A., V.J. Schoenbach, E.M. Taylor, M.R. Khan, R.J. Schwartz and W.C. Miller (2013), 'Sex ratio, poverty, and concurrent partnerships among men and women in the United States: a multilevel analysis', *Annals of Epidemiology*, **23** (11), 716–19.

Amirkhanian, Y.A. (2014), 'Social networks, sexual networks and HIV risk in men who have sex with men', *Current HIV/AIDS Reports*, **11** (1), 81–92.

Anderson, R.M., G.F. Medley, R.M. May and A.M. Johnson (1986), 'A preliminary study of the transmission dynamics of the human immunodeficiency virus (HIV), the causative agent of AIDS', *Mathematical Medicine and Biology: A Journal of the IMA*, **3** (4), 229–63.

Aral, S.O. (2010), 'Partner concurrency and the STD/HIV epidemic', *Current Infectious Disease Reports*, **12** (2), 134–9, doi:10.1007/s11908-010-0087-2.

Aral, S.O. and J.S. Leichliter (2010), 'Non-monogamy: risk factor for STI transmission and acquisition and determinant of STI spread in populations', *Sexually Transmitted Infections*, **86**, suppl. 3, 29–36, doi:10.1136/sti.2010.044149.

Aral, S.O., A.A. Adimora and K.A. Fenton (2008), 'Understanding and responding to disparities in HIV and other sexually transmitted infections in African Americans', *Lancet*, **372** (9635), 337–40, doi: 10.1016/S0140-6736(08)61118-6.

Auerbach, D.M., W.W. Darrow, H.W. Jaffe and J.W. Curran (1984), 'Cluster of cases of the acquired immune deficiency syndrome. Patients linked by sexual contact', *American Journal of Medicine*, **76** (3), 487–92.

Bearman, P.S., J. Moody and K. Stovel (2004), 'Chains of affection: the structure of adolescent romantic and sexual networks', *American Journal of Sociology*, **110** (1), 44–91, doi:10.1086/386272.

Bell, D.C., B. Belli-McQueen and A. Haider (2007), 'Partner naming and forgetting: recall of network members', *Social Networks*, **29** (2), 279–99.

Bell, D.C., I.D. Montoya and J.S. Atkinson (2000), 'Partner concordance in reports of joint risk behaviors', *Journal of Acquired Immune Deficiency Syndromes*, **25** (2), 173–81.

Berry, M., H.F. Raymond and W. McFarland (2007), 'Same race and older partner selection may explain higher HIV prevalence among black men who have sex with men', *AIDS*, **21** (17), 2349–50.

Biello, K.B., J. Malone, K.H. Mayer, A.R. Bazzi and M.J. Mimiaga (2017), 'Designing a sexual network study of men who have sex with other men: exploring racial and ethnic preferences in study design and methods', *Aids Care-Psychological and Socio-Medical Aspects of Aids/Hiv*, **29** (1), 56–60, doi:10.1080/09540121.2016.1198748.

Bingham, T.A., N.T. Harawa, D.F. Johnson, G.M. Secura, D.A. MacKellar and L.A. Valleroy (2003), 'The effect of partner characteristics on HIV infection among African American men who have sex with men in the Young Men's Survey, Los Angeles, 1999-2000', *AIDS Education and Prevention*, **15** (1 Suppl A), 39–52.

Birkett, M., L.M. Kuhns, C. Latkin, S. Muth and B. Mustanski (2015), 'The sexual networks of racially diverse young men who have sex with men', *Archives of Sexual Behavior*, **44** (7), 1787–97, doi:10.1007/s10508-015-0485-5.

Bohl, D.D., H.F. Raymond, M. Arnold and W. McFarland (2009), 'Concurrent sexual partnerships and racial disparities in HIV infection among men who have sex with men', *Sexually Transmitted Infections*, **85** (5), 367–9, doi:10.1136/sti.2009.036723.

Boily, M.-C., R. Poulin and B. Mâsse (2000), 'Some methodological issues in the study of sexual networks: from model to data to model', *Sexually Transmitted Diseases*, **27** (10), 558–71.

Boodram, B. and C.T. Williams (2013), 'Collecting whole network data for human immunodeficiency virus prevention: a review of current strategies', *Journal of AIDS and HIV Research*, **5** (4), 114–26.

Bradshaw, D., J. Raghwani, B. Jacka, R. Sacks-Davis, F. Lamoury, I. Down, et al. (2016), 'Venue-based networks may underpin HCV transmissions amongst HIV-infected gay and bisexual men', *PLOS ONE*, **11** (9), e0162002.

Brantley, M., C. Schumacher, E.L. Fields, J. Perin, A.G. Safi, J.M. Ellen, et al. (2017), 'The network structure of sex partner meeting places reported by HIV-infected MSM: opportunities for HIV targeted control', *Social Science & Medicine*, **182** (June), 20–29, doi:10.1016/j.socscimed.2017.04.006.

Brewer, D.D., S.B. Garret and S. Kulasingam (1999), 'Forgetting as a cause of incomplete reporting of sexual and drug injection partners', *Sexually Transmitted Diseases*, **26** (3), 166–76.

Brewer, D.D., J.J. Potterat, S.Q. Muth, P.Z. Malone, P. Montoya, D.L. Green, et al. (2005), 'Randomized trial of supplementary interviewing techniques to enhance recall of sexual partners in contact interviews', *Sexually Transmitted Diseases*, **32** (3), 189–93.

Brewer, D.D., R.B. Rothenberg, S.Q. Muth, J.M. Roberts Jr and J.J. Potterat (2006), 'Agreement in reported sexual partnership dates and implications for measuring concurrency', *Sexually Transmitted Diseases*, **33** (5), 277–83.

Calabrese, S.K., D.S. Krakower and K.H. Mayer (2017), 'Integrating HIV Preexposure prophylaxis (PrEP) into routine preventive health care to avoid exacerbating disparities', *American Jornal of Public Health*, **107** (12), 1883–9, doi:10.2105/AJPH.2017.304061.

Catania, J.A., D. Binson and V. Stone (1996), 'Relationship of sexual mixing across age and ethnic groups to herpes simplex virus-2 among unmarried heterosexual adults with multiple sexual partners', *Health Psychology*, **15** (5), 362–70.

Catania, J.A., D.R. Gibson, D.D. Chitwood and T.J. Coates (1990), 'Methodological problems in Aids behavioral-research – influences on measurement error and participation bias in studies of sexual-behavior', *Psychological Bulletin*, **108** (3), 339–62, doi:10.1037/0033-2909.108.3.339.

Centers for Disease Control (2016), 'HIV infection, risk, prevention and testing behaviors. National HIV behavioral surveillance system. Men who have sex with men: 20 US cities, 2011', HIV Surveillance Report, Special Report No. 8, Centers for Disease Control, US Department of Health and Human Services, Atlanta, GA.

Choi, K.H., C.S. Han, J. Paul and G. Ayala (2011), 'Strategies for managing racism and homophobia among U.S. ethnic and racial minority men who have sex with men', *Aids Education and Prevention*, **23** (2), 145–58, doi:10.1521/aeap.2011.23.2.145.

Clerkin, E.M., M.E. Newcomb and B. Mustanski (2011), 'Unpacking the racial disparity in HIV rates: the effect of race on risky sexual behavior among Black young men who have sex with men (YMSM)', *Journal of Behavioural Medicine*, **34** (4), 237–43, doi:10.1007/s10865-010-9306-4.

Cooper, H.L., S. Linton, D.F. Haley, M.E. Kelley, E.F. Dauria, C.C. Karnes, et al. (2015), 'Changes in exposure

to neighborhood characteristics are associated with sexual network characteristics in a cohort of adults relocating from public housing', *AIDS and Behavior*, **19** (6), 1016–30, doi:10.1007/s10461-014-0883-z.
Davidson, J.A., L.F. Anderson, V. Adebisi, L. de Jongh, A. Burkitt and M.K. Lalor (2018), 'Creating a web-based electronic tool to aid tuberculosis (TB) cluster investigation: data integration in TB surveillance activities in the United Kingdom, 2013 to 2016', *Eurosurveillance*, **23** (44), doi:10.2807/1560-7917.ES.2018.23.44.1700794.
De, P., A.E. Singh, T. Wong, W. Yacoub and A.M. Jolly (2004), 'Sexual network analysis of a gonorrhoea outbreak', *Sexually Transmitted Infections*, **80** (4), 280–85, doi:10.1136/sti.2003.007187.
Delva, W. and S. Helleringer (2016), 'Beyond risk compensation: clusters of antiretroviral treatment (ART) users in sexual networks can modify the impact of ART on HIV incidence', *PLOS ONE*, **11** (9), e0163159. doi:10.1371/journal.pone.0163159.
Delva, W., G.E. Leventhal and S. Helleringer (2016), 'Connecting the dots: network data and models in HIV epidemiology', *AIDS*, **30** (13), 2009–20, doi:10.1097/QAD.0000000000001184.
Doherty, I.A., V.J. Schoenbach and A.A. Adimora (2009), 'Sexual mixing patterns and heterosexual HIV transmission among African Americans in the southeastern United States', *Journal of Acquired Immune Deficiency Syndromes*, **52** (1), 114–20, doi:10.1097/QAI.0b013e3181ab5e10.
Eames, K., S. Bansal, S. Frost and S. Riley (2015), 'Six challenges in measuring contact networks for use in modelling', *Epidemics*, **10** (March), 72–7, doi:10.1016/j.epidem.2014.08.006.
Eaton, L.A., S.C. Kalichman, D.A. O'Connell and W.D. Karchner (2009), 'A strategy for selecting sexual partners believed to pose little/no risks for HIV: serosorting and its implications for HIV transmission', *AIDS Care*, **21** (10), 1279–88, doi:10.1080/09540120902803208.
Ellen, J.M., B.A. Brown, S.E. Chung, J.J. Potterat, S.Q. Muth, T.W. Valente, et al. (2005), 'Impact of sexual networks on risk for gonorrhea and chlamydia among low-income urban African American adolescents', *Journal of Pediatrics*, **146** (4), 518–22, doi:10.1016/j.jpeds.2004.11.023.
Fichtenberg, C.M., J.M. Jennings, T.A. Glass and J.M. Ellen (2010), 'Neighborhood socioeconomic environment and sexual network position', *Journal of Urban Health*, **87** (2), 225–35, doi:10.1007/s11524-009-9425-9.
Foxman, B., M. Newman, B. Percha, K.K. Holmes and S.O. Aral (2006), 'Measures of sexual partnerships: lengths, gaps, overlaps, and sexually transmitted infection', *Sexually Transmitted Diseases*, **33** (4), 209–14, doi:10.1097/01.olq.0000191318.95873.8a.
Friedman, S.R., H.L. Cooper and A.H. Osborne (2009), 'Structural and social contexts of HIV risk Among African Americans', *American Journal of Public Health*, **99** (6), 1002–8, doi:10.2105/AJPH.2008.140327.
Friedman, S.R., A. Neaigus, B. Jose, R. Curtis, M. Goldstein, G. Ildefonso, et al. (1997), 'Sociometric risk networks and risk for HIV infection', *American Journal of Public Health*, **87** (8), 1289–96.
Frost, S.D.W. (2007), 'Using sexual affiliation networks to describe the sexual structure of a population', *Sexually Transmitted Infections*, **83** (July), I37 I42, doi:10.1136/sti.2006.023580.
Frost, S.D.W. and E.M. Volz (2013), 'Modelling tree shape and structure in viral phylodynamics', *Philosophical Transactions of the Royal Society B-Biological Sciences*, **368** (1614), doi:10.1098/rstb.2012.0208.
Fujimoto, K. and M.L. Williams (2015), 'Racial/ethnic differences in sexual network mixing: a log-linear analysis of HIV status by partnership and sexual behavior among most at-risk MSM', *AIDS and Behavior*, **19** (6), 996–1004, doi:10.1007/s10461-014-0842-8.
Fujimoto, K., M.L. Williams and M.W. Ross (2013), 'Venue-based affiliation networks and HIV risk-taking behavior among male sex workers', *Sexually Transmitted Diseases*, **40** (6), 453–8, doi:10.1097/OLQ.0b013e31829186e5.
Ghani, A.C., C.A. Donnelly and G.P. Garnett (1998), 'Sampling biases and missing data in explorations of sexual partner networks for the spread of sexually transmitted diseases', *Statistics in Medicine*, **17** (18), 2079–97.
Giardina, F., E.O. Romero-Severson, J. Albert, T. Britton and T. Leitner (2017), 'Inference of transmission network structure from HIV phylogenetic trees', *PLoS Computatiuonal Biology*, **13** (1), e1005316, doi:10.1371/journal.pcbi.1005316.
Gindi, R.M., F. Sifakis, S.G. Sherman, V.L. Towe, C. Flynn and J.M. Zenilman (2011), 'The geography of heterosexual partnerships in Baltimore city adults', *Sexually Transmitted Diseases*, **38** (4), 260–66, doi:10.1097/OLQ.0b013e3181f7d7f4.
Glick, S.N., M. Morris, B. Foxman, S.O. Aral, L.E. Manhart, K.K. Holmes, et al. (2012), 'A comparison of sexual behavior patterns among men who have sex with men and heterosexual men and women', *Journal of Acquired Immune Deficiency Syndromes*, **60** (1), 83–90, doi:10.1097/QAI.0b013e318247925e.
Goodreau, S.M., E.S Rosenberg, S.M. Jenness, N. Luisi, S.E. Stansfield, G.A. Millett, et al. (2017), 'Sources of racial disparities in HIV prevalence in men who have sex with men in Atlanta, GA, USA: a modelling study', *Lancet HIV*, **4** (7), e311–e320, doi:10.1016/S2352-3018(17)30067-X.
Goyal, R. and V. De Gruttola (2018), 'Inference on network statistics by restricting to the network space: applications to sexual history data', *Statistics in Medicine*, **37** (2), 218–35.

Grabowski, M.K. and A.D. Redd (2014), 'Molecular tools for studying HIV transmission in sexual networks', *Current Opinion in HIV AIDS*, **9** (2), 126–33, doi:10.1097/COH.0000000000000040.
Grey, J.A., R.B. Rothenberg, P.S. Sullivan and E.S. Rosenberg (2015), 'Disassortative age-mixing does not explain differences in HIV prevalence between young White and Black MSM: findings from four studies', *PLOS ONE*, **10** (6), e0129877, doi:10.1371/journal.pone.0129877.
Gupta, S., R.M. Anderson and R.M. May (1989), 'Networks of sexual contacts: implications for the pattern of spread of HIV', *AIDS*, **3** (12), 807–17.
Hamilton, D.T. and M. Morris (2015), 'The racial disparities in STI in the U.S.: concurrency, STI prevalence, and heterogeneity in partner selection', *Epidemics*, **11** (June), 56–61, doi:10.1016/j.epidem.2015.02.003.
Haraldsdottir, S., S. Gupta and R.M. Anderson (1992), 'Preliminary studies of sexual networks in a male homosexual community in Iceland', *Journal of Acquired Immune Deficiency Syndromes*, **5** (4), 374–81.
Helleringer, S., H.P. Kohler, L. Kalilani-Phiri, J. Mkandawire and B. Armbruster (2011), 'The reliability of sexual partnership histories: implications for the measurement of partnership concurrency during surveys', *AIDS*, **25** (4), 503–11, doi:10.1097/QAD.0b013e3283434485.
Hethcote, H.W. and J.A. Yorke (1984), *Gonorrhea Transmission Dynamics and Control*, Berlin: and New York: Springer-Verlag.
Hippe, J. and A.M. Jolly (2012), 'STI phase and the geography of sexual partnerships: prevalence of long-distance sexual contacts among chlamydia, gonorrhea, and coinfected STI cases in Manitoba, Canada', *Spatial and Spatio-temporal Epidemiology*, **3** (3), 255–63, doi:10.1016/j.sste.2012.04.001.
Hogben, M., D. Collins, B. Hoots and K. O'Connor (2016), 'Partner services in sexually transmitted disease prevention programs: a review', *Sexually Transmitted Diseases*, **43** (2 Suppl 1), S53–62, doi:10.1097/OLQ.0000000000000328.
Hyman, J.M. and E.A. Stanley (1988), 'Using mathematical-models to understand the Aids epidemic', *Mathematical Biosciences*, **90** (1–2), 415–73, doi:Doi 10.1016/0025-5564(88)90078-8.
Janulis, P., B.A. Feinstein, G. Phillips, M.E. Newcomb, M. Birkett and B. Mustanski (2018), 'Sexual partner typologies and the association between drug use and sexual risk behavior among young men who have sex with men', *Archives of Sexual Behavior*, **47** (1), 259–71, doi:10.1007/s10508-016-0909-x.
Janulis, P., G. Phillips II, M. Birkett and B. Mustanski (2018), 'Sexual networks of racially diverse young msm differ in racial homophily but not concurrency', *Journal of Acquired Immune Deficiency Syndromes*, **77** (5), 459–66, doi:10.1097/QAI.0000000000001620.
Jennings, J.M., M.L. Reilly, J. Perin, C. Schumacher, M. Sharma, A.G. Safi, et al. (2015), 'Sex partner meeting places over time among newly HIV diagnosed men who have sex with men (MSM) in Baltimore, Maryland', *Sexually Transmitted Diseases*, **42** (10), 549–53.
Jones, J.H. and M.S. Handcock (2003), 'An assessment of preferential attachment as a mechanism for human sexual network formation', *Proceedings of the Royal Society of London B: Biological Sciences*, **270** (1520), 1123–28.
Kalmijn, M. and H. Flap (2001), 'Assortative meeting and mating: unintended consequences of organized settings for partner choices', *Social Forces*, **79** (4), 1289–312, doi:10.1353/sof.2001.0044.
Keeling, M.J. and K.T. Eames (2005), 'Networks and epidemic models', *Journal of the Royal Society Interface*, **2** (4), 295–307, doi:10.1098/rsif.2005.0051.
Kenyon, C. and R. Colebunders (2013), 'Birds of a feather: homophily and sexual network structure in sub-Saharan Africa', *International Journal of STD & AIDS*, **24** (3), 211–15, doi:10.1177/0956462412472455.
Khosropour, C.M., J.C. Dombrowski, F. Swanson, R.P. Kerani, D.A. Katz, L.A. Barbee, et al. (2016), 'Trends in serosorting and the association with HIV/STI risk over time among men who have sex with men', *Journal of Acquired Immune Deficiency Syndromes*, **72** (2), 189–97, doi:10.1097/QAI.0000000000000947.
Klovdahl, A.S. (1985), 'Social networks and the spread of infectious diseases: the AIDS example', *Social Science & Medicine*, **21** (11), 1203–16.
Kraut-Becher, J.R. and S.O. Aral (2006), 'Patterns of age mixing and sexually transmitted infections', *International Journal of STD & AIDS*, **17** (6), 378–83, doi:10.1258/095646206777323481.
Kretzschmar, M. and M. Morris (1996), 'Measures of concurrency in networks and the spread of infectious disease', *Mathematical Biosciences*, **133** (2), 165–95.
Krivitsky, P.N. and M. Morris (2017), 'Inference for Social network models from egocentrically sampled data, with application to understanding persistent racial disparities in HIV prevalence in the US', *Annals of Applied Statistics*, **11** (1), 427–55, doi:10.1214/16-AOAS1010.
Laumann, E.O. and Y. Youm (1999), 'Racial/ethnic group differences in the prevalence of sexually transmitted diseases in the United States: a network explanation', *Sexually Transmitted Diseases*, **26** (5), 250–61.
Lee, J.K., J.M. Jennings and J.M. Ellen (2003), 'Discordant sexual partnering: a study of high-risk adolescents in San Francisco', *Sexually Transmitted Diseases*, **30** (3), 234–40.
Leventhal, G.E., R. Kouyos, T. Stadler, V. von Wyl, S. Yerly, J. Boni, et al. (2012), 'Inferring epidemic contact structure from phylogenetic trees', *Plos Computational Biology*, **8** (3), doi:10.1371/journal.pcbi.1002413.
Linton, S.L., H.L.F. Cooper, R. Luo, C. Karnes, K. Renneker, D.F. Haley, et al. (2017), 'Changing places

and partners: associations of neighborhood conditions with sexual network turnover among African American adults relocated from public housing', *Archives of Sexual Behavior*, **46** (4), 925–36, doi:10.1007/s10508-015-0687-x.

Mah, T.L. and D.T. Halperin (2010), 'Concurrent sexual partnerships and the HIV epidemics in Africa: evidence to move forward', *AIDS and Behavior*, **14** (1), 11–16, discussion 34–17, doi:10.1007/s10461-008-9433-x.

McCloskey, R.M., R.H. Liang and A.F.Y. Poon (2016), 'Reconstructing contact network parameters from viral phylogenies', *Virus Evolution*, **2** (2), doi:10.1093/ve/vew029.

McPherson, M., L. Smith-Lovin and J.M. Cook (2001), 'Birds of a feather: homophily in social networks', *Annual Review of Sociology*, **27** (1), 415–44.

Mercer, C.H., C.R. Aicken, C. Tanton, C.S. Estcourt, M.G. Brook, F. Keane, et al. (2013), 'Serial monogamy and biologic concurrency: measurement of the gaps between sexual partners to inform targeted strategies', *American Journal of Epidemiology*, **178** (2), 249–59, doi:10.1093/aje/kws467.

Millett, G.A., S.A. Flores, J.L. Peterson and R. Bakeman (2007), 'Explaining disparities in HIV infection among black and white men who have sex with men: a meta-analysis of HIV risk behaviors', *AIDS*, **21** (15), 2083–91, doi:10.1097/QAD.0b013e3282e9a64b.

Millett, G.A., J.L. Peterson, S.A. Flores, T.A. Hart, W.L. Jeffries, P.A. Wilson, et al. (2012), 'Comparisons of disparities and risks of HIV infection in black and other men who have sex with men in Canada, UK, and USA: a meta-analysis', *Lancet*, **380** (9839), 341–8, doi:10.1016/S0140-6736(12)60899-X.

Morris, M. (1993), 'Epidemiology and social networks: modeling structured diffusion', *Sociological Methods & Research*, **22** (1), 99–126.

Morris, M. and M. Kretzschmar (1997), 'Concurrent partnerships and the spread of HIV', *AIDS*, **11** (5), 641–8.

Morris, M. and J. O'Gorman (2000), 'The impact of measurement error on survey estimates of concurrent partnerships', *Mathematical Population Studies*, **8** (3), 231249.

Morris, M., H. Epstein and M. Wawer (2010), 'Timing is everything: international variations in historical sexual partnership concurrency and HIV prevalence', *PLOS ONE*, **5** (11), e14092, doi:10.1371/journal.pone.0014092.

Morrison-Beedy, D., Y. Xia and D. Passmore (2013), 'Sexual risk factors for partner age discordance in adolescent girls and their male partners', *Journal of Clinical Nursing*, **22** (23–24), 3289–99.

Mustanski, B. and M.E. Newcomb (2013), 'Older sexual partners may contribute to racial disparities in HIV among young men who have sex with men', *Journal of Adolescent Health*, **52** (6), 666–67, doi:10.1016/j.jadohealth.2013.03.019.

Mustanski, B., M. Birkett, L.M. Kuhns, C.A. Latkin and S.Q. Muth (2015), 'The role of geographic and network factors in racial disparities in HIV among young men who have sex with men: an egocentric network study', *AIDS and Behavior*, **19** (6), 1037–47, doi:10.1007/s10461-014-0955-0.

Mustanski, B., D.T. Ryan, T.A. Remble, T. Richard, M.E. Newcomb and E. Morgan (2018), 'Discordance of self report and laboratory measures of HIV viral load among young men who have sex with men and transgender women in chicago: implications for epidemiology, care, and prevention', *AIDS and Behavior*, **22** (7), 2360–67.

Neaigus, A., S.M. Jenness, H. Hagan, C.S. Murrill and T. Wendel (2013), 'Reciprocal sex partner concurrency and STDs among heterosexuals at high-risk of HIV infection', *Journal of Urban Health*, **90** (5), 902–14.

Nelson, S.J., L.E. Manhart, P.M. Gorbach, D.H. Martin, B.P. Stoner, S.O. Aral, et al. (2007), 'Measuring sex partner concurrency: it's what's missing that counts', *Sexually Transmitted Diseases*, **34** (10), 801–7, doi:10.1097/OLQ.0b013e318063c734.

Newcomb, M.E., M.C. Mongrella, B. Weis, S.J. McMillen and B. Mustanski (2016), 'Partner disclosure of PrEP use and undetectable viral load on geosocial networking apps: frequency of disclosure and decisions about condomless sex', *Journal of Acquired Immune Deficiency Syndromes*, **71** (2), 200–206, doi:10.1097/QAI.0000000000000819.

Nguyen, V.-K., Z.R. Greenwald, H. Trottier, M. Cadieux, A. Goyette, M. Beauchemin, et al. (2018), 'Incidence of sexually transmitted infections before and after preexposure prophylaxis for HIV', *AIDS*, **32** (4), 523–30.

Niccolai, L.M., Farley, T. A., Ayoub, M. A., Magnus, M. and Kissinger, P. J. (2002), 'HIV-infected persons' knowledge of their sexual partners' HIV status', *Aids Education and Prevention*, **14** (3), 183–9, doi:10.1521/aeap.14.3.183.23893.

Niekamp, A.-M., L.A. Mercken, C.J. Hoebe and N.H. Dukers-Muijrers (2013), 'A sexual affiliation network of swingers, heterosexuals practicing risk behaviours that potentiate the spread of sexually transmitted infections: a two-mode approach', *Social Networks*, **35** (2), 223–36.

Nikolopoulos, G.K., E. Pavlitina, S.Q. Muth, J. Schneider, M. Psichogiou, L.D. Williams, et al. (2016), 'A network intervention that locates and intervenes with recently HIV-infected persons: the Transmission Reduction Intervention Project (TRIP)', *Scientific Reports*, **6** (December), doi:10.1038/srep38100.

Oser, C.B., E. Pullen, D. Stevens-Watkins, B.L. Perry, J.R. Havens, M. Staton-Tindall, et al. (2017), 'African American women and sexually transmitted infections: The contextual influence of unbalanced sex ratios and individual risk behaviors', *Journal of Drug Issues*, **47** (4), 543–61, doi:10.1177/0022042616678610.

Oster, A.M., C. Wejnert, L.A. Mena, K. Elmore, H. Fisher and J.D. Heffelfinger (2013), 'Network analysis among HIV-infected young Black men who have sex with men demonstrates high connectedness around few venues', *Sexually Transmitted Diseases*, **40** (3), 206–12, doi:10.1097/OLQ.0b013e3182840373.
Paul, J.P., G. Ayala and K.-H. Choi (2010), 'Internet sex ads for MSM and partner selection criteria: the potency of race/ethnicity online', *Journal of Sex Research*, **47** (6), 528–38.
Phillips, G., M. Birkett, S. Hammond and B. Mustanski (2016), 'Partner preference among men who have sex with men: potential contribution to spread of HIV within minority populations', *LGBT Health*, **3** (3), 225–32.
Phillips II, G., P. Janulis, B. Mustanski and M. Birkett (2017), 'Validation of tie corroboration and reported alter characteristics among a sample of young men who have sex with men', *Social Networks*, **48** (January), 250–55.
Phua, V.C. and G. Kaufman (2003), 'The crossroads of race and sexuality: date selection among men in internet "personal" ads', *Journal of Family Issues*, **24** (8), 981–94.
Pines, H.A., J.O. Wertheim, L. Liu, R.S. Garfein, S.J. Little and M.Y. Karris (2016), 'Concurrency and HIV transmission network characteristics among MSM with recent HIV infection', *AIDS*, **30** (18), 2875–883, doi:10.1097/QAD.0000000000001256.
Plant, A., H. Rotblatt, J.A. Montoya, E.T. Rudy and P.R. Kerndt (2012), 'Evaluation of inSPOTLA.org: an Internet partner notification service', *Sexually Transmitted Diseases*, **39** (5), 341–5, doi:10.1097/OLQ.0b013e31824e5150.
Poon, A.F. (2016), 'Impacts and shortcomings of genetic clustering methods for infectious disease outbreaks', *Virus Evolution*, **2** (2), doi:10.1093/ve/vew031.
Pouget, E.R. (2017), 'Social determinants of adult sex ratios and racial/ethnic disparities in transmission of HIV and other sexually transmitted infections in the USA', *Philosophical Transactions of the Royal Society B-Biological Sciences*, **372** (1729), doi:10.1098/rstb.2016.0323.
Prejean, J., R.G. Song, A. Hernandez, R. Ziebell, T. Green, F. Walker, et al. (2011), 'Estimated HIV Incidence in the United States, 2006-2009', *PLOS ONE*, **6** (8), doi:10.1371/journal.pone.0017502.
Purcell, D.W., D. Higa, Y. Mizuno and C. Lyles (2017), 'Quantifying the harms and benefits from serosorting among HIV-negative gay and bisexual men: a systematic review and meta-analysis', *AIDS and Behavior*, **21** (10), 2835–43, doi:10.1007/s10461-017-1800-z.
Raymond, H.F. and W. McFarland (2009), 'Racial mixing and HIV risk among men who have sex with men', *AIDS and Behaviour*, **13** (4), 630–37, doi:10.1007/s10461-009-9574-6.
Rebeiro, P.F., A.G. Abraham, M.A. Horberg, K.N. Althoff, B.R. Yehia, K. Buchacz, et al. (2017), 'Sex, race, and HIV risk disparities in discontinuity of HIV care after antiretroviral therapy initiation in the United States and Canada', *AIDS Patient Care STDS*, **31** (3), 129–44, doi:10.1089/apc.2016.0178.
Remple, V.P., D.M. Patrick, C. Johnston, M.W. Tyndall and A.M. Jolly (2007), 'Clients of indoor commercial sex workers: heterogeneity in patronage patterns and implications for HIV and STI propagation through sexual networks', *Sexually Transmitted Diseases*, **34** (10), 754–60, doi:10.1097/01.olq.0000261327.78674.cb.
Robinson, K., N. Fyson, T. Cohen, C. Fraser and C. Colijn (2013), 'How the dynamics and structure of sexual contact networks shape pathogen phylogenies', *Plos Computational Biology*, **9** (6), doi:10.1371/journal.pcbi.1003105.
Rosenberg, E.S., C.M. Khosropour and P.S. Sullivan (2012), 'High prevalence of sexual concurrency and concurrent unprotected anal intercourse across racial/ethnic groups among a national, Web-based study of men who have sex with men in the United States', *Sexually Transmitted Diseases*, **39** (10), 741–6, doi:10.1097/OLQ.0b013e31825ec09b.
Rosenberg, M.D., J.E. Gurvey, N. Adler, M.B. Dunlop, and J.M. Ellen (1999), 'Concurrent sex partners and risk for sexually transmitted diseases among adolescents', *Sexually Transmitted Diseases*, **26** (4), 208–12.
Rothenberg, R., S.Q. Muth, S. Malone, J.J. Potterat and D.E Woodhouse (2005), 'Social and geographic distance in HIV risk', *Sexually Transmitted Diseases*, **32** (8), 506–12.
Sawers, L. and E. Stillwaggon (2010), 'Concurrent sexual partnerships do not explain the HIV epidemics in Africa: a systematic review of the evidence', *Journal of the International AIDS Society*, **13** (September), 1–23, doi:10.1186/1758-2652-13-34.
Schneider, J.A., N. Lancki and P. Schumm (2017), 'At the intersection of criminal justice involvement and sexual orientation: dynamic networks and health among a population-based sample of young Black men who have sex with men', *Social Networks*, **51** (October), 73–87, doi:10.1016/j.socnet.2017.04.001.
Sewell, J., M. Daskalopoulou, F. Nakagawa, F.C. Lampe, S. Edwards, N. Perry, et al. (2017), 'Accuracy of self-report of HIV viral load among people with HIV on antiretroviral treatment', *HIV Medicine*, **18** (7), 463–73, doi:10.1111/hiv.12477.
Smith, J.A. (2012), 'Macrostructure from microstructure: generating whole systems from ego networks', *Sociol Methodol*, **42** (1), 155–205, doi:10.1177/0081175012455628.
Snowden, J.M., C. Wei, W. McFarland and H.F. Raymond (2014), 'Prevalence, correlates and trends in seroadaptive behaviours among men who have sex with men from serial cross-sectional surveillance in San Francisco, 2004–2011', *Sexually Transmitted Infections*, **90** (6), 498–504.
Stadler, T. and S. Bonhoeffer (2013), 'Uncovering epidemiological dynamics in heterogeneous host populations

using phylogenetic methods', *Philosophical Transactions of the Royal Society B-Biological Sciences*, **368** (1614), doi:10.1098/rstb.2012.0198.

Stephens, S.C., C.K. Fann, F.V. Strona, W. Wolf, S.E. Cohen, S.S. Philip, et al. (2014), 'Identifying syphilis risk networks through venue attendance in San Francisco', *Sexually Transmitted Diseases*, **41** (5), 333–7, doi:10.1097/Olq.0000000000000116.

Storholm, E.D., J.E. Volk, J.L. Marcus, M.J. Silverberg, and D.D. Satre (2017), 'Risk Perception, sexual behaviors, and PrEP adherence among substance-using men who have sex with men: a qualitative study', *Prevention Science*, **18** (6), 737–47, doi:10.1007/s11121-017-0799-8.

Sweeney, K.A. and A.L. Borden (2009), 'Crossing the line online: racial preference of Internet daters', *Marriage & Family Review*, **45** (6–8), 740–60.

Tieu, H.-V., V. Nandi, D.R. Hoover, D. Lucy, K. Stewart, V. Frye, et al. (2016), 'Do sexual networks of men who have sex with men in New York City differ by race/ethnicity?', *AIDS Patient Care and STDs*, **30** (1), 39–47.

Villandre, L., D.A. Stephens, A. Labbe, H.F. Gunthard, R. Kouyos, T. Stadler, et al. (2016), 'Assessment of overlap of phylogenetic transmission clusters and communities in simple sexual contact networks: applications to HIV-1', *PLOS ONE*, **11** (2), doi:10.1371/journal.pone.0148459.

Weir, S.S., C. Pailman, X. Mahlalela, N. Coetzee, F. Meidany and J.T. Boerma (2003), 'From people to places: focusing AIDS prevention efforts where it matters most', *AIDS*, **17** (6), 895–903, doi:10.1097/01.aids.0000050809.06065.e0.

Wertheim, J.O., S.L. Kosakovsky Pond, L.A. Forgione, S.R. Mehta, B. Murrell, S. Shah, et al. (2017), 'Social and genetic networks of HIV-1 transmission in New York City', *PLoS Pathogens*, **13** (1), e1006000, doi:10.1371/journal.ppat.1006000.

Whiteside, Y.O., S. Ruiguang, J.O. Wertheim and A.M. Oster (2015), 'Molecular analysis allows inference into HIV transmission among young men who have sex with men in the United States', *AIDS*, **29** (18), 2517–22.

Wilson, P.A., P. Valera, A. Ventuneac, I. Balan, M. Rowe and A. Carballo-Diéguez (2009), 'Race-based sexual stereotyping and sexual partnering among men who use the internet to identify other men for bareback sex', *Journal of Sex Research*, **46** (5), 399–413.

Young, L.E., S. Michaels, A. Jonas, A.S. Khanna, B. Skaathun, E. Morgan, et al. (2017), 'Sex behaviors as social cues motivating social venue patronage among young Black men who have sex with men', *AIDS and Behavior*, **21** (10), 2924–34, doi:10.1007/s10461-017-1679-8.

Zablotska, I.B., J. Imrie, G. Prestage, J. Crawford, P. Rawstorne, A. Grulich, et al. (2009), 'Gay men's current practice of HIV seroconcordant unprotected anal intercourse: serosorting or seroguessing?', *Aids Care-Psychological and Socio-Medical Aspects of Aids/HIV*, **21** (4), 501–10, doi:10.1080/09540120802270292.

Zenilman, J.M., N. Ellish, A. Fresia and G. Glass (1999), 'The geography of sexual partnerships in Baltimore: applications of core theory dynamics using a geographic information system', *Sexually Transmitted Diseases*, **26** (2), 75–81.

18. Modeling dissemination of health information and beliefs in urban social networks

Sara S. Metcalf, Harvey D. Palmer, Qiuyi Zhang and Mary E. Northridge

INTRODUCTION

Social networks play a mediating role in the dissemination of information and beliefs in which trust increases the likelihood that a person relies on a particular source (Huckfeldt and Sprague 1987; Krueger et al. 2017). Trust is often grounded in group identity, where people are more likely to receive information and accept beliefs disseminated by sources aligned with their identity group(s) than from sources aligned with groups that are in conflict with their identity group(s). Group attachment acts as an information filter, in that the beliefs of group members are shaped by an accepted information environment that is biased toward the beliefs of sources aligned with the person's identity group(s). Extant research has found the impact of trust increases with the number and variety of identity groups, that is, group heterogeneity, the average density and closeness of social links within identity groups, that is, intra-group cohesion, and the level of grievance between identity groups, that is, inter-group resentment (Fearon and Laitin 1996; Habyarimana et al. 2007). Insufficient attention, however, has been paid to how these causal factors interact within a single theoretical framework. This chapter contributes to social network theory by developing and experimenting with an agent-based model to simulate the dynamic interaction of group heterogeneity, intra-group cohesion and inter-group resentment in shaping the dissemination of beliefs within a society.

The model developed for this study simulates an information environment overlaid on an urban landscape in which particular identity group sources disseminate subjective beliefs that conflict with information disseminated by neutral or authoritative (government or scientific) sources. Whether an agent receives and accepts a message will depend on its trust in the source, which is specified as a function of group identity, and the extent to which messages received are in conflict with its current opinion, which is a function of past accepted messages. While this specification is consistent with leading theories of information processing in belief formation (Zaller 1992), it more explicitly incorporates the role of confirmation bias. The model is applied to develop theoretical expectations for the substantive context of the dissemination of information and beliefs about oral health care in low-income, Chinese American communities in New York City, in order to inform a pilot study offering oral-health outreach services to underserved adults in need of dental homes (Northridge et al. 2018).

Health care, at least relative to climate or politics, might appear to be a context in which individuals would value information more on objective grounds and be less likely to apply subjective biases. Moreover, the search costs for health-care information have arguably declined with the growth of the Internet and communication technology (Song et al.

2016). Social influence and subjective biases are evident, however, in the persistence of health-care beliefs that conflict with evidence-based knowledge, such as opposition to vaccines and community water fluoridation. We posit that subjective biases are more likely to emerge in health-care contexts involving issues that are not life-threatening, and where it is more difficult to objectively assess the relative effectiveness of alternative treatments. That is, as the individual's definition of wellness becomes more of a value judgment, the likelihood that he or she will rely more heavily on information from subjective, trusted sources, relative to governmental or scientific sources, increases. In our model, the value of truthfulness is inversely related to the impact of group identity on a person's likelihood of accepting information from a neutral or authoritative source relative to that from a source aligned with his or her identity group. Hence, when experimenting with the importance of group identity, we are also considering differences across contexts in relation to the relative value of information validated by a governmental/scientific authority.

This chapter presents a simple agent-based model of belief formation that is designed to allow for the addition of greater complexity in future extensions. It represents a proof-of-concept model both with respect to: (1) developing the central structure for a model that would apply across a wider range of substantive contexts; and (2) demonstrating the model's effectiveness in simulating health-care belief formation in a specific urban context. The focus of the model is the role of group identity as a mediating factor in belief formation. More generally, group identity would be expected to influence multiple elements of the underlying causal process of information processing, including: where people live, the heterogeneity of their social networks, how much information they receive in total, which sources of information they visit and trust, and how easily they can understand the information received. In our model, though, the impact of group identity is restricted to conditioning: (1) a person's acceptance of information from a source type, since only sources disseminate information; and (2) a person's geographic location, which is assigned in a stochastic manner so that the homes of minority group members are clustered around minority information sources (the homes of all other agents are distributed randomly throughout the landscape). The model specification is intended to facilitate future extensions that incorporate greater complexity, such as: (1) social networks that disseminate information, thereby allowing people to learn from each other; and (2) preference-based (rather than random) searches for information sources. Before specifying a more complex, that is, realistic model, we first need to validate that its central structure works as expected in simulating belief formation.

The general research question that motivated our modeling effort was how to better understand the persistence of group beliefs that diverge from evidence-based positions, even when these evidence-based positions have been endorsed by governmental agencies and professional organizations. The next section discusses this research motivation targeted at the more specific goal of exploring a potential explanation for demographic group differences in health outcomes. The subsequent section narrows the focus to differences associated with urban ethnic minority populations, and provides contextual background on the subject area for our model simulation, namely, low-income, Chinese American adults in New York City. The chapter then explicates the assumptions of the agent-based model, justifies the central assumptions through reference to theories established in the extant literature, and motivates three sets of simulation experiments that investigate the impacts of group identity, group size and spatial (geographic) constraint on information exposure. The results of these experiments are then summarized visually and interpreted relative to our theoretical expectations as a form of face validation, with the ultimate goal of improving the central model structure. Finally, the paper concludes with a discussion of future research that will incorporate greater complexity into the model, especially with respect to the role of social networks.

GENERAL RESEARCH MOTIVATION: UNDERSTANDING GROUP DIFFERENCES IN HEALTH OUTCOMES

Researchers have documented notable demographic group differences in health outcomes, including by sex, race, ethnicity and socioeconomic status (Crimmins et al. 2004; Murray et al. 2006; Olshansky et al. 2012). Avoidable and unfair group differences, termed disparities, in the quality of health care – often in the form of disparities in information received about health-care options and treatments – have been found to contribute to longstanding disparities in health outcomes (Weinick et al. 2000; Fiscella et al. 2002; Kirby et al. 2006; Mead et al. 2008). For instance, women are less likely than men to receive proper medical advice when reporting early symptoms of heart problems (Weisz et al. 2004; Wheeler et al. 2005). Researchers have also found significant racial and ethnic disparities in the level of trust in health-care providers and the quality of health care received (Lillie-Blanton et al. 2000; Johnson et al. 2004; Musa et al. 2009). While neighborhood racial and ethnic composition accounts for a large portion of disparities in health-care access (beyond insurance coverage and other institutional factors), many of the differences among racial and ethnic groups remain unexplained (Kirby et al. 2006). We believe that part of the unexplained group differences may be caused by information asymmetries that arise owing to subjective biases in how people process information and form beliefs about health-care options. The agent-based model developed in this chapter seeks to further develop a theoretical framework as the basis for investigating this potential source of racial and ethnic group disparity.

Should we expect group differences in health-care information to persist over time? The answer depends on the latent cause of these differences. If the cause is simply that members of different groups are provided with different information, we should expect this information asymmetry to dissipate over time, unless there are structural factors in society to prevent this. Hence, we should only expect group differences in information to persist if: (1) information search costs are high; (2) the value of the information is subjective at least to a degree; and (3) group membership influences the transmission of information. We would expect members of disadvantaged groups to have insufficient incentive to obtain more valuable information if at least one of these conditions exists.

Health care, at least relative to climate or politics, would seem to be a context in which individuals would value objective information, especially information which is validated scientifically by governmental or academic sources. Similarly, we might expect consumers of health-care information to be less likely to filter evidence based on its consistency with a priori preferences, rather than the reliability of the method used to gather the evidence. Moreover, the search costs for health-care information have, arguably, declined with the growth of the Internet and communication technology (Song et al. 2016). In the context of our agent-based model, a decline in the search costs can be represented by an increase in the radial distance buffer, which effectively allows people to choose among a more inclusive (less geographically bounded) set of information sources.

Nonetheless, there are situations where health-care beliefs that conflict with science-based knowledge persist over time, such as opposition to vaccines and community water fluoridation. Indeed, anti-vaccination beliefs have not only persisted in the face of governmental efforts to counter them, but have reached the point where the World Health Organization recently included this movement on the list of the top 10 threats to global health. While anti-vaccination beliefs are not obviously associated with any demographic group, the persistence of these beliefs raises questions about how the transmission of information shapes belief formation when the consequences of inaccurate beliefs are potentially lethal.

Social science research on opinion formation indicates that social networks play an intermediary role in the transmission and processing of information, and thereby indirectly shape opinion formation (Huckfeldt and Sprague 1987). In the context of politics, most people obtain most of their information from others in their social network, rather than directly from an expert source, and this has only become more so with the expansion of social media. Yet, for a personal issue such as health care, many people may only rely on their social network as the basis for choosing among information sources. Whether or not people receive and accept information depends on how much they trust the sender of that information, and people are often more likely to consider information from trusted sources when forming an opinion (Zaller 1992).

Group identity is an important basis of trust for many people (Lowry et al. 2010), and hence we should expect an individual who identifies with a group to be more likely to receive and accept information from sources that identify with the same group (Foddy et al. 2009; Metzger and Flanagin 2013). Moreover, the importance of group identity as an information filter is likely to increase with factors that influence interpersonal trust among group members and between members of different groups. Groups that are more homogeneous with stronger member ties and fewer disparate interests have greater intra-group cohesion and, presumably, greater interpersonal trust among group members (Habyarimana et al. 2007; Chen and Li 2009). In contrast, interpersonal trust among members of different identity groups ought to be lower in environments where inter-group conflict and grievance is high (Fearon and Laitin 1996).

As a foundation that is flexible enough to allow for the introduction of greater complexity, especially in respect of the role of social networks, our agent-based model restricts the impact of group identity to influencing: (1) the likelihood that a person accepts information received from a source (where exposure to the source/reception of the information is determined by a random process); and (2) the geographic distribution of households for the minority group. While this specification is restrictive, group identity is the only causal factor beyond the geographic distribution of households (relative to the fixed locations of the information sources) that drives group differences in information acquisition (consumption) and hence belief formation. Moreover, it accounts for the vast majority of variation over time in the group-specific belief distributions, even as the overall variability is conditioned by the geographic distribution of households, since this dictates the range within which the mean position of each group's distribution of beliefs will lie. The central role of group identity in our model is justified by this factor being exogenous to social network membership. Our theoretical framework seeks to capture the latent impact of social networks in their most basic form, namely, group identity influencing interpersonal trust and where members of the minority community live. A natural extension of the model for future research would incorporate social networks more directly by allowing people to acquire information from other people conditioned on shared social network membership, rather than only from the sources that originally disseminated the information.

In summary, our study seeks to develop an agent-based model of belief formation in which group identity plays a central but limited structural role. We seek to assess whether this causal factor alone can produce groups with divergent beliefs, where the macro-level behavior of the distribution for each group is consistent with established theories of how individuals process information and form opinions. To the extent that ethnicity is a form of group identity that conditions a typical person's likelihood of accepting health-care information, our model has the potential to provide theoretical insight into how information asymmetries can persist within ethnic minority groups. Toward this goal, it is informative to simulate the model by applying it to a real-world geographic landscape in which sizable minority groups reside. For our study, we apply the model to the geographic landscape of New York City with the goal of informing future interventions designed to improve the oral health and health care of low-income, Chinese Americans living in urban ethnic enclaves.

STUDY BACKGROUND: HEALTH EQUITY FOR URBAN ETHNIC MINORITY POPULATIONS

Health problems for members of racial and ethnic minority groups often result from the substantial cultural, social and economic barriers they face in obtaining high-quality health care (Mead et al., 2008). The disproportionate health burdens experienced by members of racial and ethnic minority groups are caused by many factors, including discrimination, language barriers, low socioeconomic status (Crimmins et al. 2004), inadequate access to health care (Weinick et al. 2000) and inferior quality of health care received (Lillie-Blanton et al. 2000; Mead et al. 2008). Distrust of health providers may exacerbate health disparities owing to reduced utilization of preventive and treatment services (Musa et al. 2009).

Members of particular immigrant and ethnic groups, such as Chinese Americans, arguably experience substantially larger barriers to health-care receipt than other groups, owing to language differences and concerns over citizenship status. They might also hold shared beliefs about dental and medical care that influence their propensity to seek preventive care prior to experiencing acute oral pain or discomfort. These shared beliefs may stem in part from cultural influences that emphasize homeopathic and alternative medicines and treatments.

Oral health disparities reflect fundamental social inequities, since obtaining oral health care is often expensive even with insurance coverage, and may be erroneously considered as a luxury rather than fundamental to overall health and well-being. Ethnic minority groups often have low socioeconomic status, so group differences often conflate ethnicity and socioeconomic status. In the context of information acquisition, however, socioeconomic status is expected to be far less important than ethnicity. This may be particularly

Table 18.1 Prevalence of the Chinese American population by borough, New York City

County	Borough	Total population	Chinese American population	Percentage Chinese American
Bronx	The Bronx	1 455 846	7364	0.51
Kings	Brooklyn	2 635 121	208 972	7.93
New York	Manhattan	1 653 877	108 990	6.59
Queens	Queens	2 339 280	244 232	10.44
Richmond	Staten Island	475 948	16 011	3.36
Total		8 560 072	585 569	6.84

true for information about preventive dental care, given the expansion of information technology and online access to health-care information across social strata.

In addition to contributing to the development of a general dynamic model of belief formation, our study seeks to improve understanding of how differences in health outcomes might be influenced by beliefs held by Chinese Americans about choices that influence oral health, such as regular dental visits for preventive care and oral-hygiene behaviors that include brushing with fluoride toothpaste. Cultural differences may affect whether dental care is seen as separate from medical care or whether preventive dental care is seen as necessary for the maintenance of overall health. Similarly, social networks may reinforce or counter individual beliefs about the importance of fluoride in toothpaste and water in preventing carious lesions and improving dental health, as opposed to the perception of fluoride as unnecessary or even poisonous.

Chinese Americans constitute a substantial minority population in New York City, the site of our study. Table 18.1 presents the distribution of Chinese American residents by county (borough), while Figure 18.1 provides a more detailed geographic understanding by illustrating Census-tract level differences in the Chinese American population density. The data used to construct Table 18.1 and Figure 18.1 were obtained from the US Census Bureau's 2012–2016 American Community Survey 5-year estimates (US Census Bureau 2016).

Chinese Americans account for almost 7 percent of the total New York City population. The highest concentrations of Chinese Americans are found in Queens (10 percent) and Brooklyn (8 percent), at levels that are higher than in Manhattan (7 percent), the borough where the historic Chinatown neighborhood is located. Given this geographic distribution, the landscape for our model simulation only includes these three boroughs.

AGENT-BASED MODEL OF BELIEF FORMATION

Why model belief formation using an agent-based approach? The primary reason is that social science theories of information processing and opinion formation characterize the causal process as complex and dynamic. Hence, an appropriate theoretical representation of this process should allow for dynamic interaction among the causal factors. An agent-based modeling approach enables us to consider the dynamic interaction of various factors that have been considered individually in the literature on opinion formation, that

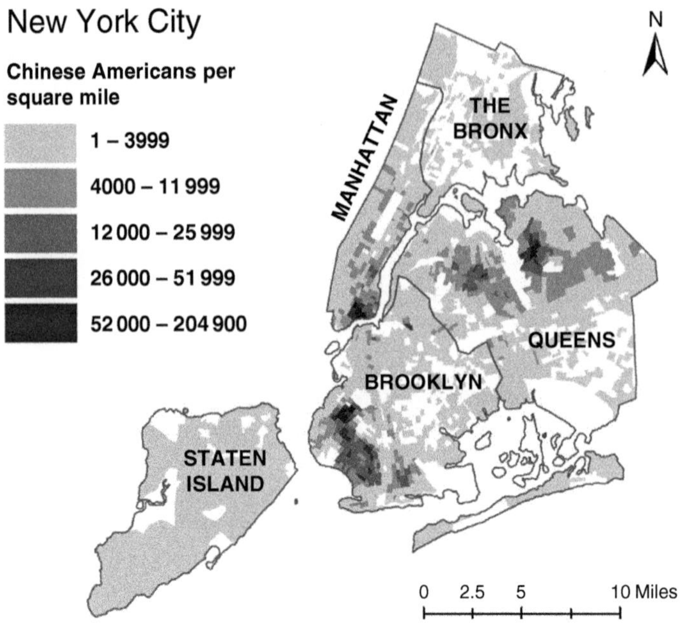

Figure 18.1 The population density of Chinese Americans by Census tract, New York City

is, confirmation bias and interpersonal trust based on group identity, conditioned by the relative size and geographic dispersion of identity groups.

This study builds upon previous efforts to simulate social networks in agent-based models of urban dynamics (Metcalfe et al. 2013). Some of these models enable consideration of the dynamics by which social ties form and break through community interaction (Metcalf and Paich 2005; Metcalf 2014; Jin et al. 2018). Agent-based models have also been used to consider transportation networks together with social networks in mediating access to oral health care (Zhang et al. 2018a). Recent developments in agent-based modeling have facilitated the explicit representation of patients' trust in dental providers and how this trust might be influenced by their social networks (Krueger et al. 2017; Zhang et al. 2018b). Figure 18.2 illustrates a simulation from the model by Zhang et al. (2018b) in the AnyLogic software platform that visualizes the spatial social networks among Chinese American adults visiting community centers in New York City. At these centers, attendees have the opportunity to participate in outreach events that involve screening for oral conditions, such as cancer and dental caries, and oral health education with demonstrations of brushing with fluoride toothpaste and flossing techniques, together with referrals for dental care to local providers as indicated.

The specification of our agent-based model adapts the theoretical foundation of the receive–accept–sample model of opinion formation advanced by Zaller (1992). In this theoretical model, citizens (members of the public) are not considered opinion holders, a priori, but form opinions by sampling relevant 'considerations at the top of their head' when asked to express or form their opinion, as when participating in a survey or presented with treatment choices by a dental-care provider (Zaller 1992, p. 62). For the

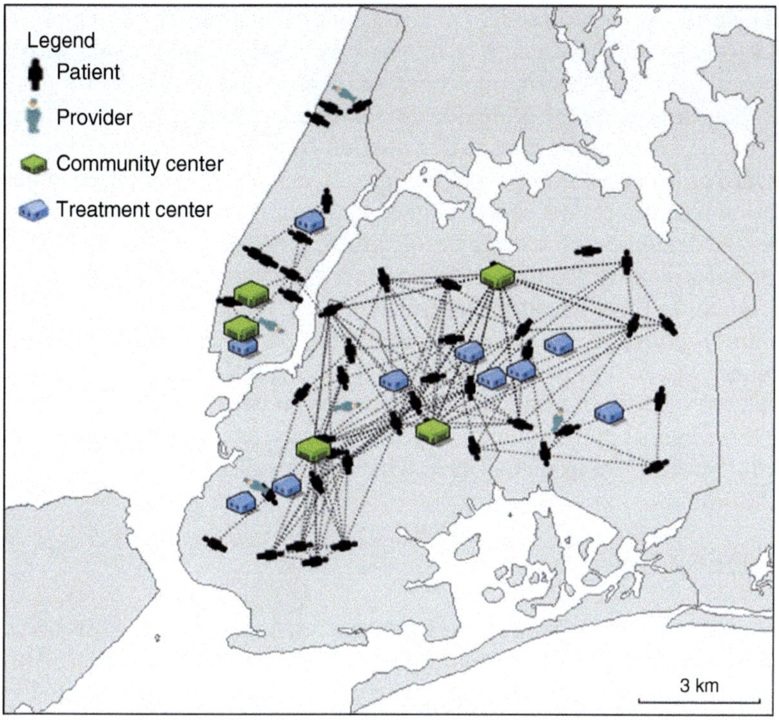

Source: Zhang et al. (2018b).

Figure 18.2 Urban social networks in an agent-based model of oral health equity

purpose of this study, we designate these individuals as 'community members' instead of 'citizens,' since the immigrant community context for Chinese Americans in New York City includes individuals who are not naturalized US citizens. In our agent-based model, these community members are agents who leave their homes to search for information.

The considerations available to community members are disseminated by elites in an information market. In our agent-based model, these elites are represented by fixed agents who disseminate information (considerations) to community members who visit them. Each community member receives a subset of these considerations based on his or her level of awareness. In our initial specification of the model, community members are homogeneous in their level of awareness, since they all have the same likelihood of leaving their homes to search for information. They then accept a subset of the received considerations based on their levels of trust (or distrust) in the source and the strength of their a priori beliefs (confirmation bias). Considerations accepted by community members are retained in memory for a limited amount of time.

Research has established that message reception is contingent on how much the receiver trusts the sender, where the likelihood of reception increases with the level of trust (Palmer and Duch 2001; Foddy et al. 2009; Lowry et al. 2010). Interpersonal trust is strongly influenced by group attachment and intergroup conflict. The level of trust for in-group senders relative to out-group senders increases with the strength of group attachment (Chen and Li 2009), and the strength of this positive relationship increases with the levels of intergroup conflict and intra-group cohesion (Fearon and Laitin 1996).

Suppose that the strength of group identity is represented by $P_{in} - P_{out}$, where P is the likelihood that a person agent accepts information from a source agent, conditioned on the congruence of his or her group type $\{in, out\}$. P_{in} denotes the likelihood of acceptance of information received from a source aligned with his or her identity group (in-group), P_{out} denotes the likelihood of acceptance of information received from a source aligned with an opposing identity group (out-group) and P_{neu} denotes the likelihood of acceptance of information from a neutral (scientific) source. We expect $P_{neu} - P_{out}$ to decrease with intra-group cohesion, to the extent that intra-group cohesion is associated with group members interacting less with information sources other than those that are aligned with their identity group. In summary, experimentation with our agent-based model can investigate how changes in the strength of group identity, owing to changes in inter-group conflict and intra-group cohesion, influence belief formation.

MODEL DESIGN

The agent-based model developed in this study was implemented in the AnyLogic simulation software (version 8.4).[1] In this basic model of belief formation, there are two types of agents: person agents who are community members, and source agents who disseminate information through messages promoting their belief position. Person agents move around the landscape searching for information, which consists of interacting with source agents (who do not move). Source agents have fixed beliefs as opinion holders, whereas person agents initially adopt a centrist (or indifferent) position, but eventually form a distinct opinion from information provided by source agents that changes dynamically over time. Person agents retain information received from a source agent, contingent on the extent to which they trust the source agent and the extent to which the information is consistent with their a priori beliefs. The latter factor constitutes the impact of confirmation bias on information processing. In our model specification, these two factors are weighted equally in determining the likelihood that a person agent accepts information received from a source agent.

Whether a person agent trusts a source agent depends on his or her group identity relative to that of the source agent. Each person agent is stochastically assigned to either a minority or majority ethnic identity group according to a specified probability (25 percent minority in the baseline scenario). Each source agent is either neutral or aligned to the minority or majority identity group. Everything else being constant, a person agent's trust in a neutral source agent does not vary with his or her group membership, but trust in other source agents will be higher or lower depending on whether the source agent is aligned with his or her identity group or with the opposing identity group.

Beliefs are represented as values ranging from 0 to 1, where 1 denotes the majority position, 0 denotes the minority position and the neutral (non-aligned) position is set at 0.8. At any specific time, a person agent's belief is the weighted sum of the information that he or she has in memory, with greater weight being placed on information retained more recently. Information retained by community members is only held in memory for a limited amount of time. Each piece of information is randomly assigned a forgetting time that ranges from seven to 11 days.

This one-dimensional conception of beliefs will better characterize issues where opinions represent how strongly someone supports or agrees with a particular belief or treatment, perhaps constituting a choice between two well-recognized alternatives, but not among a set of options that cannot be ordered obviously on a theoretically meaningful continuum. Similarly, our model will better characterize belief formation for issues where there is a true centrist opinion, which could be indifference between two options, since person agents are assumed to start with a belief at 0.5. Our focus is on health-care issues that are not immediately life-threatening, and therefore people will assess the value of information based on their subjective biases, as well as on its factual accuracy and the expertise of the source. For these issues, it simplifies the model to assume that the average person will start from an attitude of indifference. Also, while the location of the neutral position is arbitrary, placing it closer to the majority position presents a more interesting scenario, given our focus on why minority beliefs might diverge from the scientific consensus. Owing to being closer to the majority position, the effect of confirmation bias on the likelihood of accepting scientific information will differ with group identity, where agents with opinions close to the minority position will be less likely to accept scientific information than those with opinions close to the majority position, all else being equal.

In the baseline scenario, there are 15 source agents and 500 person agents. The source agents represent outreach sites, such as community centers, where health information is disseminated. Source agents are situated throughout the New York City landscape with nine sources promoting the majority position, three sources promoting the minority position and three sources promoting the neutral position. Informed by Census-tract data on the geographic distribution of Chinese Americans (see Figure 18.1), the three minority sources were placed manually in Brooklyn, Manhattan and Queens (one in each borough) at central locations where the population density of Chinese Americans was the highest. Then the remaining 12 sources were placed randomly throughout these three boroughs. While this placement was random, to minimize boundary effects, it was limited to exclude locations on the fringes of the urban area. These source agent locations were then spatially fixed for the subsequent simulation experiments, so as to better compare outcomes across different model scenarios. Figure 18.3 illustrates the information source landscape, where circles represent minority sources, triangles represent majority sources, and squares represent neutral sources.

The person agents' residences were placed on the landscape after the source locations were determined. The residence locations were set at random with the constraint that minority community members are more likely to live closer to a minority information source. Figure 18.3 presents the residences of minority community members (people icons), where the two panels contrast (a) the baseline distribution with clustering with (b) an alternative distribution that is completely random. In this scenario, 25 percent of the person agents (125) were assigned to the minority identity group, while the remaining 75 percent of the person agents (375) were assigned to the majority identity group (their residential locations are not shown in Figure 18.3). The locations of residences, as for information sources, were fixed across the simulation experiments. The clustered pattern of minority group members in proximity to minority sources, as exhibited in Figure

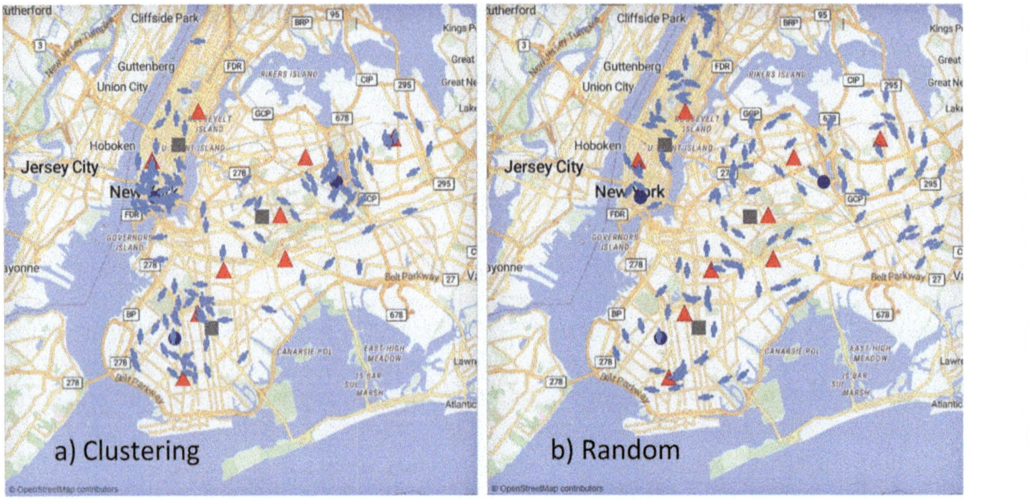

Figure 18.3 *Clustering (a) versus random (b) placement of simulated minority group members on an urban information landscape comprised of minority sources (circles), majority sources (triangles) and neutral sources (squares)*

18.3(a), provides the basis for generating the simulation experiments presented in the following section.

During the simulation, person agents move around the urban information environment; their movement is determined by a distance buffer (set to a 5-kilometer radius in the baseline scenario) and the location of the nearest source agent. Although features such as roads are visualized in the geographical information system (GIS) base maps of Figure 18.3 to provide context for the model, these features do not affect agent movement during simulation. Each day, each person agent first randomly decides whether to leave his or her residence to search for information. Note that our model specification assumes that they leave home 80 percent of the time. Person agents who choose to search will then visit a source agent within their distance buffers, if possible; otherwise, they will visit the nearest source agent. If more than one source agent is within their distance buffers, they will randomly choose one of them (weighted equally regardless of group alignment) to visit. Hence, the belief positions of person agents will change dynamically over time as they are exposed to information promoting different positions from different source agents. While their exposure to information is passive (that is, with the source agent being determined randomly when a choice is faced), person agents actively decide which information to accept based on their group identity and a priori beliefs. The distribution of source types – neutral, minority or majority – to which a person agent is exposed will differ across agents depending on their residence locations, but will become more like the total distribution (and more homogeneous across person agents) as the distance buffer increases.

The geographic clustering of minority residences can be justified both as a theoretically appropriate conception of how group identity shapes human behavior at a fundamental level, and as a more accurate empirical representation of the substantive context of interest. Yet, it is also necessary for model performance reasons, since otherwise the larger number of majority sources (and associated volume of majority belief messages) would negate the effect of group identity on the probability of acceptance, as this is conditioned on reception (that is, visiting a source). The number of majority sources is three times larger than the number of minority sources in order to match the minority–majority split of person agents specified in the baseline scenario. This difference, though, implies that minority person agents will visit majority source agents more often than minority source agents, on average, if their residences are placed completely at random or the distance buffer is set small enough that no more than the most proximate source agent is included in it for a large majority of person agents. Under either of these conditions, the probability of visiting a source agent of a particular type will be similar to the relative frequency of source agents in the space, thereby causing majority source agents to dominate minority source agents, thus exposing more people to information supporting the majority belief position.

The likelihood that a person retains information from a particular source type is the product of the probability that he or she visits a source of that type and the probability of acceptance given the source type. Group identity influences the latter but not the former, and while its effects are symmetric in respect of minority and majority group members, the unequal distribution of source types favors majority aligned sources in information dissemination. The implication of this aspect for the model simulations is asymmetry in the effect of group identity on the source type distribution of information accepted. For members of the majority group, the effect will be as specified, but for members of the minority group, group identity largely shifts the distribution to be more uniform instead of favoring majority sources. It also increases the frequency of days when the person rejects the information received, thereby creating an information gap where minority community members collect less information. Beliefs based on less information obtained from a more uniform distribution of source types will increase the variability over time of the distribution of beliefs among the minority community members.

Finally, another important aspect of the model that will structure the simulation results is the interaction between how confirmation bias is specified and the assumption that the initial belief position of person agents is 0.5. Given that the scientific position is set at 0.8, person agents will initially be more receptive to scientific information sources than their group identities would suggest, since their a priori beliefs are closer to the scientific position. Confirmation bias causes people to be resistant to new information early in the simulation runs, which can work as path dependence in limiting the range within which person agents' beliefs will vary over time as they obtain information. While a simple specification was purposely chosen as a starting point for developing a more complex model in future research, it is necessary to fully recognize the implications of this specification when assessing the representativeness of the simulated agent behavior.

EXPERIMENTAL RESULTS

Simulation experiments were performed with our model to test differences in opinion dynamics under different conditions. These opinion dynamics are compared on the basis of average opinion values for each group over a 100-day time horizon. To ensure

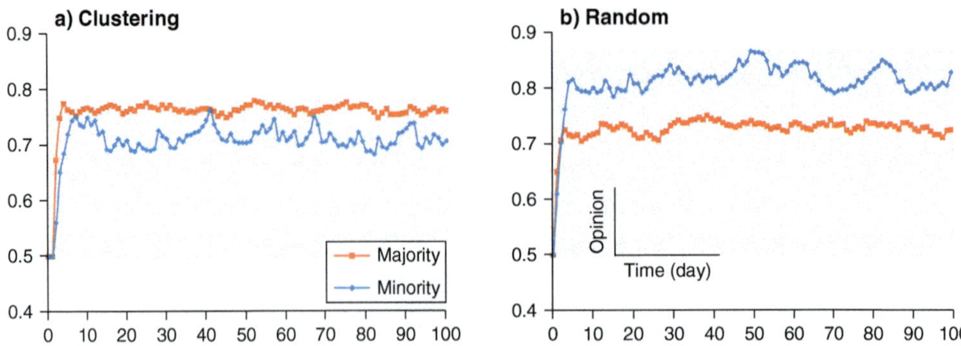

Figure 18.4 Simulation results for clustering (a) versus random (b) placement of minority agents on the information landscape as presented in Figure 18.3

reproducibility and facilitate comparison of outcomes, each scenario represents a single simulation run using a fixed seed for the stochastic mechanisms of the model. Further experimentation across different model realizations of given parameter settings would be appropriate to establish the robustness of model insights and is a topic for future research.

The contrasting patterns of behavior over time exhibited in Figure 18.4 reveal the effect of clustering minority person agents near minority source agents compared with a purely random distribution of minority residences. The latter produces an average minority opinion that is greater than that of the majority opinion, although the minority source position is 0 and the majority source position is 1. This counter-intuitive pattern illustrates that the model specification implicitly assumes a higher rate of exposure to majority sources (given their larger number), which will cause members of the minority community to be exposed to a higher percentage of majority source messages than members of the majority community are exposed to minority source messages. This comparison also demonstrates the utility of incorporating geographic clustering as a reflection of social ties among minority agents. For the clustering scenario, the average minority opinion is consistently less than the majority opinion, although it increases from the starting point of 0.5. For the simulation runs in Figure 18.4, the probability of accepting information from a source aligned with the person's group identity is set to 0.95, while the probability of accepting information from a source aligned with the opposing group is 0.05 and from a neutral source is 0.60. These parameter settings correspond to scenario D (P_{in} = 0.95, P_{out} = 0.05, and P_{neu} = 0.60) in the subsequent analyses.

Using the landscape with the clustered minority agent population, the simulation runs presented below characterize the results of three sets of experiments that vary: (1) the strength of group identity; (2) minority group size; and (3) the distance buffer for community members moving around the information landscape. First, we explore how changing the strength of group identity influences the divergence and variability of the minority community's and majority community's belief distributions. This is achieved by comparing four alternative scenarios that differ as regards the difference between the likelihood of accepting information from an in-group source agent (that is, one aligned with the person agent's identity group) and the likelihood of accepting information from an out-group source agent (that is, one not aligned with the person agent's group), or using the notation introduced above, $P_{in} - P_{out}$. In scenario A, there is no difference: $P_{in} = P_{out} = 0.60$. In the other scenarios, this difference increases by increments of 0.30: for B, $P_{in} = 0.75$ and $P_{out} = 0.45$; for C, $P_{in} = 0.90$ and $P_{out} = 0.30$; and for D, $P_{in} = 0.95$ and $P_{out} = 0.05$. In all four scenarios, the likelihood of accepting information from a neutral source is fixed at $P_{neu} = 0.60$. For this experiment, we expect greater divergence as the strength of group identity increases (that is, comparing sequentially from scenario A through scenario D) since this should increase the proportion of retained information that comes from in-group sources, all else being equal. For the majority community whose members (on average) should already be exposed more often to majority sources, belief variability should decrease with the strength of group identity, since information in memory for the average person should become more homogenous with a greater percentage of information originating from majority sources. For the minority community, however, increasing the strength of group identity will largely offset the exposure advantage of majority community sources, so belief variability is likely to increase.

Figure 18.5 illustrates the results from this first experiment with strength of group identity, by plotting the average opinions of minority and majority group members over the course of the simulation. These results are mixed in relation to how closely they match

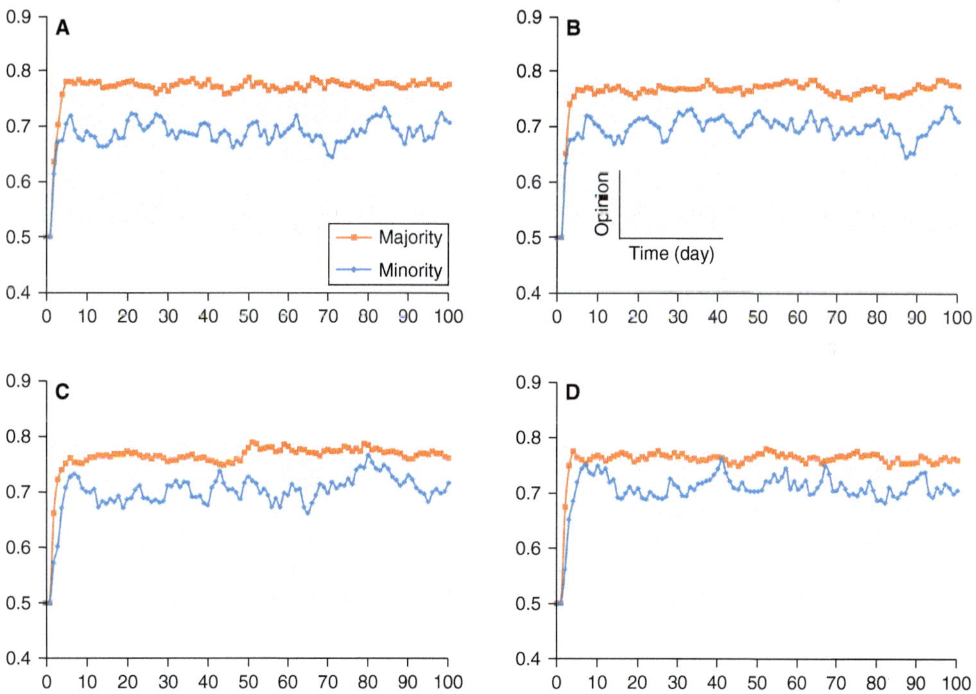

Note: The minority group size is 25 percent and the distance buffer is 5 kilometers.

Figure 18.5 Testing the effect of group identity on average minority and majority opinions simulated over time in four scenarios: A ($P_{in} = P_{out} = 0.60$); B ($P_{in} = 0.75$, $P_{out} = 0.45$); C ($P_{in} = 0.9$, $P_{out} = 0.3$); D ($P_{in} = 0.95$, $P_{out} = 0.05$)

our expectations. While beliefs among the majority community are closer to the majority position and those among the minority community are closer to the minority position, the divergence between these distributions does not increase as expected with the strength of group identity. While the variability of majority opinion does decrease slightly when comparing across scenarios A through D (as expected), the average majority opinion declines surprisingly sharply, despite already being below the neutral source position of 0.8. Similarly, contrary to expectations, the mean of the minority belief distribution is greater, on average, in scenarios C and D than in scenarios A and B, despite already being generally above 0.65. How serious is this lack of consistency between our theoretical expectations and the results of this simulation experiment? Given that our simulations fix the residential locations of the person agents (as well as those of the source agents) after the initial realization of the model, spatial contingency is a likely explanation for the lack of increased divergence between the minority and majority communities as the effect of group identity on information acceptance increases. The results of our next two experiments, especially the latter scenario that varies the distance buffer, will also help in assessing whether the inconsistencies in this experiment are symptomatic of model specification problems or unique to the particular realization of residential locations that was fixed across the simulation runs.

A more nuanced finding from the results of this first experiment, as displayed in Figure 18.5, is the greater variability of minority group opinion relative to majority group opinion in all of the scenarios. We expect that this variability is partly owing to the smaller size of the minority group relative to the majority group, since a change in individual opinion has more influence on the average for a smaller group population. Additional testing with different total population sizes confirms this general pattern of increased variability of opinion dynamics with a smaller population, even if the minority prevalence remains the same.

The second experiment considers how changes in group size influence the dynamics of the belief distribution for the minority community by separately comparing simulation results for scenarios B and D across three minority group sizes: 5 percent and 10 percent, together with the baseline size of 25 percent used in the first experiment. These two additional scenarios are closer to the actual prevalence of Chinese Americans in New York City (see Table 18.1), even as the total simulated population is just 500 agents, since representing all New Yorkers would be computationally prohibitive. In this experiment, we expect greater belief divergence and variability as the minority group size decreases from the baseline level of 25 percent to 5 percent. Figure 18.6 presents the results of this experiment, which should be compared with the right-hand panel of Figure 18.5 to contrast the results for the 25 percent baseline size with those for the 5 percent and 10 percent sizes below.

The results of the second experiment with minority group size largely confirm our expectations, especially when assessed in the context of the first experiment's results, which suggested that spatial contingency is a potential concern. As expected, the variability of the average opinion of the minority community increased as the size of that community decreased. This increased variability confounds the evidence of increased divergence between the average opinions of the minority and majority communities. While the mean difference between the averages is greater for the 5 percent than for 25 percent scenario, it is not consistently greater when considered through a narrower time

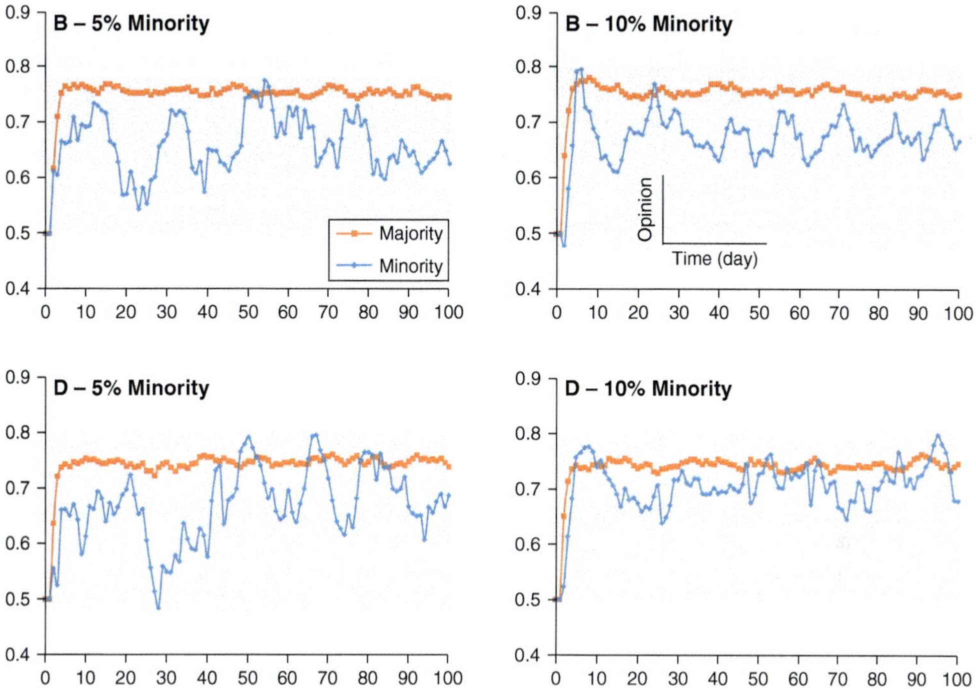

Note: The distance buffer is 5 kilometers.

Figure 18.6 *Testing the effect of smaller (5 percent and 10 percent) minority group size on average minority and majority opinions simulated over time for group identity scenarios B ($P_{in} = 0.75$, $P_{out} = 0.45$) and D ($P_{in} = 0.95$, $P_{out} = 0.05$)*

frame, such as a moving window of 20 days. If spatial contingency is accounting for the minority community being more exposed to majority and neutral sources than would occur on average across all possible landscapes (and matrices of residential locations relative to information sources), as evidenced by the minority group's higher than expected average opinion regardless of group size, then it seems probable that this type of spatial contingency would magnify the expected increase in variability associated with decreasing the group size. Similarly, Figure 18.5 also indicates that the lower than expected average opinion for the majority community is not contingent on group size and hence is probably an artifact of the spatial contingency, since greater than normal exposure of minority persons to majority sources is likely to be associated with greater than normal exposure of majority persons to minority sources.

In the third set of experiments, we investigate how sensitive the simulation results are to the size of the distance buffer. Greater mobility in the urban information landscape owing to a wider distance buffer can be viewed as representing the effects of the historical decline in search costs (increase in information accessibility) owing to advances in transportation and telecommunication technology. That is, a person agent's likelihood of interacting with an information source is less conditioned on the proximity of their household location to the locations of source agents of different types. As in the second experiment, we separately compare simulation results for scenarios B and D across a range of distance buffers from 2 kilometers to 8 kilometers that bracket the baseline scenario buffer of 5 kilometers.

Our theoretical expectations for this experiment are conditioned on group identity, even though increasing the distance buffer to cover the entire space should negate any spatial contingency in the dynamic belief distribution for both groups. For the majority community, we are uncertain how the belief distribution should change over the range between small enough that all agents have no more than their most proximate source within their buffers, and large enough to cover the entire space. At the low extreme, the belief distribution should reflect the full impact of spatial contingency associated with the distribution of types for the source closest to each of the majority person agents. This does not imply, however, that the overall impact of spatial contingency will decline monotonically as the distance buffer is increased to the upper extreme of covering the entire space, at which point the effects of spatial contingency should be completely negated. As the distance buffer increases, the importance of spatial contingency associated with the closest source declines, but the relevance of spatial contingency associated with the second-closest source increases, and as it increases further and further, spatial contingency associated with the third-closest source and then the fourth-closest source (and so on) becomes more relevant. In summary, we have no strong expectations beyond increased variability for the middle range of distance buffers and an average opinion of 0.8 (that is, the neutral position) or greater as the distance buffer approaches the level at which it covers the entire space for every person agent, since the distribution of source types to which a person is exposed should converge to the overall relative frequency distribution at this point. For the minority community, in contrast, our expectations are stronger since increasing the distance buffer will negate the effect of clustering minority residences near minority information sources, thereby increasing how often members of the minority community (as passive information gatherers) are exposed to majority opinion messages. Consequently, the expected effects of group identity for members of the minority community will be more consistently represented in simulations that have a smaller distance buffer, even though the opposite is not necessarily true for members of the majority community. Realization of our expectations in the simulations, however, could be confounded by boundary effects and unusual geographic distributions of person agent residences relative to (out-group) source agent locations (see Figure 18.3), especially given our use of the actual New York City landscape.

The results of these experiments are presented in Figures 18.7–18.10, where Figures 18.7 and 18.8 correspond with scenario B, and Figures 18.9 and 18.10 correspond with scenario D. Figures 18.7 and 18.9 plot the average belief positions of minority and majority group members over time for four alternative distance buffers (2 kilometers, 4 kilometers, 6 kilometers and 8 kilometers) that bracket the 5-kilometer baseline buffer (for which results are displayed in the right-hand panel of Figure 18.5). While these figures provide detail for belief variability, they are not easy to compare in relation to assessing how the average belief positions differ across distance buffers. To address this limitation, Figures 18.8 and 18.10 present bar graphs comparing the overall group-specific averages (together with their minimum and maximum values as whiskers) across the full range of alternative distance buffers, in increments of 1 kilometer.

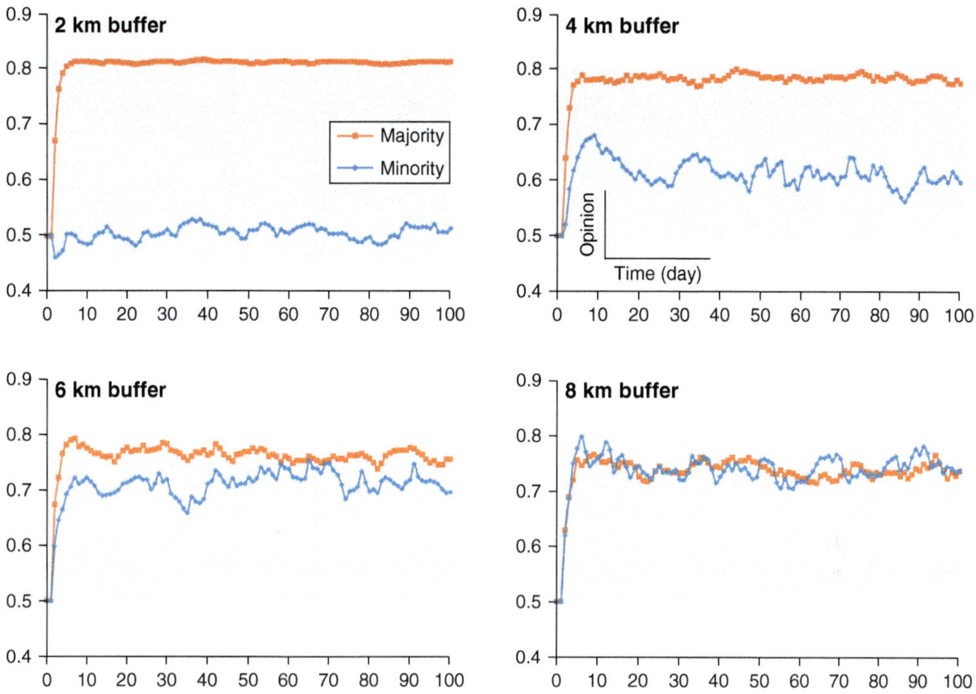

Note: The minority group size is 25 percent.

Figure 18.7 Effect of the distance buffer on average minority and majority opinions simulated over time for group identity scenario B (P_{in} = 0.75, P_{out} = 0.45)

The majority and minority opinion dynamics plotted for scenario B in Figure 18.7 illustrate a distinct difference with the smallest distance buffer of 2 kilometers. At this distance, the variability of opinion dynamics is lower than for the larger buffers. We attribute this more stable trend to the decision rule by which community members visit the nearest source if no sources are located within their buffers. As the distance buffer becomes smaller, this rule is more likely to be invoked, and removes the role of chance in source selection. In contrast, for larger distance buffers, there are likely to be more sources to choose from that fall within the buffer for a given person agent, creating more opinion variability over time for members of both minority and majority groups, until the distance buffer is large enough to cover the entire space for every agent.

Viewing these simulation results as averages across time in Figure 18.8 confirms the observations made about differences between minority and majority opinions in Figure 18.7. The results for the 2-kilometer buffer most closely match our expectations in respect of how group identity conditions the collection of information on which beliefs are formed. As the distance buffer increases, greater exposure to majority and neutral sources cause the belief distribution to become more variable as its center moves away from the minority position. With the inclusion of 3-kilometer, 5-kilometer and 7-kilometer distance buffers in the analysis, a threshold point appears to be around 5 kilometers

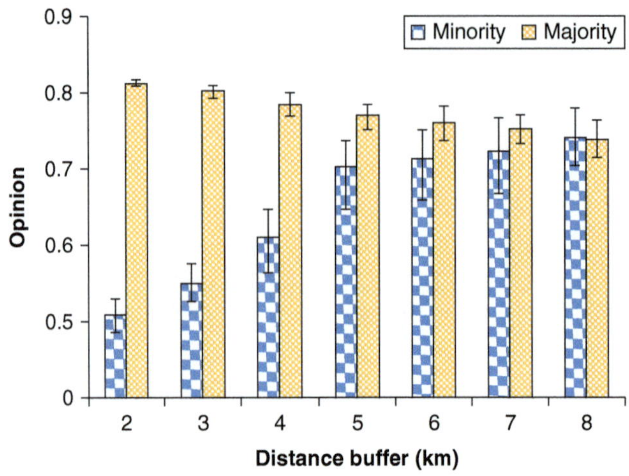

Note: Results correspond with Figure 18.7.

Figure 18.8 Effect of the distance buffer on minority and majority opinion averaged (whiskers indicate maximum and minimum opinions) over the last 70 days of simulation for scenario B

for scenario B. At this distance, the minority opinion is still lower than the majority opinion, on average, but the variability is such that their ranges almost overlap. Above this threshold, the ranges considerably overlap, and under the 8-kilometer buffer, the minority belief distribution almost matches that of the majority belief distribution. Also, as in the previous experiments, the average opinion for the majority community is surprisingly below 0.8 when the distance buffer is less than 4 kilometers, again suggesting that spatial contingency is causing members of the majority community to be exposed to minority sources at a higher rate than that which would be typical across the entire set of possible model realizations of residential locations.

Results of experimentation with the distance buffer for group identity scenario D, as shown in Figures 18.9 and 18.10, are broadly consistent with those for scenario B, as discussed previously for Figures 18.7 and 18.8. As was the case for scenario B, for the distance buffers below 4 kilometers, the belief distributions closely match our expectations as regards how group identity will produce divergence between the average opinions for the minority and majority communities. The threshold point occurs at a shorter distance, though, since the ranges in Figure 18.10 begin to significantly overlap after 4 kilometers. Another point of contrast between scenarios B and D under the largest distance buffer of 8 kilometers is that in scenario D, the minority opinion is consistently higher than the majority opinion, whereas for scenario B, the majority and minority opinions overlap.

DISCUSSION

This study proposes a basic agent-based model of belief formation whose structure is flexible enough to allow for the introduction of complexity consistent with theoretical

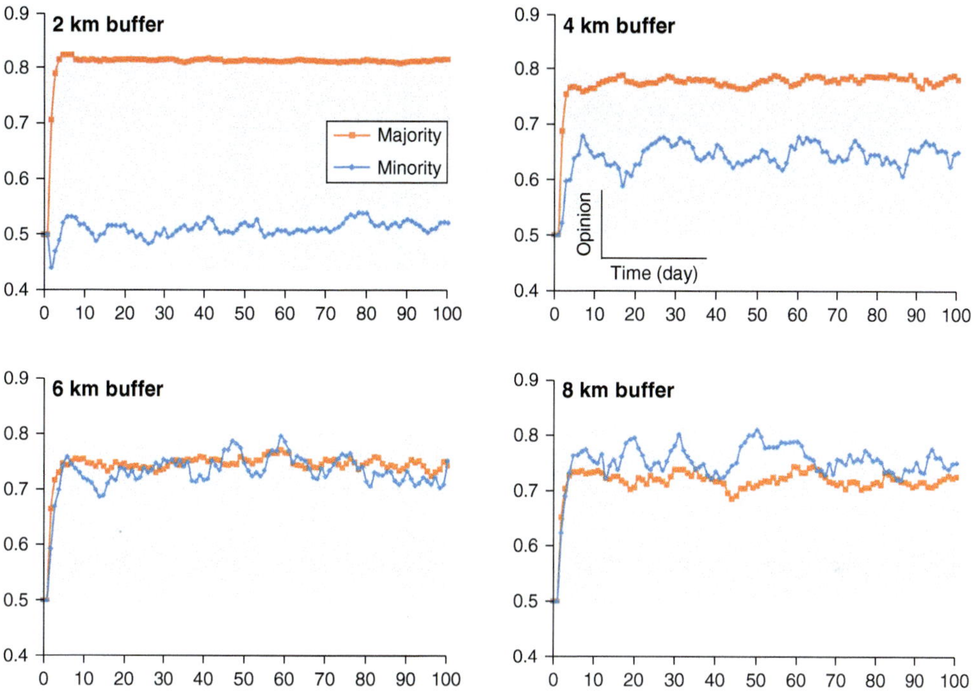

Note: The minority group size is 25 percent.

Figure 18.9 Effect of the distance buffer on average minority and majority opinions simulated over time for group identity scenario D (P_{in} = 0.95, P_{out} = 0.05)

arguments in the social science literature. Simulation of this model to characterize the dynamics of dental-care beliefs for low-income, Chinese Americans residing in New York City provides a proof-of-concept validation of the model's applicability. Preliminary results from experiments to date indicate certain parameter settings that: (1) facilitate the emergence of oscillatory minority opinion cycles; and (2) indicate sensitivity to the prevalence of the minority group in the broader population, with greater belief variability being associated with smaller group size. Changes in the probabilities of acceptance of information from source agents aligned with conflicting identity groups reveal both convergent and divergent outcomes for minority and majority groups across a range of scenarios. Finally, the experimental results demonstrate the relevance of spatial contingency in that group differences in belief dynamics are conditioned on the geographic distribution of person households by identity group relative to the locations of source agents by type. Our experimental results are conditioned on just one of many possible information landscapes, so this limits the generalizability of any conclusions drawn from the simulations.

A central aspect of our agent-based model is its simple structure that allows for extensions that introduce greater complexity, such as that associated with theories of information flow via social networks. In future research, we intend to explore three major extensions as well as several more minor adjustments that would make the model more

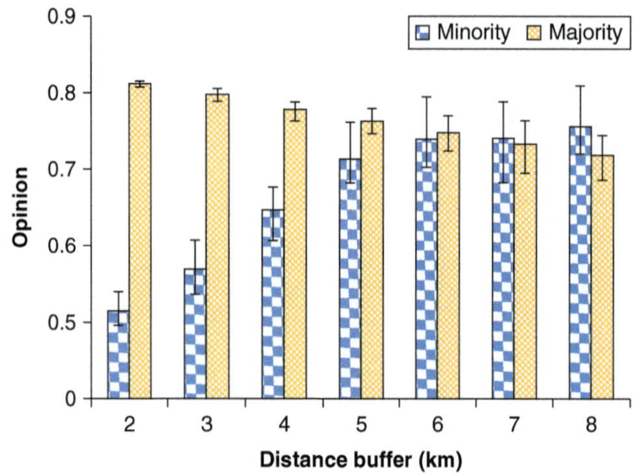

Note: Results correspond with Figure 18.9.

Figure 18.10 Effect of the distance buffer on minority and majority opinion averaged (whiskers indicate maximum and minimum opinions) over the last 70 days of simulation for scenario D

realistic and generalizable. More specifically, the major extensions will incorporate: (1) social networks that disseminate information, thereby allowing people to learn from each other; (2) alternative ways to generate geographic clustering (sorting) by group identity; and (3) preference-based (instead of random) searches for information sources.

The first extension will be implemented by differentiating among person agents to allow some to act as opinion leaders who interact with other person agents by sharing information promoting their own belief positions. The second extension will involve mechanisms such as selecting the person household locations in a manner that clusters based on identity group, locating source agents and person households of the same identity group more proximate to each other on average, and distributing agents on the landscape in accordance with GIS data from the US Census on the population density of Chinese Americans in New York City. By generating a range of alternative landscapes and varying the random number generator governing stochastic mechanisms, this second extension will also minimize the contingency of results on the specified spatial context and stochastic sequence of events realized in the simulation experiments. Finally, the third extension will be implemented by conditioning the movement of person agents based on the location of source agents from which they previously retained information. Intuitively, this extension will incorporate a different form of trust developed through repeated interaction, instead of solely on identity group attachment.

In addition to these major extensions, we plan to experiment further with the model by considering additional adjustments to the specification of the likelihood of retaining (accepting) information from a particular source agent and to how person agents form beliefs from the information in memory. In particular, we will experiment with relaxing the assumption that confirmation bias and group identity have equal weight in determining the probability of retaining information from a source agent aligned with an identity group. In addition to considering a range of different weights, we will also experiment with conditioning confirmation bias on group type such that it is only activated if the source agent is not aligned with the person agent's identity group. Moreover, we plan to adjust the specification of forgetting time so that it is not independent of the total information in memory at a given moment. In future work, the forgetting time assigned to each piece of information will be conditioned on the total number of pieces of information already held in memory. Under this formulation, a person agent's likelihood of forgetting a piece of information would increase with the total amount of information in memory and with how long ago the information was initially retained. Finally, we intend to experiment with changes to the fixed likelihood that person agents retain information from a neutral source agent. As intra-group cohesion increases, community members will interact more frequently with members of their own identity group and less frequently with members of other identity groups. A natural consequence of this less inclusive social interaction would be greater mistrust of information from all sources other than those aligned with the individual's identity group. Hence, decreasing the likelihood of acceptance from a neutral source (while keeping everything else constant) would represent the effect of stronger intra-group cohesion.

ACKNOWLEDGMENT

This research was supported in part by NIH/NIDCR U56 award DE027447, 'Implementing a Participatory, Multi-level Intervention to Improve Asian American Health' (Northridge/Trinh-Shevrin/Metcalf).

NOTE

1. The model is available upon request and may also be simulated online using the following link: https://cloud.anylogic.com/model/8a920459-b840-4017-9564-517fc0d116da?mode=SETTINGS (accessed 8 March 2021).

REFERENCES

Chen, Y. and S.X. Li (2009), 'Group identity and social preference', *American Economic Review*, **99** (1), 431–57.

Crimmins, E.M., M.D. Hayward and T.E. Seeman (2004), 'Race/ethnicity, socioeconomic status, and health', in N B. Anderson, R.A. Bulatao and B. Cohen (eds), *Critical Perspectives on Racial and Ethnic Differences in Health in Late Life*, Washington, DC: National Academies Press, pp. 310–52.

Fearon, J.D. and D.D. Laitin (1996), 'Explaining interethnic cooperation', *American Political Science Review*, **90** (4), 715–35.

Fiscella, K., P. Franks, M.P. Doescher and B.G. Saver (2002),'Disparities in health care by race, ethnicity, and language among the insured: findings from a national sample', *Medical Care*, **40** (1), 52–9.

Foddy, M., M.J. Platow and T. Yamagishi (2009), 'Group-based trust in strangers: the role of stereotypes and expectations', *Psychological Science*, **20** (4), 419–22.

Habyarimana, J., M. Humphreys, D.N. Posner and J.M. Weinstein (2007), 'Why does ethnic diversity undermine public goods provision?', *American Political Science Review*, **101** (4), 709–25.

Huckfeldt, R. and J. Sprague (1987), 'Networks in context: the social flow of political information', *American Political Science Review*, **81** (4), 1197–216.

Jin, Z., M.E. Northridge and S.S. Metcalf (2018), 'Modeling the influence of social ties and transportation choice on access to oral healthcare for older adults', *Applied Geography*, **96** (July), 66–76, doi:10.1016/j.apgeog.2018.05.005.

Johnson, R.L., S. Saha, J.J. Arbelaez, M.C. Beach and L.A. Cooper (2004), 'Racial and ethnic differences in patient perceptions of bias and cultural competence in health care', *Journal of General Internal Medicine*, **19** (2), 101–10.

Kirby, J.B., G. Taliaferro and S.H. Zuvekas (2006), 'Explaining racial and ethnic disparities in health care', *Medical Care*, **44** (5), I64–I72.

Krueger, L.K., R. Flint MacBride and N.A. Osgood (2017), 'Beyond drill and fill: modeling the impacts of risk-based care on oral health disparities', in G.A. Kaplan, A.V. Diez Roux, C P. Simon and S. Galea (eds), *Growing Inequality: Bridging Complex Systems, Health Disparities, and Population Health*, Washington, DC: Westphalia Press, pp. 229–54.

Lillie-Blanton, M., M. Brodie, D. Rowland, D. Altman and M. McIntosh (2000), 'Race, ethnicity, and the health care system: public perceptions and experiences', *Medical Care Research and Review*, **57** (1) suppl., 218–35.

Lowry, P.B., D. Zhang, L. Zhou and X. Fu (2010), 'Effects of culture, social presence, and group composition on trust in technology-supported decision-making groups', *Information Systems Journal*, **20** (3), 297–315.

Mead, H., L. Cartwright-Smith, K. Jones, C. Ramos, K. Woods and B. Siegel (2008), *Racial and Ethnic Disparities in US Health Care: A Chartbook*, New York: Commonwealth Fund.

Metcalf, S.S. (2014), 'Modeling social ties and household mobility', *Annals of the Association of American Geographers*, **104** (1), 40–59, doi:10.1080/00045608.2013.846152.

Metcalf, S.S. and M. Paich (2005), 'Spatial dynamics of social network evolution', *Proceedings of the 23rd International Conference of the System Dynamics Society*, Boston, MA: System Dynamics Society, accessed 8 March 2021 at http://www.systemdynamics.org/conferences/2005/proceed/papers/METCA487.pdf.

Metcalf, S.S., M.E. Northridge, M.J. Widener, B. Chakraborty, S.E Marshall and I.B. Lamster (2013), 'Modeling social dimensions of oral health among older adults in urban environments', *Health Education and Behavior*, **40** (1S), 63S–73S, doi:10.1177/1090198113493781.

Metzger, M.J. and A.J. Flanagin (2013), 'Credibility and trust of information in online environments: the use of cognitive heuristics', *Journal of Pragmatics*, **59** (December), 210–20.

Murray, C.J.L., S.C. Kulkarni, C. Michaud, N. Tomijima, M.T. Bulzacchelli, T.J. Iandiorio, et al. (2006), 'Eight Americas: investigating mortality disparities across races, counties, and race-counties in the United States', *PloS Medicine*, **3** (9), 1513–24, doi:10.1371/journal.pmed.0030260.

Musa, D., R. Schulz, R. Harris, M. Silverman and S.B. Thomas (2009), 'Trust in the health care system and the use of preventive health services by older Black and White adults', *American Journal of Public Health*, **99** (7), 1293–9.

Northridge, M.E., S.S. Metcalf, S. Yi, Q. Zhang, X. Gu and C. Trinh-Shevrin, for the Implementing a Participatory, Multi-level Intervention to Improve Asian American Health Research Team (2018), 'A protocol for a feasibility and acceptability study of a participatory, multi-level, dynamic intervention in urban outreach centers to improve the oral health of low-income Chinese Americans', *Frontiers in Public Health*, **6**, art. 29, 1–20, doi:10.3389/fpubh.2018.00029.

Olshansky, S.J., T. Antonucci, L. Berkman, R.H. Binstock, A. Boersch-Supan, J.T. Cacioppo, et al. (2012), 'Differences in life expectancy due to race and educational differences are widening, and many may not catch up', *Health Affairs*, **31** (8), 1803–13, doi:10.1377/hlthaff.2011.0746.

Palmer, H.D. and R.M. Duch (2001), 'Do surveys provide representative or whimsical assessments of the economy?', *Political Analysis*, **9** (1), 58–77.

Song, H., K. Omori, J. Kim, K.E. Tenzek, J. Morey Hawkins, W.Y. Lin, et al. (2016), 'Trusting Social Media as a Source of Health Information: Online Surveys comparing the United States, Korea, and Hong Kong', *Journal of Medical Internet Research*, **18** (3), e25, doi:10.2196/jmir.4193.

US Census Bureau (2016), '2012-2016 American Community Survey 5-year estimates', US Census Bureau, Washington, DC.

Weinick, R.M., S.H. Zuvekas and J.W. Cohen (2000), 'Racial and ethnic differences in access to and use of health care services, 1977 to 1996', *Medical Care Research and Review*, **57** (1) suppl., 36–54.

Weisz, D., M.K. Gusmano and V.G. Rodwin (2004), 'Gender and the treatment of heart disease in older persons in the United States, France, and England: a comparative, population-based view of a clinical phenomenon', *Gender Medicine*, **1** (1), 29–40, doi:10.1016/S1550-8579(04)80008-1.

Wheeler, D.J., R.W. Grant, J.B. Meigs, D.M. Nathan and E. Cagliero (2005), 'Sex disparities in treatment of cardiac risk factors in patients with Type 2 diabetes', *Diabetes Care*, **28** (3), 514–20, doi:10.2337/diacare.28.3.514.

Zaller, J. (1992), *The Nature and Origins Of Mass Opinion*, Cambridge: Cambridge University Press.

Zhang, Q., M.E. Northridge, Z. Jin and S.S. Metcalf (2018a), 'Modeling accessibility of screening and treatment facilities for older adults using transportation networks', *Applied Geography*, **93** (April), 64–75, doi:10.1016/j.apgeog.2018.02.013.

Zhang, Q., M.E. Northridge, S. Yi and S.S. Metcalf (2018b), 'Modeling multi-level factors to promote oral health equity for Chinese Americans', One Hundred and Fourteenth Annual Meeting of the American Association of Geographers, New Orleans, LA, 10–15 April.

19. Conflict resolution and opinion pooling in city planning

Michael Batty

INTRODUCTION

There are multiple theoretical, academic and professional perspectives associated with the study of cities. A key distinction is between urban science that reflects the city itself in relation to its physical form and function, in contrast to city planning where the main focus is on reconciling conflicting views about how the quality of life and the sustainability of cities might be improved. The development of networks in each of these domains is very different. In urban science, the concern is for the way the city functions through a multitude of networks that embody urban processes that distribute energy, materials, ideas and social interactions between its many parts. In city planning, however, the concern is largely with ways in which different views about the future of cities can be represented and then reconciled, where the network of connections between various factors that are important in this process of design often remains implicit. That is, networks in urban science are those that define the city and its component parts, whereas networks in city planning define ways in which ideas about the future are related to one another. Most of the chapters in this book are about the former – urban processes that determine how cities function and evolve – whereas in this chapter, the emphasis is on how networks can be used to represent processes associated with the actual planning of the city. These two different approaches to networks are usually developed by different constituencies of researchers and professionals, and there are few attempts at reconciling them (Batty 2013).

In city planning, there are many varieties of process that are used to explore the future; from intuitively inspired design to formal policy-making organized through different actors and stakeholders, to mathematical models that seek to find the best locations that optimize the goals and objectives that define more sustainable, equitable and efficient futures. Here we articulate these processes formally, defining various objectives that either support or conflict with one other, and which must be resolved to produce a consensus or compromise. The way in which our model of the planning process works is by defining the key relationships between actors or agents who have a stake in the outcome (the future plan), with this set of relationships represented as a network. The network provides the basis for examining conflicts and concurrences, and then resolving these if this is possible. To demonstrate its use, we define different variants of planning problem to illustrate the process of resolution that the model attempts to achieve and, in this case, we assume that the different objectives can be represented by different physical locations which imply where the best locations are for the future city. As these locations differ, the model enables a rational process of conflict resolution which homes in on the overall best location. This is achieved by altering the location or by resolving the differences in this location as actors in the process become aware of the conflicts involved and the need to compromise between them.

The model we outline is generic in that it can be applied to many different types of planning problem. Our ability to resolve conflict in this type of model is a very special case. There are many representations of the problem where the network of relationships is highly fractured and where it is not possible to reach a consensus. Arguably, many planning problems are of this nature and there is a wide class of problems that involve designing the network rather than using it to resolve a problem. That is, we can think of planning problems as involving different types of optimization and, in this context, we use the network model to achieve an optimization. Here we assume the network of relationships is akin to a social network and that actors who identify with specific objectives resolve their differences according to the objectives held by those to whom they are directly linked. If this process is achieved systematically and in an ordered way, a consensus akin in this context to an optimization is realized. We might begin with a fractured network where no consensus can be realized, but in that case we might also repair the network so that, ultimately, every actor is related to every other either directly or indirectly, thus ensuring a consensus. There are many possibilities but here our goal is to demonstrate the importance of thinking about city planning in this relational way. The model is thus more similar to a method for planning than anything that exists as a formal network in the city itself. The networks we work with here are part of the city, for planning is often seen as simply another function in the city – planning often being said to be part of the more generic problem – but this is another perspective and we will not develop it further in this chapter.

In the next section, we first outline the model using the simplest possible example which serves to introduce the type of network we work with, the objectives represented hypothetically in numerical terms, the method of compromise and the convergence towards a solution. This is sometimes termed opinion pooling, and it has a formal interpretation as a well-known process of forming weighted averages, which we can represent as a Markov chain. We explore this in a mathematical interlude in the section that follows. There are multiple applications of this linear model. The problem can be set up as one of numerical weighting of all locations to achieve a single location which is best, a binary reclassification, and involving the identification of a single best location by moving different partial best locations towards each other. We catalogue all these in the subsequent section. We then move on to show an application finding the best location for a highway, a problem we have reworked before (Batty 1971a, 1971b, 1974). This enables us to extend the formalization of the problem to one in which we can predict, or generate, the network from two different bi-partite graphs. This enriches the problem and provides us with a basis for using the model to resolve more complex problems where we formally link location – the problem of the city – with actor preferences – the problem of planning.

LINEAR OPINION POOLING BY WEIGHTED AVERAGING

Let us begin with the simplest possible model which consists of four stakeholders (which we refer to interchangeably as agents or actors). Each actor has an opinion about the solution to the problem as regards their own objectives, and these opinions differ. What is required is a pooling of opinions so that these differences might be resolved and, to this end, we need to define an appropriate network which shows how these actors might

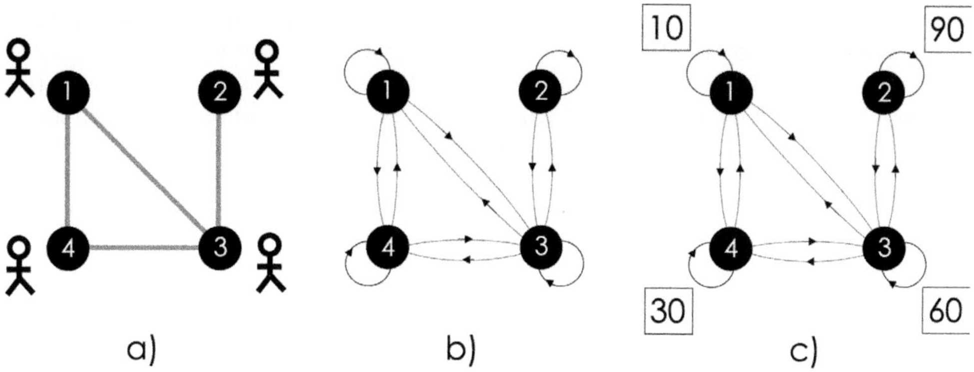

Figure 19.1 *(a) A typical network of four actors, (b) as a digraph and (c) their opinions/guesses scored as numbers*

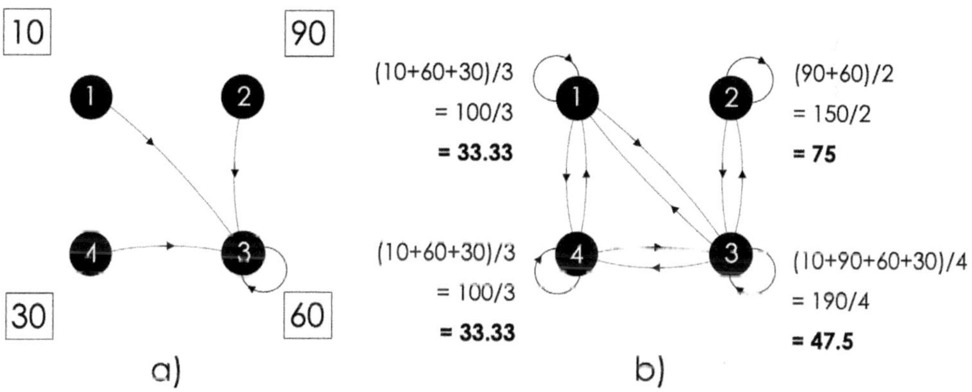

Figure 19.2 *Transmission: (a) averaging for those communicating with actor 3, and (b) averaging in the first round of communication between all actors*

be related. We define this network in Figure 19.1(a) where the communication channels between each actor are two-way. We need to show this as a directed graph (digraph) which we illustrate in Figure 19.1(b), noting that the self-arrow means that when communication takes place, the actor remembers its own opinion and transmits this to those actors to whom it is connected. To best illustrate the problem, we assign numerical values to each actor and in Figure 19.1(c), the range of these values is [10, 90, 60 and 30]. To fix ideas, it is best to think of these as first best guesses to a number which lies somewhere in the range from 10 to 90.

Opinion pooling consists of each of these actors sending its best guess to those actors to whom it is connected. In Figure 19.2(a) we show these for the links from actors 1, 2 and 4 to actor 3 with the assumption that when actor 3 receives these three values which are 10, 90 and 60, it will act rationally by taking this information and pooling it – averaging it if you like – with its own best guess which is 30. We assume that the actor is rational and believes the guesses of those who communicate their guesses to it are just as good as its own, and thus it forms the average which is 10 + 90 + 60 together with its own which adds to 190. Dividing this by 4 which is the number of actors communicating with it, including itself, its new guess becomes (10 + 90 + 30 + 60)/4 = 47.5. Now all these actors do the same on the first round of swapping their best guesses and the opinions change to those that are shown in Figure 19.2(b). This process generates the new guesses, which move from the first best guesses at the start of the process [10, 90, 60, 30] to [33.33, 75, 47.5, 33.33], and then the process begins again with these new guesses. You can see immediately that these new guesses lie between the ranges of 10 to 90 at the start to 33.33 to 75, and if you reflect a little about what the process will do in subsequent rounds, it is intuitively clear that these differences will reduce. Ultimately, if the process of transmission continues, the guesses will move to an average among all the initial opinions. In summary, the initial differences will disappear and a consensus will be reached. This is a weighted average where the average takes place according to the network. As the network varies in its links, then the consensus will also differ. If every actor is connected to every other actor, the consensus takes place in one round and becomes a true average of the initial set of guesses.

We show in Table 19.1 the succession of guesses for each actor over several iterations of the process. What is evident, although we do not demonstrate it formally, is that the convergence is very rapid at first and then moves towards an asymptote where all the best guesses are identical. Arguably, this process mirrors the kind of estimate that you might make if you were surrounded by a crowd who were pitching their best guesses. However, it is easy to see that the best guess must lie between the lowest and largest in the range, between 10 and 90, and if we were simply to average these two opinions, we would get 50 which is close to the ultimate value which from Table 19.1 is about 45. It is the network that makes a difference. If the network were very big and sprawling with many densely clustered subgraphs and various directional obstacles, assuming it was strongly connected in that every actor could reach every other either directly or indirectly, then the average would be much harder to guess. The very interesting feature, however, is when no consensus takes place.

In Figure 19.3 we show three variants of this very simple 4-node network; first in Figure 19.3(a), the network is separated into two subgraphs where 1 and 4, and 2 and 3, form their own guesses which in this instance are the simple averages of 1 and 4, and of 2

Table 19.1 Convergence of the best guesses to a stable equilibrium

Iteration no.	Actor 1	Actor 2	Actor 3	Actor 4
Best guess→	10.00	90.00	60.00	30.00
1.00	33.33	75.00	47.50	33.33
2.00	38.06	61.25	47.29	38.06
3.00	41.13	54.27	46.16	41.13
4.00	42.81	50.22	45.68	42.81
5.00	43.77	47.95	45.38	43.77
6.00	44.30	46.66	45.21	44.30
7.00	44.61	45.94	45.12	44.61
8.00	44.78	45.53	45.07	44.78
9.00	44.87	45.30	45.04	44.87
10.00	44.93	45.17	45.02	44.93

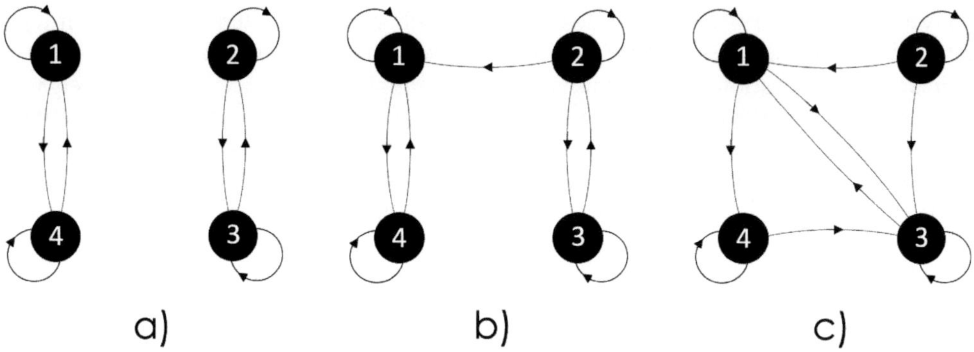

Figure 19.3 *Networks with (a) divided opinions, (b) where actors 2 and 3 dominate, and (c) where actor 2 dominates*

and 3, giving 20 and 75 respectively. This is a little similar to the system dividing into two groups where each person in the same group completely agrees with one another to form a simple compromise but never talk to other actors in the other group. We all have experience of this type of stand-off. Figure 19.3(b) mirrors a situation where 2 and 3 agree to compromise but together through actor 2, this subgroup relentlessly pushes their opinion on actors 1 and 4 with the ultimate result that actors 1 and 4 take the same opinion as the average of 2 and 3. This is a situation where actor 3 gets 2 to do its bidding but actor 2 then continues by insisting that 1 and 4 take its view. Ultimately 2 (and 3) prevail. Figure 19.3(c) is yet another situation where actor 2 by itself prevails and the other 3 relent to its opinion. In these last two examples, actors 2 and 3 form an average, which is 60, while in the last example, actor 2 prevails and everyone is converted to its initial unwavering opinion of 90. Only when we build bigger networks can we generate many different cases. However, even this simple network produces a lot of variety.

There is one final thing to present with respect to this simple model, and that involves showing specifically the pattern of influence through the successive generations. In Figure 19.4, we show the sequence of iterations by organizing the network first from top to bottom (a) and then we twist it in the horizontal direction to show it in the manner of a neural net

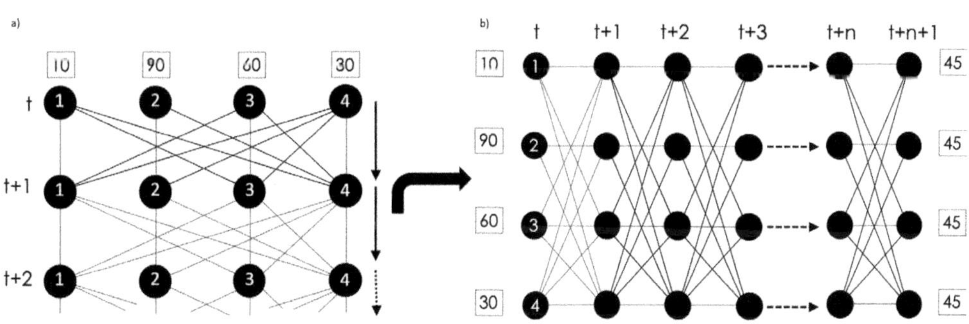

Figure 19.4 *Successive averaging: (a) the network in layers; (b) the network splayed out in neural net similar fashion*

(b), where each layer is connected to another at each subsequent iteration based on the network. This is a succession of the original network with identical weights used for the set of links at each iteration. Strictly, this is one of the simplest neural nets where the weights are compounded as the succession of opinion changes take place. To show the power of the method, we might change the weights or we might reconfigure the network at each iteration. As we start with a series of guesses at iteration 1, we then aim to reach a conclusion – a stable guess – by the end of the process, choosing weights that make that particular guess feasible. This is how a neural network actually works. You feed it some guesses and iterate and adjust the weights by a method of forward and backward propagation until the desired result is achieved. The guesses you feed it are data and if we had millions of guesses, the net could be trained to produce the required guesstimate (Gurney 1997).

This relationship to neural nets is not something we take any further in this chapter, but it does represent another dimension to this discussion and is worth exploring. However, what we need to do now is to ground the method in a more formal context. This will enable us to comment on the process and the equilibrium that results, and to generalize the method in applying it to different types of planning problem.

A MATHEMATICAL INTERLUDE

If we first examine how a single actor responds to the transmission of the best guesses from its neighbours who are directly connected to it, we can begin to generalize the process to all actors. In Figure 19.2(a), we show the communication of the opinions of actors 1, 2 and 4 in respect of how actor 3 averages these opinions with its own, and we can write these in vector form as

$$\left.\begin{aligned}[47.5] &= \left[\frac{1}{4}\right][10] + \left[\frac{1}{4}\right][90] + \left[\frac{1}{4}\right][30] + \left[\frac{1}{4}\right][60] \\ &= \begin{bmatrix}\frac{1}{4} & \frac{1}{4} & \frac{1}{4} & \frac{1}{4}\end{bmatrix}\begin{bmatrix}10\\90\\60\\30\end{bmatrix}\end{aligned}\right\}. \tag{19.1}$$

Now we can generalize equation (19.1) to all four actors in our simplest example writing the matrix-vector equation out in full for the first iteration using the values from Figure 19.2(b) as

$$\begin{bmatrix}33.33\\75.00\\47.50\\33.33\end{bmatrix} = \begin{bmatrix}\frac{1}{3} & 0 & \frac{1}{3} & \frac{1}{3}\\ 0 & \frac{1}{2} & \frac{1}{2} & 0\\ \frac{1}{4} & \frac{1}{4} & \frac{1}{4} & \frac{1}{4}\\ \frac{1}{3} & 0 & \frac{1}{3} & \frac{1}{3}\end{bmatrix}\begin{bmatrix}10\\90\\60\\10\end{bmatrix}. \tag{19.2}$$

Defining the 4 × 4 weighting matrix as $\boldsymbol{P} = [P_{ij}]$, the initial guesses as the vector $\boldsymbol{w}(t) = [w_j(t)]$ and the new vector of averaged guesses as $\boldsymbol{w}(t+1) = [w_i(t+1)]$, we can write the generalized version of equation (19.2) as

$$w_i(t+1) = \sum_j P_{ij} w_j(t), \ \sum_j P_{ij} = 1, \tag{19.3}$$

which in matrix-vector notation is

$$\boldsymbol{w}(t+1) = \boldsymbol{P}\boldsymbol{w}(t). \tag{19.4}$$

Note that $t = 1$ is the beginning of the process.

We are now in a position to demonstrate how the model is able to simulate a steady state where all the actors are reconciled with one another. Equation (19.4) is a recursive relation and if we work this through for a succession of averaging steps as portrayed in Figure 19.4, we can write it as

$$\boldsymbol{w}(t+n) = \boldsymbol{P}\boldsymbol{w}(t+n-1) = \boldsymbol{P}^n \boldsymbol{w}(t). \tag{19.5}$$

It is easy to demonstrate that equation (19.5) converges to a vector of opinions that are all identical, that is, $\boldsymbol{w} = \boldsymbol{w}(t+n)$ as $n \to \infty$. As equation (19.3) shows, $\boldsymbol{P}$ is a stochastic matrix with row sums equal to 1. A property of this matrix is that when it is multiplied by itself, it remains a stochastic matrix and its rows begin to converge to stochastic vectors which get closer and closer to a steady-state vector $\boldsymbol{z}^T$, that is $\boldsymbol{Z} = \boldsymbol{P}^n$ as $n \to \infty$ where each row of $\boldsymbol{Z}$ is this steady-state vector. Then in the limit, equation (19.5) can be written as

$$\boldsymbol{w} = \boldsymbol{Z}\boldsymbol{w}(1), \tag{19.6}$$

which generalizes to

$$\left.\begin{aligned} w &= \boldsymbol{z}^T \boldsymbol{w}(1) = \boldsymbol{z}^T \boldsymbol{w} \\ \boldsymbol{z}^T &= \boldsymbol{z}^T \boldsymbol{P} \end{aligned}\right\}. \tag{19.7}$$

Note that the vectors are column vectors in this presentation and we transform them in the usual way to row vectors by taking their transpose. Also note that the final steady-state opinion vector $\boldsymbol{w}$ can be computed directly from steady-state vector $\boldsymbol{z}^T$, and the initial opinions $\boldsymbol{w}(1)$ in equation (19.7) and this steady-state vector in turn can be generated by solving the simultaneous equations in the second row of equation (19.7). All these results are standard in linear algebra and a full presentation is in the author's book (Batty 2013).

There are two comments that need to be made about this process. First, the model is based on a well-known discrete Markov process (or chain). Second, it can be manipulated in many ways. We have assumed that the Markov or probability matrix $\boldsymbol{P}$ is strongly connected. This means that it is possible for every actor to reach every other actor either directly or indirectly. That is, the matrix cannot be partitioned into separate components for then there would be no equilibrium solution. There are several variants such as those we showed in Figure 19.3 where various actors might dominate, and these can be accommodated within the framework. The probability matrix that we have used is based on simply converting a binary matrix based on a connected and symmetric network into a matrix where each link is given equal weight at the level of the connectivity. However, the differential averaging and the probabilities are entirely dependent on a link being present or not within the original network. We could weight these links quite differently. We could, for example, assume a completely connected matrix where every actor interacts with every other actor as, for example, in spatial interaction matrix, except that the links are based on some measure of cost or distance. Actors would be swapping opinions based on their relative distance from one another. Whether the final equilibrium would be meaningful would depend on how the problem was formulated, but this indicates that the same logic as developed here could be developed for networks in city systems which do not necessarily depend on the types of social consensus assumed here but on functional relationships between different parts of a city.

APPLICATIONS OF THE MARKOVIAN FRAMEWORK

We outline three different applications of this model to problems of city planning. We begin with the simple opinion pooling model and then introduce an application to highway design originally developed by Alexander and Manheim (1962). This is in the spirit of design with nature developed by McHarg (1969) and reworked by Batty (1974). We have illustrated the opinion pooling model for a numerical example but where each actor holds an opinion which is quantified as a guess. As an abstract concept, this simply illustrates the logic of the process but in any application, this quantification has to have some meaning. There is an old story about the wisdom of crowds, told by Surowiecki (2004), which illustrates this process. In 1906, Francis Galton, the man who developed biometrics and a man of many parts, visited a livestock fair in England's West Country. There, an ox was the centre of attention and the villagers were asked to guess the animal's weight after it had been slaughtered. Some 800 people hazarded a guess but no one got near to the actual weight of 1198 pounds. When all the estimates were averaged, these gave 1197 pounds which was a much closer estimate than any single estimate by each of the 800 individuals, thus illustrating the possibility that large groups of people can converge on something which is a correct value simply by reflecting the diversity of views in a decentralized, independent group of people with no prior coordination. This is a classic example of crowdsourcing to an obvious purpose. Some planning problems of great complexity are of this nature and, in a professional group, averaging this diversity might be a good strategy for producing an optimum plan. Our model could be used in this way for appropriate problems.

However to give this some physical structure, let us assume that we have a set of n actors that each have a view about the desirability for locating a facility in a city which is divided into a series of m locations. We can define a desirability map for each actor j and each location on the map k as D_{jk}. We also have a weighting matrix in probability form as in the previous section termed $\boldsymbol{P} = [P_{ij}]$ and the process that we now initiate is where the desirability scores are averaged assuming complete rationality on the part of the actors until they all converge on an agreed set of scores. This is no more or less than applying the model in equations (19.1) to (19.4) to a set of opinions which each actor holds instead of to a single opinion. We show this in algebraic form as

$$D_{ik}(t+1) = \sum_j P_{ij} D_{jk}(t), \tag{19.8}$$

where it is clear that the scores for each location are independent of one another. After the process has converged, we can select the most desirable location from

$$D_i = \max_k D_{ik} = D_{ik}(t+n) \, as \, n \to \infty. \tag{19.9}$$

It is possible to introduce some local averaging of scores if the locations are adjacent in some way. Assuming that we identify the closest location as $D_{j\ell \in k}$, we form a local average of this which is articulated as

$$D_{ik}(t+1) = \sum_j P_{ij} \{[(D_{jk}(t) + D_{j\ell \in k}(t)]/2]\}. \tag{19.10}$$

There are many of these schemes that we might devise, but that which we have applied to identify all or nothing locations for urban development involves scoring each location as being possible or not for development. If we revert to our simple example which we show in equation (19.2), then we can assume that each of the four actors has a map of the city divided into, say, 16 zones, arranged on a 4 × 4 grid as we show in Figure 19.5. Noting that we need to scale the values by 100 so that we can represent the rounded desirability scores as integers, we can write the averaging equation for the first iteration as

$$\begin{bmatrix} 67\ 67\ 67\ 33\ 33\ 67\ 33\ 67\ 67\ 0\ 67\ 33\ 33\ 100\ 33\ 33 \\ 50\ 50\ 100\ 50\ 50\ 50\ 50\ 50\ 50\ 50\ 50\ 50\ 0\ 50\ 0\ 50 \\ 50\ 75\ 75\ 50\ 25\ 50\ 50\ 75\ 50\ 25\ 50\ 25\ 50\ 75\ 25\ 50 \\ 67\ 67\ 67\ 33\ 33\ 67\ 33\ 67\ 67\ 0\ 67\ 33\ 33\ 100\ 33\ 33 \end{bmatrix} = 100 \begin{bmatrix} \frac{1}{3} & 0 & \frac{1}{3} & \frac{1}{3} \\ 0 & \frac{1}{2} & \frac{1}{2} & 0 \\ \frac{1}{4} & \frac{1}{4} & \frac{1}{4} & \frac{1}{4} \\ \frac{1}{3} & 0 & \frac{1}{3} & \frac{1}{3} \end{bmatrix} \begin{bmatrix} 1\,1\,1\,0\,0\,1\,0\,1\,0\,0\,1\,0\,1\,1\,1\,0 \\ 0\,1\,1\,1\,0\,0\,1\,1\,0\,1\,0\,0\,1\,0\,0\,1 \\ 1\,0\,1\,0\,1\,1\,0\,0\,1\,0\,1\,0\,0\,1\,0\,0 \\ 0\,1\,0\,1\,0\,0\,1\,1\,1\,0\,0\,1\,0\,1\,0\,1 \end{bmatrix} \tag{19.11}$$

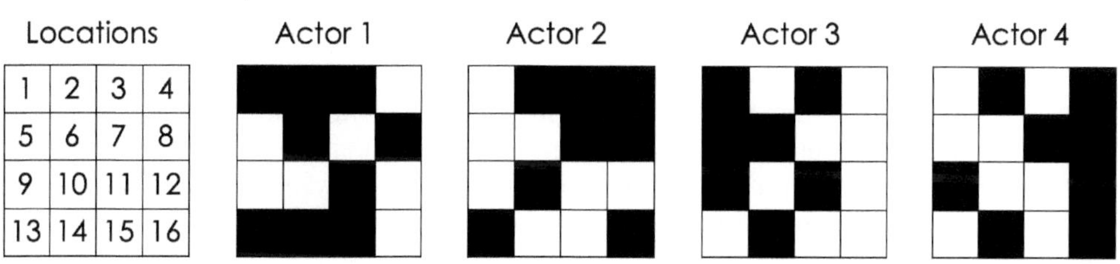

Note: Black is desirable, white is undesirable.

Figure 19.5 Actor desirability scores for locating urban land uses in a 16-zone spatial system

The four maps, one for each actor, in Figure 19.5 represent cells in the space where each actor suggests that development can take place $D_{jk} = 1$ and where it cannot $D_{jk} = 0$. There is trick in this simplified system for computing the final maps when all the actors agree with one another. Instead of iterating the equation in (19.11) according to the recursion in (19.3), we note that the steady-state vector is proportional to the indegrees (or outdegrees) of the initial digraph from which the probability matrix is derived. These indegrees, in this instance for the four actors, are [3 2 4 3] from which we compute the steady-state vector as

$$z = [1/4, 1/6, 1/3, 1/4]. \tag{19.12}$$

This vector adds to 1 as it is the set of probabilities of the actors contributing to the final solution. Applying this to the set of maps $[D_{jk}(t = 1)]$ from the first line of equation (19.7), we can produce the solution directly as

$$[58\ 67\ 75\ 42\ 33\ 58\ 42\ 67\ 58\ 17\ 58\ 25\ 42\ 83\ 25\ 42] = 100\left[\frac{1}{4}\frac{1}{6}\frac{1}{3}\frac{1}{4}\right] \begin{bmatrix} 1\,1\,1\,0\,0\,1\,0\,1\,0\,0\,1\,0\,1\,1\,1\,0 \\ 0\,1\,1\,1\,0\,0\,1\,1\,0\,1\,0\,0\,1\,0\,0\,1 \\ 1\,0\,1\,0\,1\,1\,0\,0\,1\,0\,1\,0\,0\,1\,0\,0 \\ 0\,1\,0\,1\,0\,0\,1\,1\,1\,0\,0\,1\,0\,1\,0\,1 \end{bmatrix}, \tag{19.13}$$

which we can also represent as the converged scores for all actors as

$$D_k = [58\ 67\ 75\ 42\ 33\ 58\ 42\ 67\ 58\ 17\ 58\ 25\ 42\ 83\ 25\ 42]$$
$$= \max_j D_{jk} \begin{bmatrix} 58\ 67\ 75\ 42\ 33\ 58\ 42\ 67\ 58\ 17\ 58\ 25\ 42\ 83\ 25\ 42 \\ 58\ 67\ 75\ 42\ 33\ 58\ 42\ 67\ 58\ 17\ 58\ 25\ 42\ 83\ 25\ 42 \\ 58\ 67\ 75\ 42\ 33\ 58\ 42\ 67\ 58\ 17\ 58\ 25\ 42\ 83\ 25\ 42 \\ 58\ 67\ 75\ 42\ 33\ 58\ 42\ 67\ 58\ 17\ 58\ 25\ 42\ 83\ 25\ 42 \end{bmatrix}. \tag{19.14}$$

In the vector in equation (19.14), these are the scores that the actors finally agree upon. We find the highest score by finding the maximum value of this vector, which is location 14, which has a score of $83 = \max_k D_k$.

The second application involves defining the position of each actor in some two-dimensional space. We could consider the application in equations (19.11) to (19.14) as being coordinates in a 16-dimensional space with the convergent solution being the finally agreed coordinates of the actors after they have reached a consensus about the agreed position. We can best illustrate this in two-dimensional space where the actors have an initial position and wish to move to a location where all of them are able to meet. This might be a problem where all the actors agree there is a need for a central location instead of where they are located at the start of the process, since this is similar to Hotelling's (1929) spatial problem of agglomeration. However, the power structure is such that the solution is not likely to be a true centroid but weighted according to the importance of each actor. In a situation where the actors are all able to compromise immediately, which implies that in the 4 × 4 actor matrix of probabilities the links are of equal weight, the final solution achieved in one iteration is the true centroid of the initial positions. However, for the matrix of the weighted digraph in Figure 19.2(b), we define the first iteration of the process with the matrix of coordinates $[x_j y_j]$ shown in Figure 19.6 as

$$\begin{bmatrix} 4.33 & 4.33 \\ 7.50 & 5.50 \\ 5.00 & 5.50 \\ 4.33 & 4.33 \end{bmatrix} = \begin{bmatrix} \frac{1}{3} & 0 & \frac{1}{3} & \frac{1}{3} \\ 0 & \frac{1}{2} & \frac{1}{2} & 0 \\ \frac{1}{4} & \frac{1}{4} & \frac{1}{4} & \frac{1}{4} \\ \frac{1}{3} & 0 & \frac{1}{3} & \frac{1}{3} \end{bmatrix} \begin{bmatrix} 28 \\ 79 \\ 82 \\ 33 \end{bmatrix}, \tag{19.15}$$

and using the fast solution used for the full problem, we obtain the weighted centroid expressed as the four set of coordinates

$$\begin{bmatrix} 5.084.92 \\ 5.084.92 \\ 5.084.92 \\ 5.084.92 \end{bmatrix} = \frac{\begin{bmatrix} 1111 \end{bmatrix}}{\begin{bmatrix} 4634 \end{bmatrix}} \begin{bmatrix} 28 \\ 79 \\ 82 \\ 33 \end{bmatrix}. \tag{19.16}$$

We show this convergence in Figure 19.6 where we have joined the four points to define the space in which the solution exists. In this simple problem, the convergence is quite fast,

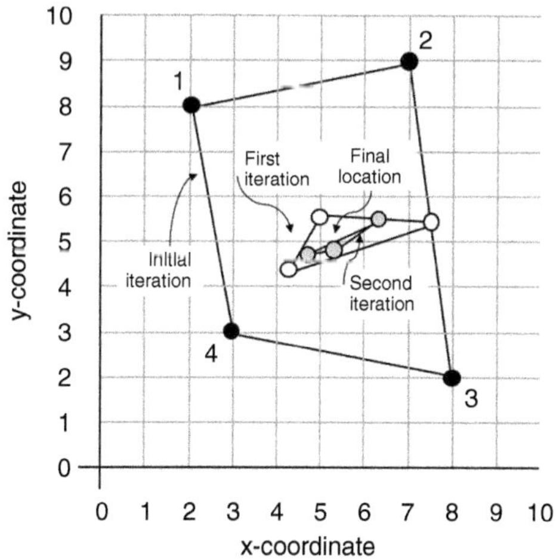

Figure 19.6 Convergence to a stable location where opinions are positions (coordinates) in a map space

but in a very big ramified network with many clusters, weak ties and asymmetric links, the process might take many iterations.

This application might be seen as an optimization problem in location theory but where the network is buried beneath the spatial solution. The two-dimensional location problem where the actors pool their opinions so they might agree on a central location somewhere within the wider space that their preferred locations define, appears in many contexts, particularly in areas such as robotics, games and problems of movement dynamics (Mesbahi and Egerstedt 2010; Ren et al. 2007). In these domains, there is a strong focus on how the dynamics of the way actors or objects move reflect other theories of how animals and humans flock and herd, usually in highly stylized ways that are characterized by well-defined physical situations (Vicsek et al. 1995). There are also many applications that have emerged in computer science where the focus on agents transmitting information by pooling has been articulated using the linear Markov model (Blondel et al. 2005, 2009). However, the origins of these ideas really lie in studies of social power, the model here being developed originally by French (1956), and then formalized by Harary (1959) and Harary and Lipstein (1962) as part of the vibrant field of group dynamics which flourished in the 1950s and 1960s. These models were then picked up in statistical and mathematical applications in sociology and political science (see de Groot 1974) and then in a succession of developments by Kelly (1981), Friedkin and Johnsen (1999) and Hegselmann (2004), and among several others working with consensus and opinion pooling.

To conclude these applications, we briefly present a more practical problem of highway location first introduced by Alexander and Manheim (1962). The problem involved the location of a highway between Springfield and Northampton in Western Massachusetts where a series of 26 factors were identified as implying different routes for the highway. The factors were mapped as desirability surfaces and all could thus be represented spatially pertaining mainly to physical constraints on the location of a highway. Many of the factors pertained to landscape amenity, topography, weather, drainage and so on, but they included a mix of socio-economic factors, such as local accessibility, urban development and transportation. The factors in question are shown in Figure 19.7 and the usual way of finding the best location (in Figure 19.7 the darker areas show more desirable locations for the highway) is to overlay these surfaces and compute an overall index of desirability for each location which would then guide the choice of the best line for the highway. If we were to do this directly, it would assume that each factor had the same weight as any other and the result would simply be a sieving or simple averaging. This is akin to methods used in professional planning and landscape design through much of the twentieth century, referred to as overlay analysis (McHarg 1969; Steinitz et al. 1976), and the layer model is still widely used in geographical information system (GIS). It is akin to a situation where each factor is associated with a single actor and actors then pool the factors or opinions as reflected in their desirability surfaces. It might be a stretch to consider factors as the opinions of actors but with judicious choice of problem and context, we can consider this to be so.

What Alexander and Manheim (1962) did was to form a matrix of differences between the factors (actors) and decompose this matrix into clusters which then produced an order for the way in which the factors could be combined. This order produced an implicit set of weights, with the most clustered based on the greatest

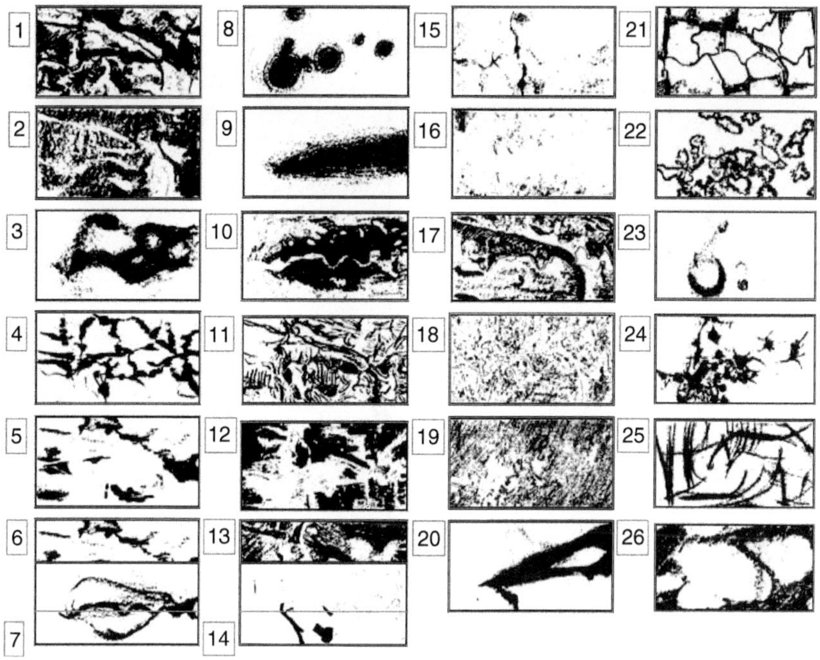

1 Earthwork costs
2 Comfort and safety
3 Regional development
4 Local land development
5 Obsolescence
6 Interference in construction
7 User costs
8 Services
9 Travel time
10 Pavement & subgrade costs
11 Drainage patterns
12 Bridge costs
13 Land costs
14 Eyesores
15 Noise
16 Air pollution
17 Weather effects
18 Public and private losses
19 Public financial losses
20 Major current traffic desires
21 Catchment areas
22 Local accessibility and integrity
23 Future transportation systems
24 Existing transportation systems
25 Duplication of facilities
26 Self-Induced congestion

Figure 19.7 Factor diagrams for the Alexander–Manheim highway location problem

differences being combined first and those on the edge of the network being the least different from the majority and thus combined last. As the order is based on differences, it becomes progressively easier to merge the factors into one another as the order progresses up the hierarchy of relationships. Alexander and Manheim should have recomputed this matrix, as it is progressively reduced in size – that is, as factors are merged – and there are many variants which could improve the process, but most have never been tested. As the factors are merged, they are strengthened in their form, for the focus is on extracting a linear form, – a highway – and reinforcing this as factors are successively merged.

The final solutions are shown in Figure 19.8. What we have done here is take the matrix of relations used in the original problem and then use Markov averaging to achieve a comparable solution. In relation to equation (19.11), the $\boldsymbol{D}$ matrix is the set of maps where i is the factor (or actor) and k is each cell of the map. In short each cell k in the ith

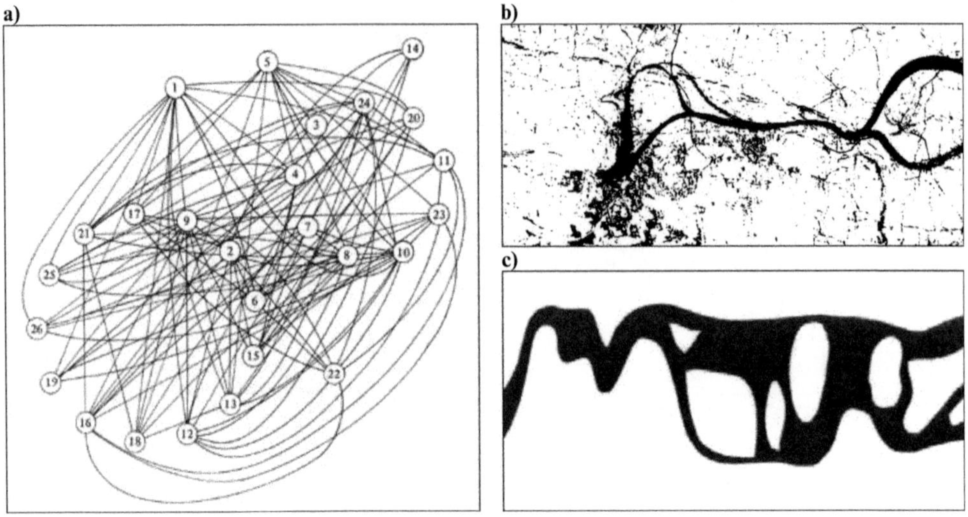

Note: (a) The relationship graph-network used for the cluster and Markov averaging methods; (b) the Alexander–Manheim cluster weighted solution; (c) the solution space from the Markov averaging.

Figure 19.8 Solutions to the highway location problem

map in Figure 19.7 is defined as $D_{ik}(t=0)$. The relationships matrix $\boldsymbol{P}_{ij}$ is defined from the graph in Figure 19.8(a) which was constructed by the author from the account given by Alexander and Manheim (1962). This led to the solution shown in Figure 19.8(b), while the Markov averaging methods associated with the examples earlier in this chapter produce a solution space in Figure 19.8(c) in which many different but similar locations for the highway might be chosen. There are many other issues involved in the Markov model and these are illustrated in Batty (2013), while the wider literature that we noted previously, particularly in sociology, shows how we might begin to ground these ideas in empirical applications.

These methods do not produce definitive solutions as regards an optimization, for they are more like tools for defining the solution space within which the designer or policy-maker or planner might make an informed choice. They act to support plan design just as participatory tools are being used to inform similar kinds of geodesign which involve pitting actors against one another in the search for a best plan (Steinitz 2012). There are many variations we can continue to make as regards models of design processes which reflect opinion pooling, and what is evidently required is a sustained attack on these variants. To conclude, we begin to explore how we can take these forward by focusing on the networks of relationships rather than on the processes of opinion pooling per se. To this end, we will sketch how we can begin to predict the network of interaction in respect of the way it can be formed.

THE CHALLENGE: DERIVING NETWORKS OF ACTOR INTERACTION

We have assumed that the network of actor relationships is observed in some way but very often these networks are not easy to define, and sometimes it is possible to see the network as a result of more indirect processes. In the example of the highway location problem, we examined the relationships between the maps in respect of how close each actor was to another in the strength of each relationship. If we take each element of the map k with for each actor, D_{ik}, where for the moment we drop the iteration index t, we can form an $n \times n$ matrix of similarities $[X_{ij}] = \boldsymbol{X}$ like correlations between the actors/factors from

$$X_{ij} = \sum_k D_{ik} D_{jk}, \text{ which in matrix terms is } \boldsymbol{X} = \boldsymbol{D}\boldsymbol{D}^T. \quad (19.17)$$

Each cell of the similarity matrix is based on a sum of how close an actor's desirability scores are to those of another actor. We then form the averaging matrix by simply summing the row elements of $\boldsymbol{X}$ and normalizing it as

$$P_{ij} = X_{ij} / \sum_j X_{ij}, \quad \sum_j P_{ij} = 1. \quad (19.18)$$

Now it is possible to compute the steady-state vector z^T which gives the relative power of each actor after a consensus has been reached. The final map could be derived in the manner we illustrated in our first example, where we defined changes in the map from successive averaging as

$$D_{ik}(t+1) = \sum_j P_{ij} D_{jk}(t), \quad (19.19)$$

which in matrix terms is

$$\boldsymbol{D}(t+1) = \boldsymbol{P}\boldsymbol{D}(t). \quad (19.20)$$

From the previous results, the final distribution of elements in the map $\boldsymbol{D}(t \to \infty)$ are computed as

$$\overline{\boldsymbol{D}} = z^T \boldsymbol{D}(t = 1). \quad (19.21)$$

It is important to note here that the original transition matrix $\boldsymbol{P}$ is computed from $\boldsymbol{D}\boldsymbol{D}^T$ and thus we are compounding the map into the process of averaging notwithstanding that the average itself depends on the original map. There is thus an element of double counting in the process of constructing the averaging.

The key insight into expressing the averaging matrix through correlation and similarity is that there is a very obvious dual problem, noting that the set of equations from (19.18) to (19.21) define a primal. In the main method, the whole process involves how n actors interact with one another, but as soon as we introduce similarities, we need to consider the similarity between any pair of actors in terms of some other dimension, in this instance factors expressed as m maps. Instead of measuring these similarities between actors, we can also compute them between the maps themselves by first forming the $m \times m$ matrix

$$Y_{k\ell} = \sum_j D_{jk} D_{j\ell}, \text{ which in matrix terms is } \boldsymbol{Y} = \boldsymbol{D^T D}. \tag{19.22}$$

This is a dual problem where, instead of finding the relative power of the actors in generating an average map, the problem is finding the relative power of each map cell/location from which we are able to compute a type of average actor. This problem is less meaningful in map factors, but it does have more relevance to situations where one set of actors is related to another, but different, set of actors. As in all the applications here, the choice of problem is critical in assessing the relevance of the method and, as we have already made clear, this is a set of tools that gives us some flexibility in handling these problems.

To complete the dual process, we form the $m \times m$ probability matrix

$$Q_{k\ell} = Q_{k\ell} / \sum_\ell Q_{k\ell}, \quad \sum_\ell Q_{k\ell} = 1. \tag{19.23}$$

with the process continuing as

$$D_{jk}(t+1) = \sum_\ell Q_{k\ell} D_{j\ell}(t), \tag{19.24}$$

which in matrix terms is

$$\boldsymbol{D^T}(t+1) = \boldsymbol{QD^T}(t). \tag{19.25}$$

If we call the steady-state vector $\boldsymbol{q}$ where $\boldsymbol{q} = \boldsymbol{qQ}$, then we can compute the averaged actors best map location as

$$\boldsymbol{\hat{D}^T} = \boldsymbol{qD^T}(t=1). \tag{19.26}$$

We conclude by taking our example based on the simple network in Figure 19.1 with the four maps in Figure 19.5, and we can simply state the primal and dual results from the steady-state equations directly as

$$\begin{aligned} &[50\,79\,75\,50\,21\,50\,50\,79\,46\,25\,50\,25\,54\,75\,29\,50] \\ &= 100\,[0.29\,0.25\,0.21\,0.25]\begin{bmatrix} 1\,1\,1\,0\,0\,1\,0\,1\,0\,0\,1\,0\,1\,1\,1\,0 \\ 0\,1\,1\,1\,0\,0\,1\,1\,0\,1\,0\,0\,1\,0\,0\,1 \\ 1\,0\,1\,0\,1\,1\,0\,0\,1\,0\,1\,0\,0\,1\,0\,0 \\ 0\,1\,0\,10\,0\,1\,1\,1\,0\,0\,1\,0\,1\,0\,1 \end{bmatrix}, \text{ and} \end{aligned} \tag{19.27}$$

$$[66\ 56\ 47\ 56] = 100 \begin{bmatrix} 0.06 \\ 0.09 \\ 0.09 \\ 0.06 \\ 0.03 \\ 0.06 \\ 0.06 \\ 0.09 \\ 0.06 \\ 0.03 \\ 0.06 \\ 0.03 \\ 0.06 \\ 0.09 \\ 0.03 \\ 0.06 \end{bmatrix} \begin{bmatrix} 1\,0\,1\,0 \\ 1\,1\,0\,1 \\ 1\,1\,1\,0 \\ 0\,1\,0\,1 \\ 0\,0\,1.0 \\ 1\,0\,1\,0 \\ 0\,1\,0.1 \\ 1\,1\,0\,1 \\ 0\,0\,1\,1 \\ 0\,1\,0\,0 \\ 1\,0\,1.0 \\ 0\,0\,0.1 \\ 1\,1\,0\,0 \\ 1\,0\,1.1 \\ 1\,0\,0\,0 \\ 0\,1\,0\,1 \end{bmatrix}. \tag{19.28}$$

We have applied the method to a problem involving key building sites and actors in the development of a small high profile neighbourhood in the City of London where the model is able to show the importance of the actors and the significance of the sites in question (Batty 2016). However, the real power of the method consists in starting with a problem involving two sets – of objects, actors, locations, and so on. The notion of devising bipartite networks or graphs as in this case for the relationships between actors and locations, always leads to primal and dual problems which involve twisting the problem one way or the other, thus gaining real insights into the nature of the networks involved and the power relations if the problem is formulated in social terms. There is a much more detailed development of this primal–dual problem where two different bipartite matrices/graphs are combined as in equations (19.17) and (19.22), respectively. If these matrices are already stochastic, the interpretations are easier in relation to the steady-state distributions. This variant of the model was first suggested by Coleman (1973) and its wider foundations are presented in Coleman's (1998) book. This has inspired many applications in social exchange theory and theories of social power, but there is still a dearth of observational data and most of these applications to date only elaborate the theoretical framework. It awaits widespread empirical testing.

CONCLUSIONS: NEXT STEPS

The message of this chapter is simple. Network models not only exist for the physical and spatial functions determining the way cities transit information, ideas, people, materials and energy, they also pertain to the ways in which we communicate information about the future form of the city itself within the wider context of design. We might think that this implies that the model in this chapter could be widely used to simulate the flows between agents that make up the entire city but here the set of agents is very specific. As designers and planners, the networks they use are to communicate ideas about the future city, rather than determining and supporting our understanding about the actual functioning of the city. These models must be taken at their face value. They are not necessarily meant to be empirically testable statements of how expert groups of designers tackle the problems they need to resolve while also resolving differences in opinions about the importance of ideas and factors. Notwithstanding that they are hard to validate mainly because the process of formulating a plan is highly convoluted and is much more than solving a simple problem in group dynamics, they are empirically applicable, although the processes of application are not well developed or understood. There are thus some important challenges in developing these applications as the conceptual ideas are widely applicable to many different kinds of problem.

The problems in this chapter are highly simplified and extremely small as regards their networks of interaction. At times we have hinted that the power of the method only comes into its own when the problem has many more actors. However, there is a limit to this as the networks are strongly purposive, in that they are appropriate only for situations where the actors have a common mission in mind. The types of network that we now find in social media are not relevant here and thus networks with more than 100 actors are not likely to be found in these types of design and planning problem. It is more likely that the model needs to be extended to deal with problems that have many more actors by developing ways in which actors can enter or leave the process as the problem alters. Planning problems involve an explicit process of learning and, even if we can conceive of planning problems that do involve thousands of stakeholders, the type of models that might explain and inform the way such a large group might move to a consensus are likely to be very different from those introduced in this chapter. The focus here is still on small-group dynamics, although the literature on opinion pooling in groups such as this has not got any closer to empirical applications in the past decade and has become ever more theoretical, concerned with the dynamics of processes moving to consensus, particularly with a focus on disruptions and disturbances from equilibrium (Dietrich and List 2016; Parsegov et al. 2017).

The most promising focus of this research is in areas associated with participatory planning and design, and the clearest statements of this have been in geodesign. Geodesign is about how we use informal but structured problem-solving among groups that articulate planning problems as problems of merging map layers in the manner shown in this chapter. The paradigm of GIS is based on these map layers, in that it originally emerged from the development of spatial overlay analysis in landscape architecture and planning. Steinitz (2012) argues that geodesign brings professional stakeholders and analysts together to solve a common problem, such as the search for development sites in a city or region, but with each stakeholder or team of stakeholders interacting with other groups who are pursuing independently their own designs. In general, groups know the opinions about the emerging solutions which other groups hold as the process proceeds. This is very close to the model developed here and, although we have considered its use as part of the International Geodesign Collaboration IGC (2020), there is much to do in first articulating the process of geodesign more explicitly, notwithstanding that there is increasing experience in its use (Pettit et al. 2019; Steinitz 2020).

To an extent this type of spatial averaging has come out of design methods where group dynamics is minimal but, as we demonstrated in relation to Alexander and Manheim's (1962) highway location problem, it is entirely appropriate to interpret 'factors influencing development' defined as maps of desirability associated with individual actors or stakeholders. Just as the work on small groups in sociology and social psychology has become less significant, formal work on design methods has fallen out of favour. To an extent, formal methods for design have become more participatory, as with geodesign. In many contexts, however, networks relating to who communicates with whom are being measured and represented in ways that enable detailed analysis of their structure to take place. However, formal processes operating on these networks are not being developed, other than the model that has been described in this chapter. As crowdsourcing is rapidly developing, there are opportunities for developing more formal processes connected to the ways in which ideas appear in this context, and a number of developments in this direction look promising (Lim et al. 2010).

Ways of extending this approach to the design of networks is another avenue to be researched (Batty 2013), but other network methods are also worth exploring. The development of the neural net model, sketched briefly in this chapter, suggests that there are new ways in which we can train the network to generate different solutions and, in approaching an ideal outcome, the neural net process encounters many less than ideal outcomes which nonetheless might be feasible. If the neural net does not converge, then less than optimal solutions are useful and, to an extent, the sensitivity of the outcomes (plans) can also be exploited by following this route. As we have emphasized, very few network models of problem-solving processes such as this exist and, although this particular model represents a baseline, we have identified many potential developments which, if pursued, will provide much richer and more relevant ways in which we might generate and evaluate plans for the future city.

REFERENCES

Alexander, C. and M.L. Manheim (1962), 'The use of diagrams in highway route location', Research Report R62-3, Civil Engineering Systems Laboratory, Massachusetts Institute of Technology, Cambridge, MA.

Batty, M. (1971a), 'An approach to rational design: part 1: the structure of design problems', *Architectural Design*, **41** (1), 436–39.

Batty, M. (1971b), 'An approach to rational design: part 2: design problems as Markov chains', *Architectural Design*, **41** (1), 498–501.

Batty, M. (1974), 'A theory of Markovian design machines', *Environment and Planning B*, **1** (2), 125–46.

Batty, M. (2013), *The New Science of Cities*, Cambridge, MA: MIT Press.

Batty, M. (2016), 'Evolving a plan: design and planning with complexity', in J. Portugali and E. Stolk (eds), *Complexity, Cognition, Urban Planning and Design*, Cham: Springer International, pp. 21–42.

Blondel, V.D., J.M. Hendrickx and J.N. Tsitsiklis (2009), 'On Krause's multi-agent consensus model with state-dependent connectivity', *IEEE Transactions on Automatic Control*, **54** (11), 2586–97.

Blondel, V.D., J.M. Hendrickx, A. Olshevsky and J.N. Tsitsiklis (2005), 'Convergence in multiagent coordination, consensus, and flocking', in *Proceedings of the Joint 44th IEEE Conference on Decision and Control, European Control Conference*, Seville, 12–15 December.

Coleman, J.S. (1973), *The Mathematics of Collective Action*, London: Heinemann Educational Books.

Coleman, J.S. (1998), *Foundations of Social Theory*, Cambridge, MA: Harvard University Press.

De Groot, M.H. (1974), 'Reaching a consensus', *Journal of the American Statistical Association*, **69** (345), 118–21.

Dietrich, F. and C. List (2016),'Probabilistic opinion pooling', in C. Hitchcock and A. Hajek (eds), *Oxford Handbook of Probability and Philosophy*, Oxford: Oxford University Press, pp. 520–42.

French, J. R. P. (1956), 'A formal theory of social power', *Psychological Review*, **63**, 181–94.

Friedkin, N.E. and E.C. Johnsen (1999), 'Social influence networks and opinion change', *Advances in Group Processes*, **16** (1), 1–29.
Gurney, K. (1997), *An Introduction to Neural Networks*, London: UCL Press Ltd.
Harary, F. (1959), 'A criterion for unanimity', in D. Cartwright (ed.), *Studies in Social Power*, Ann Arbor, MI: Institute for Social Research, pp. 168–82.
Harary, F. and B. Lipstein (1962), 'The dynamics of brand loyalty: a Markovian approach', *Operations Research*, **10** (1) 19–40.
Hegselmann, R. (2004), 'Opinion dynamics – insights by radically simplifying models', in D. Gillies (ed.), *Laws and Models in Science*, London: King's College, pp. 19–46.
Hotelling, H. (1929), 'Stability in competition', *Economic Journal*, **39** (153), 41–57.
International Geodesign Collaboration (IGC) (2020), 'International Geodesign Collaboration: changing geography by design', accessed 29 March 2020 at https://www.igc-geodesign.org.
Kelly, F.P. (1981), 'How a group reaches an agreement: a stochastic model', *Mathematical Social Sciences*, **2** (1), 1–8.
Lim, S.L., D. Quercia and A. Finkelstein (2010), 'StakeNet: using social networks to analyse the stakeholders of large-scale software projects', in *Proceedings of the 32nd ACM/IEEE International Conference on Software Engineering: ICSE'10 Volume 1*, New York: Association for Computing Machinery, pp. 295–304.
McHarg, I.L. (1969), *Design with Nature*, Garden City, NY: Doubleday/Natural History Press.
Mesbahi, M. and M. Egerstedt (2010), *Graph Theoretic Methods in Multiagent Networks*, Princeton, NJ: Princeton University Press.
Parsegov, S.E., A.V. Proskurnikov, R. Tempo and N.E. Friedkin (2017), 'Novel, multidimensional models of opinion dynamics in social networks', *IEEE Transactions on Automatic Control*, **62** (6), 2270–85.
Pettit, C.J., S. Hawken, C. Ticzon, S.Z. Leao, A.E Afrooz, S.N. Lieske, et al. (2019), 'Breaking down the silos through geodesign – envisioning Sydney's urban future', *Environment and Planning B: Urban Analytics and City Science*, **46** (8), 1387–404.
Ren, W., R.W. Beard and E.M. Atkins (2007), 'Information consensus in multivehicle cooperative control', *IEEE Control Systems Magazine*, April, 71–82.
Steinitz, C. (2012), *A Framework for Geodesign: Changing Geography by Design*, Redlands, CA: ESRI Press.
Steinitz, C. (2020), 'Re-thinking the growth of London: an interactive multi-stakeholder perspective derived in geodesign and negotiation', *Land Use Policy*, forthcoming
Steinitz, C., P. Parker and L. Jordan (1976), 'Hand-drawn overlays: their history and prospective uses', *Landscape Architecture*, **66** (1), 444–55.
Surowiecki, J. (2004), *The Wisdom of Crowds*, New York: Doubleday.
Vicsek, T., A. Czirók, E. Ben-Jacob, I. Cohen and O. Shochet (1995), 'Novel type of phase transition in a system of self-driven particles', *Physical Review Letters*, **75** (August), 1226–9.

20. Up close and impersonal: locative media and the changing nature of the networked individual in the city

Darryl A. Pieber and Anabel Quan-Haase

INTRODUCTION

By 2050, three-quarters of the world's population will live in large urban centers comprised of many municipalities of varying sizes (Castells 2002), such as the Greater Toronto and Hamilton Area (GTHA) in Canada, with a population in 2016 of almost 7 million (City of Toronto 2017) and the Pearl River delta region of China, with a population in 2010 of approximately 42 million (World Bank Group 2015). Many of the people living in these urban centers were not born there, and so will not have any historical ties to them. In the GTHA, for example, approximately 40 percent of the population was born outside Canada (City of Toronto 2013). Within these urban conurbations, people may live in one city, work in another and play in yet another, resulting in considerable mobility throughout (Castells 2002). Within this context, we see the rise of locative media, mobile technologies that capture and deliver location- and time-specific content and connections to their users. Since locative media draw on how a person moves through these urban conurbations, enhancing and transforming the experience, they can have significant impacts on the ways in which people interact with these spaces and with each other. As location awareness becomes a standard feature of increasing numbers of mobile applications (apps), the capacity for locative media to affect people's relationships with the urban spaces they inhabit, as well as with the people they inhabit these spaces with, becomes increasingly significant.

In this chapter, we examine the role that locative media apps play in cities and urban life, and the ways in which people navigate and interact in these cities. The chapter starts by tracing the origins of locative media, and demonstrating how these tools differ from other information and communication technologies (ICTs). To show how locative media apps shape current structural changes, we discuss the move from group-based interactions toward diverse, unbounded networks, and link locative media to the concept of networked individualism (Rainie and Wellman 2012). We examine how locative media apps can further complicate networked individualism through supporting ad hoc, transient relationships that further add to the fragmentation and declustering of individuals' social networks. We look at three main ways in which locative media apps insert themselves into people's daily experiences of urban space: wayfinding, accessing and sharing information in networked ways, and seeing and networking with people nearby. We also examine the social implications of locative media apps on urban sociability by discussing visibility, sensory overload, social accessibility, trust and familiarity, and finally privacy. As the field is still nascent, we are only starting to understand what privacy rights need to be in place to protect users. Our chapter also has important implications for designers of mobile technologies, as they can gain understandings of the social implications and structural transformations that locative media apps have in the context of urban life.

LOCATIVE MEDIA: APPROACHING THE PROMISE OF UBIQUITOUS COMPUTING

Locative media are a set of mobile technologies – hardware and software – that make use of an individual's spatial and temporal location to capture and deliver content specific to a particular moment in a particular place. At the device level, specific software applications or apps, make use of the device's Global Positioning System (GPS) receiver, accelerometer and compass to identify where a person is in the world, and then capture and deliver content from the Internet that is relevant to that particular place and time. A simple example of a locative media app is a public transit app that informs users when the next bus will arrive at the bus stop where they are currently located. An app that provides weather information about a specific city is not a locative media app because it obtains the weather information from a nearby weather station, and so the weather information is not based on the place where the individual is positioned. While this distinction may seem minute, it marks a shift in perspective from 'anywhere, anytime' computing to 'here and now' computing (Kitchin 2014; Villi 2015). Although locative media involve an interaction between hardware and software, in this chapter we focus on the software end – locative media apps.

Locative media apps challenge the ways we traditionally understand computing. In 1991, Weiser (1991, p. 94) wrote about efforts underway at the Xerox Palo Alto Research Center 'to conceive a new way of thinking about computers, one that takes into account the human world and allows the computers themselves to vanish into the background'. Computers could weave themselves seamlessly into the fabric of everyday life, becoming part of that fabric. Weiser (1991) was speaking of ubiquitous computing, a notion that would transform how we think about computers and mark the beginning of the current data science movement. At the center of ubiquitous computing lies the idea that, instead of having to be in a particular location to use a computer, people can access computing capabilities 'anywhere, anytime'. To highlight his point, Weiser (1991) contrasts the notion of ubiquitous computing with virtual reality. While virtual reality creates a world inside a computer, ubiquitous computing integrates into and enhances the real world, for which Weiser and his colleagues use the term enhanced virtuality. However, to approach the full potential of ubiquitous computing, the computer needs to know where in the world it is (Weiser 1991). Enter locative media.

An important consideration in understanding the role of locative media apps in everyday urban contexts is the nature of the relationship between the real and the virtual. Locative media apps reverse the notion of virtual reality. Instead, they overlay the virtual overtop of the real. That is, physical location becomes a fundamental organizing principle for digital information through locative media apps. It is not uncommon regard the relationship between the real and the virtual as a new, hybrid space. Sutko and de Souza e Silva (2011) understand the virtual as a separate space from real, physical space. Farman (2012) similarly sees this distinction, and proposes that people can have embodied experiences in both physical and virtual spaces. As an example of this, he describes an instance where a student receives a phone call during class. Instead of silencing the phone, the student takes the call in the class. The student is physically embodied in the class, but also non-physically embodied in the virtual space where the conversation is taking place. Farman (2012, p. 22) uses the term virtual in a very broad sense: 'from chatting with a loved one via text or over video conferencing to playing a multiplayer online game'. Farman (2012) sees locative media facilitating a transformation of physical space and virtual space into a hybrid space. In this hybrid space, according to Licoppe (2016), two different versions of the same here-and-now environment are available. These two different versions may present conflicting views of the same here-and-now, requiring a reconciliation on the part of the user.

In the instance of locative media, the split between real and virtual is not useful. To understand this information as existing in a virtual space that can be inhabited does not move the discussion away from notions of virtual reality. As Leszczynski (2015) notes, the digital information that is accessed through locative media is not a virtual space that we can inhabit. As she suggests, a better way to understand the relationship is that digital information is anchored to the space where it is most relevant at the time when it is most relevant (Leszczynski 2015).

Beyond reversing the notion of virtual reality, where the user enters into a virtual world through a computer, locative media also upend understandings of how ubiquitous computing works. As Farman (2012, p. 11) observes, '[i]nstead of disappearing into the fabric of our lived experience, [ubiquitous computing] is something that is consciously interacting with our environments and offering a transformative experience of space'. Instead of working away, unnoticed, in the background, locative media are visibly present through people's mobile devices, and open a type of access point to the systems working away in the background. In so doing, locative media provide additional layers of information to the spaces people inhabit at the times they inhabit them (Farman 2012).

THE EVERYDAY USES OF LOCATIVE MEDIA IN URBAN SPACE

Locative media apps can be used for a range of social and informational purposes. These apps are still a nascent technology, but already three key uses are emerging that we will focus upon in this chapter: wayfinding, accessing and sharing information in networked ways, and seeing and networking with people nearby.

Wayfinding in Urban Spaces

Wayfinding is perhaps one of the more well-known and well-established applications of locative media. A common focus of study is the use of in-car and on-foot GPS navigation. Studies have found that use of this mobile navigation technology changed how people navigate through spaces. For example, a study on the use of GPS for navigation found that its use hindered participants' cognitive mapping abilities (Willis et al. 2007; Leshed et al. 2008). Participants using GPS navigation followed the route-based commands of the system instead of relying on visual cues. Owing to their focus on the system, they did not develop a sense of familiarity with the spaces they traveled through. In another study, travelers using GPS navigation showed an increased sense of confidence during travel owing to a perceived reduced risk of getting lost (Leshed et al. 2008). This confidence also gave participants a greater willingness to explore the spaces they were traveling through. Another advantage of relying on GPS navigation was the potential to discover new landmarks that were marked on the GPS map, but not visible from the road (Özkul 2014). Wayfinding can distance people from their surroundings, hindering the development of feelings of familiarity, while also allowing them to get to know spaces in a different way and with greater confidence.

Accessing and Sharing Information in Urban Space

The ability to discover new landmarks marked on GPS maps leads to the second type of activities into which locative media apps insert themselves: accessing and sharing information. Digital information and location combine to provide individuals with enhanced ways of knowing a particular space at a particular time. Individuals and institutions (for example, governments, businesses and schools) can make available geotagged digital content that attaches to a particular location. This content becomes available for people to access in those locations using a mobile device, such as a smartphone or tablet. Spaces around locative media app users turn into information interfaces and enhance information available in a physical space (Farman 2012). Widely popular digital mapping services, such as Google Maps, combine with various information services and apps, such as Yelp and Foursquare City Guide, to provide a gateway for the crowdsourcing and delivery of context-relevant information, such as identifying nearby places to shop or eat (Farman 2012; Leszczynski 2015). These apps can identify places nearby that may go unnoticed by people who are otherwise unaware of their existence, affording people unfamiliar with the area a greater sense of what is available there (Leszczynski 2015). The information available through these apps is generated by the users themselves. Users can add, rate and review locations that they visit (Fekete 2017). These apps can also reduce individuals' willingness to explore a physical place and limit where people go within a city (van den Akker 2014) by creating what Pariser (2011) refers to as a filter bubble. As locative media apps learn more about their users, the information available through them can become increasingly tailored to their interests, thereby limiting opportunities to explore beyond the familiar.

Facilitating Social Interactions among Strangers in Urban Space

A third way in which locative media apps insert themselves into everyday activities is through facilitating various types of social interactions, such as location-based mobile gaming and dating. Both types of locative media apps involve important complications for the ways in which people present themselves and interact in urban space.

Multiplayer, location-based mobile gaming is an example of locative media apps that facilitate social interaction. Mogi is a location-aware mobile game that was launched in Japan in 2003. People played by using their smart devices to collect nearby virtual objects. A significant component for many of the players was social interaction with other players. While collecting virtual objects was a key goal of the game, for most players an important objective was to meet other players and interact with them (Licoppe and

Inada 2006). For many players, the social aspects of the game made it fun and motivated continued engagement.

Pokémon GO is a mobile gaming app that launched in various countries in 2016 (Humphreys 2017; Hamari et al. 2019). Similar to Mogi, the main purpose of the game is to collect virtual objects – in this instance, Pokémon – that can be found in various locations throughout a city. Players can also form teams and engage in virtual battles with other players. Although these battles are virtual, they nevertheless take place in physical locations designated as gyms (Zach and Tussyadiah 2017). While not specifically intended to foster social interaction, the game's team functionality has facilitated group play. Colley et al.'s (2017) study of Pokémon GO players, for example, found that as many as 70 percent of respondents play with others (Colley et al. 2017). In addition to playing with friends, Pokémon GO players regularly interact with strangers while playing, resulting in a process that Humphreys (2010) terms parochialization, that is, the development of a sense of commonality among strangers (Humphreys 2017).

Three popular locative media dating apps also facilitate social interaction. Tinder is a dating app used predominantly, though not exclusively, by cisgender heterosexuals; Grindr is a dating app used predominantly by gay men; and Scruff is a dating app used predominantly by a subculture of gay men who self-identify as bears. These apps are designed to identify people nearby who may be available and interested in meeting up with one another. All three apps are typically understood to be hook-up apps, that is, apps whose sole function is to provide connections for possible sexual encounters. However, their use is more complicated than that. Studies of all three apps have identified other uses beyond hooking up. For example, some users are looking to find friends, others are looking to find a long-term romantic relationship and yet others simply want to meet new people (Crooks 2013; Blackwell et al. 2014; Mason 2016; Sumter et al. 2017). These apps all work in relatively similar ways: they display information about other users nearby. Each app approaches this task in different ways. Tinder typically employs a user's Facebook profile to identify potential matches, and then displays information about these matches. Users can then select which users they might be interested in. With Grindr and Scruff, users set up their own profile and preferences, and nearby potential matches are displayed based upon complementary profiles.

HOW LOCATIVE MEDIA ARE CHANGING COMPUTING

Locative media become an integral part of the interface between physical space and the digital information anchored in those spaces. That is, they serve to reveal the presence of digital information that is related to a particular space at a particular time. In this context, locative media mediate among the various actors engaged in the production of place, including people, objects, images and information (Urry 2000; Farman 2012; Leszczynski 2015). In order to show how locative media reconfigures our notions of computing, McCullough (2006) outlines five central shifts in focus:

1. From virtual to embodied. Locative media make location a central organizing principle of digital information and connections, despite the many claims over the years about the irrelevance of place. Instead of visiting a website to find generalized

information, locative media apps such as Yelp and Foursquare City Guide can provide information specific to the particular location where a user is standing. Locative media apps that facilitate social interaction (such as mobile dating and gaming apps) move beyond the virtual connections of the personal computer desktop to facilitate potential physical encounters embedded in time and space.

2. From macro to micro. Focus shifts from the fast, global flows of data, information, money, goods, entertainment, and so on, to the flows of these same things in the area immediately surrounding the user. This shift is from the fast-and-far to the close-and-slow. Here, locative media information apps can move beyond information generated through global patterns and trends, to reveal the particular quirks and diversities of individual urban spaces, such as local street vendors, pop-up stores and events, and local flea markets.
3. From universal to situated. Generic, one-size-fits-all, centrally produced and distributed content is replaced with locally produced – often user-generated – content, relevant to a specific user in a specific space at a specific time. Here, economies of scale give way to economies of relevance. Locative media apps can combine an individual's interests and needs with the individual possibilities of a particular neighborhood, street or street corner of a city.
4. From behavior to intent. As situational context becomes more important, designers of locative media apps focus less on how users behave in a space and more on what they want to accomplish in a particular place at a particular time. Given the propensity of locative media apps to reconnect users with the physical spaces they inhabit, and with those around them inhabiting the same spaces, locative media app designers must consider 'sociological staples such as presentation of self, tacit boundaries, and ad hoc networks' (McCullough 2006, p. 28) in their understanding of how users will deploy their apps. A dating app, for example, must be prepared to allow for an idealized presentation of self that can withstand the realities of imminent in-person meetings.
5. From pushing to posting. The broadcast model where users consume a few large-volume generic messages is replaced by the chaos of the myriad small, diverse daily messages of day-to-day life at street level. The move here is away from the passive consumption of generalized information, to the active production and sharing of activities in the moment. In the realm of social interaction, we see a move away from the promotion of connections among like-minded people back to the messy diversity of the urban street, and interactions among people in the here and now. Apps such as Yelp and Foursquare move information-sharing away from carefully managed corporate websites to an ever-changing exchange between nearby businesses and other points of interest, and their customers and, perhaps more importantly, among existing and potential customers.

McCullough's (2006, p. 26) points of differentiation of locative media offer a good starting place for understanding how they bring a range of digital information to 'the messy multiplicity of the street'. He succinctly captures the defining features of locative media, specifically that they facilitate an embodied, spatially and temporally situated experience with digital information.

NETWORKED INDIVIDUALISM IN THE CITY

The study of social structure and its relation to modernization, including the spread of digital media, has been a topic of much concern. Simmel (1922 [1955]) analyzed social structure through group affiliation and saw in traditional societies membership of groups as largely overlapping. That is, a person who belonged to one group most likely also belonged to other groups with similar status and affiliation. The consequence is a society where groups are closely knit, and those who belong know each other and are linked in more than one way. In modern societies individuals continue to belong to various groups, but group affiliation is scattered. This leads toward sparsely knit and fragmented networks, where groups show little or no overlap (Coser, L.A. 1975; Coser, R.L. 1975).

With greater penetration, in the 1990s, of networked computers and related technologies for a range of educational, business and entertainment purposes, Wellman (1999, 2000) introduced the concept of networked individualism. Before the Industrial Revolution, people's social networks were shaped by space and a limited set of social ties, wherein they often shared similar backgrounds, status and beliefs (see Hampton and Wellman 2018). For Wellman, networked individualism represents a shift in communication, as individuals are available for interaction independent of place. Communication moves from door-to-door to person-to-person communication, as mobile technologies facilitate reaching a specific person, rather than a specific place. Building on the concept, Castells (2001, p. 132) proposes that the Internet provides the material substance for networked individualism to emerge, becoming the 'dominant form of sociability' in the future. Rainie and Wellman (2012) build on the concept and outline how society is undergoing a structural transformation.

For Rainie and Wellman (2012), three key developments have transformed communities and neighborhoods since the 1990s. First, individuals are no longer embedded in bounded, village-like settings – or urban neighborhoods – but are members of multiple, diverse social networks that are often based on shared interests (Quan-Haase et al. 2019). Secondly, connectivity via networked computers has become widespread, providing flexible communication and information exchange over great (and small) geographic distances. Finally, mobile devices have diffused widely in societies around the globe, leading to personalized and ubiquitous communication and information.

To what extent has the shift to networked individualism, with its emphasis on fragmented networks, made location obsolete? Information and communication technologies afford different network structures that can transform how people form and maintain community as well as gain access to information and social support (Hampton 2016). For example, a study of older adults in Toronto, Canada, shows that ICTs can help overcome constraints of mobility by facilitating access to social support at a distance with a range of social ties (Quan-Haase et al. 2017). This suggests that ICTs provide opportunities to form supportive social relationships in multiple contexts that do not strongly overlap: family at home; colleagues in the workplace; and friends in the neighborhood, church and voluntary associations (Rainie and Wellman 2012; Quan-Haase et al. 2019). The concept of networked individualism suggests that social ties are no longer confined to pre-existing social groupings that organize social relations, such as family, religion, ethnicity and, potentially, location. In this way, networked individualism opens up opportunities to connect individuals based on new dimensions, including through shared location.

Locative media apps add a new social dimension to networked individualism in the context of urban space, leading to additional connections that are based on colocation. ICTs, such as email, instant messaging and video chat, are mostly used for keeping in touch with existing strong and weak social ties of kin, friendship, co-workers and shared interests (Mesch et al. 2012). While locative media apps can facilitate existing connections, it favors the establishment of ad hoc, transient connections among strangers. These ad hoc connections are formed with particular aims or common interests in mind, including dating, gaming and other activities linked to a particular space. As a consequence, locative media apps can further intensify the fragmentation and declustering of social networks by adding non-overlapping ties that are of short duration and purpose, and directly linked to a place and time.

LOCATIVE MEDIA'S IMPLICATIONS ON CITY AND CITIZEN

Locative media can alter people's relationships both with the cities they inhabit and with the people around them. For networked individuals in the city, locative media reintroduce physical location as a relevant factor, but not in the same way that Simmel (1922 [1955]) noted in traditional societies. People's relationships with their cities are influenced by their use of the wayfinding and information functionalities of locative media apps. In these instances, these apps can foster a fragmented understanding of the city, but also enable a sense of security and, even, adventure. Locative media apps affect networked individuals' relationships with their fellow urban dwellers in a variety of ways: visibility; sensory overload; social accessibility; trust and familiarity; and privacy. In the next section, we discuss the relationship between cities and their inhabitants, and then we look at the effects of locative media apps on urban social relations.

IMPLICATIONS OF LOCATIVE MEDIA ON RELATIONSHIPS WITH THE CITY

Fragmented City

As Rainie and Wellman (2012) observed, networked individuals in urban spaces have become unbounded by physical location, and this remains the case with the introduction of locative media apps. Indeed, it may be accelerated. People who rely on the wayfinding capabilities of locative media tend not to develop a sense of familiarity and understanding of the urban spaces they travel through, leading to a sense of detachment from these spaces (Willis et al. 2007; Leshed et al. 2008). As Özkul (2014) discusses in her study of locative media users in London, UK, much of the city is for them an unknown space. One participant remarked that he almost never asks people for directions, because they are unlikely to know: 'nobody is from this place. So whenever you ask somebody in London "Where is this street?" they have no idea' (Özkul 2014, p. 106). Before the introduction of locative media, networked individuals in the city may have developed, over time, a cognitive map of the spaces they frequented, and therefore some degree of familiarity. Now, this activity can be relegated to the wayfinding functionality of locative media.

However, this is an incomplete picture. Both the wayfinding and information-sharing capacities of locative media afford networked individuals in the city a capacity to move through the city as if they were long-time locals (Özkul 2014). Apps such as Yelp and Foursquare City Guide provide their users with information about the businesses and other points of interest that are around them. Also, users can consult the crowdsourced ratings and reviews to obtain a sense of which businesses and points of interest to visit, and which to avoid (Fekete 2017). One area of concern here, however, is the very real potential for business owners and others to create fake reviews (Luca and Zervas 2016).

Once decided on a destination, users can make use of their smartphone's GPS navigation to find their way right to the door, irrespective of any lack of familiarity with the area they are in. Even though networked individuals may not develop a personal familiarity with their urban spaces, they can nevertheless move through them as a local, in much the same way as the inhabitants of traditional societies that Simmel (1922 [1955]) described may have done in their towns and cities. As cities continue to increase in size, and an increasing number of newcomers move into the city, these functionalities afforded by locative media have the potential to be ever more important.

Security and Adventure

While locative media use does not typically result in a sense of familiarity with urban spaces, it can foster a sense of security in unfamiliar areas (Leshed et al. 2008; Özkul 2014). Even when users are in a part of the city that they have never been in before, they do not tend to feel lost. Even when they do feel lost, or when they are genuinely lost, they feel safe because they know that they can use their mobile device to guide them home. One participant in Özkul's (2014) study remarked that, before she had a smartphone, she felt so disempowered when she got lost in London. After she got a smartphone with locative media capabilities, she felt more confident: 'I want to navigate through the space myself, because then I get empowered' (Özkul 2014, p. 106). This sense of security arises from an ability to find the way back to a familiar space, such as home or work. It can also lead to a sense of adventure within the city. In their study, Leshed et al. (2008) found that locative media app users were often more willing to explore unfamiliar spaces owing to this ability to find their way back. Some of the participants used apps, such as Serendipitor, 'to get lost willingly and explore new aspects of urban spaces' (Özkul 2014, p. 107). Some participants would use a different route to get from one place to another simply to explore new parts of the city (Özkul 2014). These adventures are not without their risk, however. In Colley et al.'s (2017) study of Pokémon GO users, for example, some players reported venturing into spaces where they felt unsafe. One player in Finland was confronted by someone with a knife (Colley et al. 2017). This sense of security and adventure has the effect of further unbounding networked individuals from the urban neighborhoods with which they might be familiar.

IMPLICATIONS OF LOCATIVE MEDIA ON URBAN SOCIALITY

Urban spaces have distinct characteristics from non-urban spaces that lead to an urban personality (Wirth 1938 [2011]), including: the size, diversity and density of a population, together with the objects, institutions and built environments. The juxtaposition of various elements makes up the key components of the urban personality (Wirth 1938 [2011]; Massey 1999; Pile 1999; Amin and Thrift 2002): social detachment, social engagement beyond family and kinship, and a tolerance for difference.

As Knopp (1998, pp. 150–51) notes, urban space occurs 'at a density and scale at once sufficiently large and complex as to feel overwhelming and almost incomprehensible, yet which remains navigable and meaningful in many particular respects from the vantage points of people's daily lives'. For Simmel (1997, p. 178), the density and scale of urban spaces bring out a greater intensity of sensory stimulus which, in turn, gives rise to a key characteristic of the urban personality – the blasé attitude: 'There is perhaps no psychic phenomenon which has been so unconditionally reserved to the metropolis as has the blasé attitude. The blasé attitude results first from the rapidly changing and closely compressed contrasting stimulations of the nerves.' This blasé attitude becomes an essential tool in contending with the difference that has long been a key characteristic of the modern city. This difference manifests through 'gender, race, ethnicity, age, life course, sexuality, or any other referent' (Fincher and Jacobs 1998, p. 5). Also, difference will be present in any of a number of combinations of these referents. The consequence is that different citizens experience the city differently from one another, even differently from their next-door neighbors.

Social relationships in urban spaces are most often relationships among strangers. Instead of face-to-face interaction, the urban social relationships of strangers are characterized by side-by-side relationships (Young 1986). The urban social relationships of strangers take place in public spaces, such as parks, streets, bars and cultural venues. In these spaces, 'the diversity of the city's residents comes together and dwells side by side, sometimes appreciating one another, entertaining one another, or just chatting, always to go off again as strangers' (Young 1986, p. 21).

A central characteristic of the urban relationships among strangers is indifference (Tonkiss 2005). In successful urban environments, where there is a great degree of difference in terms of ethnicity, gender, sexuality, and so on, there is a lack of interest or concern – an indifference – toward this difference. It is not that these urban dwellers do not notice the difference. On the contrary, as Tonkiss (2005, p. 10) notes, 'alongside an active politics that recognizes differences . . . there lies an ordinary urban ethics that looks straight past it'. She recognizes that this indifference may be 'fragile, grudging, uneven', but where it exists, it provides a greater possibility for the expression of individual difference (Tonkiss 2005, p. 10). The daily interactions of urban strangers are comprised of myriad inconsequential interactions among strangers – greetings, small talk, and so on – in public spaces such as city streets. When added up, these small interactions recognize the 'public identity' (Jacobs 1969, p. 56) of people, and serve to build a web of trust and respect among urban strangers – an indifference to difference – without demanding more intimate connections that are typically reserved for friends, colleagues and family.

Locative media use complicates the urban social relationships of networked individuals by supporting an additional layer of interactions that are ad hoc, ephemeral connections of short duration among nearby strangers. However, these connections are not straightforward. We see locative media use in the city impacting urban social relationships among networked individuals in five key areas: visibility, sensory overload, social accessibility, trust and familiarity, and privacy.

Visibility

While the main purpose of mobile dating apps is to make users aware of other users nearby, the filtering functions of these apps can make some people who are visible in the physical space disappear in the app. Some apps, such as Grindr, allow users to filter people out based upon characteristics such as ethnicity, age, and so on. As Mason (2016, p. 826) observes, 'racism often functions as "preference" and neoliberalized choice in online dating sites, especially on dating sites and apps targeted to gay, bisexual, and other men who have sex with men'. Expressions of racism are not unique to locative media apps, nor to ICTs more generally. Instead, they reinforce existing racisms. However, these manifestations of racism may be more insidious since, particularly in the case of filtering options, the erasure itself is invisible (McPherson et al. 2001; Mason 2016; Noble 2018).

By contrast, Blackwell et al. (2014) describe a case where locative media can make people visible in spaces where they might otherwise be invisible. Apps such as Grindr and Scruff effectively make all spaces 'gay spaces'. For Jim, one of the participants of Blackwell et al.'s (2014, p. 1126) study of Grindr users, Grindr provided him with 'a gay bar in my pocket'. Jim spoke of one time where he was in 'a very very straight feeling space' (Blackwell et al. 2014, p. 1126). Grindr identified another gay man at another straight bar down the road, and the two met up. Grindr, Scruff and other such apps have the capacity to make gay men visible to other gay men nearby, while remaining invisible to others who are not using these apps. As both Blackwell et al. (2014) and Roth (2014) note, it can be inadvisable and even dangerous for queer people to be visible in some spaces at some times. Being aware of other queer people nearby can be reassuring, even if the users of the apps have no particular interest in hooking up (Blackwell et al. 2014; Roth 2014).

The various filters used by locative media apps have the potential to restrict the ways in which their users express themselves through these apps. Scruff, for example, asks users to voluntarily assign themselves to various labels that are used to organize and filter Scruff's users. As Roth (2014, p. 2123) notes, 'users who do not label themselves are automatically excluded from grids that filter users on the basis of those labels'. These labels, along with the other information requested (height, weight, hairiness, and so on), constrain the many possible body expressions that do not conform to these labels.

The possibility of apps filtering people out, thereby making them virtually invisible where they are physically present, can affect their ability to participate fully in urban space. Filtering out the potentially vast amounts of stimuli in urban spaces has long been part of developing the indifference to difference that is needed to navigate these spaces. However, the type of filtering afforded through locative media apps has the potential to erase difference altogether. So, while a city's residents continue to dwell together in the same physical space, the possibility disappears for 'appreciating one another, entertaining one another, or just chatting, always to go off again as strangers' (Young 1986, p. 26). Also, the filtered-out people do not know that this filtering is happening to them, and so will not be aware of opportunities that may have existed for them to connect with strangers around them. While we have focused on dating apps in this section, any locative social media app can have the capacity to filter people in or out based upon a wide range of characteristics, including race, gender, sexuality and age (de Souza e Silva and Frith 2012).

Sensory Overload

Locative media apps can further complicate urban social relationships by increasing the already overloaded sensorial stimulation by increasing urban inhabitants' awareness of the people around them. Instead of passing through the urban space relatively anonymously, exchanging little more than pleasantries with strangers, locative social media can increase the amount of information people know about these strangers. The strangers become what Licoppe (2016, p. 108) terms pseudonymous strangers: people 'with whom one may never have interacted or talked about before . . . but who are not complete strangers either for the locative app usually makes available some info about them' through their online profiles. A person who does not want this constant stream of information about nearby strangers can turn off these apps. However, the nearby strangers may have the apps turned on still, and may therefore recognize this person, without that person's knowledge. Such a situation may amount to nothing but, equally, it may amount to unwanted attention, challenging much needed distance among urban dwellers (Licoppe 2016).

As de Souza e Silva and Frith (2012) observe, mobile technologies are often criticized for privatizing the public space. However, as they argue, mobile technologies – from the invention of the paperback novel in the nineteenth century through to the rise of locative media today – are more properly understood as being deployed in the development of the blasé attitude characteristic of navigation through urban spaces. For example, a person on public transit might read a paperback or play a mobile game to avoid unwanted conversations with strangers; a person might listen to music to drown out the noise of a busy city street (Simmel 1997; de Souza e Silva and Frith 2012). What we see with the rise of the pseudonymous stranger, however, is something of a reversal: the potential for making public what might otherwise be considered private.

Social Accessibility

Urban life is filled with opportunities for sociability, creating great relevance for understanding how people negotiate social accessibility, understood as the ability to be reached by others (Simmel 1922 [1955]; Weber et al. 1946). With the modernization of cities, potential contact among city dwellers increased (Wellman and Leighton 1979), leading to new understandings of social accessibility and norms prescribing contact (Milgram 1970; Zerubavel 1979). Milgram (1970) describes how residents of New York City experience social accessibility. He notes, for example, how overload precludes individuals from taking in all information available in a city. Individuals cope with the large number of inputs by establishing priorities as to which individuals, cues, objects and images are relevant to them at various times and locations. Milgram (1970) also identifies a series of strategies that help urban dwellers adapt, such as spending less time on each social interaction, prioritizing social interactions by formulating a set of principles based on selectivity for important relationships, and blocking unwanted encounters (see also, Quan-Haase and Collins 2008).

Similarly, the use of mobile devices adds further layers to social accessibility by making individuals potentially accessible wherever they are in a city, adding stimuli to the urban environment. Locative media apps further complicate the issue, since they do not increase people's social accessibility to their personal networks, but instead to other urban dwellers located nearby. Locative media apps expand our social accessibility in the city to strangers by facilitating new interactions. They allow individuals to connect based on shared interests that only become visible through the app. These apps also work to deal with overload as described by Milgram (1970). They make others in the environment invisible and thereby preclude social interactions. This further increases communicative selectivity and narrows the scope of individuals we encounter while navigating the city.

Trust and Familiarity

Locative media apps may well contribute a sense of trust, familiarity and safety amid the characteristic diversity of an urban environment. As Sutko and de Souza e Silva (2011, p. 819) note, locative social media apps may make urban inhabitants aware that there are 'people like you around'. This homophilic awareness may foster a sense of trust and familiarity within an otherwise unfamiliar public space full of strangers. Sutko and de Souza e Silva also suggest that this trust may spill over onto the surrounding strangers, thereby facilitating an openness to the surrounding difference. This scenario is in evidence in the situation mentioned earlier where a Grindr user was able to locate nearby gay men in a straight environment in which he felt uncomfortable. This awareness gave him a sense of security in this potentially unwelcoming space. Also, Leshed et al.'s (2008) participants experienced a similar confidence using the GPS wayfinding functionality of locative media (Leshed et al. 2008; Sutko and de Souza e Silva 2011; Blackwell et al. 2014). The effects of this sort of filtering in of 'people like you' sit in stark contrast to the effects of filtering out people who are different. Here we can see new possibilities opening up to participate more fully in unfamiliar public spaces where, otherwise, people may have felt too unsafe to engage with surrounding strangers.

Privacy

Privacy is a much discussed topic in the context of mobile technologies and is particularly relevant when it comes to locative media. Privacy is defined as a basic human need and considered an important policy and research topic, gaining much attention in the wake of the Cambridge Analytica and Facebook scandals (Bartsch and Dienlin 2016; Kezer et al. 2016; BBC 2018). Not all individuals share the same privacy concerns and there is considerable variability in terms of how much and what types of personal information people feel comfortable sharing with companies, governments and institutions (Elueze and Quan-Haase 2018). In addition to inter-individual differences in privacy attitudes, attitudes toward privacy also vary by technology and platform. Individuals often report a high degree of privacy concerns on platforms where it is difficult to control the message, such as on Facebook. By contrast, newer interactive platforms, such as Snapchat, have built-in features that help individuals protect their privacy to a greater extent not only by controlling who views a photo or message, but by also automatically removing the content permanently (Piwek and Jonson 2016).

Mobile phones pose unique privacy challenges. This is linked to two features of the technology. Mobile phones are equipped with microphones and cameras, posing difficult to evaluate risks if misused by third parties. Another risk comes from the possibility that location can be established. Shilton (2009, p. 48) discusses how surveillance via mobile phones allows companies, governments, and potentially third parties to collect personal information about a person's 'regular locations, habits, and routines'. For example, it is possible to broadcast our whereabouts to a select network of contacts throughout the day, as Sam Liam, the designer of an app named Placement, did through his phone (Farman 2013). Perhaps the greatest risk is that location data could be collected and used without the user being aware of the misuse (Shilton 2009). Google provides opportunities for users to manage their privacy by turning location information off. Yet, an investigation by the Associated Press revealed that Google continued to track users' location data even when they had turned location history off. Google misleadingly states that when location history is turned off, 'the places you go are no longer stored. When you turn off Location History for your Google Account, it's off for all devices associated with that Google Account' (Dreyfuss 2018). This demonstrates that privacy breaches in the collection and use of location-based data are widespread among technology companies. It also demonstrates that even when users actively try to protect their location-based data and turn location off on their devices, the data may still be recorded and shared without their consent.

The data collection and analysis of locative media can serve multiple purposes and is a valuable data source for companies. Companies can make use of the data to better understand purchasing behaviors as they happen in the moment. They can also improve services provided to clients or identify niche markets based on how city dwellers interact with their urban environments. Yet, the purposes of locative media are not always linked to corporate interests, nor do they necessarily serve surveillance purposes. The data recorded for research purposes is often referred to as participatory sensing. The purpose of participatory sensing is to collect and make use of data that is otherwise underutilized. Shilton (2009, p. 50) stresses that the goal is to 'avoid surveillance or coercive sensing by emphasizing individuals' participation in the sensing process'. Some of these applications collect anonymous data, while others link data to places. Some projects collect personal data on how a person moves within a city, while others collect aggregate data to learn about patterns and flows on a larger scale.

Apps based on locative media rely on detailed location-based data to facilitate experiences and interactions in the moment. There are risks associated with the use of these types of services and the broadcasting of personal information via the apps. It is important to realize, though, that risks also exist when navigating urban settings and interacting with strangers. So far there is little research assessing potential additional risks associated with the use of locative media, but this kind of research will be important as locative media evolve and become more widely used.

CONCLUSIONS

Locative media are playing an increasingly important role in the daily lives of networked individuals in urban conurbations. As increasing numbers of people move into cities, this role is also likely to increase. Even at this early stage of locative media use, it is evident that they have a profound effect on people's relationships with their cities and with the people with whom they inhabit their cities. Reliance on the wayfinding and information-sharing capabilities of locative media apps inhibits the development of a sense of familiarity with the urban spaces that app users move through. These apps can provide users with a sense of security, and even adventure within unfamiliar areas, and with information about what is going on around them. For networked individuals, locative media result in a further unbounding from the neighborhoods in which they live, work and play, becoming able to move throughout the city as a long-time local, without necessarily building a sense of familiarity with the spaces they move through.

Social relationships in urban spaces become further fragmented and declustered for networked individuals by the introduction of short, ephemeral connections with nearby strangers in complicated ways. Locative media apps make more information available to people about the strangers around them, turning them into pseudonymous strangers (Licoppe 2016). Some apps, such as Grindr and Scruff, can make people virtually visible in spaces where they might otherwise be physically invisible. They can also make people virtually invisible in spaces where they are physically visible. These affordances raise questions about who can and ought to be visible, where and when. Affordances are understood as opportunities and constraints provided by a system through features and functionalities (Wellman et al. 2003). Locative media apps can increase our accessibility to the strangers immediately around us in public spaces, but this increased accessibility can also lead to an increase in sensory overload, and add new dimensions to privacy concerns. However, within all of the information available through locative media, patterns can emerge that can reveal to users that there are people around them who are in one way or another like them, thereby facilitating a sense of trust and confidence in unfamiliar spaces. This trust can have a knock-on effect of increasing users' tolerance of the differences that surround them. In stark contrast, however, the ability to filter out people who are physically present can have the effect of limiting the ability of those filtered-out people to participate fully in the public spaces they are inhabiting. Instead of establishing an indifference to difference needed to navigate urban spaces, this filtering out can have the effect of erasing difference altogether.

As location awareness is added to increasing numbers of mobile apps, the capacity for locative media to affect people's relationships with the urban spaces they inhabit becomes increasingly significant. Licoppe (2016, p. 113) asks us to imagine 'a future in which the use of locative media becomes so commonplace that all urban denizens are digitally connected and location-aware'. In that scenario, the issues examined in this chapter have the capacity to become magnified manifold. What happens, for example, if every store we pass wants to convey information to us? What happens if everyone is sharing their profiles with everyone around them? In this sea of information, who will decide what is relevant to whom? How will these decisions be made? Who will benefit? At what cost, and to whom? These are issues and ethical questions that designers of locative media apps can and should consider when developing their apps and services. Urban planners and others concerned with the growth and development of cities can and should consider the ways in which locative media are already changing urban spaces and social relationships among urban dwellers.

ACKNOWLEDGEMENT

This chapter was informed by Darryl Pieber's Master's thesis in media studies entitled 'A place for locative media: a theoretical framework for assessing locative media use in urban environments', conducted under the supervision of Dr Quan-Haase at Western University. Funding from the Social Sciences and Humanities Research Council of Canada (SSHRC) given to A. Quan-Haase supported this research as well as an Ontario Graduate Scholarship (OGS) given to D.A. Pieber.

REFERENCES

Amin, A. and N. Thrift (2002), *Cities: Reimagining the Urban*, Cambridge: Polity.

Bartsch, M. and T. Dienlin (2016), 'Control your Facebook: an analysis of online privacy literacy', *Computers in Human Behavior*, **56** (March), 147–54.

BBC (2018), *Cambridge Analytica: The Story so Far*, 21 March, accessed 22 March 2018 at http://www.bbc.com/news/technology-43465968.

Blackwell, C., J. Birnholtz and C. Abbott (2014), 'Seeing and being seen: co-situation and impression formation using Grindr, a location-aware gay dating app', *New Media & Society*, **17** (7), 1117–36.

Castells, M. (2001), *The Internet Galaxy*, Oxford: Oxford University Press.

Castells, M. (2002), 'Local and global: cities in the network society', *Tijdschrift voor economische en sociale geografie*, **93** (5), 548–58.

City of Toronto (2013), 'Immigration, citizenship, place of birth, ethnicity, visible minorities, religion and aboriginal people's, 9 May, accessed 17 August 2018 at https://www.toronto.ca/wp-content/uploads/2017/10/9793-2011-NHS-Backgrounder-Immigration-Citizenship-Place-of-Birth-Ethnicity-Visible-Minorities-Religion-and-Aboriginal-Peoples-.pdf.

City of Toronto (2017), '2016 Census: population and dwelling counts', 9 February, accessed 17 August 2018 at https://www.toronto.ca/wp-content/uploads/2017/10/96b9-2016-Census-Backgrounder-Population-Dwellings.pdf.

Colley, A., J. Thebault-Spieker, A.Y. Liu, D. Degraen, B. Fischman, J. Häkkilä, et al. (2017), 'The geography of Pokémon GO: beneficial and problematic effects on places and movement', *Proceedings of the CHI 2017 Conference on Human Factors in Computing Systems*, May, pp. 1179–92, doi:10.1145/3025453.3025495.

Coser, L.A. (ed.) (1975), *The Idea of Social Structure: Papers in Honor of Robert K. Merton*, New York: Harcourt Brace Jovanovich.

Coser, R.L. (1975), 'The complexity of roles as a seedbed of individual autonomy', in L.A. Coser (ed.), *The Idea of Social Structure: Papers in Honor of Robert K. Merton*, New York: Harcourt Brace Jovanovich, pp. 237–62.

Crooks, R.N. (2013), 'The rainbow flag and the green carnation: Grindr in the gay village', *First Monday*, **18** (11), accessed 24 May 2016 at http://www.firstmonday.dk/ojs/index.php/fm/article/view/4958/3790.

De Souza e Silva, A. and J. Frith (2012), *Mobile Interfaces in Public Spaces: Locational Privacy, Control, and Urban Sociability*, New York: Routledge.

Dreyfuss, E. (2018), 'Google tracks you even if location history's off. Here's how to stop it', *Wired*, 8 August, accessed 19 August 2018 at https://www.wired.com/story/google-location-tracking-turn-off/.

Elueze, I. and A. Quan-Haase (2018), 'Privacy attitudes and concerns in the digital lives of older adults: Westin's privacy attitude typology revisited', *American Behavioral Scientist*, **62** (10), 1372–91.

Farman, J. (2012), *Mobile Interface Theory: Embodied Space and Locative Media*, New York: Routledge.

Farman, J. (2013), 'Site-specificity, pervasive computing, and the reading interface', in J. Farman (ed.), *The Mobile Story: Narrative Practices with Locative Technologies*, New York: Routledge, pp. 3–16.

Fekete, E. (2017), 'Foursquare in the City of Fountains: using Kansas City as a case study for combining demographic and social media data', in J. Thatcher, J. Eckert and A. Shears (eds), *Thinking Big Data in Geography: New Regimes, New Research*, Lincoln, NE: University of Nebrasaka Press, pp. 165–88.

Fincher, R. and J.M. Jacobs (1998), 'Introduction', in R. Fincher and J.M. Jacobs (eds.), *Cities of Difference*, New York: Guilford Press, pp. 1–25.

Hamari, J., A. Malik, J. Koski and A. Johri (2019), 'Uses and gratifications of Pokémon Go: why do people play mobile location-based augmented reality games?', *International Journal of Human–Computer Interaction*, **35** (9), 1–16, doi:10.1080/10447318.2018.1497115.

Hampton, K.N. (2016), 'Persistent and pervasive community', *American Behavioral Scientist*, **60** (1), 101–24.

Hampton, K.N. and B. Wellman (2018), 'Lost and saved . . . again: the moral panic about the loss of community takes hold of social media', *Contemporary Sociology*, **47** (6), 643–51.

Humphreys, L. (2010), 'Mobile social networks and urban space', *New Media & Society*, **12** (5), 763–78.

Humphreys, L. (2017), 'Involvement shield or social catalyst: thoughts on sociospatial practice of Pokémon GO', *Mobile Media & Communication*, **5** (1), 15–19.

Jacobs, J. (1969), *The Death and Life of Great American Cities*, New York: Modern Library.

Kezer, M., B. Sevi, Z. Cemalcilar and L. Baruh (2016), 'Age differences in privacy attitudes, literacy and privacy management on Facebook', *Cyberpsychology: Journal of Psychosocial Research on Cyberspace*, **10** (1), accseesd 2 January 2018 at https://cyberpsychology.eu/article/view/6182/5912.

Kitchin, R. (2014), 'The real-time city? Big data and smart urbanism', *GeoJournal*, **79** (1), 1–14.

Knopp, L. (1998), 'Sexuality and urban space: gay male identity politics in the United States, the United Kingdom and Australia', in R. Fincher and J.M. Jacobs (eds), *Cities of Difference*, New York: Guilford Press, pp. 149–76.

Leshed, G., T. Velden, O. Rieger, B. Kot and P. Sengers (2008), 'In-car GPS navigation: engagement with and disengagement from the environment', *Proceedings of the SIGCHI Conference on Human Factors in Computing Systems*, New York: ACM, pp. 1675–84.

Leszczynski, A. (2015), 'Spatial media/tion', *Progress in Human Geography*, **39** (6), 729–51.

Licoppe, C. (2016), 'Mobilities and urban encounters in public places in the age of locative media: seems, folds, and encounters with "pseudonymous strangers"', *Mobilities*, **11** (1), 99–116.

Licoppe, C. and Y. Inada (2006), 'Emergent uses of a multiplayer location-aware mobile game: the interactional consequences of mediated encounters', *Mobilities*, **1** (1), 39–61.

Luca, M. and G. Zervas (2016), 'Fake it till you make it: reputation, competition and Yelp review fraud', *Management Science*, **62** (12), 3412–27.

Mason, C.L. (2016), 'Tinder and humanitarian hook-ups: the erotics of social media racism', *Feminist Media Studies*, **16** (5), 822–37.

Massey, D. (1999), 'Cities in the world', in D. Massey, J. Allen and S. Pile (eds), *City Worlds*, London: Routledge.

McCullough, M. (2006), 'On the urbanism of locative media', *Places*, **18** (2), 26–9.

McPherson, M., L. Smith-Lovin and J.M. Cook (2001), 'Birds of a feather: homophily in social networks', *Annual Review of Sociology*, 27, 415–44.

Mesch, G.S., I. Talmud and A. Quan-Haase (2012), 'Instant messaging social networks: individual, relational, and cultural characteristics', *Journal of Social and Personal Relationships*, **29** (6), 736–59.

Milgram, S. (1970), 'The experience of living in cities', *Science*, **167** (3924), 1461–8.

Noble, S.U. (2018), *Algorithms of Oppression: How Search Engines Reinforce Racism*, New York: New York University Press.

Özkul, D. (2014), 'Location as a sense of place: everyday life, mobile, and spatial practices in urban spaces', in A. de Souza e Silva and M. Sheller (eds), *Mobility and Locative Media: Mobile Communication in Hybrid Spaces*, London: Taylor & Frances Group, pp. 101–6.

Pariser, E. (2011), *The Filter Bubble: What the Internet is Hiding from You*, New York: Penguin Press.

Pile, S. (1999), 'What is a city?', in D. Massey, J. Allen and S. Pile (eds), *City Worlds*, London: Routledge, pp. 3–50.

Piwek, L. and A. Jonson (2016), '"What do they Snapchat about?" Patterns of use in time-limited instant messaging service', *Computers in Human Behavior*, **54** (January), 358–67.

Quan-Haase, A. and J.L. Collins (2008), '"I'm there, but I might not want to talk to you": university students' social accessibility in instant messaging', *Information, Communication & Society*, **11** (4), 526–43.

Quan-Haase, A., G.Y. Mo and B. Wellman (2017), 'Connected seniors: how older adults in East York exchange social support online and offline', *Information, Communication & Society*, **20** (7), 967–98.

Quan-Haase, A., R. Zhang, B. Wellman and H. Wang (2019), 'Older adults on digital media in a networked society: enhancing and updating social connections', in M. Graham and W. H. Dutton (eds), *Society and the Internet: How Networks of Information and Communication are Changing our Lives*, 2nd edn, Oxford: Oxford University Press, pp. 96–124.

Rainie, L. and B. Wellman (2012), *Networked: The New Social Operating System*, Cambridge, MA: MIT Press.

Roth, Y. (2014), 'Locating the "Scruff guy": theorizing body and space in gay geosocial media', *International Journal of Communication*, **8**, 2113–33.

Shilton, K. (2009), 'Four billion little brothers? Privacy, mobile phones, and ubiquitous data collection', *Communications of the ACM*, **52** (11), 48–53.

Simmel, G. (1922), 'The web of group affiliations', repr 1955 in K. Wolff (trans.), *Conflict and the Web of Group Affiliations*, Glencoe, IL: Free Press, pp. 125–95.

Simmel, G. (1997), 'The sociology of space', in D. Frisby and M. Featherstone (eds), *Simmel on Culture: Selected Writings*, London: Sage, pp. 174–86.

Sumter, S.R., L. Vandenbosch and L. Ligtenberg (2017), 'Love me Tinder: untangling emerging adults' motivations for using the dating application Tinder', *Telematics and Informatics*, **34** (1), 67–78.

Sutko, D.M. and A. de Souza e Silva (2011), 'Location-aware mobile media and urban sociability', *New Media & Society*, **13** (5), 807–23.

Tonkiss, F. (2005), *Space, the City and Social Theory: Social Relations and Urban Forms*, Cambridge: Polity Press.

Urry, J. (2000), 'Mobile sociology', *British Journal of Sociology*, **51** (1), 185–203.

Van den Akker, R. (2014), 'Walking in the hybrid city: from micro-coordination to chance orchestration', in A.

de Souza e Silva and M. Sheller (eds), *Mobility and Locative Media: Mobile Communication in Hybrid Spaces*, London: Taylor & Francis Group, pp. 33–47.
Villi, M. (2015), '"Hey, I'm here right now": camera phone photographs and mediated presence', *Photographies*, **8** (1), 3–22.
Weber, M., H. Gerth and C.W. Mills (1946), *From Max Weber: Essays in Sociology*, New York: Oxford University Press.
Weiser, M. (1991), 'The computer for the 21st century', *Scientific American*, **265** (3), 94–104.
Wellman, B. (1999), 'From little boxes to loosely-bounded networks: the privatization and domestication of community', in J.L. Abu-Lughod (ed.), *Sociology for the Twenty-First Century: Continuities and Cutting Edges*, Chicago, IL: University of Chicago Press, pp. 94–116.
Wellman, B. (2000), 'Living wired in a networked world: the rise of networked individualism', paper presented at the Founding Conference, Association of Internet Researchers, Lawrence, KS, 14–17 September.
Wellman, B. and B. Leighton (1979), 'Networks, neighborhoods and communities: approaches to the community question', *Urban Affairs*, **14** (3), 363–90.
Wellman, B., A. Quan-Haase, J. Boase, W. Chen, K. Hampton, I.I. De Diaz and K. Miyata (2003), 'The social affordances of the internet for networked individualism', *Journal of Computer-Mediated Communication*, **8** (3), doi:10.1111/j.1083-6101.2003.tb00216.x.
Willis, K.S., C. Hoelscher and G. Wilbertz (2007), 'Understanding mobile spatial interaction in urban environments', *Proceedings of the Third IET International Conference on Intelligent Environments*, Ulm, Germany, 24–25 September, pp. 61–8.
Wirth, L. (1938), 'Urbanism as a way of life', repr. 2011 in R.T. LeGates and F. Stout (eds), *The City Reader*, 5th edn, New York: Routledge. pp. 96–104.
World Bank Group (2015), *East Asia's Changing Urban Landscape: Measuring A Decade Of Spatial Growth*, Washington, DC: World Bank, accessed 17 August 2018 at https://www.worldbank.org/content/dam/Worldbank/Publications/Urban%20Development/EAP_Urban_Expansion_full_report_web.pdf.
Young, I.M. (1986), 'The ideal of community and the politics of difference', *Social Theory and Practice*, **12** (1), 1–26.
Zach, F.J. and I.P. Tussyadiah (2017), 'To catch them all: the (un)intended consequences of Pokémon GO on mobility, consumption, and wellbeing', in R. Schegg and B. Stangl (eds), *Information and Communication Technologies in Tourism 2017: Proceedings of the International Conference in Rome, Italy*, Guildford: Springer, pp. 217–27.
Zerubavel, E. (1979), 'Private time and public time: the temporal structure of social accessibility and professional commitments', *Social Forces*, **58** (1), 38–58.

21. Far away ties, never so close: the geographical spread of social support resources for mobile individuals

Romina Cachia and Isidro Maya-Jariego

1. INTRODUCTION

Distinct ties offering diverse social support can be vital when moving to a new location. A move to another place implies significant changes in our social support network. Existing ties may be replaced, and new ties are expected to be established. This morphological change in the social support network is further altered with the prevalence of new communications media. Sustaining ties through virtual communication has resulted in transnational social support networks. It has been observed that tangible support (instrumental and social companionship) is strongly related to short distances, while intangible support (emotional and conflict) is more likely to stay intact despite the geographical distance (Dahinden 2010; Herz 2015; Mok and Wellman 2007; Ryan et al. 2008). The social support network is a great source of information for understanding how social support is derived in different mobility contexts and in different cities.

This chapter examines the social support networks of foreigners living in Seville. The study is based on in-depth interviews with 95 highly qualified and skilled foreigners who lived in Seville for a minimum of six months and each of their 30 ties in their personal networks ($n = 2850$). The respondents were selected from four distinct communities, namely, Erasmus students, musicians for the symphonic orchestra, partners of researchers working for the European Commission (EC) and Japanese flamenco artists. These communities provided a framework of analysis, through which four types of mobility patterns allowed us to distinguish between temporary stays, with short-term return (Erasmus students), continuous displacements in itinerancy (partners of workers of the EC), recurring temporary stays (Japanese flamenco dancers) and indefinite settlement (musicians of the symphonic orchestra). Utilizing personal networks, we explore how mobility patterns can have an influence on the provision of social support.

In the context of mobile individuals, networks are particularly insightful because they yield data on the international spread of the social support networks allowing researchers to identify social support, not merely based on geographical proximity, but also on distant ties (Castells 2000; Hampton et al. 2009; Hoete 2004; Larsen et al. 2006a; Wellman 1996). Recent literature shows an emerging interest in examining geographical mobility through informal and smaller-scale transnational networks of mobile individuals based on interpersonal connections, often as expressed through personal networks (Carrasco et al. 2008; Herz 2015; Kennedy 2004; Larsen et al. 2006b; Lubbers et al. 2007, 2010; Martinez Garcia et al. 2001; Maya Jariego 2006; Mok and Wellman 2007; Molina et al. 2007). Personal network enables researchers to study social support as detached from the geographical location, while simultaneously, the same composition of the network also sheds light on how individuals are linked to others inside or outside the city.

The chapter first looks at social support in the context of mobility. This is followed by a cross-section description of the provision of different types of social support. Data from all the respondents' ties, referred to in personal networks analysis as alters, (n = 2850) is examined. In the second part, a social support taxonomy is elaborated classifying ties into three groups: specialized, intermediate or multiplex support providers. Finally, we develop a typology derived in a clustering procedure through which four types of support ties are identified, namely: (1) diversified social companionship, (2) recent socializing ties, (3) polyvalent support by old friends and (4) co-presence support. The different categorizations are used to examine how the type of social support is related to network characteristics and the type of mobility. Finally, a discussion on the findings of this study is presented.

1.1 Measuring Social Support

During the past decades, different social support instruments have been developed to measure social support. Findings from recent studies show that studying social support as a one-dimensional phenomenon has various limitations. Different types of relationships may bestow distinct type of social support (Agneessens et al. 2002; Walker et al. 1993). Although, there is a variety of instruments to measure social support, most studies are based on three mayor types of support, namely, emotional, instrumental and informational (Barrera 1980; Burt 1984; Fischer 1982a; Marin and Hampton 2007; van der Poel 1993; Wellman 1979; Wellman and Wortley 1990).

Ties in a personal network tend to be a mixture of specialized – providers of one type of support – or multiplex – providers of different types of support (Hennig 2007). The ego–alter relationship seems to have an influence on the type of social support provided (Schweizer et al. 1998). Kin are more likely to be providers of emotional and economic support, friends act as social companions, and neighbours seem less significant but also lend instrumental help. Similar findings were previously explored by Fischer (1982a) in his analysis of how personal communities are linked to residential communities. Friends were more likely to provide companionship, relatives and neighbours provided practical help, and immediate relatives and friends gave counselling.

There are distinct methods of how the list of social support providers may be elaborated (Agneessens et al. 2002; Bidart and Charbonneau 2011; Marin and Hampton 2007; McCallister and Fischer 1978; van der Poel 1993). Social support is often studied through an ego-centred perspective, where respondents (ego) are asked about the availability of different types of support from a list of people (alters). A name generator, supplemented by additional attributes of the alters, as for instance, the type of relation they have with ego, is one of the most used tools to create a list of social support providers. Tie evaluation, consisting in identifying whether a specific relationship exists between the ties, for instance, whether ties know each other, also enables researchers to study the structure of the network, using network analysis.

1.2 Mobility and Social Support

International displacement has important repercussions on the structure of social support networks. Mobility means change, a disruption to our personal network (Sonn 2002) and the development of new demands for support in an unfamiliar environment (Litwin 1995). A residential move can lead to stressful situations and, as found by various studies based on the buffering model, social support can prevent health-related problems, such as lowering of the immune system. The simple knowledge that we have a social support network at our disposal was found to be positively related to bolster our perceived ability to cope with stressful situations (Cohen and Wills 1985) and to promote psychological well-being and quality of life.

The relationship between social support and the process of migratory integration has been explored in various studies (Aroian 1992; Berry 1997; Clarke and Jensen 1997; Leslie 1992; Lubbers et al. 2010; Lynam 1985; Martinez Garcia et al. 2001). Networks are particularly insightful in understanding how mobile individuals resort to different types of relationships to sustain social support. Aroian (1992) found that co-nationals were a major source of social support for Polish immigrants in America, especially during early resettlement. Depending on the type of mobility, family reunification and the incorporation of host individuals seem to be two basic strategies used by Africans and Latin-Americans in Spain to rebuild their personal networks (Maya Jariego 2006).

1.3 Mobility and Technology

The role of computer-mediated communication has also been highlighted in more recent studies. Chen and Choi (2011) observed that recently arrived Chinese migrants in Singapore, were more likely to rely on computer mediated social support to supplement the lack of offline social support. However, as time passes and their local networks grow, online support was more likely to be informational.

Mobility has an ambivalent nature: on the one hand, spatial dispersion from friends and family is seen as an obstacle to the nurturing of social ties; on the other hand, it also brings about opportunities for establishing new social ties (Viry 2012). Increasingly, inhabitants of urban cities sustain meaningful active ties outside their immediate locality as found in the studies carried out in East York, Ontario (Wellman et al. 1988) and London (Albrow 1997). When specific resources are not available, especially in periods of change, people will turn to media systems to obtain them (Chen and Choi 2011), as found through the media system dependency theory (Ball-Rokeach and DeFleur 1976). A recently arrived migrant may have a limited network, for instance, to look for a place to rent. Consequently, the Internet becomes an alternative source of social support. While tangible support or physical companionship is limited through new media, some types of companionship, such as playing online games, provide other forms of companionship as listed in the study of Chen and Choi (2011).

New forms of portability and connectivity, have enabled new ways of how people reconfigure social networks, by disconnecting and reconnecting them in complex ways (Larsen et al. 2006a). People travel, relocate and still remain close to emotional ties (Walker et al. 1993). This new type of connected migrant (Diminescu 2008), who does not cut ties with the social network at home, calls for new approaches in studying how the absent individuals make themselves present (Licoppe 2004) and the complexities in maintaining multiple social relations across time and space through different interactive media.

Paradoxically, although far away, emotional ties have never been so close, people travel more than ever, spending considerable amounts of money and time on the road to meet physically with their family and friends. Despite advanced communications technology, distance still matters (Axhausen and Frei 2008; Ryan et al. 2015). Respondents with migrants of Pakistani origin living in the North of England claimed to travel to Pakistan at least once a year (Mason 2004). A tie remains active through occasioned co-presence (Larsen et al. 2006a) and meetingness (Urry 2003), and it seems this presence cannot be entirely replaced by virtual presence or communication, yet. This compulsion to proximity (Larsen et al. 2006a) is important for mobility research owing to its social and emotional significance. While a wide array of literature has focused on the notion of networked individualism, less research seems to take into account the role of periodic travel and how it influences the production, sustainability and stabilization of social support networks.

Thus far we have observed how social support networks are increasingly transnational, especially in the context of mobile individuals and how the use of media and travel are crucial in sustaining such ties. These findings imply that the studying of social support networks cannot be limited by geographical limits. Moreover, we have seen how personal networks comprise ties providing distinct types of social support, and the type of social support provided is seemingly related to the type of relationship. In this study, we explore a further dimension in this relationship: the type of mobility and whether it has an influence on the composition of the social support network.

1.4 Networks and the City: The Case of Seville

Seville is a city of just over 700 000 inhabitants, making it the fourth most populated city in Spain. Taking into account the metropolitan area, the number of inhabitants amounts to just over 1 535 000. The most significant population increase occurred throughout the twentieth century, especially caused by rural immigration from other localities in the province. That has led to the urban growth of the city, mostly around the city centre.

To our knowledge, there are few studies (Maya Jariego and Armitage 2007; Maya Jariego and Holgado 2015) exploring personal networks in Seville. For this study, Seville is an interesting context since, despite its high influx of tourists, the expatriate community remains limited, especially when compared with other similar European cities. Seville is unique owing to its high level of social cohesion and it is a densely connected population, with a high perception of a sense of local belonging. Local traditions play an important role in the development of relationships and in the community integration of residents. In this context, it is interesting to explore how qualified professionals who settle in the city are integrated. On the one hand, they are incorporated into cohesive community networks. On the other, entry to these strongly articulated pre-existing networks may be a challenge. Furthermore, the high safeguarding of deeply rooted traditions by locals has limited, to a certain extent, global urban influence and internationalization of the city. Accordingly, the transnational social support networks of foreigners living in Seville play a pivotal role in their lives.

1.5 This Study

Previous research on immigrant social support networks has shown that the perception of available support is affected by international displacement. Immigrants experience a reduction in the size of support networks just after geographical relocation, initiating a process of progressive reconstruction of their personal networks. However, this impact is not uniform across all types of support; it mostly affects instrumental assistance and the provision of tangible resources. On the contrary, emotional support can be kept at a distance and remains active through periodic visits to the place of origin.

We explore how different types of support are distributed geographically according to different forms of international mobility. To do so, we examine the personal networks of four groups of qualified foreigners residing in Seville, who differ in the geographic mobility pattern: international students who make a short stay in the city; flamenco artists who make periodic visits to complete their training in dancing; relatives of workers of the EC who usually move in an itinerant way between different countries; and musicians from the symphony orchestra, settled in Spain. The four groups differ in the time of their stay, the expectation of residence and the connection with the country of origin. Erasmus students usually live less than nine months in Seville and then return to their country of origin. Flamenco students repeat several stays of a few months or a year in the city, with round trips to Japan. Family members of workers of the European Commission have lived in several different countries before arriving in Seville, where they stay for a period of three or four years before moving to a new destination. The orchestra musicians are settled in Seville, with an expectation of permanent residence, for an indefinite period.

The use of digital media and regular visits facilitates the exchange of social support resources. We analyse what type of support interactions occur in each case and to what extent they are related to the type of geographical mobility of each group.

2. METHOD

2.1 Participants

The target population of this study was mobile individuals living in Seville, for a minimum period of six months. In total, 195 respondents were contacted, of whom 100 accepted to be interviewed. For the data analysis, data from 95 interviews was used owing to uncompleted and incorrect data in five of the interviews. The data was collected between January and April 2012. Respondents participated in the study on a voluntary basis.

Participants were selected through a snowball sampling, utilizing four communities as the sampling frame. The four communities were Erasmus students ($n = 33$), partners of researchers working for the European Commission, in concrete for the Institute for Prospective Technological Studies ($n = 25$), Japanese flamenco artists ($n = 19$) and musicians from the Symphonic Orchestra of Seville ($n = 18$). These communities were selected on the basis that they framed different dimensions of mobility. In particular, we looked at time planned to stay in Seville, time spent in Seville and reason for moving. For the analysis of this work, a common classificatory framework was developed, as illustrated in Table 21.1.

Table 21.1 Common classificatory framework of mobility patterns

Type of mobility	Reason for moving		
	Study	Work	Relationship
Temporary	Erasmus		
Transit	Flamenco		EC partners
Settlement		Orchestra	

Respondents in this study were dominated by female participants (74 per cent). Erasmus students were aged under 30 years (100 per cent), respondents from Flamenco (79 per cent) and EC partners (68 per cent) were mostly aged between 31 and 50 years. Orchestra musicians were the oldest of all groups, with 61 per cent between 31 and 50 years and 33 per cent over 51 years. In terms of education, one-third of the respondents (36 per cent) had a postgraduate degree and another third (35 per cent) had a first degree.

2.2 Procedure

The data collection was designed in two phases. The first phase consisted of an electronic document sent by email, through which an electronic multiple name generator was completed based on previous work in this area (Barrera 1980; Fischer 1982b; Marin and Hampton 2007; van der Poel 1993). Respondents were presented with a list of 30 rows, with five columns, each column representing a type of support, namely, emotional, instrumental, socializing, co-presence and other, according to a series of probing questions. Respondents were requested to list 30 people starting with people providing emotional support, then instrumental support, and on. If participants had not reached 30 people, we asked them to list other people who are important for them and whom they have not listed yet in the last column, which we named as: other. All respondents in the study listed 30 alters in total. The design of the name generator served to collect data on social support multiplicity. The names of people were to be listed vertically, but any person providing more than one type of support could be listed again in the same row, in the corresponding column. In this way, multiplex alters (ties providing more than one support) were listed more than once horizontally. Multiplicity was used in this study to gauge the type and intensity of the relation between ego and alters, following findings by Marin and Hampton (2007) showing that multiple generators are more reliable than single generators when measuring size, density and measures of network compositions.

The fixed enumeration of 30 alters aimed at delineating a personal network large enough for structural network analysis (McCarty 2002), but also feasible to administer and not too complex to interact with the network visualizations. Previous research has shown that at least 30 alters are needed to capture the diversity of structures of personal networks (McCarty 2002). The fixed enumeration design allowed us to compare and contrast each and every personal network of each respondent. This is because comparison of the composition of the personal network, for instance, the proportion of compatriots living in place of origin, or the number of locals living in Seville listed in a respondent's network were more important in the context of this study than the number of ties present in each respondent's network.

The visualizations of the personal networks enabled respondents to visualize their social support network. For most, this was the first time ever to see their network and have an opportunity to discuss the different groupings within their own networks. The visualizations worked as a tool, through which respondents discussed their social support network and which added an interactive dimension to the interview.

For the analysis of this chapter, ties were divided into three groups, namely: ties providing one type of support (specialized), those providing two types of support (dual) and, finally, alters providing more than two types of intermediate support (multiple).

The second phase of the data collection consisted of an in-depth interview, elaborating a personal network visualization through the software VennMaker (Schönhuth et al. 2012), a software designed specifically for the collection, visualization and analysis of personal networks. Through this tool, respondents were asked to list the ties between all the alters and to list a number of attributes related to the ties. The interview was set up upon receipt of the first part of the data collection. All the participants who completed the first part of the data collection, participated in the second part. The personal networks elicited in this work are produced by self-reporting methods and should be regarded as a perceived social support network rather than a received support network.

3. RESULTS

3.1 Cross-section Analysis of Different Types of Social Support

The first part of the analysis provides a descriptive analysis of the social support personal networks of the respondents. A significant proportion of respondents' personal networks consists of people living abroad: two-thirds (65 per cent) of ties providing emotional support are transnational, with almost half (46.9 per cent) residing in country of origin and a small subgroup (17.5 per cent) living in other countries (Table 21.2). Instrumental and socializing

Table 21.2 Type of support across network characteristics (n = 2850)

	Emotional		Instrumental		Social companionship		Co-presence	
	Ties	%	Ties	%	Ties	%	Ties	%
Location of alters								
Seville	364	35.7	326	45.0	887	52.6	110	13.7
Country of origin	478	46.9	284	39.2	514	30.5	502	62.5
Other	178	17.5	114	15.7	286	17.0	191	23.8
Total	1020	100.0	724	100.0	1687	100.0	803	100.0
Nationality								
Spanish	210	20.1	198	26.4	455	26.3	100	12.3
Co-national	647	61.9	399	53.3	823	47.6	585	72.2
Other	189	18.1	152	20.3	452	26.1	125	15.4
Total	1046	100.0	749	100.0	1730	100.0	810	100.0

Table 21.3 Type of support across network characteristics (n = 2850)

	Emotional		Instrumental		Social companionship		Co-presence	
	Freq.	%	Freq.	%	Freq.	%	Freq.	%
Type of relationship								
Partner	50	4.7	37	4.9	47	2.7	11	1.3
Family	270	25.6	205	26.9	166	9.5	292	35.7
Friend	688	65.3	433	56.9	1343	76.6	450	55.1
Colleague	30	2.8	36	4.7	111	6.3	38	4.7
Neighbour	–	–	20	2.6	27	1.5	6	0.7
Other	15	1.4	30	3.9	60	3.4	20	2.4
Duration of tie								
In the past year	124	11.8	109	14.3	483	27.5	76	9.3
In the past 5 years	223	22.1	188	24.7	471	26.9	139	17.0
In the past 10 years	201	19.1	147	19.3	359	20.5	135	16.5
Over 10 years	469	47.1	317	41.6	441	25.2	467	57.2

support was more likely to be derived from people living in Seville. Co-nationals play a major role across different types of support, while hosts were more likely to be mentioned as potential providers of tangible support, namely, instrumental and socializing support.

As regards the role of ties listed, friends predominate across all types of support, with socializing support (77 per cent) and emotional support (65 per cent) ranking as the highest. Family ties were present across all types of support, but to a lesser extent than friends. Partners were listed mostly as providers of emotional and instrumental support. It is interesting that a small percentage of partners (1 per cent) were listed as co-presence support providers, suggesting that, in some cases, partners of our respondents in this study lived abroad. Neighbours were listed mainly as providers of instrumental and socializing support.

To understand the strength of the relationship, we also asked for the duration of the tie. Ties whom respondents knew for a longer period were more likely to be listed as providers of emotional support and instrumental support. Conversely, recently known ties were more likely to be listed as providers of socializing support. Respondents travelled mostly to visit long-known ties (Table 21.3).

3.2 Classification of Ties

The second part of our cross-section analysis is a classification of the ties listed by our respondents according to the type of support they provided. When we classify the alters according to the intensity of social support they provide, we observe that the most frequently mentioned types of ties are those providing specialized and unique type of support (50.2 per cent). Below this upper typology, we observe alters providing two types of support, mostly emotional and social (22.3 per cent). This is followed by ties providing multiplex support (three or four) to the respondents (16.9 per cent). In total, we observe that 39 per cent of the ties listed provide multiple types of support. Alters mentioned in the different types of intermediate support indicate different levels of intensity, strength

Table 21.4 Frequency of typologies of social support (n = 2850)

	Emotional	Instrumental	Social	Travel	Freq.	%
Specialized and unique support ties			X		848	29.1
				X	299	10.3
	X				185	6.4
		X			128	4.4
					1460	50.2
Intermediate support ties	X		X		239	8.2
		X	X		140	4.8
	X			X	90	3.1
			X	X	90	3.1
	X	X			65	2.2
		X		X	26	0.9
					650	22.3
Higher multiplicity ties	X	X	X		180	6.2
	X		X	X	91	3.1
	X	X		X	60	2.1
		X	X	X	20	0.7
	X	X	X	X	141	4.8
					492	16.9

of tie and different type of relationships (Maya Jariego 2006). These people are the inner layer of the network providing a combination of emotional, instrumental and social support (Fischer 1982a; van der Poel 1993).

In Table 21.4, we elaborate a qualitative typology of all the alters listed by the respondents based on which type of social support they provide. Alters who were listed as providers of one type of support, for instance, emotional support was grouped under 'Specialized and unique support ties'. Ties listed as providers of two types of support, for instance, emotional and social, were listed under 'Intermediate support ties'. While multiplex ties, providing more than two types of support, were listed under 'Higher multiplicity support ties'.

Providers of only one type of aid, people with whom respondents socialize and share their leisure time, predominate in the social support networks of our respondents. One-third of alters listed (29 per cent) are providers of socializing support. Interestingly, the second most listed alters are those living outside Spain and for whom our respondents feel the need to travel and meet periodically face to face. Providers of merely emotional support comprise only a small percentage (6 per cent).

This analysis shows that the social support networks of foreigners living in Seville studied in this work display a high level of transnationality, especially in relation to emotional support ties. Transnational ties are predominately linked to the country of origin and are mainly providers of emotional support. This suggests a shift in the location of emotional ties, whereby distance is substituted by mediated communication and periodic travel. Friends dominate in these networks, and the strength of support seems to be linked to the duration of the tie. Unique to our findings is the high number of distant ties mentioned

Table 21.5 Strength of social support by network composition characteristics

	Specialized		Dual		Multiple		
	Freq.	%	Freq.	%	Freq.	%	χ^2
Network composition: location							
Seville	686	62.4	266	24.2	148	13.5	57713*
Country of origin	496	50.7	241	24.6	241	24.6	
Other	237	54.2	111	25.4	89	20.4	
Network composition: nationality							
Spanish	360	59.7	153	25.4	90	14.9	85675*
Compatriot	715	51.8	352	25.5	312	22.6	
Other	370	62.7	135	22.9	85	14.4	
Ego-alter characteristics: duration of tie							
In the past year	450	75.4	103	17.3	44	7.4	173991*
All my life/more than 20 years	218	44.5	139	28.4	133	27.1	

Note: * p <.001.

by our respondents, with whom they sustain the relation through periodic travel. The most common multiplex ties mentioned in this study are those providing emotional and social support.

3.3 Multiplicity of Support

In this section, we examine the relationship between the type of support provided – specialized, dual and multiple – and network composition of the participants, their patterns of mobility and the way they use communications technology.

The different levels of intermediate support provides an insight into the type of relationship with ties. A positive relationship was observed across the strength of social support provided by the alters and the network composition. Ties providing specialized, dual and multiple support differed according to locality, $\chi^2_{(4,\ 2515)} = 47.70$, $p = .00$; nationality, $\chi^2_{(4,\ 2572)} = 33.08$, $p = .00$; and duration of tie, $\chi^2_{(8,\ 2602)} = 173.99$, $p = .00$ (Table 21.5).

Multiple support providers, were mainly people living in country of origin, co-nationals, partners and family members. These strong ties were more likely to have known the respondents for more than 10 years or all their lives. In contrast, specialized support was more likely to be expected from ties living in the host location, of foreign origin and who ego got to know in the past year.

As regards mobility patterns, a significant relation was observed between the community of the respondents and the type of support alters listed, $\chi^2_{(6,\ 2602)} = 40.55$, $p < .00$. Respondents from the orchestra were more likely to list multiplex support alters, while respondents from Flamenco were more likely to mention ties providing specialized support. Erasmus students and EC researchers' partners scored the highest in dual type of support alters.

From the analysis, we can observe that the type of mobility seems to have an influence on the composition of the social support network. Respondents who were more settled in the city, listed a higher frequency of multiplex social support ties. Dual ties were mostly

Table 21.6 Ranking of the eight types of support providers more frequent (average of providers per personal networks) (n = 95)

Type of providers of support	M	SD
Social companionship	8.92	5.07
Co-presence	3.15	3.80
Emotional and social	2.52	2.80
Emotional	2.03	3.28
Emotional, instrumental and social	1.89	2.47
Instrumental and social	1.47	2.37
Emotional, instrumental, social and co-presence	1.47	2.80
Instrumental	1.35	2.38

presented in social support networks of respondents with higher mobility. Ties listed as providers of multiple support in this study were mainly transnational, living in country of origin and co-nationals. In contrast, recently known ties living in the host location and of foreign origin were the main providers of specialized support.

3.4 Typologies of Social Support Providers

In this final section of the data analysis, we adopt a different level of analysis. Instead of analysing the characteristics of the alters (n = 2850), we analyse their combination in each personal network (n = 95). Specifically, we classify the types of personal networks of each respondent (n = 95) according to the type of alters (30 in each case) present in each of them.

To classify the support networks, we analyse how the 15 types of social support providers previously identified are distributed in the personal networks of the respondents (Table 21.4). First, we elaborated a ranking with the eight most frequent suppliers on average (Table 21.6). Next, we performed a cluster analysis with the quick clustering method of k-means, with 10 iterations and a criterion of convergence of 0.02 with updated means. The criterion variables were the three most frequent types of providers of support. These were alters providing only socialization support (M = 8.92); co-presence support (M = 3.15) and a combination of emotional and social support (M = 2.52) (Table 21.6).

For this work, four clusters were deemed the most apt in defining different typologies of social support. Table 21.7 presents the clusters derived based on the three main types of support, as listed previously. The figures shown in Table 21.7 are the final cluster centres on each criterion variable for each of the derived clusters. Each cluster represents a different type of respondent based on the different types of alters listed per each type of support. The first cluster includes 50 respondents (53 per cent of our sample), the second 21 (22 per cent), the third 9 (10 per cent) and the fourth 15 (16 per cent).

A descriptive analysis of the different clusters in relation to the network composition variables is presented in Table 21.8. A description of each cluster is presented in Table 21.9. The tables list the mean variable score for each cluster and the inter-cluster differences and their significance. All variables used were tested for normality. One-way Anova (Scheffé intergroup means) was used for normally distributed variables (F scores) and

Table 21.7 Cluster centroids of social support providers (n = 95)

	Cluster 1 (n = 50)	Cluster 2 (n = 21)	Cluster 3 (n = 9)	Cluster 4 (n = 15)
Type of support				
Social companionship	8	16	1	5
Co-presence	2	2	0	10
Emotional and social companionship	2	2	7	2

the Kruskal–Wallis test (*H* scores) was used for the non-parametric variables. For the non-parametric variables, follow-up tests to evaluate pair wise differences among the four clusters controlling for Type I error across tests by using the Bonferroni approach was used.

Cluster 1 (53 per cent of respondents) is mainly characterized by social companionship support. Co-presence and intermediate support (ties providing both emotional and social companionship) are also available but moderate. Respondents in this cluster listed a high number of people living in Seville and across the four clusters, and listed the highest number of people living in other locations. Their social support networks consisted mainly of friends, although family members also shared an important proportion, especially when compared with other clusters. The duration of relationship between respondents and alters in this cluster is diversified, with a slightly higher mean of alters known in recent years. Participants in this cluster communicated predominantly with their ties through face-to-face communication.

Cluster 2 (22 per cent) is similar to some degree to the previous cluster, but it reflects a higher degree of ties providing socializing support. The networks of these respondents are unique in the high listing across the four clusters of co-nationals and recently met ties. Hosts played a less important role in these networks, while friends dominated the networks. This cluster comprised Erasmus students. In this study, we found that the proportion of Spanish alters in the support network is positively related with the length of residence of the respondents, $\chi^2(8, 95) = 49.64$, $p = .00$. Respondents (100 per cent) who had lived less than a year in Seville had less than eight hosts in their network. When it comes to communication, respondents in this cluster differed significantly from other clusters in listing the highest number of alters with whom they meet weekly face to face ($M = 7.3$, $SD = 4.78$) and the highest number of alters with whom they communicated through a social networking service (SNS) ($M = 8.0$, $SD = 6.30$) also on a weekly basis.

Cluster 3 (10 per cent) is a small cluster mainly consisting of respondents providing emotional and socializing support. The ties of these respondents mainly comprised a high proportion of co-nationals and people living in the country of origin. Alters providing this type of support were more likely to have known the respondents for more than 10 years. This cluster also stands out for the lowest ranking of family members and the highest ranking of friends. Colleagues and neighbours were also at their minimal. This cluster consists of a high number of musicians from the orchestra. When it comes to media, we observe a statistically significant difference in the high number of alters mentioned with whom respondents rarely met face to face.

Cluster 4 (16 per cent) is mainly characterized by the high frequency of co-presence

Table 21.8 *Means, Oneway ANOVA and Kruskal Wallis variance of network composition characteristics per cluster ($n = 95$)*

	Cluster 1		Cluster 2		Cluster 3		Cluster 4		F	H
	M	SD	M	SD	M	SD	M	SD		
N of alters per location										
Seville	13.2	6.16	12.1	7.31	12.2	6.70	14.5	5.00	0.45	
Country of origin	10.8	5.99	13.6	5.60	13.6	6.70	12.0	4.81	1.416	
Other	6.0	5.67	4.4	4.87	4.2	4.24	3.5	2.75		1.59
N of alters by nationality										
Spanish	7.8	6.19	4.6	5.52	7.8	6.94	9.9	4.48		48.33****
Compatriot	15.0	7.18	17.8	7.82	16.8	7.86	14.5	4.19	1.001	
Other	7.0	4.07	7.6	5.78	5.4	4.93	5.6	3.81		2.86*
N of alters per duration of Tie										
In the past year	6.5	5.80	10.7	7.73	3.2	3.56	3.4	3.94		46753****
in the past five years	8.0	4.58	7.3	4.92	6.7	5.70	6.6	5.71		1.77
In the past ten years	5.7	4.03	4.1	3.40	7.2	3.15	5.6	4.45		9437*
In the past twenty years	4.4	4.30	3.1	3.65	8.2	6.40	4.9	4.85		36864***
All my life/More than 20 years	5.4	3.17	4.4	4.98	4.7	1.94	9.4	5.17		14.90***
N of alters per relationship										
Partner	0.5	0.51	0.6	0.51	0.4	0.53	0.9	0.26		16141***
Family	5.8	2.82	4.2	2.84	4.1	2.47	7.9	4.41		12.19**
Friends	19.8	4.76	20.6	5.87	22.3	3.91	16.1	5.47		9.16*
Colleagues	2.0	3.45	2.3	2.76	1.3	3.04	2.5	2.33		9390*
Neighbours	0.4	1.05	.8	1.67	0.3	0.50	1.4	2.23		10085*
Others	1.5	2.65	1.3	3.17	1.4	1.81	1.9	1.80		11920**

Note: *$p < 0.05$; ** $p < 0.01$; *** $p < 0.005$; **** $p < 0.001$.

Table 21.9 Characteristics of the social support profiles

Cluster	Description	Network composition, mobility patterns and communication
1. Diversified social companionship	• Mainly characterized by social companionship support • Low provision of co-presence and a combination of emotional and social companionship	• High number of alters living in other locations • High number of alters living in Seville • Medium number of family • High number of alters listed in weekly face-to-face communication
2. Recent socializing ties	• Highest frequency of social companionship • Low levels of co-presence and emotional and social companionship support	• Highest number of co-nationals • Highest listing of ties living in country of origin • Highest mentioning of alters from other nationalities • Support is mainly derived from recently known alters • Majority of ties mentioned have lived in Seville for a less than a year and tend to be Erasmus students • Highest listing of alters with whom they met frequently face to face and communicated through SNS
3. Polyvalent support by old friends	• Highest mentioning of combined emotional and social companionship support • No co-presence support alters • Low presence of alters solely providing social companionship	• Relative high number of Spanish alters • Have known the respondents for a long period of time • Lowest ranking of family members, neighbours and colleagues • Highest ranking of friends • Majority belong to the orchestra community • Listed the same number (an average of 5) with whom they communicate on a weekly basis face to face and through SNS
4. Co-presence support	• Mainly characterized by diversified social support alters • Highest mentioning of co-presence support alters	• Highest number of hosts • Ties who the respondents have known the longest predominate • Highest mentioning of family members • Highest mentioning of alters with whom respondent rarely communicated by mobile and lowest number of alters with whom respondents communicated by SNS on a weekly basis

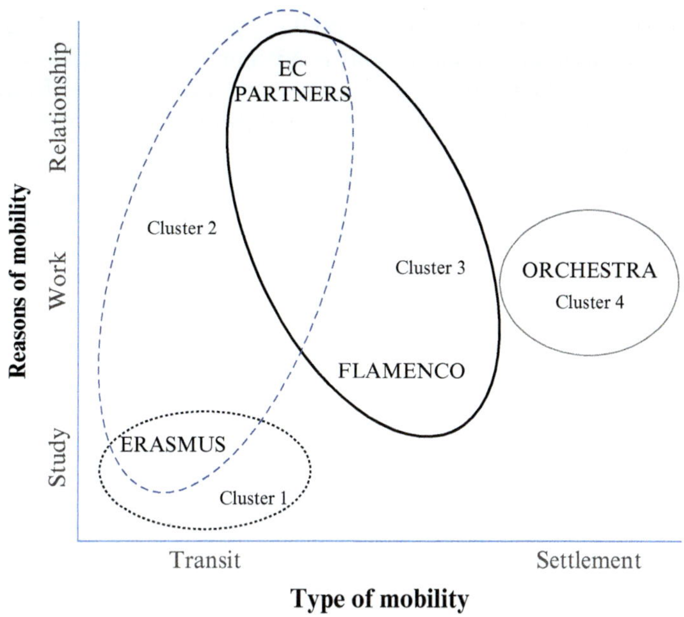

Figure 21.1 Social support clusters per subsample of respondents

support. This profile is unique in being the cluster that displays frequent mobility, through the high number of ties listed for which respondents travel to meet. The social support networks in this cluster were differentiated by the high listing of both Spanish alters and people they have known over 20 years or all their lives. Across the four categories, respondents in this cluster were more likely to have networks filled with family ties. As regards community, they were more likely to be from the Flamenco and the EC community. The majority had lived in Seville between one and five years. In general, these respondents communicated the least frequently across all the different media.

A relationship could be observed between the four clusters and the type of mobility community. Clusters dominated by social companionship support, mainly clusters 1 and 2, are characterized by respondents who are in Seville for a short stay (Erasmus) or in transit to another location in the near future (EC partners). Multiplex support seems to be associated with respondents who are more likely to be settled in Seville (orchestra). Co-presence support is more typical of respondents who have lived in Seville for a medium period, in which they have either settled in the past few years or they have been here for a couple of years and are more likely to leave soon (EC partners and Flamenco), and where short mobility of coming and going is highly present (Figure 21.1).

4. DISCUSSION

This chapter explored how social support is related to network composition in the context of mobile individuals in an urban environment. Specifically, we examined how the type and strength of social support is altered depending on the composition of the personal network and the framework within which a specific type of mobility takes place. Within urban environments, mobility of individuals is not only high, but also people tend to commute more owing to the large size of the city increasing the flux of individuals on a daily basis. People living in urban environments find themselves relying more heavily on transnational ties, as a way of bridging ties from place of origin with new social ties in the host location and in other places of mobility. In this study, we see that transnational ties play a major role in the social support network of respondents, congruent with other studies that have demonstrated the provision of support is not location bound and may be kept at a distance (Albrow 1997; Fennell 1997; Herz 2015; Horrigan 2001; Licoppe 2004; Mok and Wellman 2007; Mok et al. 2010; Neal 2012; Wellman 1996, 2002). The shift to urban environments may be one of the main reasons why social support seems to be extended globally. In rural locations, social support is more likely to be rooted to a specific location and derived from the people who live close by. However, as people move to urban cities, they find that social support is less cohesive. Effort is dedicated to extend and spread the personal network. New meaningful social relationships are created in the new location, and great effort is dedicated to maintaining contact with far-away ties. Communications technology is crucial in how mobile people derive their social support, especially intangible support.

This could indicate an important morphological change taking place in the social support network, possibly less researched and discussed in the context of migration studies. With the increase of ubiquitous and cheap social media, sustaining far-away affective ties is becoming increasingly widespread. Portability of networks has led to easy access to a wide array of social ties, through which various types of social support are sustained. Families, partners and close friends chat frequently and sustain the relation through easy to access technology such as Voice over Internet Protocol (VoIP), Face Talk and apps, such as WhatsApp. Seemingly, there seems to be less need to replace affective ties with other ones in the new residential location, given such role can be maintained at a distance.

This chapter casts light on the possible influence on the composition of the social support network, depending on the type of mobility. Settled respondents seem to rely more on emotional ties with whom they can meet face to face and on multiplex type of alters. They also felt more integrated within the city, especially through the recognition and prestige derived from their work.

In contrast, transit type of respondents seem to have more dispersed networks composed mostly of weaker ties. In the context of mobility, weaker transient ties serve specific roles related to the mobile person's need at a specific time, as well as access to other type of information, otherwise impossible to reach through strong ties (Granovetter 1973).

Hence, from our results we can argue that highly mobile respondents in Seville are hardly bowling alone, to use Putnam's (2000) expression. Mobility is often seen as a deviation to place attachment and participation in the city's community life. As yet, as seen in previous studies, mobile individuals do not have fewer social ties than mobile ties (Fischer 2011), it is just that they orientate their lives in more than one society, developing their social relationships within transnational communities (Castles 2002). During the interviews, musicians from the orchestra mentioned that through their music they felt they contributed towards the culture of the city and, hence, positively influenced the sense of belonging to the city.

This is related to the findings by Fischer, where he found that city people were as socially and psychologically integrated, as small-town people, which he attributed to selective migration, a term used to describe how the personal characteristics of the individuals play a major role in which type of community settings they opt to live. In any event, we must take into account the peculiarities of the group studied. Social support resources may be different in mobile groups with a lower socio-economic status or with less competence for intercultural communication (Briones et al. 2009; Maya-Jariego 2006; Searle and Ward 1990).

Another important dimension is the role of friendship. As people are more likely to live away from their family, friendships constitute important social bonds previously associated with the family, especially in urban cities (Bellotti 2008; Pahl 2002). Friends comprised more than half of the personal network of respondents, confirming similar results in previous studies (Fischer 1982a; Wellman and Frank 2001). Similar findings were made by Fischer (1982a) in his analysis on how personal communities are linked to the residential communities. Friends were more likely to provide companionship, relatives and neighbours provided practical help, and immediate relatives and friends gave counselling. It was also interesting that respondents in this study claimed to travel most to visit friends, followed by family. Keeping friends at a distance requires an effort, which is not necessary for staying in touch with family. Seemingly, qualified people are more likely to rely on friends, compared with low-educated people who are more likely to obtain support from family (Kennedy 2004).

Finally, the high listing of social ties in the co-presence category suggests that future studies addressing social support of mobile individuals should take into account this aspect. As the findings in this study demonstrate, despite sophisticated new communications media, distance still matters (Axhausen and Frei 2008), and people willingly go out of their way to spend quality time with significant people in their lives.

The use of a name generator with a fixed number of alters was useful to describe the multiplicity, providing practical advantages in relation to standardization and reduction of the load in data management (Maya Jariego 2018). Although the establishment of a limit of alters that can be mentioned introduces biases when aiming to estimate the size of the personal network (Neal and Neal 2017), in our study it served to efficiently compare the different forms of mobility and the consequent relational integration in the city.

5. CONCLUSIONS

Mobility is one of the defining characteristics of urban cities but, as yet, little research has looked into its diversity (Urry 2000). This study aimed to understand how the type of mobility influences the social support network by studying the support relationships of four groups of qualified immigrants in an urban city in Spain. The results show (1) the existence of a large number of distance contacts, which exert as transnational support providers; (2) frequent trips to maintain contact with relatives and friends living in other countries; and (3) a significant relationship between the type of personal support network and the migrants' geographical mobility patterns. These findings are briefly reviewed next.

Primarily, transnational support relationships are highly prevalent. Students and foreign workers residing in Spain maintain strong ties with people living in the country of origin and/or other locations where they have previously resided. Although geographic distance seems to affect adversely the accessibility of instrumental and tangible sources of help available to the individual before moving, emotional and informative support resources are more likely to be active at-a-distance.

However, travel to the country of origin is also frequent. Visits to family and friends fulfil specific needs of companionship, sociability and shared emotional connection that seem to require direct contact (or co-presence). It would be of interest to delve deeper into the factors that lead to a high frequency of international travel despite, more than ever, that affective support and advice are highly available through other means. Digital communication and frequent travel seem to be major factors in how people stay close, as regards interaction and availability, with geographically distant ties.

Third, the significant presence of active transnational support relationships can contribute to slow down the changes in the composition and structure of personal networks experienced by immigrants following international displacement. However, it appears to give rise, at least in part, to a geographical segregation of support functions. For this reason, social support multiplicity can be used as an efficient indicator of the process of acculturation and psychological adaptation. The geographical distribution of multi-purpose support providers shows how each individual responds to their needs in the transnational space.

Finally, the organization of support resources seems clearly conditioned by the pattern of geographical mobility. For example, those individuals who are settled in a stable and indefinite manner, which in our sample correspond significantly to the musicians of the symphony orchestra, have a greater relative predominance of strong ties with people residing in Spain. While those who are in transit, from country to country, or who make a short stay with an expectation of short-term return, are more likely to have weak ties (that is, specialized providers of sociability or companionship) in Spain, which they complement with a connection with providers of affective support in the sending country. We have also observed the existence of a small group of highly mobile individuals who base their exchanges of support on trips that allow them to have direct contact with family and friends with a certain frequency.

In summary, geographic mobility seems to have a different impact on the types of support available. Specifically, emotional support may be kept at a distance more easily than instrumental support and companionship. As a consequence, in situations of international displacement, it is comparatively more frequent that affective support and companionship are distributed among different providers (when in the general population these are two types of support that tend to be highly correlated with each other). In this context, periodic visits allow activating companionship from those strong ties with which the emotional bond has been maintained at a distance.

This process depends to a large extent on the type of mobility. Specifically, those people who have an expectation of temporary stay (either very short term, such as Erasmus, or medium term, such as family members of workers of the EC), are more likely to develop ties to obtain mere companionship in the place of residence. Possibly the expectation of return (or of a new displacement) leads to a reluctance in deepening relationships and, therefore, leads to fewer strong ties in the host city. However, as more time is spent in the host city, it becomes more necessary to resort to periodic visits to maintain contact with the origin. Finally, only with indefinite settlement is there greater likelihood that more multiple ties will be developed in the host city.

ACKNOWLEDGEMENT

This research has been partially funded by a research grant provided by the Joint Research Centre, European Commission, as part of the Training through Research programme.

REFERENCES

Agneessens, F., H. Waege and J. Lievens (2002), 'Social support typologies: different approaches for reducing social support data', *Developments in Social Science Methodology*, **18**, accessed 25 March 2021 at http://dk.fdv.uni-lj.si/metodoloskizvezki/Pdfs/Mz18AgnessensWaegeLievens.pdf.

Albrow, M. (1997), 'Travelling beyond local cultures: socioscapes in a global city', in J. Eade (ed.), *Living the Global City: Globalization as Local Process*, London: Routledge, pp. 37–55.

Aroian, K.J. (1992), 'Sources of social support and conflict for Polish immigrants', *Qualitative Health Research*, **2** (2), 178–207, doi:10.1177/104973239200200205.

Axhausen, K.W. and A. Frei (2008), 'Contacts in a shrunken world', paper presented at the Eighty-sixth Annual Meeting of the Transportation Research Board, Washington, DC.

Ball-Rokeach, S.J. and M.L. DeFleur (1976), 'A dependency model of mass-media effects', *Communication Research*, **3** (1), 3–21, doi:10.1177/009365027600300101.

Barrera, M. (1980), 'A method for the assessment of social support networks in community survey research', *Connections*, **3** (3), 8–13.

Bellotti, E. (2008), 'What are friends for? Elective communities of single people', *Social Networks*, **30** (4), 318–29.

Berry, J.W. (1997), 'Immigration, acculturation, and adaptation', *Applied Psychology*, **46** (1), 5–34, doi:10.1111/j.1464-0597.1997.tb01087.x.

Bidart, C. and J. Charbonneau (2011), 'How to generate personal networks: issues and tools for a sociological perspective', *Field Methods*, **23** (3), 266–86, doi:10.1177/1525822x11408513.

Briones, E., C. Tabernero, C. Tramontano, G.V. Caprara and A. Arenas (2009), 'Development of a cultural self-efficacy scale for adolescents (CSES-A)', International Journal of Intercultural Relations, 33 (4), 301–12, doi:10.1016/j.ijintrel.2009.03.006.

Burt, R.S. (1984), 'Network items and the general social survey', *Social Networks*, **6** (4), 293 339.

Carrasco, J.A., E.J. Miller and B. Wellman (2008), 'How far and with whom do people socialize? Empirical evidence about the distance between social network members', *Transportation Research Record: Journal of the Transportation Research Board*, **2076** (1), 114–22, doi:10.3141/2076-13.

Castells, M. (2000), *The Rise of the Network Society*, 2nd edn, Oxford: Blackwell.

Castles, S. (2002), 'Migration and Community Formation under Conditions of Globalization', *International Migration Review*, **36** (4), 1143–68, doi:10.1111/j.1747-7379.2002.tb00121.x.

Chen, W. and A.S.K. Choi (2011), 'Internet and social support among Chinese migrants in Singapore', *New Media & Society*, **13** (7), 1067–84, doi:10.1177/1461444810396311.

Clarke, D.E. and M.A. Jensen (1997), 'The effects of social support, life events, and demographic factors on depression among Maori and Europeans in New Zealand rural, town, and urban environments', *Journal of Community Psychology*, **25** (4), 303–23, doi:10.1002/(sici)1520-6629(199707)25:4<303::aid-jcop1>3.0.co;2-v.

Cohen, S. and T.A. Wills (1985), 'Stress, social support, and the buffering hypothesis', *Psychological Bulletin*, **98** (2), 310–57.

Dahinden, J. (2010), 'The dynamics of migrants'transnational formations: between mobility and locality', in R. Baubock and T. Faist (eds), *Diaspora and Transnationalism: Concepts, Theories and Methods*, Amsterdam: Amsterdam University Press, pp. 51–72.

Diminescu, D. (2008), 'The connected migrant: an epistemological manifesto', *Social Science Information*, **47** (4), 565–79.

Fennell, G. (1997), 'Local lives – distant ties: researching community under globalized conditions', in J. Eade (ed.), *Living the Global City: Globalization as Local Process*, London: Routledge, pp. 90–109.

Fischer, C.S. (1982a), *To Dwell among Friends: Personal Networks in Town and City*, Chicago, IL: Chicago University Press.

Fischer, C.S. (1982b), 'What do we mean by "friend"? An inductive study', *Social Networks*, **3** (4), 287–306.

Fischer, C.S. (2011), *Still Connected: Family and Friends in America since 1970*, New York: Russell Sage Foundation.

Granovetter, M.S. (1973), 'The strength of weak ties', *American Journal of Community Psychology*, **78** (6), 1360–80.

Hampton, K.N., L.F. Sessions, E.J. Her and L. Rainie (2009), 'Social isolation and new technology: how the

Internet and the mobile phones impact Americans' social networks', *Pew/Internet*, accessed 12 May 2019 at http://www.pewinternet.org/Reports/2009/18--Social-Isolation-and-New-Technology.aspx?r=1.

Hennig, M. (2007), 'Re-evaluating the community question from a German perspective', *Social Networks*, **29** (3), 375–90, doi:10.1016/j.socnet.2007.01.008.

Herz, A. (2015), 'Relational constitution of social support in migrants' transnational personal communities', *Social Networks*, **40** (January), 64–74.

Hoete, A. (ed.) (2004), *ROAM: A Reader in the Aesthetics of Mobility*, London: Black Dog.

Horrigan, J.B. (2001), 'The Internet and community networks: case studies of five U.S. cities', *Pew/Internet*, accessed at http://arxiv.org/abs/cs.CY/0109084.

Kennedy, P. (2004), 'Making global society: friendship networks among transnational professionals in the building design industry', *Global Networks*, **4** (2), 157–79.

Larsen, J., J. Urry and K. Axhausen (2006a), 'Social networks and future mobilities', report to the UK Department of Transport (Lancaster and Zurich) Department of Sociology, accessed 10 November 2018 at http://e-collection.ethbib.ethz.ch/eserv/eth:28933/eth-28933-01.pdf.

Larsen, J., J. Urry and K.W. Axhausen (2006b), *Mobilities, Networks, Geographies*, Aldershot: Ashgate.

Leslie, L.A. (1992), 'The role of informal support networks in the adjustment of Central American immigrant families', *Journal of Community Psychology*, **20** (3), 243–56, doi:10.1002/1520-6629(199207)20:3<243::aid-jcop2290200307>3.0.co;2-a.

Licoppe, C. (2004), '"Connected" presence: the emergence of a new repertoire for managing social relationships in a changing communication technoscape', *Environment and Planning D: Society and Space*, **22** (1), 135–56.

Litwin, H. (1995), 'The social networks of elderly immigrants: an analytic typology', *Journal of Aging Studies*, **9** (2), 155–74.

Lubbers, M.J., J.L. Molina, J. Lerner, U. Brandes, J. Avila and C. McCarty (2010), 'Longitudinal analysis of personal networks. The case of Argentinean migrants in Spain', *Social Networks*, **32** (1), 91–104.

Lubbers, M.J., J.L. Molina and C. McCarty (2007), 'Personal networks and ethnic identifications', *International Sociology*, **22** (6), 721–41, doi:10.1177/0268580907082255.

Lynam, M.J. (1985), 'Support networks developed by immigrant women', *Social Science & Medicine*, **21** (3), 327–33.

Marin, A. and K.N. Hampton (2007), 'Simplifying the Personal network name generator', *Field Methods*, **19** (2), 163–93, doi:10.1177/1525822x06298588.

Martinez Garcia, M.F., M. Garcia Ramirez and I. Maya Jariego (2001), 'Una Tipología Analítica de las Redes de Apoyo Social en Inmigrantes Africanos en Andalucía', *Reis*, **95** (1), 99–125.

Mason, J. (2004), 'Managing kinship over long distances: the significance of the visit', *Social Policy and Society*, **3** (4), 421–9.

Maya Jariego, I. (2006), 'Web of compatriots: relationship networks among immigrants', in J.L. Pérez Pont (ed.), *Geografías del desorden. Migración, alteridad y nueva esfera social*, Valencia: Universidad de Valencia, pp. 257–76.

Maya Jariego, I. (2018), 'Why name generators with a fixed number of alters may be a pragmatic option for personal network analysis', *American Journal of Community Psychology*, **62** (1–2), 233–8, doi:10.1002/ajcp.12271.

Maya Jariego, I. and N. Armitage (2007), 'Multiple sense of community in migration and commuting: the interplay between time, space and relations', *International Sociology*, **22** (6), 743–66.

Maya Jariego, I. and D. Holgado, D. (2015), 'Living in the metropolitan area. Correlation of interurban mobility with the structural cohesion of personal networks and the originative sense of community', *Psychosocial Intervention*, **24** (3), 185–90.

McCallister, L. and C.S. Fischer (1978), 'A procedure for surveying personal networks', *Sociological Methods & Research*, **7** (2), 131–48.

McCarty, C. (2002), 'Structure in personal networks', *Journal of Social Structure*, **3** (1), 1–11.

Mok, D. and Wellman, B. (2007), 'Did distance matter before the Internet? Interpersonal contact and support in the 1970s', *Social Networks,* **29** (3), 430–461.

Mok, D., B. Wellman and J. Carrasco (2010), 'Does distance matter in the age of the Internet?', *Urban Studies*, **47** (13), 2747–83, doi:10.1177/0042098010377363.

Molina, J.L., R. Fernandez and J. Llopis (2007), 'El apoyo social desde la perspectiva de la redes personales', in M. Zúñiga (ed.), *Redes sociales y Salud Pública*, Monterrey, Mexico: Universidad Autónoma de Nuevo León, pp. 73–98.

Neal, Z.P. (2012), *The connected city: How networks are shaping the modern metropolis*. London and New York: Routledge.

Neal, Z.P. and J.W. Neal (2017), 'Network analysis in community psychology: looking back, looking forward', *American Journal of Community Psychology*, **60** (1–2), 279–95.

Pahl, R.E. (2002), 'Towards a more significant sociology of friendship', *European Journal of Sociology*, **43** (3): 410–23.

Putnam, R.D. (2000), *Bowling Alone: The Collapse and Revival of American Community*, New York: Simon and Schuster.

Ryan, L., A. Klekowski Von Koppenfels and J. Mulholland (2015), '"The distance between us": a comparative examination of the technical, spatial and temporal dimensions of the transnational social relationships of highly skilled migrants', *Global Networks*, **15** (2), 198–216, doi:doi:10.1111/glob.12054.

Ryan, L., R. Sales, M. Tilki and B. Siara (2008), 'Social networks, social support and social capital: the experiences of recent Polish migrants in London', *Sociology*, **42** (4), 672–90.

Schönhuth, M., M. Kronenwett, M. Gamper and M. Stark (2012), 'Vennmaker 1.5. Trier: http://www.vennmaker.com', accessed 15 March 2017 at http://www.vennmaker.com.

Schweizer, T., M. Schnegg and S. Berzborn (1998), 'Personal networks and social support in a multiethnic community of southern California', *Social Networks*, **20** (1), 1–21, doi:10.1016/S0378-8733(96)00304-8.

Searle, W. and C. Ward (1990), 'The prediction of psychological and sociocultural adjustment during cross-cultural transition', International Journal of Intercultural Relations, **14** (4), 449–64, doi:10.1016/0147-1767(90)90030-Z.

Sonn, C.C. (2002), 'Immigrant adaptation: understanding the process through sense of community', in A.T. Fisher, C.C. Sonn and B.J. Bishop (eds), *Psychological Sense of Community: Research, Applications, and Implications*, New York: Kluwer Academic and London: Plenum, pp. 205–22.

Urry, J. (2000), 'Mobile sociology', *British Journal of Sociology*, **51** (1), 185–203, doi:10.1111/j.1468-4446.2000.00185.x.

Urry, J. (2003), 'Social networks, travel and talk', *British Journal of Sociology*, **54** (2), 155–75, doi:10.1080/0007131032000080186.

Van der Poel, M.G.M. (1993), 'Delineating personal support networks', *Social Networks*, **15** (1), 49–70.

Viry, G. (2012), 'Residential mobility and the spatial dispersion of personal networks: effects on social support', *Social Networks*, **34** (1), 59–72.

Walker, M.E., S. Wasserman and B. Wellman (1993), 'Statistical models for social support networks', *Sociological Methods & Research*, **22** (1), 71–98, doi:10.1177/0049124193022001004.

Wellman, B. (1979), 'The community question: the intimate networks of East Yorkers', *American Journal of Community Psychology*, **84** (5), 1201–31.

Wellman, B. (1996), 'Are personal communities local? A Dumptarian reconsideration', *Social Networks*, **18** (4), 347–54.

Wellman, B. (2002), 'Little boxes, glocalization, and networked individualism', in M. Tanabe, P. van den Besselaar and T. Ishida (eds), *Digital Cities II: Computational and Sociological Approaches*, Berlin and Heidelberg: Springer, pp. 10–25.

Wellman, B. and K.A. Frank (2001), 'Network capital in a multilevel world: getting support from personal communities', in N. Lin, K.S. Cook and R.S. Burt (eds), *Social Capital: Theory and Research*, New York: Aldine de Gruyter, pp. 233–73.

Wellman, B. and S. Wortley (1990), 'Different strokes from different folks: community ties and social support', *American Journal of Sociology*, **96** (3), 558–88.

Wellman, B., P. Carrington and A. Hall (1988), 'Networks as personal communities', in B. Wellman and S.D. Berkowitz (eds), *Social Structures: A Network Approach*, Cambridge: Cambridge University Press, pp. 130–84.

PART V

NETWORK PROCESSES BETWEEN CITIES

22. Is maritime transport an urban network? The interplay between global container flows and urban hierarchies

César Ducruet

1. INTRODUCTION

Since the 1950s and 1960s, systematic approaches to the diffusion of various flows (information, people and goods) and, especially, innovations across urban networks (or systems of cities) have emerged in geography and regional science. Both US and European (Sweden and France) scholars, through direct or indirect collaboration, developed various conceptual frameworks and empirical case studies of urban networks in such respect (see Peris 2016 for a review of this vast literature). However, these schools of thought evolved in different ways in later decades, resulting in a wider diversity of urban network studies and definitions. Yet those schools have in common that they remain relatively abstract in scope, proponing their own theories and concepts, with limited transversal dialogues among them; when operated and applied empirically, the most common geographical scale to study urban networks remained at the national level (Ducruet and Lugo 2013).

It is, perhaps, the Pour l'Avancement des Recherches sur l'Interaction Spatiale (PARIS) team in France which innovated the most in that its members gradually expanded the geographic scope of their initial French-based studies to the whole European continent, through analyses of cities connected by multinational firms (Rozenblat and Pumain 1993) or by airline and railway networks (Cattan 1995). These large-scale investigations questioned the influence of city size and specialization on the attractiveness of specific activities through a relational perspective, thereby reaching beyond the traditional analysis of technical networks and planar graphs in this discipline (Ducruet and Beauguitte 2014). It is quite common nowadays, with the explosion of computer capacity and the adoption of new methodological tools related by simulation models and big data, to find global-level studies in the literature on urban networks of all kinds. Yet, this chapter would like to insist on a remaining dark corner within urban network research, namely, cities connected by water or maritime transport.

This claim may appear surprising given the long-term historical importance of the sea to connect continents, carrying vast amounts of goods but also people and information (Stopford 2008), without which globalization would probably not have been possible (Frémont 2007). It is often recognized in the literature that ports and cities evolved from synergy to independence owing to several factors. On the urban side, cities diversified their economy through successive innovation cycles (Pumain et al. 2009) to such an extent that for many port cities, maritime transport became a minor function (Bretagnolle 2015), leading to port area dereliction, waterfront redevelopment and the construction of modern port terminals away from urban areas to mitigate environmental and land-use issues (Merk 2013). On the port side, there is a great deal of evidence that port functions have changed in ways incompatible with inner-city location, with the expansion of hinterlands (that is, the inland market area of the port), growing trade and cargo-handling volumes, and increased productivity and efficiency resulting in lack of space, reduced labor and economic impacts, increased specialization and the need for shifting towards more accessible locations to adapt ever-increasing ship sizes. In many disciplines, such as geography, history, regional sciences and economics, it is now commonly recognized that ports have ceased to be an engine for urban growth in the contemporary era.

While such trends are well documented both through spatial models of port-city evolution and numerous case studies since the 1960s, there remain two unfinished debates on (1) whether ports are still attached to cities, and vice versa, and (2) the specific role of the container revolution in this separation process. The first debate remains because cities continue to be fed by material flows to grow and prosper (for example, cement, gas, oil, various commodities and consumer goods), and maritime transport still carries about 90 percent of world trade volumes (UNCTAD 2017). Feeding cities with these flows means that physical separation does not necessarily imply a functional separation, at least for the international part of their trade (with other cities). Another argument was to say that port-city separation was not a universal phenomenon but mainly Western, given the strong permanency of these linkages in Asia for instance (Lee et al. 2008). A rise in interest for the geography of material flows led to a number of arguments in favor of maintained port-city and port-region interdependencies, either through quantitative (Ducruet and Itoh 2016) or qualitative (Hesse 2013) research. Yet, when shipping is seen through the lenses of firms' networks and maritime advanced producer services (APS) (for example, brokering, insurance and law), Jacobs et al. (2011) confirmed earlier works by demonstrating that the location of these activities is better determined by urban (population, presence of other APS, and so on) rather than port (traffic) variables.

The second debate originates from the commonly held belief that the introduction and spread of containerization greatly modified the spatial distribution of shipping networks and port hierarchies. Initially a land transport technological advance, containerization is known to have been introduced in the mid-1950s in the port and shipping sector in the US by Malcolm McLean (1935–2001), head of a truck-driving company, to speed up cargo-handling operations using a standard box, with the effect of reducing queuing time and congestion at the docks (Levinson 2008). Pallets and all various (mainly) manufactured goods were placed in these containers, thereby reducing freight transportation costs and transit times. However, containerization did not spread immediately to the rest of the world, owing to resistance among port communities and shippers, and it was only in the late 1960s that it was adopted in a few European ports, such as Antwerp (Morel and Ducruet 2015). It is only from the 1970s that containerization truly accelerated and spread globally through successive diffusion waves (Guerrero and Rodrigue 2014), from the US to Europe, Japan and, through a joint process of geographic proximity (or contagion) and hierarchy (or economic welfare), throughout the world. In a context of growing global trade, containerization was recognized to have been a catalyst of international exchanges (Bernhofen et al. 2013). It favored economies of scale in relation to port traffic concentration and increasing vessel size (Cullinane and Khanna 2000), especially with the introduction of hub-and-spokes systems in the mid-1990s in particular regions with the emergence of transshipment hubs at intermediate locations as well as inland (or dry) ports and terminals to foster freight distribution through intermodal services (Rodrigue and Notteboom 2010). Horizontal integration (mergers and acquisitions among shipping lines) and vertical integration (among different supply chain actors, for example, a shipping line such as Maersk getting involved in port operations through terminal concessions and landside logistics) reinforced this process, which still continues. The global financial crisis (2008–09) causing overcapacity in container shipping, an even more restrained handful of global actors concentrated the bulk of port and shipping operations through alliances and kept increasing vessel size while limiting vessel speed for the sake of fuel and labor cost saving. All these changes have impacted further the way port cities can handle these very selective mega-ships. Faster intermodal transfers away from urban centers accelerate the port-city spatial divide, with large ports being built and operated far away to exploit lower land rents and larger space for expansion. Very few port cities, mainly in Asia, were able to maintain this huge physical operations activity within the urban area, owing to political and spatial constraints (for example, city-states, small countries and islands, such as Hong Kong and Singapore, but also Busan and Kaohsiung, the Asian dragons (Lee and Ducruet 2009).

Nevertheless, a number of elements motivate this chapter to go one step further in understanding how containerization truly changed the longstanding ties between cities and maritime networks. This is the second debate, which is the basis of our discussion, based on three main arguments. First, containerization is not the first technological innovation in the history of shipping, bearing in mind the previous transition from oar to sail in Greek antiquity (Rivers et al. 2015) or, much later, the transition from sail to steam (Bunel et al. 2017) and then from steam to combustion and specialization. The port selection process of containerization described by so many scholars is not very different from that documented, for instance, by Marnot (2005, pp. 10–11) in his work on technological change and port competition in the nineteenth century, concluding that 'recent transformations are perhaps less different [than older ones] by their nature than by their scale'. Also, an empirical work on the topic of cities and maritime networks in general (Ducruet et al. 2018) concluded that the decreasing correlation between city size and traffic size did not coincide with the introduction of containerization, as it started much earlier, in the 1930s (see Figure 22.1).

While the transition from sail to steam did not have obvious effects on the correlation between population and number of vessel calls (all ships included), it is more the transition from steam to combustion that seems to have caused the decline. Therefore, containerization did prolong rather than cause port-city disconnection, contrary to the aforementioned belief, as part of a wider process of vessel specialization with increasing speed and size, and greater autonomy from natural conditions and combustibles (cf. coal stations replaced by fuel). Third, recent years witnessed a return of the port into the city (El Hosni 2017), especially for container terminals. Although this process does not (yet) apply to a majority of large port cities, it is perhaps symbolic that it started in London, once the world's largest port, through the construction by the global operator Dubai Ports World (DPW) of the London Gateway project within the Greater London area, thereby competing with the Hong Kong-based operator Hutchinson Whampoa (HPH) located in Felixstowe, a small town of about 3000 inhabitants situated about 300 kilometers from London, using the slogan 'ship closer, save money' to attract warehouses, shippers and related flows around the brand new container terminal. A similar phenomenon was observed in Jakarta, Taipei, Osaka and Tokyo. While the latter initiatives (that is, creating or expanding the container port of the countries' largest cities) were more based on government intervention, they have in common with London the effect of reducing trucking distance and related carbon emissions. This new phase of port-city evolution

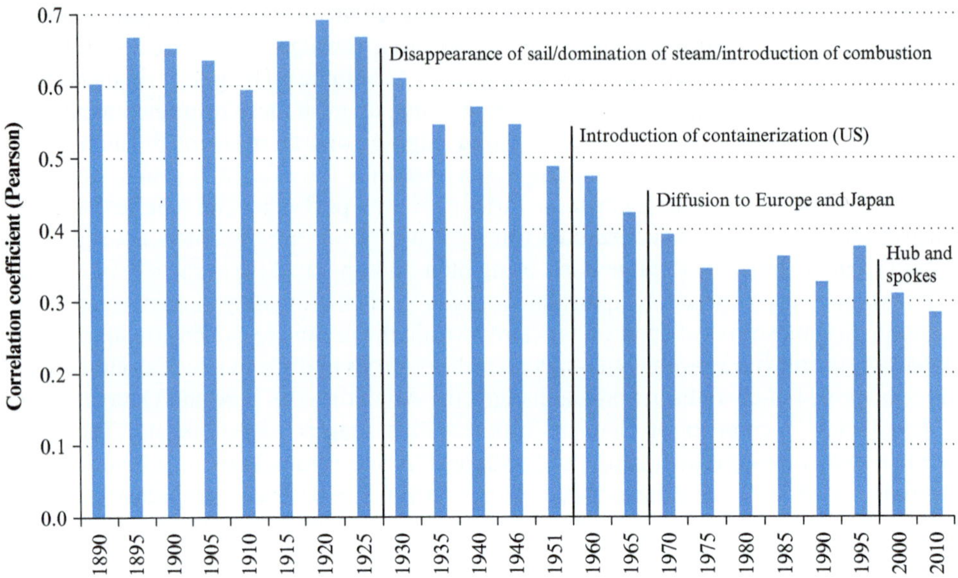

Source: Own elaboration based on Ducruet et al. (2018).

Figure 22.1 Technological change and port-city correlation, 1890–2010

might become a model for other large cities served by distant container terminals after successive phases of physical and functional disconnection. In addition, relatively large cities in comparison with the European average, such as Antwerp and Hamburg, have managed to overcome all the difficulties of their upstream location to maintain among the highest levels of port activity, being Europe's second and third largest container ports, respectively, and have been so for decades (Notteboom 2016).

In the remainder of this chapter the following section presents the data and methodology used for an empirical investigation into the spatial distribution of container shipping flows and networks across the global urban hierarchy from the 1970s to date. Then we present our main results with the main hypothesis that container shipping still remains tied to cities in one way or another, despite all the contemporary changes that apparently altered their initial symbiosis. We discuss these outcomes in the chapter's conclusion by addressing how this study can improve our understanding of transport systems, urban networks and spatial change in geography and other disciplines.

2. CONSTRUCTING A GLOBAL URBAN NETWORK MADE OF CONTAINER SHIPPING FLOWS

2.1 Methodological Specifications

While Figure 22.1 served as a long-term view of the evolution of port-city relationships, it was based on the number of vessel calls including all types of ships. In this chapter, we particularly focus on container shipping, since these flows are believed to have a specific spatial distribution compared with other types of vessels, for the reasons cited previously. A few earlier works demonstrated that container flows, which carry the most valuable goods and are operated through regular, scheduled services, continue to be concentrated in the richest and most productive port regions specialized in the tertiary sector (Ducruet 2015), while overlapping with other maritime layers (for example, passengers and vehicles, solid and liquid bulks, and general cargo) owing to a path-dependent and place-dependent process (Ducruet 2013); that is, containerization did not spread randomly but at ports already diversified in other commodity types.

This chapter delves further into the verification of how container flows maintain (or not) their ties with the global urban hierarchy. To achieve this objective, it is imperative to measure both container flows and urban variables. We collected vessel movement data from the world's leading maritime insurance company (Lloyd's List Intelligence) on the basis of two complete months of inter-port vessel flows each year (June and December) over the past four decades (1977–2016). These flows were arranged in the form of a weighted origin-destination matrix with deadweight tonnage (DWT) as the measurement unit, which corresponds to the cargo-carrying capacity of vessels. Two network topologies are used in this chapter to test our main hypothesis, usually known in the graph-theoretical literature as space-L and space-P (see Figure 22.2); the first being restrained to direct movements among ports along the voyage of each vessel (that is, Port1–Port2, Port2–Port3, Port3–Port4, and so on); the latter including indirect movements to take into account long-distance ties as well (that is, Port1–Port3, Port1–Port4, Port2–Port4, and so on). These two dimensions are interesting to compare in respect of maritime networks since they provide dramatically different results in centrality measures for nodes, such as degree centrality and betweenness centrality. For instance, a large port such as Shenzhen (China) in relation to total traffic volume – the factory of the world – has relatively fewer links to other ports (low degree) according to space-L (direct neighbors), while in space-P, its belonging to multiple global and regional trade routes multiplies its connectivity (high degree). That is, the space-L topology gives more advantage to transshipment hubs in the network, while the space-P topology give more importance to trading ports. Transshipment (or transit and redistribution) and real trade are two important facets of port activity that are sometimes combined by the same nodes, as with Rotterdam and Hamburg for instance. When it comes to urban networks, we hypothesize that space-P shall be more relevant than space-L to measure the relationship with city size.

Maritime container flows alone do not suffice to build an urban network. Still in the aforementioned matrices, nodes are ports and/or port terminals, without a proper urban dimension. This is so for most (maritime) network research, which largely ignores, to date, that such nodes are part of human settlements having their own, specific characteristics. Defining the city is a longstanding issue in geography and elsewhere. We adopted the methodology proposed by Ducruet et al. (2018) whereby each port or terminal is assigned to an urban area according to several criteria. Here also two main dimensions are taken into account to verify our hypothesis; we name them port cities (level 1) and extended city-regions (level 2). At level 1, ports and terminals are assigned to their host city based on their intra-urban location (or very close proximity) to a morphological urban area that may, in turn, contain one or more ports. These cities are those that directly handle cargo by their own infrastructural means. At level 2, ports and terminals (which may or

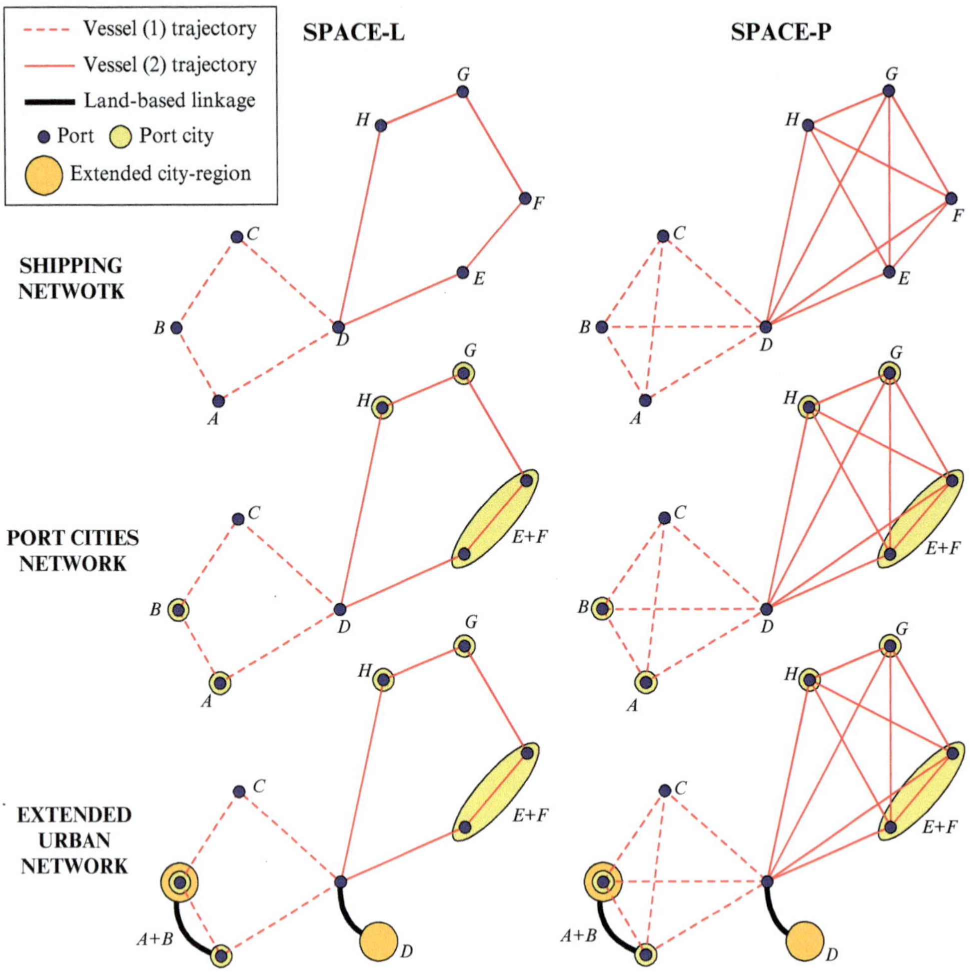

Source: Own elaboration based on Hu and Zhu (2009).

Figure 22.2 Various urban and topological perspectives on container shipping networks

not already belong to a port city) are assigned, depending on their circumstances, to a larger, nearby, landward connected urban area, whether another port city or a non-port city, either coastal or inland. Therefore, and as in Figure 22.2, the shipping network is composed of only ports and terminals as nodes; the port cities network (level 1) assigns ports to cities whenever possible (for example, A, B, H, G), some of them containing more than one port (E+F); and the extended urban network (level 2) takes into account the dependence of ports (and port cities) on larger and neighboring cities, located inland (D) or being already a port city (A+B), such as in the case of a downstream port city being in the market area of an inland city.

These two dimensions are essential because they allow us to grasp how container flows are spatially tied to cities, and check, indirectly, the role of land transport in this relationship, since many (large) cities have lost their port functions while being served by distant terminals, and others have been able to maintain this activity despite all the burdens associated with land-use conflicts and congestion. Numerous monographs and ideal-typical sequences (spatial models) describe these dynamics, but this chapter goes one step further by examining the interplay between these four dimensions.

Last but not least, assigning ports to cities remains insufficient to validate or not our hypothesis. Huge efforts were therefore made to attribute demographic features to cities, as a proxy of their economic importance. While population may not always be correlated with the true level of urban economic activity, it has the merit of being comparable across space and time and available from several urban databases. We assembled and harmonized three complementary sources – *Geopolis* (Moriconi-Ebrard 1994), Populstat (Lahmeyer 2015) for the 1975–2000 period and *World Gazetteer* (Helders 2012) – for 2010, based on the same definition of morphological urban areas and measuring their number of inhabitants every five years between 1975 and 2010. Although container flow data is available on a yearly basis, we consider in this chapter that population has been more stable and the same year can be used for testing the correlation with container flows. For instance, population numbers in 1985 are related to container flows variables for the years 1983–87, in 1990 for the years 1988–92, and so on. Container shipping variables are kept relatively simple: total traffic volume per city, degree centrality, betweenness centrality, clustering coefficient and eccentricity. Container flow variables are compared with population, taking cities either individually (for example, correlations) or grouped into six classes of demographic size (quantiles) to allow calculating the average score of each class and its evolution over time. Quantiles based on a fixed number of cities per class are preferred to arbitrary population thresholds in order to get closer to the historical context of each year or period, since being a 'large city' does not have the same meaning in the 1970s as in the 2010s. Enlarging the scope of the analysis from terminal/port to port city and extended urban area goes along with earlier work on the potential accessibility of cities, as it includes, albeit indirectly, land transport (potential) flows in the delineation of these spatial entities (see Bretagnolle et al. 1998; Rozenblat 2004). This is in line with recent efforts pushing further the empirical delineation of hinterlands, which remains a nearly unsolvable issue in geography even now (Guerrero 2014; Halim et al. 2016).

2.2 Preliminary Results

In this section, we describe the evolution of the global container shipping network according to the four dimensions explained previously. It is useful to observe how the network has changed over time and what the effect is of node aggregation (at the level of port cities or extended city-regions) on its overall topology, should it be space-L or space-P. We therefore created 10 matrices of inter-port and intercity container flows and computed the four main centrality scores for each of them. The effect of node aggregation has been discussed recently in the case of the Greek shipping network and is a promising research direction to understand the effects of shifting the study unit from port to urban area (Tsiotas and Polyzos 2018). A similar but more empirical approach was proposed by Ducruet et al. (2011) when analyzing the combined container shipping and passenger airline network from an urban perspective to study centrality determinants and functional specialization. These authors even proposed three layers of analysis, from terminal to

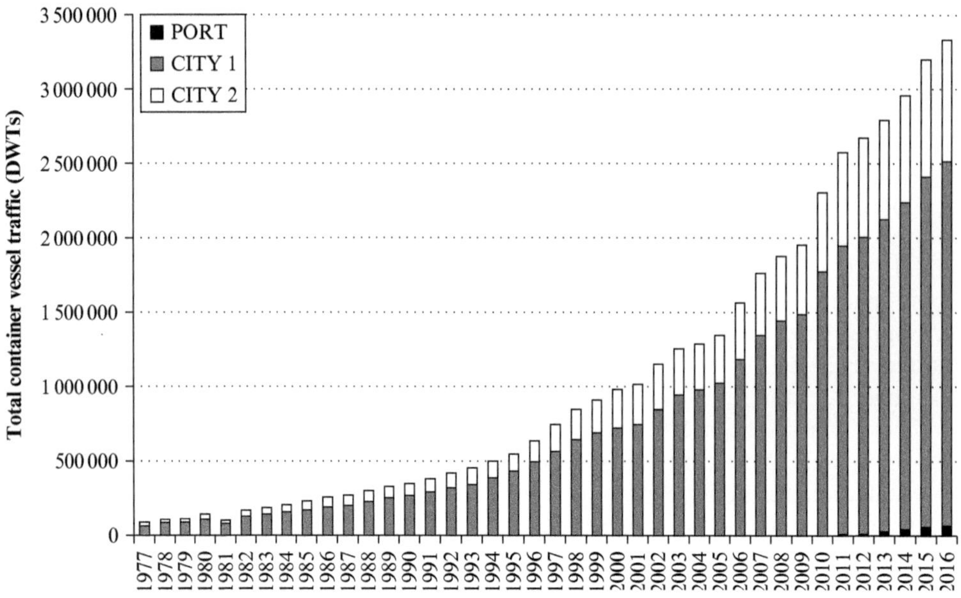

Source: Own elaboration.

Figure 22.3 Container traffic distribution by place type, 1977–2016

city and mega-region, demonstrating that higher levels of aggregation are better to understand the complementarity of networks in a multilayered system linking cities of the world (see also, Parshani et al. 2010 for an earlier attempt). It is only in the next section that we run network-analytical methods to the container shipping network to reveal the relationship between cities and flows.

First, and as seen in Figure 22.3, containership traffic has grown apace along the study period, notwithstanding a noticeable slowdown in the years 2008–09 relative to the global financial crisis causing a fall in demand, trade and economic downturn, and turmoil in the ports and shipping sector. Traffic distribution per place type is interesting as it shows a regular increase, despite fluctuations, of the weight of level 2 (extended city-regions) but also of ports (non-urban ports), the first concentrating about 20 percent of total world traffic on average, from 18 percent in 1977 to 21 percent in 2016, and the latter confirming the growing traffic outside urban areas, although with a small weight compared to port cities (level 1).

Calculating the (linear) correlation between urban population and vessel traffic shows a different picture of this evolution. Despite the huge concentration of world traffic in urban areas (especially, port cities), as seen in Figure 22.3, we observe in Figure 22.4 an opposite phenomenon, that is, the regular decline of the correlation for any city type, but especially for port cities (level 1). The correlation for port cities has never been highly significant, starting slightly above 0.4 in 1977 and regularly declining (despite a short-term increase in the years 2005–08) to above 0.3 in 2016. This moderate and relatively stable correlation shows that containership traffic is not fully explained by city size, at least at level 1. On the contrary, the same correlation for level 2 (extended city-regions),

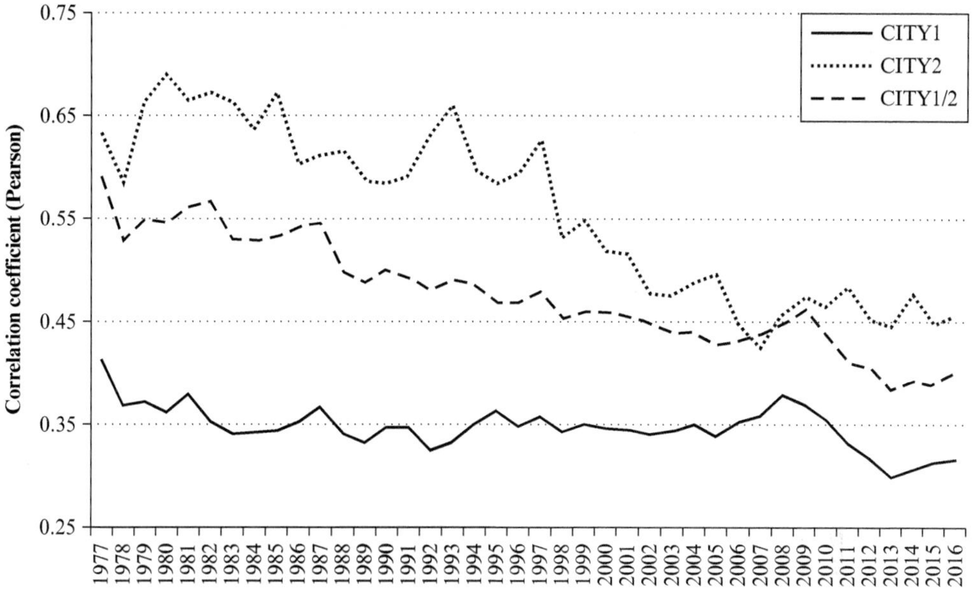

Source: Own elaboration.

Figure 22.4 Correlation between urban population and container traffic, 1977–2016

despite a similar decreasing trend, had been much more significant in the early period, oscillating around 0.65, and stabilized around 0.45 during the last decade 2006 16. This also means that urban areas have lost their power explaining container traffic, but enlarging the spatial unit had the effect of better explaining (nearly half) of this traffic. Another striking fact is that extended city-regions started witnessing a decreasing relationship between population and traffic during the mid-1990s, exactly in a phase of emerging hub-and-spokes configurations in liner shipping, dramatically increasing vessel sizes and optimization of routing schedules. The same correlation for cities altogether (combined levels 1 and 2) is more or less the average of these two tendencies, that is, a regular decline of the correlation from about 0.60 in 1977 to about 0.36 in 2016, which means a loss of nearly half of the significance of port-city relationships overtime.

Owing to the aforementioned trends perhaps being caused by the emergence of a few outliers, that is, large ports in small cities or small ports in large cities, we adopted another perspective in Figure 22.5, where urban areas are grouped into five quantiles of the same sample amplitude (20 percent), from the smallest (class 1) to the largest (class 5) cities. The class 0 corresponds to non-urban ports. The view based on a Bertin matrix intentionally ignores proportionality among the different rows, in order to better observe the behavior of each class overtime. Figure 22.5 also compares level 1 (A, top) and level 1+2 (B, bottom) to check for possible variations owning to node aggregation.

In all instances, one first result confirms the overwhelming concentration of world container traffic among the largest cities (class 5), although it had declined from 83.6 percent in 1977 to 78.1 percent in 2016 at level 1, and has increased from 93.2 percent in 1977 to 95.1 percent in 2016. This dominance confirms our intuition that the declining

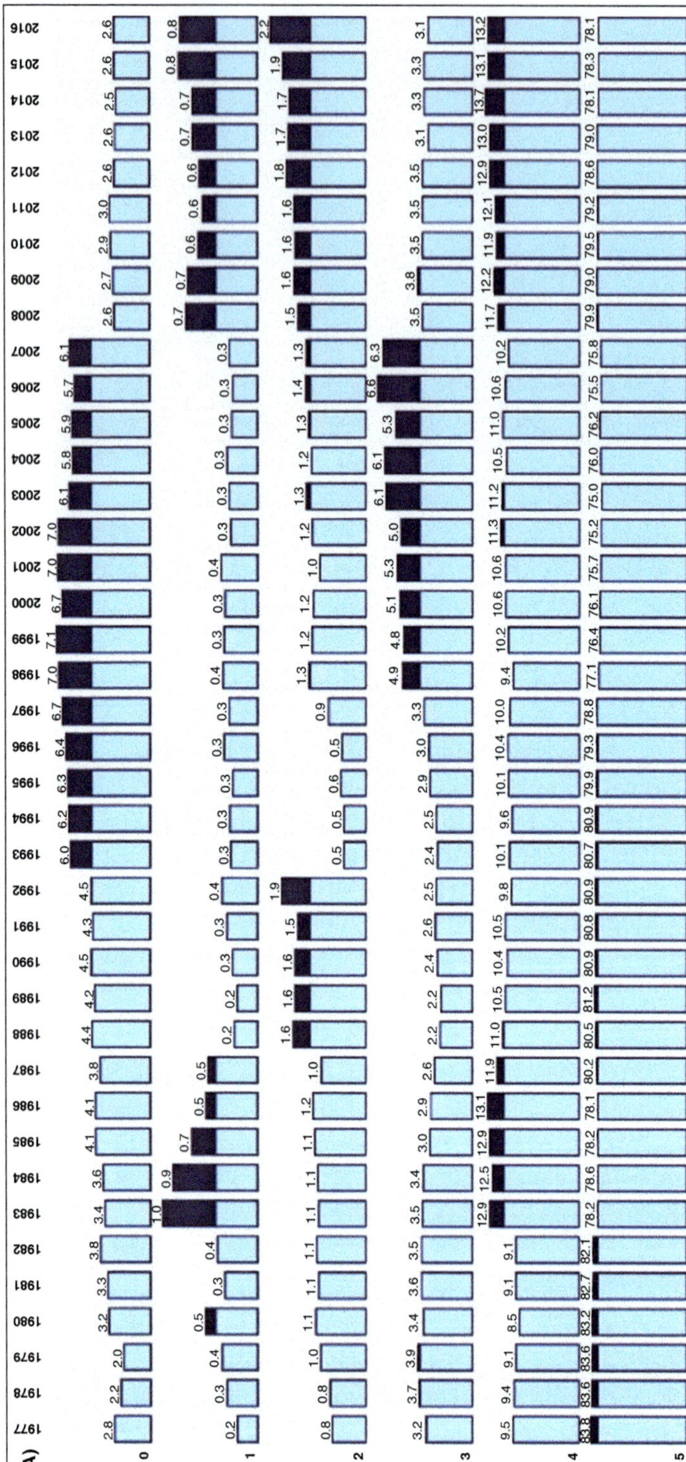
(A)
1977
1978
1979
1980
1981
1982
1983
1984
1985
1986
1987
1988
1989
1990
1991
1992
1993
1994
1995
1996
1997
1998
1999
2000
2001
2002
2003
2004
2005
2006
2007
2008
2009
2010
2011
2012
2013
2014
2015
2016
0
1
2
3
4
5

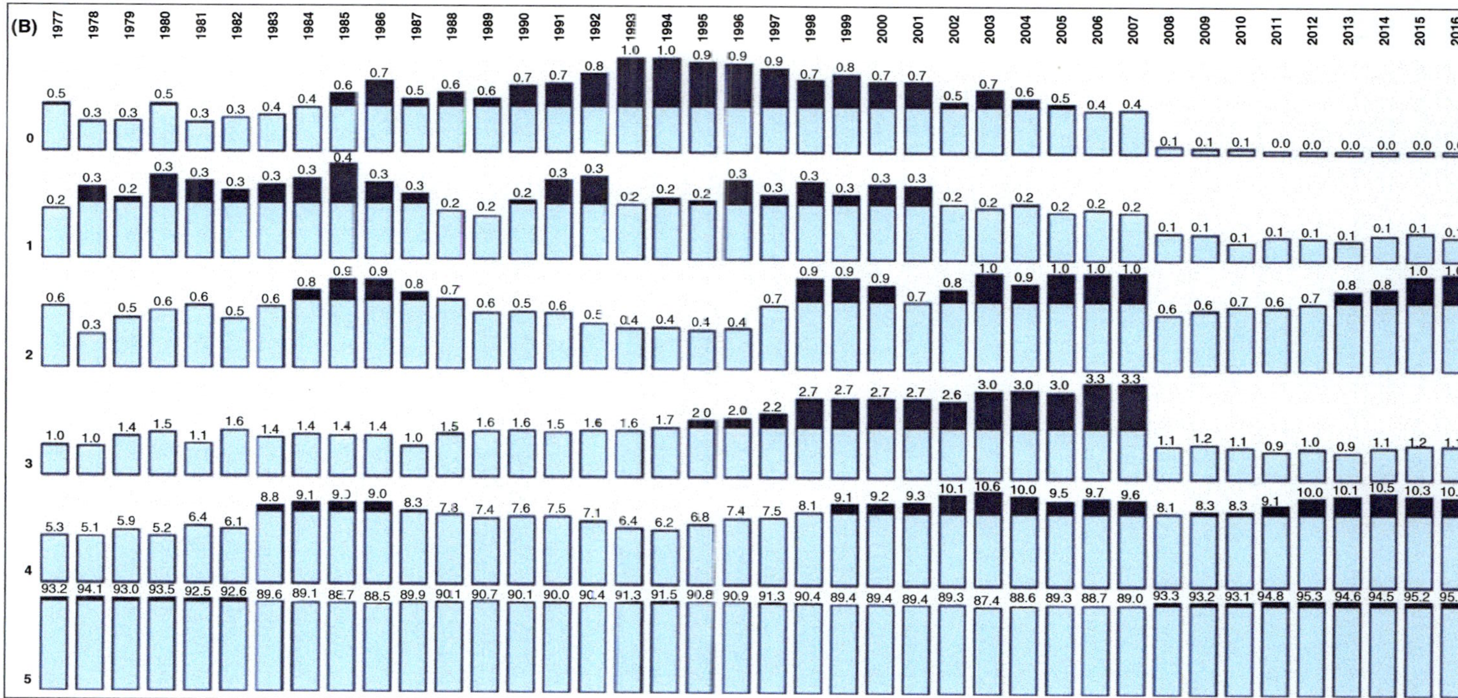

Note: Port cities (A), extended city-regions and port cities (B); non-urban ports (0), small city size (1), moderate city size (2), average city size (3), important city size (4), largest city size (5).

Source: Own elaboration.

Figure 22.5 Container traffic distribution by demographic quantile, 1977–2016 correlation observed in Figure 22.4 was caused by a few outliers rather than a structural and ineluctable decay of port-city relationships. The dominance is also caused by changes in world trade patterns whereby Western port cities have lost market share to Asian port cities during this period (see Appendix Table 22A.1), while cities in old industrial countries witnessed stable or shrinking population contrary to those in emerging countries.

The loss endured by the largest cities at level 1 still means that they lost traffic to small cities owing to well-known factors such as congestion and lack of space for further port expansion, but the loss remains relatively negligible. Traffic shifted to smaller cities of class 1 and 2, but also 4 in the late period, meaning that traffic growth has not always occurred away from large cities: class 4 gained 3 percent since 2007, while classes 1 and 2 grew less than 1 percent during the same period. The gain for the largest cities suggests that node aggregation had the effect of grasping the traffic of neighboring ports and port cities when focusing on the region's main urban center. This explains why non-urban ports (class 0), the smallest cities (class 1) and the medium-sized cities (class 3) have lost so much, given that they lie in the vicinity of a larger urban center (class 5). Yet, classes 2 and 4 witnessed substantial growth in the late period for the same reasons cited previously; not all of the largest cities have remained capable of handling ever-increasing cargo volumes so that some of their traffic, at the same time, shifted towards less-urbanized areas, as already observed by Hayuth (1981) in his work on the challenge of peripheral ports.

Level 1 (A) and level 1+2 (B) have two other common trends that need further discussion. One trend is that smaller cities witnessed a growing traffic share during a decade or so after the mid-1990s for the reasons cited previously (route optimization, port selection and competition, mergers and acquisitions in the shipping sector, growing ship sizes, and so on), especially medium-sized cities (class 3). The second common trend is the temporary growth of the traffic share of non-urban ports, despite by low numbers (class 0), in the middle of the study period, mainly owing to the construction of new ports outside urban areas. Those ports are often termed 'ports in the desert' or 'pure transshipment ports', although the transshipment function may take place in densely populated urban areas such as Singapore (Slack and Gouvernal 2015). The temporary dimension of their growth cannot obscure their absolute growth, as seen in Figure 22.3, although many of them have been incorporated into extended city-regions given their traffic share decrease since the late 2000s.

Analyzing the relationship between traffic and population distributions is meaningful in many ways, although it leads to contradictory evidence. Container shipping has been through huge technological and operational transformations during the study period, which are not reflected the same in relation to cities depending on the methodology or research angle. While correlation between traffic and population faded away, the largest cities have maintained paramount importance in global shipping flows from the late 1970s to date. Node aggregation had the effect of better understanding how (apparently) isolated ports and port cities belong to the market area of a larger and nearby city, coastal or inland, thus implying that road transport has also played an increasing role in connecting cities, directly or indirectly, to global shipping. Although larger cities maintained their prominence, this does not underestimate the role and impact of supply chain actors on traffic distribution, but it helps to balance the belief that cities and maritime transport (especially, container shipping) are no longer tied to each other. We discuss further similar issues in the light of a network perspective in the next section.

3. URBAN CENTRALITY IN THE GLOBAL SHIPPING NETWORK

As in the previous section but for selected years only, we now dive into a multidimensional perspective on how network centrality relates to the urban hierarchy in various ways. Although it has been demonstrated previously that total container vessel traffic (or weighted degree, strength index) is highly correlated with particular centrality measures (Deng et al. 2009), it is important to observe the evolution of this correlation over time and for a wider set of measures. We first calculate the linear (Pearson) correlation between urban population and the four chosen centrality measures (Figure 22.6) and we compare the average centrality score of population quantiles based on the four dimensions: level 1 (port cities), levels 1+2 (all cities), space-L and space-P. Although most previous works on the global container shipping network remained static and did not address urban issues, we did run such an analysis for only five years from 1977 to 2016, that is, providing a snapshot for every decade. We used TULIP[1] to calculate centrality indices, a graph visualization software created by the University of Bordeaux (see Rozenblat and

Measures		Year	Space-L		Space-P	
			CITY1	CITY1&2	CITY1	CITY1&2
Correlation with population	Degree centrality	1977	0.364	0.369	0.406	0.456
		1987	0.312	0.317	0.331	0.382
		1997	0.453	0.453	0.427	0.476
		2007	0.376	0.362	0.478	0.498
		2016	0.401	0.401	0.422	0.445
	Betweenness centrality	1977	0.226	0.233	0.237	0.410
		1987	0.223	0.228	0.180	0.350
		1997	0.279	0.280	0.248	0.332
		2007	0.230	0.225	0.284	0.348
		2016	0.207	0.209	0.269	0.318
	Eccentricity	1977	–0.051	–0.056	–0.310	–0.175
		1987	–0.069	–0.081	–0.238	–0.297
		1997	–0.188	–0.189	–0.308	–0.378
		2007	–0.245	–0.234	–0.321	–0.346
		2016	–0.261	–0.258	–0.343	–0.328
	Clustering coefficient	1977	0.252	0.254	0.241	0.390
		1987	0.188	0.193	0.233	0.352
		1997	0.304	0.305	0.371	0.451
		2007	0.247	0.229	0.284	0.424
		2016	0.234	0.232	0.346	0.394

Note: Based on the inverse value of the clustering coefficient.

Source: Own elaboration.

Figure 22.6 Correlation (Pearson) between centralities and population at selected years, 1977–2016

Mélançon 2013 for a wide panel of applications). Adding more years would be better to observe a continuous trend but it has the effect of exponentially increasing the number of matrices to create and manipulate (that is, $40 \times 4 = 160$). Each selected year has the merit of not being influenced by a specific event; except, perhaps, for 1997 which witnessed the Asian financial crisis, but it is an interesting year as it marks a rupture in technological and operational aspects for container shipping, in the middle of the study period. The year 2007 occurs before the global financial crisis and 2016 is the most recent data which was available at the time of this study.

3.1 Correlation between Network Centrality and City Size

Among the four centrality scores analyzed in Figure 22.6, it is important to mention that we calculated the inverse value of the clustering coefficient while keeping zero values unchanged. This operation avoided a bias in the correlation whereby peripheral nodes often exhibit zero values while important hubs have values close to zero, especially in the space-L dimension that gives advantage to hub ports over trading ports (Ducruet and Notteboom 2012). We could have done the same for eccentricity but the original measure has been kept since, unlike the previous measure, it does not suffer from any bias, although the correlation with population should be interpreted inversely, that is, values close to 1 mean peripherality and those close to zero suggest a strong (topological) proximity to other nodes.

Partly owing to the absence of a temporal continuum in the analysis, it is not always possible to find a coherent trend in the evolution of correlations. However, and as observed previously for vessel traffic, space-P correlations are (almost) always more significant than space-L correlations, just like level 1+2 compared with level 1. In the first case (topology), the space-L network is sparser than the space-P network so that space-L better reflects the service configuration of private actors than trade patterns (space-P). In the second case (node aggregation), extending the focus from the host city of the port (port city) to the extended city-region enabled the analysis of more spatially and functionally comprehensive nodes, better reflecting the extent of the hinterland, the market area of the port(s). Yet in a number of instances, even the extended city-region remains a tiny part of a vast hinterland, such as large so-named load centers or gateway ports, for example, Rotterdam and Hamburg, reaching across several countries (Merk et al. 2013) through intermodal transport; and Le Havre and Marseilles in France covering most of northwestern and southeastern regions, respectively (Guerrero 2014).

Our results (Figure 22.6) for betweenness centrality, a global centrality measure (that is, number of occurrences on shortest paths throughout the network), show similar trends to total vessel traffic (Figure 22.4). The correlation with population is fluctuating but mainly decreasing, although in the fourth dimension (space-P / levels 1+2) the decline started from relatively significant values in 1977 (0.41) to reach 0.32 in 2016, a value that is higher than any other value in other dimensions. Thus, cities as nodes in the global container shipping network have lost significance to other factors that are not tested in this chapter but can be listed intuitively, such as efficiency and productivity in terminal operations, location on optimal routes, presence of vertically integrated supply chain actors, and quality/fluidity of intermodal infrastructure (Notteboom 2004). This measure corresponds with the concept of intermediacy developed by Fleming and Hayuth (1994) to depict the ability of a transport node to become well inserted in the service networks of global players (see also Slack 1993). This explains why it is the most likely to decrease in relation to population, as it is closest to the logics and rationales of private actors' interests compared with urban and territorial factors. A perfect example is Singapore, where more than 80 percent of total container throughput is dedicated to transshipment (or transit), the rest being real-trade flows (cf. the concept of centrality or traffic generation power of Fleming and Hayuth 1994).

The other global measure, eccentricity, expresses the topological proximity to other nodes in the network. Eccentricity exhibits an opposite trend compared with betweenness, as in most cases its correlation with urban population increased dramatically, except from the fourth dimension, which lacks a clear trend despite more significant values than other dimensions. This opposite trend can be explained by the global shipping network evolving in ways beneficial to delivering the goods as close as possible to these main markets where consumers and producers concentrate, despite the world's major cities not always being at the crossroads of major shipping routes (cf. betweenness). After all, it is the *raison d'être* of maritime transport in general, and container shipping in particular, to import and export most of the world's trade volume and value, but in order to maintain and improve this objective, it had to transform itself into an ever more efficient system with its own operational logics, which do not always overlap the spatial distribution of urban settlements – the local origins and destinations of imports and exports (many of these being located further inland). This apparent paradox is well illustrated by the case of Algeria, where Algiers, the largest urban and capital city is also the country's main port, but it connects the global shipping network mainly through the transshipment hub port of Marsaxlokk (Malta), operated by the French shipping line and terminal operator CMA-CGM (Mohamed-Chérif and Ducruet 2016). Vessel movement data makes it impossible to track which part of Marsaxlokk's traffic relates to Algiers, so that the trade continuum with the rest of the world is blurred, and the correlation between urban population and centrality, at least in the Mediterranean basin, is distorted by the transit flows through low-populated hubs such as Marsaxlokk, but also Gioia Tauro, Tarento and Cagliari in Italy, Algeciras and Valencia in Spain, Tangier-Med in Morocco, and so on. Nevertheless, the eccentricity of Algiers is low (high proximity to the core of the network) as it is well connected to an important hub situated along the circumterrestrial trunk line where the largest containerships operate. The impossibility, except for a few large ports, of knowing the precise share of transshipment versus trade in total port activity is a huge limitation to address port-city relationships.

The two other centrality measures are more local in that degree centrality and clustering coefficient only consider topologically neighbor nodes. Degree centrality is the most correlated, with population among all measures in Figure 22.6, but it does not show any particular trend, or only a fluctuating stability, but in most cases the value for 2016 is slightly higher than the value for 1977, reaching a peak in the middle of the period (1997). The trend is almost the same for the (inverse) clustering coefficient, with a peak value in 1997, and slightly higher values in 2016 compared with 1977 in the case of space-P only this time. The inverse clustering coefficient expresses the ability of a node to act as a hub among its closest neighbors, but its correlation with urban population had remained fluctuating and stable, without showing any specific trend. Thus, the analysis of correlations shows that global measures are more interesting than local measures since they better express convergence or divergence between urban and port hierarchies than local measures, of which the relationship with population is fuzzier.

3.2 Centrality Distribution per City Size Classes

As in the previous section, we have calculated the average centrality score of cities for each of the four dimensions and for the same five years under investigation (Figure 22.7). We provide, in Appendix Table 22A.2, an overview of the sample size, average, minimum and maximum population of each quantile for each year. Results are strikingly straightforward, as the average centrality increases from class 1 to class 5 and this phenomenon is increasingly evident over time. This is even more true when shifting from one dimension to the other, from space-L/level 1 to space-P/levels 1+2. Yet, in particular instances there is mixed evidence as smaller cities sometimes have higher average centrality than do larger cities, especially among classes 1 to 4, while the class of the largest cities (5) nearly always surpasses the other classes. In the final dimension (space-P/levels 1+2), classes 4 and 5 surpass other classes to a large extent, with the exception of average clustering coefficient, which is superior for class 3 compared with class 4. The largest cities are thus more likely to exert hub functions than are other cities, but these functions are not linearly distributed across the urban hierarchy, since in many instances, small cities can perform this footloose activity, depending on the port selection factors and strategies of transport actors, beyond the sole spatial and functional issues faced by large port cities (Slack and Wang 2002).

These results are mainly caused by the bulk of world container traffic being concentrated in the final quantile (5), so it is not surprising to find higher centrality scores between the same thresholds, which are low compared with world average: the final quantile includes all cities with at least 200 000 to 300 000 inhabitants, thereby including most of the world's largest ports, up to the world's largest city (Tokyo) which surpasses 30 million inhabitants. The final quantile is thus characterized by a higher variance than other quantiles, which makes it unique and difficult to compare.

As a test, we have added Appendix Figure 22A.1 using more conventional urban population thresholds to check for the distribution of container traffic among those different classes than the quantiles. Still, the two largest classes (that is, over 1 million inhabitants) concentrate about 70 percent of traffic on average, while the largest cities (over 5 million inhabitants) only concentrate about 32 percent on average, so that class 4 instead of class 5 is the most likely to concentrate highest centrality measures. Additional evidence that still does not go against the quantiles method is that the largest classes have not witnessed traffic decrease overtime. Instead, their share increased up to 2007, but in 2016, when more population data was available and accessible, especially for the smallest cities, their share was artificially reduced. We therefore can conclude that city size remains a good determinant of container shipping traffic and centrality, despite all the technological and operational changes that are so often described and analyzed in transport studies in the belief that cities no longer play a role in the distribution of physical flows and supply chains.

4. DISCUSSION AND CONCLUSION

This chapter tackled for the first time a difficult challenge, that is, to verify whether maritime transport, and container shipping in particular, constitutes an urban network,

		SPACE L-CITY 1					SPACE L-CITY 1/2					SPACE P-CITY 1					SPACE P-CITY 1/2				
		1	2	3	4	5	1	2	3	4	5	1	2	3	4	5	1	2	3	4	5
Betweenness centrality	1977	1.10	122.79	5.81	77.30	488.66	0.71	106.60	38.51	62.57	547.22	326.89	106.98	397.97	435.43	2772.19	0.10	18.60	8.69	55.66	311.52
	1987	49.68	103.76	113.12	276.37	583.20	53.93	89.35	93.90	261.23	596.91	82.14	70.76	101.30	251.82	672.18	6.16	18.53	28.87	167.96	466.26
	1997	75.70	88.38	325.43	263.23	1190.08	70.34	78.00	348.72	265.41	993.03	48.56	66.43	81.75	87.75	601.59	51.04	42.51	72.68	133.81	765.89
	2007	132.16	81.80	284.60	386.68	1063.33	95.89	66.35	197.09	335.84	1175.66	7.46	29.55	186.29	196.48	665.86	6.60	32.50	129.44	262.68	863.05
	2016	125.67	238.12	160.23	403.91	1131.04	129.66	201.07	136.80	433.01	1436.21	51.49	81.34	72.86	251.94	833.79	50.59	71.15	66.79	297.10	1125.36
Clustering coefficient	1977	0.58	0.23	0.50	0.26	0.30	0.47	0.21	0.65	0.30	0.33	0.52	0.79	0.64	0.59	0.40	0.78	0.87	0.91	0.84	0.62
	1987	0.39	0.57	0.38	0.35	0.39	0.53	0.52	0.48	0.43	0.44	0.79	0.83	0.78	0.66	0.58	0.90	0.90	0.87	0.73	0.66
	1997	0.69	0.58	0.53	0.60	0.47	0.64	0.47	0.39	0.49	0.47	0.85	0.80	0.79	0.81	0.59	0.86	0.83	0.82	0.80	0.60
	2007	0.51	0.60	0.56	0.56	0.48	0.47	0.64	0.59	0.59	0.48	0.84	0.75	0.79	0.72	0.58	0.88	0.78	0.80	0.71	0.59
	2016	0.46	0.46	0.56	0.45	0.45	0.55	0.55	0.59	0.52	0.46	0.85	0.76	0.76	0.69	0.59	0.85	0.82	0.77	0.72	0.58
Eccentricity	1977	0.75	0.80	0.73	0.76	0.71	0.73	0.77	0.68	0.71	0.65	0.69	0.65	0.66	0.65	0.62	0.91	0.83	0.80	0.82	0.75
	1987	0.83	0.77	0.81	0.75	0.74	0.74	0.70	0.69	0.68	0.66	0.70	0.66	0.68	0.65	0.63	0.81	0.80	0.79	0.75	0.72
	1997	0.74	0.69	0.69	0.69	0.66	0.77	0.72	0.71	0.71	0.66	0.85	0.78	0.77	0.79	0.73	0.82	0.79	0.78	0.77	0.72
	2007	0.80	0.77	0.73	0.73	0.69	0.81	0.78	0.76	0.75	0.71	0.83	0.83	0.80	0.79	0.74	0.80	0.80	0.78	0.76	0.72
	2016	0.83	0.75	0.72	0.71	0.67	0.85	0.78	0.78	0.76	0.73	0.84	0.81	0.81	0.78	0.74	0.83	0.81	0.82	0.77	0.72
Degree centrality	1977	2.50	2.73	1.93	3.61	8.09	2.20	2.58	2.40	3.85	9.55	5.29	9.25	11.27	14.51	33.94	4.14	10.47	9.14	13.15	27.50
	1987	2.71	3.87	3.57	6.57	10.38	3.13	4.12	4.43	7.43	12.86	7.26	12.00	9.77	20.39	32.78	7.57	11.81	10.92	27.36	41.10
	1997	3.53	4.90	6.63	6.36	15.79	3.07	3.73	5.31	5.28	12.24	8.78	14.50	16.92	18.04	44.95	10.50	17.70	19.92	27.19	60.60
	2007	3.32	5.37	6.28	7.71	15.92	3.44	6.66	7.39	9.82	20.92	7.79	15.64	20.78	25.22	53.39	9.87	19.22	26.27	38.47	75.57
	2016	3.14	5.14	5.77	8.29	16.71	3.52	6.33	7.00	10.60	22.71	10.15	17.34	19.66	32.36	57.64	12.28	20.24	23.12	42.25	81.57

Source: Own elaboration.

Figure 22.7 Average centrality per population quantile at selected years, 1977–2016 similar to other networks that are so widely studied in the academic sphere and beyond, such as airlines, multinational firms, and other vectors of flows of all types whether material or immaterial. By demonstrating that technological and operational change has modified the way ports and maritime networks are tied to cities, as seen with a decreasing correlation over the past 40 years, the world's largest cities have maintained their role in relation to traffic volume, share and centrality.

This chapter provides additional evidence of the necessity for spatial network studies (and network analysis in general) not to ignore the socio-economic dimension of the connected nodes (here, population or demographic size) in order to verify the uneven territorial overlap between networks and cities. Among other results, we demonstrated that port cities, the frontlines of globalization, have lost their explaining power concerning maritime traffic volumes. However, the comparison of four different perspectives (cf. two different definitions of the city and of the network) helped us to balance this claim – that cities no longer need a port to grow, and vice versa – since we proved that expanding the spatial unit of reference (from the sole port city to the extended city-region) and adapting the network's definition to this specific context both greatly improved our results and modified the way port cities should be understood currently. Ports are still tied to cities and cities still need ports to grow and develop, but not through the same spatial and functional mechanisms as in the past, that is, before and during the spread of containerization.

Further research is underway to integrate the global road transport network with the global maritime network in order to study how cities are performing in the combined land–sea network, thereby contributing to multiplex network studies. Another research pathway is to go beyond the sole population size but other variables, such as urban gross domestic product or employment structure, are hard to obtain and often not harmonized globally across the different countries. Are richer and more productive cities more central? Before moving on to widen our analysis, we believe that this chapter contributes to both urban, transport and network studies by further bridging them together, given that they too often stand apart. This step forward towards a better understanding of the spatiality of physical, non-planar networks is also a positive response to Schwanen (2017), rightly observing that, most of the time, networks are studied in an abstract, topological space without urban attributes on nodes.

ACKNOWLEDGEMENTS

The research leading to these results has received funding from the European Research Council under the European Union's Seventh Framework Programme (FP/2007-2013)/ ERC Grant Agreement n. [313847] 'World Seastems'.

NOTE

1. http://tulip.labri.fr/TulipDrupal/ (accessed December 2020).

REFERENCES

Bernhofen, D.M., Z. El-Sahli and R. Kneller (2013), 'Estimating the effects of the container revolution on world trade', Working Paper 2013:4, Department of Economics, School of Economics and Management, Lund University.

Bretagnolle, A. (2015), 'City systems and maritime transport in the long term', in C. Ducruet (ed.), *Maritime Networks. Spatial Structures and Time Dynamics*, London and New York: Routledge, pp. 27–36.

Bretagnolle, A., D. Pumain and C. Rozenblat (1998), 'Space-time contraction and the dynamics of urban systems', *Cybergeo: European Journal of Geography*, **61**, accessed December 2020 at https://journals.openedition.org/cybergeo/373.

Bunel, M., F. Bahoken, C. Ducruet, C. Lagesse, B. Marnot, E. Mermet, et al. (2017), 'Geovisualizing the sail-to-steam transition through vessel movement data', in C. Ducruet (ed.), *Advances in Shipping Data Analysis and Modeling. Tracking and Mapping Maritime Flows in the Age of Big Data*, London and New York: Routledge, pp. 189–205.

Cattan, N. (1995), 'Barrier effects: the case of air and rail flows', *International Political Science Review*, **16** (3), 237–48.

Cullinane, K.P.B. and M. Khanna (2000), 'Economies of scale in large containerships: optimal size and geographical implications', *Journal of Transport Geography*, **8** (3), 181–95.

Deng, W.B., G. Long, L. Wei and C. Xu (2009), 'Worldwide marine transportation network: efficiency and container throughput', *Chinese Physics Letters*, **26** (11), 118901.

Ducruet, C. (2013), 'Network diversity and maritime flows', *Journal of Transport Geography*, **30** (C), 77–88.

Ducruet, C. (2015), *Maritime Networks: Spatial Structures and Time Dynamics*, London and New York: Routledge, pp. 3–26.

Ducruet, C. and L. Beauguitte (2014), 'Network science and spatial science: review and outcomes of a complex relationship', *Networks and Spatial Economics*, **14** (3–4), 297–316.

Ducruet, C. and H. Itoh (2016), 'Regions and material flows: investigating the regional branching and industry relatedness of port traffic in a global perspective', *Journal of Economic Geography*, **16** (4), 805–30.

Ducruet, C. and I. Lugo (2013), 'Cities and transport networks in shipping and logistics research', *Asian Journal of Shipping and Logistics*, **29** (2), 149–70.

Ducruet, C. and T.E. Notteboom (2012), 'The worldwide maritime network of container shipping: spatial structure and regional dynamics', *Global Networks*, **12** (3), 395–423.

Ducruet, C., S. Cuyala and A. El Hosni (2018), 'Maritime networks as systems of cities: the long term interdependencies between global shipping flows and urban development (1890-2010)', *Journal of Transport Geography*, **66** (January), 340–55.

Ducruet, C., D. Ietri and C. Rozenblat (2011), 'Cities in worldwide air and sea flows: a multiple networks analysis', *Cybergeo: European Journal of Geography*, **528**, accessed December 2020 at http://cybergeo.revues.org/23603.

El Hosni, A. (2017), *Le retour du port dans la ville*, Sarrebruck: Editions Universitaires Européennes.

Fleming, D.K. and Y. Hayuth (1994), 'Spatial characteristics of transportation hubs: centrality and intermediacy', *Journal of Transport Geography*, **2** (1), 3–18.

Frémont, A. (2007), 'Le monde en boîtes. Conteneurisation et mondialisation', Synthèse INRETS, 53.

Guerrero, D. (2014), 'Deep-sea hinterlands: some empirical evidence of the spatial impact of containerization', *Journal of Transport Geography*, **35** (February), 84–94.

Guerrero, D. and J.P. Rodrigue (2014), 'The waves of containerization: shifts in global maritime transportation', *Journal of Transport Geography*, **35** (January), 151–64.

Halim, R.A., J.H. Kwakkel and L.A. Tavasszy (2016), 'A strategic model of port-hinterland freight distribution networks', *Transportation Research Part E*, **95** (November), 368–84.

Hayuth, Y. (1981), 'Containerization and the load center concept', *Economic Geography*, **57** (2), 160–76.

Helders, S. (2012), *World Gazetteer*, accessed December 2012 at a now discontinued website.

Hesse, M. (2013), 'Cities and flows: Re-asserting a relationship as fundamental as it is delicate', *Journal of Transport Geography*, **29** (May), 33–42.

Hu, Y. and D. Zhu (2009), 'Empirical analysis of the worldwide maritime transportation network', *Physica A*, **388** (10), 2061–71.

Jacobs, W., H.R.A. Koster and P.V. Hall (2011), 'The location and global network structure of maritime advanced producer services', *Urban Studies*, **48** (13), 2749–69.

Lahmeyer, J. (2015), 'Population statistics. Growth of the population per country in a historical perspective, including their administrative divisions and principal towns', Populstat, accessed June 2018 at http://www.populstat.info/.

Lee, S.W. and C. Ducruet (2009), 'Spatial glocalization in Asia-Pacific hub port cities: a comparison of Hong Kong and Singapore', *Urban Geography*, **30** (2), 162–84.

Lee, S.W., D.W. Song and C. Ducruet (2008), 'A tale of Asia's world ports: the spatial evolution in global hub port cities', *Geoforum*, **39** (1), 372–85.
Levinson, M. (2008), *The Box: How the Shipping Container Made the World Smaller and the World Economy Bigger*, Princeton, NJ: Princeton University Press.
Marnot, B. (2005), 'Interconnexion et reclassements: l'insertion des ports français dans la chaîne multimodale au XIXe siècle', *Flux*, **59** (1), 10–21.
Merk, O. (2013), *The Competitiveness of Global Port-Cities*, Synthesis Report, Paris: OECD.
Merk, O., W.J.J. Manshanden and M.I. Dröes (2013), 'Inter-regional spillovers of seaports: the case of Northwest Europe', *International Journal of Transport Economics*, **40** (3), 401–17.
Mohamed-Chérif, F.Z. and C. Ducruet (2016), 'Regional integration and maritime connectivity across the Maghreb seaport system', *Journal of Transport Geography*, **51** (February), 280–93.
Morel, J.C. and C. Ducruet (2015), 'Interview – the man who brought containerisation to Europe', *Portus Online*, **29**, accessed December 2020 at http://portusonline.org/fr/interview-the-man-who-brought-containerisation-to-europe/.
Moriconi-Ebrard, F. (1994), *Geopolis: pour comparer les villes du monde*, Paris: Economica.
Notteboom, T.E. (2004), 'Container shipping and ports: an overview', *Review of Network Economics*, **3** (2), 86–106.
Notteboom, T.E. (2016), 'The adaptive capacity of container ports in an era of mega vessels: the case of upstream seaports Antwerp and Hamburg', *Journal of Transport Geography*, **54** (June), 295–309.
Parshani, R., C. Rozenblat, D. Ietri, C. Ducruet and S. Havlin (2010), 'Inter-similarity between coupled networks', *Europhysics Letters (EPL)*, **92** (6), 68002.
Peris, A. (2016), 'Penser les villes en réseaux: une analyse des théories sur les liens interurbains', MA dissertation, University of Paris I Panthéon-Sorbonne, Paris.
Pumain, D., F. Paulus and C. Vacchiani-Marcuzzo (2009), 'Innovation cycles and urban dynamics', in D. Lane, D. Pumain, S.E. van der Leeuw and G. West (eds.), *Complexity Perspectives in Innovation and Social Change*, Dordrecht: Springer, pp. 237–60.
Rivers, R., T. Evans and C. Knappett (2015), 'From oar to sail. The role of technology and geography in the evolution of Bronze Age Mediterranean networks', in C. Ducruet (ed.), *Maritime Networks. Spatial Structures and Time Dynamics*, London and New York: Routledge, pp. 63–76.
Rodrigue, J.P. and T.E. Notteboom (2010), 'Foreland-based regionalization: integrating intermediate hubs with port hinterlands', *Research in Transportation Economics*, **27** (1), 19–29.
Rozenblat, C. (2004), 'Les villes portuaires en Europe. Analyse comparative', final report for the Research Institute on Industrial and Territorial Strategy (IRSIT), National Centre for Scientific Research (CNRS) and GDR LIBERGEO 1559, September, accessed 8 March 2021 at https://www.mgm.fr/PUB/IRSIT.pdf.
Rozenblat, C. and G. Mélançon (2013), *Methods for Multilevel Analysis and Visualization of Geographical Networks*, Dordrecht: Springer.
Rozenblat, C. and D. Pumain (1993), 'The location of multinational firms in the European urban system', *Urban Studies*, **30** (10), 1691–709.
Schwanen, T. (2017), 'Geographies of transport II: reconciling the general and the particular', *Progress in Human Geography*, **41** (3), 355–64.
Slack, B. (1993), 'Pawns in the game: ports in a global transportation system', *Growth and Change*, **24** (4), 579–88.
Slack, B. and E. Gouvernal (2015), 'Container transshipment and logistics in the context of urban economic development', *Growth and Change*, **47** (3), 406–15.
Slack, B. and J.J. Wang (2002), 'The challenge of peripheral ports: an Asian perspective', *Geojournal*, **56** (2), 159–66.
Stopford M. (2008), *Maritime Economics*, Abingdon: Routledge.
Tsiotas, D. and S. Polyzos (2018), 'Effects in the network topology due to node aggregation: empirical evidence from the domestic maritime transportation in Greece', *Physica A*, **491** (February), 71–88.
United Nations Conference on Trade and Development (UNCTAD) (2017), *Review of Maritime Transport*, Geneva: UNCTAD.

APPENDIX

Table 22A.1 Total container traffic of the top 20 extended city-regions at selected years, 1977–2016

City	1977	City	1987	City	1997	City	2007	City	2016
Osaka	6095395	Osaka	19563383	Hong Kong	57131846	Hong Kong	116665822	Singapore	162766699
Tokyo	5839131	Tokyo	19349285	Singapore	52916822	Singapore	96641506	Shenzhen	156797205
Rotterdam	4435547	Hong Kong	17151634	Tokyo	35419105	Shanghai	72886371	Shanghai	152681263
Hong Kong	4404215	Kaohsiung	11332338	Osaka	30480647	Shenzhen	62459603	Hong Kong	119869974
New York	3996619	Rotterdam	10879412	Kaohsiung	30019133	Busan	60577372	Ningbo	103395326
Bremen	3187762	Taipei	10476247	London	22640708	Tokyo	49914010	Busan	102458199
Los Angeles	3163848	Busan	10366831	Busan	22220108	Kaohsiung	46907471	Qingdao	91026532
London	3037651	Nagoya	8651655	Rotterdam	20999826	Kuala Lumpur	42603937	Kuala Lumpur	74747413
Hamburg	2908135	London	8592861	Nagoya	16515991	Rotterdam	36931953	Kaohsiung	62353438
Le Havre	2554225	Los Angeles	8042520	Los Angeles	15770111	London	36256987	Rotterdam	55746419
Nagoya	2382368	Hamburg	7169989	Le Havre	12256797	Osaka	31220784	Tokyo	53959894
San Francisco	2309653	Le Havre	6688253	Hamburg	12227335	Ningbo	31200486	London	47488998
Singapore	2299814	New York	6294237	Kuala Lumpur	12214657	Hamburg	29139084	Dubai	46549165
Baltimore	2001996	San Francisco	5445161	Taipei	11727859	Los Angeles	28349584	Xiamen	45633545
Kaohsiung	1579028	Bremen	5403756	New York	11592911	Antwerp	27894892	Antwerp	44422867
Kuala Lumpur	1557418	Antwerp	4727480	San Francisco	10578176	Qingdao	25249995	Johore Baharu	39568214
Charleston	1537925	Seattle	3796055	Colombo	10465553	Dubai	24170231	Bangkok	38108709
Philadelphia	1365134	Jeddah	3650991	Antwerp	9787841	Nagoya	22107710	Guangzhou	36742338
Sydney	1163716	Singapore	3549919	Bangkok	9165413	Xiamen	21659488	Colombo	35703373
Marseille	1071976	Charleston	2874917	Charleston	8476277	New York	21655539	Hamburg	35051619

Note: Unit: deadweight tons (DWT).

Source: Own elaboration.

Table 22A.2 Description of population quantiles at selected years, 1977–2016

	Quantile	Nb.	Avg.	Min.	Max.
1977	1	400	10	1	18
	2	397	28	19	39
	3	451	60	40	87
	4	433	141	88	229
	5	423	1166	230	23863
1987	1	430	11	1	20
	2	425	32	21	47
	3	433	73	48	107
	4	433	176	108	293
	5	432	1434	294	27358
1997	1	439	11	1	23
	2	441	36	23	51
	3	442	81	52	119
	4	443	199	120	329
	5	442	1638	330	31594
2007	1	437	12	1	24
	2	435	39	25	55
	3	482	89	56	136
	4	406	224	136	372
	5	443	1926	373	35691
2016	1	580	3	1	9
	2	650	17	9	29
	3	621	48	30	74
	4	625	135	75	240
	5	620	1559	241	36933

Source: Own elaboration.

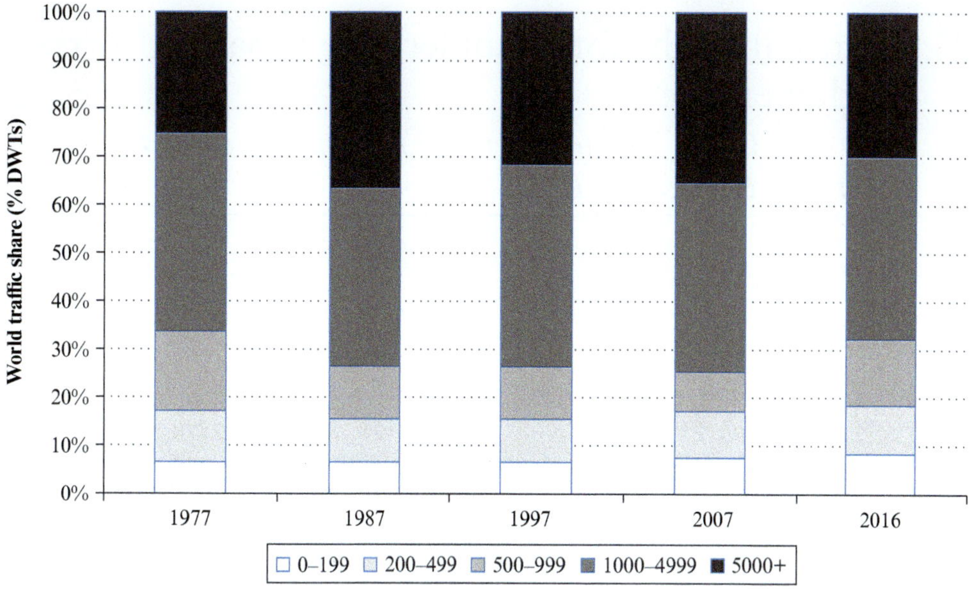

Source: Own elaboration.

Figure 22A.1 Container traffic distribution per population classes at selected years, 1977–2016

23. Unravelling the forces underlying urban industrial agglomeration

Neave O'Clery, Samuel Heroy, François Hulot and Mariano Beguerisse-Díaz

1. INTRODUCTION

Globally, a consensus has emerged that emphasises the role of cities, rather than countries, as key engines of economic growth (Jacobs 1961; Glaeser 2011). Yet, what are the drivers of urban success in the face of high costs arising from factors such as congestion, density, crime, and pollution?

Leading scholars have argued that it is the diversity of cities, and in particular the way cities facilitate and foster a diverse ecology of social interactions, that gives rise to new activities, opportunities and innovations (Jacobs 1969; Bettencourt et al. 2007). This view aligns with a growing literature that emphasises the role that larger cities have in better facilitating matching between employers and employees, knowledge spillovers between firms and innovation opportunities (Friedrichs 1993; Duranton and Puga 2001, 2004; Rosenthal and Strange 2006).

Yet, for all their diversity at an individual or worker level, cities tend to specialize at an industry level (Beaudry and Schiffauerova 2009). That is industries are geographically concentrated: examples range from the technology cluster of Silicon Valley to the automobile cluster of 'motor city' (Detroit). This phenomenon of clustering of similar firms is thought to arise mainly via three channels of cost-sharing: transport costs associated with customers and suppliers, labour costs and learning or knowledge costs (Marshall and Marshall 1881; Glaeser et al. 1992; Henderson et al. 1995).

However, until recently there has been little success in estimating the relative importance of each of these channels to the location decisions of firms. This chapter explores and extends a burgeoning literature (Ellison et al. 2010; Diodato et al. 2018) that aims to exploit the co-location patterns of industries in cities in order to unravel the relationship between industry co-agglomeration and customer–supplier, labour and idea cost-sharing.

1.1 Urban Agglomeration Patterns

The benefits of firm agglomeration have long been proposed. In 1881, Marshall argued that firms locate in close proximity in order to reduce costs: firms can save shipping costs by locating near suppliers or customers, searching and matching costs via labour market pooling, and benefit from knowledge spillovers (Marshall and Marshall 1881). For example, chemical and pharmaceutical companies might rely on access to a similar labour pool, whereas car manufacturers might be more sensitive to transport costs associated with suppliers of parts. Industries at the forefront of innovation such as biotechnology firms might benefit from a combination of skilled labour and knowledge spillovers from locating close to other similar firms. In a well-known instance of this type of location decision, Amazon was set up in Seattle to benefit from the local talent pool (that is, Microsoft and other technology firms).

Other benefits of geographical clustering of firms identified in literature are competition, which drives productivity (Porter 1990), and local demand effects (Fujita et al. 2001).

While it has been long understood that it is these 'externalities' that drive urban industry agglomeration patterns, measuring their impact, and modelling their behaviour and dynamics, remains a challenge. In particular, until recently there has been little success in estimating the relative importance of each Marshallian channel to firms' decisions – not least because each channel predicts an identical agglomeration pattern (Duranton and Puga 2004).

In response to this challenge, a pioneering study by Ellison et al. (2010a) proposed to exploit the co-location patterns of industries in cities in order to estimate the relationship between industry co-agglomeration and each of the three Marshallian channels: customer–supplier, labour and idea sharing. That is, they were interested in whether pairwise patterns of industry co-agglomeration were strongly correlated with, for example, labour pooling (which can be constructed from an estimate of skill-overlap between industries, see next section). If co-agglomeration patterns are most correlated with labour pooling (compared with the other channels), then, this study proposes, the labour channel is the most important reason for firm location decisions across all industries. Hence, Ellison et al. (2010a) argue that comparing co-agglomeration patterns with each of the three channels enables us to estimate the relative importance of each channel (across all industries).

In practice, this means constructing a pairwise industry–industry measure of co-location (for example, the propensity of firms from two industries to be located in the same cities), and corresponding pairwise estimates for each industry–industry 'proximity' or similarity for each of the three Marshallian channels. There are a variety of ways to measure these similarities as discussed below.

1.2 Industry Similarity and Networks

For each of the studies we focus on in this chapter, data on industries and employment from the USA (predominantly from 2002) is used. Technical definitions and information on data sources are provided in section 2. Following Ellison et al. (2010a), our task is to develop pairwise industry–industry measures of co-agglomeration, labour pooling, customer–supplier and knowledge sharing.

Geographic co-agglomeration of industries is typically estimated using data on the distribution of employment by industry and city (Ellison et al. 2010; Hausmann et al. 2014).

Two industries are 'proximate' in relation to the first channel, labour pooling, if they have similar skill requirements. Hence, one method to estimate shared labour needs is to compute a measure of occupational similarity (for example using data on industry employment by occupation) (Farjoun 1994; Chang 1996). Alternatively, labour flows (job switches) between industry pairs is an excellent proxy for shared skills (Neffke and Henning 2013). Unfortunately, sufficiently disaggregated data on industry–industry labour flows is currently unavailable for the USA.

Two industries are similar as regards the second channel if they share customers and suppliers. In order to capture buyer–seller relationships between industry pairs (Fan and Lang 2000), we can make use of input–output matrices, built by national statistics offices worldwide, which capture monetary flows between industries (normally with great precision).

Finally, a pair of industries is similar in relation to the third channel if they commonly exchange ideas and know-how (often via some sort of research and development collaboration). This type of knowledge sharing between industries is typically captured using patent data; more specifically, the extent to which technologies associated with industry i cite technologies associated with industry j, and vice versa. However, only a select few manufacturing industries are active in patenting (Diodato et al. 2018) and hence this third channel is omitted from much of our analysis.

It is instructive to think about each type of industry proximity (co-agglomeration, labour pooling, customer–supplier and knowledge sharing) as a network, where nodes (industries) are connected via edges with weight equal to the corresponding proximity. In section 1.5 we see that industry networks of this type are frequently used to model diversification paths, whereby places (for example, cities) branch into new economic activities that are similar to existing strengths (Frenken et al. 2007). In section 4, we propose a network-based method to estimate the relative importance of each Marshallian channel at an industry-cluster scale.

Network models are increasingly used to understand the role that interconnected structures play in economic and innovation-related processes, including research clusters (Catini et al. 2015), the inter-industry propagation of supply and demand shocks (Acemoglu et al. 2015), and banking crises (Battiston et al. 2012).

1.3 Relative Importance of Marshallian Channels

Using the pairwise proximity matrices introduced in the previous section, Ellison et al. (2010) estimated the contribution of labour, customer–supplier linkages, and knowledge spillovers to co-location using 1987 manufacturing data from the USA. The technical details of this estimation are provided in section 3.1. The authors find that each channel is a significant factor in industry co-location patterns, with a particular emphasis on customer–supplier linkages.

Diodato et al. (2018) reproduced the Ellison et al. (2010) study using data for US manufacturing industries from 2002, confirming a strong role for labour and input–output linkages in co-agglomeration patterns. Evidence for the importance of knowledge spillovers was less apparent. These studies 'however' focus on manufacturing industries and neglect services, an increasingly dominant component of advanced urban economies. Services, similar to high- and low-skilled manufacturing, may favour labour linkages (services are often labour intensive) or customer–supplier linkages (many services are non-traded). Extending the earlier study, Diodato et al. (2018) found a stronger role for labour linkages when a larger set of industries including services were considered.

Investigating further the strengthening role of labour linkages for co-agglomeration patterns, using data from 1950 to 2010 Diodato et al. (2018) showed that the importance of customer–supplier linkages has steadily decreased over the past century, while labour linkages have strengthened. This result corresponds to a well-documented shift from value-chain based manufacturing to a specialised service driven economy.

The framework proposed by Ellison et al. (2010a) assumes a homogeneous contribution of each Marshallian channel to co-agglomeration patterns for all industries. Intuitively, however, it is highly likely that individual industries will rely on each Marshallian channel in different combinations. For example, we might expect that high-technology manufacturing relies on the availability of skilled labour, whereas heavy manufacturing prioritises proximity to suppliers and customers. Diodato et al. (2018) adapt the framework proposed by Ellison et al. (2010), and estimate the relationship between co-agglomeration patterns and each of the three Marshallian channels for individual industries. Their results highlight remarkable heterogeneity across industries, with arts, media and scientific industries showing a strong preference for labour linkages, while farming and manufacturing co-locate according to customer and supplier linkages. This study is one of a number of studies to investigate heterogeneity in agglomeration forces across industries (Faggio et al. 2014; Howard et al. 2015; Beherns 2016).

1.4 Agglomeration Drivers for Industry Clusters

We can view previous approaches to measuring the relative importance of Marshallian channels as 'local' approaches, which inherently do not take into account network structure. Specifically, previous approaches have sought to relate the whole network (for example, all of the edges) or individual nodes (for example, all of the edges connected to a single node) to patterns corresponding to each of the Marshallian channels. Each of these methods has neglected to consider intermediate scales which emerge naturally from the structure of the co-agglomeration network.

Intuitively, we might expect that clusters of industries co-located across many cities may exhibit a similar dependence on each of the Marshallian channels. We can identify these clusters via an analysis of connectivity patterns in the co-agglomeration network. Then, building on the framework established by Ellison et al. (2010) and Diodato et al. (2018), we show that industry clusters exhibit heterogeneous dependence on each of the Marshallian channels.

In order to identify industry clusters, we focus on uncovering densely connected groups of nodes (known as communities) in the co-agglomeration network. This approach is connected to work on detecting and defining industrial clusters for the USA (Porter 1998; Delgado et al. 2016) and Ireland (O'Clery et al. 2019). More broadly, community detection has been used extensively to study the structure and dynamics of biological and social networks (Girwan and Newman 2002).

Although there exists a wide range of approaches to community detection (see Fortunato 2010 for a review), we employ a method term Markov Stability based on diffusion dynamics (Delvenne et al. 2010). Intuitively, we let a random walker wander on the network, jumping from node to node. If the random walker remains in the same group of nodes over a long period of time, there is high connectivity between the group of nodes, and a community has been discovered.

While the classical problem of community detection is to find a single node partition under a particular optimisation strategy (for example, modularity from Newman et al. 2004), it is more natural to consider a range of partitions on different scales (from many small node clusters to few larger clusters). Within the prism of the stability approach, this information can be extracted by analysing the patterns of random walkers on a network at different time-scales (Delvenne et al. 2010). Longer time-scales correspond to larger node aggregations, and fewer communities.

We apply the stability method to the co-agglomeration network, extracting a range of network partitions at different scales. Although not strictly nested, these partitions form a hierarchical decomposition of the co-agglomeration network into industry clusters of different sizes. For each of these partitions, we analyse the relationship between the edge weights within co-agglomeration communities and the corresponding edge weights in the customer–supplier and labour networks. We uncover regions of the co-agglomeration network – clusters of co-located industries – dominated by labour and/or value-chain agglomeration channels. This result has implications for network based diversification models, introduced in the next section, which simulate dynamics on a co-agglomeration network in order to describe economic branching processes.

As a final validating step, we investigate the relationship between the mean number of years of education (averaged across industries within a cluster) and our cluster estimates for reliance on labour and customer–supplier linkages. It is expected that labour-dominated clusters would employ workers with a higher level of education. In agreement with this hypothesis, we find that years of education correlate positively with our estimates for the labour-sharing channel, and correlate negatively with our estimates for the IO channel.

1.5 Industrial Growth and Diversification Models

An emerging perspective sees diversification as a path-dependent process, whereby the growth and appearance of new economic activities (industries) in a place is dependant on the local availability of relevant capabilities (Nelson and Winter 1982; Frenken et al. 2007; Hidalgo et al. 2007). These capabilities include, but are not constrained to, skilled labour, physical infrastructure and other necessary inputs for production.

This perspective has emerged from the (recently converging) fields of economic complexity and evolutionary economic geography, and particularly focuses on the role of local capabilities (often in the form of worker know-how) in growth and diversification process (Nelson and Winter 1982; Frenken et al. 2007; Hidalgo et al. 2007). Hence, quantifying local capabilities or know-how (termed 'economic complexity' by Hidalgo and Hausmann 2009), and identifying new activities that are 'proximate' as regards their capability requirements (Frenken et al. 2007; Hidalgo et al. 2007), is key to the development of diversification and development strategies and models.

Within this framework, cities move into new economic activities that share existing capabilities in a path-dependent manner. This process can be modelled using an industry network, where edges represent capability overlap (Hidalgo et al. 2007; Neffke et al. 2011). This network can be seen as an economic map or landscape: the position of a city (that is, the subgraph of its industries) constrains its future development path. More specifically, the diversification of a city into new industries can be modelled by a diffusion process on the industry network with initial condition governed by the initial set of industries in the city (Hidalgo et al. 2007; Neffke et al. 2011; Hausmann et al. 2014; O'Clery et al. 2016). Hence, while centrally located cities share capabilities with many potential new industries, peripheral cities have fewer options.

This network approach to modelling diversification processes was pioneered by Hidalgo et al. (2007), and focused on describing the diversification paths of countries as they develop and move into new, more sophisticated products. The underlying network of products (the product space) is a product–country analogue of the (industry–city) co-agglomeration network introduced in the previous section. That is, in the diversification literature, the co-agglomeration network has been proposed as a general measure of capability overlap or industry relatedness (Hidalgo et al. 2007; Hausmann et al. 2014), which does not distinguish between different types of linkage, and forms the 'landscape' on which to model growth and diversification dynamics. Hence, our approach to uncovering the relationship between co-agglomeration patterns and the channels of labour sharing and customer–supplier linkages at an industry cluster scale has evident implications for this class of diversification models. We expect that diversification dynamics will be better captured by skill linkages in some regions of the network (as uncovered by our analysis), and better modelled by input–output (IO) linkages in other regions of the network. Diodato et al. (2018) provide evidence that employment growth dynamics at the city-industry level can be improved by incorporating industry-level heterogeneous linkages, but there remains future work with respect to our cluster estimates.

Modelling diversification paths using industry networks is well established in the related diversification literature, and has been deployed to model growth and diversification processes across a range of spatial scales (Frenken et al. 2007; Neffke et al. 2011; Neffke and Henning 2013), and study a wide range of questions around local growth paths, including employment and formality rates, and firm and sector entry (Hausmann et al. 2014; O'Clery et al. 2016).

2. DATA AND METRICS

Here we briefly review the technical definitions of the industry similarity metrics used by Ellison et al. (2010) and Diodato et al. (2018), as introduced in section 1.2, and provide information on data sources. For the new analysis presented in section 4, we use the same data as the latter study.

2.1 Co-agglomeration

As proposed by Ellison and Glaeser (1997), we define a co-agglomeration index for industries i and j:

$$EG_{ij} = \frac{\sum_{r=1}^{R}(s_{ir} - x_r)(s_{jr} - x_r)}{1 - \sum_{r=1}^{R} x_r^2}, \qquad (23.1)$$

where s_{ir} is the employment share of industry i in region r, and x_r is the mean share of employment in region r.[1] As shown by Ellison et al. (2010), EG has the advantage of not being affected by the size distributions of firms in various industries nor by the level of spatial aggregation. This latter issue affects a similar measure of industry co-location proposed by Porter (2003) based on a correlation of employment shares across locations:

$$LC_{ij} = corr(s_{ir}, s_{jr}). \qquad (23.2)$$

For the analysis undertaken by Ellison et al. (2010), 1987 employment data from the US Census Bureau's Census of Manufacturing (122 three-digit Standard Industrial Classification, SIC, manufacturing industries) was used. In the case of Diodato et al. (2018), 2003 employment data from the US Bureau of Labour Statistics (283 four-digit North American Industry Classification System, NAICS, industries including both manufacturing and services, and 939 metropolitan areas) is used.

2.2 Labour Pooling

Next, we measure industry similarity with respect to labour pooling using occupational data as proposed by Farjoun (1994). If E_i is a vector of employment by occupation for industry *i*, we compute the Pearson correlation between E_i and E_j:

$$L_{ij} = corr(E_i, E_j). \tag{23.3}$$

Observe that this metric is symmetric with respect to *i* and *j* and does not depend on the sizes of the respective industries.

For the analysis undertaken by Ellison et al. (2010), 1987 industry-occupation employment data was taken from the National Industrial-Occupation Employment Matrix (NIOEM) published by the US Bureau of Labor Statistics (277 occupations). In Diodato et al. (2018), 2002 industry-occupation employment data was taken from the US Occupational Employment Statistics (734 occupations).

2.3 Customer–Supplier Linkages

In order to study customer–supplier relationships between industry pairs, we use an input–output matrix with entries *i, j* corresponding to the value of goods and services that industry *j* sources from industry *i*. Normalising in both directions to account for the relative importance of industries as a buyer and seller, we create a symmetric *IO* proximity matrix:

$$IO_{ij} = \max\left(\frac{IO_{ij}}{\sum_k IO_{kj}}, \frac{IO_{ji}}{\sum_k IO_{kj}}, \frac{IO_{ij}}{\sum_k IO_{ik}}, \frac{IO_{ji}}{\sum_k IO_{ik}}\right). \tag{23.4}$$

For the analysis undertaken by Ellison et al. (2010), the 1987 Benchmark Input–Output Accounts published by the Bureau of Economic Analysis (BEA) was used. In Diodato et al. (2018), analogous tables provided by the BEA for 2002 are used.

2.4 Knowledge Spillovers

We can estimate inter-industry knowledge similarity using cross-industry patent citations – specifically, the extent to which technologies associated with industry *i* cite technologies associated with industry *j*, and vice versa. Using a similar approach to equation 23.5, we compute:

$$K_{ij} = \max\left(\frac{X_{ij}}{\sum_k X_{kj}}, \frac{X_{ji}}{\sum_k X_{kj}}, \frac{X_{ij}}{\sum_k X_{ik}}, \frac{X_{ji}}{\sum_k X_{ik}}\right). \quad (23.5)$$

where X_{ij} is the number of citations from patents associated with industry i to patents associated with industry j.

Both Ellison et al. (2010) and Diodato et al. (2018) make use of the National Bureau of Economic Research (NBER) patent citations dataset (Hall et al. 2001) to compute this matrix.

2.5 Education

Finally, in Section 4.5 we use US micro-data (2002) from the Integrated Public Use Microdata Series (IPUMS), the world's largest individual-level population database, to compute the mean number of years of education for workers in US industries.

3. DEPENDENCE OF INDUSTRIAL CO-LOCATION ON MARSHALLIAN CHANNELS

In this section, we explore the methodological approach and results of Ellison et al. (2010) and Diodato et al. (2018). In summary, Ellison et al. (2010) show that input–output linkages are the dominant channel for manufacturing industries using data from 1987. Using more recent data, Diodato et al. (2018) show that labour sharing has become 'over time' an increasingly important channel, and the dominant channel in service industries. In addition, while the earlier study views these channels as homogeneous across all industries, the latter finds significant heterogeneity between industries.

3.1 Homogeneous Forces (Ellison et al. 2010)

Using manufacturing data from the US (1987), Ellison et al. (2010) sought to study the relationship between industry co-location, and shared labour, customer–supplier and knowledge requirements via estimation of the coefficients β_Z of a simple linear model:

$$EG_{ij} = \alpha + \beta_Z Z_{ij} + \varepsilon_{ij}, \quad (23.6)$$

across all pairs i, j for channels $Z \in \{L, IO, K\}$. Table 23.1 shows a much stronger (50 per cent) relationship between co-agglomeration patterns and value-chain linkages as compared to labour or knowledge linkages.

Following the approach of Ellison et al. (2010), Diodato et al. (2018) investigated the contribution of labour pooling and customer–supplier relationships to co-location using more recent data from the USA (2002). This study does not include knowledge spillovers, and natural resource availability, as was incorporated into previous work. The authors showed, using the same estimation model, that labour sharing is increasingly an equal driver of agglomeration in manufacturing industries, and a more significant driver of agglomeration in services (see Table 23.1).

Table 23.1 Ordinary least squares univariate regressions from EGK and DNO (equation 23.6)

	EGK (manu)	DNO (manu)	DNO (serv)	DNO (all)
L	0.106	0.164	0.275	0.175
	(0.016)	(0.018)	(0.037)	(0.007)
N	7381	3655	5360	16836
R^2	0.011	0.035	0.018	0.034
IO	0.167	0.177	0.171	0.138
	(0.028)	(0.032)	(0.028)	(0.015)
N	7381	3655	5360	16836
R^2	0.028	0.060	0.016	0.022
K	0.100	0.109	–	–
	(0.016)	(0.013)	–	–
N	7381	3655	–	–
R^2	0.010	0.022	–	–

Note: EGN = Ellison et al. (2010); PNO = Diodato et al. (2018).

Applying this approach to time-series data (1910–2010), the same authors showed that the influence of value-chain linkages has consistently decreased over time since before the 1940s. Furthermore, their analysis showed that while the importance of labour to firm co-location has declined in recent years, it is still significantly higher than it was in 1950. If previous trends are to persist, this study suggests that value-chain linkages will continue to lose importance, while labour pooling will remain a driving factor of co-agglomeration.

3.2 Heterogeneous Forces (Diodato et al. 2018)

It is intuitive that the relative importance of each Marshallian channel will be different for each industry. Diodato et al. (2018) explore this idea, allowing the relationship between co-location and shared labour as well as value-chain linkages to vary by industry. Specifically, they estimate the coefficients β_Z^i for each industry i via univariate regressions of the form:

$$EG_{ij} = \alpha_i + \beta_Z^i Z_{ij} + \varepsilon_{ij} \tag{23.7}$$

across all industries j for channels $Z \in \{L, IO\}$. Essentially, these coefficients capture the relationship between co-location and labour/value-chain similarities for linkages between an industry i and all other industries j.

Using this approach, it is possible to estimate labour and value-chain coefficients for each industry, enabling us to assess the relative contribution of these channels for individual industries. We plot the estimated coefficient pairs β_{IO}^i and β_L^i (averaged for 27 sectors) in Figure 23.1. This figure is a reproduction – with permission – of figure 3 in Diodato et al. (2018).

We observe a distribution of coefficients whereby knowledge-intensive industries 'such as design and accounting' rely on specific labour availability, whereas industries that rely

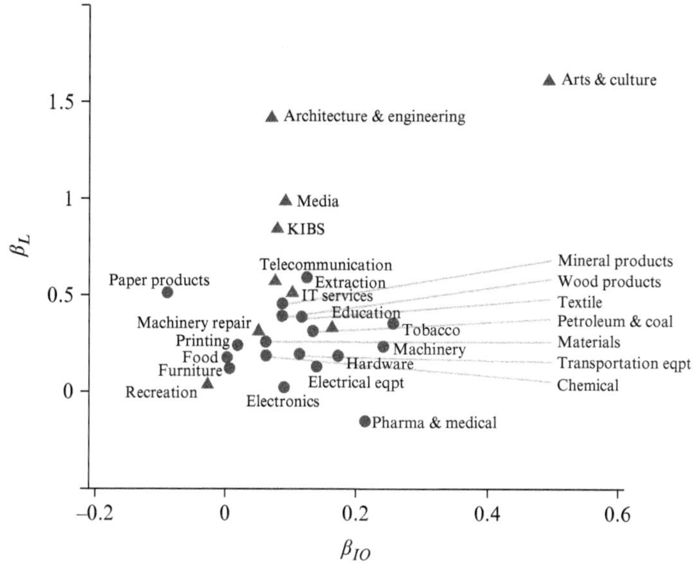

Note: Service industries (triangles) tend to be more labour pooling driven than manufacturing industries (circles).

Source: This figure is a reproduction, with permission, of figure 3 in Diodato et al. (2018).

Figure 23.1 Average estimates of labour pooling effects (β^i_L) and value chain effects (β^i_{IO}) from equation 23.7 for 27 sectors

on physical inputs such as agriculture, construction and manufacturing depend more on value-chain sharing (Figure 23.1). Moreover, we observe more generally that service industries tend to have higher labour-sharing dependence than value-chain dependence ($\beta_L > \beta_{IO}$), whereas manufacturing industries have varying ratios of β_L to β_{IO}.

Hence, by calculating industry-specific Marshallian agglomeration forces, Diodato et al. (2018) showed that labour sharing is the most important motive behind contemporary location choices of services, although value-chain linkages still explain much of the co-location patterns in manufacturing.

4. INDUSTRY CLUSTERS EXHIBIT HETEROGENEOUS DEPENDENCE ON MARSHALLIAN CHANNELS

In this section, we explore the idea that industry clusters, that is, groups of industries co-located across many cities, represent a natural intermediate scale by which to examine heterogeneous drivers of agglomeration patterns. Specifically, we identify industry clusters in the co-agglomeration network which exhibit heterogeneous dependence on each of the Marshallian channels, captured via labour and customer–supplier linkages as before.

4.1 The Co-agglomeration Network

A network is defined by a set of pairwise interactions or edges between nodes, which can represent individuals (Girvan and Newman 2002), cities (Neal 2011), or industries or products (Hidalgo et al. 2007; O'Clery et al. 2016), and so on. Analysis of the structure of those networks can elucidate important insights into complex social, economic and physical processes.

Here, we define a network describing industry–industry co-agglomeration relationships, where nodes represent industries and weighted edges represent the strength of pairwise co-agglomeration patterns. We encode this structure into a weighted adjacency matrix, in which an entry EG_{ij} (given precisely by equation 23.1) corresponds to the edge weight between nodes i and j.

Within our framework, co-agglomeration patterns (for example, the network structure) are a composite of analogous labour, customer–supplier and knowledge-sharing patterns – measured via occupation similarity, input–output linkages and patent citations, respectively. Figure 23.2 schematically illustrates this idea. In the next section we identify industry communities in the co-agglomeration network, which are clusters of tightly knit nodes that are frequently co-located across cities. We later investigate the dependence of each of these communities in respect of each of the Marshallian channels.

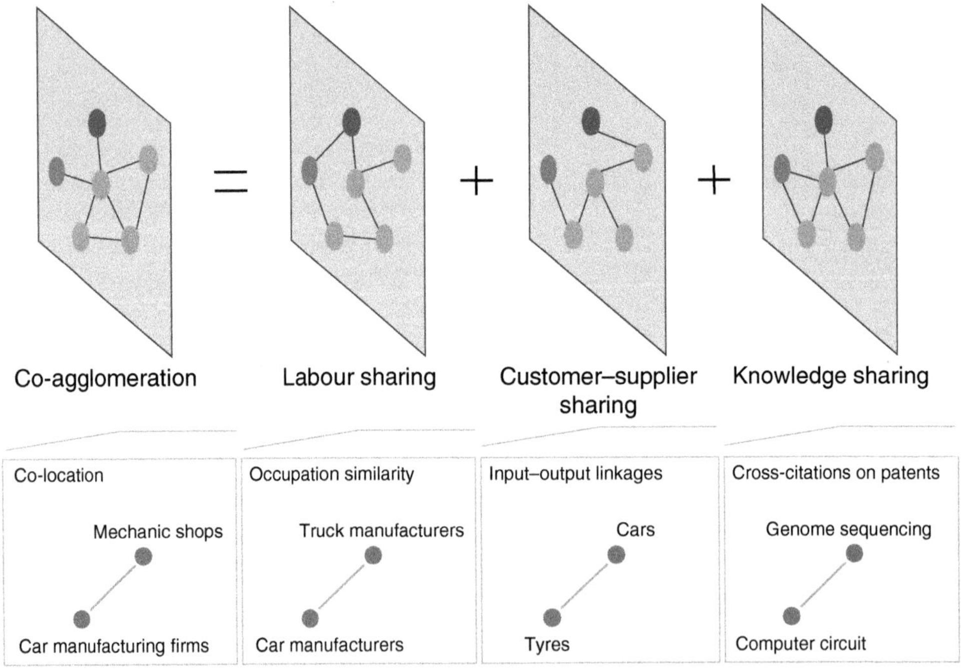

Note: These patterns are a composite of labour, customer–supplier and knowledge-sharing patterns.

Figure 23.2 Pairwise industrial co-agglomeration patterns represented as a network, with edges estimated via the co-location patterns of industries in cities

4.2 The Markov Stability Framework for Community Detection

The general task of community detection is to find a node partition (that is, a split of the nodes into communities) in that nodes in the same community have more edges than would be expected in a random graph with the same degree distribution. We use the Markov Stability algorithm (Delvenne et al. 2010) to perform this task on the co-agglomeration network. Below we briefly review the Markov stability framework, and refer readers to the original paper for more details.

In general, we can describe a (discrete) random walker diffusion process on a graph under the updating rule:

$$p_{t+1} = p_t M \tag{23.8}$$

where p_t is the (normalised) probability vector ($p_t(i)$ is the probability a random walker is on node i at time t), $M = D^{-1}EG$ is the transition matrix (D is a zero matrix with the vector of node degrees of EG – denoted $\deg(EG)$ – on the diagonal). The stationary distribution of this process is given by $\pi = \deg(EG)/\mathrm{sum}(\deg(EG))$.

We want to quantify the clustering of these dynamics for nodes within communities. We can encode a node partition in a matrix H such that $H(i,c) = 1$ if node i is in community c (and otherwise $H(i,c) = 0$). The clustered autocovariance matrix of the diffusion process is defined as:

$$R(t) = H^T[\Pi M^t - \pi^T\pi]H, \tag{23.9}$$

where $\Pi =$ diag (π). Observe that $(\Pi M^t)_{uv}$ represents the probability that a random walker who started in community u ends up in community v at time t, and $(\pi^T\pi)_{uv}$ is the probability that two independent random walkers are in u and v at stationarity. The diagonal entries of $R(t)$ therefore represent the probability that a random walker remains in their initial community after t time-steps, and hence the stability of a partition at time t is defined as:

$$r(t) = Trace(R(t)). \tag{23.10}$$

If a network's community structure is well defined, then a random walker is highly likely to remain in the community in which it started; therefore, we seek a partition matrix $\hat{H}$ that maximises $r(t, \hat{H})$ on the set of all the possible partitions. This problem is non-deterministic polynomial-time (NP)-hard, and we therefore we use the heuristic Louvain method to solve it.

Louvain's method (Blondel et al. 2008) works as follows. The method first assigns each node to its own community. Then, for each node, it considers merging it with each of its neighbours (for example, merging their communities) one by one. Merging occurs if this results in an increased value of the optimisation criteria, in this instance stability as given by equation 23.10. The process is repeated until no increase in stability can be achieved.

This algorithm is stochastic that it does not have a unique solution, and will not necessarily return the same partition on each run. Hence, for each Markov time t, we run Louvain's algorithm 1000 times. We then compare each pair of partitions found using the variation of information (Meilă 2007), and take the average across all pairs. If the mean variation of information is low at some t, the obtained partitions are similar, signalling robustness.

4.3 Community Structure in the Co-agglomeration Network

We apply the Markov stability framework for community detection on the network of industrial co-agglomeration. Communities in the co-agglomeration network correspond to groups of industries that tend to co-locate across a large number of cities.

We find (Figure 23.3A) multiple Markov times at which the computed partition is robust (for example times corresponding to local minima in the variation of information, *VI*). These correspond to node partitions on a range of scales, from many small communities to few large communities. In order to explore the network's community structure, we examine in more detail the partition corresponding to the local *VI* minimum at $t = 0.55$. This partition, which we name P_{14}, contains 14 distinct communities, of which 10 have ≥ 5 nodes (Figure 23.3C and 23.3D). At this Markov time, many communities have intermediate size, with eight having between 15 and 32 industries, while one community is quite large (75 industries).

Community B contains 38 industries, most of which are different types of services; in particular healthcare, education and finance. Owing to the tendency of service industries to require specialised labourers, we might expect that the observed co-location patterns are driven by labour linkages. Alternatively, community E contains seven industries involved in plastics manufacturing or machinery, and four industries related to motor vehicle manufacturing/repair. In this instance, we might hypothesise that customer–supplier relationships drive co-location patterns, as manufacturing/repair of motor vehicles may co-locate with plastic/metal manufacturing firms that supply necessary components. In the next section we investigate these hypotheses.

In order to visualise the multi-scale community structure of the co-agglomeration network, we choose seven representative partitions (for Markov times greater than $t = 0.55$) to construct a dendrogram (Figure 23.4). These range from the whole network (top) to the 10 community partition (leaves) discussed above (P_{14}). Since our partitions are not strictly hierarchical, we employ a simple majority voting scheme to assign communities to a parent community in a coarser partition.

In order to further demonstrate the robustness of the communities we detect in the co-agglomeration network, we compare these with communities detected in random networks derived from shuffling the edges in the co-agglomeration network. Figure 23.3A shows that for sufficient Markov time, the *VI* for the co-agglomeration network is significantly lower than for the shuffled networks. Note that in subsequent analysis, we only consider partitions for $t \geq 0.55$.

Finally, we ask whether our communities correspond to a better partition of the co-agglomeration network than administrative communities given by two-digit NAICS subsector classification. This latter comparison is particularly important as it validates to some extent the need for community detection to identify exogenous industry clusters. We see in Figure 23.3B that the stability (quality) of the detected co-agglomeration communities greatly exceeds that of the NAICS sector partitions, especially for $t \geq 0.55$.

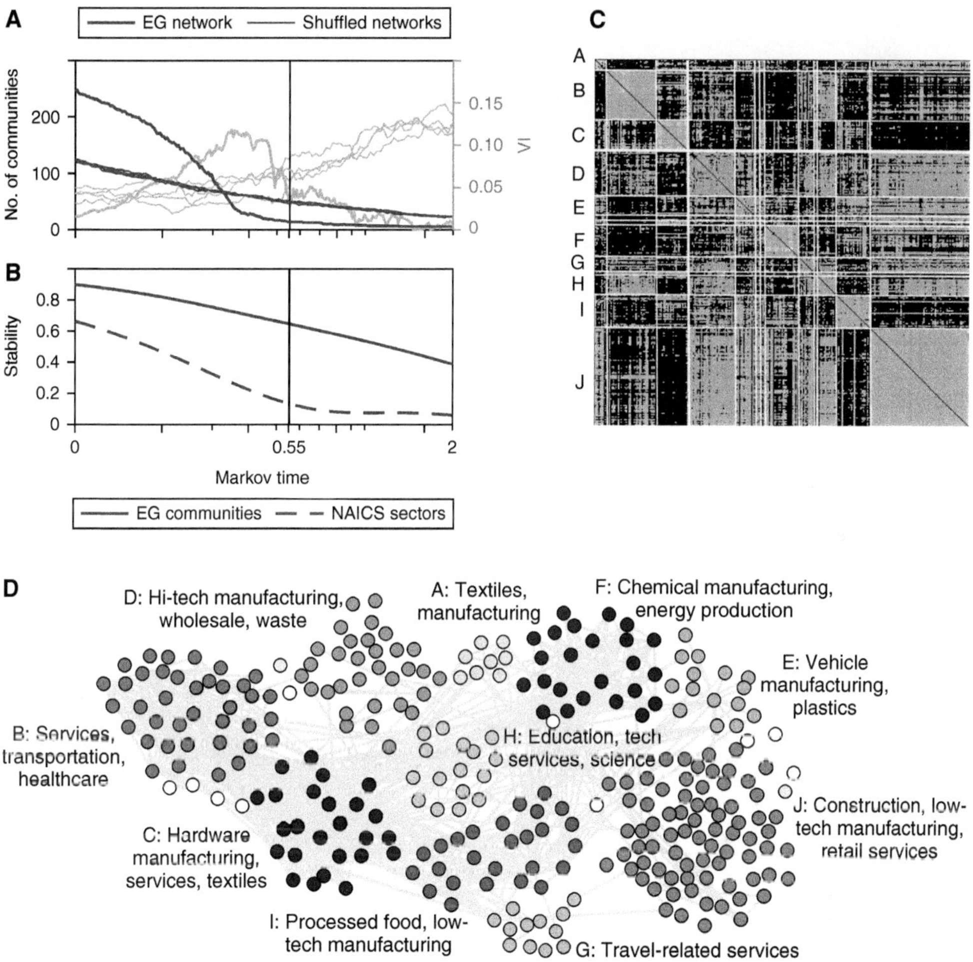

Figure 23.3 *Panel (A) The number of communities and VI as the Markov time increases (EG versus shuffled); (B) The stability (quality) of the communities (EG versus official sectors); (C) Heatmap of the co-agglomeration matrix, sorted by P_{14} and shaded by edge strength; and (D) Network visualisation, with nodes coloured according to P_{14}*

4.4 Industry Clusters Display Heterogeneous Preferences for Labour and Value-chain Agglomeration Channels

Previous analysis has been limited to measuring the relative importance of Marshallian channels either at the individual industry level or at the whole network level. Here, we argue for a meso-scale approach, with the intuition that dependence on each of the Marshallian channels varies not by individual industries, but by clusters of co-located industries, which can also be conceptualised as regions in a capability landscape (see section 1.5).

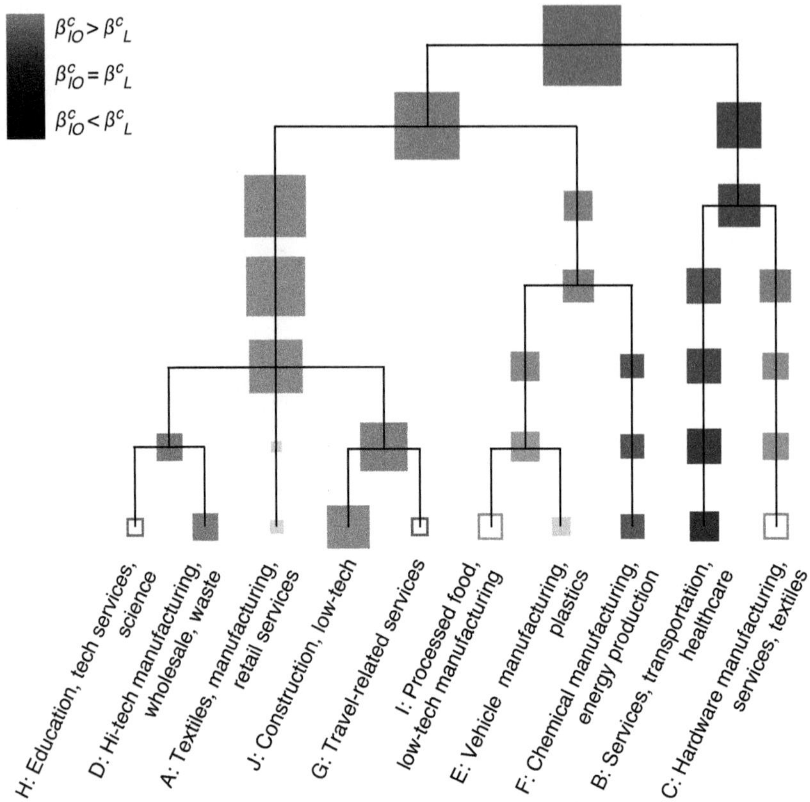

Note: Square size is proportional to community size, and colour corresponds to agglomeration channel dependence (open boxes represent communities in which neither β^c_{IO} nor β^c_L are significant).

Figure 23.4 Dendrogram of hierarchical community structure for seven partitions of the co-agglomeration network

Specifically, we seek to quantify the dependence of each community on the labour and value-chain agglomeration channels. That is, in place of equation 23.7, we let the respective channel coefficients vary by community c (instead of industry) via the regression:

$$EG_{ij} = \alpha_c + \beta^c_Z Z_{ij} + \varepsilon_{ij}, \tag{23.11}$$

for industry pairs $i, j \in c$, and channels $Z \in \{IO, L\}$. Essentially, the coefficients β^c_L and β^c_{IO} capture the relationship between co-location and labour similarities/value-chain linkages between industry pairs i and j within community c.

We estimate the coefficients in the OLS regression (equation 23.11) for various Markov times. We can visualise the progression from homogeneity to heterogeneity in the channel coefficients via the dendrogram in Figure 23.4. From this analysis, we see (from top to bottom) a shift from the simple labour versus input–output community partition to a more complex structure composed of heterogeneous network regions. At the first split, the network decomposes into two communities, one driven by input–output linkages (left) and the other driven by labour sharing (right). The input–output branch then splits into two sub-branches which are also input–output dominated, but a series of splits reveals heterogeneous channel importance at high resolution. For instance, we see that community E is dominated by input–output linkages (in line with our expectations), whereas community F (sophisticated chemical manufacturing and energy production) is dominated by labour sharing. The labour branch (top right) also splits into two sub-branches, one dominated by input–output and the other dominated by labour sharing. This latter community (driven by labour linkages), is community B, which is composed of service industries, as explored earlier.

4.5 Preference for Labour Pooling Correlates with Education Level of Workers

Here, we seek to investigate the relationship between the strength of labour pooling in co-agglomeration patterns and years of worker education at the cluster level. We expect that clusters dominated by service industries which tend to co-locate based on shared skill requirements will also tend to employ more highly educated workers.

Employing data from Integrated Public Use Microdata Series (IPUMs), we calculate for each industry the average number of years of education for workers, and then for each community the average number of years across industries (denoted ed_c). Then, we examine the relationship between ed_c and the community-level coefficients β^c_{IO}, β^c_L. Specifically, we perform regressions of the form,

$$\beta^c_Z = \alpha + b_Z ed_c + \varepsilon_c, \tag{23.12}$$

for all communities c in some partition and $Z \in \{L, IO\}$.

We perform this regression for many different partitions, corresponding to a range of Markov times. Specifically, we denote by P_k the partition corresponding to the earliest Markov time at which k communities are detected, and (independently) perform regression (23.12) for partitions $P_2, P_3, \ldots, P_{14}$. Rather than performing an OLS regression (or slope calculation in the P_2 case), we take a weighted least squares (WLS) approach to account for community size (that is, the diagonal entries of the errors covariance matrix are given by the relative sizes of the communities).

In Figure 23.5, we plot fits of the weighted regression for $Z = L$ (left) and $Z = IO$ (right). In isolation, each of the fits (pale lines) are not terribly meaningful (especially for high Markov time, given the small number of communities). However, we note that for all partitions the computed values of b_L are positive, and all computed values of b_{IO} are negative except in one case (P_7).

In order to pool the results of these partition-level fits, we perform a second set of regressions that considers all communities across all partitions. We first perform an OLS regression (column 1 of Table 23.2), in which all communities are weighted equally, and find a significant positive estimate for b_L as well as a significant negative estimate for b_{IO} ($p < .01$ for both). That is, we see a strong relationship between the strength of labour pooling in co-agglomeration patterns and years of worker education, as expected, as well as a strong negative relationship between the strength of customer–supplier relationships

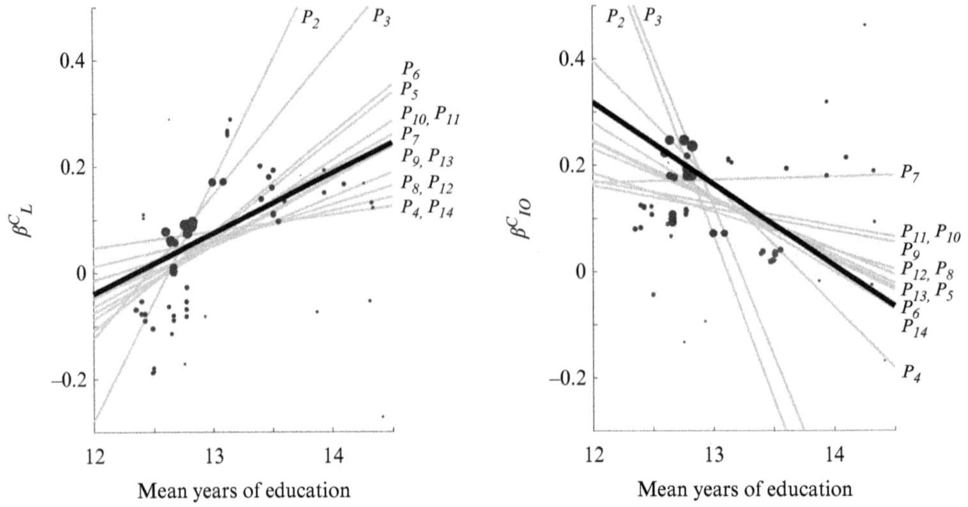

Note: Pale lines correspond to weighted least squares fits for each partition (see equation 23.12), while the dark line corresponds to a fit across all partitions (Table 23.2, col. 3).

Figure 23.5 Estimates of β^c_L and β^c_{IO} versus mean years of education for 11 different community partitions (points sized by community size)

in co-agglomeration patterns and years of education. Moreover, we find a particularly good linear fit ($R^2 > 0.4$) for the relationship between $\beta^c_L - \beta^c_{IO}$ and education level. Hence, industry communities that tend to have a very strong dependence on labour pooling relative to input–output linkages on average require more years of education.

We also perform a weighted version (WLSI) of this regression that equalizes the total error weights of each partition, and a second weighted version (WLSII) in which error weights are proportional to community size (partitions again have equal total weight), again finding significance at the 1 per cent level in both cases (columns 2–3 of Table 23.2). This latter pooled fit is shown via a dark line in Figure 23.5A and B.

Some of our estimates for β^c_L and β^c_{IO} were statistically insignificant (that is, we cannot reject the null hypothesis that these coefficients are zero). These insignificant estimates occur mainly in small communities in the finer partitions of the network (P_{14}), as seen in Figure 23.3. In order to address this issue, we perform versions of these three regressions (OLS and WLSI, WLSII) for the cases in which statistically insignificant values of β^c_{IO} and β^c_L are set to zero. For columns 4–6 (7–9) in Table 23.2, we set the coefficients that were statistically insignificant at the 10 per cent (5 per cent) level to zero. We find significant and comparable results across all cases.

5. CONCLUSION

Efforts to disentangle the relative importance of Marshallian channels to the location decisions of firms and industries in cities are building apace. In this chapter we reviewed two important studies which, in turn, sought to unravel the impact of individal

Table 23.2 Relationship between community-level L/IO estimates and mean years of education

Dependent variable: β_L	OLS	WLSI	WLSII	OLS†	WLSI†	WLSII†	OLS††	WLSI††	WLSII††
Mean yrs education	0.0715	0.0794	0.0959	0.0770	0.0780	0.1130	0.0705	0.0712	0.1037
	(0.0218)	(0.0221)	(0.0185)	(0.0184)	(0.0193)	(0.0197)	(0.0174)	(0.0183)	(0.0185)
N	80	80	80	80	80	80	80	80	80
R^2	0.1212	0.1420	0.2553	0.1837	0.1728	0.2960	0.1743	0.1629	0.2867
Dependent variable: β_{IO}	**OLS**	**WLSI**	**WLSII**	**OLS†**	**WLSI†**	**WLSII†**	**OLS††**	**WLSI††**	**WLSII††**
Mean yrs education	−0.1657	−0.1536	−0.1795	−0.1675	−0.1580	−0.0959	−0.1659	−0.1567	−0.0939
	(0.0304)	(0.0298)	(0.0304)	(0.0297)	(0.0294)	(0.0259))	(0.0301)	(0.0297)	(0.0263)
N	80	80	80	80	80	80	80	80	80
R^2	0.2757	0.2538	0.3090	0.2896	0.2707	0.1491	0.2799	0.2630	0.1401
Dependent variable: $\beta_L - \beta_{IO}$	**OLS**	**WLSI**	**WLSII**	**OLS†**	**WLSI†**	**WLSII†**	**OLS††**	**WLSI††**	**WLSII††**
Mean yrs education	0.2372	0.2330	0.1973	0.2445	0.2360	0.2089	0.2363	0.2279	0.1976
	(0.0280)	(0.0288)	(0.0244)	(0.0301)	(0.0309)	(0.0264)	(0.0310)	(0.0315)	(0.0268)
N	80	80	80	80	80	80	80	80	80
R^2	0.4797	0.4558	0.4558	0.4583	0.4279	0.4451	0.4277	0.4015	0.4098

agglomeration channels across all industries (Ellison et al. 2010) and for individual industries (Diodato et al. 2018).

Building on these studies, we constructed a hierarchical decomposition of the full set of industries into clusters based on co-agglomeration patterns, and estimated the relative importance of individual agglomeration channels for each cluster. We observe a transition from two clusters, one strongly related to the labour channel and the other to the customer–supplier channel, to a wide distribution of channel impacts at a finer partition. Finally, we find robust evidence that clusters exhibiting strong dependence on the labour channel employ more educated workers.

Our decomposition of the co-agglomeration network into regions dominated by one or another agglomeration channel has implications for diversification models using similar networks. There is evidence (Diodato et al. 2018) that those estimates can improve the predictive power of the models, and so we expect this to be a future avenue worth pursuing. From a policy perspective, a city seeking to diversify into related industries might use our cluster estimates to decide whether to focus on labour-orientated policies (for example building up the local skill base), or facilitating inter-firm transactions (for example policies aimed at promoting local value chains).

Much of this analysis focused on two of the three Marshallian channels, labour sharing and customer–supplier linkages. The third, knowledge spillovers, is more difficult to capture across a wide range of industries as patenting activity (normally used to capture research and development interaction between firms and industries) is concentrated in few industries. Future work might seek to investigate alternative methods and data sources for capturing knowledge spillovers across a broader range of industries and activities.

Finally, this analysis has been conducted using relatively recent data from the USA. Diodato et al. (2018) investigated the relative importance of labour and value-chain linkages across all industries between 1950–2010, observing a decrease in the strength of value-chain linkages over time. We would expect that over time, our framework might yield increasingly stronger relationships between labour linkages and agglomeration patterns (particularly within services clusters). Finally, while much academic study of urban economies focuses on highly developed cities (particularly US cities), little is known about whether these patterns are similar or substantially different for developing cities, where employment is often primarily informal (O'Clery et al. 2016).

ACKNOWLEDGEMENTS

NO and SH would like to acknowledge support from the PEAK Urban programme, funded by UKRI's Global Challenge Research Fund, Grant Ref: ES/P011055/1. MB would like to acknowledge support from the Oxford-Emirates Data Science Lab. The authors would additionally like to thank Dario Diodato and Frank Neffke for their contribution to the preparation of the data used in this book chapter, and for fruitful discussions.

NOTE

1. The measure becomes simply a rescaling of $cov(s_{ir}, s_{jr})$ when the sum of regional employment share is equal across regions.

REFERENCES

Acemoglu, D., A. Ozdaglar and A. Tahbaz-Salehi (2015), 'Systemic risk and stability in financial networks. American Economic Review', **105** (2), 564–608.
Battiston, S., M. Puliga, R. Kaushik, P. Tasca and G. Caldarelli (2012), 'Debtrank: Too central to fail? Financial networks, the fed and systemic risk', *Scientific Reports*, **2** (1), art. 541.
Beaudry, C. and A. Schiffauerova (2009), 'Who's right, Marshall or Jacobs? The localization versus urbanization debate', *Research Policy*, **38** (2), 318–37.
Behrens, K. (2016), 'Agglomeration and clusters: tools and insights from coagglomeration patterns', *Canadian Journal of Economics/Revue canadienne d'économique*, **49** (4), 1293–339.
Bettencourt, L.M., J. Lobo, D. Helbing, C. Kühnert and G.B. West (2007), 'Growth, innovation, scaling, and the pace of life in cities', *Proceedings of the National Academy of Sciences*, **104** (17), 7301–6.
Blondel, V.D., J.-L. Guillaume, R. Lambiotte and E. Lefebvre (2008), 'Fast unfolding of communities in large networks', *Journal of Statistical Mechanics: Theory and Experiment*, (10), P10008.
Catini, R., D. Karamshuk, O. Penner and M. Riccaboni (2015), 'Identifying geographic clusters: a network analytic approach', *Research Policy*, **44** (4), 1749–62.
Chang, S.J. (1996), 'An evolutionary perspective on diversification and corporate restructuring: entry, exit, and economic performance during 1981–89', *Strategic Management Journal*, **17** (8), 587–611.
Delgado, M., M.E. Porter and S. Stern (2016), 'Defining clusters of related industries', *Journal of Economic Geography*, **16** (1), 1–38.
Delvenne, J.-C., S.N. Yaliraki and M. Barahona (2010), 'Stability of graph communities across time scales', Proceedings of the National Academy of Sciences, **107** (29), 12755–60.
Diodato, D., F. Neffke and N. O'Clery (2018), 'Why do industries coagglomerate? How Marshallian externalities differ by industry and have evolved over time', *Journal of Urban Economics*, **106** (July), 1–26.
Duranton, G. and D. Puga (2001), 'Nursery cities: urban diversity, process innovation, and the life cycle of products', *American Economic Review*, **91** (5), 1454–77.
Duranton, G. and D. Puga (2004), 'Micro-foundations of urban agglomeration economies', in J.V. Henderson and J.F. Thisse (eds), *Handbook of Regional and Urban Economics*, Amsterdam: Elsevier, pp. 2063–117.
Ellison, G. and E.L. Glaeser (1997), 'Geographic concentration in US manufacturing industries: a dartboard approach', technical report, National Bureau of Economic Research, Cambridge, MA.
Ellison, G., E.L. Glaeser and W.R. Kerr (2010), 'What causes industry agglomeration? Evidence from coagglomeration patterns', *American Economic Review*, **100** (3), 1195–213.
Faggio, G., O. Silva and W.C. Strange (2017), 'Heterogeneous agglomeration', *Review of Economics and Statistics*, **99** (1), 80–94.
Fan, J.P. and L.H. Lang (2000), 'The measurement of relatedness: an application to corporate diversification', *Journal of Business*, **73** (4), 629–60.
Farjoun, M. (1994), 'Beyond industry boundaries: human expertise, diversification and resource-related industry groups', *Organization Science*, **5** (2), 185–99.
Fortunato, S. (2010), 'Community detection in graphs', *Physics Reports*, **486** (3–5), 75–174.
Frenken, R., F. Van Oort and T. Verburg (2007), 'Related variety, unrelated variety and regional economic growth', *Regional Studies*, **41** (5), 685–97.
Friedrichs, J. (1993), 'A theory of urban decline: economy, demography and political elites', *Urban Studies*, **30** (6), 907–17.
Fujita, M., P. Krugman and A.J. Venables (2001), *The Spatial Economy: Cities, Regions, and International Trade*, Cambridge, MA: MIT Press.
Girvan, M. and M.E. Newman (2002), 'Community structure in social and biological networks', *Proceedings of the National Academy of Sciences*, **99** (12), 7821–26.
Glaeser, E. (2011), *Triumph of the City*, London: Pan Macmillan.
Glaeser, E., H.D. Kallal, J. Scheinkman and A. Shleifer (1992), 'Growth in cities', *Journal of Political Economy*, **100** (6), 1126–52.
Hall, B.H., A.B. Jaffe and M. Trajtenberg (2001), 'The NBER patent citation data file: lessons, insights and methodological tools', technical report, National Bureau of Economic Research, Cambridge, MA.

Hausmann, R., C. Hidalgo, D. Stock and M.A. Yildirim (2014), 'Implied comparative advantage', HKS Working Paper No. RWP14-003, Harvard Kennedy School, Cambridge, MA.
Henderson, V., A. Kuncoro and M. Turner (1995), 'Industrial development in cities', *Journal of Political Economy*, **103** (5), 1067–90.
Hidalgo, C.A. and R. Hausmann (2009), 'The building blocks of economic complexity', *Proceedings of the National Academy of Sciences*, **106** (26), 10570–75.
Hidalgo, C.A., B. Klinger, A.-L. Barabási and R. Hausmann (2007), 'The product space conditions the development of nations', *Science*, **317** (5837), 482–7.
Howard, E., C. Newman and F. Tarp (2015), 'Measuring industry coagglomeration and identifying the driving forces', *Journal of Economic Geography*, **16** (5), 1055–78.
Jacobs, J. (1961), *The Death and Life of American Cities*, New York: Random House.
Jacobs, J. (1969), *The Economy of Cities*, New York: Random House.
Marshall, A. and M.P. Marshall (1920), *The Economics of Industry*, London: Macmillan.
Meilă, M. (2007), 'Comparing clusterings – an information based distance', *Journal of Multivariate Analysis*, **98** (5), 873–95.
Neal, Z. (2011), 'Differentiating centrality and power in the world city network', *Urban Studies*, **48** (13), 2733–48.
Neffke, F. and M. Henning (2013), 'Skill relatedness and firm diversification', *Strategic Management Journal*, **34** (3), 297–316.
Neffke, F., M. Henning and R. Boschma (2011), 'How do regions diversify over time? Industry relatedness and the development of new growth paths in regions', *Economic Geography*, **87** (3), 237–65.
Nelson, R.R. and S.G. Winter (1982), *An Evolutionary Theory of Economic Change*, Cambridge, MA: Belknap Press.
Newman, M.E.J and M. Girvan (2004), 'Finding and evaluating community structure in networks', *Physical Review E*, **69** (2), 026113.
O'Clery, N., E. Flaherty and S. Kinsella (2019), 'The network structure of knowledge flow', forthcoming.
O'Clery, N., A. Gomez and E. Lora (2016), 'The path to labour formality: urban agglomeration and the emergence of complex industries', HKS Working Paper No. RFWP 78, Harvard Kennedy School, Cambridge, MA.
Porter, M. (2003), 'The economic performance of regions', *Regional Studies*, **37** (6–7), 545–6.
Porter, M.E. (1990), *Competitive Advantage of Nations: Creating and Sustaining Superior Performance*, New York: Simon and Schuster.
Porter, M.E. (1998), 'Clusters and the new economics of competition', *Harvard Business Review*, **76** (6), 77–90.
Rosenthal, S. and W. Strange (2006), 'The micro-empirics of agglomeration economies', in R.J. Arnett and D.P. McMillan (eds), *A Companion to Urban Economics*, Oxford: Blackwell Publishing, pp. 7–23.

24. Cities, networks, polycentrism: examining the place of polycentrism in spaces of flows

Kathy Pain and Shuai Shi

INTRODUCTION

In a globalising world economy, cities have increasingly come to be defined not by the size of their built up area or population but by their functions in networks that connect them to capital flows that cross territorial boundaries (Rozenblat 2010). As Sassen (1991, p. 631) wrote it in her global cities book, 'massive trends toward spatial dispersion of economic activities at the metropolitan, national and global level, which we associate with globalisation, have contributed to a demand for new forms of territorial centralisation of top-level management and control operations'. As centralities for management and control functions in network enterprises operating at different spatial scales, globalising cities can be expected to be economically vibrant owing to their role in facilitating the circulation and accumulation of intellectual and financial capital. Moreover, according to protagonists of Jacobs (1984), the vibrancy of these cities spills over metropolitan boundaries to form prosperous globalising city regions (Scott 2001a, 2001b; Hall and Pain 2006). A new paradigm of city networks, functions and flows has therefore overtaken the traditional paradigm of places and territorial borders as the structure of the contemporary global space economy.

In the networked economy that has been evolving for half a century, it follows that city regions are differentiated not only by their physical pattern of urban development, but also by their functional pattern in network space. In consequence, city-region spatial form (morphology) and city-region Jacobsean economic expansion, while interrelated, are distinctive development processes that should not be confused with each other (Taylor and Pain 2007; Pain 2012). However, the concept of polycentrism has referred to morphological space without paying due regard to network space, functions and flows, for almost two decades in a policy to rebalance uneven regional economic development in Europe (European Commission 1999; Davoudi 2003; Duhr and Nadin 2005; Halbert et al. 2006; Pain 2008). Thus, European spatial planning strategy has promoted the development of polycentric urban regions (PURs) based on examples of regions populated by many similar-sized urban centres, such as the Randstad region in the Netherlands and the Rhine-Ruhr region in Germany, which are regarded as a superior spatial form to monocentric regions dominated by one centre such as South East England and the French Paris region.

In this chapter we argue for a more progressive definition of polycentrism that is informed by city-region network space patterns and how these map onto morphological space. Since capital and labour have continued to migrate to rapidly enlarging world cities that are new bases of the capitalist economy in twenty-first century economic globalisation, governments worldwide have looked to European policy on polycentrism to counter emerging regional development imbalances. Shedding fresh light on the concept of polycentrism has therefore, we argue, come to have more, not less, importance over time. Accordingly, in this chapter, we attempt to dissect the concept of polycentrism by examining interrelationships and distinctions between city-region network and morphological space development in mature and developing economy contexts. We address two questions: first, can attention to city-network space, functions and flows, usefully inform conceptualisation of city-region polycentrism; and, second, how should the concept of polycentrism inform planning policies in the context of city regions that are at different stages of connecting to economic globalisation?

In order to shed light on these hitherto under-researched questions, evidence from two city-region cases are drawn on in the chapter: The South East England city region in the densely urbanised and economically developed North Western Europe context and the Mid-Yangtze River (MYR) city region in the rapidly urbanising and developing economy context of China. The chapter first introduces the two city regions and their distinctive development contexts. Second, the concept of the city network in a globalising world economy, its relevance for the Jacobsean economic expansion process, and for polycentrism are considered. Third, the network approach to regional analysis is explained, some common methodologies for the investigation of inter-city network flows and the rationale behind different analytical approaches adopted in the case studies owing to their distinctive development stages are outlined. Fourth, the findings from the North West European study are reviewed before turning to consider evidence on recent city-region emergence in the Chinese case. Finally, distinctions between city-region economic expansion processes and polycentrism revealed by network space analysis in the two cases are reviewed, and the general implications for reference to polycentrism in planning policies are considered. It is concluded that the concept of polycentrism can only have relevance for policy to foster economic rebalancing when a theory of city-region space as an artefact of city networks is taken into account.

INTRODUCING THE CASE STUDIES: THE SOUTH EAST ENGLAND AND MID-YANGTZE RIVER CITY REGIONS

The South East England case study was undertaken as part of a larger 2003–06 research project conducted in eight North West European city regions to inform European Union (EU) policy on polycentrism aiming to rebalance the European space economy and to increase Europe's economic competitiveness in a global context (European Council 2000; European Commission 2004, 2009; Hall and Pain 2006; Pain and Hall 2008):[1] the Randstad, Netherlands; Central Belgium; Northern Switzerland; Rhine-Main, and Rhine-Ruhr, Germany; Paris Region, France; Greater Dublin, Ireland; and South East England (Hall and Pain 2006; Pain and Hall 2008). North Western Europe is one of the most urbanised and economically developed regions in the world, where city regions have become a prevailing urban form. The research was the first in the world to draw on Scott's (2001a) global city-region thesis to examine polycentrism using a network analysis lens. South East England was selected for study owing to its apparent monocentrism compared with the Randstad, Netherlands and the Rhine-Ruhr, Germany for example. South East England's primary city, London, is the most connected city in the world to dominant advanced producer services sectors and is referred to in European spatial policy as a global gateway city (European Commission 1999; Duhr and Nadin 2005) potentially capable of generating economically value-adding flows to peripheral regions with increased connectivity in the form of trans-European transportation networks (TENs), such as high-speed rail. It is therefore seen as important for Europe's global competitiveness but also a source of economic vibrancy that can contribute to rebalancing development from the European core to peripheral economies (Wallerstein 1974; Friedmann 1986).

Located in central China, the MYR region was selected 10 years after the South East England research, as a 2014–17 study[2] because, unlike the Yangtze River Delta city region that encompasses Shanghai or the Jing Jin Ji city region that encompasses Beijing, the MYR city region is seen in Chinese national spatial planning policy as rebalancing development between the coastal and the non-coastal inland regions, which is a national priority to reduce growing economic and social inequity associated with opening up China to globalisation. In particular, together with rising labour costs, competition from other emerging economies and recent global economic malaise, China initiated a crucial transition period from a labour-intensive and capital-driven economy to a more advanced and potentially resilient economy by the expansion of added-value activities. The MYR city region is thereby identified as a strategic growth region (growth pole) in China's economic transition owing to its established industrial base, well-developed infrastructure system, developed higher education, well-qualified labour, and so on, and its geographical location, which are seen as making it a conduit between the East and the West. As in the South East England case, the MYR study draws on Scott's (2001a) global city-region discourse and employs a series of network analysis methods to shed light on emergent city-network patterns in the process of regional development. However, in contrast to developed South East England, the MYR city region is a developing inland region in the context of China's rapid urbanisation and economic transition. The contrasting characteristics of the case study city regions enriches the discussion of progressive polycentrism.

CITY NETWORKS IN A GLOBALISING WORLD ECONOMY

In the globalising world, complex urban processes significant for economic development emerge from relations and flows between cities with few spatial constraints. In city-network space, economic growth is increasingly associated with connectivity to the space of flows, facilitated by advances in informational and communications technologies (ICT) and structured by the circulation of virtual flows of information, finance, and so on (Ohmae 1990, 1995; Castells 1996; Rozenblat and Pflieger 2010). However, local development outcomes in city networks remain intrinsically related to physical (geo-spatial) proximity, yet few studies have investigated the associations between network connectivity and flows, and spatial proximity that shape the space economy of city regions, and fewer still have considered the implications of the interplay between inter-city network processes and spatial proximity for city-region polycentricism.

City-region Economic Development in Inter-city Networks

The rationale behind the argument for the influence of geo-spatial proximity on economic development reflects the 'first law of geography' articulated by Tobler (1970, p. 234) as 'everything is related to everything else, but near things are more related than distant things'. Contradicting 'the end of geography' associated with increasing global financial integration (O'Brien 1992), according to this geo-spatial logic, distance remains a key determinant of economic location, since overcoming spatial constraints consumes energy and financial resources (Miller 2004). Thus, urban places that are physically close (proximate) to each other should be expected to have intensive functional interactions, for example, in employment and labour markets, that are advantaged by time-cost reductions, for example, travel to work and business travel, which will, in turn, shape the pattern of regional development as an outcome. Furthermore, spatial proximity is emphasised in new economic geography (Puga 1999; Fujita et al. 2001) and endogenous growth theories (Aghion and Howitt 1998) as a significant driver of economic growth. Further, it is argued that economic actors that are clustered in close physical proximity have access to flows of privileged information and knowledge transfer, and benefit from co-learning associated with agglomeration economies (Boschma 2004).

This geo-spatial proximity rationale appears at first sight to contrast with that of city-network theory which emphasises that, in the context of globalisation, technological developments reduce the cost of overcoming spatial constraints to a substantial degree. In particular, the rise of the knowledge economy is argued by globalists (Cochrane and Pain 2000) to generate intensive virtual flows. Information, specialised knowledge and financialised capital associated with the activities of new economy, knowledge-intensive cross-border financial and linked business network enterprises (Wood 1991), known as advanced producer services (APS), are subject to few spatial constraints. As Castells (1999 p. 295) claimed, urban spaces have become the 'material arrangements' allowing for 'simultaneity of social practices without territorial contiguity'. He asserted that new-economy activity, and specifically APS office network flows, transcend the space of places and that, in the current network society, trans-boundary flows of products, services, labour and knowledge interconnect cities and markets regardless of territorially defined place boundaries (Lizieri and Pain 2014; Pain and Van Hamme 2014; Coe and Yeung 2015). Resultant synergies between cities (Massey et al. 1999) indicate that a purely geo-spatial explanation of economic development based on city physical proximity, presents a partial view; contemporary urban spatial relations, which are now also characterised by complex virtual flows, must also be taken into account. Distinctively, this chapter considers the congruence of urban geo-spatial distance and urban economic function-related inter-city network flows in developed and developing city-region contexts to shed light on the question of polycentrism.

The Polycentric City Region in Network Space

The city-region scale of analysis provides valuable insights into the normative relevance and practical utility of the concept of polycentrism as a spatial planning tool to address uneven development. In Europe, associated with enlargement and political ambitions for territorial cohesion based on the French principle of *aménagement du territoire* (spatial planning), polycentrism has been actively promoted as a policy objective throughout the European urban system (European Commission 1999, 2005, 2008, 2012). As applied in European spatial strategy, the term polycentricity has referred to the outward diffusion of development from major commercial cores and metropolises to smaller cities in order to rebalance the uneven distribution of urban assets at city, regional, national and EU-wide scales (Davoudi 2003; Hall and Pain 2006). 'Mega-city regions are a series of anything between 10 and 50 cities and towns physically separated but functionally networked, clustered around one or more larger central cities, and drawing enormous economic strength from a new functional division of labour' (Hall and Pain 2006, p. 3). Polycentrism thereby refers implicitly to the spatial arrangement of economic development and thus to the convergence of physical proximity and network distance implied by the geo-spatial and network space theoretical rationales.

As Watts (1999) wrote in *Small Worlds: The Dynamics of Networks between Order and Randomness*, near (proximate) and distant can interact to create complex city-network interconnections and these interconnections are particularly relevant for city-region development. As Scott (2001a, p. 18) noted, 'global city-regions come to function increasingly as the motors of global economy, as dynamic local networks of economic relationships caught up in more extended worldwide webs of interregional competition and exchange'. It is argued in this chapter that understanding local (near) and distant (worldwide) interactions by adopting a network approach to city-region analysis can therefore shed valuable light on the interrelationship between functional and geo-spatial development processes, uneven development and polycentrism.

Unravelling the Concept of Functional Polycentrism

As Duranton and Puga (2001) argued, cities are increasingly recognised by their economic functions and it is the spatial patterns associated with the functional relations between cities in city networks that are the prime interest of this chapter. As first identified in Harris and Ullman's (1945) multiple nuclei model, a polycentric urban development pattern is characterised by numerous urban centres of similar size; however, as already discussed, city business network functional locations and exchanges of people, goods and information that define functional polycentricity as opposed to morphological polycentricity is critically important for space economy vitality at a city-region scale. Later, Van Oort (2010, p. 726) argued that, 'The outcome of functional polycentricity is a considerable regional cohesion in personal, occupational and corporate relationships of people, organisations and firms that transcends the boundaries of the traditional metropolitan areas'. Therefore, the pursuit of polycentrism in spatial planning policy demands a reconfiguring of the city-region urban hierarchy (Christaller 1933 [1966]) with urban service functions dispersed out from higher-order central cities to lower-order cities to reshape the space of places. However, accomplishing this aim in the contemporary world of city networks is dependent on Castells's (1996) space of flows.

Globalisation and the rise of the knowledge economy have transformed contemporary physical urban spaces into data factories where circulating labour, capital, information, commodities and human flows generate ideas, innovations, knowledge, entrepreneurships, and so on, that facilitate urban growth as an outcome. In addition to the massive scale and intensity of these flows, they are distinctive in their content, quality, value-added, distance, and directional and temporal characteristics in different local contexts. The heterogeneity of city-network flows is reflected in the characteristics of the local spaces where globalisation hits the ground. Meanwhile, the circulation of large-scale heterogeneous flows gives rise to inter-city networks in which cities have different structural positions and network roles (such as a commanding role or a broker role). That is, the circulation of heterogeneous flows results in local spaces also characterised by heterogeneity in the form of diverse city-network functions and positions and development dynamics which must inform the issue of city-region polycentrism. The South East England and MYR city-region different case settings – the former characterised by APS sectors identified by Friedmann (1986), Sassen (1991) and Castells (1996) as the drivers of a thriving knowledge economy versus the latter characterised predominantly by labour-intensive manufacturing – enable polycentrism to be considered in relation to economic transition from a capital driven to a knowledge-based economy.

CITY NETWORKS AS A LENS FOR REVEALING CITY-REGION POLYCENTRISM

With ICT technology breakthroughs and the development of high-speed modern transportation, the spatial constraints on inter-city flows and relations in economic activities have been much reduced (Friedmann 2005). The rise of the knowledge-based advanced service economy has generated intense virtual flows, capital mobility, knowledge exchange and information sharing that are dynamically interconnecting cities across territorial boundaries worldwide. Therefore, while the network argument relates to central place theory (Taylor 2009), the two theories, the Christallerian hierarchy and the concept of urban networks, are contraposed. Central place theory and Harris and Ullman's (1945) view of 'the nature of cities', which focused on hierarchical centre–hinterland relations have become less relevant to the reality of how the space economy maps onto city regions under the influence of globalisation. As Taylor et al. (2010) argued, central place thinking emphasises external urban relations as a local-hinterland process whereas city-network thinking recognises urban relations as a non-local city flow process. A network lens is thereby required to complement central place theory in the analysis of the functional polycentrism of city regions with distant as well as local economic relations. Central place hierarchies have been the traditional basis for understanding external urban relations. However, in contemporary studies of these relations, a new emphasis on urban networks has emerged. Instead of either abandoning or extending central place thinking, it is here treated as representing one of two generic processes of external urban relations. Townness is the making of 'local' urban–hinterland relations and 'city-ness' is the making of 'non-local' interurban relations. Central place theory describes the former through an interlocking hierarchical model. This chapter proposes a central flow theory to describe the latter through an interlocking network model. The key difference is the level of complexity in the two processes (Taylor et al. 2010).

The City-network Approach: Some Theoretical Reflections

As early as 1964, Berry claimed that urban geography mainly focused on 'cities as systems within systems of cities', highlighting the need to look at the relations between cities in a network perspective that pays due regard to complementary urban functions and synergistic business links. In contrast to the emphasis on the local industrial base in competitiveness theory, Burt (2009) illustrated that competition is a relationships issue as opposed to a competition between cities themselves, and this observation has been supported by interviews with APS business actors based on a city-network approach (Taylor et al. 2003; Hoyler and Pain 2001). Meijers (2007) also found that connections between cities tend to be more complementary than competitive, in order to take advantage of scale economies, knowledge exchange and synergies. Meijers (2007) further explained that horizontal links allow collaboration, while vertical links generate complementarities. Thus, it is argued by numerous authors that establishing inter-city connections is important for urban actors to develop cooperative business connections, to absorb external knowledge and to boost innovation and regional productivity as an outcome (Hakansson 1987; Powell et al. 1996; Nicolini et al. 2003; Taylor et al. 2003; Bathelt et al. 2004; Mahroum et al. 2008; Breschi and Lenzi 2014; Huggins and Thompson 2015).

The importance of spatial proximity for business connections is reflected in the significance for headquarters location decisions of access to deep infrastructure that includes an array of specialised business services suppliers, labour and skills, and international air connectivity between cities (Button et al. 1999; Taylor et al. 2003; Bel and Fageda 2008; Grubesic et al. 2008). According to Meijers (2007), simultaneous vertical and horizontal network links shape urban development processes since horizontal links enable collaboration while vertical links generate complementarities.

The new economic geography highlighted the significance of relations between cities in shedding light on urban space economy patterns (Krugman 1991; Fujita et al. 2001), however the focus is mainly on the effects of spatial proximity and the local embeddedness of economic activities, focusing particularly on trade costs and market access (Redding and Venables 2004). New economic geography perspectives have not paid attention to the multidirectionality and borderless characteristics of inter-city flows in a network context. To address the void of analysis of intangible networks constructed by multidirectional, borderless flows, the notion of calculative network capital is proposed by some authors as a means of pinning down network outcomes in corporate interactions (Huggins 2010; Huggins and Johnston 2010; Kramer et al. 2011; Smith et al. 2012), which provides an avenue for inter-city network analysis. The notion of calculative network capital argues that this capital is embedded in functional positions that are determined by network node (city) multidirectional ties or links, and the attributes of those links. It is argued that a network is not only a structure but also a source of strategic capital which can generate opportunities and constraints for city-network actors. This understanding of city networks supports Bathelt and Taylor's (2002) assertion that a city's strategic network position is not only an indicator of economic performance but also an important endogenous factor driving economic development. According to Jacobsean logic, this economic development can spill over to proximate urban places, and we argue here that this may generate functional polycentrism at the city-region scale.

Common Methodologies for the Investigation of Inter-city Network Flows

There are several ways of investigating inter-city flow networks and their contribution to economic growth and regional development methodologically. Three common approaches are global production network, interlocking office network and social network analysis. Resonating with global value chain (GVC) theory, in global production network analysis, territories are seen as integrated transnationally through value-added production processes (from decision-making, design, production, distribution, through to sales and services) that have functional production location outcomes in a globalised economy. Actors are theorised in the global production network framework not as individual agents per se, but as a constitutive part of the wider network. The approach focuses on intra-firm chained linkages that add value to final products, thereby reflecting a microeconomic industrial production perspective. In contrast, this chapter is concerned with how flows in city networks influence city-region development and their relevance for polycentrism. Thus, the case studies selected for examination here used other network analysis approaches, chosen to suit their different development stages in economic globalisation: the interlocking office network model approach in South East England and the social network analysis approach in the MYR region.

Taylor's interlocking office network model, employed in world-city network analysis (Beaverstock et al. 1999; Taylor 2001, 2004; Taylor et al. 2002), is based on Friedmann's (1986) world city hypothesis, which emphasised functionality and big cities' commanding roles in a world city hierarchy, and Sassen's (1991) global city theory, which emphasised the significance of APS sectors with high value-adding activities in forming global cities that have outstanding connectivity (Castells 1996). Since APS firms need to provide specialised services to clients, they require access and closeness to skilled labour pools, information-rich and prestigious environs, superior offices, and transport and telecommunications infrastructures, all of which are predominantly found in big cities (Harrington and Campbell 1997); second, APS firms have formed their own global service networks as a new force among multinational corporations (MNCs) (Moulaert and Daniels 1991; Aharoni and Nachum 2002; Harrington and Daniels 2006; Derudder et al. 2010). As Scott (2001a, 2001b) argued, the concentrated service functions in global cities not only represent the geography of powerful command positions in the global economy, but disseminate their power to counterparts through their international operations, and they are thereby indicators of the presence of vibrant metropolitan economies.

Drawing on social network analysis, the interlocking office network model uses data on office functions in five APS network sectors – banking/finance, accountancy, law, advertising and management consulting – to map the interlinkages between cities based on the assumption that the strength of inter-city work flows is determined by organisational network office functions – headquarters, regional offices, local offices and back offices (Taylor 2001, 2004; Taylor et al. 2002; Derudder and Taylor 2016). Computing the strategic importance of cities in all APS networks provides service values of cities hosting APS clusters as an approximation of inter-city relations, and thereby city-network structures, revealing otherwise invisible patterns of city connectivity that are an outcome of the dual spatial dynamic of office dispersal and concentration (Sassen 1991) and that generate inter-city networks.

In contrast to conventional statistical analysis, social network analysis which underpins the interlocking office network approach, is theoretically rooted in sociological enquiry that looks into the patterns of relationships between social actors. Associated with globalisation and the advent of the information age, myriad actors are involved in relations interlinking them in complex and heterogeneous ways, which allows the realm of social network analysis to be extended to complex social systems at flexible scales, from the interpersonal to the international. The approach is therefore scale-free, which is advantageous in analysing multidirectional city interactions including functional polycentrism.

Furthermore, intertwined with the rapid development of computing technology, the visualisation of social networks becomes a powerful tool for conveying complex information using graph theory. In the world-city network approach, the network has three levels: APS firms are the economic actors in the network; the cities they use in their international operations are the nodes in the network; and the world economy constitutes the supranodal network level. The global strategies of the firms that are operationalised through the nodes (cities) in which they have offices, generate the interlocking world-city network (Taylor 2001).

Rationale behind Different Analytical Approaches Adopted in the Two Case Studies

The use of inter-city APS office networks as a proxy for city-network connectivity as opposed to actual inter-city flows between cities has been subject to critical debate (see, for example, Nordlund 2004) so in the North West European study of polycentrism in which South East England was one of eight city-region cases, additional data on daily travel-to-work flows, virtual flows of information (email, telephone and electronic) and face-to-face knowledge exchange within and between offices and locations were used alongside interlocking office network analysis (Pain and Hall 2008). First, the functionally defined city region, encompassing all the areas having regular daily relationships based on commuting flows, was analysed. Second, polycentric patterns were revealed using an interlocking office network approach. Quantitative data on the office locations and importance of APS firms operating at the city-region, national, European and global scales were collected to identify regional inter-city network functional relations (Taylor 2004; Taylor et al. 2010). Third, questionnaires were used to collect data on APS firms' patterns of virtual communication to identify associated information flows (Pain and Hall 2008). Fourth, 600 in-depth interviews with senior APS actors were used to elicit data on the quality and intensity of relations between city region-located offices and others in their national, European and global networks as a proxy for the quality and intensity of inter-city flows and relations.

A pragmatic problem for the MYR case study research was that APS are not sufficiently representative of its city economic structures which are undergoing early stage economic transition to new economy enterprise in globalisation. Manufacturing remains the predominant industry and, in consequence, data on APS sectors at the firm level are neither available nor relevant metrics. At the empirically fuzzy end of the world-city network (Markusen 2003), inter-city flows more directly representative of functional polycentrism were direct inter-city capital flows in merger and acquisition (M&A) activities. First, M&A deals involve large capital flows that are free from spatial constraints, resonating with Castells's (1996) space of flows theory. Secondly, in contrast to one-off trade deals, M&A deals are characterised by control transfer and long-term interactions. Thirdly, M&A deals are characterised by intensive interactions and valuable spillovers, such as information exchange, technique sharing, elite communications and management learning (Freeman 2004). Fourthly, M&A deals not only operate between target and acquirer companies, but are also an important component of APS activity involving, in particular, financial and legal services, resonating with Sassen's (1991) global city theory. In network space, M&A projects explicitly reflect both long-term elite interactions, technology sharing, production and management modes, and capital mobility between cities across territorial boundaries.

As Böckerman and Lehto (2006) argued, M&A projects involve value evaluation, monitoring, sharing assets and synergy operations and thereby, as Rodríguez-Pose and Zademach (2003) found, M&A deals are concentrated in global and globalising city regions, driving regional economies as an outcome. Merger and acquisition projects represent capital mobility between network enterprises and are important as generators of regional aggregate capital flows and spillovers. Merger and acquisition deals grew continuously in the MYR region since 2003, peaking in 2010, and were followed by a recession period until 2014. Social network analysis from which interlocking office network analysis derives (Smith and Timberlake 1995; Taylor 2001, 2004) was the selected method used for analysis since it does not rely on data on any specific sector and can investigate any city-network relations constructed by direct flows. The approach can be adapted to illustrate complex multidirectional flows and can be used to identify specific city functions not quantified in the interlocking office network model.

Merger and acquisition data from 2003 to 2014 from the Zephyr database (Bureau van Dijk 2015) which is the most comprehensive source on worldwide M&A deals and is updated hourly, were used to establish the MYR inter-city network, based on source and target companies, and source and target cities, deal types and values, and industrial classifications including manufacturing and APS sectors (property services, banking, and insurance and financial services). In total, 1238 M&A deals from 2003 to 2014 involved 42 cities in the region, forming a city-by-city network. All deals are valued above 1 million Chinese yen and deal types include mergers, acquisitions, minority stake (greater than 2 per cent) and capital increase. The key criterion for inclusion of deals in the analysis was that they involve the transfer of a business and operations in the M&A process.

In conclusion, since the global cities literature pointed increasingly to the significance of network enterprises in the construction of inter-city capital flows and networks (Friedmann 1986; Sassen 1991; Castells 1996), investigation of the networks constructed by economic agents and circuits of inter-city flows using data relevant for a city region's development stage in economic globalisation, has become an important research area for shedding light on functional and spatial development patterns, and thereby polycentrism in the South East England and MYR region cases.

CITY-REGION EMERGENCE PROCESSES IN NETWORK SPACE: MORPHOLOGICAL OR FUNCTIONAL POLYCENTRISM?

The South East England city region is dominated by Greater London which is 25 kilometres in radius despite being bounded by a 'green belt' that has largely successfully contained the expansion of the metropolitan area since the mid-twentieth century. London's population of over 9 million represents just under half the city region's population. However, South East England is one of the densest city regions in North West Europe. Outside the London boundary, 50 smaller cities and towns ranging in size from 79 000 to 600 000 in population have shown consistent and strong service-sector economic growth in the past half century. South East England employment shrank in the 1980s, especially in London, but rebounded in the 1990s, growing by 33 per cent, with nearly a 25 per cent increase in London. Employment is heavily concentrated in the service sector, especially the APS sector; 80 per cent of all employment is in services, with banking and financial services representing more than a quarter. Significantly, financial and linked business services are strongly represented in cities and towns outside London, especially to the west of the city region.

Relative to developed South East England, the MYR city region has only recently become the location for large-scale urbanisation and economic development. Since economic reform and the introduction of the open-door policy in China which allowed its integration into the capitalist world economy, the national urbanisation rate has boomed, rising from 18.62 per cent in 1979 to 54.41 per cent in 2014. As recently as 2016, Wuhan in the MYR city region became the only city in central China with a population exceeding 10 million. With its strong manufacturing industrial base, the city region represents more than half of central China's economic activity against several economic indicators: 53.19 per cent gross domestic product, 55.05 per cent fixed asset investment and 51.22 per cent fiscal revenue (NBS 2014).[3] In contrast to the national decline in growth since the 2008 global economic crisis, the city region's economic development has maintained a double-digit growth rate (NBS 2014). However, the service economy remains relatively undeveloped in comparison with that of South East England, especially in APS. Despite these stage-of-development differences, both case studies have targeted analysis on inter-city flows associated with the activities of economic network enterprises, in order to shed light on city-region functional and associated spatial development patterns. The results therefore reveal the extent to which the regions are characterised by either functional or morphological polycentrism.

Learning from the South East England Case

Important for addressing the question, 'Can attention to city-network space, functions and flows usefully inform conceptualisation of polycentrism?', the European research identified two distinctive city-region development processes: process A, 'mega-city region economic expansion'; and process B, 'mega-city regions of proximate cities' (Taylor and Pain 2007; Pain 2008).

Process A is distinguished by a polycentric functional pattern associated with multi-sector APS clusters across South East England, which generate Jacobsean economic enveloping and upgrading of an extensive city region surrounding London. London's high multi-sector global APS network connectivity is transmitted across the larger space, forming a functionally interconnected city region with smaller centres that are growing economically at a similar rate to London. Jacobs claimed that city economic forces can be very powerful and, in the case of South East England, process A involves a simultaneous diffusion of markets, jobs, production, technologies and capital across the city-region space. Process A contrasts with process B which is distinguished by morphological polycentricity in which cities are of a similar size but are characterised by APS sectoral specialisation, and a less globally networked city region. Less globally connected first cities are failing to spread their prosperity to a wider regional scale. Jacobs (1984) argued that city regions are not defined by natural boundaries but instead they are the artefacts of cities whose economic energy dictates their boundaries. The South East England case demonstrates that, in a globalised economy, city energy comes from global city networks and it can be spread through global cities across an extensive area, producing functional complementarities between cities and towns of different sizes, with functional polycentricity as an outcome.

Related to the distinction between the two city-region development processes identified, a finding from the research relevant for the question of how polycentrism should inform planning policies in the context of city regions that are at different stages of connecting to economic globalisation, was that the morphologically polycentric city regions studied did not benefit from a balanced distribution of economic activities. Interviews with APS firms in offices across the eight city regions revealed that office networks at the regional scale are not necessarily representative of high-value information and capital flows between centres (Pain and Hall 2008). All eight city regions, whether morphologically or functionally polycentric, were found to have only one city with intense cross-border global flows, illustrating that city functional primacy versus city functional polycentricity is invisible if only morphology is considered. For example, the Rhine-Ruhr was the most morphologically polycentric city region studied, but this spatial pattern does not translate to a functional pattern and functional polycentricity. Similarly, in South East England, city-region functional polycentricity associated with London's supreme global APS network connectivity belies its morphological description in European strategy as a monocentric urban region owing to London's size. Nevertheless, the South East England geographical distribution of networked, economically buoyant centres beyond London has a distinct western orientation. Former government growth areas, that is, Milton Keynes–South Midlands and London–Stansted–Cambridge to the north-east of London, contain important networked multi-sector APS clusters which give them connectivity at different spatial scales. However, the Thames Gateway to the east and Ashford to the south-east of London, remain relatively disconnected from APS network activities and connectivity. Buoyant economic centres include the Reading–Thames Valley Corridor to the west of London which has the second-highest gross value added in England after London. Therefore, the study found that morphological polycentricity cannot reveal the underlying economic functionality and complementarities of city regions. Attention to city-network space, functions and flows is needed to inform conceptualisation of polycentrism, demonstrating the need for more network analysis. The paradigm of city-region polycentrism required to inform policy is the combination of functionality and complementarity.

Against this developed-world research backdrop, what can be learned about the concept of polycentrism from a recently emergent city region that is undergoing political economy change as China is opening up to the capitalist economy and economic globalisation at a rapid pace?

Learning from the Mid-Yangtze River Region Case

The results from the MYR case show that the number of M&A deals in the city region has been increasing between 2006 and 2014 and is gradually interconnecting its cities.The city-network patterns in 2006 and 2014 illustrate distinctive phases during the network evolution process. In the endogenous phase, most inter-city network links have been associated with city-region primary cities, Wuhan, Changsha and Nanchang, while there have been relatively few links between medium-size cities and there have been several isolated cities located at the geographical periphery of the city region. However, network links between medium-size cities have been increasing over time and the number of isolated cities has been decreasing during the endogenous phase, even though the process of change has been evolving slowly. The final pattern of the city network by 2014 showed that the whole city region had become more integrated than before, characterised by a transition from a centralised network to a balanced city-region network. Thus, Mid-Yangtze city-network development is moving from a first endogenous phase, which is characterised by the self-growth of cities, to a second interactive phase that is characterised by more multidimensional and complex inter-city relations (see also Asheim and Gertler 2005).

In addition, it is noted that although Wuhan is the most outperforming city in the region, its leading role is challenged by Changsha and Nanchang, which indicates to some extent a pattern of polycentricity. Additional investigation of the inter-city network included other city characteristics, ranging from global network to individual city-network attributes, using network centrality analysis. The findings revealed that having similar sectoral attributes is not required to initiate city interactions. Large cities with more central network positions and more intensive network links are more heterogeneous. With the evolution of the network, the economic functions and network positions of cities are changing. For example, Changsha, which was originally an indigenous hub for the Hunan sub-region, has evolved into a leading city for the whole city region with the development of more external functional connections to cities in other sub-regions. Thus, the pattern of network centrality is shifting and is dependent on 'when and at what particular relations we look' (Timberlake 2010, p. 2). In conclusion, it is found that the MYR city region is a dynamic, complex centralised network space full of multidirectional and transboundary flows where network power is concentrated on primary cities and distributed across distinctive network functions.

Although the interlocking network approach that is based on the office functions of APS firms was not adopted to estimate the city region's network connectivity owing to its stage of economic development, APS sectors are highly representative of the knowledge economy and are thereby important for the future city-region economic upgrading (the Jacobsean process A). The reciprocal association of manufacturing and APS sectors is expected to be the key to accomplishing economic transition in inland areas of China, and the MYR city region is expected to play a key part in this transition. Advanced producer services sectors provide specialised services to producers, which add value to manufacturing production, which is the foundation service market that will facilitate APS sector development in the city region and in China as a whole, therefore, APS spatial agglomeration patterns and their correlations with manufacturing were examined. A Pearson correlation matrix and principle components analysis showed that manufacturing and APS sectors (logistics and telecommunications, financial services and banking, real estate and research services) are correlated positively at the significance level.

In conclusion, based on the evidence from network analysis using data on M&A capital flows, the city region is not characterised simply by the economic functional expansion process B identified in Europe: 'mega-city regions of proximate cities'. There is evidence of an early-stage Jacobsean 'mega-city region economic expansion' (process A) and emergent functional polycentrism as an outcome. Through M&A deals, capital (the main driver of Chinese economic growth) is flowing from primary to secondary cities, stimulating an increase in APS activity beyond primary cities. However, in the MYR region inter-city network, a high proportion of deals is initiated in primary cities, particularly for high-value deals above 10 million yen, and big deals are transacted by localised APS services located in those primary cities (Taylor et al. 2003). This finding resonates with empirical work by Yang and Yeh (2013) in which they found that, in contrast to manufacturing activities, APS firms are more concentrated in Chinese high-tier cities. In contrast, the city-network links examined in South East England represent relations and information flows between APS actors and offices located in cities and towns outside London. They are therefore distinct from the MYR city-region network links represented by M&A capital flows assisted by APS actors who are concentrated in primate cities. Nevertheless, APS-assisted advances in manufacturing production could in future potentially spur the concentration and dispersion of APS offices in the city region to meet increasing market demand for specialised financial and business services. This dual concentration and dispersion space economy dynamic, first noted by Sassen (1991), could potentially advance city-region economic expansion process A and associated functional polycentrism qualitatively. Finally, we consider the two questions posed at the start of the chapter.

POLYCENTRISM EMBEDDED IN INTER-CITY FLOW NETWORKS AND POLICY IMPLICATIONS

First, based on the two case studies, what conclusions can be drawn about whether attention to city-network space, functions and flows can usefully inform conceptualisation of polycentrism?

The findings from both cases illustrate that distinctions between city-region expansion processes revealed by network analysis are relevant for polycentrism conceptualisation. Network space is dynamic, heterogeneous and complex in nature (Baum et al. 2000) making networks, functions, flows and complementarities critical for the meaning of city-region polycentrism. Increasingly, the relevant focus for policy is no longer the city; it is the city region (Kloosterman and Musterd 2001, p. 627) which is generated through Castells's space of flows and is highly networked through city-network links at different scales (Jonas 2011). Instead of focusing on morphological polycentricity, network thinking should be highlighted to shed light on functional polycentrism in the process of city-region development. The case studies demonstrate that the city networks created by cross-territorial economic activities (APS activities and M&A deals) are intrinsically embedded in city regions. Associated with deepening globalisation, these urban interactions engage advanced productive factors (human resources, knowledge and information) intangibly, representing a transformation from a two-dimensional place to a three-dimensional space. These advanced interurban flows are no longer tied to Tobler's first law of geography but overcome spatial constraints at low cost, facilitated by modern transportation and information technology. Consequently, circulating interurban flows construct a complex network space in which cities have different roles and functions that give rise to complementary inter-city relationships and to functional polycentricity at a city-region scale as an outcome. We conclude that attention to city-network space, functions and flows can usefully inform the conceptualisation of regional polycentrism.

Second, how should the concept of polycentrism inform planning policies in the context of city regions that are at different stages of connecting to economic globalisation? Corresponding implications of the results from the investigation of network space in the two city regions are that policies should be oriented to strategic network planning underpinned by cross-territorial cooperation as opposed to competition. Active institutional networks and collaboration are vital for the management of complex emerging regional spaces of flows that do not map onto established administrative areas and territorial boundaries. The multi-directionality of inter-city flows identified in both cases, points to the need for up-to-date network flow-tracking systems. Furthermore, such systems need to be populated by data that are relevant to specific city-region economic development stages in order to assist strategic planning and to inform territorial cooperation based on otherwise invisible city economic complementarities. Such systems could potentially more accurately pinpoint economic expansion processes at a granular level in different city regions in Europe than has previously been possible, as well as in other parts of the world. Policy aimed at city region rebalancing based purely on a morphological polycentric pattern contradicts actual inter-city flows generated in economic network space. In addition, intertwined with sectoral heterogeneity as an important characteristic of city interaction, network flows and functional polycentricity, the assets and capital that are embedded in urban locales will be important for network development. The findings from the case studies reviewed in this chapter therefore demonstrate that by taking into account the congruence of geo-space and network space in economic globalisation, the concept of polycentrism can better inform planning policies for city regions that are at different development stages. Given the different development stages of city regions worldwide, policies aiming to rebalance the spatial distribution of economic activity also need to take into account the characteristics of dominant city-region industries. For manufacturing-dominated regions, rebalancing should give weight to the promotion of city interlinkages in chained value-adding processes while, for APS-dominated regions, policies should be oriented towards recognition of and support for horizontal interactions across high value-added APS sectors. We conclude that the concept of polycentrism can only have relevance for policy to foster economic rebalancing when a theory of city-region space as an artefact of city networks is taken into account and an appropriate network approach is applied in analysis.

NOTES

1. 'POLYNET: Sustainable Management of European Polycentric Mega-City Regions' led by Peter Hall and Kathy Pain at the Young Foundation, London, was a three-year research project funded by the European Regional Development Fund that examined changes in functional connections and information flows in eight major city regions across North West Europe.
2. The study 'The convergence of geo-space and network space in city region development in China: taking the Mid-Yangtze River city region as an example', has been granted the award of PhD in the Department of Real Estate & Planning at the University of Reading, UK.
3. The National Bureau of Statistics (NBS) is the only agency authorised by the State Council of China to collect statistical data and economic accounting in national scope.

REFERENCES

Aghion, P. and P. Howitt (1998), 'Market structure and the growth process', *Review of Economic Dynamics*, **1** (1), 276–305.
Aharoni, Y. and L. Nachum (2002), *Globalisation of Services: Some Implications for Theory and Practice*, London: Routledge.
Asheim, B.T. and M.S. Gertler (2005), 'The geography of innovation: regional innovation systems', in J.

Fagerberg, D. Mowery and R. Nelson (eds), *The Oxford Handbook of Innovation*, Oxford: Oxford University Press, pp. 291–317.
Bathelt, H. and M. Taylor (2002), 'Clusters, power and place: inequality and local growth in time–space', *Geografiska Annaler: Series B, Human Geography*, **84** (2), 93–109.
Bathelt, H., A. Malmberg and P. Maskell (2004), 'Clusters and knowledge: local buzz, global pipelines and the process of knowledge creation', *Progress in Human Geography*, **28** (1), 31–56.
Baum, J.A., T. Calabrese and B.S. Silverman (2000), 'Don't go it alone: alliance network composition and startups' performance in Canadian biotechnology', *Strategic Management Journal*, **21** (3), 267–94.
Beaverstock, J.V., R.G. Smith and P.J. Taylor (1999), 'A roster of world cities', *Cities*, **16** (6), 445–58.
Bel, G. and X. Fageda (2008), 'Getting there fast: globalisation, intercontinental flights and location of headquarters', *Journal of Economic Geography*, **8** (4), 471–95.
Berry, B.J. (1964), 'Cities as systems within systems of cities', *Papers in Regional Science*, **13** (1), 147–63.
Böckerman, P. and E. Lehto (2006), 'Geography of domestic mergers and acquisitions (M&As): evidence from matched firm-level data', *Regional Studies*, **40** (8), 847–60.
Boschma, R. (2004), 'Competitiveness of regions from an evolutionary perspective', *Regional Studies*, **38** (9), 1001–14.
Breschi, S. and C. Lenzi (2014), 'The returns of external linkages: the role of gatekeepers for the renewal and expansion of US cities' knowledge base, 1990-2004', *Regional Studies*, **49** (5), 782–97.
Bureau van Dijk (2015), 'Zephyr M&A database', accessed 20 September 2016 at https://zephyr.bvdinfo.com/version-20171030/ZephyrNeo.HomePage.serv?_CID=1&context=BFX1F1ZSP3L0NJH.
Burt, R.S. (2009), *Structural Holes: The Social Structure of Competition*, Boston, MA: Harvard University Press.
Button, K., S. Lall, R. Stough and M. Trice (1999), 'High-technology employment and hub airports', *Journal of Air Transport Management*, **5** (1), 53–9.
Castells, M. (1996), *The Network Society*, Oxford: Blackwell.
Castells, M. (1999), 'Grassrooting the space of flows', *Urban Geography*, **20** (4), 294–302.
Christaller, W. (1933), *Central Places in Southern Germany*, English trans. 1966, Jena: Fischer.
Cochrane, A. and Pain, K. (2000), 'A globalising society?', in D. Held (ed.), *A Globalising World*, London: Routledge, pp. 5–45.
Coe, N.M. and H.W.-C. Yeung (2015), *Global Production Networks: Theorising Economic Development in an Interconnected World*, Oxford: Oxford University Press.
Davoudi, S. (2003), 'Polycentricity in European spatial planning: from an analytical to a normative agenda', *European Planning Studies*, **11** (8), 979–99.
Derudder, B. and P. Taylor (2016), 'Change in the world city network, 2000–2012', *Professional Geographer*, **68** (4), 624–37.
Derudder, B., P. Taylor, P. Ni, A. De Vos, M. Hoyler, H. Hanssens, et al. (2010), 'Pathways of change: shifting connectivities in the world city network, 2000-08', *Urban Studies*, **47** (9), 1861–77.
Duhr, S. and V. Nadin (2005), 'Spatial Vision Study no. 1 final report', 30 June, NorthWest Europe INTERREG IIIB Spatial Vision Working Group, University of the West of England, Bristol.
Duranton, G. and D. Puga (2001), 'Nursery cities: urban diversity, process innovation, and the life cycle of products', *American Economic Review*, **91** (5), 1454–77.
European Commission (CEC) (1999), *ESDP: European Spatial Development Perspective: Towards Balanced and Sustainable Development of the Territory of the European Union*, Brussels: European Commission.
European Commission (CEC) (2004), *Facing the Challenge: The Lisbon Strategy for Growth and Employment*, report from the High Level Group chaired by Wim Kok, November, Brussels: Commission of the European Communities.
European Commission (CEC) (2005), *Cohesion Policy in Support of Growth and Jobs: Community Strategic Guidelines, 2007–2013*, Brussels: Commission of the European Communities.
European Commission (CEC) (2008), *Green Paper on Territorial Cohesion – Turning Territorial Diversity into Strength*, Brussels: Commission of the European Communities.
European Commission (CEC) (2009), *Lisbon Treaty* (unconsolidated), Brussels: Commission of the European Communities.
European Commission (CEC) (2012), *Cohesion Policy 2014–2020, Investing in Growth and Jobs*, Brussels: Commission of the European Communities.
European Council (EC) (2000), *Presidency Conclusions*, Lisbon European Council, 23 and 24 March, accessed 15 March 2021 at https://www.europarl.europa.eu/summits/lis1_en.htm.
Freeman, L. (2004), *The Development of Social Network Analysis: A Study in the Sociology of Science*, North Charleston, SC: BookSurge.
Friedmann, J. (1986), 'The world city hypothesis', *Development and Change*, **17** (1), 69–83.
Friedmann, J. (2005), 'Globalisation and the emerging culture of planning', *Progress in Planning*, **64** (3), 183–234.

Fujita, M., P.R. Krugman and A. Venables (2001), *The Spatial Economy: Cities, Regions, and International Trade*, Cambridge, MA: MIT Press.
Grubesic, T.H., T.C. Matisziw and M.A. Zook (2008), 'Global airline networks and nodal regions', *GeoJournal*, **71** (1), 53–66.
Hakansson, H. (1987), *Industrial Technological Development: A Network Approach*, London: Croom Helm.
Halbert, L., J. Pain and A. Thierstein (2006), 'European polycentricity and emerging mega-city regions: "one size fits all" policy?', *Built Environment*, **32** (2), 206–18.
Hall, P. and K. Pain (2006), *The Polycentric Metropolis: Learning from Mega-city Regions in Europe*, London: Routledge.
Harrington, J.W. and H.S. Campbell (1997), 'The suburbanisation of producer service employment', *Growth and Change*, **28** (3), 335–59.
Harrington, J.W. and P.W. Daniels (2006), *Knowledge-Based Services, Internationalisation and Regional Development*, Aldershot: Ashgate.
Harris, C.D. and E.L. Ullman (1945), 'The nature of cities', *Annals of the American Academy of Political and Social Science*, **242** (1), 7–17.
Hoyler, M. and K. Pain (2001), 'London and Frankfurt as world cities: changing local- global relations', in A. Mayr, M. Meurer and J. Vogt (eds), *Stadt und Region: Dynamik von Lebenswelten*, Leipzig: Deutsche Gesellschaft für Geographie, pp. 76–87.
Huggins, R. (2010), 'Forms of network resource: knowledge access and the role of inter-firm networks', *International Journal of Management Reviews*, **12** (3), 335–52.
Huggins, R. and A. Johnston (2010), 'Knowledge flow and inter-firm networks: the influence of network resources, spatial proximity and firm size', *Entrepreneurship & Regional Development*, **22** (5), 457–84.
Huggins, R. and P. Thompson (2015), 'Entrepreneurship, innovation and regional growth: a network theory', *Small Business Economics*, **45** (1), 103–28.
Jacobs, J. (1984), *Cities and the Wealth of Nations: Principles of Economic Life*, New York: Random House.
Jonas, A. (2011), 'Scales and networks', in J.A. Agnew and S. Duncan (eds), *The Wiley-Blackwell Companion to Human Geography*, Oxford: Wiley-Blackwell, pp. 387–483.
Kloosterman, R.C. and S. Musterd (2001), 'The polycentric urban region: towards a research agenda', *Urban Studies*, **38** (4), 623–33.
Kramer, J.-P., E. Marinelli, S. Iammarino and J.R. Diez (2011), 'Intangible assets as drivers of innovation: empirical evidence on multinational enterprises in German and UK regional systems of innovation', *Technovation*, **31** (9), 447–58.
Krugman, P. (1991), *Geography and Trade*, Cambridge, MA: MIT Press.
Lizieri, C. and K. Pain (2014), 'International office investment in global cities: the production of financial space and systemic risk', *Regional Studies*, **48** (3), 439–55.
Mahroum, S., R. Huggins, N. Clayton, K. Pain and P. Taylor (2008), *Innovation by Adoption: Measuring and Mapping Absorptive Capacity in UK Nations and Regions*, London: National Endowment for Science, Technology and the Arts.
Markusen, A. (2003), 'Fuzzy concepts, scanty evidence, policy distance: the case for rigour and policy relevance in critical regional studies', *Regional Studies*, **37** (6–7), 701–17.
Massey, D.B., J. Allen and S. Pile (1999), *City Worlds*, London: Routledge.
Meijers, E. (2007), 'From central place to network model: theory and evidence of a paradigm change', *Tijdschrift voor economische en sociale geografie*, **98** (2), 245–59.
Miller, H.J. (2004), 'Tobler's first law and spatial analysis', *Annals of the Association of American Geographers*, **94** (2), 284–9.
Moulaert, F. and P.W. Daniels (1991), *Advanced Producer Services: Beyond the Micro-Economics of Production*, London: Belhaven Press.
National Bureau of Statistics (NBS) (2014), 'The National Bureau of Statistics of the People's Republic of China', accessed 1 March 2021 at http://www.stats.gov.cn/english/.
Nicolini, D., S. Gherardi and D. Yanow (2003), *Knowing in Organisations: A Practice-Based Approach*, New York: ME Sharpe.
Nordlund, C. (2004), 'A critical comment on the Taylor approach for measuring world city interlock linkages', *Geographical Analysis*, **36** (3), 290–96.
O'Brien, R. (1992), *Global Financial Integration: The End of Geography*, London: Pinter.
Ohmae, K. (1990), *The Borderless World: Power and Strategy in the Interlinked Economy*, London: HarperCollins.
Ohmae, K. (1995), *The End of the Nation State: The Rise of Regional Economies*, London: HarperCollins.
Pain, K. (2008), 'Examining "core–periphery" relationships in a global city-region: the case of London and South East England', *Regional Studies*, **42** (8), 1161–72.
Pain, K. (2012), 'Spatial transformations of cities: global city-region? Mega-city region?', in B. Derudder, M. Hoyler, P. Taylor and F. Witlox (eds), *International Handbook of Globalisation and World Cities*, Cheltenham, UK and Northampton, MA, USA: Edward Elgar, pp. 83–93.

Pain, K. and P. Hall (2008), 'Informational quantity versus informational quality: the perils of navigating the space of flows', *Regional Studies*, **42** (8), 1065–77.
Pain, K. and G. Van Hamme (2014), *Changing Urban and Regional Relations in a Globalising World: Europe as a Global Macro-Region*, Cheltenham, UK and Northampton, MA, USA: Edward Elgar.
Powell, W.W., K.W. Koput and L. Smith-Doerr (1996), 'Interorganisational collaboration and the locus of innovation: networks of learning in biotechnology', *Administrative Science Quarterly*, **41** (1), 116–45.
Puga, D. (1999), 'The rise and fall of regional inequalities', *European Economic Review*, **43** (2), 303–34.
Redding, S. and A.J. Venables (2004), 'Economic geography and international inequality', *Journal of International Economics*, **62** (1), 53–82.
Rodríguez-Pose, A. and H.-M. Zademach (2003), 'Rising metropolis: the geography of mergers and acquisitions in Germany', *Urban Studies*, **40** (10), 1895–923.
Rozenblat, C. (2010), 'Opening the black box of agglomeration economies for measuring cities' competitiveness through international firm networks', *Urban Studies*, **47** (13), 2841–65.
Rozenblat, C. and G. Pflieger (2010), 'Introduction. Urban networks and network theory: the city as the connector of multiple networks', *Urban Studies*, **47** (13), 2723–35.
Sassen, S. (1991), *The Global City: New York, London, Tokyo*, Princeton, NJ: Princeton University Press.
Scott, A. (2001a), *Global City-Regions: Trends, Theory, Policy*, Oxford: Oxford University Press.
Scott, A.J. (2001b), 'Globalisation and the rise of city-regions', *European Planning Studies*, **9** (7), 813–26.
Smith, D.A. and M. Timberlake (1995), 'Cities in global matrices: toward mapping the world-system's city system', in P.L. Knox and P.J. Taylor (eds), *World Cities in a World-System*, Cambridge: Cambridge University Press, pp. 79–97.
Smith, H.L., S. Romeo and M. Virahsawmy (2012), 'Business and professional networks: scope and outcomes in Oxfordshire', *Environment and Planning A*, **44** (8), 1801–18.
Taylor, P., J. Beaverstock, G. Cook, N. Pandit, K. Pain and H. Greenwood (2003), *Financial Services Clustering and It's Significance for London*, London: Corporation of London.
Taylor, P.J. (2001), 'Specification of the world city network', *Geographical Analysis*, **33** (2), 181–94.
Taylor, P.J. (2004), *World City Network: A Global Urban Analysis*, London: Routledge.
Taylor, P.J. (2009), 'Urban economics in thrall to Christaller: a misguided search for city hierarchies in external urban relations', *Environment & Planning A*, **41** (11), 2550–55.
Taylor, P.J. and K. Pain (2007), 'Polycentric mega-city regions: exploratory research from Western Europe', in P. Todorovich (ed.), *The Healdsburg Research Seminar on Megaregions*, New York: Lincoln Institute of Land Policy and Regional Plan Association, pp. 59–66.
Taylor, P.J., G. Catalano and D.R. Walker (2002), 'Exploratory analysis of the world city network', *Urban Studies*, **39** (13), 2377–94.
Taylor, P.J., M. Hoyler and R. Verbruggen (2010), 'External urban relational process: introducing central flow theory to complement central place theory', *Urban Studies*, **47** (13), 2803–18.
Timberlake, M. (2010), 'The kaleidoscoping world city networks: how research locates cities in global systems of cities', paper presented at the World Congress of the International Sociological Association, Gothenburg, July.
Tobler, W.R. (1970), 'A computer movie simulating urban growth in the Detroit region', *Economic Geography*, **46** (suppl. 1), 234–40.
Van Oort, F., M. Burger and O. Raspe (2010), 'On the economic foundation of the urban network paradigm: spatial integration, functional integration and economic complementarities within the Dutch Randstad', *Urban Studies*, **47** (4), 725–48.
Wallerstein, I. (1974), *The Modern World- System, Vol.1: Capitalist Agriculture and the Origins of the European World-Economy in the Sixteenth Century*, New York and London: Academic Press.
Watts, D.J. (1999), *Small Worlds: The Dynamics of Networks between Order and Randomness*, Princeton, NJ: Princeton University Press.
Wood, P.A. (1991), 'Flexible accumulation and the rise of business services', *Transactions of the Institute of British Geographers*, **16** (2), 160–72.
Yang, F.F. and A.G. Yeh (2013), 'Spatial development of producer services in the Chinese urban system', *Environment and Planning A*, **45** (1), 159–79.

25. Intracity and intercity networks of multinational firms, 2010–2019

Céline Rozenblat

1. INTRODUCTION

Globalization induces common global economic rules that are addressed by numerous various local answers (Castells 2010; Robinson and Roy 2016), but across the world, the role of cities is to leverage knowledge and information in the network of globally connected exchanges (Scott 2001; Roy and Ong 2011; Neal 2012; Scott and Storper 2015). According to Grabher and Stark (1997, p. 540), 'globalization does not displace the properties of localities but makes them all the more salient'. Firms that are organized in networks meet together in different cities in which they extend their access to resources by establishing multiple plants. At stake is the local and global coordination of transfers of value through intrafirm and interfirm networks, resulting in the concentration of wealth in specific firms in specific cities at the expense of other firms and cities, thereby contributing to the increase in interurban inequalities (Hadjimichalis 1984; Chapter 29 in this volume). Owing to the mutual construction and reinforcement between the intraurban and the interurban networks, some intraurban properties are linked to global connectiveness. In addition, between local and global scales, state and regional arrangements, such as free trade zones, are identified scales where the actors' strategies meet institutions facilitating or limiting the organizational complexity of the global value chains of firms (Sassen 2010; Dicken 2011; Rozenblat 2018). Thus, global and local networks are mutually dependent in cities, and different intermediary scales shape the actors' strategies operating at multiple spatial scales in order to ensure their stability and their power.

Many questions remain unanswered in the intraurban versus interurban networks' analysis, especially the question of how 'to make explicit an implicit and simplistic equating of spatial scale and power' (Beauregard 1995, p. 241). This question was examined by seminal papers developing reflections on the networking effects of global economic forces on cities (Amin and Thrift 1992; Castells 2000, 2010; Sassen 2000; Bathelt et al. 2004; Massey 2007; Bathelt and Glückler 2011). However, in our opinion, the absence of a distinction between scales and levels of processes avoids a better conceptualization of this intertwining between local and global networks and constituting the dynamics of cities (Pflieger and Rozenblat 2010; Pumain 2018; Chapter 1 in this volume).

In this chapter, we argue that the intertwined two-level processes – one emerging at the city level from the operations of interdependent actors and the other emerging from the interaction of these urban processes at the systems of cities' level – are not distinct processes occurring at different geographical scale or scope. Instead, if we consider that cities form systems inside systems of cities (Berry 1964), cities combine two interacting levels inside the same dynamics: the cities as systems (meso-level) and the systems of cities (macro-level) (Rozenblat 2010). The system of cities itself consists of different scales of geographic scopes and institutional organizations that are or are not nested. At stake in the mutual adaptation of these meso-levels and macro-levels are adaptive cycles permitting or not permitting individuals or firm strategies (micro-level) to blossom within the context of the lining up of local and global forces. We consider cities such as Liverpool or Detroit, which flourished in their industrial heyday but then missed out on the transformational process involving a fundamental transition to the new forms of the economy of the 1990s. The local actors' networks as well as their global linkages are in question in the collapse of these cities' economies. However, later in the 2010s, thanks to strong policies and aided greatly by a local and global network reconfiguration, the two cities rebuilt their economies and their working forces.

This chapter develops further these reflections and the empirical evidence on how global and local networks expand in mutual interdependence inside and between cities, disentangling the processes developed at the city level (meso) and at the system-of-cities level (macro) from the different geographic scales of the system of cities. The multilevel city system is presented as a useful theoretical framework able to connect local to global networks (section 2). The empirical evaluation of these local/global networks is based on the worldwide multinational firms' ownership linkages, which are considered both at their intraurban and interurban levels. This evaluation underlines the evolution of the scales of world cities' networking between 2010 and 2019 (section 3) and the uneven role of intraurban linkages in the cities' external linkages development. A focus is proposed on London's and Beijing's intraurban linkages in order to illustrate how far local and global linkages are similar in nature but develop at these two levels with some mutual adaptations (section 4). Finally, a conclusive discussion on the interpretations and implications of these intracity and intercity networks is stressed, and a future research path suggested (section 5).

2. THE LOCAL/GLOBAL CITIES' SYSTEMS AND THE SCALES OF GLOBALIZATION

Numerous reflections have conceptualized the duality of interurban and intraurban networks, stressing the critical balance and flexibility required in the intertwining between the two levels of 'local buzz and global pipelines' (Bathelt et al. 2004) for the positioning and the adaptation capacities of each city in the network society (Amin and Thrift 1992; Sassen 2000; Castells 2010). However, the issue of the network dynamics within or between cities is often addressed independently by empirical studies that are conducted either inside cities or between cities. Only few case studies have focused on the interactions between these two levels, showing in details how inter-level and intra-level networks mutually reinforce each other and how far these interactions participate in shaping the cities' development (Hall and Pain 2006; Massey 2007; Bathelt and Gluckler 2011).

2.1 The Local/Global Urban Processes

We can easily assume that local and global urban processes are closely intertwined. According to Beauregard (1995, p. 241), 'the local is no longer purely local but has incorporated global (and regional and national) interests and actors. To this extent, any spatial scale exists only temporarily as "purely" local, regional or global'. Some actors have developed local networks in a global perspective or they have mobilized global networks in a local perspective. Again quoting Beauregard (1995, p. 242): 'Local actions might have global consequences . . . .One might argue that the global is the organization of many local actors on a broader geographical canvas, and without local places the global cannot exist.' Specifically, in cities integrated in globalization, local actions may support numerous global perspectives owing to the concentration of the important control, innovation and governance of private or public institutions and private firms. Inside these cities, local connections may provide access for the global actors' connections with other places. Revisiting London, which had been built up on global and local processes, Massey (2007, p. 107) argued that 'the local is not only the product of the global, but that global itself is produced in local places'. To this mutual global/local interwoven framework in London, Amin and Thrift (1992, p. 583) introduced a temporal dimension, adding top-down or bottom up periods: 'In its old incarnation, the City [of London] was the result of the local going global. In its new incarnation, the City is the result of the global going local'. Thus, we wonder which properties of cities quickly develop these local/global processes and how the bottom-up and top-down processes occur?

2.1.1 Top-down and bottom-up processes

In a top-down way, the cities' economic advantages are traditionally attached to local agglomeration and, importantly, incorporate the economic base's interactions with other cities inside the regional/national/global cities' networks (Moulaert and Djellal 1995). Adopting a relational perspective permits the consideration of both local and global relationships as parts of the same actors' strategies. Moulaert and Djellal (1995) and, later, Bathelt and Glückler (2011, p. 115) 'extend agglomeration theory by moving beyond the issue of local externalities to examine the positive network externalities for a city as a result of its connectivity with the international city system'. International actors can be used as global connectors by other firms located in the same city, and thus the intracity networks integrate global connections as additional trans-local externalities benefiting firms located in that city: 'A city that is highly connected within the city system thus provides accessibility advantages to its firms located in that city' (Bathelt and Glückler 2011, p. 115).

In a bottom-up process, the trans-local synergies of networks reinforce the capacity of local firms to develop 'economies of overview' (Johannisson 1990), which permit the identification and exploitation of different opportunities in various markets. These economies are crucial for the most specialized activities that require a wider catchment market area. Social capital, including the know-who, is essential in very competitive markets in which innovation, regardless of whether it is industrial or service, is increasingly related to relationships with clients and consultants and is strongly based on reputation and personal trust (Bathelt and Glückler 2011). Economies of overview increase with proximity (Torre and Rallet 2005) and with the local 'buzz' (Storper and Venable 2004). They are also supported by institutions that stabilize patterns of interactions based not only on rules, norms, conventions, habits and routines (Bathelt and Glückler 2011, ch. 3; Nelson and Winter 1982) but also on culture (DiMaggio 1997) and reputation (Uzzi 1997). The reputational benefits of network transitivity increase trust and the potential economies of overview.

2.1.2 Attractiveness and power of cities

A good reputation promotes the cities' attractiveness for firms and for skilled workers. Firms attract other firms of the same or complementary activities, enabling the local complexity of subcontracting, alliances and coopetition relationships to thrive. Also, these complex local networking relationships strengthen some firms in the context of global competition, reinforcing their capacity to expand their worldwide power to other cities. To achieve local coordination, firms with linked networks must address the networks' compatibility (common standards and norms), availability (acceptation of new entrants) and malleability (open to revision) (Grewal 2008). These conditions comprise the aspects about which firms can communicate and cooperate in order to benefit from their relationships and to increase the capabilities that will reinforce the economies of overview that were evoked previously. Cities constitute the dynamic cognitive context of the functional and symbolic milieu (Crevoisier 2004) where economic agents behave with bounded rationality in an uncertain environment (Camagni 2001). Sassen (1991) deeply documented the behavior of producer services in global cities in which local and cosmopolitan businesses and the political elite created powerful social networks consolidating their updated information on global trends.

Whether controlling other firms or controlled by others, firms insert themselves in local societies, creating their own local embedded networks. Their local interactions with other firms consist in subcontracting, cooperating and competing or forging alliances. In this perspective, they often operate through a deep local labor division that is coordinated with the global labor division. For management flexibility reasons or to consolidate a cooperative or subcontracting arrangement involving strategic information that they do not want to outsource, firms also invest locally in other firms in a brownfield way (buying other firms), a greenfield way (creating a new company) or through joint ventures. In addition, they participate in local life activities, investing in sports and cultural clubs or social infrastructure. Firms with a group headquarters location specifically deploy a large set of interacting, owned firm units with different functions orientated to the control and the coordination of the global activity of the group. Long-range international control functions specifically necessitate strong coordination: numerous combined tasks are fragmented into different enterprises operating through a vertical/horizontal disintegration (Scott 2001) that comprises both intracity and intercity regions. Many of these functions remain totally or partly internalized inside the large corporate groups, facilitating the control and the adaptability of the value chain.

2.1.3 Specialization and diversification of cities

Since the 1970s, cities have experienced growing Marshallian effects of specialization thanks to the coupling of the deindustrialization of advanced capitalist countries (Peet 1987) and the deepening of the worldwide division of labor (Aydalot 1985; Storper 1997). Locally, consistent with the Jacobs's intuition of the multiplication of related activities (Jacobs 1969), numerous complementary functions were developed around specialized ones, stimulating for 15 years increased research on relatedness (Frenken et al. 2007; Hidalgo et al. 2007; Shutters et al. 2015; Content and Frenken 2016; Farinha et al. 2019). Since the 1990s, authors have remarked on the diversification of activities around some forms of specialization, in particular that for services. For example, the following comment was offered: 'Chicago is the home of the second largest number of corporate headquarters, specialized in manufacturing administration, securities, computing, and data processing (much more than New York), mailing and reproduction services, and management and public relations' (Markusen and Gwiasda 1994, p. 177). Therefore, in cities such as Chicago, the cities' specialization often leads to the apparition of complementary activities sustaining each other locally in a global production network (GPN) perspective (Henderson et al. 2002). In particular, in the knowledge economy based on technology, innovation and communication, numerous specialized services, such as those provided by information technology consulting (Moulaert and Djellal 1995), accounting, insurance, banking and finance, management, law, logistics, design and architecture, advertising and media firms (Luthi et al. 2010), contribute to the organization of activities and promote them. These specialized services form industrial or services clusters (Porter 2004) involving private firms together with education and research institutions, political governance and civil societies and constituting what are called N-helix systems (Leydesdorff 2012). Owing to their internal local relations, these N-helix systems are considered local systems, but in their dynamics, their external national or international relationships are important as well.

2.1.4 Globalization of polycentric urban regions

Global cities use their whole surrounding region of influence to develop their global scope. This ability allows them to reach a critical size and to use various local resources. City-regions extend far beyond urban cores, forming new business centers, including edge cities where financial and technological centers have expanded outside the real-estate pressure of the core and specialized subcenters that are often ancient satellite cities that are now totally incorporated inside the urban dynamic of the central metropolis (Hall 2014). It is, however, very difficult to delineate these city-regions in a global perspective. In 2006, Hall and Pain coordinated a comparative study of eight European cities in order to evaluate the cities' polycentricity stage and to what extent they could qualify them as mega-city regions (MCR). They suggested the iterative delineation of these MCR cities by first considering commuters and the advice of specialists, and then adjusting these delineations according to the local economic linkages. Following the globalization and world cities (GaWC) methodology, they considered the different locations of advanced producer services (APS) firms and assumed that intrafirm linkages between the subsidiaries of the same companies would guarantee some exchanges of information (see also Chapter 24 in this volume). According to this study, they confirmed that the subsidiary companies of APS firms spread their offices in different parts of the city-regions, contributing to their integration inside a single regional-city system. However, they observed that the majority of advanced services remained in the city cores and that owing to national policies, some strong national effects influenced the way surrounding regions interacted with the main city-core.

In the perspective of situating cities in globalization, we proposed in 2020 the delineation of 1698 large urban regions (LURs) all over the world, addressing the challenge to study the main large cities of the world in a comparative way (Rozenblat 2020a). We adopted a triple-stage approach consisting of considering first the national definitions of a city and then building upon these uneven delineation-nested definitions by using the following four main concepts:

- urban localities, the term we use for municipalities that are mostly based on political power (MUNIs);
- dense built-up areas, which are evaluated as morphological urban areas (MUAs);
- functional urban areas, which denote areas that are based on the daily flows of commuters or the density of roads (FUAs); and
- polynuclear urban regions, which represent regional systems inserted in globalization and which we term large urban regions (LURs).

The LURs are built with a third iterative stage, namely, the specific analysis of the location of the multinational firms of our database, their proximity to MUAs or FUAs, and their access to a main national or international airport, which is crucial in the context of multinational firms needing rapid access to their subsidiaries, to markets and to high-level services (Rozenblat 2020a). The detailed municipal composition of these LURs is published in open access (Rozenblat 2020b), as the LURs' delineations must be improved and updated regularly. It is only at the price of this transparency that it will be possible at the international scale to monitor and forecast the comparison of cities' global networks and their local embeddedness in city-regions.

2.2 The Scales of Global Processes

The balance between the local embeddedness of global networks and the global centrality of these local networks lies inside the numerous intermediate nested or non-nested scales (regional, national and free trade zones) that influence the effectiveness and the form of the networks' impacts on city development, regardless of whether the development is in the global North or South (Robinson 2015). In fact, we do not live in a flat world: countries or groups of countries represent institutional spaces that support firms in very uneven ways in the context of national policies, such as state market demand, taxes or monetary policies (Dicken 2011). Among numerous scales created by the multiple expansions of actors and which have tendencies to converge partly towards similar actors, these scales hold a particular attraction for network organizations. This can be explained by the facilitation properties of some geographical spaces for free circulation or institutional proximities. This situation results in a discrete set of scales that significantly override global networks.

2.2.1 National scales

Countries represent internal homogeneous spaces where global networks can preferentially expand owing to similar rules, language, regulation and control, and to the access to physical, human or institutional infrastructures that regroup diverse elements, such as roads, transport, public services, education, social welfare, culture and administration. States remain a strong regulator of labor markets, wage cost, social welfare and tax systems. In addition, the coordination between states and large domestic corporations gives the former an active role in national policies. Regarding Japanese firms, the involvement of the kereitsus (large consortia of firms) in national economy policies has existed for a long time (Dicken 2011). Currently, Chinese and American firms are becoming powerful economic influencers enabling states to dominate international trade and, consequently, control the global political order. Although this is not always so because firm groups are not always supported by particular national policies, such as those directed against monopolistic positions (and, more recently, those protecting the use of personal information), this phenomenon cannot be denied (see Dicken 2011, ch. 7).

As most of the information on multinational firms is produced at the international level between states, the role of multinationals in the international arena is very well documented through foreign direct investments (FDIs) (World Bank 2017). Owing to the lack of information at the lower scales, compared with that in the international space, the amount of information on the magnitude of multinational firms' expansion inside national spaces is very small.

The concentration of multinational firms' networks in the different national cities depends also on the inherited hierarchy of cities inside each country, and the primatial model is the most prevalent organization among the states of the world (Moriconi-Ebrard 1994; Sassen 2010; Pumain et al. 2015). Primate cities benefit from the initial advantage of being able to dominate their own national urban system, in which the national administration and stock exchanges are often concentrated in order to develop local companies and attract companies based in other cities. From the firms' network perspective, the strong embeddedness in national institutional functioning (for example, the level of decentralization and federalism) creates a preferential attractiveness of one or more cities in each country. Thus, the national average development level and the national cities' hierarchies combine to define the level of the cities' attractiveness and power (Rozenblat and Pumain 1993).

2.2.2 Supra-regional scales

Regional trade agreements (RTAs) regroup national spaces in order to facilitate network deployments and sometimes to reinforce entry barriers towards the outside. They are numerous throughout the world, and many of them overlap. The World Trade Organization (WTO) manages all the tariff barriers, which are declining in the world, but nontariff barriers (NTBs), such as technical rules and minimal standards, are increasing (Dicken 2011). Dicken (2011, p. 205, fig. 6.13) distinguishes the following types of regional trade agreements:

- free trade areas, which remove trade restrictions between member states;
- customs unions, which extend common external trade policies towards nonmembers;
- common markets, which enable free movement of factors of production between members; and
- economic unions in which the harmonization of economic policies are under supra-national control.

At the beginning of 2020, the WTO counted 303 active RTAs (WTO 2020).

Cities are positioned at these continental scales by the combination of their national average development level and their position in their own country. However, some specific functions in the RTAs (that is, the European Union headquarters functions, such as those located in Brussels, Luxemburg and Strasbourg in Europe, or the stock exchange operations, such as those in Frankfurt) constitute additional resources that firms gain by being located close to these institutions (Cattan et al. 1994; Rozenblat and Cicille 2003; Pumain and Rozenblat 2019). These RTA effects can be visible at the continental scale, which often comprises the scope of these regional agreements.

2.2.3 Other scales

In relation to the expansion of the firms' networks among world cities, other scales based on historical colonial linkages and/or on similar languages also constitute some remaining proximities from the past. The clustering approach of the global ownership networks of firms demonstrated that cities of the Commonwealth countries remain closer to Great Britain than did European cities (Rozenblat et al. 2017); Spanish and Portuguese cities are integrated with South American cities. In contrast, ex-communist countries do not exchange as much within their firms' linkages, but they are instead linked in with Western European cities and Russian cities (Di Lello and Rozenblat 2014; Zdanowska et al. 2020).

2.3 Disentangling Levels and Scales of the Multinational Firms' Networks

The balance between the cities' networks at different levels and scales is an important issue for the resilience of cities. The physical properties of the networks' structure reveal that the diversity of the cities' linkages between levels and scales and at each level and scale (in relation to origin/destination and activities) secure the system's robustness (Garas et al 2018). In addition, Hu et al. (2013) provided the evidence that the redundancy of linkages between different networks reinforces the robustness of the whole network. These results bring complementary elements to the previous fragmented approaches of linking scales and levels, enabling the further expansion of evolutionary approaches (Boschma and Martin 2010) in a multilevel and multiscale perspective (Table 25.1).

The consideration of the city as a set of interacting actors permits the identification of actors having simultaneously local multiple interactions and a large scope of global linkages (Table 25.1, cells C1, C2 and C3). These actors are identified as institutional

Table 25.1 Processes of the firms' networks in a crossed levels and scales perspective

Level (aggregation of nodes)		Geographical scale (network boundary)		
The city is considered as a level of an organization		The city is considered with different geographical scales: 1 Intralocalities	2 Intracity-region	3 Interurban
A Macro	The city as a single LUR interacting with other LURs	Degree of the localities' cohesions	Degree of intracity-region cohesion/ specialization/ relatedness	National/global cities' networks' (pipelines), reputation, economies of overview
B Meso	The city as an LUR	Localized clusters of industries inside localities Local institutions	Regional polycentrism inside the LUR; N-helix of local economic clusters; city-region's governance	LURs compared at the global scale
C Micro	The city as a set of interacting actors	Interactions of actors within municipalities	Interactions between regional actors (buzz)	Global interactions of single actors

entrepreneurs, playing the role of interfaces and facilitators of cascade spillover processes between heterogeneous social spaces (Padgett and Powell 2012).

At the meso-level, considering groups of actors in the LURs enables the determination of how far activities are concentrated in localized localities (Table 25.1, cell B1). At the city-region scale (cell B2), it is worth evaluating how far these localized clusters are embedded inside local institutions, university, research and social actors, forming the N-helix systems that only make sense at the city-region scale. Polycentric relations develop between more or less specialized places inside the city-region, strengthening the whole. Their national to worldwide linkages (cell A3) constitute another important dimension of the N-helix, causing clusters to evolve according to the global concurrency, for national and international partners and for national or supra-national regulatory policies (cell B3). At the global level, the city (LUR) as a whole reaches a cohesion, diversity and relatedness (cells A1 and A2), causing the emergence of properties of economies of overview that increase the national and international city reputation (cell A3), profiting any actors in the LUR by increasing their capacity to develop global networking.

3. LOCAL/GLOBAL CITIES' NETWORKS OF MULTINATIONAL FIRMS' OWNERSHIP LINKAGES

The consequences of attractiveness and power in the local/global interactions could be further documented in a comparative manner for all cities of the world, filling the gap in scientific literature in which these consequences are either documented separately as global or local or the local/global interactions are addressed for specific cities, such as Chicago (Markusen and Gwiasda 1994), eight European cities (Hall and Pain 2006) or the cities in the Yantze Delta region (Bathelt and Glückler 2011). The questions are numerous, and we can only address a few of them here.

The previous multilevel and multiscale framework can expound the empirical study of firms' networks within and between cities. Adopting a comparative approach for the cities of the world (section 3.1), we give an overview of the order of magnitude of the importance of each scale, the evolution of their balance over the past 10 years and the dynamics of the relationship between intracity and intercity networks (section 3.2). We question why cities experience different situations according to their national and continental contexts and how intraurban and interurban linkages adopt specific forms. Deepening some case studies of cities' intraurban networks, we discuss the importance of the specialized networks and of polycentrism in the cities' scope (section 3.3).

3.1 Multinational Firms' Networks' Evaluation Inside and Between Cities

The empirical analysis is based on multinational firms' ownership networks, the information for which we collected and organized in relation to four dates over a 10-year period (BvD 2010, 2013, 2016, 2019). At each date, the sample comprised the first 3000 firm groups in the world, and the data for all their direct and indirect subsidiaries was collected. Between the different dates, the groups partly changed and their subsidiaries changed. Another option was to consider the same sample of firms from 2010 to 2019. However, this approach would drastically reduce the size of the dataset because many firms decrease or even disappear, and it would prohibit considering newly arriving firms.

The strength of the empirical analysis (which is also a limit) is that it enables us to consider intrafirm ownership linkages. The reasons are twofold. First, it permits the use of homogeneous data based on the Bureau van Dijk (BvD) ORBIS database, certifying that these linkages are real corporate linkages and not speculative investments. Second, the linkages are intrafirm linkages, representing privileged, tangible institutionalized networks that are measured by the ownership percentage. These linkages do not represent all the firms' networks comprising interactions for trade, collaboration, services, and so on, but they reveal a large part of the firms' strategic activities provoking the sharing of participation between firms. Firm activities involving the creation of other new firms and participation in merger and acquisition (M&A) activities are a large part of the firms' strategies, contributing to capital mobility and technological spillover between firms, and thus representing a large issue in the cities' intraurban and interurban interactions (Rodrígues-Pose and Zademach 2003; see also, Chapter 24 in this volume). Recent statistics showed that the share of the economy belonging to intrafirm' networks is far from negligible: in 2014, intrafirm trade represented 50 percent of US imports (more than 60 percent for advanced countries) and 30 percent of US exports, proportions that have remained constant since 2000 (World Bank 2017). The US intrafirm trade reacted better to the 2008 crisis than the arm's length trade (trade between unrelated firms), which decreased significantly. This better resilience can be explained by intrafirm network members' different advantages, such as access to finance, their size and productivity, and a wider range of available tools permitting them to hedge against exchange-rate movements and to maintain a lower sensitivity to uncertainty (Bems et al. 2009; Bernard et al. 2009; Carballo 2015).

In addition, although the ORBIS database comprises the firms' secondary establishments since 2016, the exhaustivity of this information is very uneven between countries. Thus, from a worldwide perspective, this information is not usable in a comparative perspective, which explains why we consider only the principal establishments of each firm (a firm's official registered headquarters in the country). For our empirical study, all the firms were localized and aggregated by LURs, setting aside some firms that did not have addresses or that were outside any LUR (Table 25.2).

Table 25.2 Firms' network ownership sample and LURs' representativeness

Year	Principal establishments of direct and indirect subsidiaries			Ownership linkages		
	Total	Localized in LURs	Localized in LURs (%)	Total linkages	Localized in pairs of LURs	Localized in pairs of LURs (%)
2010	657399	440333	67	1015913	652811	64
2013	772987	551492	71	1193325	800306	67
2016	1118689	875610	78	1619536	1158429	72
2019	1303002	1067353	82	1734603	1288946	74

Source: Bureau van Dijk (2010, 2013, 2016, 2019), ORBIS – UNIL.

The LURs are defined according to the aggregation of localities that are economically dependent on the core cities (Rozenblat 2020a). The 1698 LURs defined all over the world comprise a total of 124 806 localities (Rozenblat 2020b). In total, 1186 LURs are identified as hosting some multinational firms linked to our sample of firms. We interchangeably use the terms city, urban and LUR.

3.2 Intercity and Intracity Network Construction

The method to aggregate firms and their ownership linkages by LUR and by pairs of LURs is explained in detail in Rozenblat (2018). We measured the intensity of a city relationship with another city by adding the total number of ownership relationships of the firms located in pairs of LURs: LO_{ij} denotes the number of subsidiaries located in the city *j* having their headquarters (or a minority shareholder) in the city *i*. Each orientated linkage l^e_{ij} represents an enterprise *e* having a shareholder or headquarters in city *i* and a subsidiary located in city *j*. All l^e_{ij} are summed to obtain LO_{ij}:

$$LO_{ij} = \sum\nolimits_{e=1}^{p} l^e_{ij} \tag{25.1}$$

Note that in the opposite direction, the city *j* can also have enterprises *e* hosting shareholders who invest in subsidiaries of the city *i*. Then, the non-oriented link LNO_{ij} between cities *i* and *j* is expressed as:

$$LNO_{ij} = \sum\nolimits_{e=1}^{p} l^e_{ij} + \sum\nolimits_{f=1}^{q} l^e_{ji} \tag{25.2}$$

In the following empirical study, we specify when we will use orientated or non-oriented linkages.

The total weight of each city in the network can be revealed by two different values:

> H_i denotes the number of controls from the headquarters of city *i* to other cities *j* and is evaluated by the number of outgoing ownership linkages from the city *i* towards all other cities *j*;
>
> S_j denotes the number of links of a subsidiary of city *j* from all other cities *i* and is measured by the number of ingoing linkages to the city *j* from all the other cities *i* (as these linkages are oriented from the headquarters' LUR *i* to the subsidiaries of the LUR *j*, Hi is different from Si).

H_i and S_j are computed as expressed in equations (25.3) and (25.4):

$$H_i = \sum\nolimits_{j=1}^{k} LO_{ij} \tag{25.3}$$

$$S_j = \sum\nolimits_{i=1}^{l} LO_{ij} \tag{25.4}$$

In the network analysis, the value of H_i is the city's weighted out-degree (here representing the power of the subsidiaries' control that is concentrated in the city *i* towards outside the city *i*), and the value S_j is the city's weighted in-degree (the attractiveness of the city *j* for subsidiaries controlled from outside the city *j*).

The importance of intraurban linkages is measured as:

$$I_i = \sum_{j=1}^{k} LO_{ii} = \sum_{i=1}^{l} LO_{ii} \tag{25.5}$$

Note that in this case, the links measured for the headquarters or for the subsidiaries' sums are the same and must be counted only once.

The firms' linkages (Table 25.1; cells C1–C3) were aggregated by LUR by using their origin and destination locations, which correspond to cell A3 of Table 25.1 and reveal the weight of the cities at the macro-level (Figure 25.1). This process yielded a matrix linking the 1698 LURs of the world. The intercity relationships are orientated, and they are weighted by aggregating the firms' links by pairs of cities, as described in the equations.

For intraurban linkages, the same method is adopted, and we can build intraurban networks linking the different MUAs of each LUR and the localities (MUNIs) within each MUA, revealing the intralocality linkages (Table 25.1, cell B1) or the regional polycentrism (cell B2). With this data elaboration, we are able to consider the same networks with different levels of dynamic processes at the global and local scales.

In addition to the attention we paid to the delineation of LURs, we also considered the different geographical territories of 247 countries (UN 2018) aggregated by six large continents (not considering the Arctic, where there are no LURs, and merging Oceania and Asia). This allows us to distinguish the intranational, intracontinental and intercontinental linkages (Table 25.1, cell A3), considering therefore different scales between the intraurban regional linkages and interurban global ones. In these scales, it is important to not count the same networks twice (for example, owner or subsidiary firm networks inside cities are the same; the same holds for national intercity networks inside countries).

3.3 Evolution of Multinational Firms' Networks Between Cities

Figure 25.2 displays the graph built by the sum of inter-LUR linkages (Figure 25.1c). The position of each city-LUR is determined by its networks' proximities with other cities. The continents of the cities are represented by colors, underlying the preponderance of the intracontinental systems of cities. We applied the same thresholds for every year in order to clarify the figure and to avoid too much noise:

- the LURs must have more than 30 total linkages; and
- there must be over 10 linkages between the LURs.

In 2010, European and North American cities constituted the core of the worldwide system dominated by London, then Paris and New York (each regrouping half of London's linkages), and Tokyo was lower (with also half of Paris's linkages). Other cities surrounded them. The former-USSR and Ukrainian cities remained isolated, and this isolation existed over the period 2010–19. The Middle East cities, such as Kuwait, Bahrain, Doha, Amman, Beirut and Riyadh, also form an isolated system. The Asian cities clustered around the domination of Tokyo, and other Japanese cities, such as Osaka and Nagoya, and Seoul formed a distinct subsystem as well.

Asian or Oceanian cities, such as Hong Kong, Kuala Lumpur, Taipei, Singapore, Bangkok, Sydney, Beijing and Shanghai, have multiple linkages with core cities and thus

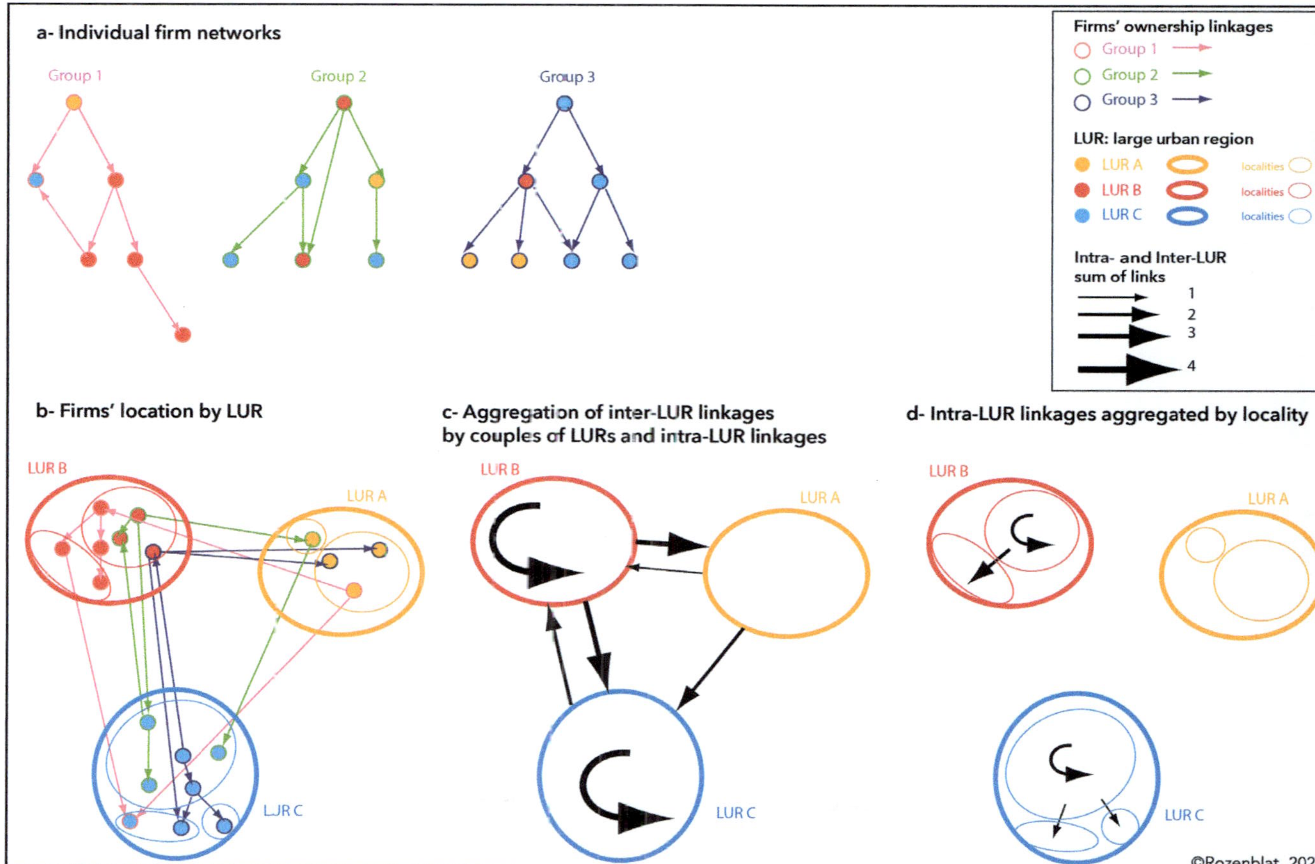

Figure 25.1 *Aggregation of the firms' linkage inside and between LURs*

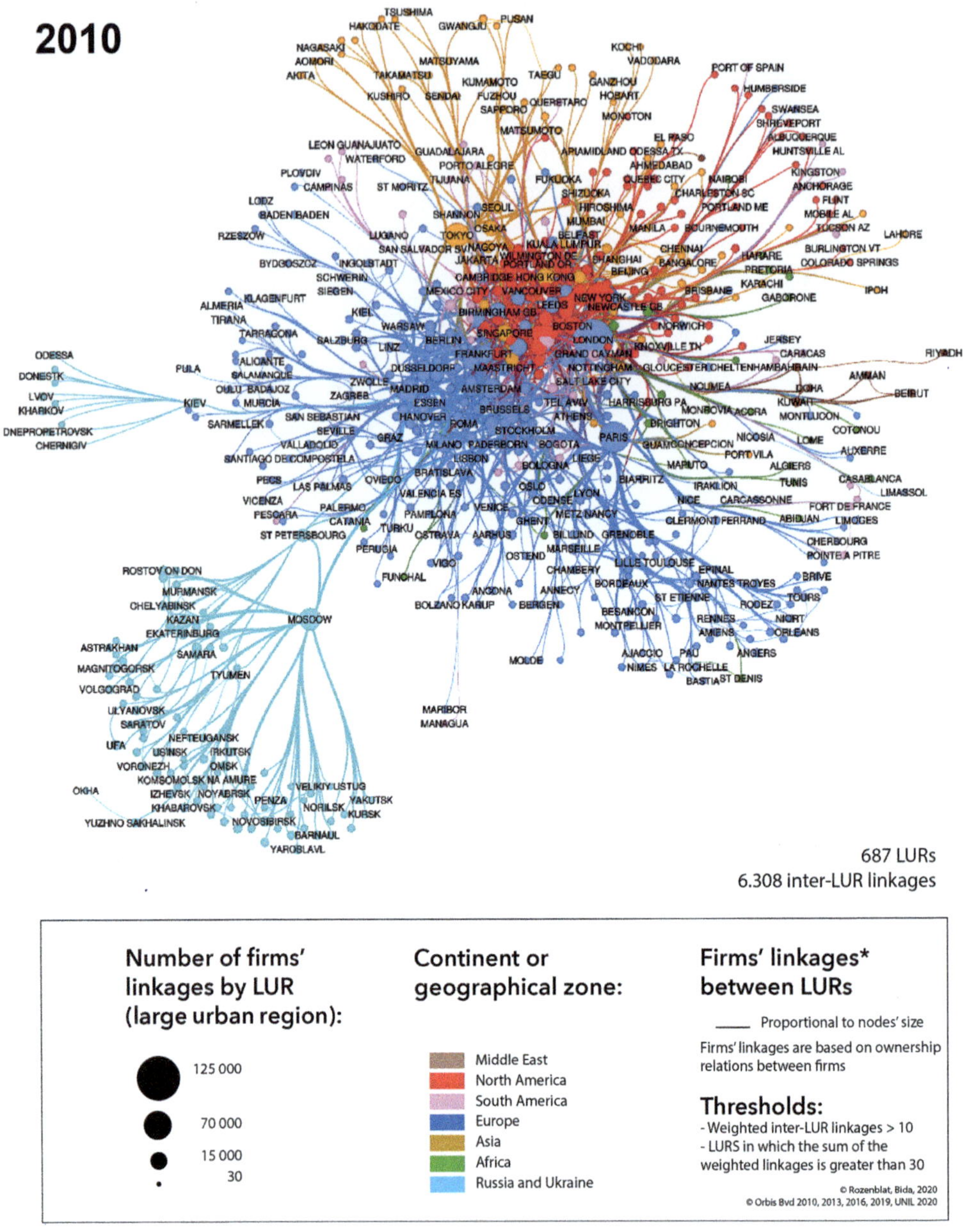

Figure 25.2a Cities in the globalization of firms in 2010

are positioned in between them in the core area. African and South American cities are either linked to the core system through European cities or through North American cities. This attachment mostly depends on their colonial period legacy, from which remains a special bond often reinforced by language (a good example of this type of legacy is the still strong ties between Spain and Portugal and South American cities; Rozenblat et al. 2017).

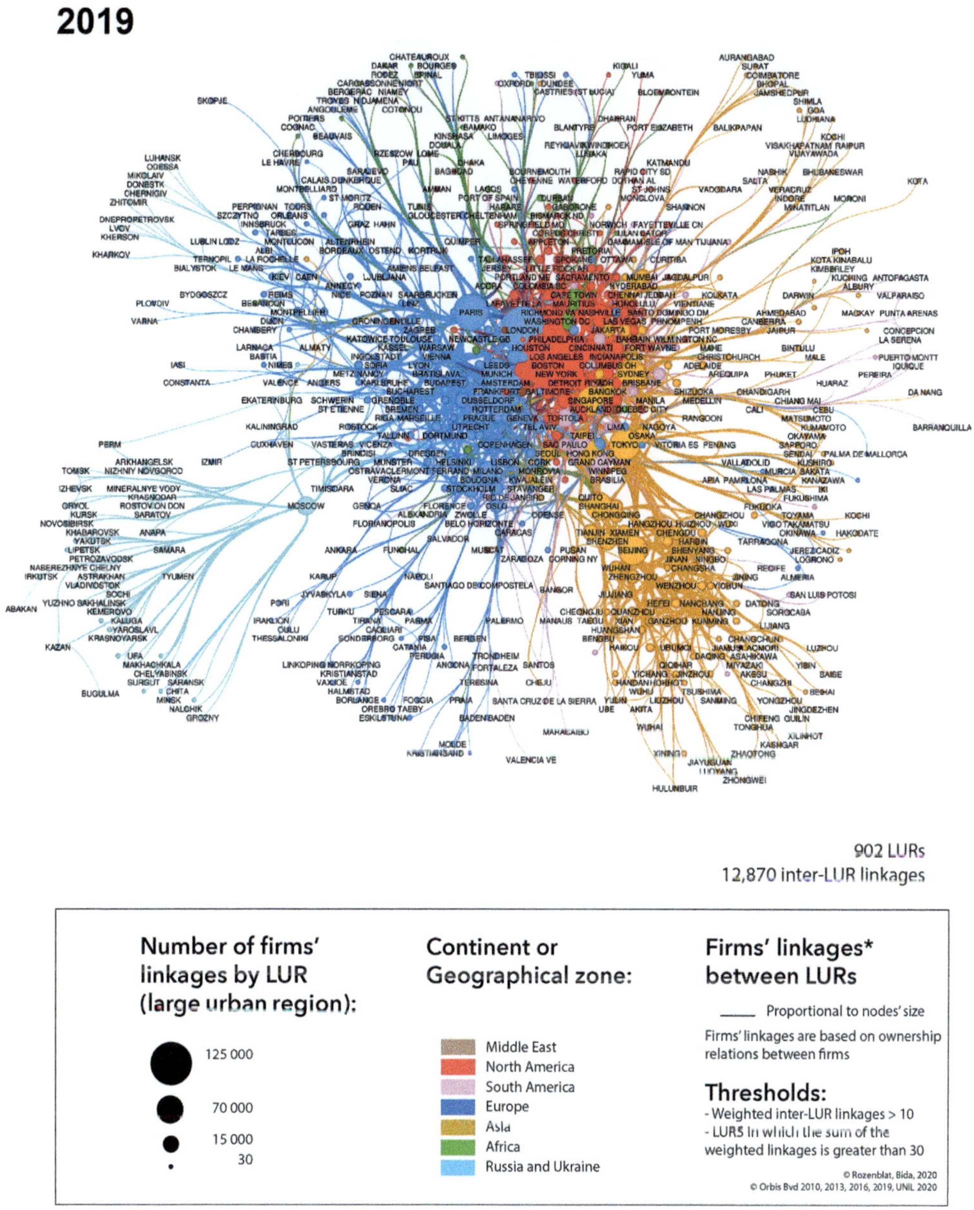

Figure 25.2b Cities in the globalization of firms in 2019

The evolution between 2010 and 2019 is principally characterized by the expansion of the cities' network. The number of cities above the threshold of 30 firm external linkages grew notably between 2010 and 2013, and the network of intercity linkages grew faster between 2013 and 2016, demonstrating a complexification of the cities' networks. In Figure 25.3 this expansion is more visible by geographical areas for all LURs.

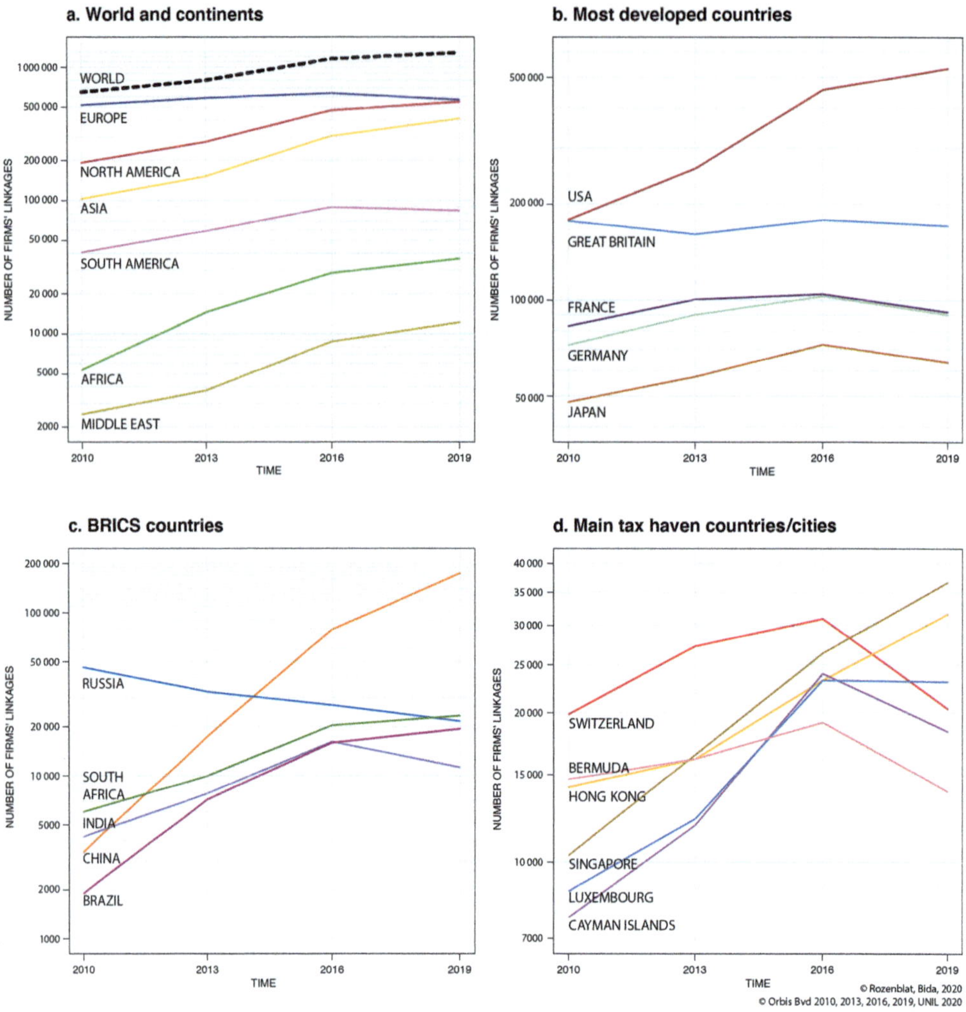

Figure 25.3 Evolution of the firms' linkages by areas

While most developed countries remained stable during the period 2010–19 and even decreased during 2016–19, the US's linkages continuously increased. Middle Eastern, Asian and North American firms' linkages (intraurban or interurban) grew the most from 2010 to 2019. The US (Figure 25.3b) and China (Figure 25.3c) had the most impressive total increase of their firms' networks in cities. Specifically, Chinese cities' firm networks expanded very fast. In Figure 25.2, Chinese cities began to constitute a distinct national Chinese system around Beijing in 2013, and this system was reinforced until 2019. In parallel, Shanghai and Hong Kong remained closer to the international core of the network.

Other BRICS countries in general gained importance in the worldwide city firms' system, except Russia, which regularly decreased from 36000 total linkages in 2010 to 22000 in 2019 (Figure 25.3c). For the tax haven countries, Singapore grew at the same speed as China, supplanting Hong Kong, which grew slower. After large increases until 2016, Switzerland, Luxembourg, Bermuda and the Cayman Islands decreased during the 2016–19 period.

3.4 Evolution of the Cities' Networks' Scopes

To better understand these different national trajectories, the geographical scope of the cities' networks can be summed up by intraurban, national, intracontinental and intercontinental scales (Figure 25.4).

The geographical scope of the firms' linkages at the world scale during the period 2010–19 shows a decreasing proportion of international linkages (intracontinental and intercontinental) coupled with an intraurban linkages' decrease for the benefit of intranational linkages. On average, at the world scale, intraurban linkages represented 40 percent of the total firms' linkages in 2010 and still represented 36 percent in 2019.

This worldwide average hides some very uneven evolutions according to the initial level of each country, as a dependence on statistics based only on categories of countries may not show evidence of very diverse country situations. Of the BRICS member countries, China, Brazil, India and South Africa, whose firms' networks increased a great deal during the period, decreased their intercontinental share of flows to the benefit of intraurban and intranational flows. However, India increased more its intracontinental linkages with other Asian cities. The US also experienced a large increase in firm networks at the national scale. One of the US cities that benefited more from this national consolidation was Wilmington (Delaware), which offers very low taxes to firms operating in its jurisdiction, reducing their tax burden. Russia is the country with a lower level of intercontinental linkages; however, during the period, its links with Europe increased not only proportionally, from 4 percent to 17 percent, but also in absolute value, from 2200 to 3600, despite the Russian multinational linkages' general decrease when considering all scales combined.

3.5 Interrelations Between Intraurban and Interurban Linkages

The role of cities in the global interurban system is based on a very stable proportion of intraurban linkages, and there is a log-linear relationship between the two numbers (Figure 25.5).

The urban scaling analysis in general 'reveals the *general* non-linear properties of how cities work, manifested as economies of scale, when certain quantities grow more slowly than city size ("sub-linearly") or as *increasing returns to scale*, when quantities grow faster than city size ("super-linearly")' (Bettencourt et al. 2020, p. 2, original emphases). The scaling properties can be interpreted as elasticity by economists, but also as network effects reflecting the way the agglomerated individuals may produce different quantities through a better and more complex coordination of their social networks, their institutional organization or their innovative level (Pumain et al. 2006; Batty 2006; Bettencourt et al. 2007; West 2017; Lobo et al. 2020).

In this instance, the scaling approach directly addresses two network quantities: the total firms' linkages that mostly comprise national and global networks and the local networks remaining inside the LURs. Thus, it presents a vision of the complexity of the

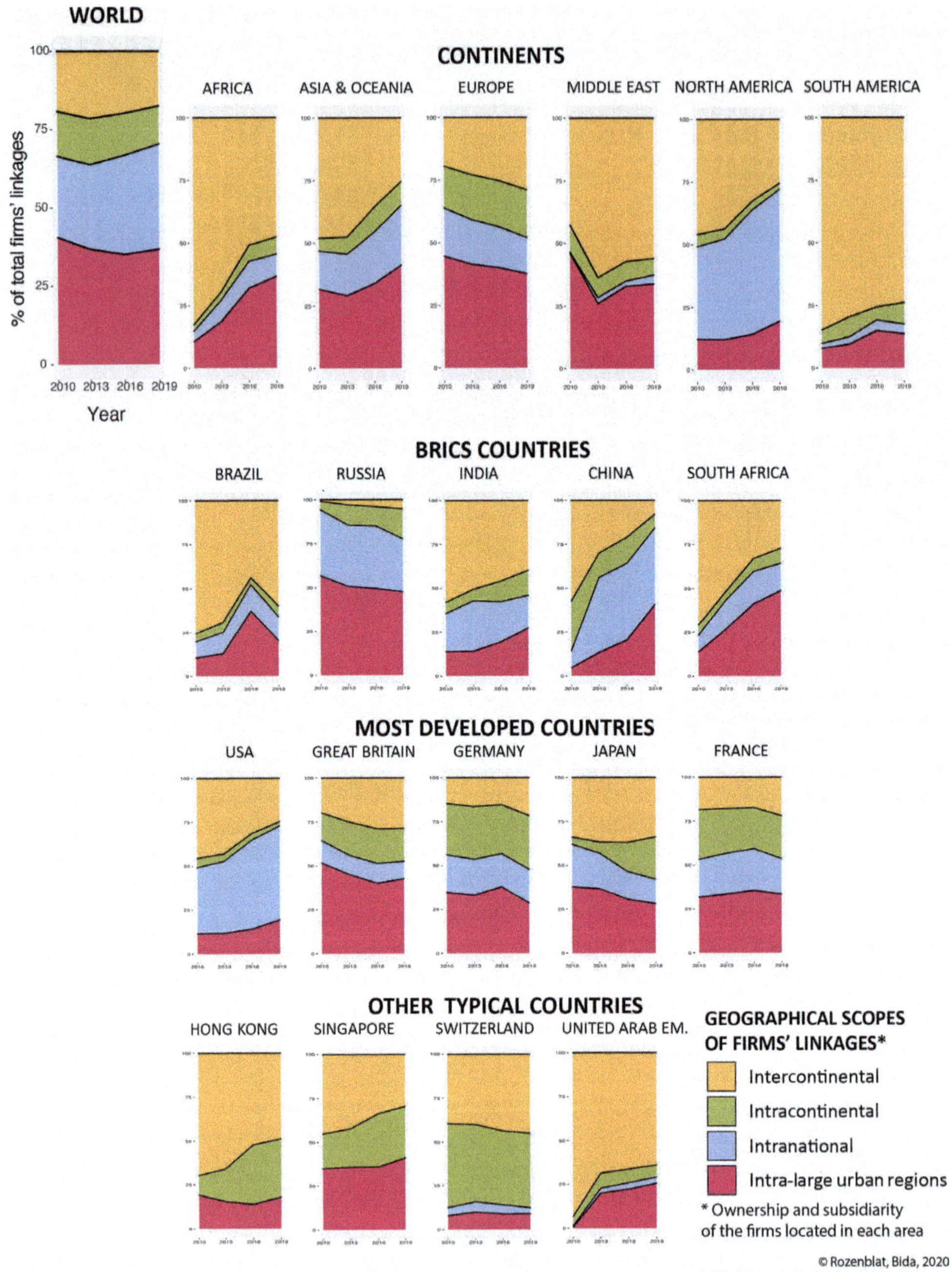

Figure 25.4 Evolution 2010–19 of the geographical scope of the firms' linkages by areas

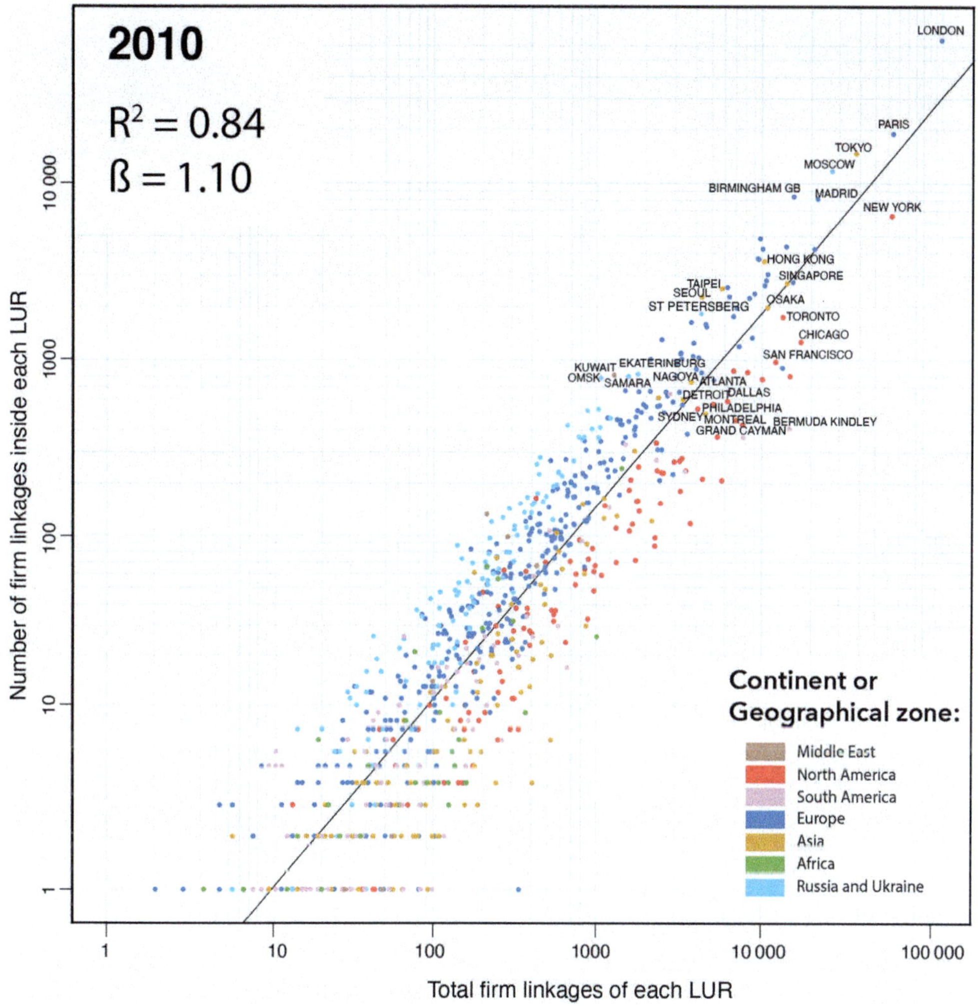

Figure 25.5 Intrafirm and interfirm linkages by LUR (2010–19)

local networks, while the total networks at every scale are varying. A stable ß parameter over the 2010–19 period means that the intraurban linkages remain, on average, at the same proportion in relation to the total linkages. A ß value that is always significantly over 1 means that local linkages are more than proportional to the total linkages and, therefore, they are relatively more numerous and part of larger connected cities.

This superlinear scaling-law demonstrates that local firm linkages are more complex in the most connected cities. It is no surprise that the most connected cities are London, Paris, New York and Tokyo. They did not change their positions during the period, but some cities changed considerably. In particular, Sydney and Singapore joined the leading pack in 2016, leaving Hong Kong behind, as it increased much less. Beijing and Shanghai increased in 2019. Moscow's linkages decreased during this period.

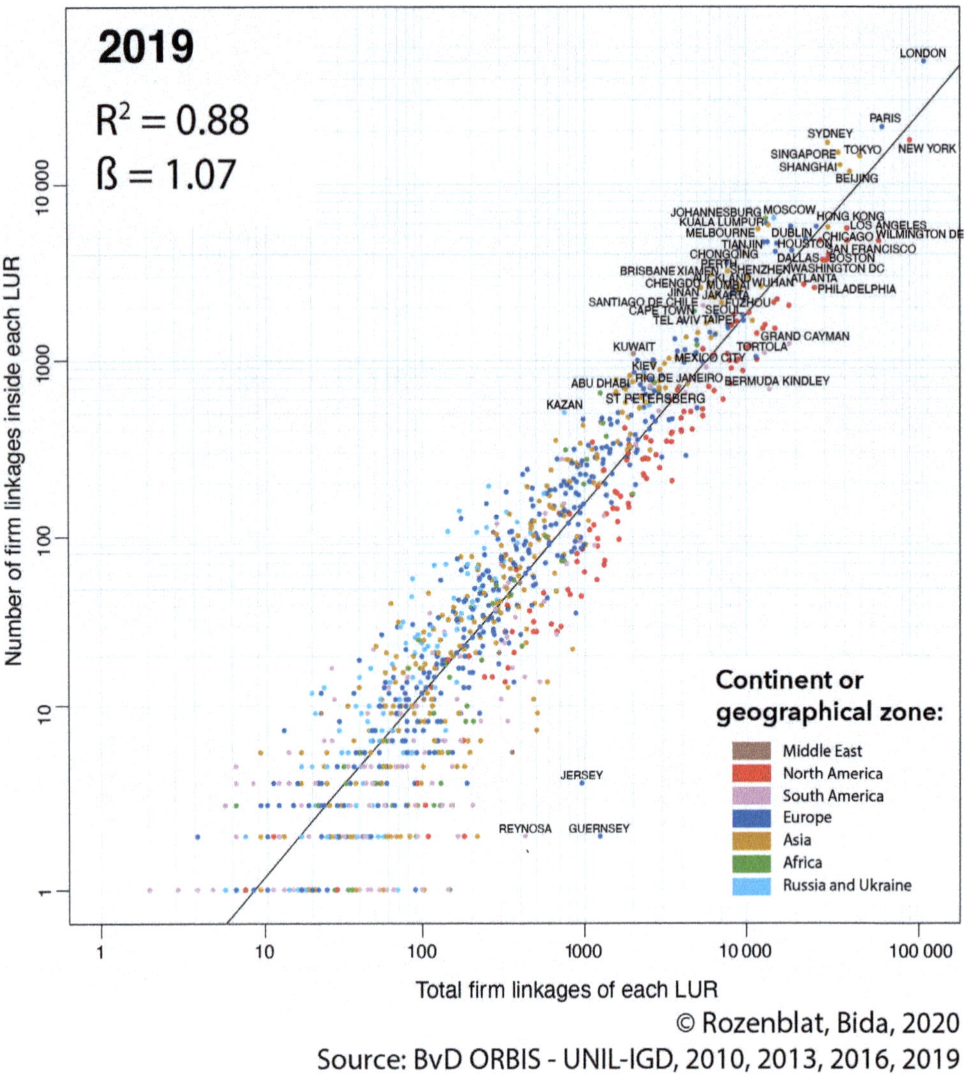

Figure 25.5 (continued)

In Figure 25.5, the continental colors of the cities (LURs) also underline a significant difference regarding this proportion: North American cities (red dots below the line) developed fewer local linkages in proportion to their total linkages than did European cities (dark blue dots above the line). For the cities of other continents, there is no systematic logic regarding this difference. The tax havens, such as Grand Cayman, Bermuda Kinsley, Tortola, lower Guernsey and Jersey, are systematically positioned below the line, demonstrating their weak local activity in relation to their global role.

The superlinear scaling between intraurban and total linkages hides a difference between out-reach linkages (power towards other cities) and the in-reach ones (attractiveness from outside). The distinction between these two directions is a higher rate of attractiveness for local linkage complexity (ß ≈ 1.10 during all the periods) than for power linkages (ß ≈ 0.80). This means that firms develop preferentially a larger number of local linkages when they arrive in a local place than when they develop their coordination of worldwide external activities. This is of great importance for the comprehension of their embeddedness in local places.

Most of the cities increased along the diagonal during 2010–19, so there is an allometric growth of the local linkages in relation to the total linkages (all the system and its elements grow at the same speed). This is a very common feature of this evolutionary distribution (Bettencourt et al. 2020).

The trajectories do not depend on the initial conditions in 2010 nor on the size of the cities. Some national trends reflect the importance of states' policies for cities, but these trends are not sufficient enough to affect each position of city evolution inside global networks and their local developments. A graph on the trajectories of the 15 first LURs according to their total number of linkages in 2019 (Figure 25.6) illustrates the diversity of trajectories at the top of the hierarchy of the urban system.

The colors of the cities' trajectories in Figure 25.6 correspond with their types according to a classification we implemented for whole cities (we do not show here the whole classification). London, Paris and Tokyo observed an absolute decrease of their firms' linkages but with different speeds. At the opposite end, New York increased its position a great deal until 2016, as its linkage growth was reinforcing specifically by its intraurban linkages, which increased from 10 percent of the total in 2010 to 18 percent in 2019. Boston evolved from 7 percent in 2010 to a 12 percent level for intraurban linkages in 2019, experiencing a trajectory very similar to that of New York. Singapore shared the same trajectory with the US cities of Los Angeles, Chicago, Washington, DC and San Francisco. The relative weight of their intraurban linkages was not reinforced much, which is different from Boston and New York.

The very impressive growth of Shanghai and Beijing demonstrates a difference between the two cities: at the end of the period, Shanghai had a 34 percent level of intraurban linkages, while Beijing had only 27 percent. Hong Kong, compared with the two leading Chinese cities, had slower growth and was bypassed by both of them. Sydney and Wilmington, DE, shared the same type of trajectory of growth, and Sydney had a higher absolute level.

4. LOCAL LINKAGES INSIDE LARGE URBAN REGIONS (LURS)

According to these very uneven trajectories of cities, we propose to deepen the examination of two LURs that have experienced very opposite evolutions: London and Beijing. Despite a small increase of external linkages, London saw its total number of linkages decrease from 129 000 total linkages in 2010 to 123 000 in 2019 owing to a strong decrease of internal linkages from 65 000 to 50 000, decreasing the share of intraurban linkages from 50 percent to 41 percent. Alternatively, Beijing increased its networks in our sample from fewer than 900 in 2010 to more than 43 000 in 2019. In the meantime, Beijing's intraurban linkages grew from 30 to 11 800, representing an evolution of the share of intraurban linkages from 4 percent to 27 percent.

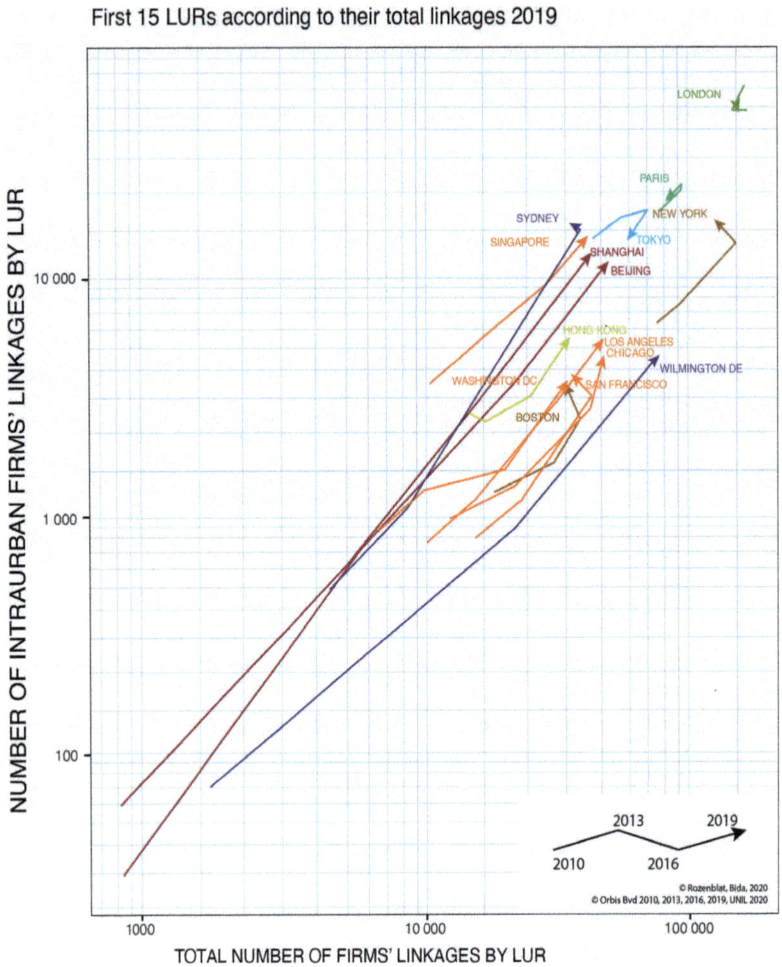

Figure 25.6 Trajectories of 15 cities according to their total and intraurban linkages

These two case studies illustrate the extent to which multinational firms use intraurban networks to maintain and expand their global activities in two very different economic and political contexts, inducing differences that are outlined in this section.

4.1 London's Intraurban Linkages

London is a place characterized by a long heritage of global management of the development of a worldwide market economy. Containing a myriad of firms all linked together, the 2008 crisis revealed the extent of this intertwined network, demonstrated by the cascading failure of a number of firms (London still comprised 50 percent of the total linkages in 2010). Owing to this decrease, since 2010, the weight of these intraurban linkages has been highly deflated but has remained at the worldwide cities' average, comprising 41 percent of all London firms' linkages.

4.1.1 London's network components of local linkages

There is a great deal of interest in analyzing London's intraurban linkages that are linked to each other inside the LUR and that constitute what is termed a main network component. In 2019, the main network component of London contained more than 20 000 local firms connected together by more than 30 000 ownership linkages that represent 40 percent of all the firms involved in London's intra-LUR linkages and comprise 60 percent of their intra-LUR linkages (Figure 25.7).

The main local component of firms that are interlinked by financial-ownership relationships reveal the strong city system of firms that acquire individual strength by their direct access to the financial and competency resources of the whole system. This main London component is centered on financial and insurance firms that support a myriad of other specialized branches. The most central financial firms of that system (in relation to betweenness centrality) are Premier Asset Management Holding (PAM), which is a company that specializes in investment management, and the City of London Investment Management Company Limited (CLIM), which specializes in closed-end funds (CEFs). These two companies are very interlinked with each other and are even linked socially, as some of the directors of CLIM come from PAM as well as from other important rival companies, such as The Vanguard Group. Together with more than 3200 other financial and insurance services companies, such as other investment management companies (for example, Aviva Group, Legal and General group, Schroders, Prudential, and the Janus Henderson group), they demonstrate the diversity of London's numerous financial firms, each of which specializes in very narrow financial sectors. Together with them, the main banks (for example, Lloyds, Barclays and HSBC) are also linked. All together, they form the financial core of this main component system (in light pink in Figure 25.7), permitting the flow of complex investments between them and with other companies specialized by linked branches around this core. The role of real-estate companies is particularly central (for example, Banner Homes Group, the British Land, Land Securities Group and Intu Properties); these companies are situated in different parts of the central network and constitute specific branches. The presence of real-estate linkages in the network confirms the role of real estate in the consolidation of parts of portfolio investments and the intimate relationship between capitalization and real estate (Aalbers 2019). Similarly, construction companies (that is, the Berkeley group, Taylor Wimpey, Balfour Beatty, Canary Wharf developments and Kier Building Limited) provide direct investments in real construction not only in London but also in the rest of the country and all over the world.

Other companies of this main network component are involved in wholesale trade (for example, Tesco, J. Sainsbury, Tyson Foods, Universal Meat, Mark & Spencer, BCA Market Place, Diploma, and Kingfisher), transportation and storage (for example, the First Property Group, Clarkson, Braemar Shipping, British Airways, John Swire & Sons, Big Yellow, and Gama Aviation), and mining and quarrying (for example, Royal Dutch Shell, BG Group, Rio Tinto, Tullow oil, BHP Group, Ensco Rowan, Sirius Minerals, ENI Lasmo and Esso). Companies involved in professional, scientific and technical activities include WPP Jubilee Limited, which specializes in advertising, media and public relations and invests in more than 400 London companies involved in artwork, film and design (the orange pack at the top of Figure 25.7).

In addition to this main network component, London's intraurban linkages comprise more than 2600 independent network components. Each of these secondary components

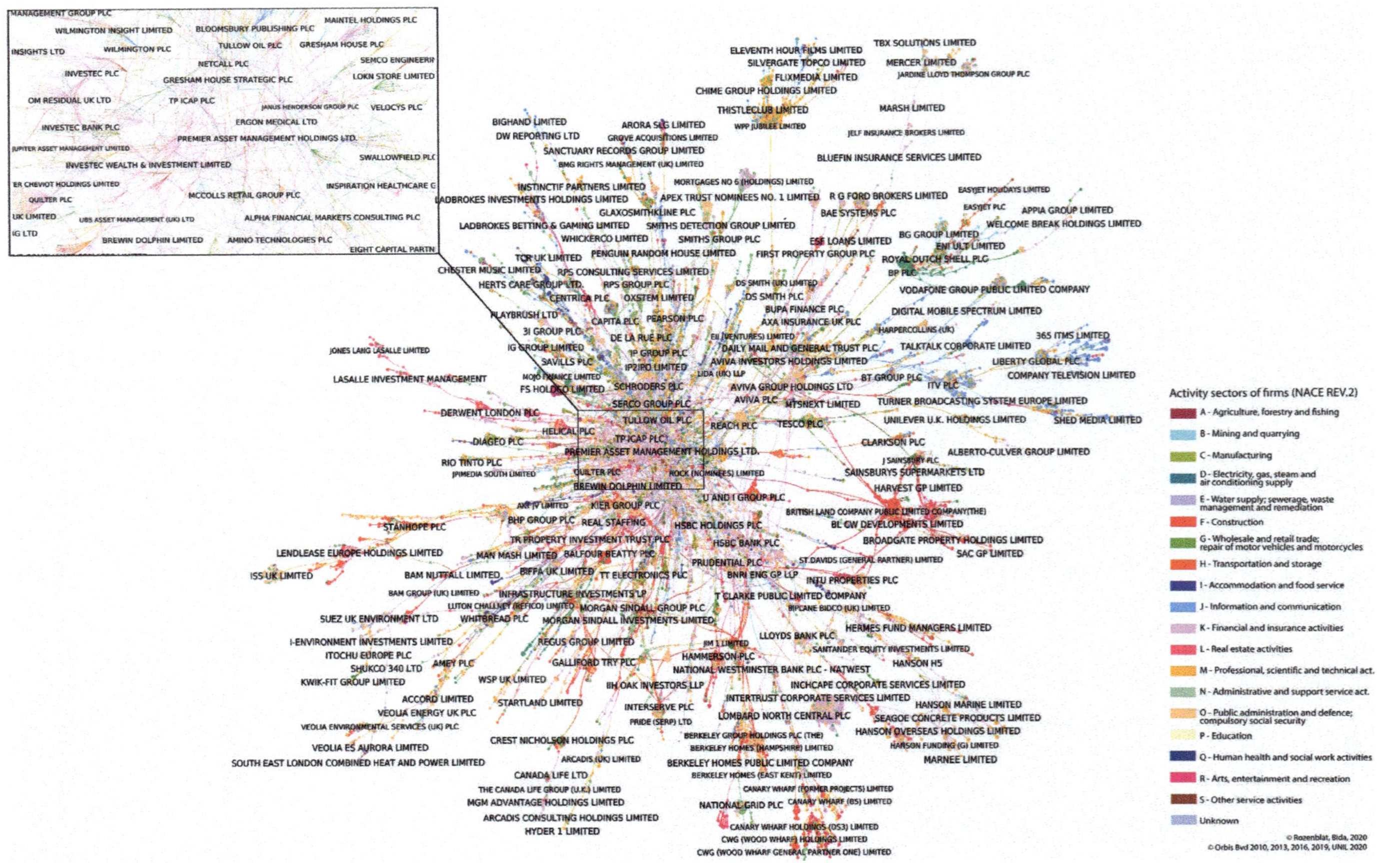

Figure 25.7 London's intra-LUR main network component in 2019

also contains a mix of several activities. We propose some examples of these smaller networks of firms in order to better understand the arrangement of the activities' complementarities (Figure 25.8).

The second main component contains 'only' 144 firms linked by 203 ownership relations (Figure 25.8a) and demonstrates well the way in which a foreign firm could insert itself into London's game. Mitsui is the core of one of the largest Japanese keiretsus. Created in 1876, this company is now known as the Sumitomo Mitsui Banking Corporation (SMBC). Specializing in raw material trade, Mitsui is also active in rail material and construction.

This scheme demonstrates clearly how SMBC, represented by Mitsui Power Venture Limited, entered London and established financial contact with International Power Limited through both IPM Holdings and the IPM Eagle LLP, which operate worldwide as investment management firms specifically in the raw materials and energy sectors. On the other side of this relationship, International Power Limited is the international part of the National Power firm, which was created from the privatization of the nationalized electricity industry of England and Wales in 1990s. It was acquired in totality in 2012 by the GDF-Suez French group (now ENGI) and diversified its competencies to all kinds of solar, nuclear, wind electric power generation and power grid management. Its activities in London are conducted through productive companies of electricity generation (in green in Figure 25.8a) and by exploiting a diverse portfolio of energy sources – renewables, natural gas and coal (similar in operation to Trustenergy BV, based in the Netherlands but mostly active in Portugal, and to Moray Offshore), electric installation and repair (Swindon Power Technical Services Limited) and administrative and headquarters activities (in orange). Investment holding companies raise funds (IPM International BV, International Power Group Trustee Limited and Rugeley Power Generation Limited) and sometimes save taxes by being based in Jersey, Gibraltar or Luxembourg (International Power Luxembourg Holding).

Mitsui is also linked with one of the biggest British rail group of companies, namely, the First Rail Group, which encompasses five main companies: Avanti West Coast (a joint venture with Trenitalia), Great Western Railway, South Western Railway, TransPennine Express and Hull trains. This link was made possible by taking control of West Midlands Holdings Limited in 2017 (less than 50 percent ownership but more than 25 percent), consequently placing Japanese individuals as directors of the firm and reinforcing Japan's participation in the West Midlands Railway as part of a joint venture between Abellio, the East Japan Railway Company and the Japanese conglomerate Mitsui. Included also in this local network is Mass Transit Railway Corporation (MTR), a Hong Kong company, which manages the Hong Kong metro-system and which built the Guangzhou-Shenzhen-Hong Kong High Speed Rail and many other recent projects in mainland China, in Sweden and in Australia. The double insertion of Mitsui in energy, on the one hand, and in transportation, on the other, enables all the companies to strengthen their position in numerous worldwide projects, for example, projects from China, Dubai or Australia.

Another incoming example can be seen in the edition sector: the Hachette Group, a French company, acquired a large set of British companies that included some British patrimonial publishing companies, such as J.M. Dent and Company, fiction publishers, such as the Orion Group, and nonfiction publishers, such as the Octopus group. Unlike Mitsui, the Hachette group remains in the unique specialized sector of book editions.

INTERNATIONAL POWER - MITSUI - FIRST RAIL

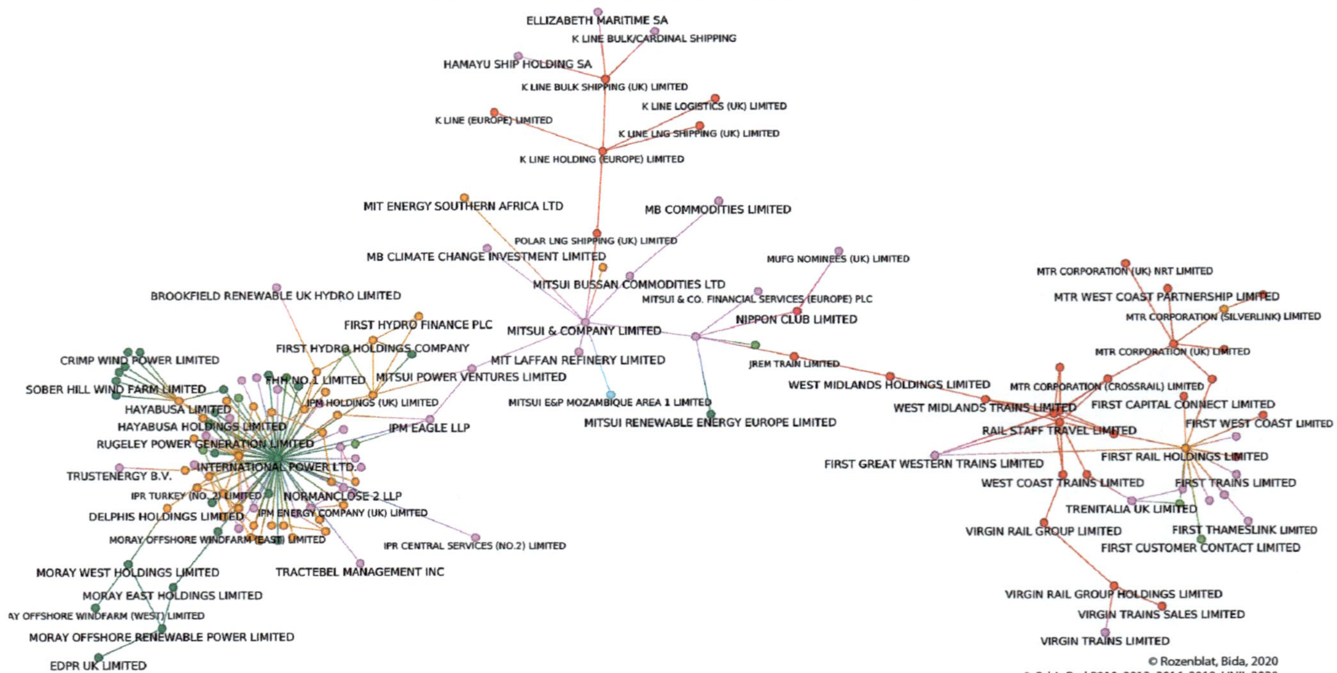

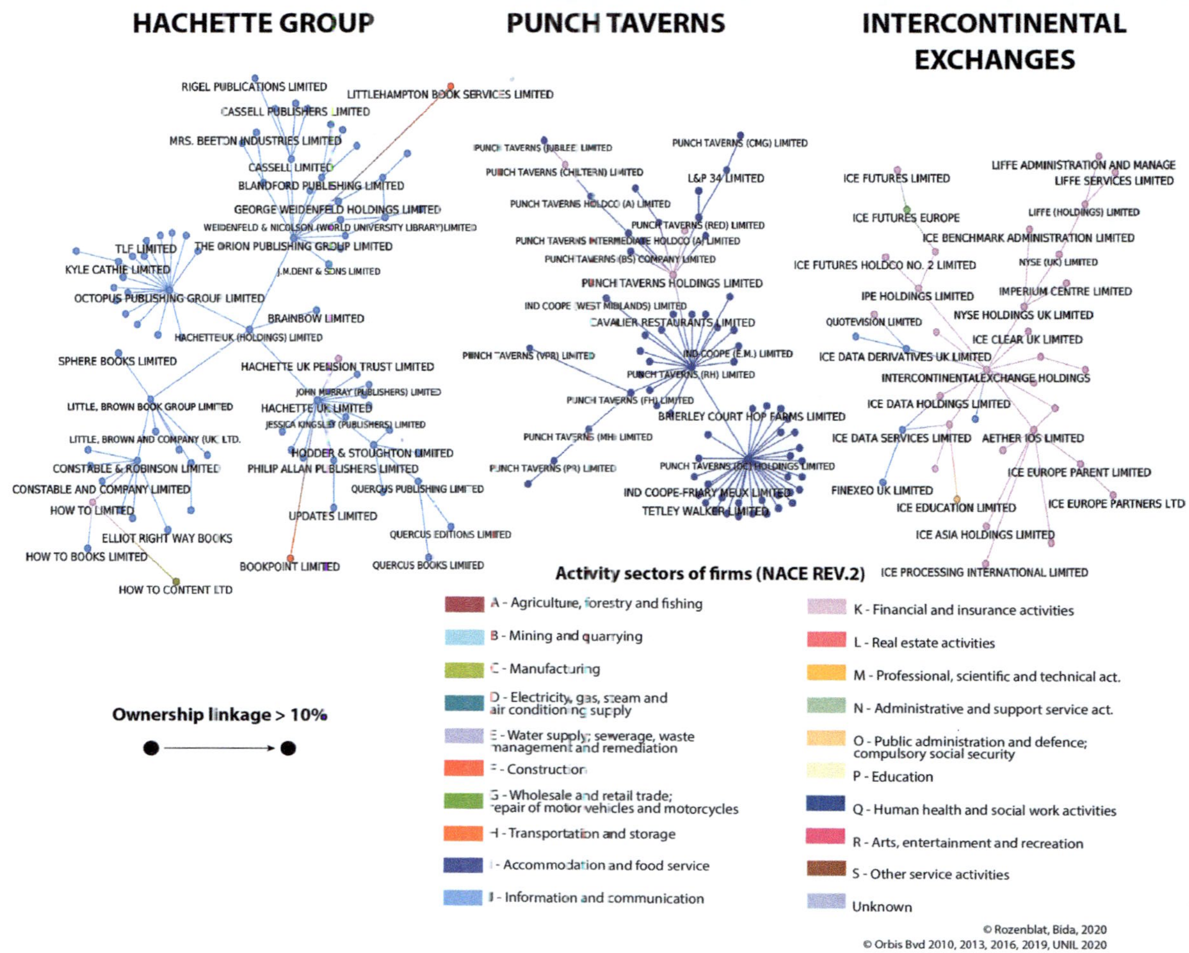

Figure 25.8a Examples of London's other intra-LUR network components in 2019

FAIRVIEW NEW HOME

CHAUCER CAPITAL INVESTMENTS

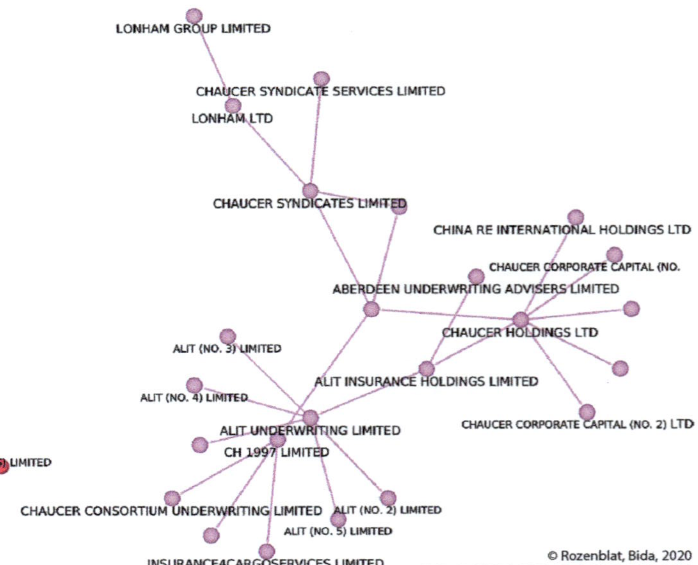

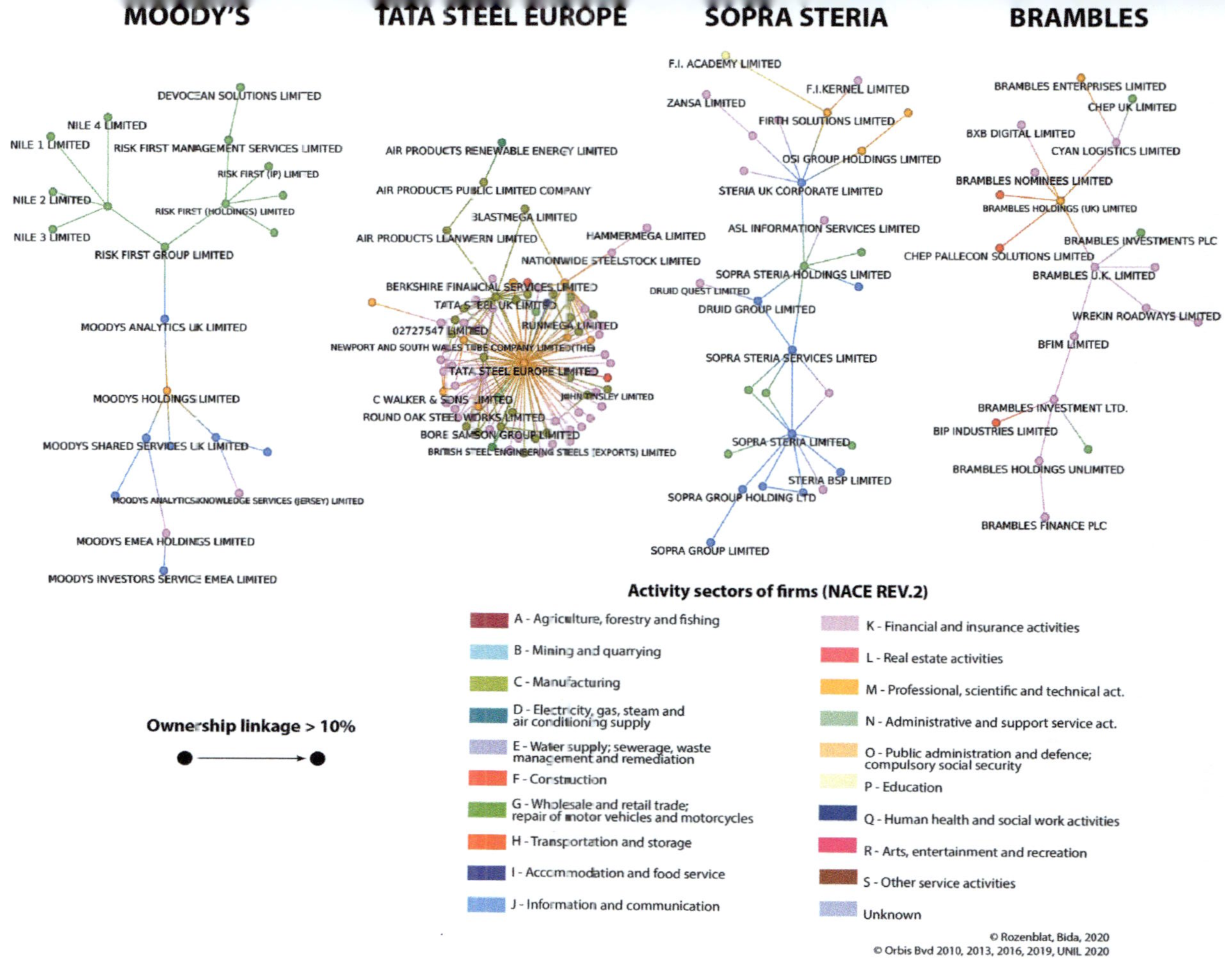

Figure 25.8b *Examples of London's other intra-LUR network components in 2019 (continued)*

A local firm, Punch Tavern, which comprises 69 companies operating inside London, operates the traditional English public house (pub) and has more than 1300 pubs and breweries all over England, Wales and Scotland (including brands, such as Pubmaster, Inspired Group and Avebury Holdings). It was registered on the stock exchange in 2002 and was acquired by Patron Capital and May Capital in 2017.

The Intercontinental Exchange Group (ICE) provides electronic trading platforms that operate in global markets through three main components: ICE provides market strategic designs; the New York Stock Exchange (NYSE) provides trading for equities, options, Exchange Traded Funds (ETFs) and listings; and ICE Data service provides analytics to enable risk management. Represented in Figure 25.8b, another financial service company, Chaucer Capital Investment, specializes in real-estate investments. The Moody's Group provides more advanced services in information analytics for firm monitoring. Based in Brussels, the Bureau van Dijk, which provided all the data used in this chapter, was acquired by Moody's in 2016 (which from then on complicated much of our data acquisition procedures).

Figure 25.8b shows other examples illustrating the diversity of the patterns of these local firm networks in London; this pattern of diversity is demonstrated as much by the networks' shape as by their activity combination. Fairview New Home combines strong construction and real-estate companies with financial companies; Tata Steel Europe together with Corus management interlinks 81 companies, combining engineering, finance, human health, trade, manufacturing, real estate and administration to support the financial aspects of London's steel activities; Sopra Steria edits specialized software required by scientific companies (in orange in Figure 25.8b), information and communication companies (in blue), and financial service companies (in light pink); the company also support training with the FI Academy (in yellow). Brambles is a logistic company transporting goods all over the globe. Its London plants combine only two logistic companies, which are complemented by scientific, administrative and financial support firms.

4.1.2 Local firm activities in London

In addition to looking at the components, we consider London's total intra-LUR linkages in order to analyze their activities (Table 25.3).

According to Table 25.3, London's firms as a whole participated in activities involving intra-LUR linkages, although since 2010 the number of firms decreased in total and there was a transformation in the firms' intraurban linkages' activity composition. The firms mainly were in the finance sector, insurance sector and administrative services in 2010, but the share of these sectors dramatically decreased during 2010–19: these sectors together represented 47 percent of all firms in 2010, while they represented only 33 percent of all firms in 2019. In addition to the decrease of financial and administrative services, there were other activities lost owing to the decrease in the number of local network companies in construction, manufacturing, art and entertainment or agriculture. In contrast, there was an increase in the number of companies engaged in professional, scientific, technical and information and communication activities; the number of firms engaged in these activities grew both in absolute numbers and in percentages.

Considering their linkages in 2019, 25 percent of the firms' linkages were owned by financial firms, 20 percent by firms in professional, scientific and technical activities, and approximately 8 percent by firms in administration, real estate, manufacturing and trade.

Table 25.3 Activities of the firms participating in intra-London linkages 2010–19

2010 – activity of firms deploying local linkages				2019 – activity of firms deploying local linkages			
NACE REV.2		Number of firms	%	NACE REV.2		Number of firms	%
K	Financial and insurance activities	9977	26.3	K	Financial and insurance activities	7558	19.9
N	Administrative and support service activities	8215	21.6	M	Professional, scientific and technical activities	5688	15.0
F	Construction	4221	11.1	N	Administrative and support service activities	5091	13.4
J	Information and communication	3492	9.2	J	Information and communication	4503	11.9
C	Manufacturing	2516	6.6	L	Real estate activities	2972	7.8
L	Real estate activities	2089	5.5	F	Construction	2381	6.3
M	Professional, scientific and technical activities	2066	5.4	C	Manufacturing	1974	5.2
G	Wholesale and retail trade; repair of motor vehicles and motorcycles	1946	5.1	G	Wholesale and retail trade; repair of motor vehicles and motorcycles	1901	5.0
S	Other service activities	1246	3.3	D	Electricity, gas, steam and air conditioning supply	1118	2.9
H	Transportation and storage	924	2.4	H	Transportation and storage	888	2.3
R	Arts, entertainment and recreation	862	2.3	S	Other service activities	721	1.9
B	Mining and quarrying	698	1.8	B	Mining and quarrying	717	1.9
I	Accommodations and food service act.	615	1.6	I	Accommodations and food service act.	705	1.9
Q	Human health and social work act.	340	0.9	Q	Human health and social work act.	668	1.8
D	Electricity, gas, steam and air conditioning supply	302	0.8	R	Arts, entertainment and recreation	453	1.2
E	Water supply; sewerage, waste management and remediation activities	100	0.3	T	Activities of households as employers; undifferentiated goods- and services-producing activities of households for own use	209	0.6
P	Education	96	0.3	E	Water supply, sewerage, waste management and remediation activities	164	0.4
A	Agriculture, forestry and fishing	83	0.2	P	Education	163	0.4
O	Public administration and defense; compulsory social security	36	0.1	O	Public administration and defense; compulsory social security	80	0.2
T	Activities of households as employers; undifferentiated goods- and services-producing activities of households for own use	22	0.1	A	Agriculture, forestry and fishing	42	0.1
U	Activities of extraterritorial organizations and bodies	0	0.0	U	Activities of extraterritorial organizations and bodies	4	0.0
No information		16706		No information		11174	
Total		56552		Total		49174	
With information		39846	100	With information		38000	100

Source: © Rozenblat, Bida, 2020. © ORBIS BvD – UNIL 2010–2019.

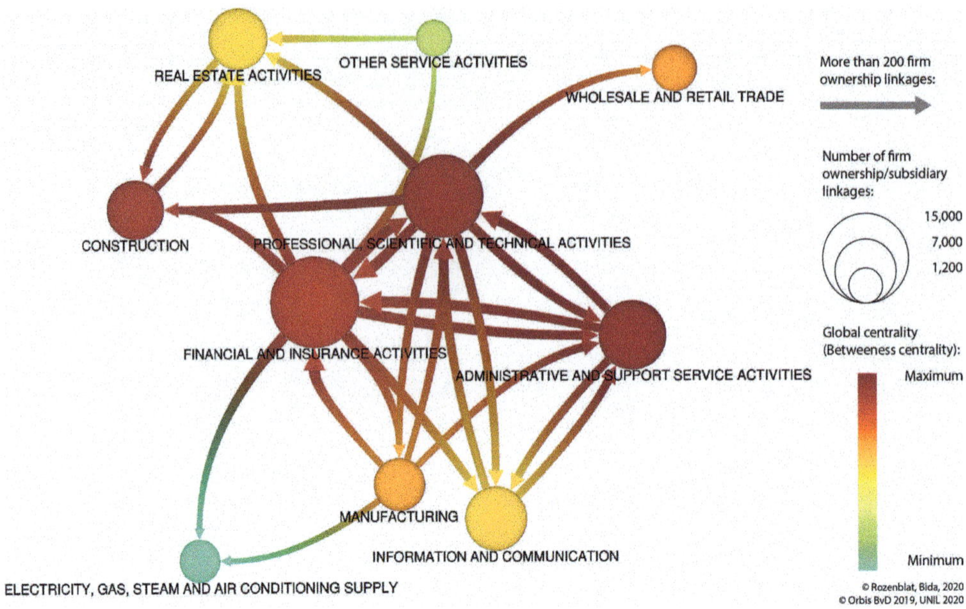

Figure 25.9 Ownership system between activities of London's intra-LUR firms' network in 2019

The local ownership system is centered on finance, scientific, technical and administrative activities, while other activities formed their main periphery (Figure 25.9).

4.1.3 London's LUR polycentricity

As regards the spatial distribution inside the LUR of London, there is a high concentration inside London's MUA (the equivalent of Greater London) – concentrating 98 percent of all the firms' linkages (Figure 25.10).

Around the London MUA of Greater London, forming a star, most of the other MUAs of the LUR are only directly linked with firms of the core MUA. The direction of the linkages slightly benefits the owners in London possessing subsidiaries in the periphery (200 links), but the linkages in the opposite direction are balanced (130 links). The main component that was observed in Figure 25.7 is partly distributed in all the surrounding London satellites. This can be seen in services industry companies such as Personal Assurance (insurance) (24 linked companies) or industrial companies, such as Design Group UK (cards and gift wrap) (18 linked companies), Essentra (plastic and fiber products) (31 linked companies) or Domino's Pizza (19 linked companies), which are all located in Milton Keynes and belong to the main London component. This is also demonstrated by Venture Life Group (healthcare) in Bracknell Ascot and Sureserve Compliance Services (repair services of gas, fire and electrical, water and air conditioning) in Reading.

Other than this star that the MUAs constitute around London city, only Milton Keynes, Bracknell Ascot and Reading form a weak polycentric network together with Chelmsford, Colchester, Slough, Guilford and Burgess Hill. Outside London, we count approximately 500 firm links that are located 90 percent inside the MUAs.

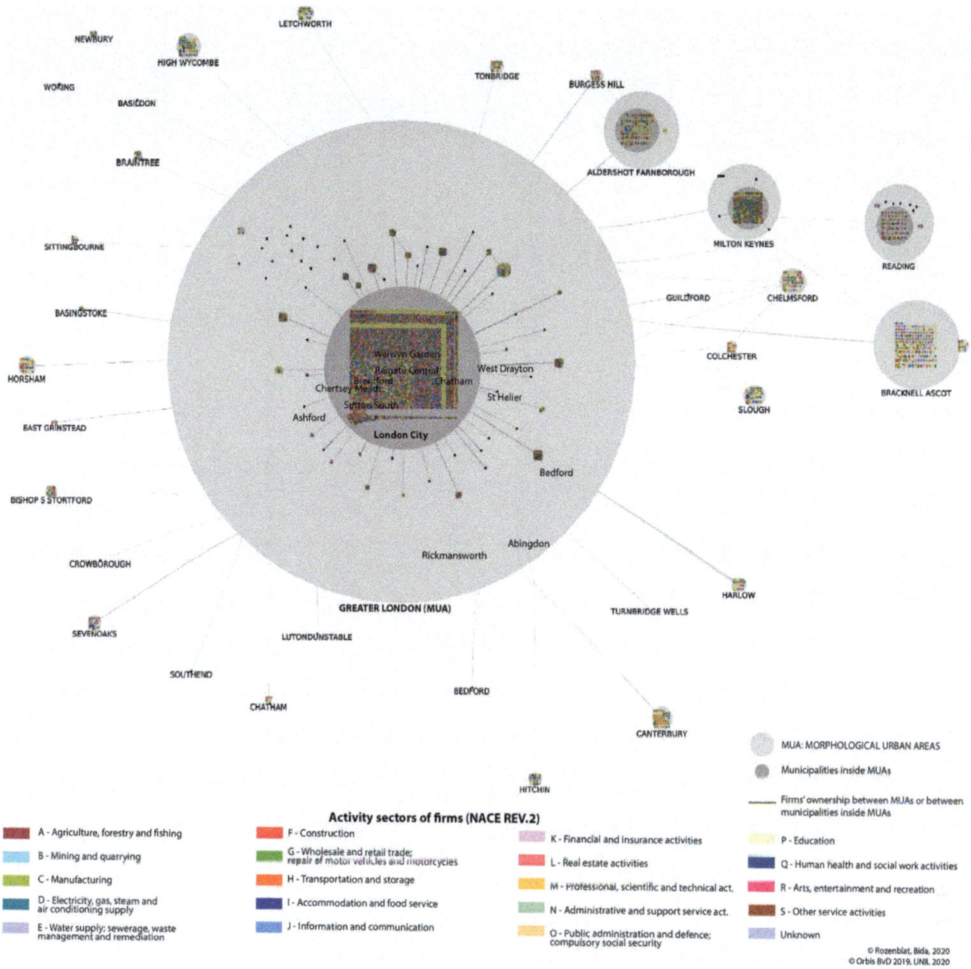

Figure 25.10 *London's intra-LUR firms' network polycentrism in 2019*

4.2 Beijing's Intraurban Linkages

Unlike London, Beijing experienced a huge increase in the weight of its global networks, as well as a high level of complexification in its intraurban linkages, which increased from 4 percent in 2010 to 27 percent in 2019.

4.2.1 Beijing's network components of local linkages

In 2019, Beijing's main component remained much smaller than that in London, with 8626 firms and 9725 linkages (Figure 25.11).

This main component comprises more than 75 percent of the total 11 416 firms involved in the intraurban firms and 82 percent of the total intraurban linkages. This is much larger than London's main component, which comprises 40 percent of all the firms involved in

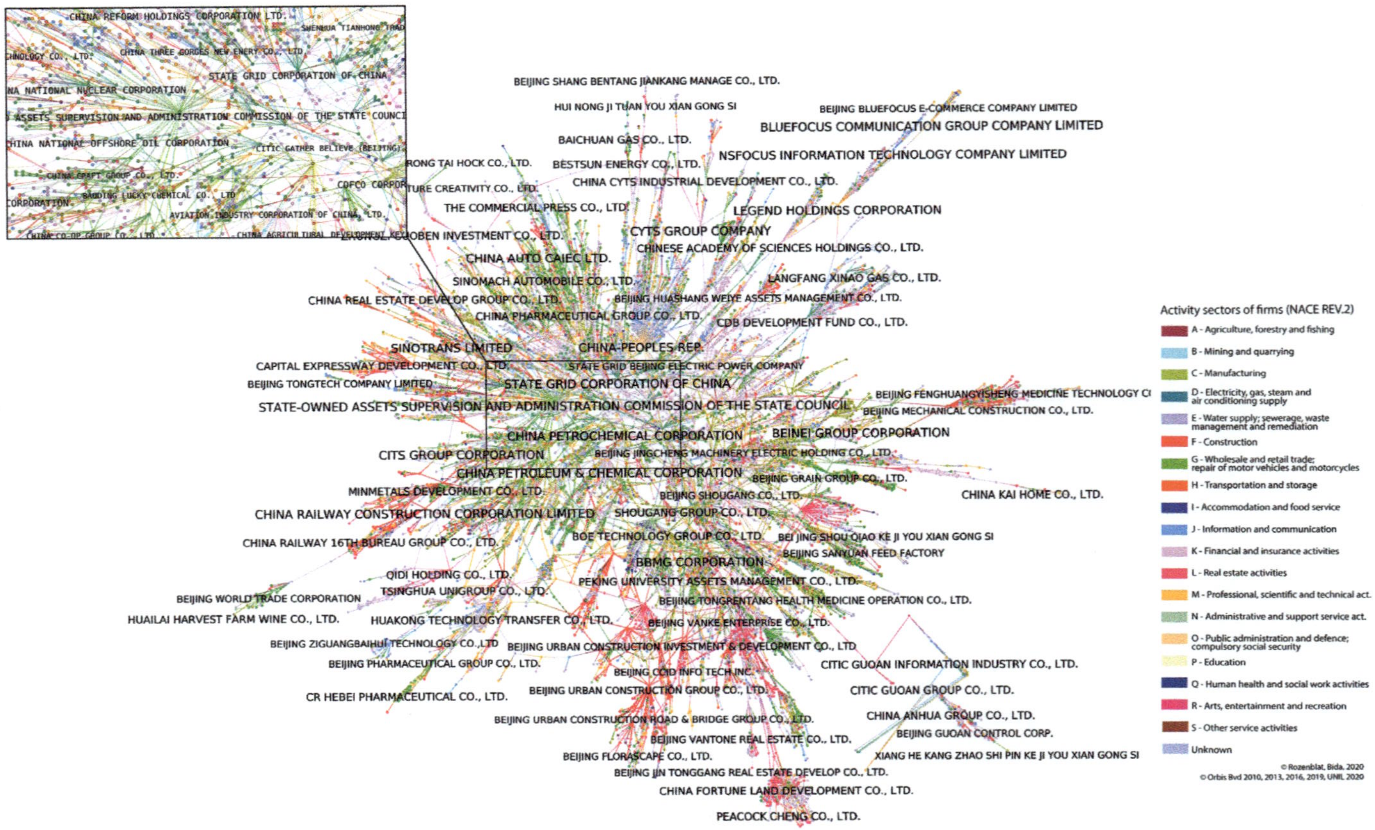

Figure 25.11 Beijing's intra-LUR main network component in 2019

London's intra-LUR linkages and 60 percent of their intra-LUR linkages. Beijing's main component reflects the organization of the Chinese state, which provides leadership in a very concentrated and hierarchical way.

Beijing's main component is totally centered around the administrative support functions of the State-owned Assets Supervision and Administration Commission of the State Council (green star in Figure 25.8), affirming the importance of the state in financial and economic planning in cooperation with financial state companies, such as the CDB Development Fund Company, the Citic Group Corporation, Beijing Huashang Weiye Assets Management and China Chengtong Holdings Group. Other numerous firms also classified as administrative and support services firms occupy the center of this network, each of them orientated to specific activities (for example, China Agricultural Development Key Construction Fund, Beijing Energy Holding, Aviation Industry Corporation of China, Beijing Jingcheng Machinery Electric Holding, Beijing Urban Construction Investment & Development, Beijing Uni-Construction Group and Peking University Assets Management Company). Each of these companies directs the activities of specialized manufacturing or trade companies that are close to the central government. This relationship is revealed by green linkages (manufacturing or trade) in the center of this main component.

Beijing's intraurban firm linkages are characterized by a very dominant power arrangement in which all firms are close to the central head, with only one or two intermediate edges. It is structured by a dominance of hierarchical linkages and a lack of complexity: on average, Beijing's main intraurban component comprises 1.12 linkages per firm, while that of London comprises 1.5 linkages. Thus, Beijing is much closer to 1, denoting the extreme case in which every firm has only 1 linkage, demonstrating a pure tree structure.

In addition to this huge main component, Beijing has 760 other much smaller local components, the largest of which is that for Beijing Haidian Science and Technology, which contains 28 firms (Figure 25.12a). This company, located in the Haidian district near the former Beijing Steel and Iron Institute (today, named the University of Science and Technology of Beijing), has supported for more than 20 years the development of the Haidian Science Park, also called Zhongguancun. With approximately 100 Fortune 500 companies that have located their research and development centers in Haidian Science Park and including several innovative Chinese corporations, such as Lenovo, Founder and Baidu.com, it is one of China's largest and most important high-technology bases. Dedicated to healthcare services, it supports diversified activities not only in medicine but also in robots and, more surprising, in jewelry. Its structure is also organized in a star shape and includes financial, manufacturing and trade companies.

This star shape has also been adopted by other components, such as the Vantone Real Estate component that mixes financial and real-estate companies together with trade companies and hotels, or the component for Pony Roll Media Culture, an entertainment company located in Dongcheng and engaged in providing e-tickets for concerts. This company was acquired in 2017 by the famous Chinese Alibaba group Holding, based in Hangzhou and located 200 kilometers from Shanghai; it is not located in Beijing.

Moreover, focusing on foreign companies entering the Beijing arena (Figure 25.12b), we found that Siemens and Volkswagen are two German groups that have formed small local components in Beijing. They each appear with two independent components. Siemens has had a presence in China since 1994 and produces electronic components, as

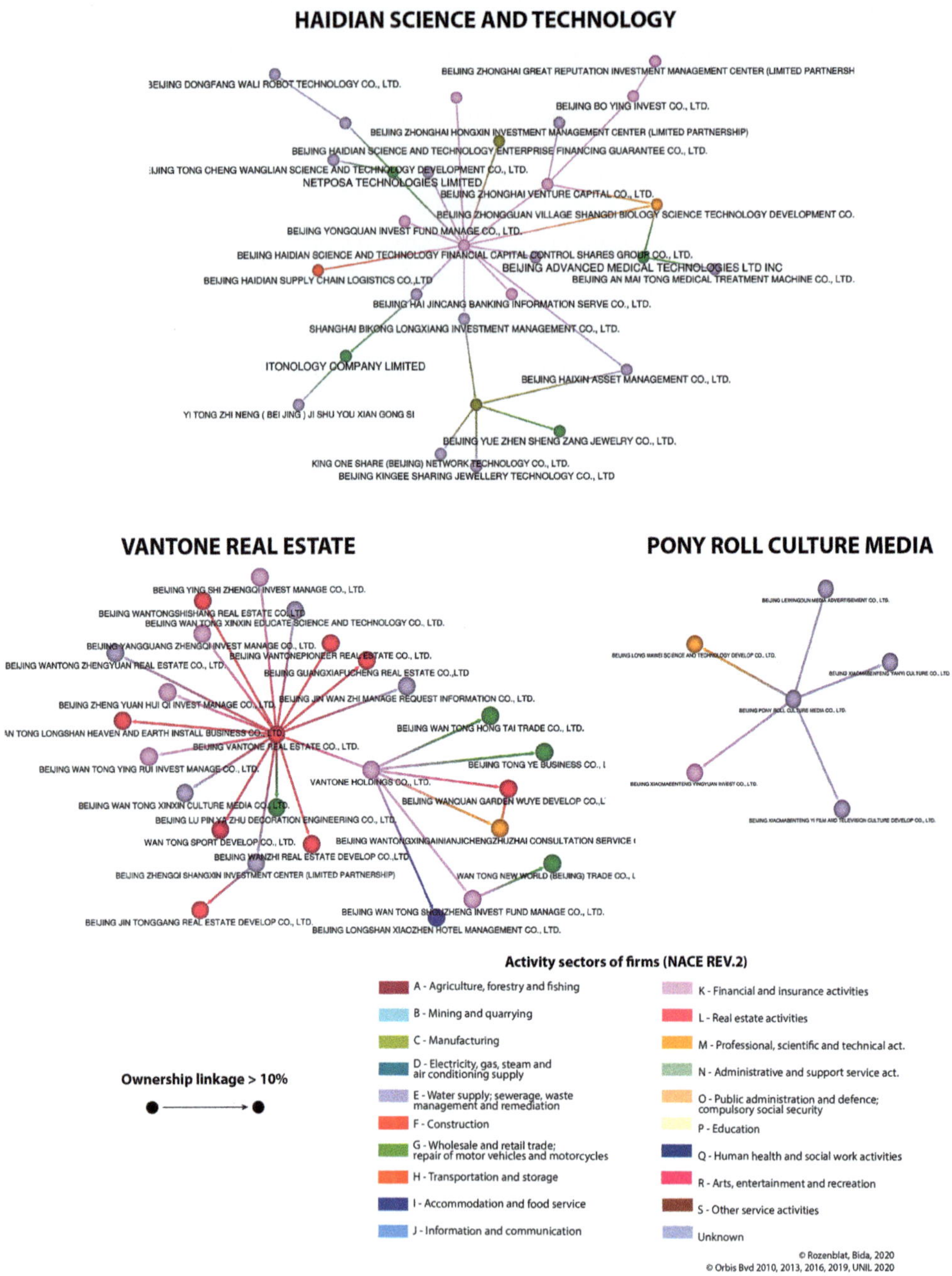

Figure 25.12a Examples of Beijing's other intra-LUR network components in 2019

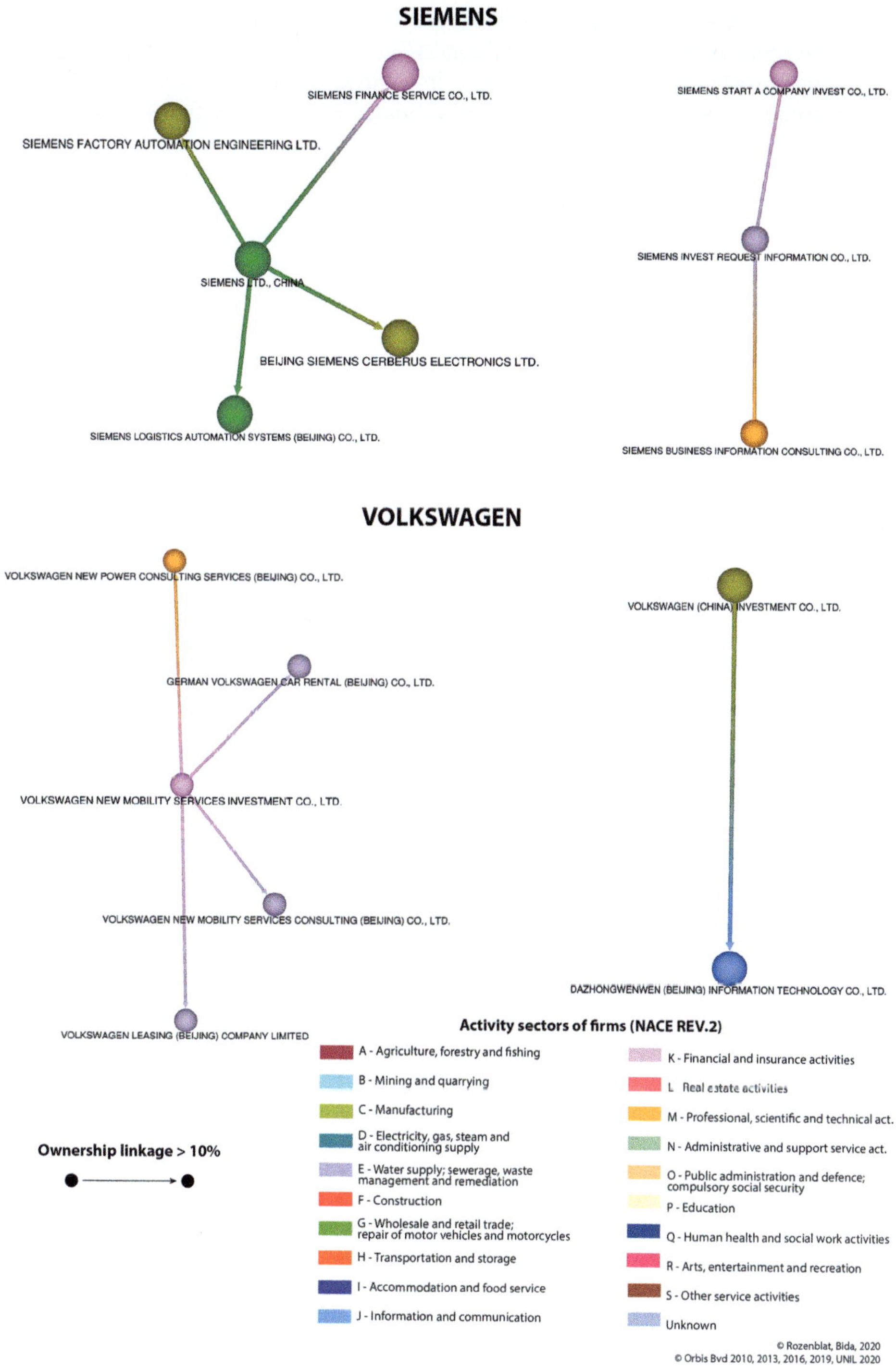

Figure 25.12b *Examples of Beijing's other intra-LUR network components in 2019*

shown in the left-hand side component in Figure 25.12b. Through its second component, namely, the network of Siemens Invest Request/Business Information Consulting/Start a Company Invest, which is headquartered in Beijing and has branches in Shanghai, Siemens offers access to the know-how and financial expertise for finding investments in technologies.

Similarly, Volkswagen is present through two independent components, and even a third component which supports a company belonging to the main Beijing component: Beijing Shangqi Shouchuan Shanghai Volkswagen Maintenance is partly owned (30 percent) by the Beijing Capital Group. Initially located in 1985 in Shanghai (automobiles were built in nearby Anting), Volkswagen is now present in all parts of the country (joint ventures and wholly owned enterprises exist in Beijing, Changchun, Shanghai, Chengdu, Nanjing, Dalian and Hong Kong). The Beijing plants mostly provide a financial integration function and include more than 2300 employees who were previously spread out in four different places in Beijing; these employees were regrouped in a unique headquarters (the Vspace in the Chaoyang district of Beijing along the third ring road) in 2019.

4.2.2 Local firms' activities in Beijing

Considering all the firms using intraurban linkages in Beijing, there were very few that participated in activities using intraurban linkages in 2010; most of these firm activities were in finance, insurance and manufacturing, and they expanded dramatically up until 2019, completely transforming the composition of the local firms' ownership networks (Table 25.4).

The most central activity in the ownership network inside Beijing is manufacturing, followed by financial and insurance activities, professional, scientific and technical activities and wholesale trade (Figure 25.13). Note that the network in Figure 25.13 is not directly comparable to that of London (Figure 25.9) as we had to adapt the thresholds to the scale of Beijing.

The main difference between Beijing and London in 2019 is the weight in Beijing of manufacturing and wholesale trade, as London is dominated by finance, insurance, administrative and support activities. In both cities, professional, scientific and technical activities are positioned at the center of the system, revealing the place of innovation in each capital city. In London, the administrative and support services seem much stronger (in weight and centrality) than in Beijing owing to their nature being very different. In London, these are privately provided services that comprise numerous companies that are spread everywhere in the productive processes, whereas Beijing's administrative and support services are mostly state companies that direct a fewer number of companies in a hierarchical way.

4.2.3 Beijing's LUR polycentricity

Similarly to London, in Beijing, the intra-LUR linkages are mostly concentrated in Beijing's core (Figure 25.14): a total of 87 percent are in Beijing city (corresponding to the official city definition). The number was even higher for London's core, which concentrated 98 percent of all linkages of the LUR. Therefore, the Beijing urban region is less polarized in the core area, with privileged linkages from the core to the periphery (755 links) rather than in the opposite direction (65 links). The most important link goes from Beijing to Langfang, situated in the southeast at the middle way to Tianjin. In

Table 25.4 Activities of the firms participating to the intra-Beijing linkages 2010–19

2010 – activity of firms deploying local linkages				2019 – activity of firms deploying local linkages			
	NACE REV.2	Number of firms	%		NACE REV.2	Number of firms	%
K	Financial and insurance activities	25	35.2	G	Wholesale and retail trade; repair of motor vehicles and motorcycles	1618	18.7
C	Manufacturing	21	29.6	M	Professional, scientific and technical activities	1206	13.9
F	Construction	5	7.0	C	Manufacturing	1176	13.6
J	Information and communication	4	5.6	K	Financial and insurance activities	1164	13.4
G	Wholesale and retail trade; repair of motor vehicles and motorcycles	3	4.2	L	Real estate activities	1059	12.2
H	Transportation and storage	3	4.2	J	Information and communication	777	9.0
L	Real estate activities	3	4.2	F	Construction	469	5.4
N	Administrative and support service activities	3	4.2	H	Transportation and storage	271	3.1
D	Electricity, gas, steam and air conditioning supply	2	2.8	N	Administrative and support service activities	239	2.8
E	Water supply; sewerage, waste management and remediation activities	1	1.4	D	Electricity, gas, steam and air conditioning supply	231	2.7
I	Accommodation and food service activities	1	1.4	I	Accommodation and food service activities	177	2.0
No information		1		B	Mining and quarrying	58	0.7
Total		72		A	Agriculture, forestry and fishing	55	0.6
With information		71	100	E	Water supply; sewerage, waste management and remediation activities	55	0.6
				O	Public administration and defence; compulsory social security	30	0.3
				R	Arts, entertainment and recreation	29	0.3
				Q	Human health and social work activities	21	0.2
				P	Education	18	0.2
				S	Other service activities	11	0.1
				No information		2745	
				Total		11 409	
				With information		8664	100

Source: © Rozenblat, Bida, 2020. © ORBIS BvD – UNIL 2010–2019.

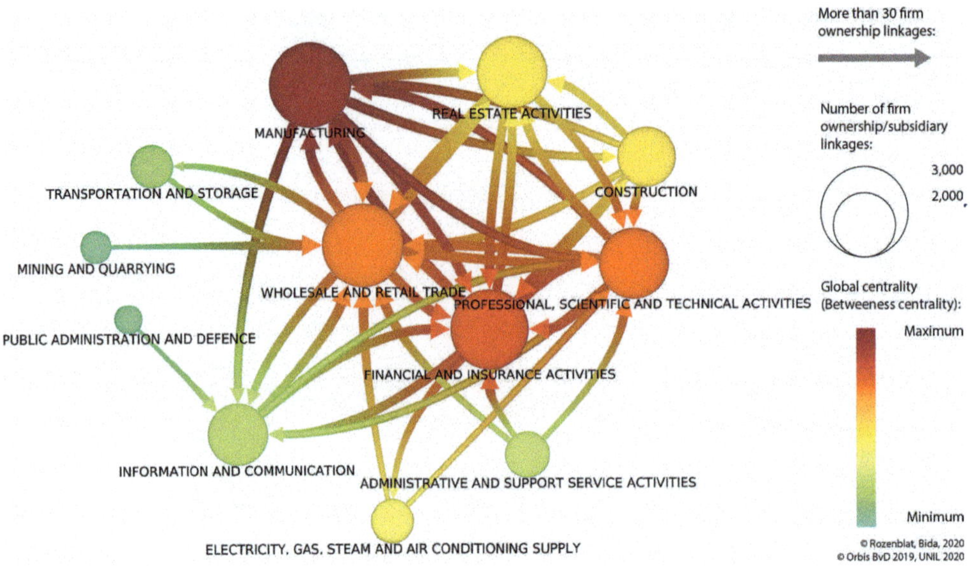

Figure 25.13 Ownership system for the activities of the Beijing's intra-LUR firms' network in 2019

Langfang, there are many extensions of Beijing's main component, especially those linked to Langfang Jingyu Real Estate Development, which comprises 117 firms primarily involved in real estate but also includes import/export companies of metallic products. Langfang also hosts the China Petroleum company and numerous branches of pipelines and logistics.

The second linked satellite to Beijing is Baoding, the capital of the Heibei province, situated farther to the southwest of Beijing. With approximately 1 million inhabitants, the city is more industrial, producing electric power equipment (Baoding Tianwei Baobian Electric, 21 firms) and cars (Great Wall Motor, 33 firms). The city also includes offices for the China Lucky Film Corporation (13 firms), which is the largest photosensitive materials and magnetic recording media manufacturer. Baoding has the same number of firms deploying intraurban linkages as Langfang (450), but the firms are more linked together inside Baoding and less linked to Beijing.

The city of Zhangjiakou (situated to the northeast of Beijing towards Inner Mongolia) is also very industrial, hosting a very important thermic central power plant and numerous steel, iron and coal companies. Chengde, a historical city and the home of the Imperial Summer Palace of the Qing dynasty, is situated 200 kilometers northeast of Beijing. In addition to tourism, the city specializes in steel production, as provided by the Chengde Iron and Steel Group, which is directly supervised from the Beijing's main component. The city also has mining and blasting equipment companies.

The deconcentrating activities of Beijing benefit all these surrounding cities of Hubei province and have a network rebalancing towards this province; however, the control still comes from the center of the network. Note that we did not include Tianjin in the Beijing LUR because Tianjin, an important coastal city with more than 15 million inhabitants, has developed its own regional system in its LUR. However, examining the exchanges

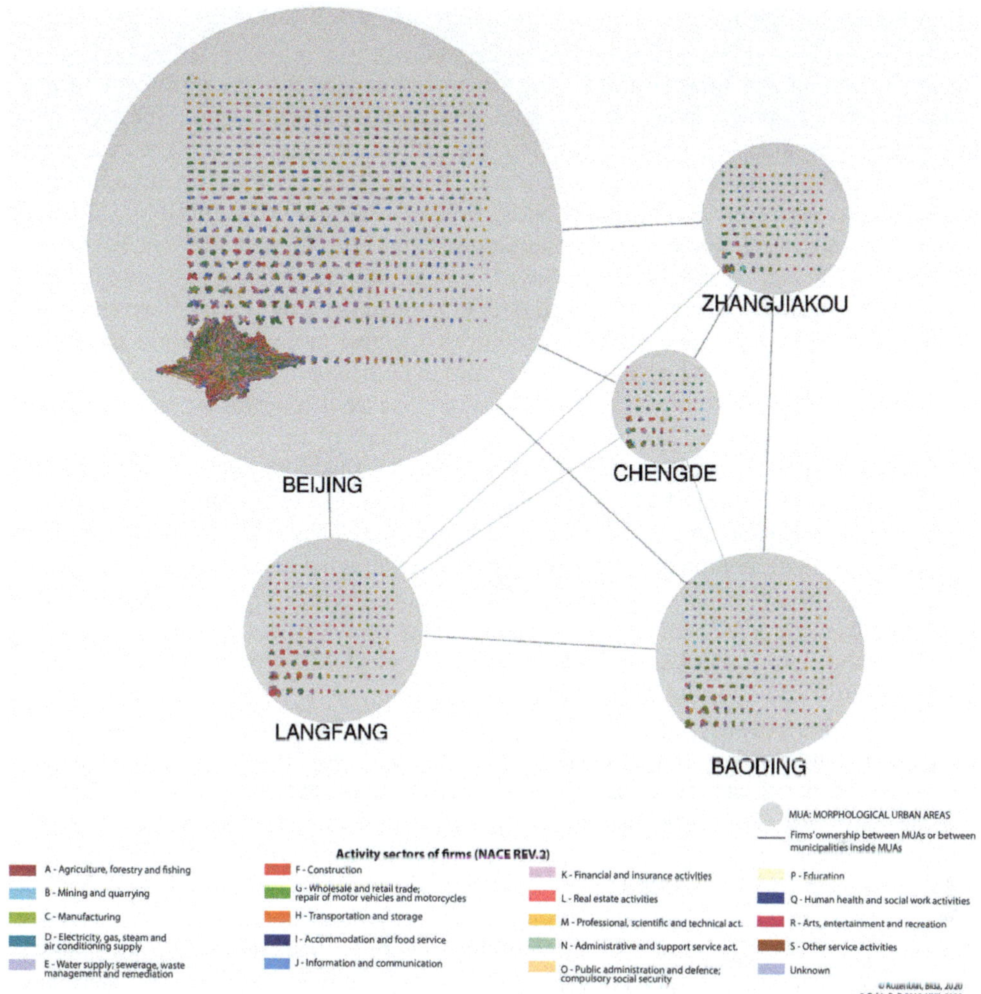

Figure 25.14 *Ownership system for the activities of the Beijing's intra-LUR firms' network in 2019*

between the Beijing LUR and the Tianjin LUR, we note the same type of disequilibrium between the firms' linkages: there are eight times more firms' linkages from the Beijing LUR to the Tianjin LUR than in the opposite direction, and the linkages are orientated mainly towards manufacturing, trade, professional, scientific and technical activities. Thus, the centrifugal movement of the linkages in the Chinese capital is a local, regional and national movement towards the largest regional capitals.

The Chinese main development during 2010–19 has been revealed as the expansion of its intraurban and national networks, strengthening the international networks that have developed so dramatically. The Chinese urban networks' dynamic also reveals the intertwining of the intraurban networks' densification and the interurban expansion of firms.

5. CONCLUSION

These two examples of London and Beijing are two extreme cases of the transformation of intracity networks in the cities' global expansion: London, which has been the archetype of the global city since the 1990s, reduced its internal linkages (which were too vulnerable and experienced cascading effects) between 2010 and 2019. However, to strengthen its worldwide governance value, it increased its level of diversity of financial services and private administrative support for the numerous firms coming from all over the world. At the opposite end of the spectrum, the Chinese economic global expansion has been sustained by Beijing's strong hierarchical structure, in which international firms are used to reinforce the country's own firms' capabilities. Chinese national networks expanded around Beijing into its surrounding satellite cities in Hubei and into the whole country in order to scale up its firms in its national diversified market. The linkage complexification is far from that which we find in London, and the shape of the linkages is forged by a controlling regime. However, the focus of the country is the leveraging of many Chinese companies to enable them to face North American and European competition.

We must also consider the national scale expansion that is very important in the Chinese case. The multinational linkages of Beijing expanded nationally (outside Beijing's LUR itself) during 2010–19, from 11 percent in 2010 to 63 percent in 2019, and the intra-Beijing LUR linkages comprised 90 percent of all its links in 2019. In comparison, London decreased the share of linkages in Great Britain (outside the London LUR) from 13 percent in 2010 to 10 percent in 2019.

With much less efficient political structures and smaller national capacities to strengthen the national firms by their own territorial capacities, other BRIC countries' cities have followed the pattern of Chinese cities. Most of the cities in the increasing number of groups owe their internal development to the meeting of external actors and global conditions with local human or natural resources, and applies to the growing number of US cities in this group, as well as to the cities of Vietnam, Liberia or central Asia.

Another extreme case is that concerning tax havens. Wilmington, DE, as well as Grand Cayman, Bermuda Kinsley or Tortola, does not truly base its network scope on a strong internal complex game of actors. These cities do not have large national economies and are part of the game played by actors located in other cities and who only take advantage of the cities' intermediation ability. However, these cities are not major players. The first part of the chapter on the global expansion of the multinational firms' networks at different scales showed how much national scale matters and the extent to which this scale sustains the expansion of their cities abroad. The evidence of the global trend demonstrates the tendencies of cities in which intraurban linkages are low (such as the US cities) to relatively increase their local complexities, while the evidence also reveals that cities where local firms were too much intertwined (such as London, as well as other European, Russian and Japanese cities) tend to relax their internal linkages for a higher diversity of networks, thus supporting their resilience. Local and national networks also adopt to the shape of the institutions and the regime. This was very clear in China, where most of networks form star shapes. The embeddedness of foreign networks within Beijing also adopted this form, as they were obliged to adapt to the local society. In this perspective, network and hierarchy must not be opposed, as they intrinsically reflect the organizational structure of the society. Hierarchies (trees or concentrated networks on certain nodes) are specific types of mathematical networks that are more centralized and not just circuits. Despite the claim of greater equality brought by the networks, the development of complex network studies in the past 20 years showed how these hierarchized structures (or scale-free networks) were much more dominant than the egalitarian structures.

The result of disentangling the different scales and levels of urban networks appears in this chapter through the systematic comparison of the cities' networks at local/global levels and at different scales. The trajectories are diverse and would deserve in many instances to be examined in greater depth. In particular, it would be worth specifying to what extent each city plays the role of a network governance node in the context of the duality of embedded governance in their services and economies of scope for their innovation. For example, in a study on Marseilles-Aix en Provence metropolis (Rozenblat et al. 2015), we demonstrated the extent to which inside this urban area the intramunicipal firms' linkages supported specialized clusters of innovation, while intermunicipal linkages were mostly financial flows coming from the two poles of Marseilles and Aix-en-Provence. It was much easier to underline this double logic in such a medium-sized city than in large metropolises where industrial and financial linkages are dense and intertwined.

The local flows linked to global flows that we report here are different in nature than those described in the Christaller (1933) model. We recall that the Christaller model described local and regional dependences of inhabitants for daily services. The industrial and financial networks that are at stake here do not replace local services flows but are juxtaposed and intertwined with them to constitute the thickness of urban societies: These networks are supported by inherited local spatial structures developed previously owing to the local service hierarchies, the legacy of local savoir-faire and the specificities of the reputation of places. Globalization has caused the dual emergence of the intercity and intracity networks, and it would be wrong to say that local networks would be less complex than global networks. In the interactive processes of financial functions and information flows, firms' network dynamics find a balance between intraurban and interurban levels, demonstrating the adaptive organizational capacities of the economy within the political and institutional national and international transforming frameworks. The future evolution of this balance towards a greener economy would certainly be helpful for facilitating the development and resilience of cities.

REFERENCES

Aalbers, M.B. (2019), 'Financial geography II: financial geographies of housing and real estate', *Progress in Human Geography*, **43** (2), 376–87.

Amin, A. and N. Thrift (1992), 'Neo-Marshallian nodes in global networks', *International Journal of Urban and Regional Research*, **16** (4), 571–87.

Aydalot, P. (1985), *Economie régionale et urbaine*, Paris: Editions Economica.

Bathelt, H. and Glückler, J. (2011), *The Relational Economy: Geographies of Knowing and Learning*, Oxford: Oxford University Press.

Bathelt, H., A. Malmberg and P. Maskell (2004), 'Clusters and knowledge: local buzz, global pipelines and the process of knowledge creation', *Progress in Human Geography*, **28** (1), 31–56.

Batty, M. (2006), 'Rank clocks', *Nature*, **444** (7119), 592–6.

Beauregard, R.A. (1995), 'Theorizing the global–local connection', in P. Taylor and B. Derruder (eds), *World Cities in a World-System*, Cambridge: Cambridge University Press, pp. 232–48.

Bems, R., R.C. Johnson and K.M. Yi (2009), 'The collapse of global trade: update on the role of vertical

linkages', in R.E. Baldwin (ed.), *The Great Trade Collapse: Causes, Consequences and Prospects*, Geneva: Center for Economic and Policy Research.
Bernard, A.B., B.J. Jensen, S.J. Redding and P.K. Schott (2009), 'The margins of US trade', *American Economic Review*, **99** (2), 487–93.
Berry, B.J. (1964), 'Cities as systems within systems of cities', *Papers in Regional Science*, **13** (1), 147–63.
Bettencourt, L.M., J. Lobo, D. Helbing, C. Kühnert and G.B. West (2007), 'Growth, innovation, scaling, and the pace of life in cities', *Proceedings of the National Academy of Sciences*, **104** (17), 7301–6.
Bettencourt, L.M., V.C. Yang, J. Lobo, C.P. Kempes, D. Rybski and M.J. Hamilton (2020), 'The interpretation of urban scaling analysis in time', *Journal of the Royal Society Interface*, **17** (163), 20190846, doi:10.1098/rsif.2019.0846.
Boschma, R. and R. Martin (eds) (2010), *The Handbook of Evolutionary Economic Geography*, Cheltenham, UK and Northampton, MA, USA: Edward Elgar.
Bureau van Dijk (BvD) (2010, 2013, 2016, 2019), 'Orbis database', Bureau van Dijk – Moodys, Brussels, accessed 26 March 2021 at https://www.bvdinfo.com/en-gb/our-products/data/international/orbis.
Camagni, R. (2001), 'The economic role and spatial contradictions of global city-regions: the functional, cognitive and evolutionary context', research report, Politecnico di Milano, Milan.
Carballo, J.R. (2015), 'Essays in trade and uncertainty', PhD dissertation, University of Maryland, College Park, MD.
Castells, M. (2000), *The Rise of the Network Society*, 2nd edn, Oxford: Blackwell.
Castells, M. (2010), 'Globalisation, networking, urbanisation: reflections on the spatial dynamics of the information age', *Urban Studies*, **47** (13), 2737–45.
Cattan, N., D. Pumain, C. Rozenblat and T. Saint-Julien (1994), *Le système des villes européennes*, Paris: Anthropos-Economica.
Christaller, W. (1933), *Die zentralen Orte in Suddeutschland: Eine okonomisch-geographische Untersuchung uber die Gesetzmassigkeit der Verbreitung und Entwicklung der Siedlungen mit stadtischen Funktionen*, Jena: Gustav Fischer.
Content, J. and K. Frenken (2016), 'Related variety and economic development: a literature review', *European Planning Studies*, **24** (12), 2097–112.
Crevoisier, O. (2004), 'The innovative milieus approach: toward a territorialized understanding of the economy?', *Economic Geography*, **80** (4), 367–79.
Di Lello, O. and C. Rozenblat (2014), 'Les réseaux des firmes multinationales dans les villes d'Europe centre-orientale', *Cybergeo – European Journal of Geography*, art. 678.
DiMaggio, P. (1997), 'Culture and cognition', *Annual Review of Sociology*, **23** (1), 263–87.
Dicken P. (2007), *Global Shift: Mapping the Changing Contours of the World Economy*, London: Sage.
Dicken, P. (2011), *Global Shift: Mapping the Changing Contours of the Global Economy*, 6th edn, New York: Guilford Press.
Farinha, T., P.A. Balland, A. Morrison and R. Boschma (2019), 'What drives the geography of jobs in the US? Unpacking relatedness', *Industry and Innovation*, **26** (9), 1–35.
Frenken, K., F. Van Oort and T. Verburg (2007), 'Related variety, unrelated variety and regional economic growth', *Regional Studies*, **41** (5), 685–97.
Garas, A., C. Rozenblat and F. Schweitzer (2018), 'Economic specialization and the nested bipartite network of city-firm relations', in S. Battiston, G. Caldarelli and A. Garas (eds), *Multiplex and Multilayer Networks*, Oxford: Oxford University Press, pp. 75–85.
Grabher, G. and D. Stark (eds) (1997), *Restructuring Networks in Post-Socialism: Legacies, Linkages, and Localities*, Oxford: Clarendon Press.
Grewal, D.S. (2008), *Network Power: The Social Dynamics of Globalization*, London: Yale University Press.
Hadjimichalis, C. (1984), 'The geographical transfer of value: notes on the spatiality of capitalism', *Environment and Planning D: Society and Space*, **2** (3), 329–45.
Hall, P. (2014), *Cities of Tomorrow: An Intellectual History of Urban Planning and Design since 1880*, Somerset, NJ: John Wiley & Sons.
Hall, P.G. and K. Pain (eds) (2006), *The Polycentric Metropolis: Learning from Mega-City Regions in Europe*, London: Routledge.
Henderson, J., P. Dicken, M. Hess, N. Coe and H.W.C. Yeung (2002), 'Global production networks and the analysis of economic development', *Review of International Political Economy*, **9** (3), 436–64.
Hidalgo, C.A., B. Klinger, A.L. Barabási and R. Hausmann (2007), 'The product space conditions the development of nations', *Science*, **317** (5837), 482–7.
Hu, Y., D. Zhou, R. Zhang, Z. Han, C. Rozenblat and S. Havlin (2013), 'Percolation of interdependent networks with intersimilarity', *Physical Review E*, **88** (5), 052805, doi:10.1103/PhysRevE.88.052805.
Jacobs, J. (1969), *The Economy of Cities*, London: Vintage Books.
Johannisson, B. (1990), 'Economies of overview – guiding the external growth of small firms', *International Small Business Journal*, **9** (1), 32–44.

Leydesdorff, L. (2012), 'The triple helix, quadruple helix, . . ., and an N-tuple of helices: explanatory models for analyzing the knowledge-based economy?', *Journal of the Knowledge Economy*, **3** (1), 25–35.
Lobo, J., L.M. Bettencourt, M.E. Smith and S. Ortman (2020), 'Settlement scaling theory: bridging the study of ancient and contemporary urban systems', *Urban Studies*, **57** (4), 731–47.
Lüthi, S., A. Thierstein and V. Goebel (2010), 'Intra-firm and extra-firm linkages in the knowledge economy: the case of the emerging mega-city region of Munich', *Global Networks*, **10** (1), 114–37.
Markusen, A. and V. Gwiasda (1994), 'Multipolarity and the layering of functions in world cities: New York City's struggle to stay on top', *International Journal of Urban and Regional Research*, **18** (2), 167–93.
Massey, D. (2007), *World City*, Cambridge: Polity.
Moriconi-Ebrard, F. (1994), *Geopolis*, Paris: Anthropos.
Moulaert, F. and F. Djellal (1995), 'Information technology consultancy firms: economies of agglomeration from a wide-area perspective', *Urban Studies*, **32** (1), 105–22.
Neal, Z.P. (2012), *The Connected City: How Networks Are Shaping the Modern Metropolis*, London: Routledge.
Nelson, R.R. and S.G. Winter (1982), *An Evolutionary Theory of Economic Change*, Cambridge, MA: Belknap Press.
Padgett J.F. and W.W. Powell (2012), 'The problem of emergence', in J.F. Padgett and W.W. Powell, *The Emergence of Organizations and Markets*, Princeton, NJ: Princeton University Press, pp. 1–29.
Peet, R. (ed.) (1987), *International Capitalism and Industrial Restructuring: A Critical Analysis*, Crows Nest, NSW: Allen & Unwin.
Pflieger, G. and C. Rozenblat (2010), 'Introduction. Urban networks and network theory: the city as the connector of multiple networks', *Urban Studies*, **47** (13), 2723–35.
Porter, M.E. (2004), *Building the Microeconomic Foundations of Prosperity: Findings from the Microeconomic Competitiveness Index*, London: Palgrave Macmillan.
Pumain, D. (2018), 'An evolutionary theory of urban systems', in C. Rozenblat, D. Pumain and E. Velasquez (eds), *International and Transnational Perspectives on Urban Systems*, Singapore: Springer, pp. 3–18.
Pumain, D. and C. Rozenblat (2019), 'Two metropolisation gradients in the European system of cities revealed by scaling laws', *Environment and Planning B: Urban Analytics and City Science*, **46** (9), 1645–62.
Pumain, D., F. Paulus, C. Vacchiani-Marcuzzo and J. Lobo (2006), 'An evolutionary theory for interpreting urban scaling laws', *Cybergeo: European Journal of Geography*, art. 343.
Pumain D., E. Swerts, C. Cottineau, C. Vacchiani-Marcuzzo, C.A. Ignazzi, A. Bretagnolle, et al. (2015), 'Multilevel comparison of large urban systems', *Cybergeo: European Journal of Geography*, art. 706.
Robinson, J. (2015), 'World cities, or a world of ordinary cities?', in F. Miraftab and N. Kudva (eds), *Cities of the Global South Reader*, London: Routledge, pp. 66–72.
Robinson, J. and A. Roy (2016), 'Debate on global urbanisms and the nature of urban theory', *International Journal of Urban and Regional Research*, **40** (1), 181–6.
Rodríguez-Pose, A. and H.M. Zademach (2003), 'Rising metropoli: the geography of mergers and acquisitions in Germany', *Urban Studies*, **40** (10), 1895–923.
Roy, A. and A. Ong (eds) (2011), *Worlding Cities: Asian Experiments and the Art of Being Global*, Hoboken, NJ: John Wiley & Sons.
Rozenblat, C. (2010), 'Opening the black box of agglomeration economies for measuring cities' competitiveness through international firm networks', *Urban Studies*, **47** (13), 2841–65.
Rozenblat, C. (2018), 'Urban systems between national and global: recent reconfiguration through transnational networks', in C. Rozenblat, D. Pumain and E. Velasquez (eds), *International and Transnational Perspectives on Urban Systems*, Singapore: Springer Nature, pp. 19–49.
Rozenblat, C. (2020a), 'Extending the concept of city for the delineation of large urban regions (LUR) for the cities of the world', *Cybergeo – European Journal of Geography*.
Rozenblat, C. (2020b), 'Large urban regions of the world', Database Zenodo, accessed 12 March at 10.5281/zenodo.3944181.
Rozenblat, C. and P. Cicille (2003), *Les villes européennes*, Paris: La documentation française.
Rozenblat, C. and D. Pumain (1993), 'The location of multinational firms in the European urban system', *Urban Studies*, **30** (10), 1691–709.
Rozenblat, C., A. Bellwald and P. Gomez (2015), 'L'international est l'affaire de toute la métropole', in V. Fouchier and N. Frachon, '06-Interpellation stratégique Internationalisation: l'affirmation euroméditerranéenne et mondiale d'Aix-Marseille-Provence', report, Mission interministérielle pour le projet métropolitain Aix-Marseille-Provence, pp. 35–48.
Rozenblat, C., F. Zaidi and A. Bellwald (2017), 'The multipolar regionalization of cities in multinational firms' networks', *Global Networks*, **17** (2), 171–94.
Sassen, S. (1991), *The Global City*, New York: Princeton University Press.
Sassen, S. (2000), 'Spatialities and temporalities of the global: elements for a theorization', *Public Culture*, **12** (1), 215–32.

Sassen, S. (2010), 'Global inter-city networks and commodity chains: any intersections?', *Global Networks*, **10** (1), 150–63.
Scott, A. (2001), *Global City-Regions: Trends, Theory, Policy*, Oxford: Oxford University Press.
Scott, A.J. and M. Storper (2015), 'The nature of cities: the scope and limits of urban theory', *International Journal of Urban and Regional Research*, **39** (1), 1–15.
Shutters, S.T., R. Muneepeerakul and J. Lobo (2015), 'Quantifying urban economic resilience through labor force interdependence', *Palgrave Communications*, **1** (1), 1–7.
Storper, M. (1997), *The Regional World: Territorial Development in a Global Economy*, New York: Guilford Press.
Storper, M. and A.J. Venables (2004), 'Buzz: face-to-face contact and the urban economy', *Journal of Economic Geography*, **4** (4), 351–70.
Torre, A. and A. Rallet (2005), 'Proximity and localization', *Regional Studies*, **39** (1), 47–59.
United Nations (UN) (2018), 'Methodology: Standard country or area codes for statistical use (M49)', United Nations, Statistics Division, accessed 26 March 2021 at https://unstats.un.org/unsd/methodology/m49/.
Uzzi, B. (1997), 'Social structure and competition in interfirm networks: the paradox of embeddedness', *Administrative Science Quarterly*, **42** (1), 35–67.
West, G.B. (2017), *Scale: The Universal Laws of Growth, Innovation, Sustainability, and the Pace of Life in Organisms, Cities, Economies, and Companies*, London: Penguin.
World Bank (2017), *World Bank Annual Report 2017*, Washington, DC: World Bank.
World Trade Organization (WTO) (2020), 'World trade report 2020', World Trade Organization, Geneva.
Zdanowska, N., C. Rozenblat and D. Pumain (2020), 'Evolution of urban hierarchies under globalization in Western and Eastern Europe', *Regional Statistics*, **10** (2), 1–24.

26. Uneven ties! The imposition of inequality through interscalar networks

Ronald Wall and Umakrishnan Kollamparambil

INTRODUCTION

The Emergence of Uneven Geographies

Notwithstanding the fast globalizing world we live in, academics and policymakers continue to view economic geography as a science of particularly urban and sometimes regional scales (Grabher 2006). This has led to an over-territorialized notion of development (Dicken and Malmberg 2001; Hess 2004; Crescenzi et al. 2016). Within our contemporary context, a need for the smart coupling of regional assets to global production networks, across geographic scales, is evident (Henderson et al. 2002; Coe et al. 2004; Hess 2004). This type of approach is important when considering that globalization can be characterized by the integration of international labor markets, capital markets and commodity markets (Bordo et al. 2005), through which corporations intensify their networks globally, in their pursuit of profit maximization, and driven by natural and human resource scarcity (Maddison 1995).

Combined expansion and integration has enabled the increased development of international networks and the splintering of markets and production across various spatial scales (Dicken 2003; Taylor 2004; Alderson and Beckfield 2004; Wall 2009). Urban and regional economic development does not depend particularly on local production and knowledge characteristics, but instead requires 'global pipelines' (Bathelt et al. 2004), and 'local buzz' (Storper and Venables 2004). The emergence of these linkages does not necessarily depend on geographic proximity (Alderson and Beckfield 2004; Boschma 2005; Wall and van der Knaap 2011), and where attributes that are non-spatial, such as cognitive, organizational, institutional and social, play a critical role in the formation of corporate networks (D'Este et al. 2013).

Transnational corporations are important in the governance and establishment of global pipelines, which stems from their vital positions within worldwide production networks (Dicken and Henderson 2003; Coe et al. 2004; Wrigley et al. 2005; Hobday et al. 2005; Yeung 2009). The conceptions discussed above lead from theories on the evolution of theorized hierarchic central place structures (Christaller 1933) towards dual hierarchic and heterarchic systems, through which cities hold increasingly local and global dualities, owing to the economic linkages within and between them (Pred 1973; Hohenberg and Lees 1985; Castells 1989; Neal 2011). This has enabled an uneven dissemination of technology, capital and labor across differing hierarchic, economic and political domains (Wallerstein 1999; Graham and Marvin 2002; Harvey 2006), by which transnationals propagate a labor division among nations that mirrors the labor division between different levels of corporate hierarchy (Hymer 1982). This division originates from top-level centralized decision making and cutting-edge production in a limited range of nations, while the rest of the world is subjected to lesser levels of economic activity and income. These mechanisms confine the world into three hegemonic zones, that is, periphery, semi-periphery and core (Myrdal 1957; Wallerstein 1974). Similarly, it has been postulated that the economic networks between cities do not take place uniformly (Camagni 1993), but consist of horizontal and vertical levels of a corporate system, defined as (1) network (cooperation), (2) competitive (hierarchical), and (3) territorial (state), which are dependent on global, regional and local geographic scales. Subsequently, the evolution of the world economic system into a complex duality of globally integrated and spatially dispersed urban hubs, in which centralized activities are located and which control global production through advanced producer services, is explained (Sassen 2001). In a similar manner to Friedmann (1986), Sassen argued that the globalization of these services would enable a hegemonic world by which these hubs control a multitude of producer cities, in which the majority of people would be gradually excluded from the global economy (Sassen 1994). This has led to the rise of multinational complexity in which these firms carry out global value-chain activities across various spatial scales (Gereffi et al. 2005), revealing diverse structures and multiple globalizations (Wall and van der Knaap 2011; Krätke 2014).

Seven conceptions concerning world city networks were revealed in Friedmann's (1986) *The World City Hypothesis* which can be generalized into the following observations. First, a city's level of development depends on the degree to which it is integrated functionally into the global economy. Second, transnational firms use important cities as access points to worldwide production and markets, which, thirdly, allows for the configuration of global cities into elaborate hierarchic networks. From this, Friedmann (1995) argued that cities of the world are ranked hierarchically in accordance with their economic ties, where highest ranked cities form the control centers of the global economy, followed at a lesser level by cities that oversee regional economies and, at an even lower level, cities that form a nexus between national and subnational economies. Two of Friedmann's seven postulations are particularly important to this chapter. The first, which concerns the sixth postulation, argues that the formation of world cities reveals the contradiction of industrial capitalism, that is, class and spatial polarization, which concerns the growing divide between periphery and core in relation to power, wealth and income (global, regional and local). This divide exacerbates the polarization of class owing to enormous income gaps between elites and low-skilled workers, as well as the proliferation of rural–urban migration. In his seventh postulation he argues that unevenly developed cities have generated change rates that tend to exceed state fiscal resources (welfare, transportation, health and education). It leads to a struggle for resources between the rich and poor, for instance, firms that are often exempted from paying taxes and are often also subsidized. Consequently, that the burden of accumulated capital is shifted towards the most politically and economically disenfranchised parts of the population.

In view of this, notwithstanding the considerable advances made in integrating space-specific locational assets and institutional structures, that is, 'strategic coupling attributes' (Coe et al. 2004; Yeung 2016), with transnational networks, there remain severe gaps in literature concerning the nexus between the global and the local, in reference to functional causality, socio-economic impacts, situational determinants and geographic gravitation (Crescenzi and Rodríguez-Pose 2011; Burger et al. 2014; Wall and Stavropoulos 2016). For example, in comparison to geographically well-dispersed global interactions, transnational flows tend to be more concentrated between closely proximate economies (Rugman 2005; Guy 2015), hence stimulating spiky geographies and asymmetrical regional development. The primarily adjacent economic activity influences business functions, competences, skills and talents, thereby having a general impact on society, in this way accentuating the eminence of global corporate spillovers to host locations (Javorcik and Spatareanu 2008). Through this, it is necessary to be aware that regional advantages strongly influence the evolution of firms located in particular economic zones, shaped by the coexistence of various scales of flows. Therefore, local institutional capacity is needed which can combine external and internal knowledge and resources, in order to enable systemic integration and resilience, diminish spatial hierarchies and promote interregional equality, in relation to economic inclusiveness and wage distribution.

Globalization and Inequality

Since the nineteenth century, studies on the growth of the economy and inequality have existed, however, these studies neglected the regional and international context of capital and trade flows (Lindert and Williamson 2001). In 1976 Arthur Lewis was key to showing that the openness of trade would ultimately lead to development and inequality enclaves. In contrast to this, the Heckshe–Ohlin model (1930; see Ohlin 1935) conjectured that trade openness would enable relative returns to labor and thereby minimize developing country inequality. The cogency of this model has been scrutinized owing to rising worldwide inequality (Feenstra and Hanson 1997). Instead, more recent studies show differing results in the context of trade openness, ranging from the reduction of inequality (Herzer and Nunnenkamp 2013), its insignificance to inequality (Sylwester 2005) or the tendency to increase inequality (Mahutga and Bandelj 2008). Some studies have divided the inequality of nations into within-region and between-region constituents, postulating that since average incomes differ geographically within and beyond regions, this determines regional inequality (Zhang and Kanbur 2001; Kanbur and Venables 2005).

Wage inequality is the result of the wage premiums paid by transnational corporations, and therefore these corporate network structures will have a strong effect on equality (Poon et al. 2015). The findings substantiate a wage premium for the internationally networked city, implying that the economic linkages between cities affect wage differences (Poon et al. 2015). Where financial flows are generally international, Marshallian externalities are mostly localized, the former arising from expansive geographical reach. Hence, a strong possibility exists that international networks accommodate in-migration, innovation and capital exchanges, by which spatial connections determine the destinies of world cities (Glückler 2013). Consequently, the wages of skilled labor are expected to be higher in the hub cities, and those cities that are less internationally connected will form part of global periphery and have lower wage returns. This outcome reminds us of what Sassen (2001) posited as the polarization of income distribution and what was argued by Friedmann (1986) to be the increase of social costs of immigrants and poor workforces that are marginalized in the contention for services, housing and jobs. This leads to emergent disparity based on the increase of well-paid jobs in developed countries and the diaspora of middle-income manufacturing jobs owing to rising real-estate costs (Dasgupta and Osang 2002). Based on this, rich residents and commuters with elaborate lifestyles surface in the core areas, as well as an even larger group of low-wage providers of services and goods (Sassen 2001). Contrary to this, it has been argued that the polarization of income cannot be seen as the consequence of global city connectivity, but of the dependency on welfare policies and immigration, where world cities are subjected to professionalization instead of polarization (Monaghan and Ikeler 2014).

With reference to the income inequality and foreign investment nexus, studies are typically organized into dependency theory and neoclassical theory studies, in which the former notes that the reliance of developing economies on foreign direct investment (FDI) has a negative social and economic impact, which increases inequality between low-skilled and high-skilled workers (Firebaugh and Beck 1994). The latter depicts the opposite. More contemporary research reveals that FDI tends not to affect income inequality, when instrumental variables are included in the study, or alternatively that FDI does reduce significantly the inequality levels of a developing country (Im and McLaren 2015). Furthermore, Johannson et al. (2017), revealed that investment, in addition to directly impacting the creation of jobs, positively influences wage spillover effect on domestic firms, and has an agglomeration effect on unskilled and skilled labor. Wei et al. (2009) contradicts this finding, by showing that the geographic distribution of FDI in China has been primarily influential in increasing regional inequality in Chinese regions. It is argued by Wei et al. (2009) that the disproportionate distribution of FDI, not FDI itself, is responsible for regional growth differences. In light of this, Hymer's (1976) and Krugman's (1991) new economic geography reveals that agglomeration economies are the force that drives corporate decision making, by which these economies contribute to the reduction of transaction costs for investors (Guimaraes et al. 2000) and that, when the investment concentration of a city reaches a specific threshold, it will exhibit a reduction of comparative advantages and a rise in agglomeration diseconomies emerging from negative externalities, such as rising labor costs and transportation bottlenecks (Chan et al. 2008). Similarly, a study by Timberlake et al. (2012), on cities in the US, shows that a city's positon within air-traffic flows can determine its level of income inequality.

METHODOLOGY

Investment and Inequality Data

In the theoretical section, the crossover between inequality literature and world city network theory has been investigated, as well as the relationship between the uneven distribution of foreign investment and income inequality. However, these are case studies on particular countries, for example, the USA or China. Our study goes beyond this, and explores for the first time the effect of FDI networks at the combined city and country level, for the entire world, and showing regional differences. Based on this approach, the structural influence of FDI networks upon inter-regional and intra-regional inequality is explored (and not domestic investments), as well as the indirect and direct effects on host countries. The analysis includes key cities and countries of the world and is executed using different network centrality measures, derived from social network analysis literature, and is carried out at country and city aggregation levels.

The underlying data is from fDi Markets data (greenfield investments), used to capture the globalization of investment, between countries and cities of the world. The source of this data is the *Financial Times*, but the raw data has been supplemented and elaborated using the Bureau van Dijk's ORBIS data. Greenfield investments relate to those in which headquarters launch new projects in host countries by means of the development of new operational facilities. This form of FDI includes investments in research and development (R&D) facilities, regional headquarters, production plants, and sales and marketing offices, and has a measurable direct effect on countries and cities. Also, greenfield FDI typically concerns high-profile company factors, as it relates to transactions by multinational corporations, and is particularly known to support the growth of capital formation, employment and local productivity, including the transfer of prime technology and processes of production from source countries to destination countries. Unlike other investment databases, fDi Markets can be aggregated uniquely at city and country scales, serving the need of this analysis. Initially, the raw data had 20 percent missing values, which were complemented by means of Internet searches. The benefit of this data is that cities are not pre-delineated or predefined beforehand, that is, prior to collecting the investment data for the cities. Instead, fDi Markets provides all the major global investments that took place during 2005–16 regardless of which cities or towns they refer to. The binary cities of each investment were identified afterwards, leading to a well-represented sample of roughly 1200 international cities and towns.

Country and city locations were geocoded to XY coordinates, by means of batch engines, then mapped by means of ArGiS. Since many identical cities were erroneously named differently, or where sub-cities of larger municipalities had different names, a spatial joining technique was executed to ensure that the cities were assigned the correct names. It is a standard procedure in a geographic information system (GIS) to assign the correct name to any investments that fall within a particular municipal boundary. About 200 000 investments were correctly aggregated to approximately 1200 international cities. As regards inequality determinants, we used country and city socio-economic indicators from Euromonitor Passport, including Gini coefficients and control variables, such as unemployment rate and gross domestic product (GDP) growth.

Network Measures

Based on the constructed database, various network centrality measures were calculated for country and city levels. These measures relate to calculations in which countries or cities have different network path relationships with particular cities or countries, or groups of countries or cities. City structural holes serve as the first measure, and rely on the conception that economic information similarities are stronger within a cluster of cities that receive the same types of investment, than those between two or more city clusters that receive dissimilar types of investment. Hence, a city with a high structural hole value serves as an intermediary between two or more dissimilar clusters, and is therefore seen to hold comparative advantages because it permits a city to gate-keep valuable information between clusters of cities. It can therefore be seen as a measure of a city's innovation ability when that city situates itself between clusters of similar cities. In our study, we rely on the interpretation that investment information similarities are stronger within distinct clusters of cities, and that cities that serve as gateways between two or more clusters will connect different types of investment information, and thereby hold a relative innovation advantage.

Next, the city clustering coefficient is used as a measure of how strongly a city clusters with other cities which receive similar types of investment. This relates to cities that create tightly bound clusters dependent on a high density of investment ties. The probability of these types of linkages occurring is said to be higher than the likelihood of linkages being randomly established between cities. In this context, when a city has a large clustering coefficient, the possibility of attracting similar future FDI is high. In turn, city closeness forms the next measure used, and expresses how proximate a city is to all pairs of cities based on their shortest paths between each other. This is a measure of the steps it takes to spread information from a city to all other cities in a network, and is sometimes seen as a measure of network regionalism. These measures are calculated similarly for country level, and include country structural holes, country clustering coefficient and country closeness. For the computations used, see the appendix at the end of this chapter.

Geographic Sample Representation and Summary

The used dataset has a complete set of indicators for 85 destination cities spread over 29 countries (Table 26.1). There are about 800 source cities in the database. Therefore it should be clear that the sample is limited by data availability at the city level, partly owing to the Gini coefficient. Furthermore, adequate regional coverage exists, with developing countries having fair representation in the sample used. Table 26.2 shows the standard deviations and means of the variables used.

Exploratory Analysis and Model Conceptualization

To ascertain the coincidence between network and inequality measures, an exploratory cluster analysis has been undertaken to assist with the hypothesis development, using the K-means algorithm with five Gini clusters at both the city and country levels. The averages of the network measures and Gini are traced using line graphs over the five clusters at

Table 26.1 Region-wise sample distribution and Gini averages

World region	City			Country		
	No.	Percentage	Gini	No.	Percentage	Gini
Africa	4	4.71	50.23	4	13.79	49.86
Asia and Pacific	24	28.24	39.9	7	24.14	41.05
Latin America	6	7.06	43.69	3	10.34	48.23
Middle East	1	1.18	42.63	1	3.45	38.6
North America	22	25.88	47.1	2	6.90	46.91
Europe	28	32.94	37.2	12	41.38	34.225
Total	85	100.00	39.67	29	100.00	41.04

Source: Calculated from the Euromonitor Passport database.

Table 26.2 Summary statistics: mean and standard deviation averages variable

Summary statistics	City	Country
Economic growth	3.73 (5.20)	3.15 (3.79)
Unemployment rate	8.26 (6.86)	8.20 (5.24)
Structural hole	0.64 (0.19)	0.53 (0.16)
Cluster coefficient	0.43 (0.30)	0.52 (0.24)
Closeness	0.16 (0.016)	1.50 (0.16)

Note: Standard deviation in parentheses.

Source: Calculated from Euromonitor Passport database.

city and country levels. The measures of clustering coefficient and structural holes follow very similar patterns, with quite a bit of variance between clusters, while closeness proves to have a consistent variance across the five clusters. Most importantly, this enables us to derive the hypothesis that there is a negative relationship between Gini coefficient and network measures. The averages within each cluster, however, hide the variation within city and country measures, therefore it is useful to take a closer look to capture these differences within each cluster. This allows us to identify the cities and countries with extreme Gini values and how their network measures compare. These results are explained in detail in the results section. The importance of introducing the cluster results and scatterplots in this section is to clarify that they are needed in order to conceptualize a hypothetical model on the relationships between the variables (Figure 26.1). The relationships depict the direct effect of FDI on inequality as well as the indirect effect of FDI on

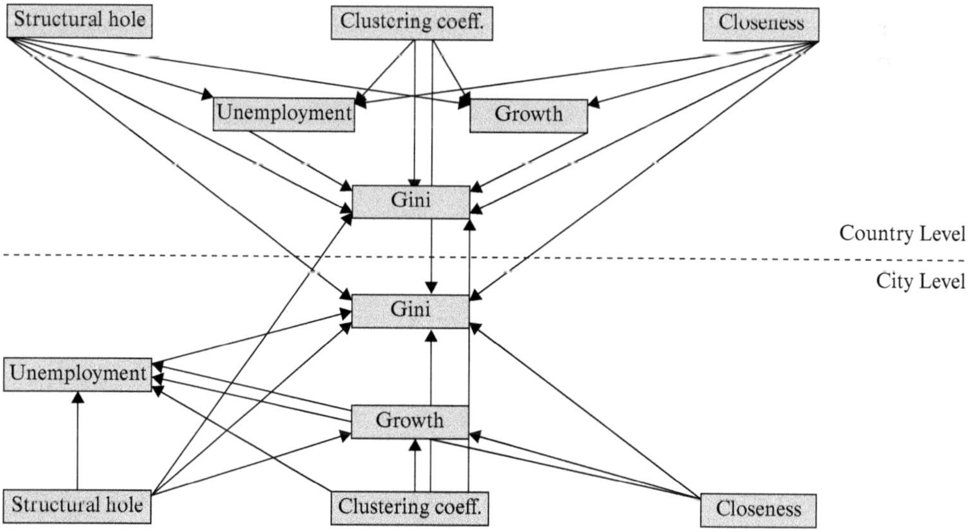

Figure 26.1 Conceptualized path relationships of centrality measures at city and country levels

inequality via economic growth and unemployment. This hypothetical construct forms the foundation for a structural equation model (SEM) that enables not only the testing of the hypothesis, but also the path determination using confirmatory path analysis. It identifies which network measures impact on inequality at the city and country levels and through which paths. The SEM model is chosen for estimation over single equation regression analysis to understand the paths through which FDI impacts inequality at the city and country levels, as conceptualized in Figure 26.1. Since all utilized variables are observables and the paths are determined by a priori theory, the structural equation model comes down to the confirmatory path analysis (Kline 2005). Confirmatory path analysis (CPA) is suitable for this study because there are probably multiple dependent variables, and CPA assists in testing competing hypotheses about complex relationships. In this way, the total impact of FDI on inequality at the city and country levels was captured, through direct and indirect effects, using CPA. Given that this analysis seeks to analyze the impact of city- and country-level network measures on country inequality, as well as the impact of city- and country-level network measures on city inequality, the data structure is nested or hierarchical in nature. Variance clustering of cities within the same countries is factored into the analysis by allowing for a multilevel CPA using generalized SEM estimation. The final model for selection is based on Akaike information criteria (AIC) yielding the lowest value (Shipley 2013).

RESULTS

GIS Maps of Investment Networks

City level

The first map (Figure 26.2) shows the geographic distribution of FDI (2005–16) between 800 source cities (white nodes) and 125 destination cities (black nodes). The size of the node represents the number of investments (count) sent or received by cities. Most source nodes are located in the Northern Hemisphere, particularly North America, Europe, the Middle East and Asia Pacific. Very few source nodes (white) are found in the Southern Hemisphere, and their sizes are relatively small. This disparity illustrates the hegemony of the outward investment held by the developed regions of the world, which control the majority of production and markets in both the developed and developing worlds. The 125 destination nodes (black) are more abundant in the Southern Hemisphere and would be even more so if the data were not limited by the matching of investment and indicator (inequality) databases. Underneath the cloud of source nodes in Europe and North America, there are also many hidden destination nodes. Most destination investment occurs here, showing that the developed world invests most in itself. It is also evident that Eastern Europe and Asia Pacific are premier investment destinations. The thick black linkages represent the highest amount of FDI between source and destination nodes (cities). These powerful capital flows occur strictly between cities in the developed world (trans-Atlantic, trans-Asian and trans-Pacific connectivity) and traverse huge global distances. The companies represented here are top multinationals that hold a very extensive global reach. The white linkages represent different strengths of medium- and small-sized investments between cities. The thicker linkages tend to be located in the

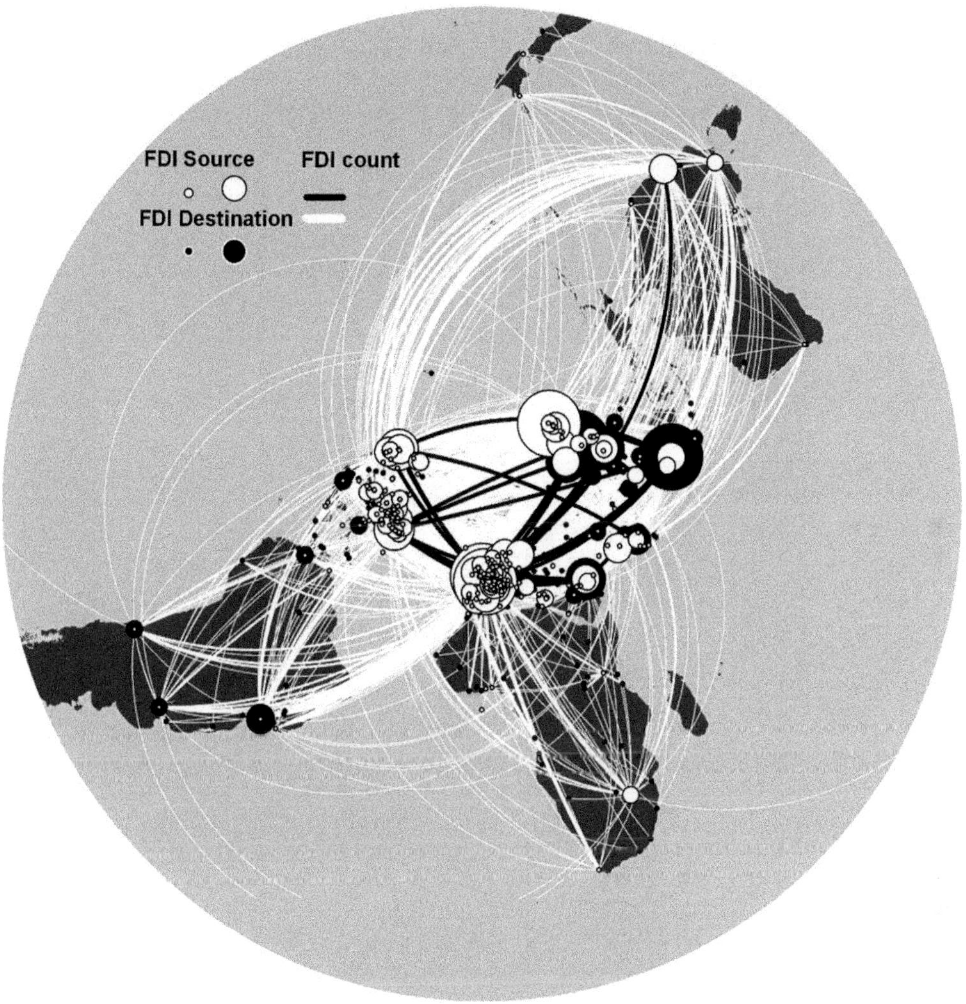

Source: fDi Markets database; mapped in ArcGIS.

Figure 26.2 FDI count between global cities (polar projection)

Northern Hemisphere, while the smaller linkages are represented more in the Southern Hemisphere. Africa, South America and South East Asia receive the least FDI, while those regions are known to have some of the highest levels of inequality. It is also observed that South–South FDI is much weaker than is commonly proclaimed in the media and some literature.

Country level

The next map (Figure 26.3) represents exactly the same data as Figure 26.2 but aggregated to the country level. Instead of the investments crossing between city coordinates, the flows are now between the middle points of countries (centroids). For source countries (white

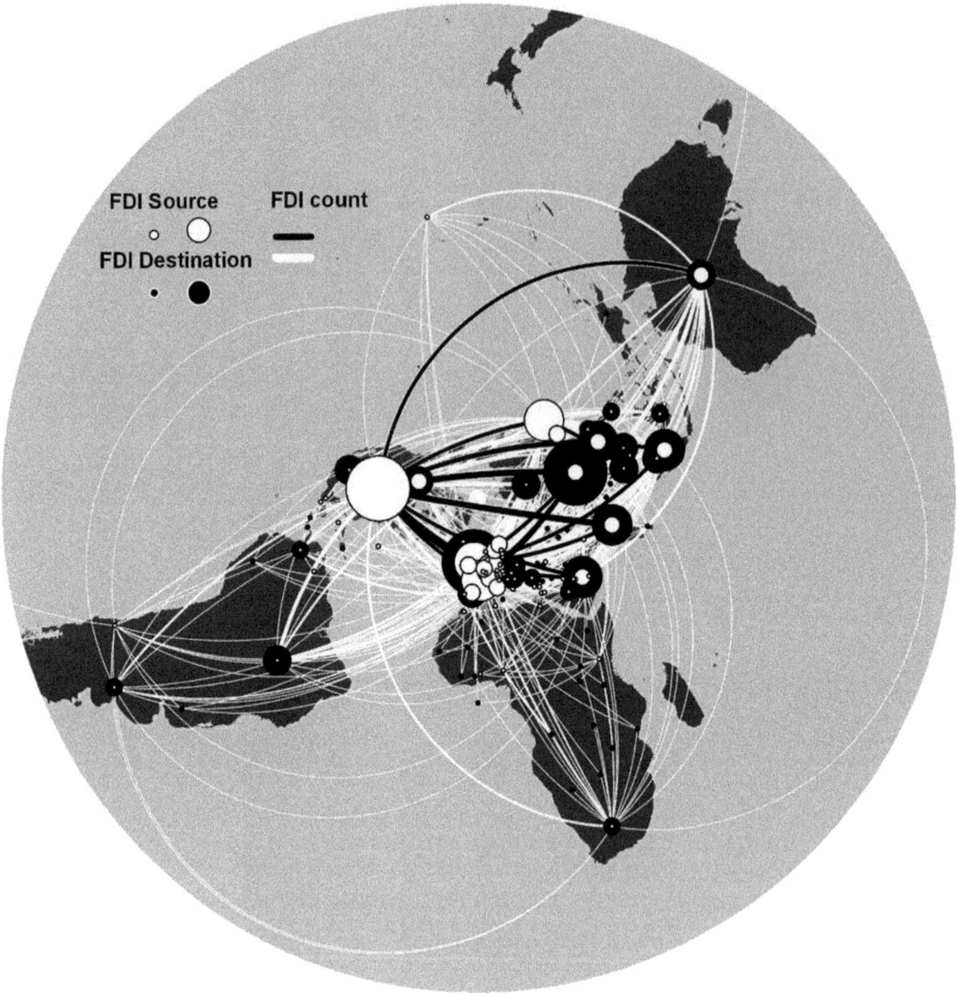

Source: fDi Markets database; mapped in ArcGIS.

Figure 26.3 FDI count between countries (polar projection)

nodes), we again see the dominance of investments emanating particularly from countries in the developed world. The USA, the UK and Japan hold center stage as regards outward investment. Source countries are found far less frequently in the developing regions. The map in Figure 26.3 reveals the disproportionate power of outward investment held by the developed parts of the world, as the firms in these countries control production and markets in both the developed and developing worlds. Again, in relation to inward investment (black nodes), we see a higher presence of countries in the developing world. Strong investment linkages between countries of the developed world are also observed, with weaker ties being more in the Southern Hemisphere. The presence of strong South–South ties between countries in this region is, once again, not evident at this scale.

GIS Maps of Network Measures

The previous maps (Figures 26.2 and 26.3) represent the world investment network (2005–16). The nodes are simple measures of indegree and outdegree. However, based on this data, the previously discussed complex nodal measures have been calculated and mapped for both city and country aggregations; that is, clustering coefficient, structural holes and inward closeness. These maps have been omitted to save space. However, it is important to note that completely different geographic distributions of network data are seen for each measure and aggregation level.

Cluster Analysis Results

City level

Figure 26.4 summarizes the output of the cluster analysis with Gini on the secondary Y-axis and the three network measures on the primary Y-axis. The X-axis indicates the five clusters based on the Gini of 125 cities. The negative relationship between inequality and network measures is brought out at the city level regarding structural holes and cluster coefficients, especially through the extreme clusters, 1 and 5. The variation of inward closeness is limited and hence the relationship with inequality is visually less evident. The city variation within the two extreme clusters is evident regarding network measures and Gini. However, the negative relationship between network measures and Gini is evident within each cluster, as depicted within all six scatterplots. This supports the

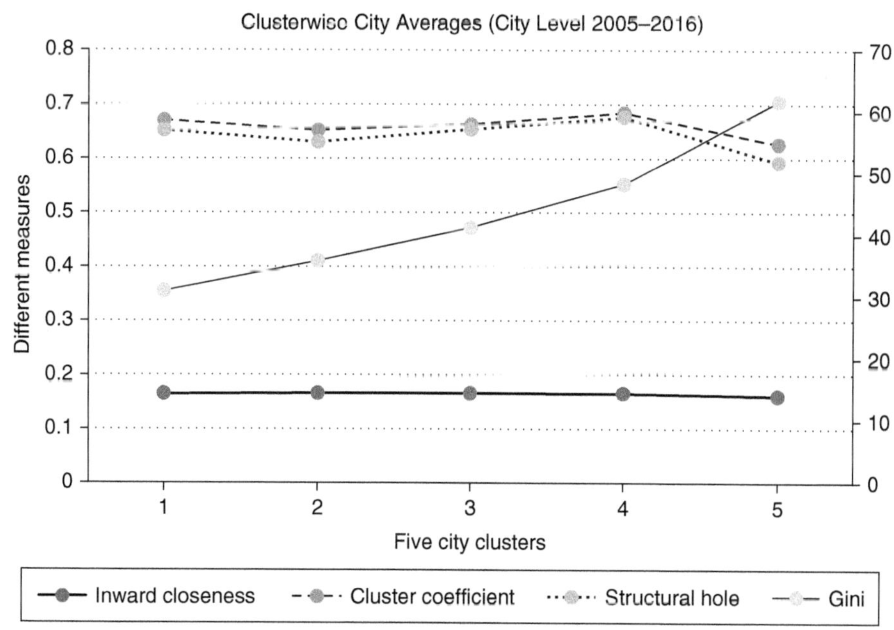

Figure 26.4 Averages of the five city clusters according to network measures and Gini coefficients

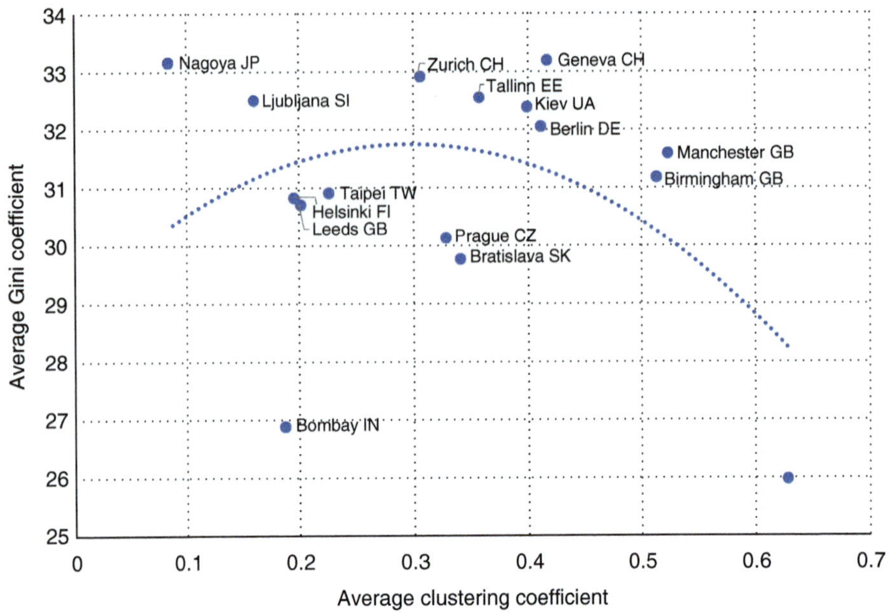

Figure 26.5 Clustering coefficient of lowest Gini cluster (city level 2005–16)

hypothesis that a negative relationship exists between network measures and inequality. Furthermore, these scatter plots reveal the relationship between inward closeness and Gini more clearly than that of the cluster average earlier on. Moreover, the possibility of non-linear relationships are also indicated by the shape of the trend lines in some scatter plots.

Next, in the form of scatterplots, the three city network measures for the lowest and highest inequality clusters is discussed. In the first plot (Figure 26.5), we see the city relationship between clustering coefficient and Gini coefficient, for the lowest inequality cluster. Even within this equitable cluster, cities such as Taipei, Nagoya, Ljubljana, Zurich, Helsinki and Leeds have lower average clustering coefficients, which tend to increase wage inequality in these cities. This means that these cities are less capable of forming tight complete bonds with multiple other cities, which has an adverse effect on their wage distribution. Cities such as Almaty, Manchester, Birmingham, Prague and Bratislava have a higher clustering coefficient, so they form complete bonds with many more cities, which explains their exponentially lower inequality levels. When looking at the highest Gini cluster (Figure 26.6), similar evidence is seen. These are cities exhibiting some of the highest inequality levels in the world. Even in this high-inequality cluster there are differences. Lagos, Johannesburg and Nairobi have very low clustering coefficients, which explains their extremely high levels of wage inequality; while Cape Town and Rio de Janeiro reveal the opposite effect, where high clustering coefficients determine exponentially low inequality levels. In the subsequent scatterplot (Figure 26.7), the relationship between city structural hole scores and city inequality at the lowest Gini cluster level are seen. Although the relationship is weaker than the previous measure, it

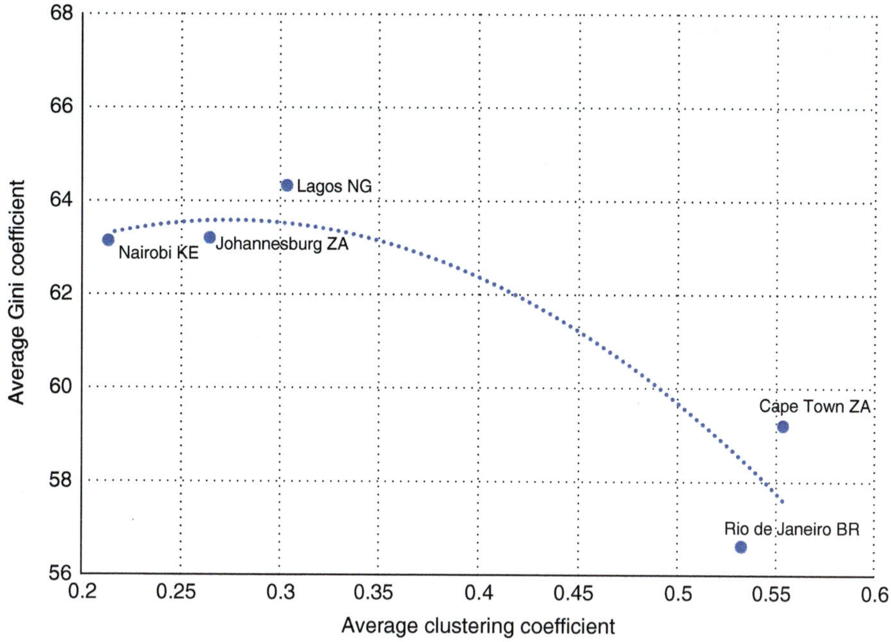

Figure 26.6 Clustering coefficient highest Gini cluster (city level 2005–16)

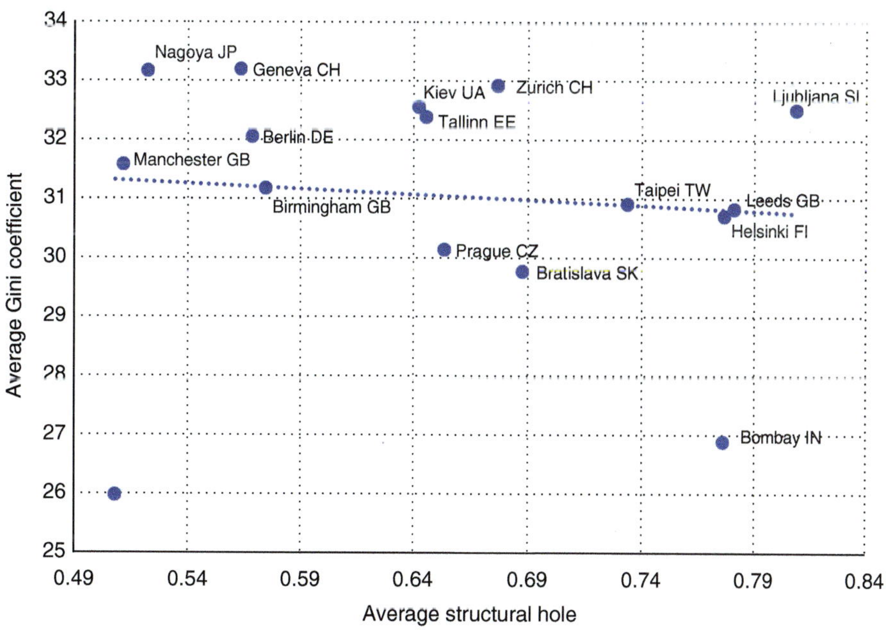

Figure 26.7 Structural hole lowest Gini cluster (city level 2005–16)

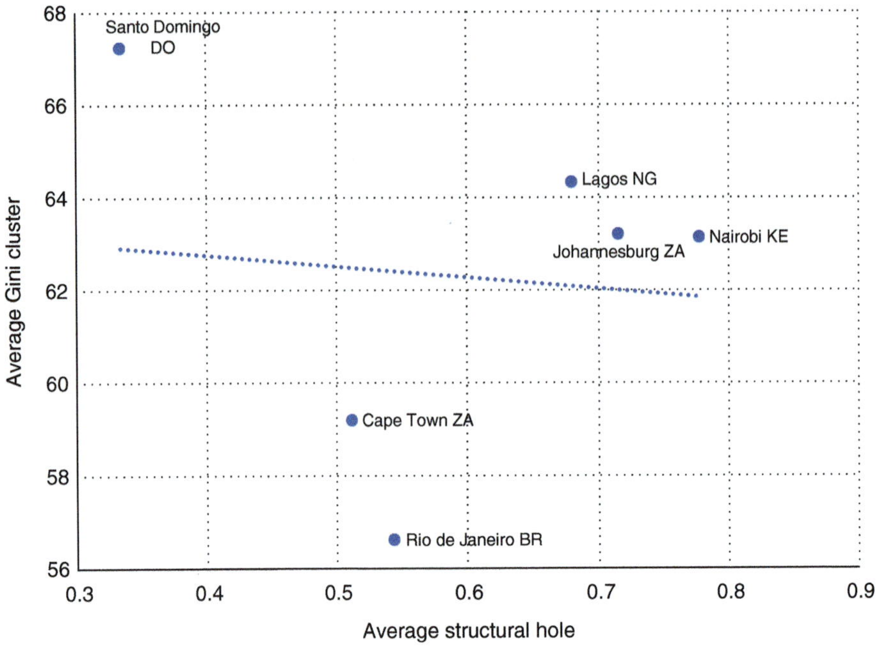

Figure 26.8 Structural hole highest Gini cluster (city level 2005–16)

is evident that the higher the structural hole score, the less inequality there is. This means that cities that form strong gateways between clusters of other cities (that is, hold strong innovation characteristics) tend to, on average, have lower levels of inequality. As regards the highest Gini cluster (Figure 26.8), we see similar evidence. The next plot (Figure 26.9) shows the interdependence between inward closeness and wage inequality for the lowest Gini cluster. Here the effect is stronger than the previous measure. At this equitable level there is visually not much variance in inward closeness, which means that prominent world cities all tend to have high closeness values that are functionally proximate to many other cities in the world. This high level of global economic integration leads to a strong linear decrease in wage inequality. At the lower end of the equity spectrum, we see that low inward closeness determines high Gini coefficients (Figure 26.10).

Country level

The country-level cluster analysis undertaken on the same lines as the city level, reveals much the same pattern but with larger variations observed in the network measures. For instance, the negative relationship of inward closeness together with the other network measures and inequality is evident, especially considering the two extreme outer clusters (Figure 26.11). The scatter plot of the low- and high-inequality countries further indicates that a non-linear relationship possibly exists between network measures and inequality. For the relationship between clustering coefficient and inequality at the lowest Gini cluster level, we see that low clustering coefficients lead to higher Gini coefficients (Figure 26.12). This means that counties such as Britain (GB), France (FR), Japan (JP), Spain (ES) and Germany (DE) form less tightly woven structures

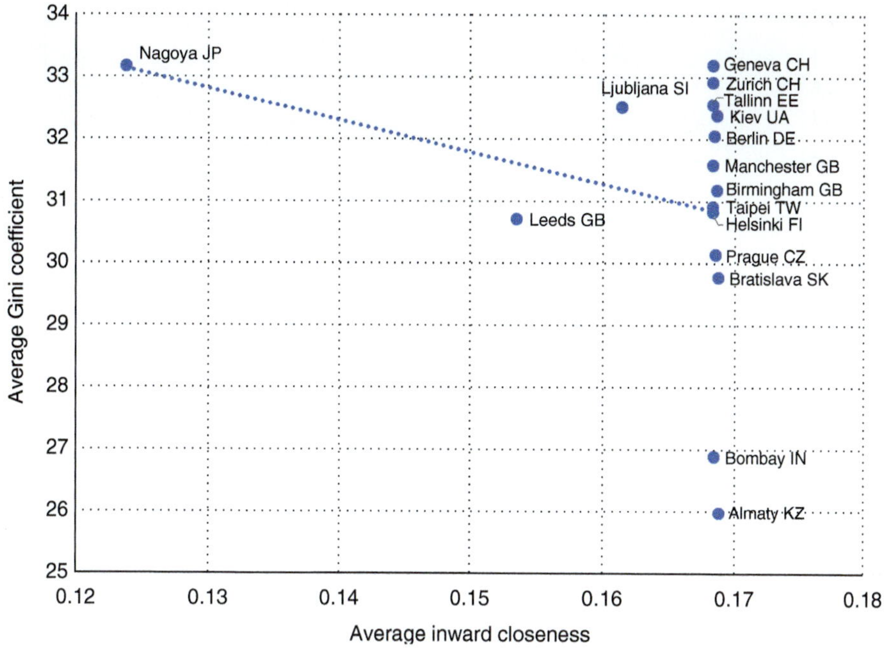

Figure 26.9 Inward closeness lowest Gini cluster (city level 2005–16)

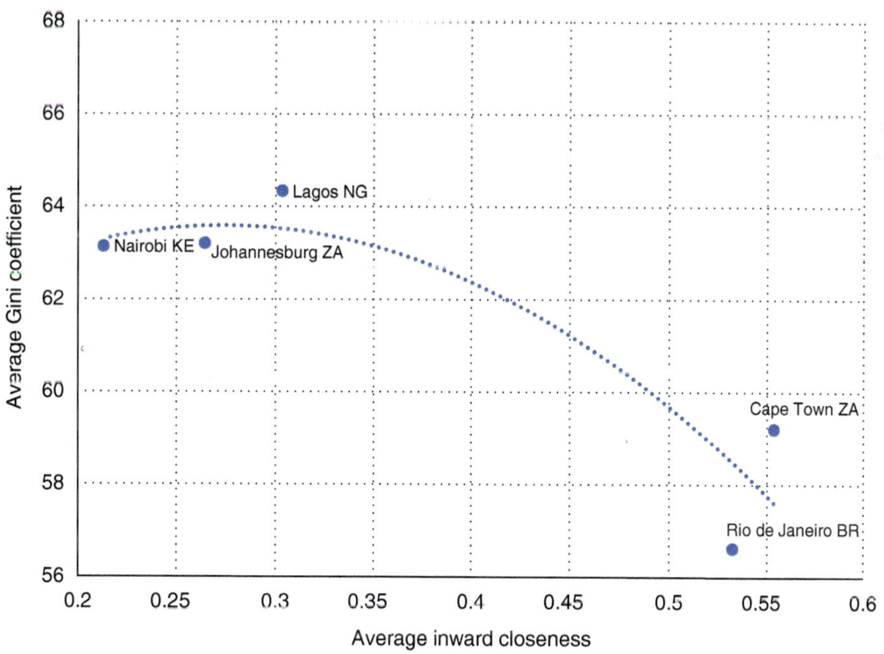

Figure 26.10 Inward closeness highest Gini cluster (city level 2005–16)

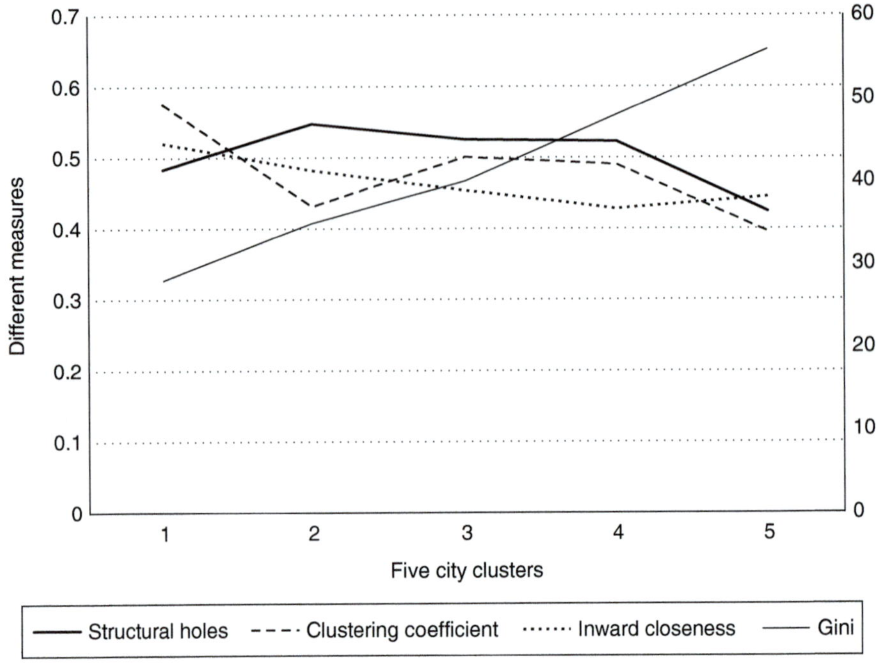

Figure 26.11 Cluster-wise country averages (country level 2005–16)

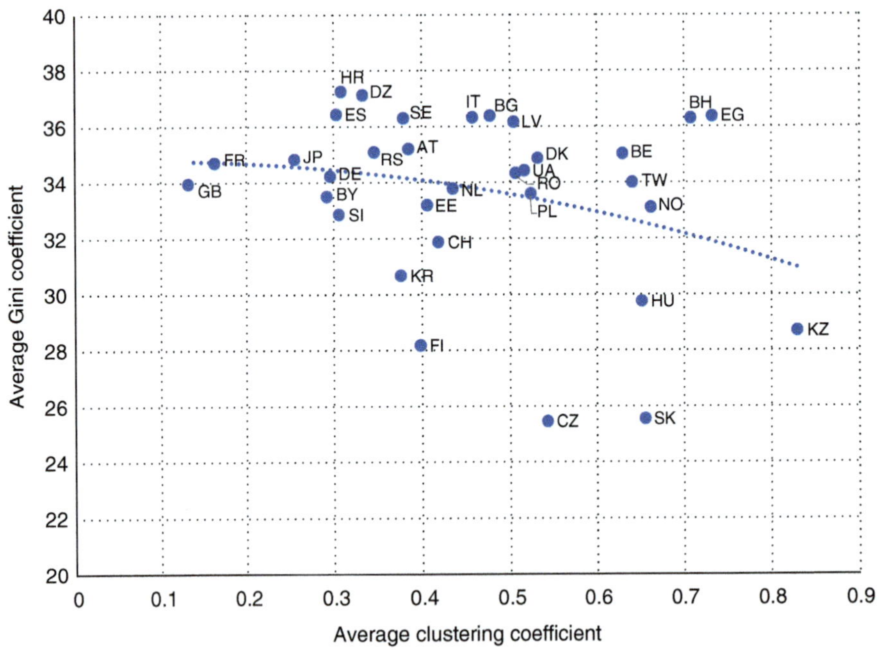

Figure 26.12 Clustering coefficient low Gini cluster (country level 2005–16)

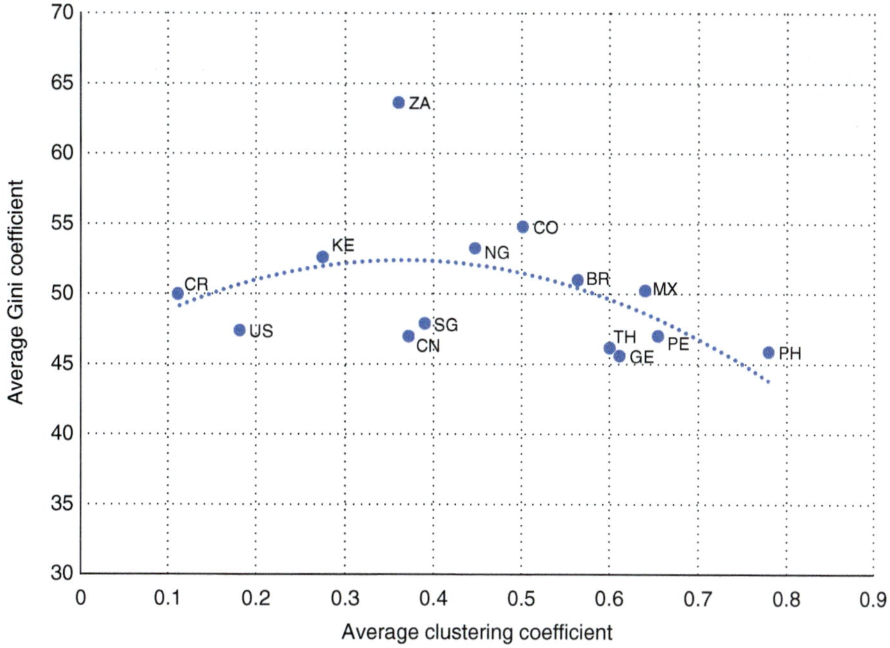

Figure 26.13 *Clustering coefficient high Gini cluster (country level 2005–16)*

with other countries, than do, for instance, Kazakhstan (KZ), Slovakia (SK), Hungary (HU), Czech Republic (CZ), Croatia (KR) and Switzerland (CH). In the case of the high Gini cluster (Figure 26.13), we see similar evidence, where countries such as South Africa (ZA), Costa Rica (CR), the USA (US) and Kenya (KE) have very high wage inequality, stemming from their low clustering coefficient scores. The inverse is seen for countries such as Peru (PE), Philippines (PH) and Georgia (GE). These countries tend to form more tightly knit bonds with other countries. Next, in relation to structural holes versus income inequality, we see a more modest relationship (Figure 26.14). Here, too, counties with higher levels of structural hole values will be less unequal. Countries with high network innovation levels, such as Britain (GB), France (FR), Belarus (BY), Slovenia (SI), Germany (DE) and Serbia (RS), will exhibit modestly lower Gini levels. For the higher Gini country cluster level, similar evidence is shown, in Figure 26.15. Countries with high inequality, for example, South Africa (ZA), Colombia (CO), Kenya (KE), Nigeria (NG) and Guatemala (GT), also have lower structural hole values, that is, their innovation capacity within networks is lower than that of other countries. As regards inward closeness and inequality at the lower Gini cluster level, it is again evident that the more integrated a country is with other countries of the world, the less the inequality levels will be (Figure 26.16). A similar relationship is found at the higher level of Gini (Figure 26.17).

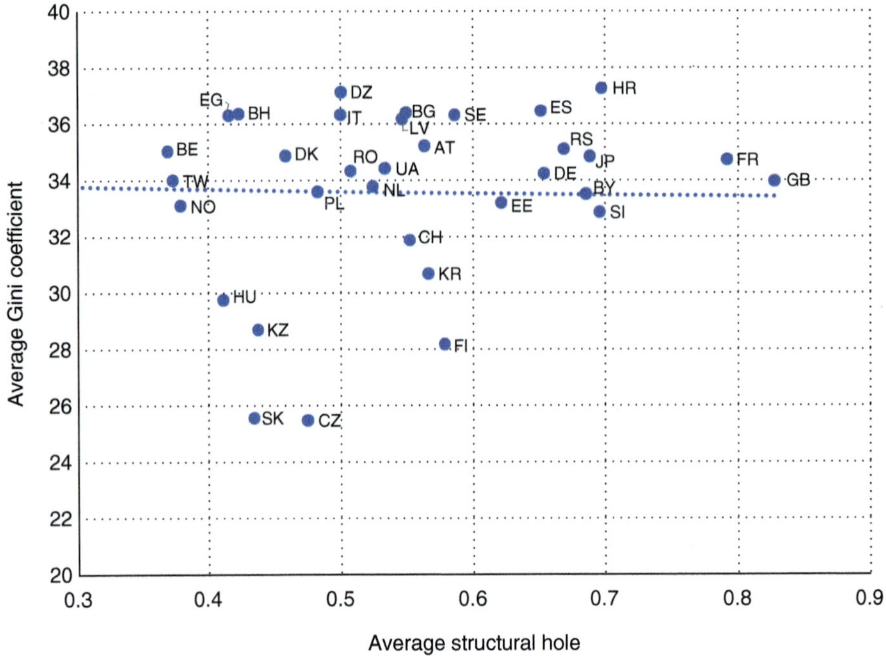

Figure 26.14 Structural hole low Gini cluster (country level 2005–16)

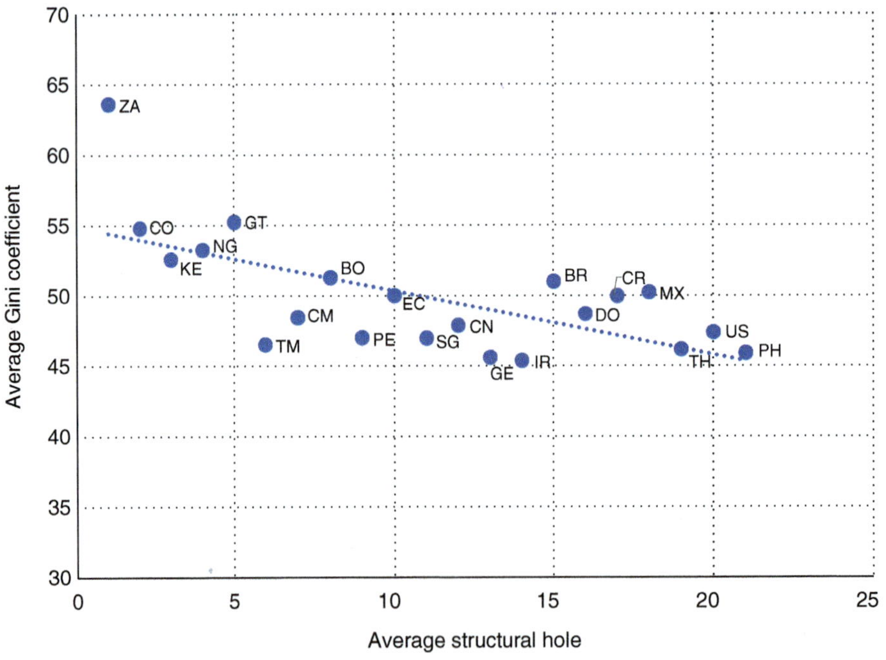

Figure 26.15 Structural hole high Gini cluster (country level 2005–16)

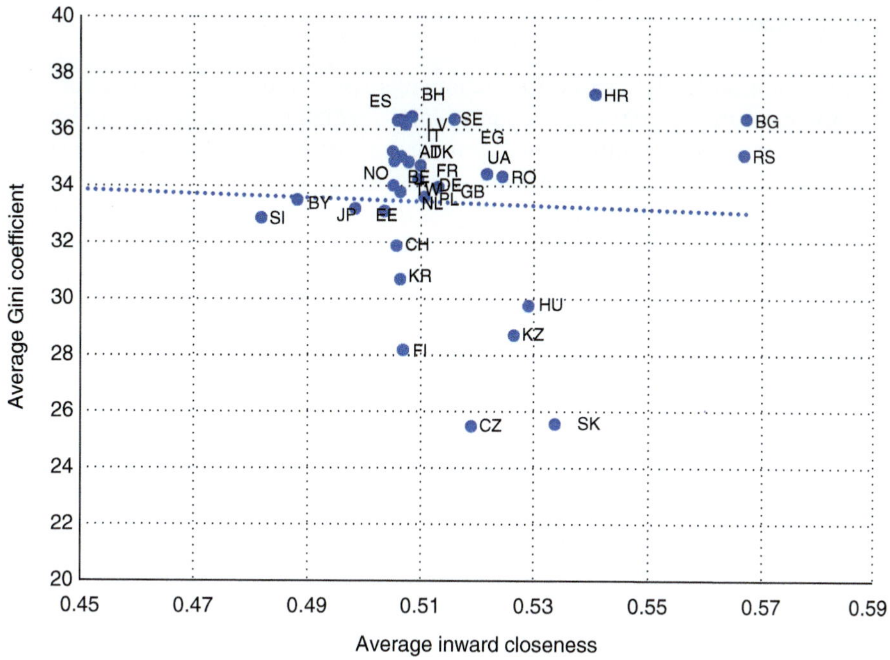

Figure 26.16 Inward closeness low Gini cluster (country level 2005–16)

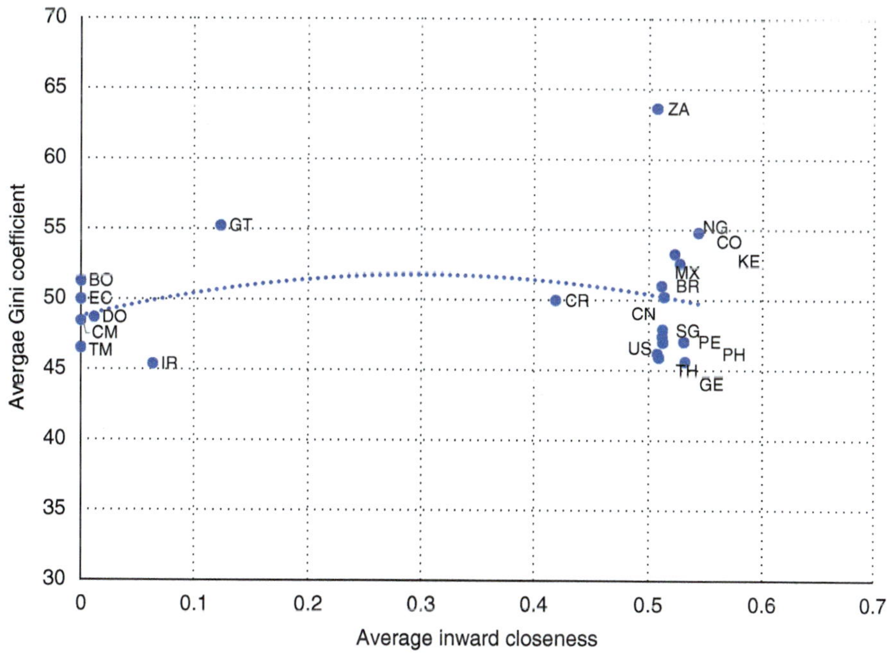

Figure 26.17 Inward closeness high Gini cluster (country level 2005–16)

Econometric Results

The results of multilevel structural equation regression models analyzing the impact of network measures on city- and country-level inequality indicate that the network measures impact predominantly inequality at the same level (Table 26.3). That is, city-level network measures impact city inequality, while country-level measures impact country inequality. Closeness is a measure of the degree to which a city or country is functionally close to all other cities or countries in a network (Bavelas 1950) in which functional gravity or regionalism is expressed, and which, in turn, enables the city or country to more easily access information from functionally proximate cites or countries. City closeness is found to not directly influence city inequality, arguably because only hub cities in a country have access to globally accessible information. This is in line with our cluster analysis where we found that variation of city closeness was limited. Unlike city closeness, country closeness is found to directly reduce both country and city inequality. This reduction in inequality is owing to the ready availability of information and being functionally close to many other countries. The results contribute to filling the literature gap concerning the intersection of global and regional economic networks (Crescenzi and Rodríguez-Pose 2011). Considering that most countries have low closeness values, those countries face a disadvantage in accessing vital information, which in turn propagates inequality and contributes to the spurring of spiky geographies and irregular development (Javorcik and Spatareanu 2008). The impact of country closeness on city as well as country inequality is found to be non-linear, indicating that inequality is conditional on certain thresholds of closeness.

Similarly, country closeness has a non-linear relationship with city and country GDP growth; too much or too little information can determine the effect on inequality. While initial increases in country closeness reduce both city and country inequality, a sustained increase in the closeness quotient increases inequality at both the city and country levels. These findings are in line with Pham (2014), that is, whereas regionalism between countries at the same level of economic development tends to reduce inequality, integration between developed and developing countries worsens the within-country income inequalities of developing countries. Therefore, assuming that the initial regionalism develops between countries at the same level of economic development, a country's income inequality can be expected to decrease at initial levels of regionalism but thereafter increase as the country integrates with other countries at different levels of development. Therefore, as noted by Wall et al. (2018), regionalism on a larger scale leads to higher inequality if it entails integration of countries and cities at different levels of development.

For high city structural hole values, it is evident that cities have significantly lower inequality levels. A city with high structural hole strength acts as a bridge between two or more city clusters of differing qualities and is said to gain comparative advantages because it is able to transfer and absorb valuable information from one city cluster to another. Structural hole strength expresses a city's innovation capacity. The results reveal that highly innovative cities are more likely to experience reductions in inequality.

Similarly, a country with high country structural hole values is seen to have lower country-level inequality, indicating that innovative countries tend to have lower inequality. This is expected, since the city or country acts as a mediator between two or more differing geographic clusters and is therefore said to gain comparative advantages, as being a

Table 26.3 Multilevel structural equation model results: inequality and network measures

	City Gini	Country Gini	City GDP growth	Country GDP growth	City unemployment	Country unemployment
City structural hole	–13.25*** (3.978)	–0.771 (2.364)	1.304 (1.579)	–0.405 (1.172)	4.142 (4.259)	–1.142 (1.614)
City clustering coefficient	–4.840** (2.456)	–0.797 (1.497)	0.178 (1)	–0.0446 (0.742)	1.363 (2.603)	–1.109 (1.022)
City closeness	120.2 (564.8)	544.8* (321.8)	101 (214.4)	159.8 (159.1)	–1198** (593.7)	–466.2** (219.2)
City closeness squared	–493.5 (1901)	–1704 (1083)	–264.4 (721.6)	–510.7 (535.5)	4045** (2001)	1489** (737.7)
Country structural hole	–3.643 (4.001)	–18.78*** (2.466)	2.773* (1.597)	2.570** (1.185)	6.339 (4.299)	–9.479*** (1.636)
Country clustering coefficient	3.511 2.475)	–7.789*** (1.634)	0.715 (1.071)	0.802 (0.795)	5.186* (2.651)	–3.226*** (1.095)
Country closeness	–103.4*** (18.52)	–62.57*** (11.73)	24.02*** (7.791)	17.82*** (5.781)	–19.83 (19.83)	9.489 (8.001)
Country closeness squared	33.01*** (7.46)	21.04*** (4.61)	–14.15*** (3.043)	–10.62*** (2.258)	4.819 (7.997)	–4.595 (3.145)
City GDP growth	–0.0256 (0.079)				–0.0912 (0.0838)	
City unemployment rate	0.447*** (0.0585)					
Country GDP growth		–0.117* (0.0693)				–0.187*** (0.0444)
Country unemployment rate		0.0990** (0.0479)				
Developing region	6.457*** (0.822)	3.732*** (0.489)				
C.Hub#M1[city>isocode]	1 (0)					
Constant	117.3*** (43.4)	56.71** (24.87)	–12.8 (16.53)	–13.82 (12.27)	103.6** (45.21)	48.83*** (16.9)
Observations	960	960	960	960	960	960
AIC	26282.96					

Note: Standard errors in parentheses; *** $p < 0.01$, ** $p < 0.05$, * $p < 0.1$.

mediator allows the city or country to transfer or gate-keep valuable information from one cluster to another. The impact of the city or country structural hole quotient, however, is limited to city and country inequality, respectively. This underlines Glückler's (2013) reasoning that the probability that extraterritorial networks keep a city open to innovation will determine the fate of these cities across regions. The spillover effects of city structural hole quotient on country inequality, or vice versa, are not observed.

The final network centrality measure, that is, clustering coefficient, proves to decrease both city and country inequality. The clustering coefficient is a measure of the degree to which cities in the network tend to cluster together. Of interest are cities that tend to create tightly knit cohesions, characterized by a relatively high density of linkages (Holland and Leinhardt 1971; Watts and Strogatz 1998). For the city clustering coefficient, it is evident that it directly and significantly reduces city inequality. This means that the more a city forms closed integrated networks with other cities, the less inequality is experienced. Being well connected to others increases opportunities, which in turn ensures equity. Increasing a city's inclusion in complete network clusters will contribute to reducing inequality. Similarly, the city clustering coefficient reduces country-level inequality.

Although both unemployment and GDP growth reveal the expected effect on inequality, only unemployment is seen to exert a significant impact. As expected, the developing region dummy variable shows that the developing world has higher levels of inequality at both city and country levels compared with their counterparts in the developed world.

The results of the path analysis indicate the indirect effect of network measures on inequality through economic growth and unemployment is limited (Table 26.4). This

Table 26.4 Direct, indirect and total impact of network measures on city and country inequality

Complex measures	Direct	Indirect	Total
City inequality			
Country clustering coefficient	3.511 (2.475)	2.298* (1.225)	5.809 (4.730)
Country structural holes	–3.643 (4.001)	2.761 (1.971)	–0.882 (4.406)
Country closeness	–103.4*** (18.52)	–9.474 (9.128)	–112.83*** (20.373)
Country closeness squared	33.01*** (7.460)	2.5151 (3.752)	35.528*** (8.176)
City clustering coefficient	–4.840*** (2.456)	0.604 (1.166)	–4.235*** (2.714)
City structural holes	–13.25*** (3.978)	1.817 (1.92)	–11.429*** (4.403)
City closeness	120.2 (564.8)	–5.380** (2.745)	–4.178 (6.218)
City closeness squared	–493.5 (1901)	18.142** (9.250)	1.321 (2.093)
Country inequality			
Country clustering coefficient	–7.789*** (1.634)	–0.413* (0.215)	–8.202*** (1.639)
Country structural holes	–18.78*** (2.466)	–1.238** (0.521)	–20.018*** (2.453)
Country closeness	–62.57*** (11.73)	–1.142 (1.702)	–63.71*** (11.706)
Country closeness squared	21.04*** (4.610)	0.7855 (0.883)	21.821*** (4.57)
City clustering coefficient	–0.797 (1.497)	–0.104 (0.143)	–0.901 (1.502)
City structural holes	–0.771 (2.364)	–0.066 (0.219)	–0.837 (2.373)
City closeness	544.8* (321.8)	–64.799* (37.416)	479.89 (322.403)
City closeness squared	–1704 (1083)	206.9* (123.293)	–1497.157 (1085.16)

Note: Standard errors in parentheses; *** $p < 0.01$, ** $p < 0.05$, * $p < 0.1$.

is indicative of the absence of spillover benefits to host economies through forward/backward linkages limiting the potentially larger role that FDI could play in reducing inequality. At the city level only city incloseness is seen to have significant indirect impact on inequality, while at the country level only country structural hole has a negative and significant impact. However, it is interesting that the non-linear direct impact of incloseness carries through to indirect impact, indicating that while some level of regional integration is beneficial to reducing inequality, over-committing to regionalism is not beneficial. Nevertheless, the total impact of network measures follows the direct effect, indicating that indirect effects are limited and merely compound the direct effect, in the instances were indirect effect is found to be significant.

CONCLUSION

Existing studies on financial globalization have not taken into account how various scales of investment impact countries or cities (IMF 2007). In this context, our study attempts to bridge world city network literature with inequality theory, thereby integrating modern-day economic globalization studies with development economics (Alderson and Beckfield 2004; Coe et al. 2008; Wall 2009). The study accomplishes this by bridging inequality theories with world city network literature in order to address the complexity of international capital flows and its impact on territorial development at the city and country levels. It enables the exploration of nuanced effects of FDI on economic inequality at inter-regional and intra-regional levels. The study covers the world's major cities and their respective countries using a unique, comprehensive database, and contributes to the literature by providing a comparative analysis at the city and country levels through a multilevel structural equation model estimation.

The findings indicate that the income distribution of a city is defined by its position within transnational networks, measured through different scales of FDI network centralities (closeness, clustering coefficient and structural holes). The study shows that apart from agglomeration factors, such as GDP growth and unemployment rates, income inequality is also significantly affected by network centrality measures. The results of multilevel structural equation models, indicate that network measures predominantly impact inequality at their respective levels, where city-level network measures impact intra-city inequality, and country-level measures impact country inequality.

The results reveal that reductions in inequality are more likely to be experienced by highly innovative cities/countries or in other words cities/countries with high structural hole values. For city and country clustering coefficients, it is evident that these directly and significantly reduce intra-city and country inequality, respectively. As regards country closeness, a non-linear relationship is observed, suggesting that inequality is conditional on particular thresholds of closeness.

The results of the path analysis indicate that the indirect effect of network measures on inequality through economic growth and unemployment is limited. This is indicative of the absence of spillover benefits to the host economy through forward/backward linkages limiting the potentially larger role that FDI could play in reducing inequality except for city closeness and country structural holes. However, it is interesting to note that the non-linear direct impact of city closeness carries through to indirect impact as well, indicating that while some level of regional integration is beneficial to reducing inequality, over-committing to regionalism is not beneficial. This finding indicates the existence of an optimal level for globalization, which needs to be taken up for future research.

REFERENCES

Alderson, A.S. and J. Beckfield (2004), 'Power and position in the world city system', *American Journal of Sociology*, **109** (4), 811–51.

Bathelt, H., A. Malmberg and P. Maskell (2004), 'Clusters and knowledge: local buzz, global pipelines and the process of knowledge creation', *Progress in Human Geography*, **28** (1), 31–56.

Bavelas, A. (1950), 'Communication patterns in task-oriented groups', *Journal of the Acoustical Society of America*, **22** (6), 725–30.

Bordo, M., M. Taylor and J. Williamson (2005), *Globalization in Historical Perspective*, Chicago, IL: University of Chicago Press.

Boschma, R. (2005), 'Proximity and innovation: a critical assessment', *Regional Studies*, **39** (1), 61–74.

Brautbar, M. and M. Kearns (2011), 'A clustering coefficient network formation game', in G. Persiano (ed.), *International Symposium on Algorithmic Game Theory*, Berlin and Heidelberg: Springer, pp. 224–35.

Burger, M.J., B. van der Knaap and R.S. Wall (2014), 'Polycentricity and the multiplexity of urban networks', *European Planning Studies*, **22** (4), 816–40.

Burt, R.S. (2009), *Structural Holes: The Social Structure of Competition*, Cambridge, MA: Harvard University Press.

Camagni, R.P. (1993), 'From city hierarchy to city network: reflections about an emerging paradigm', in T.R. Lakshmanan and P. Nijkamp (eds), *Structure and Change in the Space Economy*, Berlin and Heidelberg: Springer, pp. 66–87.

Castells, M. (1989), *The Informational City: Information Technology, Economic Restructuring, and the Urban-Regional Process*, Oxford: Blackwell.

Castells, M. (1996), *The Rise of the Network Society*, Oxford: Blackwell.

Chan, K.W., J.V. Henderson and K.Y. Tsui (2008), 'Spatial dimensions of Chinese economic development', in L. Brandt and T. Rawski (eds), *China's Great Economic Transformation*, Cambridge and New York: Cambridge University Press, pp. 776–828.

Christaller, W. (1933), *Die Zentralen Orte in Suddeutschland*, Jena: Gustav Fischer.

Coe, N.M., P. Dicken and M. Hess (2008), 'Global production networks: realizing the potential', *Journal of Economic Geography*, **8** (3), 271–95.

Coe, N.M., M. Hess, H.W.-C. Yeung, P. Dicken and J. Henderson (2004), '"Globalizing" regional development: a global production networks perspective', *Transactions of the Institute of British Geographers*, **29** (4), 468–84.

Crescenzi, R. and A. Rodríguez-Pose (2011), *Innovation and Regional Growth in the European Union*, Berlin: Springer Science & Business Media.

Crescenzi, R., D. Luca and S. Milio (2016), 'The geography of the economic crisis in Europe: national macroeconomic conditions, regional structural factors and short-term economic performance', *Cambridge Journal of Regions, Economy and Society*, **9** (1), 13–32, doi:10.1093/cjres/rsv031.

D'Este, P., F. Guy and S. Iammarino (2013), 'Shaping the formation of university–industry research collaborations: what type of proximity does really matter?', *Journal of Economic Geography*, **13** (4), 537–58.

Dasgupta, I. and T. Osang (2002), 'Globalization and relative wages: further evidence from US manufacturing industries', *International Review of Economics & Finance*, **11** (1), 1–16.

Dicken, P. (2003), *Global Shift: Reshaping the Global Economic Map in the Twenty-First Century*, London: Sage.

Dicken, P. and J. Henderson (2003), 'Making the connections: global production networks in Britain, East Asia and Eastern Europe', final report on ESRC Research Project R000238535, Economic and Social Research Council, Swindon.

Dicken, P. and A. Malmberg (2001), 'Firms in territories: a relational perspective', *Economic Geography*, **77** (4), 345–63.

Feenstra, R. and G. Hanson. (1997), 'Foreign direct investment and relative wages: evidence from Mexico's maquiladoras', *Journal of International Economics*, **42** (3–4), 371–93.

Firebaugh, G. and F.D. Beck (1994), 'Does economic growth benefit the masses? Growth, dependence, and welfare in the third world', *American Sociological Review*, **59** (5), 631–53.

Friedmann, J. (1986), 'The world city hypothesis', *Development and Change*, **17** (1), 69–84.

Friedmann, J. (1995), 'Where we stand: a decade of world city research', in P.L. Knox and P.J. Taylor (eds), *World Cities in a World-System, World City Research*, Cambridge: Cambridge University Press, pp. 27–47.

Gereffi, G., J. Humphrey and T. Sturgeon (2005), 'The governance of global value chains', *Review of International Political Economy*, **12** (1), 78–104.
Glückler, J. (2013), 'Knowledge, networks and space: connectivity and the problem of non-interactive learning', *Regional Studies*, **47** (6), 880–94.
Grabher (2006), 'Trading routes, bypasses, and risky intersections: mapping the travel "networks" between economic sociology and economic geography', *Progress in Human Geography*, **30** (2), 163–89.
Graham, S. and S. Marvin (2002), *Splintering Urbanism: Networked Infrastructures, Technological Mobilities and the Urban Condition*, London: Routledge.
Guimaraes, P., O. Figueiredo and D. Woodward (2000), 'Agglomeration and the location of foreign direct investment in Portugal', *Journal of Urban Economics*, **47** (1), 115–35.
Guy, F. (2015), 'Globalisation, regionalization and technological change', in D. Archibugi and A. Filippetti (eds), *The Handbook of Global Science, Technology and Innovation*, Oxford: Wiley-Blackwell, pp. 575–96.
Harvey, D. (2006), *Spaces of Global Capitalism: Towards a Theory of Uneven Geographical Development*, London: Verso.
Henderson, J., P. Dicken, M. Hess, N. Coe and H.W.-C. Yeung (2002), 'Global production networks and the analysis of economic development', *Review of International Political Economy*, **9** (3), 436–64.
Herzer, D. and P. Nunnenkamp (2013), 'Inward and outward FDI and income inequality: evidence from Europe', *Review of World Economics*, **149** (2), 395–422.
Hess, M. (2004), '"Spatial" relationships? Towards a reconceptualization of embeddedness', *Progress in Human Geography*, **28** (2), 165–86.
Hobday, M., A. Davies and A. Prencipe (2005), 'Systems integration: a core capability of the modern corporation', *Industrial and Corporate Change*, **14** (6), 1109–43.
Hohenberg, P.M. and L.H. Lees (1985), *The Making of Urban Europe, 1000–1950*, Cambridge, MA: Harvard University Press.
Holland, P.W. and S. Leinhardt (1971), 'Transitivity in structural models of small groups', *Comparative Group Studies*, **2** (2), 107–24.
Hymer, S.H. (1976), *The International Operations of Multinational Firms: A Study of Foreign Direct Investment*, Cambridge, MA: MIT Press.
Hymer, S. (1982), 'The multinational corporation and the law of uneven development', in J.M. Letiche (ed.), *International Economics Policies and their Theoretical Foundations*, New York: Academic Press, pp. 325–52.
Im, H. and J. McLaren (2015), 'Does foreign direct investment raise income inequality in developing countries? A new instrumental variables approach', working paper, School of Economic Finance, Yeungnam University.
International Monetary Fund (IMF) (2007), 'An empirical investigation of globalization and inequality', in A. Kumar (ed.), *World Economic Outlook: Globalization and Inequality*, Washington, DC: International Monetary Fund, pp. 46–54.
Javorcik, B.S. and M. Spatareanu (2008), 'To share or not to share: does local participation matter for spillovers from foreign direct investment?', *Journal of Development Economics*, **85** (1), 194–217.
Johansson, A.C., D. Liu and M. Zhen (2017), 'Foreign direct investment and urban inequality: evidence from Chinese cities', Working Paper No. 2017-45, Stockholm China Economic Research Institute, Stockholm School of Economics.
Kanbur, R. and A.J. Venables (2005), *Spatial Inequality and Development*, Oxford: Oxford University Press, pp. 3–11.
Kline, R.B. (2005), *Principles and Practice of Structural Equation Modeling*, 2nd edn, New York: Guilford.
Krätke, S. (2014) 'How manufacturing industries connect cities across the world: extending research on "multiple globalizations"', *Global Networks*, **14** (2), 121–47.
Krugman, P. (1991), 'Increasing returns and economic geography', *Journal of Political Economy*, **99** (3), 483–99.
Lewis, W.A. (1976), 'Development and distribution', in A. Cairncross and M. Puri (eds), *Employment, Income Distribution and Development Strategy: Problems of the Developing Countries*, New York: Holmes & Meier, pp. 26–42.
Lindert, P.H. and J.G. Williamson (2001), 'Does globalization make the world more unequal?', NBER Working Paper No. 8228, National Bureau of Economic Research, Cambridge, MA.
Maddison, A. (1995), *Monitoring the World Economy 1820–1992*, Paris and Washington, DC: Organisation of Economic Co-operation and Development.
Mahutga, M.C. and N. Bandelj (2008), 'Foreign investment and income inequality: the natural experiment of Central and Eastern Europe', *International Journal of Comparative Sociology*, **49** (6), 429–54.
Monaghan, D. and P. Ikeler (2014), 'Global centrality and income inequality in U.S. metropolitan areas: a test of two hypotheses', *Sociological Focus*, **47** (3), 174–93.
Myrdal, G. (1957), *Economic Theory and Underdeveloped Regions*, London: Duckworth.
Neal, Z. (2012), 'Structural determinism in the interlocking world city network', *Geographical Analysis*, **44** (2), 162–70.

Neal, Z.P. (2011), 'From central places to network bases: a transition in the US urban hierarchy, 1900–2000', *City & Community*, **10** (1), 49–75.
Ohlin, B. (1935), *Interregional and International Trade*, Cambridge, MA: Harvard University Press.
Pham, T.H.H. (2014), 'Intra-regional trade and income inequality: where do we stand?', working paper, LEMNA, University of Nantes, hal-01087366.
Poon, J.P., G.K. Tan and W. Yin (2015), 'Wage inequality between financial hubs and periphery', *Applied Geography*, **61** (4), 47–57.
Pred, A. (1973), 'The growth and development of systems of cities in advanced economies', in A. Pred and G. Törnqvist (eds), *Systems of Cities and Information Flows: Two Essays*, Lund Studies in Geography, series B, no. 38, pp. 9–82.
Rugman, A.M. (2005), *The Regional Multinationals*, Cambridge: Cambridge University Press.
Sassen, S. (1994), *Cities in a World Economy*, Thousand Oaks, CA: Pine Forge Press.
Sassen, S. (2001), 'Global cities and developmental states: how to derail what could be an interesting debate: a response to Hill and Kim', *Urban Studies*, **38** (13), 2537–40.
Shipley, B. (2013), 'The AIC model selection method applied to path analytic models compared using a d-separation test', *Ecology*, **94** (3), 560–64.
Storper, M. and A.J. Venables (2004), 'Buzz: face-to-face contact and the urban economy', *Journal of Economic Geography*, **4** (4), 351–70.
Sylwester, K. (2005), 'Foreign direct investment, growth and income inequality in less developed countries', *International Review of Applied Economics*, **19** (3), 289–300.
Taylor, P.J. (2004), *World City Network: A Global Urban Analysis*, London: Routledge.
Timberlake, M., M.R. Sanderson, X. Ma, B. Derudder, J. Winitzky and F. Witlox (2012), 'Testing a global city hypothesis: an assessment of polarization across US cities', *City & Community*, **11** (1), 74–93.
Wall, R., J. Maseland, K. Rochell and M. Spaliviero (2018), 'State of African cities report 2018: the geography of African investment', United Nations Human Settlements Programme (UN-Habitat), Nairobi.
Wall, R.S. (2009), 'Netscape: cities and global corporate networks', PhD thesis, Erasmus Research Institute on Management (ERIM), Haveka.
Wall. R.S. and G.A. van der Knaap (2011), 'Sectoral differentiation and network structure within contemporary worldwide corporate networks', *Economic Geography*, **87** (3), 267–308.
Wall, R.S. and S. Stavropoulos (2016), 'Smart cities within world city networks', *Applied Economics Letters*, **23** (12), 875–9.
Wallerstein, I. (1974) 'The rise and future demise of the world capitalist system: concepts for comparative analysis', *Comparative Studies in Society and History*, **16** (4), 387–415.
Wallerstein, I. (1999), *The End of the World as We Know It: Social Science for the Twenty-First Century*, Minneapolis, MN: University of Minnesota Press.
Watts, D.J. and S.H. Strogatz (1998), 'Collective dynamics of "small-world" networks', *Nature*, **393** (6684), 440–42.
Wei, K., S. Yao and A. Liu (2009), 'Foreign direct investment and regional inequality in China', *Review of Development Economics*, **13** (4), 778–91.
Wrigley, N., N.M. Coe and A. Currah (2005), 'Globalizing retail: conceptualizing the distribution-based transnational corporation (TNC)', *Progress in Human Geography*, **29** (4), 437–57.
Yeung, H.W.-C. (2009), 'Regional development and the competitive dynamics of global production networks: an East Asian perspective', *Regional Studies*, **43** (3), 325–51.
Yeung, H.W.-C. (2016), *Strategic Coupling: East Asian Industrial Transformation in the New Global Economy*, Ithaca, NY: Cornell University Press.
Zhang, X. and R. Kanbur (2001), 'What difference do polarisation measures make? An application to China', *Journal of Development Studies*, **37** (3), 85–98.

APPENDIX: COMPUTATION OF NETWORK MEASURES

1. Structural Holes (Innovation Advantage of Cities in FDI Networks)

According to Burt (2009), social structures comprise dense clusters of strong connections, where the homogeneity of information, new ideas and behavior tends to be higher within a cluster. However, an individual that mediates between two or more clusters is said to gain important comparative advantages, because the individual can gate-keep and transfer different types of information between clusters. Also, an individual can merge and absorb different ideas received from surrounding clusters, and thereby hold a relative innovation advantage. In relation to our study, we rely on the interpretation that economic information similarities are stronger within distinct clusters of cities, and that cities that serve as intermediaries between two or more clusters will connect different types of investment information, and thereby hold a relative innovation advantage. Hence, a city with a high structural hole value holds comparative advantages, because it permits a city to gate-keep valuable information between clusters of cities.

To measure a structural hole in a network, Burt (2009) developed a measure of network redundancy, which estimates to what extent nodal contact j is redundant in relation to other nodal contacts of node i, in which redundancy is measured as an investment in a relationship with another node q, with whom node j is strongly connected – *redundancy* = *piq mjq* – and where *piq* is a portion of *i*'s investment in relation to q, and *mjq* is calculated as *j*'s interaction with q divided by *j*'s biggest relationship with any others. The network's redundancy is calculated by the sum of this product across all nodes of q. A negative factor is that this output expresses the non-redundant part of this relationship. The effective size of *i*'s network is determined by the sum of *j*'s non-redundant contacts.

$$Structural\ hole_i = \sum_j [1 - \sum_q p_{iq} m_{jq}], \quad q \neq i,j \tag{26A.1}$$

The more disconnected each node is from other primary contacts, the greater will be the effective size of the network. This measure varies from 1 (where the network has a single link) to a total number of links N (in which each contact is non-redundant).

2. Clustering Coefficient (Proximity Advantage of Cities in FDI Networks)

In network theory, the clustering coefficient measures the degree to which nodes in a network gravitate towards each other. It has been shown that social networks create tightly knit clusters that are characterized by relatively high densities of connections, and that this likelihood is generally larger than the probability that a connection is randomly established between nodes (Holland and Leinhardt 1971). Through this, social networks are seen to exhibit high-linkage clustering; that is, the density of local neighborhoods, as expressed by the clustering coefficient, is greater than the overall linkage density of the entire network. In social networks, the desire for these social cliques is seen as an advantage of proximity within networks (Brautbar and Kearns 2011). Interpreted in our study, the clustering coefficient is a measure of how strongly a city clusters with other cities, which receive similar types of investment. This relates to cities that create tightly bound clusters dependent on a high density of investment ties. The probability of these types of linkages occurring is said to be higher than the likelihood of linkages being randomly established between cities. In this context, when a city has a large clustering coefficient, the possibility of attracting similar, future FDI is high. It is sometimes considered to be a measure of network regionalism.

The clustering coefficient can be measured for an entire network, that is, global, or for each node in a network, that is, local (Watts and Strogatz 1998). The clustering coefficient represents a node's neighborhood, indicating the level of cohesion with its neighbors. This coefficient is the fraction of nodal pairs that are neighbors of the given node and are connected by linkages (see equation 26A.2).

$$Clustering\ coefficient_i = \frac{2|e_{jk}|}{k_i(k_i - 1)} : v_j v_k \in N_i, e_{jk} \in N_i, e_{jk} \in E \qquad (26A.2)$$

where *Ni* is the neighborhood of node *Vi*, *ejk* represents the edge that connects node *Vj* to node *Vk*, where *ki* is the degree of node *Vi*, and |*ejk*| indicates the proportion of links between the nodes within the neighborhood of node *Vi*.

3. Closeness (Accessibility Advantage of Cities in FDI Networks)

This measure relates to the closeness of an individual to all others in a network. In social networks, individuals are seen to have a high level of closeness, when they can quickly interact with others. A central individual holds the shortest path distance to all others and can thereby rapidly communicate with them without utilizing many intermediaries. Hence, if two individuals are not directly tied, then the smallest number of steps to reach one another is important. It is an index which measures the communication accessibility to others, which is vital to the successful performance of the individual. For our study, the measure expresses how proximate a city is to all pairs of cities, based on the shortest investment paths between each other. It is a measure of the steps taken to spread investment information from a city to all other cities in a network.

Closeness is an approximate measure of the general position of a node in a network, revealing the number of steps it will take to reach other nodes from a node. It is formally the average length of the total shortest paths of one node to all the other nodes in a network. This measure is generally only computed for nodes within the largest component of a network (see equation 26A.3), defining how fast a given node can reach all others in the network. In the equation, closeness centrality is the reciprocal sum of the shortest distances from v to all $n - 1$, nodes. As the sum of distances depends on the number of nodes in the graph, the measure is normalized by the minimum sum of likely distances $n - 1$, where $d(u, v)$ represents the smallest path distance between u and v, and n indicates the number of nodes.

$$Closeness_v = \frac{n - 1}{\sum_{u \in V(G) v} d(u,v)} \qquad (26A.3)$$

27. Research progress of Chinese city networks

Fenghua Pan, Cheng Fang and Xiande Li

INTRODUCTION

Chinese city networks have attracted tremendous research attention in recent years (Timberlake et al. 2014). Since the beginning of China's reforms, cities in China have achieved sustained and rapid growth, with a significant increase in their global influence. Beijing, Shanghai, Guangzhou, Shenzhen and other Chinese cities have become important nodes and hubs in the world city network (WCN). More importantly, they have been gradually reshaping the very structure of the WCN (Derudder et al. 2013, 2018; Taylor et al. 2014b). In addition, China possesses a vast territory with significant regional differences. These conditions not only form complex city networks at the national level, but they also create unique regional city networks in China (Zhao 2011; Zhen et al. 2013; Zhao et al. 2015a). Moreover, the development of cities in China has been strongly influenced by the background settings in China. Therefore, the dynamics of the formation and evolution of Chinese city network structure are unique (Ma and Timberlake 2008; Zhao et al. 2015a; Pan et al. 2017).

The perception of cities as network nodes in academia has evolved over time. The external connections of a city are considered its second nature (Taylor 2004), and each city is seen as a relationship phenomenon (Scott 2012). The establishment of multiple inter-city connections can be important to the development of a city (Amin and Thrift 2002). Globalization and informatization have greatly changed the spatial organization patterns of cities typically manifested by the increasing spatial differences of cities and the continuous linkages between them (Dicken et al. 2001; Sheppard 2002; Florida 2003). Thus, cities have formed extensive and complex interconnections at different spatial scales. These interconnections have presented a network-based spatial organization structure, namely, the city networks (Camagni and Capello 2004). The city network offers an important perspective for analyzing external connections of cities. As a dimension of social spatial relations, the city network can well analyze the spatial flows of capital, information and other essential elements. In city network research, size is no longer the sole determinant of function, and geographical proximity does not necessarily mean that linkages between cities are closer (Wang 2007; Li 2015).

Although academia has long realized the importance of a city's external connections in theory, the breakthrough of empirical research on city networks has come from the theoretical and methodological innovation of world city research. Since the 1960s, world city research has gone through three distinct stages. The progression of these stages has been from world cities with attributes but no relationships, to world cities in the network society and to the research of WCN (Yang et al. 2011). Inter-city linkages were largely ignored during the first two stages. Castells (1996, pp. 407–59) put forward the concept of 'space of flows' and stated that, in the society of the information age, the core control of the global economy is 'the generation of knowledge and flows of information' within the advanced producer service (APS) sectors and cities are 'nodes and hubs in the space of flows'. The key to connecting these nodes and hubs is social practice, namely, the exchange and communication between social objects (Castells 1996, p. 442). The theory of space of flows has spurred the shift of the world city research to WCN research. However, it is very difficult to collect city relationship data, in particular, at the global scale.

Entering the twenty-first century, the Globalization and World City Research Group (GaWC), had broken the limitations of previous research. The GaWC used a variety of indicators to measure the strength of inter-city links and study the patterns and evolutions of WCN. The breakthrough of methodology has pushed world city research to a new stage (Taylor 2004; Derudder et al. 2010; Pereira and Derudder 2010; Hanssens et al. 2013; Taylor et al. 2014a, 2014b; Liu et al. 2014). In particular, the inter-locking network model (INM), proposed by Taylor and others, based on the inheritance of Sassen's emphasis on the concept of APS and 'space of flows' theory of Castells (Castells 1996; Sassen 2013), has realized the transformation from firm links to city relations (Taylor et al. 2002). Among these studies, APS intra-firm linkages have been widely applied to empirical studies on various scales (Taylor et al. 2003; Zhao and Liu 2010; Zhang 2018).

Inspired by the GaWC's studies, Chinese city network research has been booming. In earlier years, city network research on Chinese cities applied the same data and methods used by the GaWC. Later, more research used various data and methods of building city linkages at different spatial scales. More importantly, scholars have realized that city networks in China are not just a simple extension of WCN. It is necessary to examine the formation and evolution of Chinese city networks under China's unique social, economic and institutional context (Li 2014; Li et al. 2015; Zhao et al. 2015a). In addition, with the rise of China's status within the global economy, scholars have begun to study how Chinese factors affect and reshape WCN from a Chinese perspective. These studies have opened up a new perspective into the research of WCN (Pan et al. 2018). That is, a large amount of city network research on Chinese cities has promoted worldwide progress in this field. Existing research has enhanced the understanding of the evolutionary process and driving mechanism of the Chinese city network structure. It has also promoted the development of urban geography and economic geography research in China. However, there are also limitations and challenges to contemporary research. There are deficiencies in theoretical innovation, and original ideas are limited (Ma 2016). Some empirical studies have utilized similar data and perspective, and their findings and conclusions tend to be homogenous.

This chapter summarizes and reviews the research trends of the Chinese city networks from the aspects of data and methods, core issues and major contributions. We also discuss the limitations in existing literature and, accordingly, propose directions for future research.

DATA AND METHODS IN CHINESE CITY NETWORK RESEARCH

Chinese city network research had adopted a variety of data and methods to build city networks. Many early studies directly used GaWC data and methods or followed the GaWC approach. In recent years, many studies have applied new data sources and perspectives to investigate city networks in China.

Data of Inter-city Linkages

Various data and approaches are used in Chinese city network research to measure inter-city linkages (see Table 27.1). In traditional urban system research, attributes of cities, such as population and gross domestic product (GDP), were used to estimate and

Table 27.1 Different ways to measure inter-city linkages based on different data in Chinese city network research

Inter-city linkage types	Network-building method	Data
Estimated linkages	Model-transforming based on gravity or radiation model	Attribute data of cities (e.g., GDP and population)
Intra-firm linkages	Model-transforming based on inter-locking network model	Geographical distribution data of different level branches of producer service firms (Ni et al. 2011; Zhao and Liu 2012; Zhen et al. 2013; Yin et al. 2011; Pan et al. 2018)
	Directly building	Ownership relationships and geographical distribution data of headquarters and branches (Zhao 2011; Li 2014; Wu and Fang 2015) or some specific industries (Zhao et al. 2014)
Inter-firm linkages	Directly building	Inter-firm cooperation relationship data based on questionnaires (Ma and Fang 2011; Yeh et al. 2015), venture capital investment, inter-firm service relationship data in IPOs (Pan et al. 2016, 2017), and so on
Infrastructure linkages	Directly building	Telecommunication link and Internet flow data (Yu 1988; Wang and Ning 2004), air passenger flow data (Wang and Ning 2011), highway passenger flow data (Chen et al. 2017), high-speed rail stations and shift data (Jiao et al. 2017), comprehensive link data for various modes of transportation with a weighted calculation (Zhu et al. 2018) and so on
Other inter-city linkages	Directly building	Link data in the co-authorship of scientific research papers (Liefner and Hennemann 2011; Lu and Huang 2012), co-application for patents (Ma et al. 2015), mutual notice data of microblog users (Zhen et al. 2012), buyers' and sellers' links in Internet shopping behavior (Xi et al. 2018), resident mobile data based on Weibo positioning (Zhang et al. 2018), and so on

simulate interactions between Chinese cities based on the gravity model (Gu and Pang 2008). However, this attribute data cannot sufficiently reflect real flows between cities. There exist obvious limitations in this strand of literature.

With the increasing availability of data, firms and infrastructure are used to build inter-city connections. Among all studies, those that used firm linkages to build city networks have become the mainstream method. The firm linkages can be divided into intra-firm links and inter-firm links. Intra-firm link data, in particular APS intra-firm linkage, was widely used to characterize Chinese city networks (Derudder et al. 2013, 2018; Taylor et al. 2014b; Zhao and Liu 2012; Zhang 2018). In addition, some studies use headquarter–branch linkages of large manufacturing firms to explore urban networks in China (Zhao 2011; Li 2014; Wu and Fang 2015). While intra-firm linkages cannot fully reflect the inter-city connections in the real world, inter-firm linkages are important in modern society (Yeung 2005). Existing urban networks studies based on intra-firms linkages were criticized for the limitation of their methodology (Csomós and Derudder 2014; Rozenblat and Pumain 2008). Thus, more studies started to apply inter-firm linkages to examine Chinese city networks. For example, Ma et al. (2011) built the city network based on the upstream and downstream links of inter-firm production activities. Pan et al. (2016) applied venture capital firms' inter-city investment data to study city connections. Yeh et al. (2015) conducted a study based on questionnaire surveys on inter-firm service linkage in China. Pan et al. (2017) used the inter-firm service linkage between ABS firms and publicly listed companies among the initial public offerings (IPOs) in China to study urban networks. These studies based on inter-firm linkages in China have significantly advanced this research field.

As well as using firm linkages to build city networks, some earlier studies employed inter-city communication and transportation flows to capture inter-city relationships (Yu 1988; Zhou and Hu 2002; Wang and Ning 2004; Jin 2001), as these flows can well reflect the material and information exchanges between cities. In recent years, inter-city population movements (Zhang et al. 2018) and commodity flows (Xi et al. 2018) have been used to study urban networks (Ma 2016). In addition, co-authorship of scientific papers (Liefner and Hennemann 2011; Lu and Huang 2012), co-application data of patents (Ma et al. 2015) and social media data (Zhen et al. 2012; Chen et al. 2013) have been used to construct urban linkages in China.

Methods of Building City Networks

There are two ways to build city networks: model transforming and directly building (see Table 27.1). The gravity model, radiation model and inter-locking network model are used in model transforming. Both the gravity model and radiation model use attributes of cities to estimate urban linkages, but the two models are rarely used now. The inter-locking network model, first proposed by Taylor et al. (2002), is very widely used, and transforms firm distributions into inter-city linkages. Based on the database of multi-location APS firms, the firm–city matrix $V_{m \times n}$ is built (see Figure 27.1). V_{aj} and V_{bj} refers to the service values of APS firm *j* in any two cities *a* and *b*. The value of V_{aj} and V_{bj} are determined by the firm *j*'s branch level in city *a* and *b*. Usually, the service value of a branch is a number between 0 and 5. If there is no branch in the city, the value is 0. If there is a headquarters, the value will be 5. Based on the service values of a specific firm *j* in both cities, we can obtain the link value ($C_{ab, j}$) between city *a* and *b*. Equation

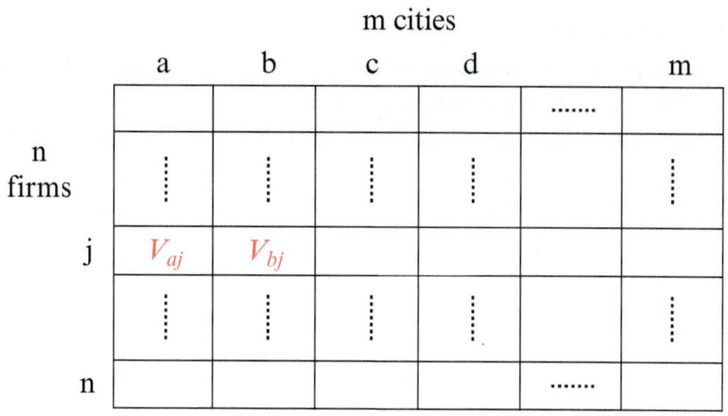

Figure 27.1 *The city–firm matrix in an interlocking network model*

(27.1) shows how $C_{ab,\,j}$ is calculated. If we have *n* firms, we can calculate the link value C_{ab} through formula (27.2).

$$C_{ab,\,j} = V_{aj} \times V_{bj} \tag{27.1}$$

$$C_{ab} = \sum\nolimits_{j=1}^{n} C_{ab,j} \tag{27.2}$$

The interlocking network model has been widely adopted in studies on Chinese city networks. In some studies, the original model was modified for specific purposes. For example, the original method cannot adequately reflect the huge differences in firm size between different APS firms, Zhang and Kloosterman (2016) chose to re-establish the assignment rule of service values. Zhao et al. (2015b) introduced an improved partition core algorithm, which identified the core cities in the region according to the service values of APS firms in each city. Moreover, some scholars argue that we should comprehensively examine the two-mode network, formed by both cities and firms, to reduce the information loss (Liu and Derudder 2013).

Another method of network-building uses specific types of inter-city relationship data that can be directly quantified (see Table 27.1). In studies of directly built city networks with link data, the measurement of link strength is the key issue. Some direct linkages can be assigned values of 0 or 1, for example, ownership linkages between the corporate headquarters and branches (Zhao 2011), co-authorship linkages of scientific papers (Liefner and Hennemann 2011; Lu and Huang 2012) and service linkages of APS firms with their clients (Pan et al. 2017). Some linkages can be quantified very easily. For example, air passenger flows (Wang et al. 2011) and venture capital flows (Pan et al. 2016) can measure the linkages between cities. Given that some inter-city linkages are directional, such as venture capital investments (Pan et al. 2016) and transportation linkages (Jin 2001), some studies will build a directed network which can provide more analytical information (Zhao et al. 2015b).

Analysis of City Networks

The social network analysis (SNA) method and complex network analysis (CNA) method are widely used in the analysis of city networks. Social network analysis is a mathematical analysis method originating from graph theory. It is used to measure and explain the topological structure of social relations (Scott and Carrington 2011). Some analytic indicators and concepts include centrality, network density and network concentration cohesion subgroups. Social network analysis is well suited to analyze the linkages between cities, therefore scholars have introduced SNA into urban research (Derudder and Taylor 2005). Complex network analysis is also based on graph theory, but it is more conducive to depicting network structure, function and the internal dynamic interaction mechanism of complex systems. As well as analyzing the static geometry and its statistical properties, complex network theory can also be used to explore the dynamic properties of the networks in combination with small world networks, scale-free networks and other evolutionary models. The spatial organization of cities is a giant complex system, so CNA has great potential in city network research (Wu and Fang 2015). Based on the results of SNA and CNA, the overall structure and characteristics of networks, as well as the status and roles of specific node cities, can be analyzed further.

In addition, city-dyads analysis is also an important method (Taylor et al. 2014b). By calculating the city-dyads with the strongest connections, the core cities and city-pairs within the network and their characteristics can be examined. Owing to the diversity of data, city networks based on different linkages usually show different structures (Burger et al. 2014a). Therefore, some studies try to make a comparative analysis of different city networks (Zhao and Chen 2011; Li et al. 2015; Lao et al. 2016; Wang and Jing 2017). The comparative analysis of different networks can reveal the characteristics of spatial organization of cities from different aspects (Derudder and Taylor 2018). Chinese city network research aims to use more complicated methods; as Neal (2013) noted, the methodology of city network research should be compatible with the key research questions.

PATTERNS OF CHINESE CITY NETWORKS AND DRIVING FORCES

Current research on Chinese city networks has mainly investigated the structure and formation mechanism of networks at three different spatial scales: global, national and regional. In this section, we introduce the research progress at the three scales, respectively.

Global Scale: Chinese Cities in the WCN

With the rapid growth of China's economy, Chinese cities have been playing an increasingly important role in the WCN (Ni et al. 2011; Wei et al. 2012; Derudder et al. 2013; Taylor et al. 2014a; Timberlake et al. 2014). Derudder et al. (2018) studied Chinese cities in the WCN from 2010 to 2016 and it was found that all Chinese cities' connectivity had increased in the WCN within that period. This strand of literature has been focusing on several large cities, such as Beijing, Shanghai and Hong Kong. These three global cities of China have collaborated and become important gateways for other Chinese cities to connect with the world. Taylor et al. (2014b) noted that Beijing, Shanghai and Hong Kong, are constantly improving their global status. Lai (2012) and Taylor et al. (2014a) believed that Beijing has had an unparalleled advantage in human capital, scientific research and state-owned enterprises. As an intermediary between the global financial market and China's mainland (Wójcik and Camilleri 2015), Hong Kong plays a key role in the global expansion of Chinese financial enterprises (Pan et al. 2018). As regards global connectivity, Taylor et al. (2014b) found that Shanghai is closely linked to the world's most important global cities, while Beijing tends to establish stronger ties with important political cities.

In addition to these three lead cities, other large cities have also become more integrated into WCN (Ni et al. 2011). The Chinese cities with the highest global connectivity are mostly located in the coastal areas, while the inland hub cities are closely connected to those coastal cities (Zhou and Hu 2002). For example, Zhang and Kloosterman (2016) found that Guangzhou and Shenzhen have been regarded as the global workshop behind Hong Kong. However, while the global connectivity of top cities continues to increase rapidly, regional disparity has been rising. As some studies have found, there exists a hierarchical structure with Beijing and Shanghai holding the dominant positions within the networks (Hennemann et al. 2011; Ni et al. 2011; Zhao and Liu 2012; Zhu et al. 2018).

Continuous opening and reforming have accelerated the integration of Chinese cities into the WCN. Since its accession to the World Trade Organization (WTO), China has been promoting the development of global cities as an important national strategy (Timberlake et al. 2014). Under this strategy, Beijing, Shanghai and other cities are equipped with the most superior development resources. As Wu and Ning (2012) noted, multinational corporations tend to choose more developed cities and regions in which to make investments, and this leads to the rapid growth of these selected cities. In addition, the institutional contexts of the Chinese economy and market forces also helped to shape the urban networks in China (Ma and Timberlake 2008). In particular, the concentration of state-owned firms' headquarters in a few large cities and the agglomeration economies driven by market forces have determined the overall pattern of Chinese cities' integration into the WCN (Derudder et al. 2013).

More studies are needed to explore how those relatively undeveloped cities and regions are connected with the WCN. In addition, Chinese cities have the potential to reshape the WCN (Pan et al. 2017), and so it is necessary to pay more attention to the driving forces and consequences of this process.

National Scale: The Chinese National[1] City Network

Existing research has found there exists a diamond-shaped structure, with the four vertices of Beijing, Shanghai, Guangzhou-Shenzhen and Chengdu within Chinese national city network. This diamond-shaped structure was confirmed in a large number of studies using different data, including financial firms (Yin et al. 2011), producer services companies (Zhao and Liu 2012), micro-blogging mutual notice (Zhen et al. 2012), cooperation in patent application (Wu and Fang 2015) and bank branches (Ma and Li 2017). This structure covers China's four major urban agglomeration areas, which are the backbone of the Chinese national city network. Owing to differences in data and methods, the network structures in some studies have unique characteristics. For example, the city network structure is more centralized, based on the data of financial service firms. Beijing, Shanghai and Shenzhen are at the core of the network and have the closest ties with each other, and the three cities have overwhelming advantages over other cities (Pan et al. 2016, 2017). Within the transportation network of aviation, railways and highways,

Table 27.2 Key node cities in the Chinese national city network

Cities	Network position and functions
Beijing	Beijing has key resources and decision-making power, with an asymmetric advantage in information (Yin et al. 2011). Beijing plays the role of the substantive management and control center (Zhen et al. 2013). Beijing retains lead status in the national city network (Lu and Huang 2012; Zhao et al. 2015a; Pan et al. 2016, 2017).
Shanghai	Shanghai has a well-developed headquarters economy and is an important national APS center (Zhen et al. 2013). Its radiating effect on the surrounding area is stronger than that of Beijing (Cheng et al. 2013).
Guangzhou and Shenzhen	These are regional central cities of the Pearl River Delta region, and are well connected in the Chinese national city network (Yeh et al. 2015; Zhang and Kloosterman 2016; Zhang 2018).
Wuhan, Chengdu and Chongqing	These are the main node cities in the Midwest of China's national city network (Zhao and Liu 2012).

Source: Compiled by the author.

the structure of city networks tend to be more dispersed (Chen et al. 2017; Wang et al. 2017), as the investment of infrastructure is more balanced owing to concerns of reducing regional disparities (Lao et al. 2016). The positions and functions of some key node cities in the national network have been discussed in previous studies (see Table 27.2).

It is not surprising that the city connectivity in the whole country has increased. For instance, Wang et al. (2014) observed China's aviation network for a long period and found that cities are better connected over time. However, there are significant regional differences in the density of city networks. The three major regions, namely, the Bohai Rim, the Yangtze River Delta and the Pearl River Delta region, have the highest city connectivity. As noted by Leng et al. (2011), the city connectivity in some of the more developed areas has increased significantly. Zhao et al. (2015a) and Liu et al. (2018) found that the city networks in the eastern coastal areas are polycentric, but the networks in the Middle and West parts of the country are monocentric. Wu and Fang (2015) found that there exists a rich-club phenomenon in the national city network of China, which indicates that well-connected cities tend to interact more with each other.

Market forces and institutional factors have shaped the patterns of Chinese national city networks. On the one hand, market factors play a fundamental role in generating flows between cities, which determine the city network structures. For example, firms' linkages have been used in a great deal of previous research to build city networks, while firms' activities are largely affected by market forces. The benefits of agglomeration economies in large cities can attract firms and talented workers. The economic development of a city is crucial for the location choices of firms, and these location decisions will in turn affect the status of the city within the network (Wu and Fang 2015). It was observed that APS firms are highly concentrated in eastern coastal cities with a higher economic and social development level (Zhen et al. 2013). The concentration of high-end factors continuously strengthens the control status for some important node cities within the network.

On the other hand, the institution contexts in China are also very important to understanding the formation of city networks in China. The administrative level of cities will affect their status in the city network. Usually, cities of high administrative level hold a more important position in the network. Lu and Huang (2012) found that most of the regional innovation centers are provincial capital cities, and Wu and Ning (2012) found that electronic information firms are often concentrated in the provincial capital cities. Meanwhile, some strategic resources are located in specific cities, which in turn improves the importance of cities in the network. For example, Shanghai and Shenzhen have attracted many APS firms and the two cities are very well connected in the city networks shaped by APS firms, while Beijing has been benefitting as the financial regulation center (Pan et al. 2017). In particular, the China Securities Regulatory Commission (CSRC) is headquartered in Beijing, which has significantly improved Beijing's role in China's venture capital investment network (Pan et al. 2016). Moreover, the administrative boundary will also affect the structure of Chinese national city networks. For example, Hennemann et al. (2011) studied the knowledge network of China and found that the proportion of co-authored papers by scientists from different provinces was relatively low. Similarly, Li (2014) found that urban networks shaped by publicly listed companies in mainland China have been significantly affected by provincial boundaries.

In addition, path dependence also affects, and even shapes, the structure of Chinese national city networks. Zhao et al. (2015a) found that the special path dependence caused by informal relationships (*guanxi*) in China has made the city networks of China more complex. Since Beijing has a large number of universities and research institutes, Li et al. (2015) found that the knowledge network in China has been significantly shaped by this pattern. The center of knowledge flows within China's city network, Beijing's position is partly a result of path dependence (Hennemann et al. 2011).

Regional City Networks in China

The Yangtze River Delta, Pearl River Delta and Bohai Rim region are the most developed areas in China, and the urban networks within these three areas are very dynamic, which has attracted a lot of research attention. The city networks in the Yangtze River Delta area have been transforming from a monocentric structure to a polycentric networked structure, although Shanghai is still in a very dominant position within the network (Zhao 2011). The key city-pairs in this region are Shanghai–Nanjing, Shanghai–Hangzhou and Hangzhou–Ningbo, which has shaped a Z pattern. The administrative level and market forces are the main influencing factors for the formation of the spatial structure of the city network in this area (Li 2014). In addition, the provincial administrative boundary in the Yangtze River Delta area still seems to be a hindrance to the process of regional integration from the city network research (Feng et al. 2011; Zhang et al. 2018). With the economic transformation and industrial upgrading, the inter-city linkages within and outside the Pearl River Delta region have been increasing (Zhao et al. 2017). Guangzhou and Shenzhen are the two key hubs in the regional city networks (Yeh et al. 2015; Zhang 2018). The study based on service linkages found that Shenzhen is the hub of intra-region connectivity for the Pearl River Delta area, while Guangzhou is the gateway for the region to connect with cities outside of this region (Zhang and Kloosterman 2016). In contrast, the Bohai Rim region has a hierarchical pattern of urban networks. Beijing and Tianjin hold dominant positions in the networks, while all other cities are marginal (Zhao et al.

2014). The three major urban agglomerations have attracted most research attention, and there are only a few studies on other parts of China (Ma and Fang 2011; Tang et al. 2013; Dong et al. 2014).

MAJOR CONTRIBUTIONS OF CHINESE CITY NETWORK RESEARCH

Existing research on Chinese city networks might have contributed in several ways. First, city network research on Chinese cities is multi-scale. Camagni (1993) noted that city networks are multi-scale nested. Figure 27.2 shows the nested nature of city networks at varied spatial scales. Given China's large territory size, city networks at different spatial scales might present different patterns and have different driving forces and implications, thus existing studies have examined Chinese city networks at varied geographical scales (Li 2018). In particular, the comparison results on the patterns of Chinese city networks at the global and local scales have provided rich information from which to understand the ongoing evolution of the Chinese economic landscape (Ma and Timberlake 2013; Zhang 2018; Zhou et al. 2017).

Second, Chinese city network research avoids Western centralism views on the study of globalization. Since Friedmann (1986) proposed the 'world city hypothesis', most studies have focused on established global cities in western countries (Douglass and Friedmann

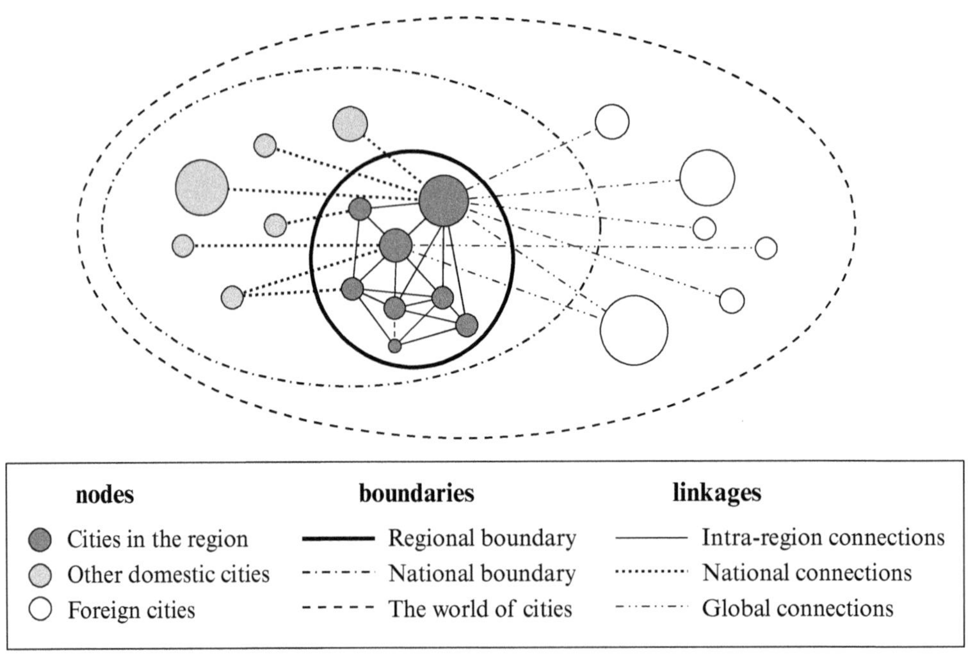

Source: Drawn by the author.

Figure 27.2 Geographic boundaries and inter-city linkages at different geographical scales

1998). In earlier research, Chinese cities are very marginal in WCN, although the rise of Chinese cities in world city networks are significant (Derudder et al. 2013; Taylor et al. 2014b). Western centralism studies mainly explore how Chinese cities are passively integrated into WCN, and largely neglect the domestic factors in shaping the networks. While scholars have realized that the city network in China is not a simple extension of the WCN (Zhao et al. 2015a; Pan et al. 2017), China's institutional contexts should be taken into account in selecting data and explaining results. More recently, scholars have begun to study the WCN formed by Chinese firms' overseas expansion activities (Pan et al. 2018).

Third, the use of novel perspectives and data to measure linkages between cities has advanced the progress of WCN research. In particular, a large number of empirical studies on Chinese city networks have built city linkages by using inter-firm linkages, which has overcome the limitation of being overly reliant on intra-firm data in previous studies. For example, Yeh et al. (2015) used a questionnaire survey to obtain the upstream and downstream industrial links of firms. In addition, the investment links driven by venture capital firms (Pan et al. 2016), service links between APS and publicly listed companies in the IPO process (Pan et al. 2017), and cooperation links between APS firms in the IPO process (Pan et al. 2020) were used to build inter-city linkages.

Finally, Chinese city network research has strengthened the network perspective and helped promote the research in urban and economic geographies in China. The status of modern cities depends more on their linkages with other cities than on their absolute size (Beaverstock et al. 2000). Thus, network perspective has become increasingly important in urban studies (Batty 2013). Owing to research progress in this area, it has been widely accepted that it is important to study a city's position as a node within the urban network (Zhou and Yang 2014). This shift in the research paradigm has huge impacts on urban studies in China. In addition, using firm links to build city networks has expanded the analysis methods and perspectives of firm production network research in economic geography. It shows the possibility for a quantitative analysis on the structure of global and local production networks at the inter-city level.

LIMITATIONS AND FUTURE RESEARCH ON CHINESE CITY NETWORK RESEARCH

Although Chinese city network research has made great achievements, it also has limitations and faces significant challenges. First, most existing studies on Chinese city networks are empirical orientated, while theoretical and methodological contributions are rare. A great deal of research is very descriptive of the network structures but is not analytical enough, while the mechanisms of the formation and evolution of networks are not well explored. Second, existing literature uses a variety of link data in the Chinese city network research, but it is becoming increasingly convergent on data sources and research methods. Additional novel data sources and perspectives are needed in this research area.

We have the following suggestions for future research on Chinese city networks. First, the research perspectives of the city network should be expanded. On the one hand, with the increasing influence of Chinese cities worldwide, it would be necessary to conduct the research of the WCN from China's perspective. For instance, the 'the Belt and Road initiative' might provide a great opportunity to explore how Chinese cities are involved in WCN and how WCN will be impacted by this initiative. On the other hand, more link data between cities generated by important economic and social activities should be applied in Chinese city network research. There already exists rich information on inter-firm service and cooperation linkages, which could be transformed into inter-city connections. In addition to the links between firms from service sectors, we should also pay attention to those city networks formed by production links between manufacturing firms (Li 2016). Moreover, with the rise of the knowledge economy, it is also important to explore knowledge flows between cities (Ma et al. 2015).

Second, the mechanisms of the formation and evolution of Chinese city networks need more elaboration, therefore more theoretical contribution is expected. To date, empirical analysis of Chinese city network research has been weak in explaining the formation and evolution mechanisms of the network structure. It might be possible to introduce the perspective of network evolution from evolutionary economic geography (Boschma and Lambooy 1999) into Chinese city network research. By studying the long-term evolution of city networks, more in-depth interpretations of the formation and evolution mechanism of the city networks in China can be generated. The city network should be used as a research tool to explore important theoretical issues concerning urban and regional development (Burger et al. 2014b; Cirilli and Vereri 2014). In particular, it might be helpful to combine city network research with urban agglomeration, regional integration, regional governance and other important research questions. Thus, we may explore the impacts of city networks on urban, economic and social development.

Third, more policy-relevant research on city networks in China is needed and the cities in less developed regions deserve more research attention. Existing research on Chinese city networks mainly focuses on describing the network patterns and explaining the driving forces, however, the policy research of city networks is still rare. Revealing the structure and driving factors of city networks has important policy implications for urban and regional development. Researchers on Chinese city networks should get involved in policy discussions. In addition, more research attention should be paid to smaller cities in less developed areas, which are also in very marginal positions in the city networks. World city network research has been neglecting smaller cities, and these cities are therefore described as 'off the map' (Robinson 2002). Considering the vast territory and huge regional differences across China, more research on less developed areas and small or medium-sized cities will help to understand the full map of the city landscape (Hino 2013), and the efforts might be helpful to promote development of those areas and cities.

NOTE

1. Hong Kong, Macao and Taiwan are not included in this section.

REFERENCES

Amin, A. and N. Thrift (2002), *Cities: Reimagining the Urban*, Cambridge: Polity Press.
Batty, M. (2013), 'The new science of cities', *Building Research & Information*, **38** (1), 123–6.

Beaverstock, J.V, R.G. Smith and P.J. Taylor (2000), 'World-city network: a new metageography?', *Annals of the Association of American Geographers*, **90** (1), 123–34.

Boschma, R.A. and J.G. Lambooy (1999), 'Evolutionary economics and economic geography', *Journal of Evolutionary Economics*, **9** (4), 411–29.

Burger, M.J., E.J. Meijers and F.G. van Oort (2014b), 'Regional spatial structure and retail amenities in the Netherlands', *Regional Studies*, **48** (12), 1972–92.

Burger, M.J., B. van der Knaap and R.S. Wall (2014a), 'Polycentricity and the multiplexity of urban networks', *European Planning Studies*, **22** (4), 816–40.

Camagni, R.P. (1993), *From City Hierarchy to City Network: Reflections about an Emerging Paradigm*, Berlin and Heidelberg: Springer.

Camagni, R. and R. Capello (2004), 'The city network paradigm: theory and empirical evidence', in R. Capello and P. Nijkamp (eds), *Urban Dynamics and Growth*, Amsterdam: Elsevier, pp. 495–529.

Castells, M. (1996), *The Rise of the Network Society*, Oxford: Blackwell.

Chen, W., W. Liu, W. Ke and N. Wang (2017), 'The spatial structures and organizational patterns of China's city networks based on the highway passenger flows', *Acta Geographica Sinica*, **72** (2), 224–41 (in Chinese).

Chen, Y., F. Zhen, B. Wang and W. Zou (2013), 'Chinese city network structure in the cyberspace based on social network analysis', *Economic Geography*, **33** (4), 56–63 (in Chinese).

Cheng, J., J. Fan and D. Chen (2013), 'Discernment of Chinese financial center system: gravity model approach', *Economic Geography*, **33** (3), 8–14 (in Chinese).

Cirilli, A. and P. Veneri (2014), 'Spatial structure and carbon dioxide (CO) emissions due to commuting: an analysis of Italian urban areas', *Regional Studies*, **48** (12), 1993–2005.

Csomós, G. and B. Derudder (2014), 'European cities as command and control centres, 2006–11', *European Urban & Regional Studies*, **21** (3), 345–52.

Derudder, B. and P. Taylor (2005), 'The cliquishness of world cities', *Global Networks*, **5** (1), 71–91.

Derudder, B. and P.J. Taylor (2018), 'Central flow theory: comparative connectivities in the world-city network', *Regional Studies*, **52** (8), 1029–40.

Derudder, B., Z. Cao, X. Liu, S. Wei, D. Liang, W. Zhang, et al. (2018), 'Changing connectivities of Chinese cities in the world city network, 2010-2016', *Chinese Geographical Science*, **28** (2), 183–201.

Derudder, B., P.J. Taylor, M. Hoyler, P. Ni, X. Liu, M. Zhao, et al. (2013), 'Measurement and interpretation of connectivity of Chinese cities in the world city network, 2010', *Chinese Geographical Science*, **23** (3), 261–73.

Derudder, B., P.J. Taylor, P. Ni, A. Vos, M. Hoyler, H. Hanssens, et al. (2010), 'Pathways of change: shifting connectivities in the world city network, 2000–08', *Urban Studies*, **47** (9), 1861–77.

Dicken, P., P.F. Kelly, K. Olds and H.W.C. Yeung (2001), 'Chains and networks, territories, and scales: towards a relational framework for analyzing the global economy', *Global Networks*, **1** (2), 89–112.

Dong, C., C. Xiu and Y. Wei (2014), 'Network structure of "space of flows" in Jilin province based on telecommunication flows', *Acta Geographica Sinica*, **69** (4), 510–19 (in Chinese).

Douglass, M. and J. Friedmann (1998), *Cities for Citizens: Planning and the Rise of Civil Society in a Global Age*, London: John Wiley & Sons.

Feng, F., L. Zhang, Y. Du, L. Ma and M. Fu (2011), 'Research on scientific and technological cooperation network of cities in Pan-Yangtze river delta of China: based on social network analysis', *International Journal of Business and Social Science*, **2** (20), 116–20.

Florida, R. (2003), 'Cities and the creative class', *City & Community*, **2** (1), 3–19.

Friedmann, J. (1986), 'The world city hypothesis', *Development & Change*, **17** (1), 69–83.

Gu, C. and H. Pang (2008), 'Study on spacial relations of the Chinese urban system: gravity model approach', *Geographical Research*, **27** (1), 1–12 (in Chinese).

Hanssens, H., B. Derudder and F. Witlox (2013), 'Are advanced producer services connectors for regional economies? An exploration of the geographies of advanced producer service procurement in Belgium', *Geoforum*, **47**, 12–21.

Hennemann, S., T. Wang and I. Liefner (2011), 'Dynamics of center-periphery patterns in knowledge networks – the case of China's biotech science and technology system', arXiv preprint, accessed 27 March 2021 at https://arxiv.org/ftp/arxiv/papers/1112/1112.2498.pdf.

Hino, M. (2013), 'Individual-city-centered networks for cities' self-sustainability: lesson from the Great East Japan Earthquake', *Geography*, **59** (1), 1–12.

Jiao, J., J. Wang and F. Jin (2017), 'Impacts of high-speed rail lines on the city network in China', *Journal of Transport Geography*, **60** (C), 257–66.

Jin, F. (2001), 'A study on network of domestic air passenger flow in China', *Geographical Research*, **20** (1), 31–9 (in Chinese).

Lai, K. (2012), 'Differentiated markets: Shanghai, Beijing, and Hong Kong in China's financial centre network', *Urban Studies*, **49** (6), 1275–96.

Lao, X., X. Zhang, T. Shen and M. Skitmore (2016), 'Comparing China's city transportation and economic networks', *Cities*, **53**, 43–50.

Leng, B., Y. Yang, Y. Li and S. Zhou (2011), 'Spatial characteristics and complex analysis: a perspective from basic activities of urban networks in China', *Acta Geographica Sinica*, **66** (2), 199–211 (in Chinese).
Li, D., Y.D. Wei and T. Wang (2015), 'Spatial and temporal evolution of urban innovation network in China', *Habitat International*, **49**, 484–96.
Li, X. (2014), 'Spatial structure of the Yangtze River delta urban network based on the pattern of listed companies network', *Progress in Geography*, **33** (12), 1587–600 (in Chinese).
Li, X. (2015), *Structure and Evolution of Urban Network*, Beijing: Science Press (in Chinese).
Li, X. (2016), 'Measuring the in-and-out degrees of industrial networks in Shanghai: beyond the myth of "post-industrial society"', *Geographical Research*, **35** (11), 2185–200 (in Chinese).
Li, Y. (2018), 'Difference between Chinese and Western studies on urban networks and its reflection', *Urban Planning International*, **33** (2), 61–7 (in Chinese).
Liefner, I. and S. Hennemann (2011), 'Structural holes and new dimensions of distance: the spatial configuration of the scientific knowledge network of China's optical technology sector', *Environment & Planning A*, **43** (4), 810–29.
Liu, X. and B. Derudder (2013), 'Analyzing urban networks through the lens of corporate networks: a critical review', *Cities*, **31** (2), 430–37.
Liu, X., B. Derudder and P.J. Taylor (2014), 'Mapping the evolution of hierarchical and regional tendencies in the world city network, 2000–2010', *Computers, Environment, and Urban Systems*, **43**, 51–66.
Liu, X., B. Derudder and M. Wang (2018), 'Polycentric urban development in China: a multi-scale analysis', *Environment and Planning B: Urban Analytics and City Science*, **45** (5), 953–72.
Liu, Z., S. Wang, M. Zhao and K. Wu (2013), 'Exploratory analysis of directed weighted network of city', *Geographical Research*, **32** (7), 1253–68.
Lu, L. and R. Huang (2012), 'Urban hierarchy of innovation capability and inter-city linkages of knowledge in post-reform China', *Chinese Geographical Science*, **22** (5), 602–16.
Ma, H. (2016), 'Research progress and prospects of city networks based on knowledge flow in China', *Economic Geography*, **36** (11), 207–13 (in Chinese).
Ma, H. and C. Fang (2011), 'Spatial organization of production networks in city-region based on enterprises relationship: a case of apparel production in urban agglomeration of Eastern Guangdong Province', *Scientia Geographica Sinica*, **31** (10), 1172–80 (in Chinese).
Ma, X. and L. Li (2017), 'Network spatial connection structure of Chinese cities based on bank branches data', *Progress in Geography*, **36** (4), 393–403 (in Chinese).
Ma, X. and M.F. Timberlake (2008), 'Identifying China's leading world city: a network approach', *Geojournal*, **71** (1), 19–35.
Ma, X. and M.F. Timberlake (2013), 'World city typologies and national city system deterritorialisation: USA, China and Japan', *Urban Studies*, **50** (2), 255–75.
Ma, H., C. Fang, P. Bo and S. Wang (2015), 'Structure of Chinese city network as driven by technological knowledge flows', *Chinese Geographical Science*, **25** (4), 498–10.
Neal, Z. (2013), 'Does world city network research need eigenvectors?', *Urban Studies*, **50** (8), 1648–59.
Ni, P., K. Liu and P.J. Taylor (2011), 'Zhongguo Chengshi Lianxidu: Jiyu Liansuo Wangluo Muxing de Cedu', *Comparative Economic & Social Systems*, (6), 96–103 (in Chinese).
Pan, F., W. Bi, J. Lenzer and S. Zhao (2017), 'Mapping urban networks through inter-firm service relationships: the case of China', *Urban Studies*, **54** (12), 3639–54.
Pan, F., W. Bi, X. Liu and T. Sigler (2020), 'Exploring financial center networks through inter-urban collaboration in high-end financial transactions in China', *Regional Studies*, **54** (2), 162–72.
Pan, F., Z. He, T. Sigler, K. Martinus and B. Derudder (2018), 'How Chinese financial centers integrate into global financial center networks: an empirical study based on overseas expansion of Chinese financial service firms', *Chinese Geographical Science*, **28** (2), 217–30.
Pan, F., S.X.B. Zhao and D. Wójcik (2016), 'The rise of venture capital centers in China: a spacial and network analysis', *Geoforum*, **75**, 148–58.
Pereira, R.A.O. and B. Derudder (2010), 'Determinants of dynamics in the world city network, 2000-2004', *Urban Studies*, **47** (9), 1949–67.
Robinson, J. (2002), 'Global and world cities: a view from off the map', *International Journal of Urban & Regional Research*, **26** (3), 531–54.
Rozenblat, C. and D. Pumain (2008), 'Firm linkages innovation and the evolution of urban systems', in P.J. Taylor, B. Derudder, P. Saey and F. Witlox (eds), *Cities in Globalization Practices Policies & Theories*, London: Routledge, pp. 130–56.
Sassen, S. (2013), *The Global City: New York, London, Tokyo*, Princeton, NJ: Princeton University Press.
Scott, A.J. (2012), *A World in Emergence: Cities and Regions in the 21st Century*, Cheltenham, UK and Northampton, MA, USA: Edward Elgar.
Scott, J. and P.J. Carrington (2011), *The SAGE Handbook of Social Network Analysis*, London: Sage.
Sheppard, E. (2002), 'The spaces and times of globalization: place, scale, networks, and positionality', *Economic Geography*, **78** (3), 307–30.

Tang, F., H. Tang, Q. Sun and D. Tang (2013), 'Analysis of the economic network structure of urban agglomerations in the Middle Yangtze River', *Acta Geographica Sinica*, **68** (10), 1357–66 (in Chinese).

Taylor, P.J. (2004), *World City Network: A Global Urban Analysis*, London: Routledge.

Taylor, P.J., G. Catalano and N. Gane (2003), 'A geography of global change: cities and services, 2000-2001', *Urban Geography*, **24** (5), 431–41.

Taylor, P.J., G. Catalano and D.R.F. Walker (2002), 'Measurement of the world city network', *Urban Studies*, **39** (13), 2367–76.

Taylor, P.J., B. Derudder, J. Faulconbridge, M. Hoyler and P. Ni (2014a), 'Advanced producer service firms as strategic networks, global cities as strategic places', *Economic Geography*, **90** (3), 267–91.

Taylor, P.J., B. Derudder, M. Hoyler, P. Ni and F. Witlox (2014b), 'City-dyad analyses of China's integration into the world city network', *Urban Studies*, **51** (5), 868–82.

Timberlake, M., Y.D. Wei, X. Ma and J. Hao (2014), 'Global cities with Chinese characteristics', *Cities*, **41**, 162–70.

Wang, J. and Y. Jing (2017), 'Comparison of spatial structure and organization mode of inter-city networks from the perspective of railway and air passenger flow', *Acta Geographica Sinica*, **72** (8), 1508–19 (in Chinese).

Wang, J., H. Mo and F. Wang (2014), 'Evolution of air transport network of China 1930–2012', *Journal of Transport Geography*, **40** (40), 145–58.

Wang, J., H. Mo, F. Wang and F. Jin (2011), 'Exploring the network structure and nodal centrality of China's air transport network: a complex network approach', *Journal of Transport Geography*, **19** (4), 712–21.

Wang, M. (2007), *The Production and Consumption of Urban Cyberspace*, Beijing: Science Press (in Chinese).

Wang, M. and Y. Ning (2004), 'The Internet and the rise of information network cities in China', *Acta Geographica Sinica*, **59** (3), 446–54 (in Chinese).

Wei, Y.H.D, Y. Zhou, Y. Sun and G.C.S. Lin (2012), 'Production and R&D networks of foreign ventures in China: implications for technological dynamism and regional development', *Applied Geography*, **32** (1), 106–18.

Wójcik, D and J. Camilleri (2015), '"Capitalist tools in socialist hands?" China mobile in global financial networks', *Transactions of the Institute of British Geographers*, **40** (4), 464–78.

Wu, K. and C. Fang (2015), 'The spatial organization and structure complexity of Chinese intercity networks', *Geographical Research*, **34** (4), 711–28 (in Chinese).

Wu, Q. and Y. Ning (2012), 'China's urban network based on spatial organization of electronic information enterprises', *Geographical Research*, **31** (2), 207–19 (in Chinese).

Xi, G., F. Zhen, J. He and Y. Gong (2018), 'City networks of online commodity services in China: empirical analysis of T-mall clothing and electronic retailers', *Chinese Geographical Science*, **28** (2), 231–46.

Yang, Y., B. Leng, Y. Tan and T. Li (2011), 'Review on world city studies and their implications in urban systems', *Geographical Research*, **30** (6), 1009–20 (in Chinese).

Yin, J., F. Zhen and C. Wang (2011), 'China's city network pattern: an empirical analysis based on financial enterprise layouts', *Economic Geography*, **31** (5), 754–9 (in Chinese).

Yeh, A.G.O, F.F. Yang and J. Wang (2015), 'Producer service linkages and city connectivity in the mega-city region of China: a case study of the Pearl River Delta', *Urban Studies*, **52** (13), 2458–82.

Yeung, H.W.C. (2005), 'The firm as social networks: an organizational perspective', *Growth & Change*, **36** (3), 307–28.

Yeung, H.W.C and W. Liu (2008), 'Globalizing China: the rise of mainland firms in the global economy', *Eurasian Geography and Economics*, **49** (1), 57–86.

Yu, W. (1988), 'A systematic analysis of information interaction among the major cities of China', *Acta Geographica Sinica*, **43** (2), 141–9 (in Chinese).

Zhang, W., B. Derudder, J. Wang and W. Shen (2018), 'Regionalization in the Yangtze River Delta, China, from the perspective of inter city daily mobility', *Regional Studies*, **52** (4), 528–41.

Zhang, X. (2018), 'Multiple creators of knowledge-intensive service networks: a case study of the Pearl River Delta city-region', *Urban Studies*, **55** (9), 2000–2019.

Zhang, X. and R.C. Kloosterman (2016), 'Connecting the "workshop of the world": intra-and extra-service networks of the Pearl River Delta city-region', *Regional Studies*, **50** (6), 1069–81.

Zhao, M. (2011), 'Evolution of network and spatial structure in the Yangtze River Delta', *Geographical Research*, **30** (2), 311–23 (in Chinese).

Zhao, M. and C. Chen (2011), 'A comparison between the airline passenger network and the advanced producer service network in the urban system of China', *Urban Planning Forum*, (2), 24–32 (in Chinese).

Zhao, M. and Z. Liu (2012), 'Research on China's city network based on production service industry', *Urban Planning*, (9), 23–8 (in Chinese).

Zhao, M., B. Derudder and J. Huang (2017), 'Examining the transition processes in the Pearl River Delta polycentric mega-city region through the lens of corporate networks', *Cities*, **60**, 147–55.

Zhao, M., X. Liu, B. Derudder and Y. Zhong (2015a), 'Mapping producer services networks in mainland Chinese cities', *Urban Studies*, **52** (16), 3018–34.

Zhao, M., J. Wei and K. Wu (2014), 'Functional linkages in the Beijing-Tianjin–Hebei conurbation region and the evolution of the complex networks', *Urban Planning Forum*, (1), 46–52 (in Chinese).
Zhao, M., K. Wu, X. Liu and B. Derudder (2015b), 'A novel method for approximating intercity networks: an empirical comparison for validating the city networks in two Chinese city-regions', *Journal of Geographical Sciences*, **25** (3), 337–54.
Zhen, F., B. Wang and Y. Chen (2012), 'China's city network characteristics based on social network space: an empirical analysis of Sina Micro-blog', *Acta Geographica Sinica*, **67** (8), 1031–43 (in Chinese).
Zhen, F., X. Wang, J. Yin and M. Zhang (2013), 'An empirical study on Chinese city network patterns based on producer services', *Chinese Geographical Science*, **23** (3), 274–85.
Zhou, C., G. Zeng, F. Wang, et al. (2017), 'Innovation network structure and innovative performance: a study of China's electronic information industry', *Scientia Geographica Sinica*, **37** (5), 661–71 (in Chinese).
Zhou, Y. and Z. Hu (2002), 'Looking into the network structure of the Chinese urban system from the perspective of air transportation', *Geographical Research*, **21** (3), 276–86 (in Chinese).
Zhou, Y. and Q. Yang (2014), 'Changes in the urban hierarchy in China and the territorial types by province', *Chinese Sociological Review*, **19** (3–4), 137–59.
Zhu, Z., A. Zhang and Y. Zhang (2018), 'Connectivity of intercity passenger transportation in China: a multi-modal and network approach', *Journal of Transport Geography*, **71** (C), 263–76.

28. The GaWC perspective on global-scale urban networks

Ben Derudder and Peter J. Taylor

1. INTRODUCING GAWC: AN 'INVISIBLE COLLEGE'

The Globalization and World Cities (GaWC) research network was founded at Loughborough University in 1998 by Peter Taylor and Jon Beaverstock as an invitation to urban scholars across the world to help understand the changing worlds of cities under conditions of contemporary globalization. These changes were so dynamic, so unusual and so worldwide that they were beyond the comprehension of any one group of researchers in one place to even begin to adequately research this then-emerging field of study. GaWC's core business has been to more narrowly focus on one conspicuous topic in research on globalized urbanization: the external connections of world cities. Just as world/global cities research as a whole never intended to provide a comprehensive account of globalized urbanization (Parnreiter 2014; van Meeteren et al. 2016; Harrison and Hoyler 2018), so too the formal analysis of worldwide networks between cities is not necessarily the main activity of world/global cities research. Nonetheless, as the latter literature is premised upon the existence of interactions between cities, these should not be neglected as they so commonly were in earlier global/world city literature (Taylor 2001a); GaWC was thus formed to aid in rectifying this situation.

Although world/global city research in general and GaWC in particular is best described as an 'invisible college' (Acuto 2011; Bassens and van Meeteren 2015) – that is, a continuously evolving group of authors in a particular research field who constitute a social circle, but have varying degrees of involvement and pursue different research questions on the basis of shared interests – the research ethos of GaWC at large has been critical realist (Sayer 1992). In practice this implies that we firmly believe that these extensive quantitative studies should exist alongside, and be connected to, intensive qualitative investigations of world city-formation (see Watson and Beaverstock 2014). Examples of a fruitful combination of formal quantitative analysis and qualitative research strategies can be found in studies on the connections between London and Frankfurt at the time of the euro currency launch (Beaverstock et al. 2001) and on the connections between New York, Los Angeles and Detroit in light of changing advertising practices (Faulconbridge et al. 2011). Literature that contextualizes, challenges and provides formative interpretations of extensive quantitative studies remains a vibrant strand of research (Watson and Beaverstock 2014), and includes Lai's (2012) analysis of the Beijing–Shanghai–Hong Kong triad and Wójcik's (2013) analysis of the commonality, complementarity and connectivity between New York's and London's financial markets. In addition, the broader question of the nature of world cities as sites through which economies flourish is also being addressed (Taylor 2013; Bassens and van Meeteren 2015).

Within this diverse set of approaches, the purpose of the quantitative research is above all to specify, measure and generalize some of the insights derived from qualitative research to arrive at a more systematic appraisal of how cities are connected across the globe. An example of how GaWC's quantitative research can complement the qualitative research can be found in the work of Faulconbridge et al. (2011), who use as an example the changing advertising industry to explore how urban economies across the globe are increasingly being interconnected. Advertising as we know it today has its roots in archetypal American consumer culture. As an initially US service industry, New York, with its famous Madison Avenue cluster, was the main national hub throughout much of the twentieth century. However, as advertising gradually diffused to other parts of the world, New York became a world hub, leading to global campaigns designed and executed in this one city for the rest of the world to consume. However, based on interviews with advertising practitioners, Faulconbridge et al. (2011) argue that in the twenty-first century an increasingly global business emerged, which implied the need for more nuanced messages reflecting changes in audiences and specific geographical contexts. In practice, this has meant the implementation of cooperative arrangements between offices across cities: different offices are understood to be composed of different varieties of skills and local knowledge that need to be brought together for the client. Currently, new creative ideas are accepted and expected from offices in non-US cities, such as Bangkok and São Paulo, in part deriving from their very different cultural contexts. In addition, New York is no longer the automatic lead office: it might be part of a project team led, say, from Los Angeles. The end result is a multi-nodal network approach to the advertising business that harnesses internal and network assets from around the world. This results in advertising firms setting up multiple offices in multiple cities across the globe, which are interconnected by flows of people, information, knowledge, capital, and so on. The rationale behind GaWC's quantitative research stream is that it seeks to turn these qualitative insights into a measurable understanding of global inter-city relationships, by explicitly expressing these as a world city network. In the remainder of this chapter, we focus on this quantitative research as it is the part of the research that is most closely aligned with research on the city/networks nexus.

Following early inventories of world cities based upon their level of advanced producer services (Beaverstock et al. 1999), most of GaWC's quantitative research has been premised upon the application of the interlocking network model (INM). Initially presented in Taylor (2001b), this model continues to be applied, both in its original guise (Taylor et al. 2014; Derudder and Taylor 2018) and in adapted forms (Derudder et al. 2013; Hennemann and Derudder 2014). The purpose of this chapter is to review this perspective on the analysis of global-scale urban networks, and to this end it is organized in three parts. Section 2 offers a brief description of the research context in which the quantitative side of GaWC operates, situating our approach in the broader research agenda detailing city networks at the global scale. In section 3 we present the basic model and some straightforward results based on an operationalization of the model using the latest available data for 2016. In section 4, we discuss a number of model extensions and uses: (1) its application to other urban network agents and/or at other geographical scales; (2) complementary specifications of city-dyads and connectivity; and (3) its theoretical integration into wider urban network research questions.

2. RESEARCH CONTEXT

Critiques of existing ways through which research is conducted are the means by which we move forward to produce better – that is, more reliable, more relevant, more focused, more critical, more integrated, more credible, more inclusive or more efficient – knowledge. In this section we consider quantitative GaWC as deriving from a critique of research on world/global cities as practised throughout the late 1980s and the 1990s. The lack of empirical research on cities in globalization, in general, the global connections between these cities, in particular, was one of the great paradoxes of this literature, which was thus vulnerable to criticism for its dearth of evidence backing up its key propositions. For instance, disquiet for the paucity of evidence behind Friedmann's (1986) world city hypothesis surfaced in the initial published discussion of the paper (Korff 1987). Over the past two decades there have been an increasing number of efforts to address this lacuna, and GaWC provided one of the earliest and arguably most systematic responses to it.

There are two potential solutions for specifying global-scale urban networks, involving direct and indirect measures of inter-city relationships, respectively. These alternative specifications include radically different answers to the questions of how cities are connected. Direct measurement solutions can be found in two different sets of data. First, and most directly, there is secondary information on the provision and execution of movements between cities, both material and virtual, by companies providing transport and communication services in the manner pioneered by Taaffe (1956) and later extended to the continental and global scale by O'Connor and Scott (1992) and Smith and Timberlake (2001) (for example, Mahutga et al. 2010; Ducruet 2017). Second, there is information on corporate command structures, specifically headquarter–subsidiary relationships indicating flows of directive between cities in the manner pioneered by Pred (1978) for the USA and later extended to the continental and global scale by Cohen (1981) and Rozenblat and Pumain (1993) (for example, Alderson et al. 2010; Rozenblat et al. 2017). Both solutions constitute key infrastructures that have enabled global inter-city networks to develop, and there is therefore a clear-cut rationale for the often refined analysis of these infrastructures to map how cities are connected in globalization.

In our research, we have taken a different route. The principal reason why we believe these direct measures need to be complemented with other approaches is that these empirical analyses can be viewed as chaotic from a theoretical viewpoint: some of these analyses combine very different city processes, and some do not necessarily address key processes in city formation. Thus small urban places can be both global headquarters (HQ) centres and sites of globally renowned universities, just as tourist destinations can be well connected in airline networks. Therefore, our view of this is to specifically produce measures of inter-city connections that are integral to the current drivers of city development, and thereby revert to indirect but process-focused measurement of business flows. These indirect measures are based upon a theory of the process that can be modelled in such a way that other data can be used to provide estimated measures of the process. In a great deal of our research the theory is drawn from the foundation studies that treat cities as interconnected global service centres (Sassen 1991). We cannot know all the information and knowledge flows between many cities that the service firms in this business generate, but we can easily find where they have offices in cities across the world. Combining these data with a model of the everyday inter-office work practices in firms – the interlocking city network model – provides a specification of inter-city networks. The measures will not reflect real flows but guesstimates of potential flows, however they do have the advantage of providing a theorized empirical basis for network studies of cities to complement the comparative methodology by Sassen (1991) from which this idea is originally derived. In the next section, we introduce the interlocking network model and its most recent operationalization, for 2016.

3. THE INTERLOCKING NETWORK MODEL

3.1 Model Specification

Given that GaWC takes its cue from Sassen (1991), the interlocking network model (INM) was initially specified from the perspective of how producer services firms organize their global networks through strategic presences in cities. The INM provides a specification of Allen's (2010, p. 2898) observation that city powers are mobilized through inter-city networking in financial and business services firms. To this end, a universe of m advanced producer service firms located in n world cities is defined. The elemental measure is a service value, v_{ij}, with information on the importance of the presence of firm j in city i. These observations can be arrayed as a $m \times n$ service value matrix V. A very straightforward measure of a city's position would be its total level of services, but this is also an inadequate measure as it does not provide an understanding of how the offices providing these services fit into a broader office network and the exchanges this entails between offices. Instead, these presences are most important because of the practices employed by the financial and business professionals distributed in the corporate networks that flow through cities. Thus we need to convert V so that it gives us insight into the connections between cities through firms, instead of simply taking stock of firms' presence in cities.

In network analysis, the service value matrix V is commonly termed a two-mode network (Neal 2008; Liu and Derudder 2012). In contrast to one-mode networks, where actors are directly interlinked, a two-mode network is characterized by connections between two separate sets of nodes. In this case V would be a network connecting cities with firms. In the initial specification there is no direct linkage within the same set of nodes; we simply know which firms are in which cities, and which cities house which firms. However, it is possible to infer one-mode networks from the two-mode dataset by applying a projection function. The INM is this type of projection function; it entails applying a method of converting the service value matrix V into a relational matrix R of city interactions, and ultimately draws on seeking out co-presences of firms in cities.

The crux of the INM projection function is the definition of city-dyad connectivity CDC_{a-b} between cities a and b for each pair of cities and firms based on V:

$$CDC_{a-b} = \Sigma_i v_{ai} \cdot v_{bi} \quad a \neq b \tag{28.1}$$

Two-mode to one-mode projections come with a set of assumptions. In the INM, the conjecture behind conceiving CDC_{a-b} as a surrogate for actual flows of inter-firm information and knowledge between cities is that the more important the office, the more connections there will be with other offices in a firm's network. A city's overall connectivity, which we term global network connectivity GNC_a, can then be calculated by aggregating all its possible links:

$$GNC_a = \Sigma_b \, CDC_{a-b} = \Sigma_{bi} v_{ai} \cdot v_{bi} \qquad a \neq b \tag{28.2}$$

In network-analytical terms, global network connectivity (GNC) is a measure of degree centrality: the total number of connections incident upon a city-as-node. This measure is hard to interpret on its own terms, as it depends on the number of firms and cities in the data. To make GNC more readably interpretable, we therefore convert it into a proportion of the maximum value. Based on operationalizations of this approach, which entail systematically gathering information that allows assigning service values to city/firm pairs, we can explore the connectivity patterns of individual cities, city-dyads, and the network at large.

3.2 Model Operationalization

The INM specification guides data collection: data are required on the (importance of the) presences of the producer services firms interlocking cities to devise a two-mode network matrix V of service values v_{ij}. Here we summarize the latest exercise in data collection for producer services carried out in 2016; a more detailed elaboration of the rationale and specifics of the data collection can be found in Taylor and Derudder (2016, ch. 4).

The first step entails the selection of firms. Following Daniels (1995) and Sassen (1991), we define producer services firms as firms providing knowledge-based (expert/profession/creative) services to other corporations to facilitate their business activities. The selection of producer services firms is based on rankings of leading firms in five sectors: 75 financial services firms, 25 management consultancy firms, 25 advertising firms, 25 law firms and 25 accountancy firms. The information on the location strategies of these 175 firms used in this research exercise was gathered in July and August 2016. Firms were selected based on sectoral rankings for 2015, which tended to be based upon 2014 data. For each sector, the top-ranked firms were chosen based on sectoral rankings (for example, financial services firms were chosen based on their ranking in the Fortune ranking). This selection mechanism, using a predefined number of firms from each sector selected via independent rankings, puts the data collection on a footing that enables coherent longitudinal analysis. However, given ongoing changes and disruptions in both urban economies and the service sector (for example, the impact of FinTech and block chain on the financial services industry), using such a consistent definition of 'firms' in 'service sectors' may at some time challenge the conceptual purchase of our framework, and call for revisions.

The second step entails the selection of cities. In addition to the criteria for city inclusion in and of themselves, we try to adopt a consistent definition and bounding of cities. In principle, our definition of a city is aligned with that which in the USA would be termed a metropolitan statistical area (MSA), consisting of a core city and the adjacent areas in so far as these are firmly centred on that core city. However, given the lack of a coherent comparative framework at the global scale, we take a pragmatic approach and combine judgement calls with information on how firms tend to self-identify on their websites.

In practice, we devised a roster of cities based on a number of overlapping criteria. For example, we included all cities covered in earlier research, with a population of more than 1.5 million inhabitants in 2014, all capital cities of countries with a population of more than 1 million and every city with a headquarter office of one of our selected firms. This led to the selection of 707 cities. It is this roster of cities that is used in recording information on the global service networks of the 175 firms.

The third and final step in the data gathering is to allocate standardized service values to each city/firm pair. To this end, we draw on the self-presentation of firms on their corporate websites. A putative global presence, no matter how uneven it is in practice, has become an integral part of leading firms' public marketing and recruitment policies (for example, Beaverstock et al. 2010). Hence among producer service firms, locational strategy is of necessity transparent. The corporate websites of these firms therefore customarily showcase the geographical range of the services on offer. Advantage is taken of this geographical transparency for information gathering. For each of the 175 firms, two types of information have been gathered. First, information about the size of a firm's presence in a city is obtained, ranging from the number of offices the firm has in a city for accountancy/financial services firms to the areas of expertise for law firms. Second, the extra-locational functions of a firm's office (for example, headquarter functions) in a city are recorded. The end result is that for each of the 175 firms, information is available to create standardized service values in each of the 707 cities. This standardization involved assigning scores ranging from 0 to 5 as follows. The city housing a firm's headquarters was scored 5, a city with no office of that firm was scored 0. An ordinary or typical office of the firm resulted in a city scoring 2. With something missing (for example, no partners in a law office), the score reduced to 1. Particularly large offices were scored 3 and those with important extra-territorial functions scored 4. The end-result was a 707 × 175 matrix V where v_{ij} ranges from 0 to 5. Applying equations 28.1 and 28.2 to this dataset produces an overview of the overall connectivity of a city (equation 28.2) and where it is connected (equation 28.1). In this chapter we restrict discussion to what is arguably the most well-known output of this approach, that is, an overview of the geographies of global network connectivity; alternatives and complements to this approach and some of the corresponding geographies they produce are discussed in section 4 of this chapter.

3.3 Geographies of Global Network Connectivity

Figure 28.1 shows our operational roster of world cities as those 172 cities with at least one-fifth of the connectivity of the most connected city (London). This cartogram illustrates the most connected cities (with darker shades reflecting stronger connectivities) as this mode of presentation solves the problem of illustrating a very uneven distribution of cities across the world. Relative city concentration in some regions (for example, in western Europe) coupled with relative sparseness of cities elsewhere (for example, in Central Asia) makes depiction of results on orthodox maps, with extremes of overlaps and empty spaces, sometimes difficult to perceive and interpret. In the cartograms, each city is given its own equal space in approximately its correct relative position. Cities are indicated by intuitive two-letter codes, for example, 'NY' for New York and 'JB' for Johannesburg.

Our most basic finding is that the world city network reproduces the old North–South divide: higher-connected cities tend to be in the North and lower-connected cities in the

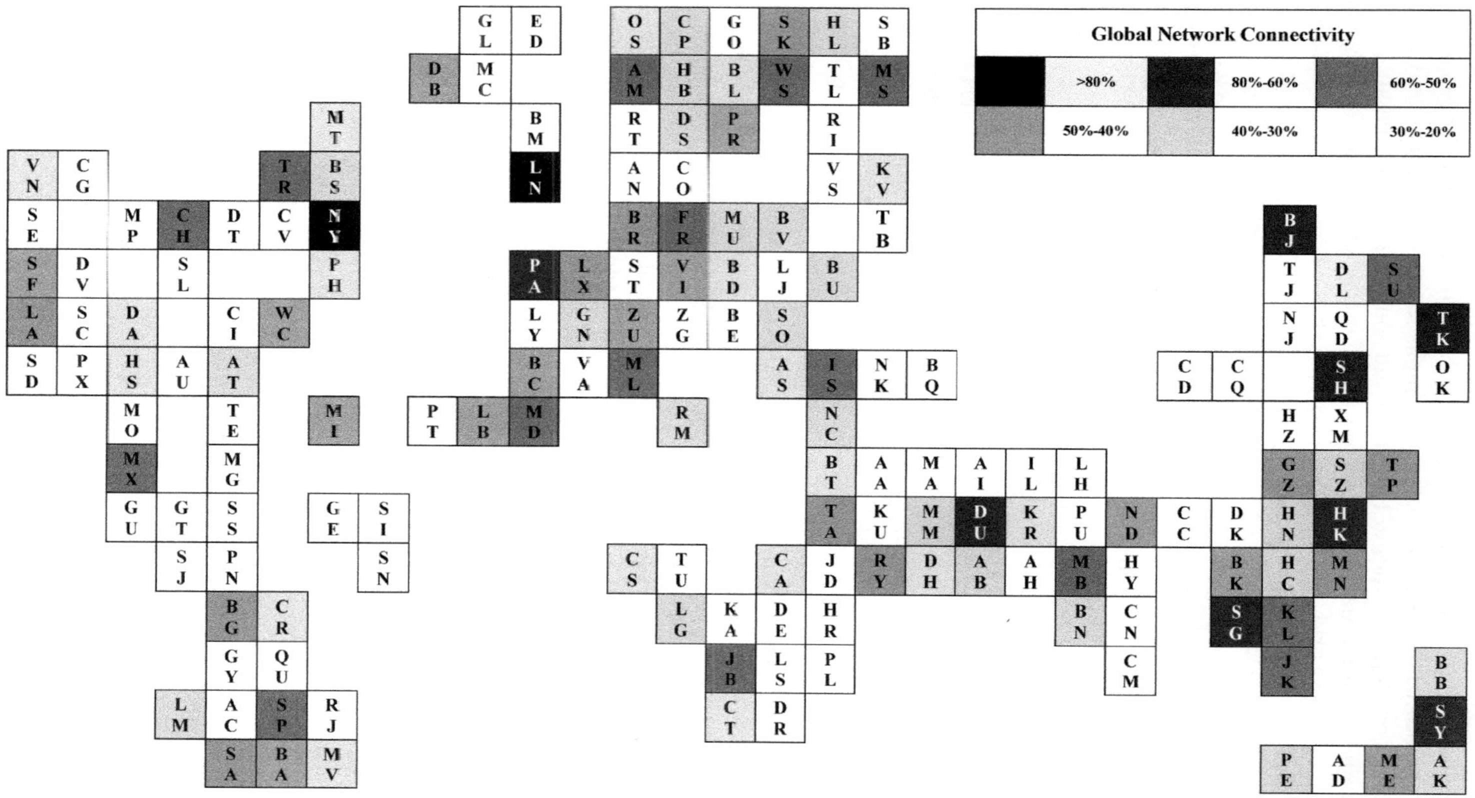

Source: Derudder and Taylor (2018).

Figure 28.1 Cartogram with GNC of cities with a GNC

South, with the western Pacific Rim firmly bucking this trend. However, the geography is much more complicated: this simple, not to say simplistic, interpretation is only a trend with many lower connectivity cities in the North and some higher connectivity cities in the South beyond Pacific Asia. Figure 28.1 clearly illustrates the three contemporary zones of the core of the world economy: northern America, western Europe and parts of Pacific Asia. This is not a homogeneous core; the three zones have very different histories associated with their trajectories to core status in contemporary globalization.

The oldest, indeed original, core zone is western Europe and this is reflected in two features. First, this region has more world cities (49) in Figure 28.1 than the other regions. Second, there is a wide variety of levels of connectedness among the region's cities, ranging from London and Paris to Leeds and Rotterdam. That is, in this region, there is a mixture of cities of varying importance all linking into the world city network. This is the opposite of Pacific Asia in which the connectivity levels of the cities is generally top heavy. In spite of booming and widespread urbanization – there are now more than 200 cities in Asia that have more 1 million inhabitants, about half of which are located in China – few of these cities in the most recent of the core zones have made the threshold. Thus this region has far fewer world cities (24) than does western Europe, although the number increases to 30 if we add Australasian cities and Auckland to create a Western Pacific Rim region. The third core zone, northern America (USA and Canada), is in between the other two in numbers of world cities identified (26). In this instance the range of levels of connectedness is very similar to western Europe, with numerous less important cities, such as Tampa and Charlotte, joining the world city network. However, there is a regional difference in that in northern America the more connected cities tend to be in the east and west of the region, leaving the centre bereft of well-connected cities, apart from the major exception of Chicago.

Beyond the core there are no regions with concentrations of highly connected cities. In eastern Europe (the former Communist states), the most common pattern is for capital cities to take on the world city role; apart from St Petersburg, the only cities (4) that feature are capital cities. Having lost its political and economic distinctiveness, this region has become an appendage to the western European core, albeit that unlike western Europe there is only one well-connected city (Moscow) and there is much greater concentration, with generally one (capital) city per state. The same can be said for Latin America (23 cities) where, again, capital cities dominate, although here São Paulo, despite being neither a former nor current Brazilian capital city, has become a highly connected world city in its own right. Together with Mexico City, this city stands out in Latin America. This pattern is also found in South Asia (13 cities) and the Middle East/North Africa (17 cities), where Mumbai and Dubai as non-capital cities have become highly connected world cities. Sub-Saharan Africa has only nine cities but it does sport a clear regional leading city in relation to connectivity: Johannesburg.

What does all this locational detail mean? Uneven globalization has spawned an uneven world city distribution, but most certainly not a simple one. First, the political influence can be seen in that nearly half the cities are capital cities (85), mostly outside the core (especially eastern Europe, Latin America and the Middle East) but also in parts of the core where countries form relatively small markets linked to the world city network through one city, their capital (for example, Scandinavian countries and South East Asian countries). Equally, in some larger countries capitals do not feature in this economic network role: the countries of which Abuja, Brasilia, Canberra, Ottawa and Pretoria are capitals are represented by other cities in Figure 28.1. Secondly, there are a large number of cities outside the core in regions that have become much more integrated into the world city network. A simple comparison of the number and the distribution of cities with the 2000 world city network (Taylor 2004) hints at the basic contours of this change. Overall, the number of cities meeting the threshold for inclusion has risen from 123 to 172. Significantly, this change has been most marked in regions that were previously 'off the map' as identified by Robinson (2002). For instance, there are now far more cities from South Asia (up from six to 13) and Latin America (up from 11 to 23). The world city network is even less constituted as an exclusionary club of the world's major cities than it was a decade ago, and increasingly has numerous linkages into regions beyond major world cities.

4. EXTENSIONS AND MODIFICATIONS

The previous two sections have been self-centred in that the discussion focused on those measures, operationalization and findings that have arguably been at the core of the quantitative research of GaWC. However, in addition to this research being part of a much broader world/global cities research agenda in general (see Harrison and Hoyler 2018 for the most recent overview) and of the quantitative analysis of the worldwide connections of cities in particular, note that there have been extensions and modifications to this approach. In this section, we discuss some of the most important extensions and modifications.

4.1 Extensions: Other Uses of the INM

Extensions to the INM detailed in the previous sections come in different guises. We collectively term these 'extensions' because the basic remit and assumptions of equation 28.1 are not challenged (in contrast to the research discussed in section 4.2), but instead are applied to other selections of agents and/or at other scales.

Lüthi et al. (2018) have recently argued that in the selection of producer services firms two approaches can be distinguished: a global top-down approach that studies the world city network from the perspective of the largest firms, and a macro-regional bottom-up approach that starts with the most important knowledge-intensive firms located within specific territorial boundaries. The research presented up until now would be representative for the top-down approach, while an example of the bottom-up perspective can be found in the work of Schmitt and Smas (2012), who use a number of northern European metropolitan areas (Copenhagen, Helsinki, Oslo and Stockholm) as geographic focal points to study inter-city connections using place-based information of financial and market service firms.

This bottom-up perspective highlights the importance of the pre-hoc definition of the scale of analysis. Although the INM was initially devised for studies of city networks at the global scale, it has been adopted in urban networks studies at other scales. Zhao et al. (2015), for example, analyse the geographies of urban networks created by producer services firms in China. They argue that owing to regulation of the Chinese state-led space economy, the position of Chinese cities as encapsulated in Figure 28.1 needs to be complemented with empirical studies drawing on leading producer series firms in the Chinese market. They explore the spatial differentiation in the connectivity of Chinese cities based on the location strategies of national firms, and conclude that these networks are not a simple extension of urban networks at a global scale but are also an embodiment of economic activities at other scales (see also Bentlage et al. 2013). One conspicuous scale is that of polycentric mega-city regions (Hall and Pain 2006). Developed around one or more cities of global status, a polycentric mega-city region is characterized by a cluster of cities and towns, physically separate but intensively networked in a complex spatial division of labour. Taylor et al. (2008) use the INM to uncover these regional networks and use this to contrast the monocentric spatial structure of the Paris and Rhine-Main metropolitan regions (dominated by Paris and Frankfurt) with the polycentric spatial structure of the Rhine-Ruhr and Randstad metropolitan regions (consisting of multiple important nodes).

This observation of the 'scalar specificity' of urban connectivity has become a key theme in the INM literature, and is comprehensively developed by Zhang (2018), who explores how producer firms with headquarters in the Pearl River Delta (PRD), in Mainland China or overseas, differentially impact both the PRD's urban system and its external relationships through their business networks. Zhang shows that the variegated urban networks created by different types of producer services firms within and outside China not only reflect the firms' different development histories, client orientations in specific markets and home regions' economic conditions, but also are significantly shaped by China's unique regulatory environment and complex state–market relationships (for other multiscalar analyses, see Schmitt and Smas 2012; Lüthi et al. 2013).

A final set of extensions is found in the observation that producer service firms are by no means the only world city network makers. The explicit recognition of the 'multiplexity of urban networks' (Burger et al. 2014) has led to research on the 'multiple globalizations' of cities (Krätke 2014). Some of this research uses the INM to analyse city networks through the lens of other agents, including global media firms (Hoyler and Watson 2013), maritime producer services (Jacobs et al. 2011), higher education (Chow and Loo 2015) and non-governmental organizations (Taylor and Derudder 2016, ch. 5). In the latter, the focus is on city networks of new social movements as reflected in the organization of the 25 largest environmental, development, humanitarian and human rights non-governmental organizations (NGOs) as defined in a ranking in the *Global Journal* in 2013. Using a similar measurement methodology and applying equations 28.1 and 28.2, the global network connectivity of cities in the networks of NGOs can be measured and mapped. The cartogram in Figure 28.2 shows the connectivity of the 100 most connected cities using four simple categories for cities ranked 1–10, 11–25, 26–50 and 51–100.

Figure 28.2 reveals a very different distribution of global connectivity cities across the world compared to that in Figure 28.1. The most notable feature is the widespread and relatively even pattern of cities across most world regions – if globalization is about network organizations that are worldwide in scope and operation, NGO connectivities show these institutions to be the globalizers par excellence. This relatively more even geography implies that regions previously identified as having few or no world cities are now well represented: the most obvious example is sub-Saharan Africa which now has almost 20 per cent of the cities on the cartogram but notice also the appearance of Central Asia

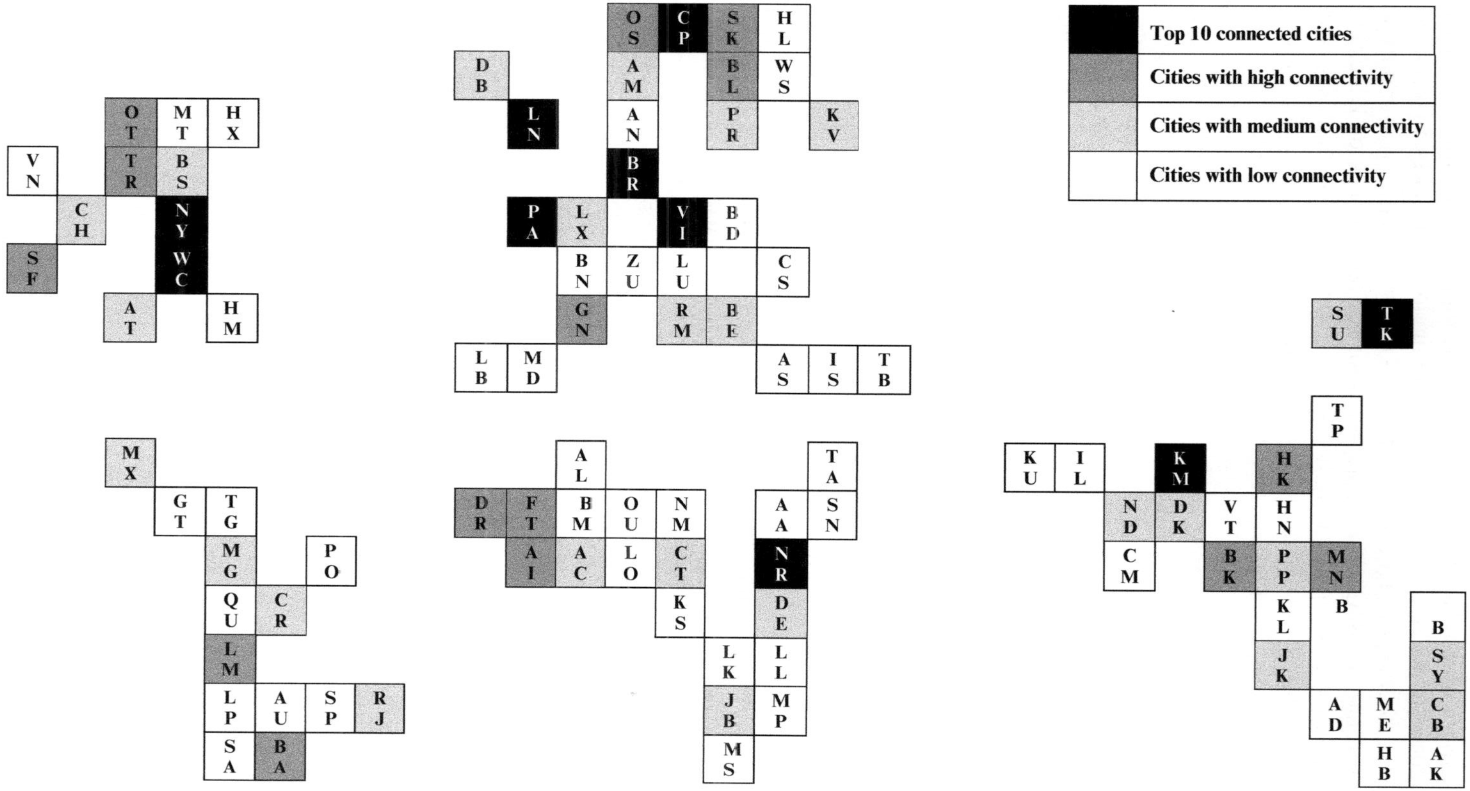

Source: Derudder and Taylor (2018).

Figure 28.2 Cartogram with 100 cities with largest GNC for NGOs and Central America which are both almost wholly missing from the world city network created by producer services firms. In contrast, the reduction of the USA to just a couple of cities on this cartogram is startling; evidently, US cities are relatively unimportant to the global networking of NGOs. The explanation for the relative lack of US cities has partly to do with the political nature of NGO business. Figure 28.2 is dominated by capital cities, locales where NGOs work with and through national politicians and government; an example of how states do indeed continue to matter in the geographies of the world city network. Unsurprisingly, then, and similar to the office networks of global law firms, Washington, DC, and Brussels as 'EU capital', feature very prominently in NGO networks. Although New York and London remain the dominant cities in this network, there are major differences at the apex of this ranking. The most interesting result is the presence of dedicated NGO centres, which emerge as specialist centres in the organization of securing environmental, development, humanitarian and human rights. These include Geneva and Brussels in Europe, but above all, cities such as Nairobi and Dakar in Africa, Kathmandu and Phnom Penh in Asia, and Lima and Managua in Latin America.

4.2 Modifications: Changing the City-dyad Specification in the INM

The analyses in sections 2 and 3 are not only confined to a specific scale and a specific set of actors, they are also idiosyncratic in relation to network specification and the ensuing analytical framework: there are other bipartite network specifications than the INM, and there are other centrality measures than degree centrality. However, the key point here is that things can be done differently, but that these alternative specifications and measures can provide complementary insights in the global connectivity of cities.

Some of these alternatives remain broadly in the spirit of the INM. Hennemann and Derudder (2014), for example, re-specify equation 28.1 by geographically restricting connections, while Derudder et al. (2013) aggregate different subsets of dyads to explore the relative strength of a city's connectivity with specific selections of cities (for example, leading world cities and cities in specific regions). More fundamental modifications can be found in a series of publications by Neal, who has put together a robust set of alternative specifications of city-dyads as per equation 28.1 (Neal 2011a, 2013a, 2013b, 2017) as well as centrality measures that diversify our understanding of cities' prominence in urban networks. Here we focus on one specific example, that of Neal's (2013a) alternative specification of city-dyads. In his paper, Neal proposes to take another approach to specifying strong city-dyads. Rather than defining these in absolute terms – the co-presence of a relatively large number of relatively important offices – Neal proposes specific strong city-dyads as those city-pairs that share more offices with different firms than could be expected by pure chance. A concrete example serves to illustrate. In her qualitative research on world city formation in China, Lai (2012) reveals that there is a division of labour between Shanghai (as Mainland China's premier gateway for international business) and Beijing (as the capital city of a state with a strongly regulated economy). As a consequence, producer services firms with a presence in Mainland China tend to have a major presence in both cities. Thus the Shanghai–Beijing city-dyad can clearly be thought of as being strong; firms with an office in Shanghai tend to also have an office in Beijing (and the other way round), leading to major service connections for this city-dyad. This pattern is revealed in by the INM, as Shanghai features prominently in the ranking of

Beijing's city-dyad connectivity (CDC) values (and the other way round). Yet both cities appear to have stronger connections with London and New York, but this is simply a reflection of the latter cities' large number of offices. This leads each and every city to be relatively strongly connected to New York and London, with any city's CDC distribution broadly replicating the GNC ranking. A way out of this is to come up with a strategy that allows revealing the relative statistical strength of city-dyads across the world city network.

Neal's (2013b) solution is to formally specify the level of co-presence of firms in city-pairs that can, with a given probability, be associated with non-random sorting of firms in cities. Lai's (2012) observation of producer services firms quasi-systematically having a major office in both Beijing and Shanghai is an example of this sorting process. Regionality, connections between specialist centres, strong connections between major world cities, and so on would be other examples of sorting processes. Importantly, in Neal's (2013b) approach these patterns are treated as hypotheses that need statistical validation. Although this is not strictly needed and other and more refined methods have been developed (for example, Neal 2017), the method in Neal (2013b) is perhaps the most straightforward way to explain this approach, as it first simplifies the original valued data into simpler binary data. The transformed service value matrix V thus merely reflects a presence (1) or absence (0) of a given firm in a given city, so that CDC_{a-b} equals the number of firms operating offices in cities a and b.

Formally answering the question of non-randomness requires consideration of two parameters: the capacity of cities to host firms, and the total number of firms being sorted into them. First, each city has a specific capacity, C, to serve as the site of firms' branch offices. Some cities, such as New York or London, have the capacity to host many firms, while other cities, such as Raleigh and Leeds, have the capacity to host only a few firms. Secondly, the world city network contains a finite number of producer services firms, N, that are being sorted into cities. With these two parameters defined, the issue of distinguishing inter-city linkages arising from random versus non-random sorting processes can be framed as a question of probabilities: what is the probability of observing an interlocking network model-estimated linkage between city i and city j that is stronger than X, if C_i out of N firms are randomly sorted into city i, and C_j out of N firms are randomly sorted into city j? The answer to this question is given by the cumulative distribution function:

$$1-\int_0^X \frac{\binom{N}{X}\binom{N-X}{c_i-X}\binom{N-c_i}{c_j-X}}{\binom{N}{C_i}\binom{N}{C_j}} \tag{28.3}$$

This function is constructed from a series of binomial coefficients, where each element of the form

$$\binom{A}{B} \tag{28.4}$$

indicates the number of ways in which B items can be chosen from a population of A, and is read as 'A choose B'. In this case, the denominator indicates the total number of ways

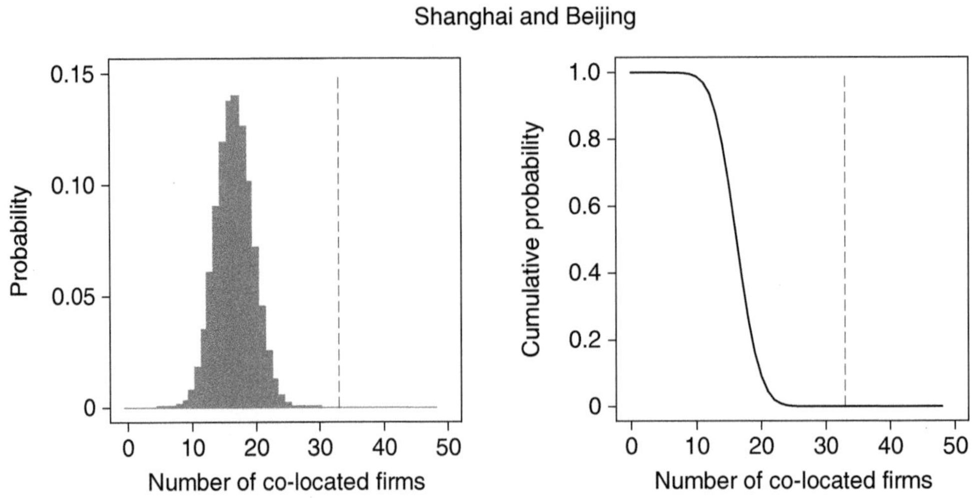

Source: Neal (2013b).

Figure 28.3 Cumulative distribution function for all possible values of X

that C_i of N firms could be sorted into city i and C_j of N firms into city j. The numerator indicates the total number of ways that, if C_i of N firms were sorted into city i and C_j of N firms were sorted into city j, they would host exactly X of the same firms. Thus, the entire fraction indicates the probability of observing exactly X firms co-located in cities i and j, given their respective capacities (C_i and C_j) and the total population of firms (N), if the firms-into-cities sorting process is random. Aggregating the probabilities across all values of X (which in our case ranges from 0 to 175) reveals how unlikely the observed value of X is.

The example of Beijing and Shanghai serves to illustrate. In our data, the three parameters of the cumulative distribution function are $N = 175$, $C_{Shanghai} = 61$, $C_{Beijing} = 48$, while the observed value for $X = 33$. That is, of the 175 firms that are assumed to be producing the world city network, 61 and 48 of them have a major office in Shanghai and Beijing, respectively, with 33 of them having a presence in both cities. Figure 28.3 shows the values of the cumulative distribution function for all possible values of X in this particular case, thus indicating the probability of having an interlocking network model-estimated linkage between Shanghai and Beijing that is stronger than the observed value of 33 (indicated by a dashed vertical line). The figure shows that, given Shanghai's and Beijing's total number of important offices, we would expect anywhere between 10 and 22 shared offices if global service firms would randomly open offices. This is clearly not the case here: the fact that Shanghai and Beijing share 33 offices is highly unlikely in statistical terms (the chance that this occurs is in fact almost zero), suggesting that there is a non-random – that is, meaningful – link between both cities as predicted by Lai (2012). As with other new empirical approaches to measuring global inter-city connections, this alternative specification (as well as the development of other alternatives in Neal 2014 and Hennemann and Derudder 2014) begs the question as to which empirical road to follow in world city network analysis. However, we agree with Neal (2013b, p. 1286) that the particular choice of network specification (or specific measures) should not be framed as a simple matter of preference or choice; these different perspectives are not simply right or wrong but 'Instead, they provide different and potentially complementary understandings of how cities might come to participate in the network and of how the network is structured. Viewing these . . . perspectives, and the networks they imply, in relation to one another can serve to draw on the strengths of each.'

5. CONCLUSIONS

Thinking through city networks has gained broad appeal in urban studies. Meijers (2007) and Neal (2011b) identify an empirically verifiable paradigm change from location-based to network-based renderings of city-systems. GaWC's approach implicitly draws upon this argument to underpin its specification of a world city network. However, analysis of world city networks can take very different forms in ontological, epistemological and methodological terms than that advanced here. Drawing on a parallel stream of qualitative research into the globalization of service industries, the central aim of our quantitative research has been to contribute to the literature on global inter-city relationships through the development and application of a network model that can be flexibly applied, transformed and modified in line with the research question to hand. In this chapter, we have probed how this research fits into the broader literature on cities in globalization, on the one hand, and on the quantitative analysis of urban networks on the other. It is only relatively recently that research that is broadly co-inspired by our approach (and critiques thereof) has started constructively engaging and subsequently taking it in new, often very different, and sometimes surprising directions. This fast-evolving set of applications, transformations and modifications has led to a more diverse and a richer understanding of global inter-city relationships, which in turn complements parallel approaches as developed by, *inter alia*, Mahutga et al. (2010), Wall and van der Knaap (2011), Alderson et al. (2010) and Rozenblat et al. (2017). We hope that the review presented in this chapter opens up spaces for further developing and diversifying this literature.

REFERENCES

Acuto, M. (2011), 'Finding the global city: an analytical journey through the "invisible college"', *Urban Studies*, **48** (14), 2953–73.

Alderson, A.S., J. Beckfield and J. Sprague-Jones (2010), 'Intercity relations and globalisation: the evolution of the global urban hierarchy, 1981–2007', *Urban Studies*, **47** (9), 1899–923.

Allen, J. (2010), 'Powerful city networks: more than connections, less than domination and control', *Urban Studies*, **47** (13), 2895–911.

Bassens, D. and M. van Meeteren (2015), 'World cities under conditions of financialized globalization: towards an augmented world city hypothesis', *Progress in Human Geography*, **39** (6), 752–75.

Beaverstock, J.V., M. Hoyler, K. Pain and P.J. Taylor (2001), *Comparing London and Frankfurt as World Cities: A Relational Study of Contemporary Urban Change*, London: Anglo-German Foundation.

Beaverstock, J.V., R.G. Smith and P.J. Taylor (1999), 'A roster of world cities', *Cities*, **16** (6), 445–58.

Beaverstock, J.V., P. Taylor, J.R. Faulconbridge and C. Nativel (2010), *The Globalization of Advertising: Agencies, Cities and Spaces of Creativity*, London: Routledge.

Bentlage, M., S. Lüthi and A. Thierstein (2013), 'Knowledge creation in German agglomerations and accessibility – an approach involving non-physical connectivity', *Cities*, **30** (February), 47–58.

Burger, M.J., B. van der Knaap and R.S. Wall (2014), 'Polycentricity and the multiplexity of urban networks', *European Planning Studies*, **22** (4), 816–40.
Chow, A.S. and B.P. Loo (2015), 'Applying a world-city network approach to globalizing higher education: conceptualization, data collection and the lists of world cities', *Higher Education Policy*, **28** (1), 107–26.
Cohen, R.B. (1981), 'The new international division of labor, multinational corporations and urban hierarchy', in M. Dear and A.J. Scott (eds), *Urbanization and Urban Planning in Capitalist Society*, New York: Taylor and Francis, pp. 287–318.
Daniels, P.W. (1995), 'Producer services research in the United Kingdom', *Professional Geographer*, **47** (1), 82–7.
Derudder, B. and P.J. Taylor (2018), 'Central flow theory: comparative connectivities in the world-city network', *Regional Studies*, **52** (8), 1029–40.
Derudder, B., P.J. Taylor, M. Hoyler, P. Ni, X. Liu, M. Zhao, et al. (2013), 'Measurement and interpretation of connectivity of Chinese cities in world city network, 2010', *Chinese Geographical Science*, **23** (3), 261–73.
Ducruet, C. (2017), 'Multilayer dynamics of complex spatial networks: the case of global maritime flows (1977–2008)', *Journal of Transport Geography*, **60** (April), 47–58.
Faulconbridge, J.R., P. Taylor, C. Nativel and J. Beaverstock (2011), *The Globalization of Advertising: Agencies, Cities and Spaces of Creativity*, London: Routledge.
Friedmann, J. (1986), 'The world city hypothesis', *Development and Change*, **17** (1), 69–83.
Hall, P.G. and K. Pain (2006), *The Polycentric Metropolis: Learning from Mega-city Regions in Europe*, New York: Routledge.
Harrison, J. and M. Hoyler (2018), *Doing Global Urban Research*, London: Sage.
Hennemann, S. and B. Derudder (2014), 'An alternative approach to the calculation and analysis of connectivity in the world city network', *Environment and Planning B*, **41** (3), 392–412.
Hoyler, M. and A. Watson (2013), 'Global media cities in transnational media networks', *Tijdschrift Voor Economische En Sociale Geografie*, **104** (1), 90–108.
Jacobs, W., H. Koster and P. Hall (2011), 'The location and global network structure of maritime advanced producer services', *Urban Studies*, **48** (13), 2749–69.
Korff, R. (1987), 'The world city hypothesis: a critique', *Development and Change*, **18** (3), 483–93.
Krätke, S. (2014), 'How manufacturing industries connect cities across the world: Extending research on "multiple globalizations"', *Global Networks*, **14** (2), 121–47.
Lai, K. (2012), 'Differentiated markets: Shanghai, Beijing and Hong Kong in China's financial centre network', *Urban Studies*', **49** (6), 1275–96.
Liu, X. and B. Derudder (2012), 'Two-mode networks and the interlocking world city network model: a reply to Neal', *Geographical Analysis*, **44** (2), 171–3.
Lüthi, S., A. Thierstein and M. Bentlage (2013), 'The relational geography of the knowledge economy in Germany: on functional urban hierarchies and localised value chain systems', *Urban Studies*, **50** (2), 276–93.
Lüthi, S., A. Thierstein and M. Hoyler (2018), 'The world city network: evaluating top-down versus bottom-up approaches', *Cities*, **72** (February), 287–94.
Mahutga, M.C., X. Ma, D.A. Smith and M. Timberlake (2010), 'Economic globalisation and the structure of the world city system: the case of airline passenger data', *Urban Studies*, **47** (9), 1925–47.
Meijers, E. (2007), 'From central place to network model: theory and evidence of a paradigm change', *Tijdschrift Voor Economische En Sociale Geografie*, **98** (2), 245–59.
Neal, Z. (2008), 'The duality of world cities and firms: comparing networks, hierarchies, and inequalities in the global economy', *Global Networks*, **8** (1), 94–115.
Neal, Z. (2011a), 'Differentiating centrality and power in the world city network', *Urban Studies*, **48** (13), 2733–48.
Neal, Z. (2011b), 'From central places to network bases: a transition in the US urban hierarchy, 1900–2000', *City & Community*, **10** (1), 49–75.
Neal, Z. (2013a), 'Does world city network research need eigenvectors?', *Urban Studies*, **50** (8), 1648–59.
Neal, Z. (2013b), 'Brute force and sorting processes: two perspectives on world city network formation', *Urban Studies*, **50** (6), 1277–91.
Neal, Z. (2014), 'Validity in world city network measurements', *Tijdschrift Voor Economische En Sociale Geografie*, **105** (4), 427–43.
Neal, Z. (2017), 'Well connected compared to what? Rethinking frames of reference in world city network research', *Environment and Planning A*, **49** (12), 2859–77.
O'Connor, K. and A. Scott (1992), 'Airline services and metropolitan areas in the Asia-Pacific Region 1970-1990', *Review of Urban & Regional Development Studies*, **4** (2), 240–53.
Parnreiter, C. (2014), 'Network or hierarchical relations? A plea for redirecting attention to the control functions of global cities', *Tijdschrift Voor Economische En Sociale Geografie*, **105** (4), 398–411.
Pred, A. (1978), 'On the spatial structure of organizations and the complexity of metropolitan interdependence', in L.S. Bourne and J.W. Simmons (eds), *Systems of Cities*, New York: Oxford University Press, pp. 115–42.

Robinson, J. (2002), 'Global and world cities: a view from off the map', *International Journal of Urban and Regional Research*, **26** (3), 531–54.

Rozenblat, C. and D. Pumain (1993), 'The location of multinational firms in the European urban system', *Urban Studies*, **30** (10), 1691–709.

Rozenblat, C., F. Zaidi and A. Bellwald (2017), 'The multipolar regionalization of cities in multinational firms' networks', *Global Networks*, **17** (2), 171–94.

Sassen, S. (1991), *The Global City*, Princeton, NJ: Princeton University Press.

Sayer, A. (1992), *Method in Social Science: A Realist Approach*, London: Hutchinson.

Schmitt, P. and L. Smas (2012), 'Multi-scalar "intercity connectivities" from a Northern European perspective', GaWC Research Bulletin 41, 1 October, accessed 6 April 2018 at http://www.lboro.ac.uk/gawc/rb/rb411.html.

Smith, D.A. and M.F. Timberlake (2001), 'World city networks and hierarchies, 1977-1997: an empirical analysis of global air travel links', *American Behavioral Scientist*, **44** (10), 1656–78.

Taaffe, E.G. (1956), 'Air transportation and United States urban distribution', *Geographical Review*, **46** (2), 219–38.

Taylor, P.J. (2001a), 'Being economical with the geography', *Environment and Planning A*, **33** (6), 949–54.

Taylor, P.J. (2001b), 'Specification of the world city network', *Geographical Analysis*, **33** (2), 181–94.

Taylor, P.J. (2004), *World City Network: A Global Urban Analysis*, London: Routledge.

Taylor, P.J. (2013), *Extraordinary Cities: Millennia of Moral Syndromes, World-Systems and City/State Relations*, Cheltenham, UK and Northampton, MA, USA: Edward Elgar.

Taylor, P.J. and B. Derudder (2016), *World City Network: A Global Urban Analysis*, 2nd edn, London: Routledge.

Taylor, P.J., B. Derudder, J. Faulconbridge, M. Hoyler and P. Ni (2014), 'Advanced producer service firms as strategic networks, global cities as strategic places', *Economic Geography*, **90** (3), 267–91.

Taylor, P.J., D.M. Evans and K. Pain (2008), 'Application of the interlocking network model to mega-city-regions: measuring polycentricity within and beyond city-regions', *Regional Studies*, **42** (8), 1079–93.

Van Meeteren, M., B. Derudder and D. Bassens (2016), 'Can the straw man speak? An engagement with postcolonial critiques of "global cities research"', *Dialogues in Human Geography*, **6** (3), 247–67.

Wall, R.S. and G.A. van der Knaap (2011), 'Sectoral differentiation and network structure within contemporary worldwide corporate networks', *Economic Geography*, **87** (3), 267–308.

Watson, A. and J.V. Beaverstock (2014), 'World city network research at a theoretical impasse: on the need to re-establish qualitative approaches to understanding agency in world city networks', *Tijdschrift Voor Economische En Sociale Geografie*, **105** (4), 412–26.

Wójcik, D. (2013), 'The dark side of NY LON: financial centres and the global financial crisis', *Urban Studies*, **50** (13), 2736–52.

Zhang, X. (2018), 'Multiple creators of knowledge-intensive service networks: a case study of the Pearl River Delta city-region', *Urban Studies*, **55** (9), 2000–2019.

Zhao, M., X. Liu, B. Derudder, Y. Zhong and W. Shen (2015), 'Mapping producer services networks in mainland Chinese cities', *Urban Studies*, **52** (16), 3018–34.

29. Global cities, centripetal wealth transfer and uneven development

Christof Parnreiter

INTRODUCTION

While much of global or world city research is framed as urban studies, in this chapter I apply an economic geographical perspective to scrutinize the role of these cities in uneven economic development.[1] A concern with the role of specific cities in the organization of economic globalization marked the beginning of global city research (Hymer 1972; Cohen 1981; Friedmann and Wolff 1982; Friedmann 1986; Sassen 1988), and Sassen (2001, pp. 359 and 361) recapped in the second edition of *The Global City* that '[t]he key indicator of global city status is whether a city contains the capabilities for servicing, managing, and financing the global operations of firms and markets. . . . The question is whether coordination and specialized servicing of global firms and markets is taking place'.

Over time, however, this economic geographical perspective faded into the background. Instead of examining global cities as 'highly concentrated command points in the organization of the world economy' (Sassen 1991, p. 3), their analysis was increasingly moved to the field of urban studies (Parnreiter 2013). Together with the increasing dominance of quantitative studies on the form and changes of the world city network (Taylor 2001; Taylor and Derudder 2016), this shift has caused an unfortunate move away from what had originally been a central interest of global city research, namely, the critical analysis of the geographies of power in globalization processes. Sassen, who describes herself as a 'political economist interested in the spatial organization of the economy and in the spatial correlates of economic power' (Sassen 1998, p. 182), reiterated in the second edition of *The Global City* that the notion of global cities has been elaborated as a critical inquiry into 'questions of power and inequality' (Sassen 2001, p. 351). She contends that global cities represent spatial correlates of economic power; they are a 'spatialization of inequality' (Sassen 1998, p. 182).

It is important to remember that global city scholarship does not relate these power functions to formal command power (as expressed, for example, in the number of corporate headquarters in a city). Instead, Sassen focused on the activities behind transnational corporate power, on the inputs, that make the exercise of power possible, on 'the *practices* that constitute what we call economic globalization and global control' (Sassen 2001, p. xxii, original emphasis). Building on the service literature that emerged in the 1980s, Sassen argued that management, coordination and control functions of firms operating in or orchestrating entire global commodity chains have become so complex that they have been increasingly outsourced to specialized service firms.[2] That is, a key argument of global city scholarship is that 'the capabilities for global operation, coordination, and control . . . need to be produced' (Sassen 2010, p. 158), and that this production occurs outside the big transnational corporations, namely, in legal, financial, accountancy and other consulting firms.

From that follow two related arguments. First, producer services (PSs) are more than just technical support. They combine practical assistance with guidance. Articulation functions, which make global commodity chains run, and control functions, which govern them, are interconnected and interwoven. Global producer service firms (PSFs) are contracted because they are (or are believed to be) the providers of the knowhow necessary for the relatively assured realization of capital (Bassens and van Meeteren 2015), and that is why their counsels are usually followed. Since advice which is known to be taken becomes more than an advice, PSFs, even though they have no formal autonomy in decision making, pre-structure corporate decisions in that they shape a decision in advance. Sassen (2010, p. 158) terms this influence as by means of the services provided 'embedded governance – embedded in the lawyering, the accounting and the investment choices of the firm'.

Secondly, PSFs are not just providers of inputs for their clients' commodity chains. They have also become strategic partners of their clients, who in addition to facilitating the global operations of their clients make available elements for, and even pre-structure, governance processes (Parnreiter 2015a). The relationships between global corporations and PSFs are therefore best understood as being horizontal (Bryson and Daniels 2007). Accordingly, PSFs' power is not as exercised over clients, but with them, in order to achieve an empowerment vis-à-vis business partners, unions or governments. Therefore, the even and mutual relationships between PSFs and their clients are embedded into, and serve, the hierarchical relationships which PSFs' clients maintain in their global commodity chains or production networks (Parnreiter 2019).

In this perspective, global city research is a key theoretical and methodological instrument with which to examine uneven development as it is being produced and reproduced in globalization processes. Providing critical insights on by whom, how and from where global commodity chains are organized and governed, the global city concept draws attention to the practical work of organizing centripetal wealth transfers as well as to its geographies. Global city research helps, thus, to achieve a more nuanced account of the geographies of global economic governance and the production of uneven development. This is important, not at least in order to correct all too simple notions of a clear-cut North–South power divide, which we find, for example, in the postcolonial studies critique of global city research.

Yet, more than 25 years after the publication of Sassen's *The Global City* (1991) we have to admit that the call to understand the novel geographies of uneven economic power relationships and the functions which global cities assume in them is still more of a programmatic claim than lived research practice. Although studies substantiating that global cities are organizational and power nodes in globalization exist (for example, Rossi et al. 2007; Parnreiter 2010, 2015a; Jacobs et al. 2011; Hanssens et al. 2012; Jacobs 2014; Bassens and van Meeteren 2015), we have not advanced satisfactorily in understanding the role of global cities in the exercise of economic power. To further develop this research agenda, we need to understand global cities as critical governance nodes in global commodity chains or production networks (Parnreiter et al. 2007; Brown et al. 2010; Parnreiter 2010, 2014, 2019), and to link the notion of global cities' governance functions to Hadjimichalis's (1984, 1987) concept of the 'geographical transfer of value'.

GLOBAL CITIES AND THE GOVERNANCE OF GLOBAL COMMODITY CHAINS

Turning to the literatures on global commodity chains (GCCs), global value chains (GVCs) and global production networks (GPNs) (for example, Gereffi and Korzeniewicz 1994; Henderson et al. 2002; Gereffi et al. 2005; Gibbon et al. 2008; Dicken 2011; Bair 2014; Yeung and Coe 2015; Werner 2016) and integrating them with global city scholarship is apt, since questions of economic power and of uneven development are central to both. Moreover, Gereffi's (1994, p.97) widely used conceptualization of economic governance as the 'authority and power relationships that determine how financial, material and human resources are allocated and flow within a (commodity) chain' provides an actor-focused and explicitly relational perspective on asymmetrical power relationships that is suitable to substantiate the notion of command and control, which is only vaguely defined in the global city concept.[3]

Yet, despite commitments to analyze governance as multi-actor and multi-scalar relationships, many studies in the fields of GCCs, GVCs and GPNs have prioritized the link between lead firms and key suppliers, thereby ignoring other possible actors (Yeung and Coe 2015). Producer service firms, in particular, have barely been examined, though their general importance has been recognized (Rabach and Kim 1994; Henderson et al. 2002; Kaplinsky and Morris 2002; Dicken 2011). In the past few years, however, the significant influence that PSFs, in general, and financial firms, in particular, exercise on corporate strategies has been put center stage in GPN research (Coe et al. 2014). For example, the consequences of lead firms' financialization have been discussed as regards the shape and extension of GCCs (Gibbon 2002; Milberg 2008) and as to the distribution of value among different actors (Palpacuer 2008). Wójcik (2013) and Bassens and van Meeteren (2015) indicate the critical role of PSFs in creating an offshore world, thereby enabling their clients to escape governmental or intergovernmental control and to keep profits through tax evasion. As regards non-financial PSFs, Coe (2014) has drawn attention to logistic firms, while the Institute for Global Law and Policy (IGLP) Law and Global Production Working Group (2016) is launching an initiative to focus more on the role of law in GVC-governance.

These studies echo the findings of the broader literatures on financialization and transnational private governance which confirm that PSFs supply key inputs for economic governance. For example, Flood and Sosa (2008) have shown that legal firms successfully support their clients in resolving tax issues. As regards financialization, it has been corroborated that PSFs are functional for the implementation of shareholder values (for example, Froud et al. 2006; Milberg and Winkler 2010; Baud and Durand 2012). Producer service firms are also key players in developing standards, which commit other firms to specific ways of doing business, for example, as regards the harmonization of accountancy principles (for example, Djelic and Sahlin-Anderson 2006; Graz and Nölke 2008).

All these studies confirm that PSs combine practical assistance with strategic guidance. The importance of an aspect is case-specific, and depends on factors such as the nature of the business case, the service provided, the size and firm culture of the client, and the client's market position. Yet, while there are many routine services where pure technical skills, such as proficiency in legal affairs or bookkeeping, matter and where strategic guidance remains a minor factor, the reason why big, global PSFs have gained their market position (and can charge extremely high fees) is that they offer more than just those technical advices. Many PSs are customized, tailored to specific customers and business cases, and therefore include strategic elements which serve to drive clients' businesses in a desired direction. Accordingly, PSFs are more than providers of inputs. From the perception of clients, PSFs enjoy authority in Arendt's (1961) or Sennett's (1980) sense: they master knowledge which is seen to satisfy the desire and need for leadership in very complex economic settings. Therefore, they allow PSFs to steer their own behavior and to pre-structure governance processes. Producer service firms thus exercise significant influence on how their clients develop their business and on how value is created and distributed in GCCs, GVCs or GPNs. When Sassen (2010) refers to 'embedded governance', what she means is that financial, legal or accountancy services play an active role in firms' decision making (including firm location).

The consequences for the geographies of the world economy are evident: If PSFs are important actors in governance processes, then global cities, where these firms are clustered, have to be understood as central governance nodes in the world economy (Parnreiter et al. 2007; Brown et al. 2010; Parnreiter 2010, 2014). It is from there that strategic inputs for economic decision-making processes in GPNs are provided.

UNEVEN CITY NETWORKS

As the interlocking network model (INM) to measure relationships between world cities (Taylor 2001; Taylor and Derudder 2016; for an assessment see Derudder and Parnreiter 2014) became dominant in the global city literature, the notion that inter-city networks were even became increasingly prevalent. While Taylor originally grounded the elaboration of the INM in the need to substantiate the notion of a hierarchy of cities both theoretically and empirically (Taylor 1997), he later dismissed the idea of a hierarchy in favor of a world city network. This shift implied much more than just using a new vocabulary. The clarification of the subject matter as being urban network studies had a great deal to do with Taylor turning to the work of Jane Jacobs (1984 for example). Specifically, Taylor draws strongly on Jacobs's idea that cities need the exchange with each other to be economically dynamic, whereby external relationships of cities are characterized by cooperation rather than by competition: 'For Jacobs, cities come in groups and they need each other; this mutuality indicates (horizontal) network relations rather than (vertical) hierarchical relations' (Taylor 2011, p. 53).

Yet, if we define a hierarchy as a set of social relations where 'control up and down different levels' (Taylor 2004, p. 39) is the norm, it seems strange, against the backdrop of a world characterized by enormous economic and social differences at various levels, to describe the city networks that form the backbone of the world economy as being flat. On the contrary, I suggest that most inter-city relations are asymmetrical, marked by differences in economic and/or political power. Thus, while relationships between world cities are even because the division of labor between the offices of PSFs in global cities are even (for Germany, see Parnreiter 2015a; for the relationships between London and Frankfurt, see Beaverstock et al. 2005), the many inter-city connections on which the world city network is built and that link global cities to the countless non-global, yet globalized cities, where production for global markets is carried out, are far from being even. The world city network and the inter-city ramifications on which it rests form the operational framework of the world economy through which uneven development is organized. The inter-city connections serve as pipelines through which value is syphoned off and transferred to the centers.

The idea of a world/global city, as introduced by Friedmann and Sassen, only makes sense if these cities are seen in a relational perspective as being hierarchically connected to all the globalized 'ordinary cities' (Robinson 2006) whose work has to be articulated with the world market. There is, thus, no contradiction between Sassen's (2016, p. 103) 'analytic tactic . . . to focus on the extreme ends', such as the financial sector in New York, and Robinson's (2002, p. 547) assertion that export processing zones in poor countries constitute 'quite the other end of the command and control continuum of global city functions'. It is precisely this relational approach to center–periphery interactions in GVCs and in urban networks that justifies the selection of PSs in global cities as key analytical lenses. The often disapproved of narrowing down of the focus on PSs (for example, Robinson 2002) is thus a means of a 'critical exploration of exceptional power' (Peck 2015, p. 164), conceptually intended to draw attention to the ugly truth that most people in most cities (Southern and Northern alike) are off the map of economic decision making. Since the world economy is so spiky, it is neither analytically accurate nor politically progressive to speak homogeneously of 'globalizing cities' (Marcuse and Kempen 2000).

Let me exemplify this by one example. In Mexico's automotive and electronics industries, which belong to the leading economic sectors of that country, there is a pronounced division of labor between cities where manual labor – most of it assembling imported parts, largely accomplished by female workers – is carried out, and parts of Mexico City, from where the high-end services come (often provided by men), which make this global assembly-line work. Just three maquiladora cities on Mexico's northern border – Ciudad Juárez, Tijuana and Reynosa – account for 43 percent of the value-added in the electronics industry, and just five cities – Cuautlancingo (Puebla), Hermosillo, Ramos Arizpe, Ciudad Juarez and Silao – provide 41.4 percent of the value-added in the automotive industry. However, none of these cities has a significant production of PSs – together they reach 1.4 percent of the value-added of all services to the producer, and only 0.2 percent of financial services. Mexico City accounts for more than two-thirds of the national value-added in PSs. In financial and insurance services, the percentage of Mexico City over the national added value rises to 81.5 percent. Moreover, PSs are also highly concentrated within Mexico City: just four *delegaciones* (districts) – Cuauhtémoc, Álvaro Obregón, Benito Juárez and Miguel Hidalgo – achieve 71 percent of all Mexican added value (Instituto Nacional de Estadística y Geografía 2016). Therefore, service flows connect areas in Mexico City, such as Santa Fe, Paseo de la Reforma, Polanco and Avenida de los Insurgentes Sur, with manufacturing cities in the country's central region and in the North (Parnreiter 2010, 2011, 2015b).

The data quoted on the functional and geographical division of labor within commodity chains in Mexico indicate that there are deep inequalities within city networks. This unevenness is evident at first glance of the basic economic data: Value-added per employed person in Álvaro Obregón's financial sector is, for example, nearly 10 times higher than in Ciudad Juárez's transport equipment sector, while payment is about four times higher (own calculations, based on Instituto Nacional de Estadística y Geografía

2016). These inequalities in value creation, wages and (presumably) profits not only signal uneven development within Mexico's cities, but also imply hierarchical relationships in that manual manufacturing work – often conducted by women – in cities such as Ciudad Juarez is controlled from (often male) service workers in Mexico City.

As regards these hierarchical relationships, or power asymmetries, critics of global city research have objected that the concept is 'unnecessarily focused on urban places and spaces at the expense of transnational firms and most especially key social actors' (Jones 2002, p. 340). Conflating 'resources and power' (Allen 2003, p. 2), global city scholars fail 'to get to grips with theorising power' (Jones 2002, p. 340). While I share the notion that '[o]ne cannot infer interaction, coordination, command, control, domination and subordination from the mere existence of office networks' (Smith and Doel 2011, p. 27), it is important to remember that global city scholarship does not relate power functions to formal command power but focuses on the activities behind the exercise of command and control. I therefore maintain that the lack of information on how specific firms in particular cities accomplish strategic control of distanciated economic activities is an empirical rather than a conceptual problem. Where global city researchers have endeavored to investigate the mechanism of economic command and control, they have found it (for example, Parnreiter 2010, 2015a).

One example of how PSFs are involved in the governance of global commodity chains refers to labor relations. According to interviews conducted in Mexico City with partners and lawyers in global legal firms, these professionals see themselves as strategic actors in shaping labor relations in their clients' production networks. A recurring issue addressed in the interviews relates to the payment of social benefits. For example, the level of insurance benefits for employees depends on the risk class in which a company is classified by the Mexican Institute for Social Security. In order to reduce payments for insurance protection of the employees of one client, an interviewed global law firm has successfully claimed in the courts so that the client is re-classified in a less risky business field. Another issue mentioned repeatedly by labor lawyers concerns the configuration of employment contracts. If workers, for example, belong directly to the hiring company, they normally receive higher wages and more social benefits than if they were employed by a subsidiary founded for this purpose.

Another frequent topic is the restructuring of labor relations, for example, through switching to a different union with which the collective bargaining agreement is signed. This is a critical issue, owing to the closed-shop principle according to which companies in Mexico have to accept only one union as the workers' legitimate representation. As unions differ significantly in strength and political orientation, the closed-shop principle boosts high selectivity as regards with whom to sign the collective bargaining agreement. Lawyers therefore recommend their clients to 'actively hunt for a trade union', as the German Association for Foreign Trade and Location Marketing informs (Buerstedde and Päffgen 2015). This chasing of convenient unions is particularly important owing to the geographical dimension of unions' strength. Since unions are more employer-friendly in some federal states than in others, the proactive selection of workers' representation involves localization decisions: Owing to unions' impacts on the working environment, wages, the payment of social benefits and hence on profits, companies decide where to establish a plant also with regard to the geography of union power. However, these spatialities of labor relations are difficult to grasp for foreign companies and especially for newcomers to the Mexican market. Lawyers from global law firms describe it therefore as one of their tasks to guide the client in this respect. As one lawyer underscores:

> [the partners of his firm] have conferences with the clients to plan a deal, to structure a deal. . . . The one who takes the decision how things should be done, rather would be here, from the [law] firm. . . . I do believe that the one who makes the strategy, it's the partners of the law firm. (Parnreiter 2019, pp. 89–90)

A partner of another law firm corroborates:

> [the client] contracts you precisely for that. That you tell him which union is the most convenient one. . . . I think there's much influence one has. A good example, as to labor issues, is . . . the firm x. X has closed plants, and has put other plants, only and exclusively because of the advice we have given . . . [They faced] problems with the union, because the union is very aggressive. . . . Because [in that federal state] a contract was common, they had to pay certain, hm, benefits that were above market standards. . . . And the decision to relocate the plants, it was a joint decision, the client and you? Well, . . . the client decides because it is his money, (but) based on the advice you give him. . . . If you tell him the best place to establish a plant, then it is unlikely that the client will decide for anything else. (Parnreiter 2019, pp. 89–90)

These examples of the strategic involvement of PFSs in economic governance fit both the literature on the importance of barriers to entry, and labor relations in shaping the structure, geography and governance of global commodity chains (Gereffi 1994; Kaplinsky and Morris 2002; Smith et al. 2002; Mahutga 2012; Selwyn 2012) and the literature on the role of financial and legal firms in GPN governance processes (Coe et al. 2014; IGLP Law and Global Production Working Group 2016). To emphasize: PSFs are not only enablers of production networks through 'greasing' assembly lines, they are also drivers, shaping how and where value is produced and accumulated. And they are involved in the production and re-production of uneven development, helping to direct value flows towards regional and global centers. I contend that this is valid also for Southern global cities, such as Mexico City, in line with the quantitative Globalization and World Cities (GaWC) research that proves that the world city network is formed by more than 500 cities around the world (Taylor and Derudder 2016).

Without claiming that the cases quoted here draw a complete picture, I think they do support Sassen's (2010, p. 158) contention that governance is being 'embedded' in the services provided. The cases show that services such as patenting, financing or drawing up labor contracts are far more than technical devices. They 'entail command functions that are distributed across those operations' (Sassen 2010, p. 158). The Mexico City lawyer, for example, who in his search for convenient unions pre-structures clients' strategies for plant relocation, is an active agent in GCC governance in two ways. Suggesting to clients a geographical reorganization of their GPNs implies that the lawyer functions as a disarticulator (Bair and Werner 2011) of plants, cities or federal states with combative (or otherwise unfriendly) trade unions. Since these exclusions are as constitutive (Werner 2016) for the formation and restructuring of GVCs as their opposite, namely, the integration through 'strategic coupling' (Yeung 2015), lawyers can be identified as important shapers of uneven economic and urban network. Also, plant relocation in order to reduce wages increases local value transfer from workers' households to the firm: Weaker unions imply fewer direct and indirect payments and therefore more value to be siphoned off from the workers' households. Yet, workers compensate the loss of income among other things through externalization of costs to their households, which feed values into GPNs because (mainly female) household members shoulder more unpaid work for producing and reproducing the labor force (Dunaway 2014).

THE GEOGRAPHICAL TRANSFER OF VALUE

The dynamic underlying the production and reproduction of uneven inter-city networks is what Hadjimichalis (1984, 1987) has termed the geographical transfer of value (GTV). In general, this GTV refers to a process through which value produced at one location is transferred to and realized in another, with the consequence that the latter location's accumulation base grows at the expense of the first. The GTV thus leads to and intensifies differences in localized accumulation processes, which in turn increases and accelerates inter-regional inequality. This connection between the geographical reallocation of wealth, regionally differentiated growth rates and uneven development becomes most obvious in respect of the 'urbanization of capital' (Harvey 1985).

According to Hadjimichalis (1984, p. 340), value is transferred geographically either indirectly through the market, or directly through violent methods and 'more sophisticated policies such as state taxation policies, price policy, public transfer payments, multinational profit repatriation, and transfer-pricing'. This suggests that PSFs are strategic agents facilitating the GTV.

As to the market-mediated shifting of value, GCC, GVC and GPN analysis consistently indicate entry barriers are a key instrument for achieving rents and hence an uneven distribution of surplus: 'The GCCs approach explains the distribution of wealth within a chain as an outcome of the relative intensity of competition within different nodes' (Gereffi et al. 1994, p. 4). For the purpose of this chapter, two things are important. First, erecting entry barriers is a function of governance: 'It is this role . . . of identifying dynamic rent opportunities and apportioning roles to key players which reflects the act of governance' (Kaplinsky 2000, p. 124). Secondly, for creating and managing entry barriers, lead firms have to rely on PSFs. These are particularly important for erecting strategic barriers, which the Organisation for Economic Co-operation and Development (2005) defines as being intentionally created to deter entry of competitors. Prime examples are the creation and subsequent protection of intellectual property or establishing brands, firm reputation and customer loyalty, which would be impossible without the involvement of patent lawyers or advertisers. However, PSFs are also important as regards some structural entry barriers (which are related to industry conditions), such as financing, which requires engagement with banks and other financial institutions.

Producer service firms are also key agents in many direct mechanisms of value transfer. It is hard to imagine that policies related to taxation, pricing, foreign direct investment or the repatriation of profits could be put into practice without the involvement of bankers, lawyers, accountants and other consultants (for example, Flood and Sosa 2008; Sikkaa and Willmott 2010; Wainwright 2011; Clark and Monk 2013; Wójcik 2013; Bassens and van Meeteren 2015). Another field in which PSFs are involved in value transfer are labor relations. These are critical to GCC governance not least because, under competitive conditions, companies must reduce costs without impairing performance (Yeung and

Coe 2015). Labor costs and regulations are therefore essential criteria for whether and how companies in a particular region become integrated into GPNs (Henderson et al. 2002; Smith et al. 2002; Selwyn 2012). Finally, PSFs are key agents in the GTV for another reason. As 'gatekeeper(s) of an enormous stock of sought-after knowledge' (Bassens and van Meeteren 2015, p. 763), they draw massive fees from their clients, thereby appropriating an increasing part of the wealth produced along GVCs.

While all this confirms once again that PSs are not just technical inputs, considering PSFs as strategic actors in the GTV carries consequences for the geographies of wealth transfer, in particular as regards its organization. Contrary to Hadjimichalis (who developed his arguments more than 30 years ago, when the new geographical contours of the world economy were just emerging), I contend that the social agents active in transferring value are not located 'usually . . . *outside* the donor region' (Hadjimichalis 1984, p. 340, emphasis added). Current urban theorizing stresses that knowledge-intensive activities require constant face-to-face interactions and, thus, co-presence (for example, Storper 2013). This certainly applies to high-end bankers, lawyers or accountants. As distance-sensitive knowledge workers, professionals of PSFs had to follow their clients in globalization processes (Bryson and Daniels 2007). The global cities where these services cluster are necessarily spread over the world. Remember that Sassen (1991) contended that New York, London and Tokyo are just leading examples of a much broader process of global city formation which cuts across the traditional North–South divide. Sassen's (2001, p. 348) wording that global cities form 'the operational scaffolding of . . . the global economy' is unambiguous – scaffolding typically covers the whole area under construction, not just key intersections. Accordingly, Taylor and Derudder (2016) identify 525 cities creating the world city network, almost half of them in poorer countries (including classical mega-cities such as Mumbai, São Paulo, Mexico City or Jakarta). While the resulting geography of economic governance looks far more decentralized than that deduced from a focus only on lead firms, the theoretical insight drawn from these spatialities of PSFs' influence on GCC governance is that wealth transfer is organized from outside the donor regions as well as from within, and it is accomplished through inter-city networks. Centre–periphery relationships unfold thus as successive inter-city connections on various scales; an idea that Frank (1969) suggested with his spatialized account of the 'development of underdevelopment', in which Latin American capital cities were conceived of as bridgeheads for the interests of the dominant centers. Alternatively, as John Friedmann (1995, p. 42) put it: 'We must understand global cities in relation to their respective peripheries.'

The principle identified here of how the capitalist global or regional divisions of labor are organized is anything but novel. The city network of the Hanseatic League not only served the interests of the merchants in leading cities such as Lübeck or Cologne, but Kontor cities such as Novgorod or Bergen, or secondary Hanse cities such as Danzig obtained economic stimuli not abstractly through their privileges, but because they functioned as nodes for the transfer of resources from the East and North to the West, namely, to the rising economic centers such as Flanders, Brabant and the Northern Netherlands (Dollinger 2012).[4] Similarly, Mexico City's role in the Spanish viceroyalty was to 'to suck capital . . . out of its own satellites and to channel part of this surplus to the world metropolis of which all are satellites' (Frank 1969, p. 6). While the key economic activity of Spanish colonialism was silver mining, the silver excavated by wage and forced labor in the North and the central region of what is now named Mexico was 'funneled into Mexico City' (Kandell 1990, p. 180). It was there where the Crown Mint was established, founded in Mexico City in 1535 as the first mint in the colonies. Minting was not only spatially highly centralized – in all Spanish possessions in the Americas there were only six mints – but also held a monopoly. Mexico City's elite merchant families who ran the Crown Mint controlled the emergent world monetary system that linked the viceroyalty of New Spain to both the colonial power and other Spanish possessions, namely, the Philippines. Local mine-owners, however, could keep only a small percentage of the silver output. Thus, Mexico City's prosperity during the viceroyalty of New Spain was based on Spanish colonialism and exploitative transatlantic and transpacific relationships it created and lived off. Finally, on the West African coast, the economic growth of cities was spurred by their pivotal role in the triangular slave trade, which enriched, for example, Lagos's nobility and turned the city into an important port (Mann 2007). Therefore, some mutuality existed between Lagos's kings and the merchants from Lisbon or London. Yet, this mutuality was deeply embedded in hierarchical relationships between the transatlantic slave traders and their local partners. Furthermore, the win–win situation for slave traders in Lisbon and Lagos was sustained by very uneven relationships between Lagos and the towns and villages in its hinterland that had to supply the slaves.

CONCLUSION

The intention of this chapter has been to re-orientate global city studies towards their original economic geography focus and, in particular, to re-draw attention to the role of global cities in making and reproducing uneven development. My key contention is that since the direct as well as the indirect mechanisms through which value is transferred geographically require the intervention of PSFs, global cities can be theorized as governance nodes for centripetal wealth transfers along GCCs. Yet, this form of governance does not directly correspond to the often quoted notion of command and control, nor does it fit the dominant approach in GCC, GVC and GPN studies, which until very recently has focused on bipolar governance relationships between lead firms and key suppliers. I think that economic governance is best understood as a multi-actor, multi-scale and multi-local process. Paraphrasing Jane Jacobs (1970, p. 121), PSFs agency resembles the 'little movements' that turn 'the great wheels' of the GTV. Mexico City is just one of the many global cities where these wheels are turned, and that is why the global city concept is a meaningful approach to study the role of cities all over the world in uneven development. Thus, the challenge for urban studies is not to theorize the city 'from the Global South' (Parnell and Robinson 2012), but to comprehend the role of global cities all over the world in the shaping of the uneven relationships between cities that guarantee the centripetal GTV.

NOTES

1. While some authors use the terms 'global' and 'world city' either without further justification or interchangeably, others have stressed the conceptual differences behind the different terms (Derudder 2006). In

this chapter I adopt the respective terminology of the authors I refer to. Where I allude to the general debate, I use 'global cities', since my own work draws strongly on Sassen's (1991) *The Global City*.
2. I am aware of the conceptual differences between the literatures on global commodity chains, global production networks and global value chains. For the purpose of this chapter, however, these differences are of less importance. Therefore, the terms are used synonymously.
3. Today a narrowed view of governance is dominant, in which the focus on uneven power relations has been replaced by discussions of transaction costs and of win–win situations in the context of upgrading of firms. There are, however, many calls (Selwyn 2012; Bair 2014; Werner 2016) to recover the original critical impetus of GCC analysis, which aimed to uncover the structural inequalities inherent to capitalism's cross-border division of labor (Hopkins and Wallerstein 1986).
4. A Kontor was a trading post of the Hanseatic League abroad. The Kontors formed the key nodes of the Hanseatic trade network.

REFERENCES

Allen, J. (2003), *Lost Geographies of Power*, Oxford: Blackwell.
Arendt, H. (1961), *Between Past and Future: Six Exercises in Political Thought*, New York: Viking Press.
Bair, J. (2014), 'Editor's introduction: commodity chains in and of the world system', *American Sociological Association*, **20** (1), 1–10.
Bair, J. and M. Werner (2011), 'Commodity chains and the uneven geographies of global capitalism: a disarticulations perspective', *Environment & Planning A*, **43** (5), 988–97.
Bassens, D. and M. van Meeteren (2015), 'World cities under conditions of financialized globalization: towards an augmented world city hypothesis', *Progress in Human Geography*, **39** (6), 752–75.
Baud, C. and C. Durand (2012), 'Financialization, globalization and the making of profits by leading retailers', *Socio-Economic Review*, **10** (2), 241–66.
Beaverstock, J., M. Hoyler, K. Pain and P. Taylor (2005), 'Demystifying the euro in European financial centre relations: London and Frankfurt, 2000-2001', *Journal of Contemporary European Studies*, **13** (2), 143–57.
Brown, E., B. Derudder, C. Parnreiter, W. Pelupessy, P. Taylor and F. Witlox (2010), 'World city networks and global commodity chains: towards a world-systems' integration', *Global Networks*, **10** (1), 12–34.
Bryson, J. and P. Daniels (eds) (2007), *The Handbook of Service Industries*, Cheltenham, UK and Northampton, MA, USA: Edward Elgar.
Buerstedde, P. and C. Päffgen (2015), 'Nach wie vor Fachkräftemangel in technischen Berufen', accessed 11 June 2018 at http://www.gtai.de/GTAI/Navigation/DE/Trade/Maerkte/Geschaeftspraxis/lohn-und-lohnnebenkosten,t=lohn-und-lohnnebenkosten--mexiko,did=1289472.html.
Clark, G. and A. Monk (2013), 'The scope of financial institutions: in-sourcing, outsourcing and off-shoring', *Journal of Economic Geography*, **13** (2), 279–98.
Coe, N. (2014), 'Missing links: logistics, governance and upgrading in a shifting global economy', *Review of International Political Economy*, **21** (1), 224–56.
Coe, N., K. Lai and D. Wójcik (2014), 'Integrating finance into global production networks', *Regional Studies*, **48** (5), 761–77.
Cohen, R. (1981), 'The new international division of labor, multinational corporations and urban hierarchy', in M. Dear and A. Scott (eds), *Urbanization and Urban Planning in Capitalist Society*, New York: Methuen, pp. 287–315.
Derudder, B. (2006), 'On conceptual confusion in empirical analyses of a transnational urban network', *Urban Studies*, **43** (11), 2027–46.
Derudder, B. and C. Parnreiter (2014), 'The interlocking network model for studying urban networks: Outline, potential, critiques, and ways forward', *Tijdschrift voor Economische en Sociale Geografie*, **105** (4), 378–86.
Dicken, P. (2011), *Global Shift: Mapping the Changing Contours of the World Economy*, London: Sage.
Djelic, M. and K. Sahlin-Anderson (eds) (2006), *Transnational Governance. Institutional Dynamics of Regulation*, Cambridge: Cambridge University Press.
Dollinger, P. (2012), *Die Hanse*, Stuttgart: Kröner.
Dunaway, W. (2014), 'Bringing commodity chain analysis back to its world-systems roots: rediscovering women's work and households', *American Sociological Association*, **20** (1), 64–81.
Flood, J. and F. Sosa (2008), 'Lawyers, law firms, and the stabilization of transnational business', *Northwestern Journal of International Law & Business*, **28** (3), 489–526.
Frank, A. (1969), *Latin America: Underdevelopment or Revolution: Essays on the Development of Underdevelopment and the Immediate Enemy*, New York: Monthly Review Press.
Friedmann, J. (1986), 'The world city hypothesis', *Development and Change*, **17** (1), 69–83.

Friedmann, J. (1995), 'Where we stand: a decade of world city research', in P. Knox and P. Taylor (eds), *World Cities in a World System*, Cambridge: Cambridge University Press, pp. 21–47.
Friedmann, J. and G. Wolff (1982), 'World city formation: an agenda for research and action', *International Journal of Urban and Regional Research*, **6** (3), 309–44.
Froud, J., S. Johal, A. Leaver and K. Williams (2006), *Financialization and Strategy. Narrative and Number*, London: Routledge.
Gereffi, G. (1994), 'The organization of buyer-driven global commodity chains: how U.S. retailers shape overseas production networks', in G. Gereffi and M. Korzeniewicz (eds), *Commodity Chains and Global Capitalism*, Westport, CT: Praeger, pp. 95–122.
Gereffi, G. and M. Korzeniewicz (eds) (1994), *Commodity Chains and Global Capitalism*, Westport, CT: Praeger, pp. 95–122.
Gereffi, G., J. Humphrey and T. Sturgeon (2005), 'The governance of global value chains', *Review of International Political Economy*, **12** (1), 78–104.
Gereffi, G., M. Korzeniewicz and R. Korzeniewicz (1994), 'Introduction: global commodity chains', in G. Gereffi and M. Korzeniewicz (eds), *Commodity Chains and Global Capitalism*, Westport, CT: Praeger, pp. 1–14.
Gibbon, P. (2002), 'At the cutting edge? Financialisation and UK clothing retailers' global sourcing patterns and practices', *Competition and Change*, **6** (3), 289–308.
Gibbon, P., B. Jennifer and P. Stefano (2008), 'Governing global value chains: an introduction', *Economy and Society*, **37** (3), 315–38.
Graz, J.C. and A. Nölke (eds) (2008), *Transnational Private Governance and its Limits*, London: Routledge.
Hadjimichalis, C. (1984), 'The geographical transfer of value: notes on the spatiality of capitalism', *Environment and Planning D: Society and Space*, **2** (3), 329–45.
Hadjimichalis, C. (1987), *Uneven Development and regionalism: State, Territory and Class in Southern Europe*, London: Croom Helm.
Hanssens, H., B. Derudder and F. Witlox (2012), 'Managing organizational and geographical complexity: The "positionality" of advanced producer services in the globalizing economies of metropolitan regions', *Erdkunde*, **66** (1), 45–55.
Harvey, D. (1985), *The Urbanization of Capital*, Oxford: Blackwell.
Henderson, J., P. Dicken, M. Hess, N. Coe and H. Yeung (2002), 'Global production networks and the analysis of economic development', *Review of International Political Economy*, **9** (3), 436–64.
Hopkins, T. and I. Wallerstein (1986), 'Commodity chains in the world-economy prior to 1800', *Review*, **10** (1), 157–70.
Hymer, S. (1972), 'The mulitnational corporation and the law of uneven development', in J. Bhagwati (ed.), *Economics and World Order*, New York: Free Press, pp. 113–140.
Institute for Global Law and Policy (IGLP) Law and Global Production Working Group (2016), 'The role of law in global value chains: a research manifesto', *London Review of International Law*, **4** (1), 57–79.
Instituto Nacional de Estadística y Geografía (2016), 'Censos Económicos 2014 Resultados Definitivos', accessed 11 June 2018 at http://www.inegi.org.mx/est/contenidos/proyectos/ce/ce2014/.
Jacobs, J. (1970), *The Economy of Cities*, New York: Vintage.
Jacobs, J. (1984). *Cities and the Wealth of Nations: Principles of Economic Life*, New York: Vintage Books.
Jacobs, W. (2014), 'Rotterdam and Amsterdam as trading places? In search of the economic-geographical nexus between commodity chains and world cities', *Tijdschrift voor de Economische en Sociale Geografie*, **105** (4), 483–91.
Jacobs, W., H. Koster and P. Hall (2011), 'The location and global network structure of maritime advanced producer services', *Urban Studies*, **48** (13), 2749–69.
Jones, A. (2002), 'The "global city" misconceived: the myth of "global management" in transnational service firms', *Geoforum*, **33** (3), 335–50.
Kandell, J. (1990), *La Capital. The Biography of Mexico City*, New York: Henry Hold.
Kaplinsky, R. (2000), 'Globalisation and unequalisation: what can be learned from value chain analysis?', *Journal of Development Studies*, **37** (2), 117–46.
Kaplinsky, R. and M. Morris (2002), 'A handbook for value chain research', accessed 11 June 2018 at https://www.ids.ac.uk/ids/global/pdfs/VchNov01.pdf.
Mahutga, M. (2012), 'When do value chains go global? A theory of the spatialization of global value chains', *Global Networks*, **12** (1), 1–21.
Mann K. (2007), *Slavery and the Birth of an African City: Lagos, 1760–1900*, Bloomington, IN: Indiana University Press.
Marcuse, P. and R. van Kempen (eds) (2000), *Globalizing Cities: A New Spatial Order?* Oxford: Blackwell.
Milberg, W. (2008), 'Shifting sources and uses of profits: sustaining US financialization with global value chains', *Economy and Society*, **37** (3), 420–51.
Milberg, W. and D. Winkler (2010), 'Financialization and the dynamics of offshoring in the USA', *Cambridge Journal of Economics*, **34** (3), 275–93.

Organisation for Economic Co-operation and Development (2005), *Policy Roundtables: Barriers to Entry. Directorate for Financial and Enterprise Affairs Competition Committee*. accessed 11 June 2018 at https://www.oecd.org/regreform/sectors/36344429.
Palpacuer, F. (2008), 'Bringing the social context back in: governance and wealth distribution in global commodity chains', *Economy and Society*, **37** (3), 393–419.
Parnell, S. and J. Robinson (2012), '(Re)theorizing cities from the Global South. Looking beyond neoliberalism', *Urban Geography*, **33** (4), 593–617.
Parnreiter, C. (2010), 'Global cities in global commodity chains: exploring the role of Mexico City in the geography of global economic governance', *Global Networks*, **10** (1), 35–53.
Parnreiter, C. (2011), 'Formación de la ciudad global, economía inmobiliaria y transnacionalización de espacios urbanos. El caso de Ciudad de México', *EURE*, **37** (111), 5–24.
Parnreiter, C. (2013), 'The global city tradition', in M. Acuto and W. Steele (eds), *Global City Challenges. Debating a Concept, Improving the Practice*, Houndmills: Palgrave Macmillan, pp. 15–32.
Parnreiter, C. (2014), 'Network or hierarchical relations? A plea for redirecting attention to the control functions of global cities', *Tijdschrift voor Economische en Sociale Geografie*, **105** (4), 398–411.
Parnreiter, C. (2015a), 'Managing and governing commodity chains: the role of producer service firms in the secondary global city of Hamburg', *Die Erde*, **146** (1), 1–15.
Parnreiter, C. (2015b), 'Strategic planning, the real estate economy, and the production of new spaces of centrality. The case of Mexico City', *Erdkunde*, **69** (1), 21–31, doi:10.3112/erdkunde.2015.01.02.
Parnreiter, C. (2019), 'Global cities and the geographical transfer of value', *Urban Studies*, **56** (1), 81–96.
Parnreiter, C., K. Fischer and K. Imhof (2007), 'El enlace faltante entre cadenas globales de producción y ciudades globales: el servicio financiero en Ciudad de México y Santiago de Chile', *EURE*, **33** (100), 135–48.
Peck, J. (2015), 'Cities beyond compare?', *Regional Studies*, **49** (1), 160–82.
Rabach, E. and E.M. Kim (1994), 'Where is the chain in commodity chains? The service sector nexus', in G. Gereffi and M. Korzeniewicz (eds), *Commodity Chains and Global Capitalism*, Westport, CT: Praeger, pp. 1–14.
Robinson, J. (2002), 'Global and world cities: a view from off the map', *International Journal of Urban and Regional Research*, **26** (3), 531–54.
Robinson, J. (2006), *Ordinary Cities: Between Modernity and Development*, London: Routledge.
Rossi, E., J. Beaverstock and P. Taylor (2007), 'Transaction links through cities: "Decision cities" and "service cities" in outsourcing by leading Brazilian firms', *Geoforum*, **38** (4), 628–42.
Sassen, S. (1988), *The Mobility of Labor and Capital. A Study in International Investment and Capital Flow*, Cambridge: Cambridge University Press.
Sassen, S. (1991), *The Global City. New York, London, Tokyo*, Princeton, NJ: Princeton University Press.
Sassen, S. (1998), *Globalization and its Discontents*, New York: New Press.
Sassen, S. (2001), *The Global City. New York, London, Tokyo*, 2nd edn, Princeton, NJ: Princeton University Press.
Sassen, S. (2010), 'Global inter-city networks and commodity chains: any intersections?', *Global Networks*, **10** (1), 150–63.
Sassen, S. (2016), 'The global city: enabling economic intermediation and bearing its costs', *City & Community*, **15** (2), 97–108.
Selwyn, B. (2012), 'Beyond firm-centrism: re-integrating labour and capitalism into global commodity chain analysis', *Journal of Economic Geography*, **12** (1), 205–26.
Sennett, R. (1980), *Authority*, New York: Alfred A. Knopf.
Sikkaa, P. and H. Willmott (2010), 'The dark side of transfer pricing: Its role in tax avoidance and wealth retentiveness', *Critical Perspectives on Accounting*, **21** (4), 342–56.
Smith, A., A. Rainnie, M. Dunford, J. Hardy, R. Hudson and D. Sadler (2002), 'Networks of value, commodities and regions: reworking divisions of labour in macro-regional economies', *Progress in Human Geography*, **26** (1), 41–63.
Smith, R. and M. Doel (2011), 'Questioning the theoretical basis of current global-city research: structures, networks and actor-networks', *International Journal of Urban and Regional Research*, **35** (1), 24–39.
Storper, M. (2013), *Keys to the City: How Economics, Institutions, Social Interaction, and Politics Shape Development*, Princeton, NJ: Princeton University Press.
Taylor P. (1997), 'Hierarchical tendencies amongst world cities: a global research proposal', *Cities*, **14** (6), 323–32.
Taylor P. (2001), 'Specification of the world city network', *Geographical Analysis*, **33** (2), 181–94.
Taylor P. (2004), *World City Network. A Global Urban Analysis*, London: Routledge.
Taylor P. (2011), 'The interlocking network model', in B. Derudder, M. Hoyler, P. Taylor and F. Witlox (eds), *International Handbook of Globalization and World Cities*, Cheltenham, UK and Northampton, MA, USA: Edward Elgar, pp. 51–63.
Taylor, P. and B. Derudder (2016*), World City Network: A Global Urban Analysis*, 2nd edn, New York: Routledge.

Wainwright, T. (2011), 'Tax doesn't have to be taxing: London's "onshore" finance industry and the fiscal spaces of a global crisis', *Environment and Planning A*, **43** (6), 1287–304.
Werner, M. (2016), 'Global production networks and uneven development: exploring geographies of devaluation, disinvestment, and exclusion', *Geography Compass*, **10** (11), 457–69.
Wójcik, D. (2013), 'Where governance fails. Advanced business services and the offshore world', *Progress in Human Geography*, **37** (3), 330–47.
Yeung, H. (2015), 'Regional development in the global economy: a dynamic perspective of strategic coupling in global production networks', *Regional Science Policy & Practice*, **7** (1), 1–24.
Yeung, H. and N. Coe (2015), 'Toward a dynamic theory of global production networks', *Economic Geography*, **91** (1), 29–58.

Index

3H-city triangle 11, 274, 276–8, 280, 286

abstraction 253–5
Abu-Lughod, J. 81
access networks 49–51
accessibility
  complexity of accessibility trajectories 178–9
  rank correlations for 178, 179
  social 420–21, 423
accessing and sharing information 412, 417
Achibet, M. 168
active spaces 11
activity spaces 51
  segregation and 323–4, 332–3
  ties through place 13, 196, 197, 198–9
actor interaction
  deriving networks of 403–5
  opinion pooling 389–408
adjacency matrix 128
administrative level of cities 593
advanced producer services (APS) 515, 558, 586
  centripetal wealth transfer and uneven development 618–31
  Chinese city networks 588–9, 592, 593, 595
  GaWC perspective on global-scale urban networks 604, 605–10, 612–15
  polycentrism 496, 500, 501, 503–4, 505–6
adventure 417
advertising industry 602
Aelius Aristides 88
affordances 423
Africa
  South Africa 23
  sub-Saharan 610, 611
  West African coast 627
age
  and sexual networks 352–3
  and travel 315
agent-based model of belief formation 367, 369–70, 371–87
agglomeration, industrial, *see* industrial agglomeration
agglomeration economies 16, 20, 560
aggregation 3
agrarian economy 34–5
airline networks 248–9
Alaska Natives 335, 340–41, 346, 347
Albert, R. 131, 133
Alderson, A. 234
Alexander, C. 400–402
Alexandria 98, 100
Algiers 463
Alice Springs 249
Alinsky, S. 296
Alonso, W. 255–7
alpha world cities 79, 80
Amaral, L.A.N. 131, 133
Amazon 473
American Indians 335, 340–41, 346, 347
Amin, A. 207, 513
Anchorage 245
ancient urbanism, *see* Roman urbanism
Andersson, C. 31
Anthropocene 36
anthropological framework of comparison 63–4
Antioch 98
antiretroviral therapy (ART) 353–4
anti-vaccination beliefs 369
Antwerp 265, 266, 452
appreciative city 276–7
archaeological material 96
Archaic nomads 65, 66
archipelago metaphor 264
area of city block faces 138–40, 141, 145, 147
  typology of planar graphs 140–44
area X density method 96–7
aristocracy 112
Aroian, K.J. 429
Asia
  Central 73, 610–12
  East 71
  Pacific 607, 608
  South 608, 609
Asian Americans 334–47
Association of Community Organizations for Reform Now (ACORN) 298
associational life 107–25
  American studies 108–10
  British studies 110–12
  Newcastle 113–24
assortativity 245, 339
Athens 142–3
attractiveness of cities 514, 530–31
author citation analysis 203–7

average closeness centrality in time 176–7
average degree 128–9, 155, 338, 344
average increase in connectivity per person 45
average shortest path 133–5
avocational associations 13, 196, 197, 198

backwash effects 262, 263
Ban, Y. 169
Baoding 550, 551
Baptiste, H. 170
Barfield, T. 73
Barthelemy, M. 129, 130, 136, 148, 151, 168
Bathelt, H. 513
Bearman, P.S. 354
Beauregard, R.A. 512–13
Beaverstock, J.V. 79, 80, 601
Beckfield, J. 234
Beijing 529, 590–91, 592, 593
  city-dyad with Shanghai 612–13, 614
  intraurban linkages 531, 532, 543–51, 552
    local firms' activities 548, 549, 550
    LUR polycentricity 548–51
    network components of local linkages 543–8
Belgium 265, 266
beliefs
  agent-based model of belief formation 367, 369–70, 371–87
  dissemination of health information and 366–88
Belt and Road initiative 595–6
Berger, T. 167
Berry, B.J. 3, 21, 25–6, 498
Bess, K.D. 297
beta centrality 221, 222–3
beta world cities 79, 80
Bettencourt, L.M. 527
betweenness centrality (BC) 244–5, 248, 265, 266
  container shipping networks 461, 462–3, 465
  macro-comparative social network analysis 220, 222, 223
  road networks 148–51
biases 208
  confirmation bias 377, 386–7
Biello, K.B. 358
big data 286
  socio-material network analyses 198–9, 208
Bigotte, J.F. 169–70
bijection approach 144
biomed sorting 353–4
Black and African Americans
  segregation and mobility 334–47
  sexual networks and health 351, 352, 357
Blackwell, C. 419

blasé attitude 418
block models 224, 231–2
blocks 49–51
  shape and area 138–40, 141, 145–8
  typology of planar graphs 140–44
Blokland, T. 201, 207
Blokland-Potters, T. 195
Blumenfeld-Lieberthal, E. 170
Bohai Rim region, China 592, 593–4
Bonacich, P. 221
Bonnafous, A. 166
Borgatti, S. 221–2
Boston 109, 531
Botero, G. 23–4
bottom-up perspective 609–10
bottom-up processes 513
boundary specification problem 4
Bourdieu, P. 207
Bouttier, J. 144
Brambles 539, 540
Brelsford, C. 49–50
Bretagnolle, A. 167, 184
BRICs countries 526, 527, 528, 552
Browning, C. 203
Brussels 265, 266, 612
budget constraint 56
Buenos Aires 142–3
Buhl, J. 128, 129
Burgess, E.W. 166
Burt, R. 207, 583
business regionalism 113–14
Butler, T. 207
Butts, C.T. 3

calculative network capital 499
California
  Southern California power grid 131, 133
  urban mobility and segregation 334–47
California Household Travel Survey (CHTS) 334–5, 336
Calthorpe, P. 283–4
Camden, New Jersey 296
Canada 409
capability theory 279
capital cities 18
capitalism 76–8
  contradiction of industrial capitalism 558
capitalist city-states 74, 77–8
Cardillo, A. 128, 129, 135, 136
Carrasco, J.A. 333
Castells, M. 240, 496, 585–6
Cattell, V. 200
causality regimes 180–82, 182–4
cellular automatons 169–70
census data 209, 321

Center for Popular Democracy 298
Centner, R. 322
Central America 611, 612
Central Asia 610–12
  steppe nomads 73
central place theory (CPT) 18–22, 53, 254, 257–61, 498
centrality 252–71
  abstraction 253–5
  centre–periphery theory 261–3
  FDI and inequality 561–2, 562–3, 567–79, 583–4
  global shipping network 455–6, 461–4, 465
  and land-use theory 255–7
  macro-comparative social network analysis 219–23, 234
  proximity and urban network structures 244–6, 248–9
  urban network theory 264–7
centrifugal tendencies 252
centripetal tendencies 252
centripetal wealth transfer 618–31
Certeau, M. de 282
chaining 227
Changsha 504, 505
Chapron, P. 32–3
Chase-Dunn, C. 64, 65, 71, 73, 75
Chaucer Capital Investments 538, 540
Chaudhuri, G. 168
Chebyshev distance 241
Chen, W. 169, 429
Chengde 550, 551
Chengdu 592
Chesapeake region 71
Chicago 81, 322, 515
  community organizing and interorganizational network changes 293–312
  Right on Justice (ROJ) Network 299–311
  Southwest Organizing Project (SWOP) 293, 299
Chicago School 110, 196
chiefdoms 68–9, 71–2
China 75, 273, 609–10
  Belt and Road initiative 595–6
  Bohai Rim region 592, 593–4
  multinational firms' networks 526, 527, 528
  MYR city region 494, 495, 498, 501–2, 503, 504–6
  Pearl River Delta region 409, 592, 593–4
  Shanghai–Beijing city dyad 612–13, 614
  see also under individual cities
China Securities Regulatory Commission (CSRC) 593
Chinese Americans 370–86
Chinese city network research 585–600
  data and methods 586–90
  major contributions 594–5
  network-building methods 587, 588–9
  patterns and driving forces 590–94
    global scale 590–91
    national scale 591–3
    regional city networks 593–4
Choi, A.S.K. 429
Chongqing 592
Christaller, W. 19–20, 21, 23, 26, 257–61, 553
Christens, B.D. 295, 296
citations, quantitative analysis of 203–7
cities4people 284–5
citizenship 110–11
city blocks, *see* blocks
city-dyad analysis 590
city-dyad connectivity 604–5
  changing the city–dyad specification in the INM 612–15
city lights 81
city of opportunity 11, 273–92
  cities as sources of opportunities 277–9
  microcosmic city 274–7, 280
  public space and cultural urbanism 282–4
  shared spaces and networks 280–82
  social interactions and networks 284–5
  urban analytics 285–6
  urban management dashboards 287–8
city planning 389–408
  applications of the Markovian framework 396–402
  deriving networks of actor interaction 403–5
city regions 81–3
  economic development in inter-city networks 495–6
  emergence processes in network space 502–6
  Mid-Yangtze River (MYR) 494, 495, 498, 501–2, 503, 504–6
  polycentrism 493–510
    implications 506–7
  South East England 494–5, 498, 501, 502–3, 504, 506
city size distribution 26–7, 52–3, 77–8, 82
Civil Rights Movement 331
Clark, K. 168
class, social 112, 240
classification of nation-states 233
clergy 118, 120, 122
Cloke, P. 278
closed-shop principle 623–4
closeness centrality 245, 264–5, 266
  average closeness centralities in time 176–7
  FDI and inequality 562, 563, 567, 570, 571, 573, 575, 576, 577, 578, 579, 584

macro-comparative social network analysis 220, 222, 223
cluster analysis 562–4, 567–75
clustering 247, 248
hierarchical 224–5, 226–32
industrial agglomeration 472–92
modelling dissemination of health information and beliefs 375–7, 378, 386
clustering coefficient 128, 136–7
container shipping networks 461, 462, 463, 465
FDI and inequality 562, 563, 567–8, 569, 570–73, 577, 578, 583–4
urban mobility and segregation 338, 341, 344
co-agglomeration index 477–8
co-agglomeration network 475–6, 481–8, 490
community structure in 484–5, 486
coalitions 296–7
Chicago's ROJ Network 299–311
regional 298
co-evolution
inferring from comparative socio-economic change 27–9
sedentary–nomadic 72–3
urban systems, complexity and 25–33
co-evolution models 166–93
application to French system of cities 182–5, 186, 187
distance-based network growth specification 173–4
family of 172–5
generic formulation 172–3
microscopic and mesoscopic scales 168–70
model with a physical network 185–8, 189
synthetic system of cities 176–82
urban systems modelling 170–71
cognitive overload 44–5
cognitive proximity 241–2, 243
Colby, C.C. 252, 261, 263
Coleman, C. 207
Coleman, J.S. 405
Colley, A. 413, 417
colonization 22–3
commercial/economic FCAs 116–17, 121
community
defining 195–8
typology of ties that form 196–8
community centres 372, 373
community detection
Markov stability framework and industrial agglomeration 475, 483–5, 486
socio-material network analyses 194–214
community organizing 293–312
Chicago's Right on Justice (ROJ) Network 299–311
cyclical process 294
interorganizational networks 296–7
interpersonal networks 294–5
interurban networks 297–8
companionship 428, 433–44
comparative evolutionary world-systems perspective 64–5
competition, interurban 23–5
complementary regions (*umlands*) 257–61
complex network analysis (CNA) 589–90
complex networks 41–61
intercity flows 42–3, 51–4
socio-economic networks in the city 42, 43–9
urban infrastructure networks 42, 43, 49–51
complexity
of accessibility trajectories 178–9
urban systems, co-evolution and 25–33
computer-mediated communications 429–30
computing
impact of locative media 413–14
ubiquitous 410–11
concurrent sexual partnerships 355
confirmation bias 374, 375, 377, 386–7
confirmatory path analysis (CPA) 564, 578–9
conflict-based organizing 296, 306–8
conflict resolution 389–408
confrontation 306–8
connectivity
city-dyad 604–5, 612–15
and distance 241
latent and associational life 120–22, 123
socio-economic networks 43–9
urban mobility and segregation 337–8, 341, 344
consensus organizing 296, 307–8
constraints 56
container shipping 449–71
evolution of the network 455–60
global urban network of flows 452–60
urban centrality in the network 461–4, 465, 469
containerization 450–51
continuous coreness 223
conurbation 24
co-presence support 428, 433–44
Copus, A.K. 263
core and branch structure 152–6
core and periphery 261–3, 626
discrete and continous structures 222
macro-comparative social network analysis 215, 216–18, 221–3, 233
settlement networks 12, 65

core–periphery differentiation 65, 73
core–periphery hierarchy 65, 73
coreness 221–3
correspondence analysis (CA) 224, 227–32
cost–benefit analysis 157–62
  application to subways 158–62
cost-sharing 472–92
Cottineau, C. 32–3, 170
coverage of a subway system 158–9
Covid-19 pandemic 36
crises 111–12
critical realism 601
critical success factors (CSFs) 287
cross-industry patent citations 478–9, 482
crowds, wisdom of 396
Crown Mint 627
Cullercoats artists' colony 113
cultural ecology school 278
cultural/recreational FCAs 116–17, 121
cultural urbanism 282–4
Curitiba, Brazil 83
customer/supplier cost-sharing 472–92
  industrial clusters' preferences 485–7, 490

dashboards 286, 287–8
data analytics 285–6
dating apps 413, 419
Gay, G. 195
de Souza e Silva, A. 420, 421
De Vries, B. 110
defence 17–18
degree centrality 245, 264–5, 266, 605
  container shipping networks 461, 463, 465
  macro-comparative social network analysis 219–20, 222, 223
degree disassortativity 354
degree distribution 127
  non-scale-free degree distributions 129–33, 134
demand 105
demographics of neighbourhood ties 201–2, 202–3
densification 149, 150
density 338, 341, 344
Denton, N.A. 316
dependency theory 216, 560
Derudder, B. 590, 626
design methods 406–7
desirability scores 396–8, 400–402, 403
determinism 278–9
Detroit 245, 512
diameter, network 338, 341, 344
Dicken, P. 517
Dickinson, R. 19
diet 319
diffusion of innovations 28–9
digital technology 14, 239–40
  locative media, *see* locative media
  online communities 13, 196, 197, 198
  sexual networks 357
  smart cities 275–6
  social interactions 284–5
Dimaggio, P. 109
Ding, R. 169
Diodato, D. 474, 475, 477, 478, 479–81, 490
direct network-building 587, 589
disassortative mixing 352–4, 356
discrete core/periphery structures 221–2
dissemination of health information and beliefs 366–88
distance
  measuring 241
  role in interurban networks 239–51
    length of links and 246–8
distance-based network growth 173–4
distance buffers 369, 376, 381–4, 385, 386
diversification 474, 476–7, 490, 514–15
diversity 2
divided opinions 393
Djellal, F. 513
Dollinger, P. 626
dominance 261–2, 263
double-embeddedness 195
Drennan, R.D. 102
Dresden road network 130
dual graphs 50, 131
dual–primal problem 403–5
Dubai 242, 608
Dubai Ports World (DPW) 451
Ducruet, C. 451, 452, 453, 455–6

East Asia 71
Eastern Europe 608
eccentricity 461, 463, 465
economic co-evolution models 169
economic/commercial FCAs 116–17, 121
economic development
  ancient urbanism 104–5
  city regions in interurban networks 495–6
  uneven, *see* uneven development
economic globalization 493–4
  city networks in 495–8
economic governance 619, 620–27
economic growth 44
  city regions 495–6
  GDP growth 577, 578, 579
economic rebalancing 507
economic segregation 324
economic urban networks 4–6
economies of overview 513

edge lists 336
edge weights 337
edges 2–3, 264
education 479
  labour pooling and level of 487–8, 489, 490
ego name generators 208
eigenvector centrality 221, 222–3, 245–6
elites 109
Ellison, G. 473, 474–5, 477–8, 479–80
embedded governance 619, 621, 624
emergent properties 30
emotional social support 428, 433–44
Empire of the Great Khan 18
empires
  power and size 73–6
  growth and decline sequences 75
Enflo, K. 167
entertainment structures 96
entropy of populations 177–8
entry barriers 625
environmental determinism 278–9
environmental transition 35, 36
epidemic processes 46–8
eponymous relationships 13, 196, 197, 198, 199–201
equivalence criteria 224, 225–32, 233
Erasmus students 427, 431–44
Erdös–Rényi graphs 128, 130, 136
ethnicity, *see* race/ethnicity
Euclidean (geodesic) distance 241
Europe 74
  Eastern 608
  urbanism from AD 800 to AD 1900 104
  western 607, 608
  world city system centred on 78
European Commission workers' partners 427, 431–44
European Union (EU) 494–5, 496–7
Everett, M.P. 221–2
evolution mechanisms of networks 596
evolutionary population dynamics 56
evolutionary theory of urban systems 26–36, 167
  reconstructing evolutionary trajectories 29–33
exchanges of capital 4
exclusion
  social 314–16
  spatial 316–18
  see also residential segregation
exploration 149, 150
exploratory spatial data analysis (ESDA) 337, 340–41
extended city-regions 453–66, 469
Fairview New Home 538, 540
Faith in Action National Network (FIA) 297
familiar strangers 13, 196–7, 197–8, 198–9
familiarity 421
Farman, J. 410–11
Faulconbridge, J.R. 602
Faust, K. 207, 223
fDi Markets data 560–61
Feagin, J.R. 321–2
Feinberg, M.E. 296–7
Felixstowe 451
field observations 208–9
filtering by locative media 419–20
financial firms 533, 620–21
financialization 620
First Rail Group 535, 536
Fischer, C.S. 48, 207, 428, 442–3
flamenco artists 427, 431–44
Flammini, A. 129, 130, 136, 168
Flap, H. 202, 203, 356
Fleming, D.K. 462–3
Fletcher, R. 76, 82
flood plain settlements 71–2
flow networks
  interurban 42–3, 51–4
  methodologies for investigating 499–501
  polycentrism 593–510
food buying 320
food-deserts literature 319, 320
foreign direct investment (FDI) 557–84
  cluster analysis 562–4, 567–75
    city level 567–70, 571
    country level 570–75
  GIS maps of investment networks 564–7
    city level 564–5
    country level 565–6
  greenfield 560–61
  structural equation model 564, 576–9
forgetting time 387
former colonies 22–3
fortification 17–18
Foursquare City Guide 417
fragmented city 416–17
France 18, 24–5
  Marseilles-Aix en Provence metropolis 553
  Paris, *see* Paris
  railway network 167, 168, 172, 189
    co-evolution model applied to 182–5, 186, 187
Francis, J. 200–201
Frank, A. 626
French Revolution 18
Friedmann, J. 79, 255, 263, 500, 558, 603
friends 428, 434, 435, 438, 440, 443
Frith, J. 420

Fujita, M. 170–71
functional communicative associations (FCAs) 107–8, 114–24
functional polycentrism 497–8, 502–6
functional urban areas (FUAs) 516
fundamental urban processes 41–61
future city of humans 82–3

Galen 95–6
Galster, G. 207
Galton, F. 396
Galtung, J. 217
Gamaliel Foundation 297–8
gaming, mobile 412–13
gamma world cities 79, 80
GDP growth 577, 578, 579
Geddes, P. 24
gender, and travel 314–15
general systems theory 53
Genoa 77
geodesic (Euclidean) distance 241
geodesign 406
geographic information systems (GIS) 336, 406
  maps of FDI networks 564–6
  maps of network measures 567
geographical (spatial) proximity 243
geographical scales 9–10, 11
geographical transfer of value (GTV) 625–7
geometrical co-evolution models 168–9
Gereffi, G. 620
Germany's Rhine-Ruhr region 493, 610
Gibrat, R. 26–7, 28–9
Gini coefficient 562–3, 567–75, 577
Glaeser, E. 477
Glick, S.N. 355
global cities, *see* world/global cities
global commodity chains (GCCs) 618–31
  geographical transfer of value 625–7
  governance of 620–27
global financial crisis 451
global/local interactions
  multinational firms' ownership linkages 519–31
  urban processes 512–16
Global Navigation Satellite Systems (GNSS) 241
global network connectivity (GNC) 264, 605
  geographies of 606–9
  NGOs 610–12
global production networks (GPNs) 240, 499–500, 620–27
global scale 4, 5–6, 12
  Chinese cities in the WCN 585, 590–91, 594–5
  GaWC perspective on global-scale urban networks 601–17
  global urban networks in daily life 239–40
global settlement system 79–81
Global South 82, 83
global value chains (GVCs) 620–27
global warming 82
globalization 493–4
  city networks in a globalizing world economy 495–8
  and inequality 559–60
  polycentric urban regions 515–16
  scales of global processes 516–18
Globalization and World Cities (GaWC) Research Group 586
  perspective on global-scale urban networks 601–17
    interlocking network model 602, 604–15
    research context 603–4
Glückler, J. 513
GMP 161
Golledge, R.G. 21
goods, flows of 53–4
Google 422
Google Scholar 194
Gottmann, J. 25
Goulias, K. 316
governance 83
  embedded 619, 621, 624
  global commodity chains 620–27
GPS technology 323–4, 410
  navigation 411–12, 417
Granovetter, M. 207, 442
graph edit distance (GED) 141
gravity model 587, 588
Greater Toronto and Hamilton Area (GTHA), Canada 409
Greek urbanism 104
greenfield FDI 560–61
Grindr 413, 421
Groane region, Italy 144, 145–8, 150
Gromov-Hausdorff distance 141
group identity 366–88
  strength of 374, 378–80
group size 380–81, 385
growth and decline sequences 75
Guangzhou 591, 592, 593
Guimera, R. 245

Hachette Group 535, 537
Hadjimichalis, C. 625, 626
Hägerstrand, T. 28
Haidian Science and Technology 545, 546
Hall, P.D. 109
Hall, P.G. 497, 515

Hall, T.D. 64, 65, 71, 73
Hamburg 452
Hanna, M.G. 294–5
Hanseatic League 22, 626
Hansen, M.H. 104
Hanson, J.W. 91–2
happiness 276–7; *see also* 3H-city triangle
Haraldsdottir, S. 354
harbours 93
Harris, C.D. 262, 497, 498
Harvey, D. 253
Haussmann works in Paris 144–5, 148, 150, 151
Hayuth, Y. 462–3
health 276–7
  dissemination of health information and beliefs 366–88
  equity for urban ethnic minority populations 370–71
  impact of urban social life 350–65
  residential segregation and the distribution of services 319–20
  understanding group differences in health outcomes 368–70
  see also 3H-city triangle
Hecksher–Ohlin model 559
Herculaneum 98
heterogeneous forces 480–81
  industrial clusters and heterogeneous dependence on Marshallian channels 481–8, 489
heterophily 352–4, 356
Hickman, P. 201
hierarchical clustering 224–5, 226–32
hierarchical diffusion of innovation 28–9
hierarchies 552–3
  city infrastructure networks 50
  settlement-size 26–7, 28–9, 68–70
  street networks 50
  urban, *see* urban hierarchies
Hidalgo, C.A. 476–7
hilly flanks 68
hinterlands 99
Hipp, J. 202, 207
Hispanic Americans 334–47
historical contingency 55–6
historical materialism 110
HIV/AIDS 350–65
Hołyst, J.A. 131
home 276–7; *see also* 3H-city triangle
homo urbanus 273–4
homogeneous forces 479–80
homophily 246
  sexual networks 352, 356
Hong Kong 242, 529, 531, 590–91
horticulture 68, 70–71
household characteristics, and travel 316
hub-and-spoke highway systems 321
hubs 127–8, 247–8
human ecology perspective 44
Hutchinson Whampoa (HPH) 451

idea cost-sharing 472–92
ideal range 258, 259
immigrants
  enclaves 317
  social support networks 427–47
  status and travel 315, 317
impactful collaborations 308, 309, 310
impedance 241
implicit network 17–18
incarceration 357
inclusion 4
income, and travel 316
income inequality 559–60
  FDI and 560–80
indifference 418
individualism, networked 409, 415–23
industrial agglomeration 472–92
  agglomeration drivers 475–6
  agglomeration patterns 472–3
  industrial growth and diversification models 476–7
  industry similarity and networks 473–4
  Marshallian channels 472–90
    heterogeneous dependence on 481–8, 489
Industrial Areas Foundation (IAF) 297–8
industrial capitalism, contradiction of 558
Industrial Revolution 34–5, 196
inequality
  globalization and 559–60
  imposition through interscalar networks 557–84
  residential segregation and spatial inequality 317–21
information
  accessing and sharing using locative media 412, 417
  dissemination of health information and beliefs 366–88
  flows of 53–4
information and communication technologies (ICTs) 415–16
informational social support 428, 433–44
infrastructure networks 42, 43, 49–51, 587, 588
  evolution of 144–56
  residential segregation and transportation infrastructure 320–21
  see also under individual types of network
innovation ability 561–2

innovations, diffusion of 28–9
Inoue, H. 76
input–output linkages 478, 482, 487
institutional proximity 242, 243
institutional ties 13, 196, 197, 198, 201
instrumental social support 428, 433–44
insurance firms 533
intent 414
inter-city networks, *see* interurban networks
Intercontinental Exchange Group (ICE) 537, 540
interdependencies 43
inter-firm linkages 587, 588, 595, 596, 603
intergovernmental organizations 235
inter-group conflict 366, 369, 373–87
interlocking network model (INM) 8, 500, 586, 587, 588–9, 621
  advanced producer services, *see* advanced producer services (APS)
  GaWC perspective on global-scale urban networks 602, 604–15
    extensions to the model 609–12
    geographies of global network connectivity 606–9
    model operationalization 605–6
    model specification 604–5
    modifications of the model 612–15
intermediacy 462–3
intermediate social support 428, 434–6
International Monetary Fund 82
international nongovernmental organizations 235
international organizations 235
International Power Limited 535, 536
interorganizational networks 296–7
  changes in Chicago's ROJ network 299–311
interpersonal relationships 294–5
interstation distance 159–61
interurban competition 23–5
interurban interactions 29–33
interurban networks
  centrality 264–7
  Chinese cities 585–600
  city-region economic development in 495–6
  community organizing 297–8
  evolutionary theory of systems of cities 34–5
  flow networks, *see* flow networks
  multinational firms' intraurban networks and 511–56
  role of proximity and distance 239–51
interviews 207–8
intra-firm linkages 587, 588, 595; *see also* multinational firms
intra-group cohesion 366, 369, 373–87

irrigation 68
isolated state 19
Italy
  Groane region 144, 145–8, 150
  power grid 131, 132
Ito, K. 113

Jacobs, J. 44, 55, 107, 503, 621
Janulis, P. 357
Japan 75, 566
  energy companies 249
  Tokyo, *see* Tokyo
Japanese flamenco artists 427, 431–44
Jeffrey, P. 131, 134
Jennings, J.M. 357
Jim Crow laws 331–2
Johannesburg 608
Jones, C. 44
judicial/political FCAs 116–17, 121

Kalapala, V. 131
Kalmijn, M. 356
Katifori, E. 144
Kegler, M.C. 297
key performance indicators (KPIs) 287
Kick, E. 217, 225
Kim, S. 218
Knopp, L. 418
knowledge economy 34–5, 496, 497, 498
knowledge spillovers 472–92
Kohl, J.G. 22
Kourtit, K. 275, 276, 280, 281, 284
Krätke, S. 245
Kretzschmar, M. 355
Krivitsky, P.N. 355
Kurant, M. 135
Kwan, M.-P. 321

labour division 557–8
labour pooling 472–92
  industry clusters' preferences 485–7, 490
    and workers' education level 487–8, 489, 490
labour relations 623–5, 625–6
lagged correlations 180–82, 182–4
Lagos 627
Lai, K. 612–13, 614
Lalanne, L. 23
Lämmer, S. 130
land rent 19
land-use theory 255–7
Land-use Transport Interaction (LUTI) models 171
Langfang 548–50, 551
large urban regions (LURs) 515–16, 518, 519

local linkages inside 531–51
Beijing 531, 532, 543–51, 552
London 531, 532–43, 548, 552
multinational firms' ownership linkages 521–31
Larsen, J. 200
latent connectivity 120–22, 123
Latin America 608, 609
Lattimore, O. 73
Laumann, E.O. 354
law firms 623–4
Lefebvre, H. 278
Lepetit, B. 18
Leshed, G. 417
Leszczynski, A. 411
Levant, the 64, 68, 70
levels 2, 3, 6–9
city in inter-level interactions 9
economic urban networks 4–5, 5–6
multi-level approaches 10–14
non-nested levels 8–9
and scales of multinational firms' networks 518–19
see also under individual levels
Levinson, D. 157, 169
Lewis, W.A. 559
Li, Y. 187
Licoppe, C. 420, 423
lines, number of in subways 161, 162
links/ties
centrality of 248–9
length of and distance on a network 246–8
number of 145, 146
proximity and urban network structures 243–4
social support networks for mobile individuals 427–47
classification of ties 434–6
ties through place 12–13, 194–214
Literary and Philosophical Society of Newcastle upon Tyne 109, 113, 114
Liverpool 512
Lloyd, P. 223
Lo Cascio, E. 101
local capabilities 476
local/global interactions
multinational firms' ownership linkages 519–31
urban processes 512–16
local scale 4, 5–6, 12
Chinese regional city networks 593–4
locative media 409–26
everyday uses in urban space 411–13
impact on computing 413–14
implications on city and citizen 416
implications on relationships with the city 416–17
implications on urban sociality 417–22
networked individualism in the city 409, 415–23
ubiquitous computing 410–11
Logan, J.R. 317–18
London 24, 513, 529, 531, 612, 613
intraurban linkages 531, 532–43, 548, 552
local firm activities 540–42
LUR polycentricity 542–3
network components of local linkages 533–40
locative media 416, 417
South East England city region 494–5, 498, 501, 502–3, 504, 506
world city 77, 79, 218, 242, 522, 626
London Gateway Project 451
Los Angeles 81, 334
Los Angeles County 334–47
Lösch, A. 21, 257
Louf, R. 159, 160, 161
Louvain's method for community detection 483–4
LSE Library Database 194
Lucas, R. 44
Lüthi, S. 609
Lynch, K. 240

Ma, H. 588
Mackinder, H. 24
macro level 3, 5–6, 8, 9, 11, 12
macro–micro shift of locative media 414
multinational firms' networks 518–19
macro-comparative social network analysis 215–38
methods 219–32
centrality 219–23, 234
roles and positions 223–32
nature of 216–19
state of the field 233–5
macroscopic geographies 255
Magnasco, M.O. 144
Mahutga, M. 218, 234
maize cultivation 71
majority opinions 378–84, 385, 386
Manheim, M.L. 400–402
Mapping L.A. 336
Marcuse, P. 316
maritime transport, *see* container shipping
MARIUS model 32–3
market economy 34–5
Markov process 394–407
applications to city planning 396–402

Markov stability framework for community detection 475, 483–5, 486
Marnot, B. 451
Marsaxlokk 463
Marseilles-Aix en Provence metropolis 553
Marshall, A. 16, 44, 472
Marshallian channels 472–90
  dependence of industry co-location on 479–88
    heterogeneous dependence 481–8, 489
  relative importance 474–5
Marxism 19
Mason, C.L. 419
Massey, D. 282, 283, 316, 513
Masucci, A. 135
Maya Jariego, I. 429
Mayhew, H. 112
McCarty, C. 432
McCullough, M. 413–14
McKenzie, R. 261, 263
McLean, M. 450
McNeill, J. 71
McNeill, W.H. 71
McPherson, J.M. 112
McPherson, M. 207
measurement error 357–9
Measuring Community in an Urban Age (MCiUA) study 194–214
medical service 19–20
megacities 273
mega-city region economic expansion 503, 505–6
mega-city regions (MCRs) 497, 515, 610
mega-city regions of proximate cities 503, 505
Meijers, E. 499
men who have sex with men (MSM) 351–9
Mercer, C.H. 355
merger and acquisition (M&A) deals 501–2, 504–6
meso level 3, 8–9, 11, 12
  economic urban networks 4–5, 5–6
  multinational firms' networks 518–19
Mesopotamian world-system 72
mesoscopic scale 168–70
Mexico 622–4, 626–7
Mexico City 608, 622, 623, 626–7
Meyer, T. 282–3
micro level 3, 7–8, 9, 11, 12, 33–4
  economic urban networks 4–5, 5–6
  shift from macro to micro by locative media 414
  multinational firms' networks 518–19
microcosmic city 274–7, 280
microscopic scale 168–70
middle class 112
Mid-Yangtze River (MYR) city region 494, 495, 498, 501–2, 503, 504–6
migration 52–3, 99–100
  immigrants' social support networks 427–47
  nomadic migration circuits 65–6
  selective 443
  see also mobility
Milan 245
Mileyko, Y.H. 144
Milgram, S. 420
military superiority 73
Milne, G.J. 113–14
minority opinions 378–84, 385, 386
Mitsui 535
mobile applications (apps) 357
  locative media, *see* locative media
mobile gaming 412–13
mobile phones 421–2
mobility 603
  networked travel activity, urban mobility and segregation 331–49
  racial/ethnic residential segregation and 313–30
  and social support 429
  social support networks of immigrants 427–47
  and technology 429–30
modelling, abstraction and 254
Modelski, G. 73, 75–6
modern world city network 76–8
modernization theory 216
modifiable areal unit problem (MAUP) 3
Mogadishu 142–3
Mogi 412–13
money, flows of 53–4
monogamy 354–5
monuments 91
Moore, J. 200
morphological urban areas (MUAs) 516, 542
Morris, M. 355
Morris, R.J. 110, 111–12
Moscow 264, 529
Moulaert, F. 513
multi-agents systems (MAS) 29–33
multi-level approaches 10–14
  evolution of urban systems 33–5
multinational firms 511–56
  disentangling network levels and scales 518–19
  local/global urban processes 512–16
  local linkages inside large urban regions 531–51
    Beijing 531, 532, 543–51, 552
    London 531, 532–43, 548, 552
  networks of ownership linkages 519–31

evaluation of networks inside and between cities 519–21
evolution of cities' network scopes 527, 528
evolution of interurban networks 522–7
interrelations between intraurban and interurban linkages 528–31
network construction 521–2, 523
scales of global processes 516–18
multiplex social support 428, 432, 434–7, 441, 444
multiscale approaches 10–14
Chinese city network research 590–94
community organizing 298–9
Mumbai 608
mundanity 200–201
musicians 427, 431–44
Myrdal, G. 262, 263

name generators 208–9, 428, 432, 443
Nanchang 504, 505
Nast, J. 201
national scales
Chinese national city network 591–3, 594
multinational firms' networks 516–17
Native Hawaiians 335, 340–41, 346, 347
Natufian culture 64
Natural History Society of Northumbria 112, 113
Neal, Z. 5, 108, 612–13, 614–15
neighbourhood effects 48
neighbourhoods 143–4
characteristics and sexual networks 356–7
ties through place 13, 196, 197, 198, 201–3
urban mobility and segregation 335–6
neighbourhood-interaction potential 338–40, 341, 344, 345
neoclassical theory 560
neo-determinism 278
neoliberalism 79–81
Neolithic revolution 16
nested levels 8
Netherlands, the 110
Randstad region 493, 610
network-building methods 587, 588–9
network compositional features 351–4
network flow-tracking systems 507
network metaphor 216
networked individualism 409, 415–23
neural nets 393–4, 407
new economic geography 499, 560
New Orleans 142–3
New Urban World 273
New York 531, 612, 613, 626
advertising industry 602
dissemination of health information and beliefs 370–86
road network 139–40, 141, 144
social accessibility 420
world city 77, 79, 81, 218, 242, 522, 529, 626
Newbold, B.K. 315
Newcastle 107–8, 109
associational life 113–24
data structure 116–20
latent connectivity 120–22, 123
previous studies 113–14
Newcastle Antiquarian Society 112, 113
Nielsen, T.H. 104
Nijkamp, P. 279
Nissen, H.J. 69, 71
nodality 25
nodalization 265, 266
node partitions 475, 483–4, 486
nodes 3, 264
number of 145, 146
subway systems 155
nomadic polities of foragers 63–4, 65–7
nomadic–sedentary co-evolution 72–3
non-geographical scales 10
non-governmental organizations (NGOs) 610–12
non-nested geographical scales 10
non-nested levels 8–9
non-profit organizations 301, 303
non-random sorting 613–15
non-scale-free degree distributions 129–33, 134
North America 607, 608
power grid 131, 133
see also United States of America (USA)
Nuceria 96

occupation similarity 478, 482
occupational diversity 46
Ockham's razor 254
Offner, J.-M. 166
Oldenburg, R. 197
one-to-one meetings 294–5
online communities 13, 196, 197, 198
OpenMOLE simulation platform 32, 33, 176
opinion leaders 386
opinion pooling 389–408
applications of the Markovian framework 396–402
deriving networks of actor interaction 403–5
linear 390–94
opportunity, city of 273–92
oral health care 366–88
Orange County 334–47
ORBIS database 520
orchestral musicians 427, 431–44

organicist metaphor 24–5
organizational proximity 242, 243
organized complexity 55
organizing principles 294
Östh, J. 324
Ostia 98
outcast segregation 316, 317
overlay analysis 400
ownership linkages of multinational firms 519–31
Özkul, D. 416, 417

Pacific Asia 607, 608
Pacific Islanders 335, 340–41, 346, 347
Pain, K. 497, 515
Paleo nomads 65, 66
Pan, F. 588, 595
Panopeus 90
Pareto fronts 185, 186
Paris 218, 245, 522, 529, 531, 610
  infrastructure network evolution 144–5, 148, 150, 151
  subway system 153, 159
Park, R. 44
participation lists 114–17
partitions 475, 483–4, 486
Pastor, M. 298
patent citations 478–9, 482
path dependence 593
Pausanias 88, 90
Pavé, A. 3
Pearl River Delta region, China 409, 592, 593–4
Pergamum 95–6
Perrin, A. 202
person agents 374–8, 386
personal networks
  mobility and social support networks 427–47
  socio-economic 43–9
  sexual, *see* sexual networks
Peterson, C.E. 102
Pham, T.H.H. 576
phylogenetic transmission trees 358
physical network co-evolution model 185–8, 189
piazzas 280–82, 284
PICO California 297
Pitts, F.R. 264
place
  space and 282
  ties through 12–13, 194–214
place-based discrimination 321–2
place-name generators 208–9
planar graphs 50
  typology of 140–44
planning, city, *see* city planning
Plassard, F. 166
point-integral measures 321
Pokémon Go 413, 417
polarization of income distribution 559–60
policy
  implications of polycentrism 506–7
  implications of research on mobility and segregation 346
  need for Chinese city network research to include 596
political/judicial FCAs 116–17, 121
Polo, Marco 18
polycentric land-use model 256, 257
  polycentric urban regions (PURs) 493
    globalization of 515–16
polycentrism 493–510
  Beijing's LUR polycentricity 548–51
  city networks in a globalizing world economy 495–8
  city networks as a lens for revealing city-region polycentrism 498–502
  city-region emergence processes in network space 502–6
  functional 497–8, 502–6
  London's LUR polycentricity 542–3
  policy implications 506–7
Pompeii 96, 98
Pony Roll Media Culture 545, 546
population density 73
  ancient urbanism 96–8
population ergodic theorems 52
population growth 67
population size
  and container flows 455, 456–60, 471
    network centralization and 462–4, 465
  and power 74–5
  Roman urbanism 95–102
Port Moresby 245
Portes, A. 207
ports 22–3, 248, 264
  global container flows and urban hierarchies 449–71
  port and city relationships 449–52
Portugali, J. 170
position of an actor in two-dimensional space 398–400
position and role analyses 223–32
positional power 234–5
possibilism 278–9
posting 414
potential capacity 24
Pounds, N.G. 88, 89
Pour l'Avancement des Recherches sur l'Interaction Spatiale team (PARIS) 449

poverty 316–17
power
  and attractiveness of cities 514, 530–31
  and settlement size 73–6, 78
  world cities 234–5
power grids 131, 132, 133, 137
power laws 129–30, 247
Prebisch, R. 262
Pred, A. 26
predictability 54–5
pre-exposure prophylaxis (PrEP) 353–4
Preston, R.E. 21
primal–dual problem 403–5
primal graphs 50, 131
privacy 421–2
producer services firms, *see* advanced producer services (APS)
propinquity 201–2
proto-agriculture 65
proximity 284
  city-region economic development 495–6
  role in interurban networks 239–51
    types of proximity 241–3
    and urban network structures 243–6
proximity constrained interactions 32
public health departments 359
public spaces 200, 418
  city of opportunity 280–84
    and cultural urbanism 282–4
Pumain, D. 26, 167
Punch Taverns 537, 540
Purcell, D.W. 353
Putnam, R. 207

Quan-Haase, A. 415

race/ethnicity
  dissemination of health information and beliefs 370–86
  health disparities 370–71
  racial/ethnic residential segregation 313–30
  sexual networks and HIV 351, 352, 356, 357
  urban mobility, segregation and networked travel activity 331–49
racial homophily 352, 356
radiation model 587, 588
railway networks 23, 135, 167
  France 167, 168, 172, 189
    co-evolution model applied to 182–5, 186, 187
Raimbault, J. 169, 188
Rainie, L. 415, 416
Rallet, A. 243
random geometric graphs 137
random sorting 613–15
Randstad region, the Netherlands 493, 610
range 258–9
  real and ideal 258, 259
rank correlations for accessibility 178, 179
rank-size distributions 70, 102–3
rank-size rule 26, 27, 52–3
rationality assumption 255
real-estate firms 533
real range 258, 259
recall errors 358
receive–accept–sample model of opinion formation 372–4
Reclus, E. 24
recreational/cultural FCAs 116–17, 121
regional city networks 593–4
regional coalitions 298
regional scale 5–6, 12
regional trade agreements (RTAs) 517
regular equivalence 225–32
relationships 2–3
  economic urban networks 4
  interpersonal 294–5
  urban sociality 417–22, 423
relative space 253
relocation 356–7
reproductive rate 46–7
residential segregation 332
  racial/ethnic segregation and travel 313–30
    distribution of services 318–20
    social exclusion 314–16
    spatial exclusion 316–18
    stigmatization 321–2
    and transportation infrastructure 320–21
resourceful region 279
restorative justice 299–311
restorative justice court 309
reticular principle 16–25
Reynaud, J. 20, 22, 23, 25
Rhine-Ruhr region, Germany 493, 610
rice cultivation 71
ridership (subways) 158–9, 160
Right on Justice (ROJ) Network 299–311
rivers 93, 94, 261
roads 23, 130, 131, 248
  ancient world 93, 95
  evolution of road networks 144–51
    general properties 144–8
    structural changes revealed by the BC 148–51
  location and opinion pooling 400–402
  typology of road networks 137–44
Robinson, B. 294–5
Robson, B. 27
role and position analysis 223–32
Roman urbanism 88–106

structure and scale 91–103
hierarchies and networks 102–3
numbers of cities 91–5
numbers of inhabitants 95–102
Rome 98, 99, 100
Romer, P. 44
Roth, C. 152, 153, 154, 155, 156
Roth, Y. 419
rounding out 107–8, 111; *see also* associational life
Rozenblat, C. 515–16, 553
Rui, Y. 169
Russia 526, 527, 528
Moscow 264, 529

sampling error 357–9
Sampson, R.J. 207
San Diego Organizing Project (SDOP) 297
São Paulo 608
Sassen, S. 6, 79, 493, 500, 558, 618, 619, 624, 626
scale-free networks 31, 129–30, 247–8
scales 2, 4, 6–7, 9–10
economic urban networks 5–6
extension of the INM 609–10
geographical 9–10, 11
of global processes 516–18
and levels of multinational firms' networks 518–19
multiscale approaches, *see* multiscale approaches
non-geographical 10
non-nested geographical 10
Roman urbanism 91–103
scaling laws 29
Schlapfer, M. 199
Schmitt, C. 33, 170
scope of multinational firms' networks 527, 528
Scott, A. 494, 495, 497, 500
Scranton, Pennsylvania 107
Scruff 413, 419
seaports, *see* ports
second moment 127, 130
security 417
sedentism 64, 65, 66–8
sedentary–nomadic co-evolution 72–3
segment length 135–6
segregation
networked travel activity, urban mobility and 331–49
racial/ethnic residential segregation and travel 313–30
selective migration 443
self-organizing open systems 28
self-reinforcing networks 188, 189
semiperipheral capitalist city-states 74, 77–8
semiperipheral development 72, 73–4
semipheripheral marcher states 76
Sen, A. 279
sensory overload 420, 423
serial partnerships 354–5
serosorting 353
service value matrix 604, 605, 613
services
advanced producer services, *see* advanced producer services (APS)
residential segregation and the distribution of 318–20
settlement networks 11–12, 63–87
city regions 81–3
first villagers 66–7
hilly flanks 68
modern world city network 76–8
rise of cities and states 70–76
settlements before the rise of states 65–6
world cities and the global settlement system 79–81
settlement-size hierarchies 26–7, 28–9, 68–70
Seville, immigrants in 427–47
sexual networks 350–65
network mechanisms 351–5
compositional 351–4
structural 354–5
population disparities and HIV/AIDS 350–51
social life, venues, geography and online/mobile applications 356–7
sexually transmitted infections (STIs) 350–65
Shanghai 529, 531, 590–91, 592, 593
city-dyad with Beijing 612–13, 614
shape of city blocks 138–40, 141, 145–8
typology of planar graphs 140–44
shared philosophy 304–8
Shen, C. 108
Shenzhen 591, 592, 593
Shilton, K. 421–2
Shin, E.-H. 218
shipbuilding 113–14
shops 200–201
Siemens 545–8
Sienkiewicz, J. 131
Silva, J. de A. e 315
Simmel, G. 415, 418
Simões Aelbrecht, P. 201
Simpop models 30–33, 34–5
Simpop2 167
SimpopLocal 33, 35
SimpopNet 170
Singapore 463, 529, 531

singular value decomposition (SVD) 224, 227–30
Skocpol, T. 109
slave trade 627
smart cities 11, 275–6
Smith, D. 217–18, 225
Smith, D.H. 108
Snijders, T. 207
Snyder, D. 217, 225
social accessibility 420–21, 423
social anchor points 333
social capital 513
social class 112, 240
social companionship 428, 433–44
social exclusion 314–16
social FCAs 116–17, 121
social interactions
  city of opportunity 284–5
  locative media and 412–13, 417–22, 423
  sexual networks 350–65
social media 208, 284–5, 286, 323–4
social network analysis (SNA) 34, 216, 311, 500–501, 589–90
  macro-comparative 215–38
  proximity and urban network structures 244–9
social networks 33–4, 246–7, 248
  as metaphor 216
  modelling dissemination of health information and beliefs 366–88
social proximity 242–3
social support networks 427–47
  classification of ties 434–6
  measuring 428
  mobility and 429
  multiplicity of support 428, 432, 434–7, 441, 444
  typology of social support providers 437–41
social typologies of relations 196–8
socialising frequency 200
socially desirable responding 358
socio-cultural evolution 11–12, 63–87
socio-economic change, comparative 27–9
socio-economic networks 41–2, 43–9, 51
socio-economic status
  low-income Chinese Americans and health information dissemination 370–86
  and travel 315
socio-material network analyses 12–13, 194–214
  comparison of data collection instruments to type of analysis 209
  data collection and sources 207–9
  neighbourhoods 13, 196, 197, 198, 201–3
  quantitative analysis of citation sourcing 203–7
  studies combining discrete sets of social and spatial data 199–201
  studies using large data sets 198–9
socio-materiality 195
Solé, R. 131, 132
Sopra Steria 539, 540
source agents 374–8, 386
South Africa 23
South Asia 608, 609
South East England city region 494–5, 498, 501, 502–3, 503–4, 506
Southern Californian power grid 131, 133
Southwest Organizing Project (SWOP) 293, 299
space 7–8
  place and 282
space of flows theory 585–6
space-L topology 453, 454, 461–4, 465
space-P topology 453, 454, 461–4, 465
Spain
  colonialism 626–7
  social support networks for immigrants in Seville 427–47
span of control 100
spatial capital 322
spatial exclusion 316–18
spatial framings of Newcastle 117–18, 119, 121
spatial inequality, segregation and 317–21
spatial Keynesianism 262
spatial mismatch hypothesis 318
spatial network structure 127–65
  average degree 128–9, 155
  average shortest path 133–5
  clustering coefficient 128, 136–7
  connecting to the economics of the substrate 157–62
  evolution of infrastructure networks 144–56
  non-scale-free degree distributions 129–33, 134
  segment length and total length of the network 135–6
  typology of road networks 137–44
spatial proximity 243
spatial typologies of relations 196, 197–8
specialization 21–2, 24, 32, 33–4, 45–6, 514–15
specialized social support 428, 434–7
Speer, P.W. 294, 296
spread effects 262, 263
Springfield–Northampton highway 400–402
standardized service values 606
states, rise of 70–76
stations in subway systems
  interstation distance 159–61

number of 152, 153, 159–61
status 112
socio-economic, *see* socio-economic status
Steinitz, C. 406
stigmatization 321–2
stops on long-distance routes 22–3
Strano, E. 144–5, 146, 147, 150
strategic entry barriers 625
strong ties 196
structural centrality 264
structural entry barriers 625
structural equation model (SEM) 564, 576–9
structural equivalence 225–32
structural holes 561–2, 563, 567, 568–70, 573, 574, 576–8, 579, 583
structure
proximity and urban network structures 243–6
Roman urbanism 91–103
sexual networks 354–5
spatial network structure, *see* spatial network structure
subgraph isomorphism 339
sub-optimal services 320
sub-Saharan Africa 610, 611
substrate 128
connecting structure of networks to the economics of 157–62
subway systems
cost–benefit analysis 158–62
evolution 151–6
successive averaging 393–4
Sumitomo Mitsui Banking Corporation (SMBC) 535, 536
Sun, L. 198–9
superlinear scaling 527–31
supra-regional scales 517
surface area of city blocks, *see* area of city block faces
Surowiecki, J. 396
surplus production 71
surveillance 421–2
surveys 207–8
survival rates of ancient cities 101
Susiana Plain 71
Sutko, D.M. 421
Sweden 167, 324
Swiss railway network 135
Sydney 242, 529
synthetic system of cities 176–82
systems of cities, *see* urban systems

Tata Steel Europe 539, 540
tax havens 526–7, 530, 552
Taylor, P.J. 83, 498, 590, 591, 601, 626
interlocking network model 500, 586, 588, 602, 610, 621
Taylor, T.G. 278
technology
digital, *see* digital technology
locative media, *see* locative media
mobility and 429–30
technological change and maritime networks 451, 452
Tell Ouelli 71
territorial principle 16–25
territorially constrained interactions 32
Tertullian 95
Texas Organizing Project (TOP) 298
Thévenin, T. 172
Thiran, P. 135
third places 13, 196, 197, 198, 200–201
threshold 258–9
Thrift, N. 513
Tianjin 550–51
ties, *see* links/ties
Tinder 413
Tobler, W.R. 243
Tokyo 153, 245, 529, 531
road network 138, 139–40
world city 79, 218, 242, 522, 626
Tonkiss, F. 418
Tönnies, F. 196
top-down perspective 609
top-down processes 513
Torre, A. 243
total network length 135–6, 145, 146
tourism 243
trade
ancient urbanism 105
cities as stops on trade routes 22–3
container shipping, *see* container shipping
macro-comparative social network analysis 217–18, 227, 229, 233–4
openness 559
settlement networks 67, 71
trade blocs 242
traffic routes 22–3
trait-based discrimination 321–2
transnational corporations (TNCs)
FDI and inequality 557–84
headquarters 218
transnational support ties 435–6, 437, 438–41, 442, 443–4
transport model (Christaller's) 21
transportation networks 247–8
co-evolution of cities and 166–93
maritime, *see* container shipping
network structure connected to economics of substrate 157–62

railways, *see* railway networks
residential segregation and transportation infrastructure 320–21
roads, *see* roads
structuring effects 166, 190
subways, *see* subway systems
travel
racial/ethnic residential segregation and 313–30
social support networks 430, 434, 435–6, 441, 443, 444
urban mobility and segregation studied through networked travel activity 331–49
trust
dissemination of health information and beliefs 366–88
locative media 421, 423
Twitter 323–4
two-dimensional location problem 398–400
two-mode networks 604

Ubaid period 71–2
ubiquitous computing 410–11
Ullman, E.L. 262, 497, 498
umlands 257–61
underdevelopment 216–18
undetectable viral load (UVL) 353
unemployment 577, 578, 579
unequal terms of trade 217–18, 262
uneven city networks 621–5
uneven development 72, 73–4, 83, 262–3, 557–9
global cities, centripetal wealth transfer and 618–31
unions 623–4
United Kingdom (UK) 566
associational life 109, 110–12
Newcastle, *see* Newcastle
evolution of cities 24
London, *see* London
South East England city region 494–5, 498, 501, 502–3, 504, 506
United Nations (UN) 83, 273
United States of America (USA) 23, 77–8, 566
California, *see* California
Civil Rights Act 1964 331
community organizing 293–312
Fair Housing Act 1968 316, 331
GNC for NGOs 612
history of segregation 331–2
Jim Crow laws 331–2
multinational firms' linkages 526, 527, 528
New York, *see* New York
power grids 131, 133, 137
racial/ethnic residential segregation and urban spatial networks 313–30
railway network 167
segregation and mobility 331–49
Springfield–Northampton highway 400–402
studies of associational life 108–10
Voting Rights Act 1965 331
urban development locations 396–8
urban dominance 261–2, 263
urban economics 44, 56
urban growth
Botero's theory of 23–4
distribution of city sizes and 26–7
Gibrat's model 26–7, 28–9
urban heat islands 82
urban hierarchies 53, 261, 621
central place theory 18–22
global container flows and 449–71
Roman empire 102–3
urban management dashboards 286, 287–8
urban network theory of centrality 264–7
urban piazzas 280–82, 284
urban scaling theory 29, 56
urban science 389
urban sociology 110
urban systems 20, 25–36
complexity and co-evolution 25–33
evolutionary theory of 26–36, 167
models 170–71
multi-level urban networks 33–5
urbanism, defining 90–91
urbanization rate 101–2, 104
Uruk 72

value-chain agglomeration channel 472–92
industrial clusters' preferences 485–7, 490
value transfer 618–31
geographical 625–7
Van Oort, F. 497
Vantone Real Estate 545, 546
Vauban 18
Venice 77
venues 356–7
Vidal de la Blache, P. 25, 278
villages, first 66
virtual reality 410–11
virtual social interactions 284–5
visibility 419, 423
Völker, B. 202, 203
Volkswagen 545, 547, 548
voluntary associations, *see* associational life
Von Ferber, C. 131
Von Thünen, J.H. 19, 255, 256, 257

wage inequality 559–60

Walhström, M. 284
Wang, D. 324
Wang, Q. 314, 323–4
Washington, DC 612
Wasserman, S. 223
water distribution networks 131, 134
Watts, D.J. 497
wayfinding 411–12, 416–17
weak ties 196
wealth transfer, centripetal 618–31
Weber, M. 284
weighted averaging 390–407
Weiser, M. 410
Wellman, B. 207, 415, 416
West African coast 627
western Europe 607, 608
White, D. 217–18, 225
White Americans 334–47
Whyte, H.W. 283
Wilson, A.I. 96–7
Wilson, W.J. 322
wisdom of crowds 396
Wolf, L.J. 3, 4
women
  in associational life 118, 120, 122
  and travel 314–15
work segregation 332
world city network (WCN) 240, 501, 585–6, 621
  Chinese cities in 585, 590–91, 594–5
  modern 76–8
  researching from China's perspective 595–6
world/global cities 32, 76–81, 218, 263, 500, 558
  centripetal wealth transfer and uneven development 618–31
  and the global settlement system 79–81
  and the governance of global commodity chains 620–21
  macro-comparative social network analysis 234–5
  stages in research on 585–6
world historical framework of comparison 63–4
world-systems 64–5
  of city states 72
  co-constitution of world cities and their countries 234
  comparative evolutionary world-systems perspective 64–5
World Trade Organization (WTO) 517
Wright, H.T. 71, 72
Wu, J.R. 169
Wuhan 503, 504, 505, 592

XXQ city, *see* city of opportunity

Yangtze River Delta region 592, 593–4
Yazdani, A. 131, 134
Yeh, A.G.O. 588, 595
Yelp 417
Youm, Y. 354
Young, I.M. 418
youth organizing 295

Zaller, J. 372
Zhang, Q. 372, 373
Zhang, X. 610
Zhangjiakou 550, 551
Zipf, G.K. 26
Zipf's law, *see* rank-size rule
Zozman, G. 75